Temperature Conversion Scale

For conversion of Fahrenheit to centigrade, the following formula can be used:

$$°C = 5/9 \ (°F - 32)$$

For conversion of centigrade to Fahrenheit, the following formula can be used:

$$°F = 9/5 \ °C + 32$$

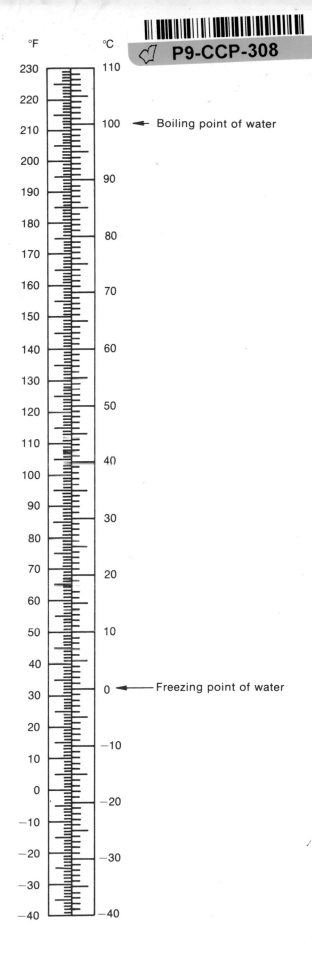

P9-CCP-308

Contemporary Biology

concepts and implications

MARY E. CLARK *California State University, San Diego*

W. B. SAUNDERS COMPANY
PHILADELPHIA / LONDON / TORONTO

W. B. Saunders Company: West Washington Square
Philadelphia, Pa.   19105

12 Dyott Street
London, WC1A   1DB

833 Oxford Street
Toronto 18, Ontario

Contemporary Biology   ISBN   0-7216-2597-5

© 1973 by W. B. Saunders Company. Copyright under the International Copyright Union. All rights reserved. This book is protected by copyright. No part of it may be reproduced, stored in a retrieval system, or transmitted in any form or by any means, electronic, mechanical, photocopying, recording, or otherwise, without written permission from the publisher. Made in the United States of America. Press of W. B. Saunders Company. Library of Congress catalog card number 72–86447.

Print No.:   9   8   7   6   5   4   3

*This book is dedicated
to the memory
of a most human man,
my father.*

PREFACE

Biology, the study of life, is a subject which has an intrinsic fascination for a certain fraction of people, including myself. For such people it is no chore to grapple with the vocabulary which surrounds the subject nor with the abstract concepts which are the basis of scientific thinking. It is easy to learn about that which attracts one. But, fortunately for mankind, not all are alike in their individual tastes and interests. This book is addressed mainly to those of you who may not find anything especially intriguing about the intricacies of a spider's web, the life cycle of a fungus or the structure of the hemoglobin molecules in your red blood cells—who prefer to absorb yourself in other matters such as the building of bridges (be they substantive or communicative) or other social and economic problems of man. Many of you may agree with the English poet, Alexander Pope, who stated over two hundred years ago that "the proper study of mankind is man."

Yet mankind, through application of scientific principles, has developed a technology which affects himself not only directly, as he intended it to, but indirectly in ways that were unforeseen—in explosive population growth, drug abuse and environmental deterioration. Man today threatens the very health of the biosphere which supports him.

These are biological problems, and they affect us all. To correct them requires not merely study and advice by scientists but, since the problems are created by all of us, a willingness on the part of society as a whole to redirect itself. Such change in direction will occur, however, only if there is general agreement, and this can come about only if people understand the nature of the problems. It is clear that biology has now become a "proper" subject for study by all mankind—it is a subject relevant to all of us.

This book attempts to present the subject matter of biology in a way which relates it to every individual. In thinking about how best to do this it first occurred to me to present only the relevant aspects—those parts of the subject easily seen to have practical applications to everyday life. But this approach proved impossible. How can one meaningfully discuss the problems raised by alcohol, drugs and pollution if the fundamental information by which these problems can be understood is not known to the reader? It would be like explaining the significance of Shakespeare's puns to someone who couldn't read English, or the importance of a carburetor to someone who had no idea of how an internal combustion engine works. Some background information is essential for the useful application of any piece of information, and this is as true of biology as of any other subject.

In this book, therefore, a considerable amount of basic biology is presented, although the specialized vocabulary has been kept to a minimum. The coverage of the subject is not comprehensive. Those areas of basic information which are most important for an understanding of current biological problems have been emphasized. As each new topic is introduced it is followed by a chapter which applies the principles just learned to some specific contemporary problem. Although it has not been possible to cover all of the potential applications of biology in today's world, I hope that the examples given in the relevant "a" chapters will enable the reader to gain experience in applying the principles he learns to other matters which become important in his future life—in understanding news articles, essays and lectures about environmental and health topics and in voting on issues related to them.

Biology is an enormous and complex subject whose various facets are interrelated. A newcomer to the subject is likely to feel frustrated at first because to understand fully one part of the discipline he often must have information about another part not yet covered. The sequence of subject matter has been chosen to minimize this difficulty as much as possible, but the reader may find it worthwhile to glance ahead in anticipation of a fuller explanation of some points and to refer back to earlier chapters for review when appropriate. Important concepts which may be useful at several points in the book have been set into boxes that can easily be referred to as the need arises (see end of Contents, page xiv). New terms which appear in the text are in **boldface,** and definitions for these will be found in the glossary. A list of symbols and abbreviations, and conversion tables from English to metric units and from Fahrenheit to centigrade temperature scales, are located inside the covers.

Much of the chemical background generally thought to be required for an understanding of biology has been placed in an appendix. There is also an appendix reviewing the use of graphs and exponents, both of which are met with extensively in the text. A third appendix covers the classification of organisms.

Although this book is intended for non-science majors, it is by no means an "easy" book. On many pages the reader will discover that almost every word counts and will find himself suitably challenged intellectually, as a college student should be. In fact, a student who suddenly finds himself "turned on" to biology after reading this book should, with but little further effort, be able to go on to more advanced courses in the subject.

Any writer of a general biology textbook recognizes his deep obligation to many others—to the scientists about whose work he writes, to other authors whose ideas he uses, to his own teachers who opened up the subject to him and most of all to his students and colleagues who, through their suggestions and criticisms, help to shape the book in its final form. All of these people I thank, and especially those many personal friends in the field of biology who willingly gave so much time to reading parts of the manuscript in its various stages. Appreciation is also due the cooperative and patient staff of W. B. Saunders Company. And finally, a special word of gratitude belongs to Wanda Humphrey, who not only typed the many drafts of the manuscript but also offered innumerable suggestions from the student reader's point of view. For the errors which remain, I alone am responsible.

MARY E. CLARK

La Jolla, California

CONTENTS

7a

8

8a

9

9a

10

U.S. Forest Service photo

BIOLOGY: A RELEVANT SUBJECT

Our current environmental and human dilemmas demand understanding and application of biological principles by the average person if they are to be resolved.

Biology, of all the sciences, is perhaps most closely associated with our everyday life, and we shall attempt not only to survey biological principles but also to apply them to contemporary problems. A great many of these problems are of a biological nature—the population explosion, cancer and other diseases, pollution, to name but a few—and many of them are man-made. Biology is thus not merely an esoteric subject to be studied by the curious; it has relevance for the non-scientist as well. In this introductory chapter, therefore, we shall not talk about the scope of biology itself, but instead shall consider some of the ways in which biology is relevant to all and so set the stage for the rest of our text.

THE BIOLOGICAL NATURE OF MAN

Mankind has been around for over one million years. Like all other living organisms, man has struggled to survive and maintain himself by reproducing and by utilizing his environment. This is a purely biological goal, no different from that of any other species of plant or animal on Earth. Other species, however, have never threatened the total survival of life as man now has the potential for doing. With his manual dexterity, his language and his ability to form abstract concepts, man is able to modify his surroundings in ways that no other organism can.

These same attributes make man readily adaptable. Unlike plants and animals, which must generally undergo gradual genetic changes to meet changing environmental conditions, man is able to change his habits and utilize new environments within a single lifetime. He can survive in the arctic or in the tropics—or even briefly on the moon—but there has been little genetic adaptation to these varied surroundings. Instead of passing on adaptational information by inheritance, man passes it on by teaching his children.

It is true that animals other than man can teach their offspring many things. For instance, a few years after the second world war, dairies in England began to cap milk bottles with thin aluminum tops. Soon a tit, somewhere in the Midlands, discovered that by pecking through this shiny lid he could obtain a drink of cream. During the ensuing months the knowledge of this trick spread amongst the population of these little birds in ever-widening circles, and within less than two years tits all over England had learned how to find their morning drink of cream. The English are inveterate bird watchers, so this minor phenomenon is well documented.

Man's ability to transmit information, however, is both qualitatively and quantitatively far greater than that of any other living organism. Relatively recently in his history man learned to write down his ideas and thus to transmit information without the necessity of personal contact between individuals. Writing has added enormously to the potential for storing and communicating observations and experiences, and was an essential prerequisite for today's technological explosion.

Man also differs from all other organisms in the degree of his curiosity about the world around him. He tries to understand how things work and to account for such things as his own origins, the meaning of life and the necessity of death. Man's need to explain himself originated far back in prehistoric times, and all known primitive cultures have an oral tradition of the origin of man. This curiosity was also a prerequisite for the advent of science and technology.

MAGIC AND RELIGION

Awareness of death and the struggle to survive induced primitive societies to attempt to manipulate nature in ways favorable to the public good. According to Sir James Frazer, this was the origin of magicians—witch doctors, rainmakers and other functionaries invested with magical powers. Although magicians served an important psychological function, their contribution toward improving the physical condition of man was minimal and their impact on the environment virtually nonexistent.

The powers of the magicians were gradually replaced by those of super-beings or deities. In some cultures, the soul of the immortal being retained its contact with society through a priesthood, in which the powers once held by magicians were invested. In these primitive religions, the functions of the omnipotent deity in regard to human survival were fundamentally the same: to protect and prolong life, increase fertility, relieve suffering, and provide food and water. Early religion was man's second attempt to control nature and to improve his condition; like magic, it served a psychological and sociological function, but its direct effects on the environment were inconsequential.

THE ADVENT OF SCIENTIFIC RATIONALISM

As time passed, primitive man began to learn by trial and error to modify his environment and improve his condition. He kept livestock and he farmed; he learned to breed plants and animals, to till, to fertilize and to irrigate. Folk medicine was sometimes effective in curing disease; in some cultures, even, medicated tampons were used for birth control.

These activities led in some places to local changes in the landscape. Overgrazing in the Near East and North Africa may have contributed to the formation of the present deserts along the once fertile Mediterranean shores. Plato described how the removal of forests and overgrazing on the Attic peninsula caused a drying up of springs and loss of fertile land because the water ran off the barren ground to the sea. The countryside of most of Europe was totally altered from wilderness to farmland and cultivated forests long before the industrial revolution (Fig. 1.1). But these changes did not threaten to destroy mankind. In fact, those in Europe permitted an even larger population to survive.

Until about 200 years ago, the environmental changes brought about by man were gradual, almost imperceptible. Then the Age of Enlightenment, which began to flower in the eighteenth century, ushered in a new era, and scientific rationalism quickly gained ground. With it came discovery, invention and rapid change. Technology seemed the promised answer to the biological goals of mankind.

FIGURE 1.1

A man-made countryside. Once covered with trees, this farmland in County Tyrone, Ireland, is divided into small fields separated by hedgerows. (Courtesy British Ministry of Defence [Air Force Department]. Crown Copyright reserved.)

"PURE" AND "APPLIED" SCIENCE

With the advent of the scientific age, the application of scientific ideas to the solution of human problems became widely accepted. At first science was deemed worthwhile only if it pursued useful ends; the study of science for its own sake was regarded as a frivolous exercise. Gradually attitudes underwent a reversal; "pure" science became an intellectual pursuit of the leisure class, and thus attained an aura of superiority — *useless* came to equal *good.* "Applied" science, carried out for a specific purpose, was equated with trade and the necessity of earning a living. To a certain extent, this distinction persists today, although pure science now is generally called "basic research," and is often conducted in the hope of uncovering that which may later prove useful. We shall return later to the role of "pure" and "applied" research in solving human problems.

It is not generally recognized that the methods of scientific inquiry are the same whether or not the subject under investigation has immediate practical application. The difference is simply that the results of applied research can be immediately exploited by technological development.

At first, most of the outstanding inventions of the Industrial Revolution were devised by men with little or no formal scientific training. Only later were the basic principles of science studied and utilized for the development of new technological goals. This was as true in the field of biology as in the physical sciences—while the leisure class debated the origins of man, Claude Bernard, Louis Pasteur and others were making great discoveries in the areas of physiology and medicine. Only in the past few decades has there been a wide application of general biological principles to the solution of specific problems. However, the lag in time between explanation of "useless" phenomena and their development into "useful" applications is constantly decreasing. To a large extent, this continuing emphasis on rapid technological application of scientific discoveries has created many of the new biological problems which face us today.

SOME RESULTS OF TECHNOLOGY

Knowledge, of itself, is neither "good" nor "bad"—it lies outside the domain of morality. It is, rather, the use to which knowledge is put that determines whether it brings blessings or evils. What we shall attempt to show here is that the application of scientific ideas for the benefit of mankind has been carried out in a haphazard way. Most technological advances have been directed toward solving specific demands of society without considering their possible repercussions—a "hasty opportunism," to use the words of the distinguished English biologist P. B. Medawar. Much of the disrepute into which science has recently fallen is due to the backlash from the application of both biological and physical technology; the biological consequences are what concern us here.

THE POPULATION EXPLOSION

Among the biological goals of man are the desire to live longer, to be free from pain and disease and to see all his children survive him. To serve these ends the fields of medicine, dentistry and public health have evolved and, in the past hundred years, have made remarkable advances. Although the maximum human life span has not been prolonged, the average life expectancy has increased enormously. Death from disease has sharply declined, especially among infants and children. Only a hundred years ago it was common to find in a churchyard recent gravestones of people who died before they were thirty. Now, in most of Europe and North America, this is rare. There has not been a compensating drop in birthrate, however. The result, shown in Figure 1.2, has been a rapid upsurge in the world population. In the year 1930, there were about two billion people on Earth; today there are more than 3.7 billion. If the present trend continues, in the year 2000 there will be about 7.5 billion, or twice as many as now. In just 30 years, the world population will have doubled!

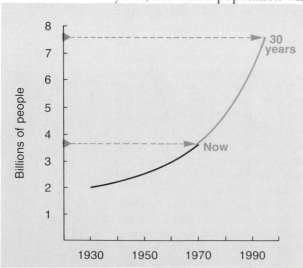

FIGURE 1.2

Projected growth of human population in the world, based on the United Nations constant fertility projection. (From Population Bulletin 21, No. 4).

The curve in Figure 1.2 is known as a geometric growth curve. That is, the absolute increase in numbers is greater each year. The present **doubling time** for the world's population is about 30 years. Were this rate to be maintained, when today's pre-school children were ready to retire in 2030 there would be *four* times the number of people on Earth as when they were born. When their children retired, there would be *ten* times as many. It is unlikely that this rate of increase will continue, however, for reasons to be discussed shortly. Doubling times are not evenly distributed in the world. Some countries are growing much faster than others. In North America the doubling time is about 63 years, and it appears to be slowing down. In India the doubling time is 28 years; in Costa Rica, only 18 years. Although in 1969 the United States population was still growing, and would have doubled in 70 years, the recent decline in birth rate has led to a leveling off — we are now at zero population growth.

We have all heard Earth referred to as a spaceship. What is meant by this is that our Earth has a limited supply of the air, water and sunlight necessary to produce our food. We cannot increase any of these things. Our planet, like the spaceships recently sent to the moon, has physical limitations on the number of poeple it can support. It is likely that the director of NASA would be fired for incompetence if he suggested doubling the number of astronauts in a space capsule the night before takeoff. Although our larger spaceship, Earth, may be able to accommodate somewhat more people than presently exist, our current rate of increase will, in but a few generations, create the same problems that doubling the number of astronauts in a spaceship would. In just 30 years, the world will have to supply twice as many individual goods and services as are now consumed — food, water, electricity, houses, cars, schools, hospitals, churches and mortuaries — just to keep even. The chances of accomplishing this all over the world in such a short time are remote. The likelihood of widespread famine, disease and decreased life expectancy is considered high by many scientists and demographers. These consequences have not yet been seriously contemplated by most of the world's people. The success of medical technology in meeting one set of goals of mankind has created a new problem — overpopulation!

FOOD PRODUCTION

Another biological goal of mankind is to obtain food. Man began farming about 10,000 years ago, and until the Industrial Revolution introduced mechanization, the majority of the populace lived in rural communities and were engaged in agriculture. The advent of technology meant the production of more food with considerably less labor, and the migration of large numbers of rural workers (some 75 per cent in the United States) to towns and cities.

Technological aids introduced into farming are legion. During a drive through California's rich central valley one sees advertised "Rain for Rent," "Perma-rain," "Farm-Builder Biochem," "Probiotic" and numerous other aids to the farmer. There are also billboards urging the farmer to borrow the cash necessary to buy such commodities, from companies with names like "Production Loan Association." Farm machinery of all descriptions is for sale or hire everywhere. There are huge farms in California given over to one particular crop — a practice known as **monoculture.** It is truly an agro-industry. These same things can be observed throughout much of rural America and, indeed, most industrialized countries.

But technology designed specifically to increase productivity and minimize costs has produced a myriad of problems in its wake. Single crops tend to wear out the soil more rapidly, while the use of heavy tractors and frequent plowing pulverizes the soil, causing it to lose nutrients. Formerly productive land becomes impoverished and requires the use of more and more artificial fertilizers, a practice which tends to pollute lakes and streams. Irrigation of poorly drained land may cause waterlogging of the soil and accumulation of salts which poison crops.

One of the greatest problems resulting from monoculture is that large fields of the same crop invite rapid multiplication of destructive insects, since there are no suitable habitats for the predators which normally keep pests in check. Various insecticides have been developed by technology to control these pests. But there is a major difficulty: these chemical poisons are not specific; they kill not only the harmful target insects, but beneficial insects and useful soil organisms as well. Moreover, because the target insects are widely dispersed, far more pesticide must be sprayed on crops or forests than is required to kill only the harmful individuals, and many unintended targets are hit. The effect is similar to that found when a whole city is bombed to destroy one munitions factory. It is a non-selective, non-specific way of killing.

The best known of these pesticides is DDT—one of a group of chlorinated hydrocarbons which are characterized by their virtual indestruc-

FIGURE 1.3

Our vanishing national emblem, the bald eagle, a victim of the accumulation of DDT in the environment. (From Walker, M. J., and Lowen, J. 1967. The vanishing American. American Forests 73(10). Photo by Bureau of Sport Fisheries and Wildlife.)

tibility in the environment. It was Rachel Carson's revealing book, *Silent Spring*, which first alerted the public to the dangers of these chemicals. Since then, much evidence has been amassed which indicates that DDT and its relatives accumulate in the fatty tissues of organisms, especially those at the end of the food chain. DDT from water or soil gets into plants which are eaten by small animals which are in turn eaten by larger animals. Eventually, those at the top of the chain accumulate very high concentrations of the pesticide in their tissues — a process known as **biological magnification.** The effect of DDT, at least in many species of birds, is to interfere with reproduction, so that few eggs are laid and few offspring are hatched. The brown pelican, once the state bird of Louisiana, has all but disappeared from the Gulf Coast, and will soon disappear from California's coast as well. In 1970, only three young were hatched on the offshore islands where these birds breed; in 1971, only one. The osprey, the peregrine falcon and other birds of prey are also disappearing. Our national emblem, the bald eagle (Fig. 1.3), may soon be only a figure on flagpoles and official stationery.

To some, while the loss of a few bird species is regrettable, it is of no real consequence; to the biologist it signals a polluted global environment. DDT has been found in penguins as far off as the Antarctic. Of more practical importance are findings of traces of DDT in almost all fish, meats and dairy products. The story of Coho salmon in Lake Michigan, for example, is a cautionary tale. Due to the activities of man, the original commercial whitefish and trout fisheries were declining when, early in the 1960's, Coho salmon were introduced in an attempt to restock the lake. They proved enormously successful photographs were published showing Michigan's governor at that time holding a 20 pound salmon; canneries were built or reconditioned. Only five years later, however, DDT levels in the salmon were found to exceed Food and Drug Administration tolerance limits, and the commercial fisheries were closed. Although the acute toxicity of DDT for humans is low, no one yet knows what its long-term effects on human health may be. What is certain is that no technology can retrieve DDT or other toxic substances once they are let loose into the environment.

In addition to DDT, mercury compounds, used to treat seeds to inhibit fungal growth, have found their way into wildfowl in parts of western Canada and the United States. Hunters in some areas have been warned not to eat the game they shoot because of mercury contamination. DDT and mercury are but two examples of the many chemicals used in agriculture to increase productivity which have resulted in far-reaching and unexpected side effects. We shall consider others later in the text.

RESOURCE CONSUMPTION

Besides good health and food, creature comforts have always been a biological goal of man — clothing, shelter, warmth, and the goods and chattels of domestic life. In furnishing himself these material items, man consumes both living and non-living resources that supply the raw materials and energy necessary to their manufacture. The search for new sources of minerals played an important role in dispersing mankind in prehistoric times. The Phoenicians from the Near East, for example, ventured to the shores of Britain in search of tin. With time, the increasing importance of manufacturing began to affect the landscape: during the Middle Ages many indigenous forests in Europe were consumed as fuel for smelting. Compared to those occurring at present, however, these effects were relatively minor and gradual.

Since the Industrial Revolution, the basic "needs" of man, at least in developed countries, have expanded to include such things as refrigerators, washing machines, automobiles and garbage disposals — the accouterments of a high standard of living. The rate of resource consumption has consequently risen geometrically, and continues to increase.

The minerals mined today from concentrated ore deposits are being dispersed as solid wastes in a thin layer over the Earth. These deposits are non-renewable resources—a geologic bank account from the Earth's past history which we are rapidly exhausting. Exploitation of the sea bed (or, as some suggest, the moon) is only a temporary stopgap and may prove more costly than recycling the minerals we have already unearthed. Sooner or later we shall be forced to change our habits.

Our present manufacturing processes are also a source of environmental contamination. Noxious fumes are released into the air (Fig. 1.4) and toxic wastes are discarded into rivers. These substances are being produced at rates which exceed the capacity of natural restorative processes to handle them. In some instances the very organisms which we rely upon to restore the purity of our air, soil and water are being destroyed by the poisonous by-products of our industry.

The energy needs for maintenance of our comfort and production of our material blessings have also skyrocketed. Primitive man burned wood, peat, animal fat and sea-coal—coal lumps cast up fortuitously on the shore. Extensive mining of coal and pumping of oil and natural gas coincided with the Industrial Revolution. These new energy sources greatly expanded man's manufacturing potential. They also brought pollution. Queen Elizabeth I deplored the smoke-filled air of London—then a city of but a few hundred thousand. Today's smoggy urban skies are more than irritating; they endanger the health of both man and his environment. Energy production also results in the generation of waste heat, which further affects the quality of our atmosphere and our water.

Our present energy comes mainly from our fossil fuel bank account.

FIGURE 1.4

TVA steam plant in western Kentucky, generating electricity from combustion of fossil fuel. Many of the gases released are toxic to plants, animals and man. (Reprinted with permission from The Courier-Journal and Louisville Times.)

Coal, oil and natural gas are forms of energy that were stored in the past. As miners of this energy bank we are like someone trying to make an inheritance stretch indefinitely in the absence of any income from interest or dividends—we are living on capital which is gradually disappearing.

Meanwhile the demand for power increases. The Joint Committee of Congress on Atomic Energy predicts that in North America the *per capita* consumption of electric power alone will increase four times by the year 2000. Taking the expected local population increase into account (33 per cent), 5.3 times as much electricity as we now use will need to be generated in the United States in just 30 years. Whether these demands will be made and met depends upon the priorities adopted by tomorrow's citizens.

The development of nuclear power has much to commend it, for it does not pollute the air and, if thermonuclear fusion reactors become practical, there will be virtually no radioactive contaminants released to the environment. Nevertheless, no matter how energy is produced, there is always the problem of waste heat which warms the environment and thus has biological repercussions. No technology can overcome this law of nature, so there exists an upper limit on how much energy mankind can generate.

OTHER TECHNOLOGICAL PROBLEMS

As man learns more about nature, he is tempted to use his knowledge for specific ends. The power of the atom provides us the means not only of generating energy and of curing disease but also of destroying ourselves in a nuclear war. The development of chemistry has produced potent pharmacological tools which can be either a boon to medicine or a source of drug abuse leading to human degradation. Our skills in the field of genetics may soon make it possible for man to produce genetic replicas of any living person—the future world could be peopled with facsimiles of men like Einstein, Hitler or Gandhi. But who is to decide which genetic characteristics are of value? Who is to be in control? Science provides the tools of technology, but the uses made of scientific discoveries cannot be scientifically determined. These are the responsibility of society.

HOW DID IT HAPPEN?

Never before has mankind been faced so squarely with such great moral decisions. We now have the potential to destroy ourselves or to create a brave new world, although not necessarily according to the vision of Aldous Huxley. Many of us are perplexed, pondering the magnitude of it all, and are wondering what decisions one individual can make to affect the future. Perhaps the first questions to ask are "How did it all come about?" and "Why has society allowed these crises to befall us?"

One explanation is that proposed by Professor Lynn White, Jr., who lays most of the blame at the feet of the Judeo-Christian ethic. In Genesis (1:26) it says that man was given by God "dominion over the fish of the sea, and over the fowl of the air and over the cattle, and over all the earth." Dr. White argues that although the same basic scientific knowledge was available in most civilized parts of the world just prior to the Industrial Revolution, only in Europe and North America—where the Judeo-Christian ethic predominates—was this knowledge exploited for technological developments which could affect the age-old lot of man and which, incidentally, had a great environmental impact. The spread of technology to non-Europeans was secondary. It is Dr. White's thesis that the Judeo-Christian ethic is more anthropocentric than other religions, regarding man as the controller of nature rather than as a part of nature. According to him, this ethic was interpreted in the early nineteenth century by Western civilization as meaning that scientific knowledge implies (and justifies) technological power over nature. In other words, Dr. White is saying that Judeo-Christian traditions urged the exploitation of nature by man rather than the harmonious interaction of the two.

It is perhaps possible to find equally plausible alternative explanations for the initial rapid rise of technology in Western culture. As we have already seen, all primitive peoples have attempted by magic or religion to control natural phenomena, and there seems to be no reason to suppose that Europeans were more motivated than others to do so. In addition to their anthropocentric ethic, Western thinkers were exposed to a wide array of earlier knowledge, accumulated in the writings of their circum-Mediterranean predecessors, which was not widely known elsewhere in the world. A great variety of minerals and other natural resources were also available to them. Communication among peoples in Europe was remarkably good compared to that in other parts of the world, facilitating dissemination of ideas and information. These factors could all have contributed to the origin of the Industrial Revolution in Western society.

Whether one subscribes to Dr. White's thesis or not is perhaps of little importance. The point is that technological development, with its environmental consequences, has spread from Europe and North America to all parts of the world as fast as educational conditioning and a supply of capital would permit. Most people are ready to improve their condition by technological means, and there is little indication that the underdeveloped countries are reticent about catching up with their technologically developed neighbors. This seems to be true despite widespread awareness of the ecological dangers being created daily as the result of today's technological approaches.

SOLVING THE PROBLEM

The following brief discussion is an attempt to describe the present state of our technological society and to suggest a basis for change. It is offered as a stimulus to further discussion and not as an all-encompassing explanation and solution. Extensions and alternatives to the views presented here abound. The writings of people like John Kenneth Galbraith, Barry Commoner, Margaret Mead, René Dubos, Paul Ehrlich and Charles Reich will quickly introduce the concerned student to current thinking about these matters.

THE STATUS QUO

Technology, as we have seen, has developed to serve the historical demands of society, but despite its increasing impact, there has been little change in approach. Goals are still sought individually without regard to possible unlooked-for consequences. The great forward strides in the life-saving aspects of medical technology have only recently begun to be accompanied by parallel advances in birth control. The industrialization of agriculture has proceeded without an understanding of ecological principles. Our manufacturing plants have treated air, water and minerals as though they were infinitely available. And the military has been largely unconcerned about either the biological or cultural consequences of its technology—a fact implicit in its headlong development of nuclear, chemical and biological warfare. The absence of a unified or **holistic approach** in the application of technology is thus at the root of many of our current biological problems.

The Technology Equation. A recent consequence of this piecemeal, **reductionist approach** to achieving human goals has been the widespread acceptance of technology for its own sake. Technology—*any* technology—has come to be equated with "good." This equation derives largely from the important role of technology in economics. According to British economist E. F. Schumacher,* the rise of affluence has placed economics at the center of public concern. But it is in the very nature of contemporary economics to be fragmentary: it excludes all criteria except those based on monetary values.

We live today with a myth of growth—that bigger means better, that growth equals progress. The meaninglessness of our gross national product as a measure of human value is recognizable in the kinds of things which are added together to calculate the annual GNP. Bombers and hospitals which happen to cost the same are given the same value; a basketball star may contribute more than the President, salary for salary; a dollar's worth of comic books is the same as a dollar's worth of textbooks. The absence of any human values from these calculations is obvious. The term "standard of living" has lost its meaning, for it is no longer related to the quality of life.

Nor is this "economocentricity" limited to capitalist countries; it applies equally to socialist and communist systems, which consequently suffer problems identical to ours. The average person in Western society has lost control of the technology which is supposed to serve him. Somewhere the system has broken down.

BREAKDOWN OF THE SYSTEM

For convenience let us use the greatly simplified model sketched in Figure 1.5 to reveal some of the weak points in the system which exists in the real world. In this model, the world is peopled by citizens of all sorts; our "citizen" represents not only the individual, but also the collective citizen, acting through public organizations such as governments, unions, cooperatives and so forth. The technologists are a smaller group who meet the citizen's demands with goods and services, and are in theory regulated by the citizen. Technology utilizes natural resources and generates wastes. The scientist, working in universities and public or private research laboratories, makes discoveries about nature, some of which are of immediate use to the technologist. The scientist is supported by the citizen as taxpayer and also by industry, which supplies both money and tools. The scientist also plays a role in educating the individual citizen, through schools and universities, and in advising the collective citizen by serving on government committees, giving expert testimony and so forth. Our spaceship, Earth, is the biological support system for the whole hierarchy, and also provides those esthetic amenities that the psychic nature of man requires. It also is the source of natural resources and the repository for discarded waste generated by technology.

There is an elaborate network of feedbacks in this system, far more complex than we have shown. We cannot survive without the system, so destroying industries is no solution. What is necessary is to correct the weaknesses of the system and restore a balance to our spaceship. Where are the weak links?

All Passengers and No Crew. One link is missing altogether. As pointed out by Paul Ehrlich, our spaceship, Earth, has only passengers; it has no crew concerned with keeping it running smoothly and looking after its functioning. It is as though, instead of trained astronauts, recent spacecrafts to the moon were manned only with sightseers "going along for the ride." Primitive man, who had no real need of a crew, at least *tried* to insure the continuance of his world by performing rites to make the sun come up and the seasons follow one upon the other. Today, when the need is great,

*Reprinted in *Omega*, by Paul K. Andersen. 1971. W. C. Brown, Dubuque, Iowa.

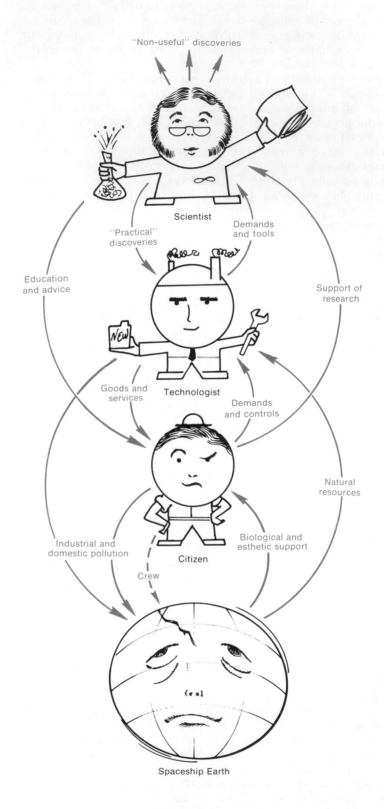

FIGURE 1.5

only a relative handful seem to be concerned about the continuing state of our Earth. Even scientists, who might be expected to be aware of the need, have given far too little concern to such a practical problem. Those few people, scientists and others, who have spoken up, however, have made little headway; they have no power of their own to change things, and those with power have chosen not to listen. Somewhere a crew must be found.

Technology and the Citizen.   A major point of breakdown is in the link between technology and the citizen. The original, rather simple demands of man have been blown up out of all proportion. Most of today's excessive consumption does not fill legitimate biological goals; rather it meets artificial needs created by technology and advertising.

The control of technology by the citizen at large is another weak spot in the system. The citizen, through passivity and apathy, has largely abdicated his powers. Technology, both of private industry and of governmental departments such as the Department of Defense and the Atomic Energy Commission, is virtually autonomous. James D. Reilly, a Vice President of Consolidated Coal Company, said in a speech in 1969:

The conservationists who want strip miners to restore land are stupid idiots, Socialists, and Commies who don't know what they are talking about. I think it is our bounden duty to knock them down and subject them to the ridicule they deserve.*

Representative Craig Hosmer, during the 1969 fiscal hearings of the Joint Congressional Committee on Atomic Energy, said:

Somebody should start to talk about power plant siting in terms of reason instead of emotion, because if this continuous badgering about thermal enhancement and esthetics and a lot of other things continues there is not going to be any place to put new power plants down.

The choice is not between some thermal enhancement and none. The choice is between some thermal enhancement and blackouts and brownouts. We ought to be thinking in those terms and the *public ought to be educated in those terms.*** (italics mine)

To which Chairman Chet Holifield added:

I agree with my colleague's comments and would add to it in this way, that because of this tremendous increase in need of electricity the American people are going to demand that electricity.†

(The "need" referred to here is that projected by the committee itself, which we mentioned earlier; the speakers seem to be implying that because they foresee a need it will therefore of necessity become a fact. No alternatives are considered.)

These quotations, by no means unique or even unusual, indicate the depth of the rift between technology and the governmental bodies charged with its control on the one hand, and the general public on the other. It is virtually impossible to discover a technologist or a politician who advocates educating the public to reduce its demands for consumption as a solution to any problem!

Public Functions of Scientists.   Another weak link in the system is the failure of scientists to effectively advise the government. Martin Perl of Stanford University has analyzed the situation and finds that, although the governmental scientific advisory system functions well on specific technical problems, scientists have little effect on policies. Reports are kept confi-

---

*Quoted by Harry M. Caudill in, Are conservation and capitalism compatible? *In* H. W. Helfrich, Jr. 1970. *Agenda for Survival: The Environmental Crisis-2.* Yale University Press, New Haven, p. 177.

**From *Selected Materials on Environmental Effects of Producing Electric Power.* Joint Committee on Atomic Energy, Congress of the United States, August 1969. U.S. Government Printing Office, Washington, D.C., pp. 26–27.

†Ibid.

dential, and a scientific advisor who breaks the rules and publicly objects is soon ousted for having "rocked the boat."*

Scientists could do far more in the arena of public education, however. Universities, colleges and the public lecture forum offer excellent opportunities, yet to be maximally utilized, for informing the citizen.

Support of Research. The short-term gain has been a major determinant in man's application of technology. Until quite recently, it seems not to have occurred to more than a few individuals that any other considerations were necessary. Without our noticing it, we have let our technological cleverness outstrip our basic understanding of the world we live in. What is needed is not a return to the rustic past but a greater understanding of how nature works. The recent cutbacks in public support of research are a shortsighted economy; basic research needs vigorous support, especially in those areas where insight is needed to correct present errors.

Pollution. We have already noted some of the effects of pollution on ourselves and our environment. During the course of this text we shall refer to specific cases in detail. The production and dispersal of pollutants is, quite obviously, a major weakness in our present system; it derives from our habit of treating Earth as an infinite resource. It is time to begin counting, in hard cash, the cost of restoring the health of our spaceship.

REGAINING CONTROL

We must change direction, set new priorities, find new goals. It seems unlikely that either technological or governmental institutions will take the initiative—nor, in a democratic society, should they. In theory, they are instruments of the public will; in practice, due to our abdication of control, they are currently autonomous. It is therefore up to the large, amorphous mass of citizenry to take the lead and bring about change.

Individual Action. Unless enough individuals want a change, nothing will happen. Redirection of cultural attitudes away from mindless consumption and toward an awareness of man's role in the world as a whole, requires an education to alternative life styles and purposeful planning (see, for example, Ian McHarg's *Design with Nature*). Once alternatives become apparent, rational choices are possible. Technology today offers little choice: there is really no significant basis upon which to choose among 25 almost identical cars, or among a half-dozen commercial television channels all showing equally poor programs. A real choice would consist of a distinct variety of transportation facilities—walking, cycling, rapid transit—and a variety of leisure activities—film clubs, dance festivals, painting classes—and so forth. To a limited extent, such choices are already available, and public demand could greatly increase them. Specific suggestions on what to do to improve the environment on an individual basis are to be found in paperbacks such as *The Environmental Handbook* and *Ecotactics*.

Collective Action. Despite the apparent autonomy of technology and government, the possibility of public action exists. In the early 70's, consumers' grape boycotts in California and throughout the country forced growers to recognize farm-workers' unions and so ended an exploitation of a virtually captive labor force. (Although this may seem a purely sociological problem, there are environmental ramifications. Many grape-pickers suffered poisoning from toxic pesticides, a situation they can now protest through exercise of collective action.) As a result of the boycott of earlier detergents by housewives, almost all soap manufacturers now market "non-polluting," phosphate-free products. Although some of these

---

*Perl, Martin J. 1971. The scientific advisory system: some observations. Science *173*:1211–1215.

contain chemicals which may prove to be almost as environmentally un-desirable as phosphates, the example demonstrates the power of the public to effect technological change. Ehrlich in his books, and Nader and Ross, in *Action for a Change,* offer further suggestions on opportunities for extra-governmental organization to bring about desired changes.

Even today's governments are not totally unresponsive to public pres-sure when it is strong enough, although the initial time-lag can be exasper-atingly long. For example, legislation regarding automobile exhaust emis-sions, now enacted in several states, has been a political response to collective action. So, too, is the gradual liberalization or abolishment of anti-abortion laws in various states. Governments, however, cannot legis-late, even for desirable changes, unless there is public support. India's fail-ure to control her population growth, despite widespread government campaigns, is a case in point: sufficient cultural demand for such a program to be successful simply does not yet exist there. Even the most enlightened government cannot, by itself, bring about change.

TOWARD NEW GOALS

We are caught upon the horns of a dilemma: much of today's technol-ogy seems to produce bad consequences, but we can no longer live without it. Although science is a primary prerequisite of technological innovation, it is not itself the villain. We are inclined, as Professor Medawar says, to "give too much thought to the material blessings or evils that science has brought with it, and too little to its power to liberate us from the confinements of ig-norance and superstition."* Science has taught us that it is not necessary to pray for the sun to come up every morning. It could also teach us, would we but let it, that man can live a rich, varied and rewarding life without exterminating himself in the process, and further, that this life can be sup-ported by the intelligent use of technology.

We exist, at present, in a cultural trap of our own creation. It is impos-sible to correct all our errors in one fell swoop, since to do so would bring about our total collapse — we all depend too much on contemporary tech-nology. Fortunately, in a world society of 3.7 billion, an immediate change-about is impossible; the inertia is too great. Yet our survival demands a reasonably rapid and concerted change. Technology, even with the best of intentions, cannot at the moment supply all the answers. But answers will come with the application of new technologies which are coupled with an increased understanding of how our spaceship functions. We must learn to pay the same respect to our "biotic constitution" as we do to our political constitution. Eventually, the answers to our problems must be found through the application of biological principles which treat our environ-ment as a total system rather than as a series of unconnected compartments.

In this text we shall not only present biological principles, but point out how they relate to some of our current problems — both those which per-tain to the environment and those affecting the health and well-being of man. The guidance of future technology must be determined by rational decisions which can only be made by an informed public. It is to this end that this book is directed.

Carson, R. 1962. *Silent Spring.* Houghton Mifflin, Boston. A highly readable layman's account of the environmental impact of pesticides.
Commoner, B. 1971. *The Closing Circle.* Alfred A. Knopf, New York. One ecologist's views on environment, technology, economics and politics.

*Medawar, P. B. 1967. *The Art of the Soluble.* Methuen, London, p. 15.

De Bell, G. 1970. *The Environmental Handbook.* Ballantine Books, New York. A collection of essays and useful ideas, prepared for the First National Environmental Teach-In, April 22, 1970.

Dubos, R. 1968. *So Human An Animal.* Charles Scribner's Sons, New York. A well-known bacteriologist interweaves the biological and spiritual needs of man in a plea for humanity.

Ehrlich, P. R. and Ehrlich, A. H. 1970. *Population, Resources, Environment: Issues in Human Ecology.* W. H. Freeman & Co., San Francisco. A comprehensive account of our major biological problems.

Ehrlich, P. R. and Harriman, R. L. 1971. *How to Be a Survivor: A Plan to Save Spaceship Earth.* Ballantine Books, New York. Describes global problems in terms of a lack of human controls and suggests solutions.

Frazer, Sir J. G. 1922. *The Golden Bough* (2 volumes). MacMillan, New York. A monumental treatise cogently arguing the succession of magic by early religion.

Galbraith, J. K. 1971. *The New Industrial State.* 2nd ed. Houghton Mifflin, New York. Describes the technological system in America, including creation of demand.

McHarg, I. L. 1969. *Design with Nature.* Natural History Press, New York. A stimulating handbook which explains how urban planning can bring harmony between man and nature.

Mead, M. 1970. *Culture and Commitment: A Study of the Generation Gap.* Natural History Press, New York. A foremost anthropologist shows that today's generation gap is qualitatively different from that in the past and can lead to rapid cultural change.

Medawar, P. B. 1967. *The Art of the Soluble.* Methuen, London. A collection of thoughtful essays about science and scientists.

Nader, R. (ed.) 1970. *Ecotactics.* Pocket Books, New York. Suggestions for environmental activists.

Nader, R. and Ross, D. 1971. *Action for a Change.* Grossman, New York. A student's manual for organizing public interest groups.

Reich, C. A. 1970. *The Greening of America.* Random House, New York. Describes the emergence of a new culture with ideals beyond technology and the corporate state.

White, L., Jr. 1967. The historical roots of our ecological crisis. Science *155*:1203–1207. One historian's view that the Judeo-Christian ethic encourages technological exploitation of nature.

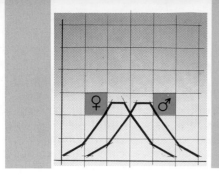

SCIENTIFIC PREDICTIONS
AND HUMAN AFFAIRS

In the process of explaining nature, scientists discover principles on the basis of which predictions can be made. Such predictions must be used in making decisions about human affairs far more frequently than they are today if we are to steer our spaceship on a sane course.

The word science comes from the Latin *scire,* "to know," and may be broadly applied to any field of systematized knowledge. In its more restricted sense, it refers to the natural sciences, of which biology is one. The scientist's subject is the study of the natural world; his job is to explain its phenomena. In this chapter we shall attempt to describe how scientists arrive at statements about the world of nature and how these statements, or principles, are useful in making predictions in the realm of human affairs.

HOW SCIENTISTS FIND OUT THINGS

Much has been written about how scientific discoveries are made—about what a scientist does when he is being a scientist. There is probably no one answer which covers all cases, although detailed descriptions purporting to explain the Scientific Method are common enough. Scientists come in all shapes and sizes and are found in diverse cultures. There is no such thing as The Scientific Mind, and the varied methods used by most scientists in making discoveries are often as obscure to themselves as to the non-scientist. To some extent we all use scientific methods in our daily lives.

In studying some part of the natural world, a scientist often comes across a problem which intrigues him and which he believes he can solve. To formulate the problem better, he may make further observations and may read the published findings of others. At some point he attempts an explanation of the phenomenon at hand. It is at this stage that imagination plays a primary role. The attempted explanation is clearly a guess, or, put more formally, a **hypothesis.**

Hypotheses are among a scientist's more useful intellectual tools. They are not merely random guesses, but are likely descriptions of "what might be true," to use an expression of the English biologist P. B. Medawar. A hypothesis is a possible model or picture of the real world. Without a hypothesis, a scientist could only amass facts which would serve no purpose. In formulating his hypothesis, a good scientist—one who may eventually succeed in explaining some aspect of nature—is careful to see that it meets certain criteria:

1. The hypothesis must explain what is known about the phenomenon at the time—it must be consistent with the facts.
2. The hypothesis must be capable of predictions about observations not yet made.
3. The hypothesis must be capable of being tested—it must have the potential for being proved false.

A hypothesis which cannot, by its nature, be tested and perhaps proved false does not belong to the realm of science. For example, the hypothesis that God exists is not amenable to testing—it cannot be proved, one way or the other, and hence lies outside of scientific inquiry.

The function of a hypothesis is to direct the scientist's next efforts. The story of what might be true is only the beginning. If the hypothesis is on the right track, it should predict that certain things will be consistent with it. At this point, the scientist sets about testing his hypothesis, hopefully realizing that it is only one of many possible explanations. He tries to obtain proof concerning the correctness of his idea, using experiments or observations designed to test its predictions. The strongest support a hypothesis can have comes from the failure of tests consciously designed to prove it wrong to do so. During this stage, the scientist employs deductive argument—if such-and-such (the hypothesis) is true, then it should follow that thus-and-so (the predictions) will also be true. If his further observations or experimental results do not agree with the predictions of the hypothesis, it must be discarded or modified. If, after rigorous testing, the hypothesis stands up, it may become accepted as a theory or scientific principle. Usually this occurs only after a number of scientists in different laboratories have subjected the same idea to a variety of tests, during which it may be modified and molded to fit new facts.

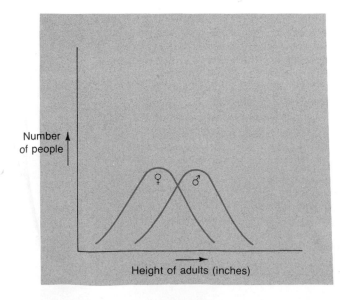

FIGURE 1a.1

The hypothesized distributions of heights among adult men and women. Men as a group tend to be taller than women.

It is important to realize that no scientific theory can ever be proved with 100 per cent certainty. All that is absolutely certain in science is that body of ideas which have been shown to be wrong. Nevertheless, it is convenient to assume that explanations which in human experience always appear to be true, are, in fact, very close to the truth; these we accept as "laws of nature."

Science is thus a mixture of imagination, on the one hand, and of rigorous proof and self-criticism, on the other. Its strength lies in the fact that, no matter who does a particular experiment, the results should be the same. It is for this reason that, in publishing their findings, scientists give details of their experimental methods; others can then repeat the experiments and confirm, or deny, their conclusions. Secrecy does not lead to scientific truth, a point to which we shall return later in the chapter.

HOW SCIENTISTS HANDLE DATA

Although we cannot at present explain how scientists come by hypotheses, we can describe some of the problems they face in interpreting observations or experiments designed to test their ideas. Let us, by way of example, take a very simple hypothesis and follow it through.

"HEIGHT IS RELATED TO SEX"

It may have struck you, from casual observation, that men generally seem to be taller than women—there appears to be a correlation between sex and height. You might formulate the hypothesis that, in human populations, height is related to sex. We can diagram your hypothesis to indicate that, in terms of height there are two separate populations, one male (σ) and one female (φ). See Figure 1a.1.

An alternative hypothesis is that there is only one population. See Figure 1a.2.

How might you go about testing your hypothesis? First of all, it would probably occur to you that a great many things could affect height. You are not asking, however, "What are *all* the factors which affect height?" but rather, "Is it dependent upon sex?" In this case, the **dependent variable** is height, and the **independent variable** is sex. If your original hypothesis is

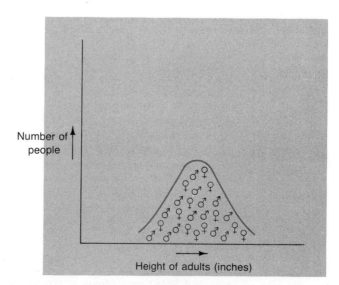

FIGURE 1a.2

An alternative hypothesis: There is no difference between men and women with respect to height.

true, then, given that a person is male (♂), he will be taller than if he had been female (♀).

A good test of a hypothesis excludes, as far as possible, all extraneous factors (or variables) which could affect the outcome. In your experiment, you wish to examine only the effects of sex on height. Let us consider some of the other factors which may affect a person's height. A deficiency of vitamin D leads to rickets, a disease characterized by improper calcification of bones. Until recently, this was a common disease in Scotland, and many older people living there today are shorter than they would otherwise have been, owing to their bowed legs. Diets low in protein can cause retardation of growth and small adult size. A deficiency of iodine in food affects the functioning of the thyroid gland and, if severe, leads to a type of dwarfism known as cretinism. Still another factor is the general health of an individual. Excessive illness in childhood may lead to short stature; so may developmental abnormalities (e.g., hunchback). Accidents may affect height—the severe fracture of both legs suffered as a child by the painter Toulouse-Lautrec resulted in his being a dwarf. A most important factor affecting height is the genetic variable. Tall parents tend to produce tall children, and short children are likely to come from short parents. Japanese men are seldom stars of basketball teams, and Swedish women are known for their majestic stature.

Were it possible to measure the height of everyone in the world in order to test your hypothesis, these differences would cancel out and you would be able to form a true conclusion about the effect of sex on height. But since this is not possible, you must select a **sample** of people to measure.

How should you choose the sample which will provide the data necessary for you to decide whether to accept or reject your hypothesis? Suppose you decide initially to measure a sample of 200 students drawn from a large population on a university campus. In choosing which students to measure, you will want to eliminate bias, or the tendency to measure only those people who are likely to fit your hypothesis. To do this, you might employ a lottery system. Each male student on campus is assigned a number; the same is done for the females. The numbers are placed in a drum and the first 100 drawn for each sex determine the students who will form the sample.

After choosing your sample and measuring your subjects, you will need to analyze and interpret the data. Suppose the data you obtain produce the points shown in Figure 1a.3A. On drawing smooth curves to fit as nearly as possible the points for males and for females, you observe

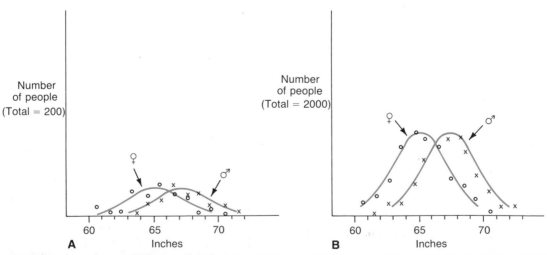

FIGURE 1a.3   Frequency data of heights of men and women on a college campus. *x*, number of males falling within each one-inch interval; *o*, number of females falling within each one-inch interval. Smoothed curves drawn to fit data.
   A. Data from a sample of 200 people (100 ♂ + 100 ♀). Peaks hard to distinguish.
   B. Data from a sample of 2000 people. Peaks easy to distinguish.

that, although there are a few very short men and a few very tall women, the two curves have peak (average height) values which appear to be separated. The question is: Are these two curves sufficiently distinct to allow you to accept your hypothesis that males are taller than females? The range of height within each group ($\male$ and $\female$) is so great that the answer is in some doubt. It is possible that the smoothed curves are not really separated at all — that in the smoothing process a difference has been introduced which is not really there.

CONFIDENCE IN THE RESULTS

At this stage, you may find yourself in a quandary: "Shall I accept or reject my hypothesis? The data are not quite conclusive and I do not have 'confidence' in them, one way or the other." Several possibilities may then occur to you to increase your level of confidence in the conclusion you draw from the data.

More Measurements. You might immediately decide that more measurements are called for. Were you to increase your sample size from 200 to 2000, you might obtain results such as those of Figure 1a.3B. The range of heights of males and females is almost the same as before, but the curves you draw now require much less smoothing. You have now obtained a more accurate estimate of precisely what the peak (average height) values of males and females are. Thus, the larger the sample size, the more accurate is your basis for accepting or rejecting the hypothesis. But perhaps you still lack confidence in the conclusions you can draw from your data.

Controlling for Extraneous Variables. We noted earlier that factors other than sex may affect height. These may be both genetic (inherited) and environmental (nutrition, culture, disease, accidents and so on). Since such extraneous factors will increase the variability in your sample, it is desirable to eliminate them if possible. It may occur to you to eliminate many of them by selecting only brother-sister pairs in your sample, since siblings usually have a similar genetic and environmental background. By doing so, you are "controlling" for these extraneous variables.

The sort of data you might obtain from measuring 1000 pairs of randomly chosen brothers and sisters, compared to your earlier measurement of 2000 unrelated students, are shown in Figure 1a.4.

The two peak (average height) values are now more sharply separated because much of the extraneous variability has been eliminated. You now have even greater confidence in your conclusion that, on the average, men are taller than women. It should be mentioned here that a scientist would not simply look at these curves, as we have done, and decide whether they are in fact the same curve or two separate curves. Instead, he would apply a formal mathematical treatment to his data, known as statistical analysis, to tell him whether to accept or reject his hypothesis.

Thus, both increasing sample size and controlling for extraneous variables gives a result in which greater confidence may be placed; your hypothesis may, in fact, be an accurate estimate of the true state of affairs. Upon achieving a certain level of confidence, you may decide to accept your original hypothesis as a scientific principle: Height is a sex-dependent characteristic and males tend to be taller than females.

There are, of course, many other ways in which scientific data are expressed and interpreted, but our example will give the reader a feeling for the way scientists handle data and draw conclusions. Before a principle is established and accepted by the scientific community as a whole, however, it is usually demonstrated several times by various investigators acting independently of one another.

USEFULNESS OF SCIENTIFIC PRINCIPLES

Once established, a scientific principle is used to make further predictions. For example, your conclusion, based only on data obtained from a student population at one university, might be extended to predict that men everywhere tend to be taller than women. Without further testing, however, such an extrapolation would be unwise—there still may be, after all, populations of Amazons extant in the world!

When a principle is accepted, it leads to further questions, requiring new observations, hypotheses and tests, which lead to still more predictions. One question among the many that may arise from our example is, Why are men taller than women? To satisfy the curious, since we shall not pursue this topic further, the answer follows: The reason for the height difference of men and women lies primarily with the effect of sex hormones, released in large quantity during adolescence, on the growth of bone and muscle. The bones of women stop elongating at an earlier age than those of men, owing to the presence of female estrogens. The male hormone testosterone, on the other hand, causes an increased rate of bone and muscle development during puberty.

The usefulness of a scientific principle thus lies in its ability both to make accurate predictions about the world and to serve as a basis for further inquiry into natural phenomena. Virtually all of the scientific principles discussed in the text have developed from the repeated application of processes similar to those we have been considering.

SCIENTIFIC PREDICTIONS ABOUT MAN

Let us briefly consider some of the problems which arise in the scientific study of human beings. In studies on animals it is relatively easy to control extraneous variables such as heredity, nutrition, age and sex. The experiments on the induction of bladder cancer in rats fed cyclamates—which resulted in the withdrawal of that sugar-substitute from the market by the Food and Drug Administration—used inbred, genetically homogeneous strains of animals, all born at the same time and raised under identical conditions. The animals were thus genetically and environmentally homogeneous. The scientist cannot manipulate human popula-

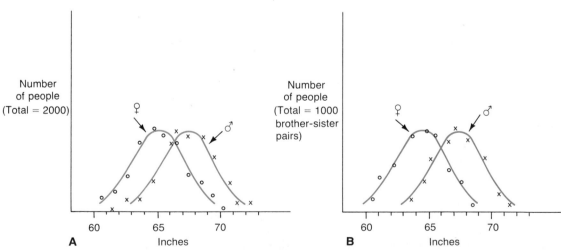

FIGURE 1a.4   Frequency data of heights of unrelated (A) and related (B) men and women on a college campus. x, number of males falling within each one-inch intervals; o, number of females falling within each one-inch interval. Smoothed curves drawn to fit data.
   Note that the two peaks move apart slightly and are more readily distinguished when brothers and sisters are compared (B).

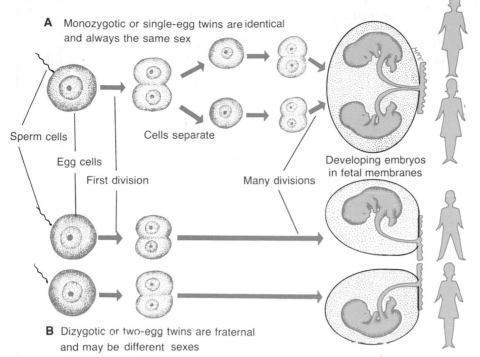

A Monozygotic or single-egg twins are identical
and always the same sex

Sperm cells

Egg cells

First division

Cells separate

Many divisions

Developing embryos
in fetal membranes

B Dizygotic or two-egg twins are fraternal
and may be different sexes

FIGURE 1a.5

Monozygotic versus dizygotic twins.
  A, Monozygotic (identical twins) are produced from a single egg and sperm.
  B. Dizygotic (fraternal twins) are produced from two separately fertilized eggs.

tions to suit himself, however. Even if it were possible to perform such dangerous experiments on human subjects, extraneous variables would require far larger sample sizes than are needed for studies on laboratory animals.

Human populations are always affected by cultural patterns over which the investigator has no control. Even such apparently homogenous groups as the Eskimos or Australian aborigines have strict totem rules about who may marry whom. These social factors affect reproductive patterns and, hence, the distribution of genetic attributes in the population. It is also not easy to control for such other, non-genetic differences between individuals as nutrition, amount of exercise, history of disease and so on, all of which may introduce extraneous variation and thereby affect the results. Therefore, when studying human populations, much larger sample sizes are needed than for laboratory animals, to take into account the uncontrolled variables. Unfortunately, the reverse is usually the case; most investigations on humans involve fewer individuals than do parallel studies on laboratory animals. Human behavior is especially difficult to study precisely because of the multitude of uncontrolled cultural variables involved.

It is possible to compare the relative contribution of genetic factors on the one hand and cultural factors on the other to the incidence of a particular characteristic by studying differences between identical (monozygotic) and fraternal (dizygotic) twins. Identical twins arise from the union of one egg and one sperm cell (Fig. 1a.5A). At the first division, the two cells do not stay together as usual, but separate to form two individuals who are genetically identical. Fraternal twins, however, arise from two separately fertilized egg cells (Fig. 1a.5B), and such twins are genetically no more alike than any pair of siblings (children having the same parents). Differences between monozygotic (one-egg) twins with respect to a given trait are therefore due only to cultural factors, whereas differences between dizygotic (two-egg) twins are due to both genetic and cultural factors. By comparing differences between a group of monozygotic and a group of dizygotic twins it is possible to assess the contribution of genetic factors to the trait in question—for example, frequency of cavities in teeth, or the ability to sing a given note when asked.

In certain types of situations where cultural and behavioral factors are insignificant, it is possible for scientists to make predictions about individual people based on information gained from previous cases. For example, certain diseases follow a similar pattern, no matter what the environmental background of the person. This is especially true of genetically inherited diseases, such as diabetes, Tay-Sachs syndrome (a degenerative brain disease in children) and so on. But for most human problems past experience offers at best only a vague guide as to the predicted outcome of a particular situation. In most illnesses, for instance, a doctor cannot predict accurately if and when a sick person will get well. There are too many variables, mostly unknown to the doctor, which determine this—and too many unknown variables existed in the past cases with which he is comparing the present one. Medicine is still largely an inexact science because of the great variation among patients. At best, it can offer only a rough prediction for a particular case.

SCIENTIFIC PREDICTIONS AND PUBLIC DECISIONS

What role do scientific predictions play in making political, social and economic decisions today? It is probably safe to say, very little. In many instances, not enough information exists to make reasonably accurate predictions; in other cases, predictions are ignored. A few examples will illustrate the point.

DISREGARD FOR ADVISORY SYSTEMS

As we saw in the last chapter, the usefulness of governmental scientific advisory committees, whose function is to make predictions about the results of political decisions, is often vitiated because advice is ignored. The secrecy which all too frequently surrounds the decision-making process increases the chance of a wrong decision being made. One such case is the decision taken in 1962 to detonate nuclear devices in the Van Allen radiation belt. The events are well detailed by Dr. Barry Commoner in his book, *Science and Survival.*

In the upper atmosphere there are two naturally occurring regions of radiation-emitting particles held in position by the Earth's magnetic field. These regions were first discovered by scientific instruments carried on board ballistic rockets developed after World War II; they are known as Van Allen belts. It soon became obvious that these belts were important in long-range radio communications and therefore were of military significance. The United States government proposed to explode nuclear devices within these belts to discover how they could be disrupted if military need arose. Many scientists at research institutes throughout the world protested, claiming that the projected explosions would have a long-lasting effect on the belts and might harm astronauts and research satellites. The government asked a team of its own classified scientific advisers for their secret opinion, which was that the effects would last only a few weeks at most; the nuclear devices were duly exploded in 1962. It is now public knowledge, admitted by the United States government, that the effects will in fact last for at least 30 years, interfering with further scientific study of these belts and their natural radiation during that time. Not only will astronauts and research satellites passing through the belts be affected; this is also near the region where supersonic transport planes will be flying, so there will be a possibility of effects to passengers and crew.

The government, by forcing its advisers to hold their debates in secret, deprived them of discussions with better informed scientists throughout the world. Although information was available at the time which could have provided a more accurate prediction, secrecy caused that information to be ignored.

The conscious exclusion of information and ignoring of predictions is practiced not only by the American government but by others as well. Governments, the military organizations they support and industries are all guilty of ignoring the environmental and public health consequences of their actions.

DECISIONS BASED ON INSUFFICIENT PREDICTIONS

In many cases, decisions are made before sufficient information is available to provide an accurate prediction about the outcome. Frequently, the right questions are not even asked before the decision is taken. For example, the decision to use DDT was made by many governments shortly after its insecticidal properties were discovered in the late 1930's; its potency against insects and low toxicity for man were clearly demonstrated and this was all the information thought necessary for deciding that it was safe to use. No one even thought to ask the question, Will this substance have environmental side effects?

A similar oversight was made in the case of the tranquilizer drug thalidomide. Its low toxicity and lack of side effects in adults caused many governments to license its use. Because this substance was also shown to be safe when used at low concentrations on pregnant laboratory animals, it was assumed that it would have no effect on unborn babies. No one thought to ask, Is the safety margin great enough? The debilitating deformities came as an ugly shock.

One could cite numerous other instances where decisions have been made which predicted one outcome and where, because information was incomplete, a different outcome actually occurred. As more holistic approaches are used in making decisions, failure to ask the right questions should decline in frequency. But there will still be times when decisions are made where it is known that not enough information exists to accurately predict the outcome. We shall return to this point later.

IMPROVING FUTURE DECISIONS

Much more accurate predictions could be obtained about the likely outcome of decisions if both government and industry were required to act publicly rather than secretly. This would serve two functions: first, it would insure that the long-term public interest was not being sacrificed to special interests through suppression of information and deliberate disregard of predictions; second, it would allow a far greater quantity of information to be made available in the course of public debate.

Although at first glance this might seem to make the decision-making process unwieldy, the application of computers offers a powerful tool for analyzing massive amounts of information. Such computer models, as they are called, are also capable of determining how accurate a prediction is likely to be and of assessing what further information would be of greatest value in improving the accuracy of a prediction.

A recent study by an international group of scientists at the Massachusetts Institute of Technology will serve as an illustration of the potential value of computer modeling. Using currently available information about trends in world population, resource utilization, technological development and pollution, the team constructed a global model designed to predict what will happen on Earth in future years. Their admittedly grim prediction is that, without a major change in direction, there will be virtually total economic collapse in just 70 years, leading to worldwide famine and disease (Fig. 1a.6). By making hypothetical decisions which are alternatives to simply following present trends, the model is able to show what actions will affect the predicted outcome. The conclusion is that only by stopping both population *and* economic growth can our present progress towards destruction be halted. The model thus poses the not very cheerful alternatives of global destruction, on the one hand, and the possibility of war and revolution if economic growth is curtailed, on the other.

Although this model may not be entirely accurate because it is based on today's incomplete information, it illustrates not only the magnitude of the decisions which face us but also the need for using such powerful models to determine our future course of action. The M.I.T. group, incidentally, did test different possible versions of the model to determine how errors in their available information might affect the outcome and found discouragingly little effect.

FIGURE 1a.6　Computer modeling allows predictions to be made about the future outcome of complex courses of action. Predictions based on available information about current trends are not overly encouraging.

It is all too evident that the magnitude of the decisions which face mankind is too great to leave decisions in the hands of special interest groups. The public must make its own decisions between alternative courses of action.

Any decision is taken because it is expected to have a certain predicted outcome. The likelihood of that outcome taking place is only as good as the probability that the predictions on which the decision is based are correct. It is an axiom of science that no prediction can be 100 per cent accurate, but, as we have seen, the more information there is available, the more accurate a prediction will be.

It thus follows that, in order to improve accuracy in predicting an outcome, more data are necessary. But the acquisition of data—the designing, carrying out and interpretation of experiments—requires money. And when the public is asking for a scientific prediction it must pay for obtaining the data. Since the scientist cannot provide absolute accuracy, it is up to the citizen to decide what degree of accuracy he is willing to pay for. In some cases, the citizen might be willing to pay the financial costs, but not the time costs involved in postponing a decision until the required experiments are made.

Once the public has obtained a scientific prediction about the probable outcome of a particular course of action, it must then go about making its decision, weighing the possible benefits against the possible risks. Decision-making rests with the public and not with the scientist. The scientist provides information and prediction, sometimes given as advice—but it is the citizen's responsibility to know the limits of accuracy of such predictions before acting on them.

Finally, what happens when scientists disagree? How will the uninformed layman, the average citizen, know which one to believe? This situation frequently arises during the course of expert testimony before a public body when the two sides present conflicting predictions. Those charged with the decision can do either of two things: they can go ahead and make the decision anyway, trusting the evidence of one expert over another (in which case, those making the decision are presuming themselves to have the ability to evaluate scientific facts); or they can seek further predictions, preferably from disinterested experts (which may delay the decision and add to its cost). Despite its drawbacks, the latter course will probably provide the citizen with a more accurate prediction.

In the course of this book, many new instances of lack of evidence about decisions which now face the citizen will come up. It will be the citizen's responsibility to determine how much evidence is "necessary" for him to make the decisions of the future.

SUMMARY

A scientist's job is to explain natural phenomena. After formulating an imaginative hypothesis of what might be true, he tests the hypothesis by observation and experiment, noting whether his results confirm the predictions of his tentative explanation. If they do, and if others agree, the hypothesis becomes a principle which permits further predictions and leads to new questions.

The decision about whether to accept or reject a particular hypothesis is made on the basis of factual information. In obtaining this information, the scientist attempts to exclude extraneous variables—he estimates the degree of confidence he places in the conclusions drawn from his data, which largely determines whether he accepts or rejects, or perhaps modifies, his original idea.

Scientific predictions about man himself are made difficult by the numerous uncontrolled variables inherent in studying human populations. On the other hand, application of scientific predictions to public decisions is of the utmost importance for the future survival of man. Errors of the past, resulting from ignorance or willful disregard of scientific predictions, are no longer acceptable. It is the citizen's responsibility to determine what degree of accuracy in predictions is necessary before decisions are made, as well as to weigh the benefits and risks to mankind in making a particular decision. The magnitude of our present and future decisions is so great and their component parts so complex that computer technology must be applied. The best way to obtain accurate predictions of the outcome of alternative courses of action is to seek information from demonstrably disinterested experts.

READINGS AND REFERENCES

Commoner, B. 1963. *Science and Survival.* Viking Press, New York.

Meadows, D. H., Meadows, D. L., Randers, J. and Behrens, W. W., III. 1972. *The Limits to Growth.* Universe Books, New York. The M.I.T. team's report of its predictions if present growth trends continue, with the interesting conclusion that life could be far better when growth stops.

Medawar, P. B. 1967. *The Art of the Soluble.* Methuen, London. The last two chapters, especially, give some very thoughtful insights into the nature of scientific processes.

Scientific American. September, 1958. An issue devoted to "Innovation in Science."

OUR PHYSICAL SURROUNDINGS

Our spaceship Earth is the stage upon which life is enacted. Each environment has its own physical and chemical properties which set the scene for a different group of actors, but the play itself is an integrated whole.

Life occurs almost everywhere on the surface of the Earth, even in the most improbable places. There are worms living at 9000 meter depths in the oceans, bacteria in snow banks at high altitudes and in the hot springs of Yellowstone Park. Living organisms everywhere interact with their non-living surroundings, and the nature of the inanimate world largely determines which organisms live where. The earth's surface can be conveniently divided into four major types of environments: marine, freshwater, brackish water and terrestrial. Brackish waters occur in bays and estuaries where fresh and salt waters mix, producing a fluctuating salinity. It is the physical environments within which life is lived that are the subject of this chapter.

ESSENTIAL CHEMICALS

All environments, whatever their special properties, have one common function—they are the ultimate source of the chemical *elements* which compose living organisms. The most important of these elements, together with their chemical symbols, are listed in Table 2.1.

Green plants are the main link between the inanimate world and all other living organisms; they take up minerals from water, soil and air, and

TABLE 2.1 COMMON ELEMENTS IN LIVING MATTER AND THEIR CHEMICAL SYMBOLS		
Carbon : C	Phosphorus : P	
Oxygen : O	Sodium : Na (for Latin *Natrium*)	
Hydrogen : H	Magnesium : Mg	
Nitrogen : N	Calcium : Ca	
Potassium : K (for Latin *Kalium*)	Chlorine : Cl	
	Sulfur : S	

convert them into the organic molecules characteristic of all living things. In a particular environment, the availability of the common elements in utilizable forms thus sets a limit on the rate at which green plants can grow and provide food for other organisms; this growth rate of plants within an environment is known as its **productivity.** One purpose of this chapter, then, is to examine the availability of essential elements in the major environments of the world.

CHEMICAL FORMS OF THE ELEMENTS

The elements necessary to life are found in a variety of chemical forms in nature. Some occur mainly as gaseous molecules, while others exist in the form of simple or complex **ions** — soluble forms of elements carrying positive or negative electrical charges. (If the reader is unfamiliar with chemistry, he will find a short summary of the nature of atoms, chemical bonds and simple molecules and ions in Appendix I.)

The mineral elements K, Na, Mg, Ca and Cl are all obtained by plants as water soluble ions: K^+, Na^+, Mg^{++}, Ca^{++} and Cl^-. Although sodium (Na) is often present in plants, it is not essential for the growth of most of them. Carbon is obtained by terrestrial plants as gaseous carbon dioxide (CO_2) and by aquatic plants as bicarbonate ion (HCO_3^-). Carbon dioxide gas readily dissolves in water (H_2O) and combines chemically with some of the water molecules to form a new compound, carbonic acid (H_2CO_3). The reaction is reversible, as shown by the two-way arrows:

$$CO_2 + H_2O \quad \rightleftarrows \quad H_2CO_3$$
$$\text{gas} \quad \text{liquid} \qquad \text{molecule in solution}$$

Carbonic acid in turn can split into two parts, hydrogen ion (H^+) and bicarbonate ion (HCO_3^-):

$$H_2CO_3 \rightleftarrows H^+ + HCO_3^-$$

The element oxygen must also be available to green plants; although they produce oxygen during daylight, they require it at night. It is obtained as molecular oxygen (O_2), either directly from air or dissolved in water. Animals also require oxygen in the same form. The hydrogen needed by plants is obtained from water itself, as we shall see in Chapter 8.

The two remaining major elements required by plants, nitrogen and phosphorus, are obtained almost exclusively in the form of complex ions — nitrate (NO_3^-) and ammonium (NH_4^+) ions, and phosphate ($PO_4^\equiv$) ion, respectively. Since the ionic forms of these two elements are often available only in very small quantities, they may limit the productivity of an environment.

THE ATMOSPHERE

The atmosphere is in contact with all the major types of environment, interacting with them and greatly affecting their ability to support life. It filters sunlight reaching the Earth, affects climate, and is a reservoir of several elements essential for life.

Of the four main gases in dry air (Table 2.2), all of which readily dissolve in the Earth's waters, oxygen and carbon dioxide are used directly by

TABLE 2.2 COMPOSITION OF DRY AIR

Nitrogen	: N_2	78	per cent
Oxygen	: O_2	21	per cent
Argon	: A	0.9	per cent
Carbon dioxide	: CO_2	0.03	per cent

plants. Although the nitrogen of the air does not react directly with most living organisms, it serves as a constant source of nitrogen which a few special bacteria can convert to forms useful to plants. Argon, however, is inert. The atmosphere also contains varying amounts of water vapor.

If all air on Earth were compressed to the same density as occurs at sea level, the atmosphere would be only five miles high. In fact, air gets thinner with altitude, and extends to a height of about 150 miles, as shown in Figure 2.1.

The atmosphere is divided into two major regions. The **troposphere,** which extends to a height of about seven miles, has a variable temperature. Over the oceans, the air temperature ranges from 0° to 16° C; the range over land is much greater — from −50° to +50° C or more. (For a rough conversion from the centigrade to Fahrenheit scale, multiply the temperature by two and add 30. The two scales are directly compared inside the front cover.) The troposphere is thus a region of strong air movements and cloud formation, in which all our "weather" occurs. The upper region, the **stratosphere,** is much more stable, with a narrow temperature range between −50° and −60° C; there are no clouds.

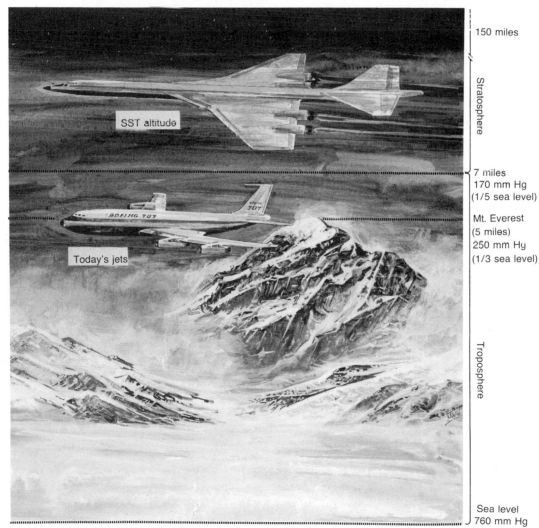

Divisions of the atmosphere. The density of air decreases with altitude. At sea level, air pressure is sufficient to force a column of mercury to a height of 760 millimeters (760 mm Hg). At the top of Mt. Everest, the air pressure is only one-third that at sea level (250 mm Hg). A man breathing at the summit obtains only one-third the oxygen he would at sea level with each breath.

FIGURE 2.1

The ultimate source of virtually all the energy on Earth is the sun, and this energy must pass through the atmosphere. Not only are weather and climate, which both affect living organisms, related to the energy received from the sun; life as we know it is directly dependent on sunlight. Solar energy drives the chemical reactions of **photosynthesis,** which make the growth of green plants possible—a subject to be covered in detail in Chapter 8.

MAN'S IMPACT ON THE ATMOSPHERE

Since changes in the atmosphere have effects on weather, climate and organisms, it is worth considering here a few of the ways in which man may affect the atmosphere. Other examples will become apparent later on.

The amounts of atmospheric dust or particulate matter are thought to be increasing because of man's activities (plowing, bulldozing, industry, logging, bombing). Even though volcanoes temporarily increase atmospheric dust, their effects usually disappear in four or five years. After the great 1963 eruption of Mt. Agung in Bali, however, atmospheric dust did not return to its earlier level as expected. In the intervening years, increasing amounts of man-made pollutants had presumably accumulated. A persistent rise in air particulates is expected to have a long-term effect on climate. By increasing the reflectivity of the atmosphere and decreasing the amount of sunlight reaching the Earth's surface, a cooling effect will result.

The stratosphere is the region in which it is proposed to fly the supersonic transport (SST) "above the weather." One objection to the SST is that the contrails (visible trails of condensed water vapor) it leaves behind may produce long-lasting clouds in the stratosphere, which, like dust, will decrease sunlight at the Earth's surface and produce a cooling effect. We know too little about the causes of climatic change on Earth to predict how great the effects will be; in addition, insufficient experiments have been carried out on contrails to permit prediction of their rates of formation and disappearance.

In Chapter 7a we shall consider some of the ways in which man may be having an opposite effect—that is, warming the atmosphere. Since the likely magnitude of each of these effects is not known, no one can predict what the net outcome will be.

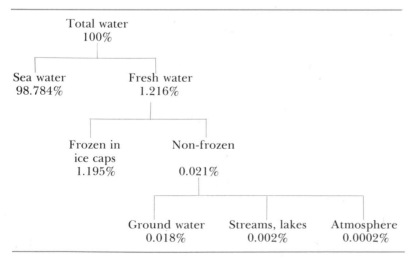

TABLE 2.3 DISTRIBUTION OF WATER ON EARTH* (Expressed as Per Cent of Total Water)

*Calculated from G. E. Hutchinson. 1957. *A Treatise on Limnology.* John Wiley and Sons, New York, Volume 1, Table 15, p. 223.

DISTRIBUTION OF WATER ON EARTH

Water, which composes 60 to 98 per cent of the weight of living organisms, is essential for the survival of life, and its quality and availability strongly affect the distribution of plants and animals. Despite what people living in the northwestern part of the United States or in tropical rain forests may think, very little of the Earth's water occurs in the atmosphere. Nor, relative to the total amount of water, is there very much fresh water; most of what there is, moreover, lies frozen in the ice caps or stored under the surface as ground water. Only a tiny fraction, as shown in Table 2.3, exists as fresh water in lakes and rivers.

THE HYDROLOGIC CYCLE

Water continuously circulates between the atmosphere and the Earth's surface in what is known as the *hydrologic cycle* (Fig. 2.2). The energy for driving this cycle and thus insuring a constant supply of fresh water on land comes from the sun. Solar heat evaporates water from the ocean. A lesser amount of water is also evaporated from the surface of the land and from plants, a process known as *evapotranspiration.* This process, as we shall see later, is of great importance to the survival of terrestrial plants. There is a daily net transfer of about 100 cubic kilometers of water vapor through the air from the oceans to the land, which is balanced by a return of the same volume, via rivers and ground water, back to the sea. When the volume turned over each day is considered, one becomes aware of why so much rain can fall even though the air itself contains very little water vapor.

ACIDITY AND ALKALINITY

One property of natural waters is their acidity or alkalinity. About one water molecule in every 10,000,000 splits into two, producing a hydrogen ion (H^+) and a hydroxyl ion (OH^-):

$$H_2O \rightleftharpoons H^+ + OH^-$$

In pure water, there are equal numbers of H^+ and OH^- ions; it therefore has a *neutral* reaction. Some natural waters, however, acquire an excess of H^+ and are *acidic*, while others, with an excess of OH^-, are *alkaline.* The particular units used by chemists to describe the degree of acidity or

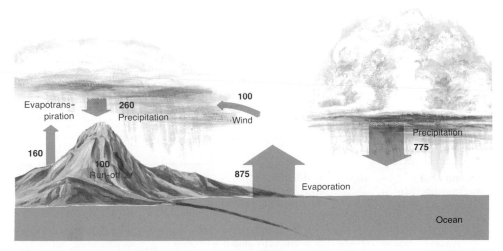

The hydrologic cycle. Units are in cubic kilometers of water per day (1 cubic kilometer = 0.24 cubic mile).

FIGURE 2.2

alkalinity of a sample of water, blood or any other solution are given in Appendix I.

THE MAJOR TYPES OF ENVIRONMENTS

Our interest in the various environments of the world lies mainly with those physical and chemical properties which are of greatest importance to the plants and animals which inhabit them—the availability of the various chemicals needed for life, the degree of saltiness, the temperature, and the acidity and alkalinity.

THE MARINE ENVIRONMENT

The oceans are vast, covering more than two-thirds of the Earth's surface; they are deep; and they are salty. Some geologists think the oceans may have always been very nearly as salty as they are now because of continual leaching of minerals from exposed rocks during freshwater run-off from land. Eventually, the minerals precipitate out onto the ocean floor, forming sedimentary rocks. This slow cycle of erosion and deposition keeps the oceans' mineral composition constant (Table 2.4).

Chemistry and Physics of Sea Water. The salts found in the ocean are all dissociated into positive and negative ions, of which sodium (Na^+) and chloride (Cl^-) ions, the components of table salt, are most abundant. Other elements essential to life are also present among the ions found—potassium, magnesium, calcium, sulfur and carbon. Phosphorus and nitrogen do not appear in the table, for the ions containing them—$PO_4^=$ and NO_3^-—are present only at very low concentrations in the sea. Thus, the oceans, despite their vastness, are relatively deficient in some of the elements essential for plant growth. It is for this reason that, compared to the land, the oceans as a whole have a low productivity.

The salinity or saltiness of the oceans is almost constant at all depths, except near river mouths. Many organisms living in the sea have a salt content in their body fluids similar to that of their surroundings and therefore experience no difficulties in maintaining a balance between salt and water in their body. Man and many other terrestrial and freshwater animals, however, have only one-third as much salt in their blood as is found in sea water. Such organisms are unable to utilize sea water for drinking, since they are unable to excrete the excess salt—a problem we shall consider further in Chapter 11.

Sea water is only slightly alkaline, and on this account poses no problems to the organisms living in it.

Dissolved oxygen is another biologically important component of sea water, since it is required by most animals and plants. The concentration of dissolved oxygen is expressed in milliliters of O_2 per liter of water; a liter is about one quart, and a milliliter, or 1/1000 liter, is about one-half teaspoon. (More exact equivalents are given inside the front cover.) Near the surface of the ocean, water is nearly saturated with oxygen from the atmosphere,

Sodium (Na^+)	10.6	Chloride (Cl^-)	19.0
Magnesium (Mg^{++})	1.3	Sulfate ($SO_4^=$)	2.6
Calcium (Ca^{++})	0.4	Bicarbonate (HCO_3^-)	0.1
Potassium (K^+)	0.4		

TABLE 2.4 MAJOR MINERALS FOUND IN SEA WATER (In Parts Per Thousand)*

*Parts per thousand may be expressed in any units one chooses: grams per kilogram, pounds per thousand pounds and so forth. For American equivalents of metric units, see inside front cover.

the concentration being 5 to 6 milliliters of O_2 per liter of sea water. Air, on the other hand, contains about 200 milliliters of O_2 per liter, or about 40 times as much per unit volume as in sea water. This means that animals living in the ocean must have either lower oxygen requirements or more efficient respiratory structures than their terrestrial relatives. In very deep waters, greater than 3000 meters, the oxygen concentration is only 3 to 4 milliliters per liter. In between the surface and the deepest regions there is an oxygen minimum zone, where O_2 may fall to 0.5 milliliters per liter, or 1/400 the concentration in air. Mixing in the ocean is very slow, and these zones are quite stable. Each zone has organisms specially adapted to live under the conditions found there.

The temperature of the oceans, compared to that of air, varies relatively little. The range in the open ocean is from 0° C (freezing point of fresh water) to 25° C (room temperature). Shallow tropical seas may reach 30° C, while the temperature near the poles and at great depths may fall to −1° C. The waters of the ocean have an almost unlimited capacity to exchange heat with the overlying atmosphere, thus producing a moderating effect on world climate.

Subdivisions of the Ocean. Despite the general uniformity of sea water, the marine environment is by no means the same everywhere. Some of its major subdivisions are shown in Figure 2.3. Scientists distinguish "blue water," or the open ocean, from the shallow coastal regions overlying the continental shelf. The latter may drop off sharply as a steep submarine cliff. There are also vertical divisions in the ocean. Light penetrates the upper layer of water only to a depth of about 200 meters. Green plants, which require sunlight, are limited to this region; in the open ocean only microscopic, single-celled plants, the **phytoplankton** (Fig. 2.4) are to be found floating in the water, while along the coastal margins the large, attached algae or seaweeds also occur. In the darkness below, a great variety of animals exist; living on the bottom or within the sediments, they are supported by food drifting down from the surface (Fig. 2.5). Those at great

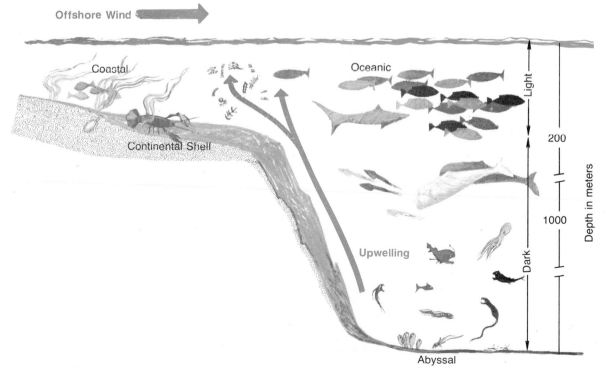

Divisions of the marine environment. Note that light penetrates only to depths of about 200 meters (a meter is slightly longer than one yard). Upwelling movements of water near coastlines are shown by solid arrows; the offshore wind which causes upwelling is shown by a broad arrow.

FIGURE 2.3

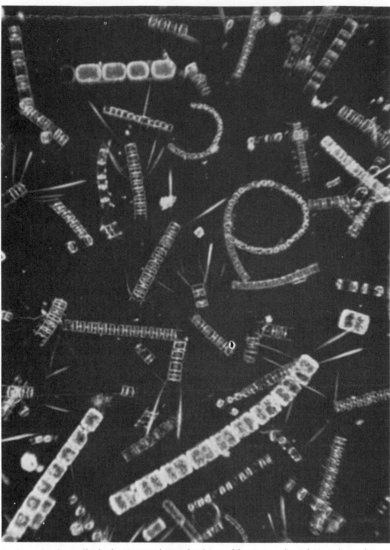

FIGURE 2.4

Living single-celled plants or phytoplankton. Most species shown here have mineralized skeletons. In several cases the single cells are linked together to form chains. (Courtesy of D. P. Wilson.)

depths, some of which resemble plants in appearance, are adapted to the high hydrostatic pressures.

The coastal regions are far more productive than the open ocean. Although they comprise only one-tenth of the ocean surface, more than a fifth of the sea's organisms are found there. The reason for this is that the nutrients phosphate ($PO_4^=$) and nitrate (NO_3^-), in scarce supply in the open ocean, are fed into coastal waters by run-off from the land. In some regions a phenomenon known as **upwelling** also occurs. As dead organisms fall to the bottom they are decomposed by bacteria, which release $PO_4^=$ and NO_3^-. Whenever offshore winds push surface waters out to sea, the deeper coastal waters upwell to replace them (see arrows in Fig. 2.3), bringing with them the nutrients required for plant growth.

THE BRACKISH WATER ENVIRONMENT

Although bays and estuaries are a small part of the total Earth surface, they are of great importance to man (Fig. 2.6). Like the coastal zone of the ocean, they are regions of high productivity, being rich in dissolved nutrients from river waters; they are thus a major source of seafood. Bays and estuaries are also utilized as a recreational amenity by large human populations, at least where they are not so polluted as to be a menace. They undergo daily salinity fluctuations, the magnitude of which depends on the rate of freshwater inflow and tidal exchange. There may also be seasonal fluctuations, with low salinities in the winter and spring when freshwater run-off is greatest. Only certain groups of plants and animals can tolerate these fluctuations—those which have evolved special mechanisms for withstanding them.

THE FRESHWATER ENVIRONMENT

The composition of various bodies of fresh water on the Earth's surface is by no means identical. Of prime importance is their content of dissolved solids or salts, leached from soil and rocks. Most fresh water passes through soil as ground water before reaching streams, so its salt content depends both on the time the water is in the soil and on the type of soil through which it flows. The total amount of dissolved solids found in fresh water affects its quality. Low values fall in the 100 to 200 ppm range; high values, as found in the lower Colorado River, reach 700 to 1000 ppm, or five times as much. (ppm = parts per million, usually expressed as milligrams per kilogram—or liter—of water.)

Hard, Soft and Alkaline Waters. The type of salt dissolved in fresh water is also important and depends on the geology of the drainage area. Granite soils and soils containing much humus, as the floors of forests, release excess amounts of hydrogen ions (H^+) into the waters draining them. These waters are markedly acid, and contain low levels of Ca^{++} and Mg^{++} ions; since they do not precipitate soaps, they are known as "soft" waters. Rain water is also soft because of the absence of Ca^{++} and Mg^{++} ions. Soft waters are preferred by housewives because they taste better and make suds for washing; but because they are usually poor in plant nutrients they are not favored by farmers.

There are two types of alkaline waters. Those which have an excess of Ca^{++} and Mg^{++} ions, leached out from limestone, are known as "hard" waters. Those which contain excessive amounts of Na^+ and K^+ ions, on the other hand, are the familiar alkali or "bad" waters of desert regions. Moderately alkaline waters usually contain most of the elements required by plants, and are thus useful for agriculture. Highly alkaline waters, however, are toxic and unsuitable either for irrigation or for drinking.

Nutrients in Fresh Waters. Given that a freshwater environment, such as a lake or river, is neither too salty nor too acid or alkaline, its ability to

FIGURE 2.5

Photograph of ocean bottom at 1200 meters off the southern California coast, taken with a special camera. No plants occur in these dark waters. Animals visible include brittle starfish and sea cucumbers; the conical humps are probably made by burrowing worms. (From Odum, E. P. 1971. *Fundamentals of Ecology.* W. B. Saunders Co., Philadelphia.)

FIGURE 2.6

An estuary in Georgia, with typical tidal creeks, mudflats (middleground) and salt marshes. Marsh grass, *Spartina,* is especially adapted to this type of environment. (From Odum, E. P. 1971. *Fundamentals of Ecology.* W. B. Saunders Co., Philadelphia.)

support life depends on the amount of nutrients dissolved in it. As in the ocean, the limiting elements in freshwater systems are usually nitrogen (N) and phosphorus (P). Nitrogen, as we shall see in Chapter 18, comes mainly from the air by a circular route involving bacteria; it does not originate from the rocks of the Earth's crust. Phosphorus, however, comes from the soil in the form of phosphate ion ($PO_4^=$) and is often the limiting nutrient in freshwater environments. It is usually low in mountain lakes and high in enriched lakes; it is commonly a factor in the over-enrichment or **eutrophication** of lakes polluted by man (see Chapter 2a).

As in the oceans, nutrients are not uniformly distributed in bodies of fresh water, nor is the amount of dissolved oxygen nor the temperature uniform. In lakes, mixing of waters at various depths is slow, although not nearly as slow as in the ocean. In temperate lakes, for example, more or less complete mixing occurs twice a year, in spring and fall (Fig. 2.7). Mixing rates in rivers and streams are far more variable and usually somewhat faster than in lakes.

A study of Figure 2.7 reveals the cyclic nature of the productivity of freshwater lakes in temperate regions. In the summer, warm surface waters of the **epilimnion** float on top of the cold, dense waters of the **hypolimnion.** At first the plants grow rapidly, but gradually the nutrients required for their growth, especially NO_3^- and $PO_4^=$, are used up. Some of the microscopic phytoplankton near the surface and the larger shore plants die and sink to the bottom, together with wastes excreted by animals; the nitrogen and phosphorus bound within them are temporarily stored in the sediments, eventually to be recycled by bacterial decomposition. During the fall overturn, however, the presence of oxygen in the water keeps most of the nutrients bound to the sediments, and only during winter are they gradually released into the water. Following the spring overturn, both nutrients and oxygen are dispersed throughout the lake, setting the stage for a renewed spurt of growth.

Problems Faced by Freshwater Animals. A freshwater lake poses special problems for animals. In addition to the cyclic nature of both temperature and food supply, the changes in oxygen concentration affect animals. Cold water, 0° C, can dissolve but nine milliliters of O_2 per liter, and warm water, 15° C, can dissolve only seven milliliters per liter; thus animals and bacteria living in the stagnant hypolimnion gradually deplete its oxygen supply during the summer. This deeper part of a lake may become restricted to only those animals with low oxygen requirements. This is particularly true of polluted lakes, as we shall see in Chapter 2a. Only the epiliminion, whose oxygen is renewed from the air and by its plant life, maintains a high O_2 level throughout the summer.

Animals are also affected by the low salt level in fresh water. The blood of most freshwater animals contains about 30 to 75 times more salt than occurs in their environment, and this salt tends to be leached out through the skins of the animals. It is therefore necessary for such animals to perform work to pump salts back into their blood, concentrating them from the dilute environment. This adaptation is a characteristic property of animals living in fresh water.

An interesting corollary to this is that "soft" waters, low in calcium ion (Ca^{++}), are often impoverished in animal life. The reason is that Ca^{++} is necessary to prevent leaching of salts from the body of an animal and also may be required for the pumping process.

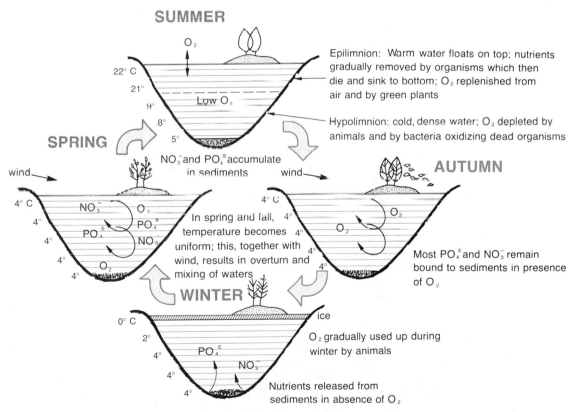

Layering and mixing of waters in a temperate lake. Temperatures are given in degrees centigrade. The nutrient cycles have been considerably simplified for clarity. (After Kormondy, E. J. 1969. *Concepts of Ecology.* Prentice-Hall, Inc., Englewood Cliffs, N.J.)

FIGURE 2.7

THE TERRESTRIAL ENVIRONMENT

The terrestrial environment supports both organisms living entirely on the surface or in large underground burrows and organisms which come in intimate contact with the soil. Surface-dwelling organisms have air as their major environment, and air has two properties which directly affect them. First, oxygen is readily available, since one-fifth of the atmosphere is O_2. Second, the low water content of the atmosphere poses a major problem, since plants and animals tend to lose water and dry out. Water constitutes about 70 per cent of the weight of animals and about 90 per cent or more of the weight of plants. Preventing desiccation is a major problem. Terrestrial plants have waxy coverings on their leaves to slow down water loss, and they draw water from the soil to balance their losses to the air. Animals have also evolved water-impermeable coverings, such as those found in insects and lizards, for example. Human skin is covered by a hard, water-impermeable protein called **keratin**; the greatest danger to extensively burned patients, who have lost much of this protective covering, is desiccation.

Chemistry and Physics of Soil. Organisms living in intimate contact with soil face rather different problems from those living on the surface, and indeed, it is the soil itself which is the most important physical part of the terrestrial environment and which determines its productivity. We tend to think of soil as being just "dirt," but it is in fact a very complex system, composed of three phases: solid, liquid and gas (Fig. 2.8).

The solid phase of soil contains both minerals, derived from the erosion of rocks, and organic matter, the decaying remains of plants and animals. The liquid phase is composed of water, in which the minerals and organic substances dissolve. The gaseous phase is air, whose oxygen must

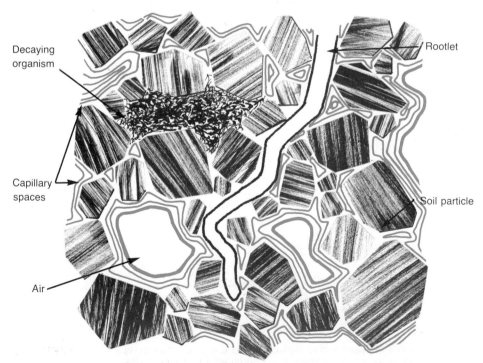

Decaying organism

Capillary spaces

Air

Rootlet

Soil particle

FIGURE 2.8   Diagram of microscopic nature of soil. Capillary spaces between the soil particles contain water and air. Minerals from soil particles and decaying organisms (humus) are eventually taken up by plant roots.

be present in soil for the metabolism of plant roots and other organisms. The rate at which oxygen can dissolve in and diffuse through soil water is far too slow to supply the oxygen needs of most terrestrial organisms. It is thus possible for plants to die in a flooded field because their roots literally become asphyxiated—most plants require a porous, well-drained soil. A reference summary of the elements required for the growth of terrestrial plants, the form in which they are supplied and their source is given in Table 2.5.

As shown in the table, almost all the mineral nutrients derived by plants from the solid phase of soil are products of the rock itself, and hence the mineral composition of a soil largely affects the types of plants it will support. The one exception is nitrogen; the peculiar cycle of this element is considered in detail in Chapter 18.

The abilities of a soil to hold water, to prevent excess leaching of minerals and to provide air spaces are all important in determining the productivity of a particular terrestrial environment. We can recognize three layers in most soils (Fig. 2.9). The quality of the topsoil is of great importance to plants. It varies from a fraction of an inch to several inches in depth. If it is too fine, it will hold too much water and not enough air in the capillary spaces between the solid particles. This explains the practice of adding coarse particles like sand to lighten heavy soils. Excessive plowing, especially with heavy machinery, pulverizes and compresses topsoil, making it more compact and less porous. Humus is also important to the quality of topsoil, since it provides a natural adsorbent for some of the mineral nutrients and prevents their being washed away by rain or irrigation water. Concentrated farming may cause a decrease in the humus content of soils and a loss of ability to retain nutrients.

Soil Aging. The gradual aging of soil is a natural phenomenon. Young soils, freshly formed by erosion of bedrock, readily release their nutrient minerals. These nutrients are either taken up by plants or else lost in the ground water, eventually reaching the ocean. In natural environments, most nutrients are taken up by plants, which eventually die, decay, and return most of the nutrients to the soil. Only a small fraction is lost in surface run-off or ground water. With time, the soil gives up its nutrients at a slower

TABLE 2.5 PRIMARY SOURCES OF ELEMENTS REQUIRED BY TERRESTRIAL PLANTS*

	Element	Symbol	Form Supplied	Primary Source
Air	Carbon	C	CO_2	atmosphere
	Oxygen	O	O_2	atmosphere
Soil				
Gaseous	Oxygen	O	O_2	atmosphere
Liquid	Hydrogen	H	H_2O	rain, ground water
Solid	Nitrogen	N	NO_3^-	recycled from decayed organisms
Major	Calcium	Ca	Ca^{++}	rock
nutirents	Magnesium	Mg	Mg^{++}	rock
(required	Potassium	K	K^+	rock
in large	Phosphorus	P	$PO_4^{\equiv}$	rock
amounts)	Sulfur	S	$SO_4^=$	rock

*Several minor nutrients, not shown here, are obtained from soil particles.

and slower rate. The most readily soluble fraction of minerals, present in young soils, disappears, and the more tightly bound minerals are only slowly released.

Agriculture hastens the rate of soil aging. After harvesting, the ground is temporarily bare; this increases the rate of run-off of rainwater, which carries with it valuable dissolved minerals. The harvested plants themselves represent another loss; nutrients and humus which would have returned to the soil are removed, and unless organic wastes are returned as compost and manure, the soil rapidly ages. Artificial fertilizers, applied to soils depleted of natural nutrients, are often used in quantities far in excess of those actually needed by the crops. The absence of humus precludes binding of the inorganic ions in the soil, and much fertilizer is lost in the run-off of rainwater or irrigation water. Only by taking steps to maintain humus is this problem overcome.

Soil Water. Water is distributed in several compartments within the soil. The water directly utilized by plants is that held in the capillary spaces, the tiny pores, between the soil particles (see Figure 2.8). Gravitational water is the excess water which cannot be held in the capillaries and runs down into the water table; it is only present in topsoil after a rain or after irrigation. The residual capillary water which remains behind is removed only by evaporation from the surface or by the roots of plants. The coarse, sandy soils characteristic of deserts are quite dry near the surface but retain considerable moisture underneath. After rain falls on the desert, water rapidly penetrates deep into the sand. During the day, many desert animals utilize burrows, where the air is cool and the soil moisture maintains a high humidity, thus slowing their rate of water loss.

So far we have not considered the ***water table*** and its role in the terrestrial environment. At some depth beneath the surface, virtually all soils are saturated with ground water. The upper level of this underground reservoir is known as the water table. We have already seen in Table 2.3 that about 90 per cent of liquid fresh water exists as ground water. Capillary water is only a small part of this. Most ground water occurs in a diffuse underground reservoir which moves exceedingly slowly through the subsoil, eventually entering rivers and flowing to the sea. It is the water which percolates into wells. It can also move upward into the topsoil, again exceedingly slowly, replacing capillary water lost by evaporation. Some of man's uses—and misuses—of this freshwater resource are discussed in our next chapter.

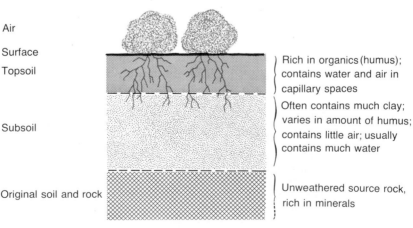

Air

Surface

Topsoil

Rich in organics (humus); contains water and air in capillary spaces

Subsoil

Often contains much clay; varies in amount of humus; contains little air; usually contains much water

Original soil and rock

Unweathered source rock, rich in minerals

FIGURE 2.9

Stratification of soil.

SUMMARY

For the survival of life on Earth, certain physical and chemical conditions must be met. Light, carbon dioxide, oxygen, water and a variety of mineral nutrients must all be present for the growth of green plants, upon which virtually all other life forms ultimately depend. Temperature also affects growth rates and hence productivity. Some environments are relatively deficient in one or another of these factors, and this becomes limiting to their productivity: oceans and fresh waters are low in mineral nutrients; much of the land is deficient in water; polar regions have very low temperatures.

It is further apparent that the various environments are linked with one another, especially by the hydrologic cycle. There are also significant interactions, which we have so far touched upon only briefly, between the physical environments of Earth and the organisms which inhabit them.

READINGS AND REFERENCES

Atwater, M. A. 1970. Planetary albedo changes due to aerosols. Science *170*:64–66. A research report calculating the amount of sunlight that is reflected into space because of dust in the atmosphere.

Clark, M. E., Jackson, G. A. and North, W. J. 1972. Dissolved free amino acids in the coastal waters of southern California. Limnology and Oceanography *17*:749–758. A research paper dealing with the addition of organic wastes into the coastal marine environment.

Hutchinson, G. E. 1957. *A Treatise on Limnology* (2 volumes). John Wiley and Sons, New York. A basic reference on the physical and biological properties of freshwater environments.

Kormondy, E. J. 1969. *Concepts of Ecology*. Prentice-Hall, Englewood Cliffs, New Jersey. A short and easily understandable introduction to ecological principles.

Peterson, J. T. and Bryson, R. A. 1968. Atmospheric aerosols: Increased concentrations during the last decade. Science *162*:120–121. A research paper describing recent increases in dust in the global atmosphere.

Sverdrup, H. U., Johnson, M. W. and Fleming, R. H. 1942. *The Oceans*. Prentice-Hall, Englewood Cliffs, New Jersey. A standard compendium of information about the marine environment.

2a

OUR FRESHWATER RESOURCE

Men really know not what good water's worth.

This apt phrase penned by Lord Byron in the early nineteenth century is even more true today.

Man, like almost all terrestrial organisms, requires a supply of fresh water to survive. If, like other animals, man used fresh water only for drinking and perhaps an occasional bath, there would be no water shortage in most inhabited areas today—streams, lakes and wells easily could supply these needs. But man's present-day water requirements involve many other uses which consume prodigious quantities of fresh water. For millennia, fresh water has been diverted from streams to irrigate fields. As cities grew, water was transported to them by aqueducts to serve not only for drinking and bathing but for removal of household wastes. It became the custom to dilute human excreta with many volumes of water to flush them away.

With the Industrial Revolution, the per capital consumption of water increased even further. In some instances, water is required at a particular stage in a manufacturing process, such as in pulp mills; in other industries, such as food processing, it is used to wash away unwanted by-products. In mines, water is contaminated unintentionally by the release of acids from minerals exposed at the mine face; the acids enter the ground water and render it unsuitable as a freshwater supply. Power plants also alter fresh water by using it for cooling purposes; although the water released is chemically unchanged, its increased temperature often adversely affects organisms living near the site of discharge which are physiologically adapted to lower temperatures. This special case will be considered further in Chapter 7a.

Several interacting factors must be considered in a discussion of man's use of fresh water. Obviously, as world population increases, the amount of water used will increase; also, as more and more countries become industrialized the per capita consumption throughout the world will go up. Moreover, rainfall is not distributed uniformly throughout the world; it varies from only a few inches per year in desert regions to several feet in rain forests (Fig. 2a.1). As populations continue to expand in dry climates, such as the southwestern part of the United States and the countries bordering the eastern Mediterranean, the disparity between water demand and natural water supply will continue to increase in these regions. Nor is rainfall constant throughout the year in most areas, though people use the same amount of water at all seasons, or more in dry periods if they have fields and gardens to water. Thus, not only the total supply of fresh water, but its distribution in relation to population is important. As more and more people concentrate in large cities, distribution will become increasingly more difficult.

There are at least three things to consider in determining priorities for use of our freshwater resource: the total quantity of water required for all purposes; the quality of the water to be used for each purpose; and the values man puts on the esthetic properties of the water in his environment. All three overlap to a considerable extent, and although the last factor is of least biological importance, it will probably play the major role in fashioning future decisions about how our water resource is used. It is the aim of this chapter to point out at least some of the biological principles which also should be taken into account in making such decisions.

THE QUANTITY OF WATER

It has been argued that there is plenty of fresh water in the world for all of man's needs for a long time into the future. All that is necessary is to move the people to the water (for example, the residents of Los Angeles to Vancouver Island—a possibility likely to be so unpopular that no govern-

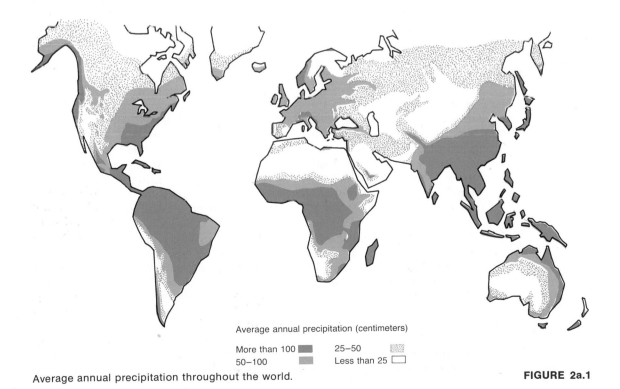

Average annual precipitation (centimeters)

More than 100   25–50

50–100   Less than 25

Average annual precipitation throughout the world.

FIGURE 2a.1

ment would seriously consider it), or to transport water to the people. Before considering the latter alternative, it is first necessary to know something of the distribution of the world's freshwater supplies.

THE DISTRIBUTION OF FRESH WATER

As noted in Table 2.3, only about *0.02 per cent* of the water on Earth exists in the form of liquid fresh water, and most of this lies underground. In fact, our knowledge of the actual amounts and rates of movement of underground water is very sketchy for most parts of the world. Yet we are likely to become more and more dependent on ground water as our water requirements increase.

The Fate of a Raindrop. Let us briefly consider the possible routes taken by a molecule of rainwater between the time it falls on land and the time it eventually returns to the sea. As soon as a raindrop forms, it begins to lose its purity. Dust and gaseous molecules in the air dissolve in the liquid drop. But unless it has fallen through the heavily polluted air of a big city, a raindrop reaches Earth in a reasonably pristine condition.

Where and when a raindrop falls plays a large role in its immediate fate. If it falls during a heavy storm on land already water-logged, it forms part of the surface run-off, finds its way to a stream and reaches the sea fairly soon. It may dissolve relatively few chemicals from the soil and remain nearly pure during its sojourn on land.

But if our raindrop is part of a brief summer shower, it is likely to be absorbed into the soil, where it may meet one of several fates. If the sun comes out quickly it may evaporate back into the atmosphere without reaching the sea. Or it may fall near the root of a plant, which quickly absorbs it and transports it to its leaves. There, if it is not incorporated into plant tissue, it again evaporates, through transpiration. **Evapotranspiration** — the evaporation from the surface layers of the soil and transpiration from leaves of plants — causes a significant loss of rainwater, which never reaches the sea. The hydrologic cycle is said to be "short-circuited."

If our raindrop is not returned to the air by evapotranspiration it gradually seeps through the surface layers of soil to join the ground water. There are two sorts of ground water — that found in underground streams or channels, called **aquifers**, which are usually very deep and transport water quite rapidly; and the interstitial ground water, which resembles water held in the interstices of a fine-meshed sponge. The "sponge" is formed by particles of soil which are far enough below the surface and the roots of plants so that evapotranspiration cannot occur. Interstitial ground water is present below the surface of virtually all dry lands. If one digs a deep enough well almost anywhere, water will be found. The upper level of ground water, as noted earlier, is known as the water table.

Just as streams and subterranean aquifers return water to the sea, so does a raindrop in the interstitial ground water eventually reach the sea. However, the rate of movement is much slower; if our raindrop enters the ground water of a humid region, it may move five feet per day, but in the desert it may move only five feet per year! Eventually, of course, if man does not intervene, it reaches the sea, but it dissolves out salts from the soil as it moves, and is far less "pure" than surface water.

Underground Water. It is at once apparent that movements of water underground are exceedingly slow, and that the rate of renewal is also very slow. Incorporated in the hydrologic cycle is a huge amount of ground water, which recycles at rates which have not yet been carefully measured. Ground water acts as a large reservoir in the cycle, helping to stabilize it.

Consider the diagram in Figure 2a.2. Part of the water flowing from precipitation and snow-melt at high altitudes flows rapidly via surface streams to the sea. Part of it enters the ground water reservoir and moves slowly downhill. Underground rivers or aquifers, where the water is channeled between impermeable layers of rock, supply springs and artesian wells which flow spontaneously when the water pressure is high enough. Water in aquifers, because it contacts few small particles, remains fairly "pure" and salt-free. Most underground water, however, enters the interstitial ground water, and dissolves out salts from the soil as it flows.

In the absence of man-made wells, there is a natural exchange between surface waters and ground water. A stream above the water table (stream A) will lose water to the interstitial reservoir, whereas a stream whose bed lies below the water table (stream B) will be fed by the water table. In fact, such a stream may run all summer, despite lack of surface run-off into it.

HISTORICAL METHODS OF OBTAINING FRESH WATER

Primitive man, like other animals, depended mainly on surface waters. As population pressures increased, people moved to areas where the water supply varied seasonally and built dams or dug wells.

Dams, Old and New. Dams act both to store water for use during the dry season and to prevent flooding during the rainy season. Unfortunately, like natural lakes, dams have relatively short lives. They become filled with silt. As we shall see in Chapter 19a, the silt which is deposited behind a dam deprives the alluvial flood plain downstream of its natural source of nutrients, and its soil becomes less fertile.

The numerous dams of modern man release clear, silt-free water downstream. In some cases, this results in erosion of the river banks, since the water, as it flows, tends to re-acquire a normal silt load. To prevent this,

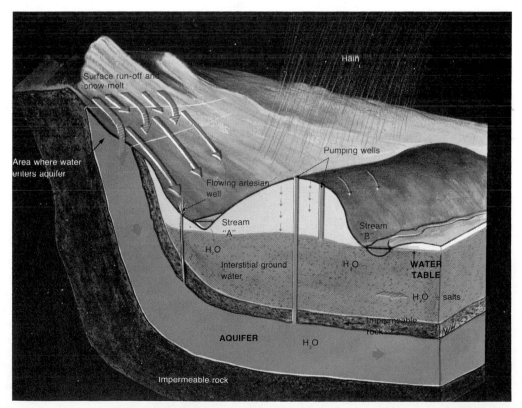

Relations between surface waters and ground waters, shown in geographic profile. Details of water movements are given in the text.

FIGURE 2a.2

concrete banks are sometimes poured to "stabilize" the river. However, if flood conditions occur, there is still an increase in water flow downstream despite the dams. Natural river banks, being composed of porous soil, absorb surprisingly large amounts of flood water, but concrete banks cannot perform this function. It is evident that porous levees are more desirable than concrete banks for containing flood waters. Moreover, levees do not prevent replenishment of ground water as concrete banks do. (In this connection, paving large urban areas—Los Angeles County in California is about one-third paved—also increases flood problems and decreases replenishment of local ground water. Fresh water from storms, which once went partly to ground water, is confined to flood-control sewers, whence it flows unused out to sea.)

In arid regions about one-quarter of the water stored in reservoirs behind dams evaporates (Fig. 2a.3). It is estimated that the potential irrigation water lost through evaporation from Lake Mead has a greater dollar value than does all the electricity generated by Hoover Dam. Water which should have replenished ground water is lost and the concentration of salts in the water left behind is increased. The natural balance between surface and ground water levels is thus put out of equilibrium by the water management practices of man. This is also true of another source of fresh water historically used by man, namely, wells.

Wells: The Mining of Ground Water. Wells have long been used as a source of fresh water, but because of the vastness of the underground reservoirs, wells long had little effect on the water table in most parts of the world. When engine-driven pumps were introduced, however, enormous quantities of ground water were removed. In many arid regions the water table has now dropped so much that pumping is no longer economical. In these areas it will take thousands of years to replenish the supply, owing to the slow rate of movement of underground water. This practice is known as "mining" of ground water.

Land developers in semi-arid regions such as southern California all

FIGURE 2a.3

The water evaporated from Lake Mead behind Hoover Dam, if it could be sold for irrigation, would be worth more than the electricity produced by the generators. (Courtesy of U.S. Dept. of the Interior Bureau of Reclamation, Boulder Canyon Project.)

too often promise "unlimited supplies" of well water to unwary buyers. The mined water supply, in fact, is usually sufficient to last only a few decades, at the end of which the entire existing community will either have to import water from elsewhere or perish. If the community is large enough or many such communities occur in one area, political pressure will result in importation of water. Since more and more communities are being developed in areas where the natural water supplies — rainwater and ground water — are insufficient for future needs, it is necessary to examine alternative sources of fresh water for such regions.

MODERN METHODS OF OBTAINING FRESH WATER

Technology has introduced three new ways of obtaining more fresh water in areas where demand now exceeds supply: cloud-seeding, desalination and interbasin water transfer. Each has at present certain limitations, either environmental or practical.

Cloud-Seeding. The idea of changing the weather, making it rain in the deserts and stop raining in tropical and northern regions, has always been attractive to man. Modern rainmakers have abandoned the homeopathic or imitative magic of primitive peoples for more scientific methods of inducing rain. Cloud-seeding involves dispersing tiny crystals of silver iodide in a storm cloud; these tend to cause the condensation of water vapor into raindrops. So far, cloud-seeding has proved only marginally successful, for frequently the rain falls not on the drought area over which seeding occurred but many miles away or perhaps not at all. Rainmaking still seems to be mostly magic.

Desalination. On the other hand, desalination is a likely means for obtaining fresh water from the sea, and will no doubt become more important in coming years. A variety of ingenious methods is now available, although none yet compete by our current economic standards with other sources of fresh water. To separate water from salt requires energy, and therein lies the rub. If energy for desalination is supplied by conventional means, such as burning fuel or using electricity, the polluting effects from the power plants may vitiate the advantages gained from the increased supplies of fresh water. Air pollution generated by conventional power plants is already a problem, and the excess heat released by all types of power plants, including nuclear reactors, poses a large question for the future, as we shall see in Chapter 7a. Production of useful amounts of fresh water from the sea using non-polluting energy sources, such as wind, tides and solar radiation, however, is a possibility which technology may soon make feasible.

Interbasin Transfer. The most commonly suggested solution for increasing freshwater supplies, however, is interbasin water transfer. For centuries man used aqueducts for transporting water downhill within a single watershed, but transfer of water over mountains has become technically possible only in recent decades. The Los Angles basin is a prime example of an importer of water (Fig. 2a.4).

First, in the 1920's, the Owens River was tapped via the Los Angeles aqueduct, and once productive ranch land in that valley is now desert through loss of its water supply. Then, in the 1930's, the Colorado River was tapped, but even this supply of relatively impure water became insufficient; an aqueduct from the Feather River at Lake Almanor in sparsely populated northern California has just been constructed. Such precedents have led to the general acceptance by many politicians and engineers of interbasin water transfer as a solution to freshwater deficits in dry regions. An enormous project, proposed by the North American Water and Power Alliance, is being formulated to bring water from Alaska and western Canada to dry regions of central Canada, the southwestern United States

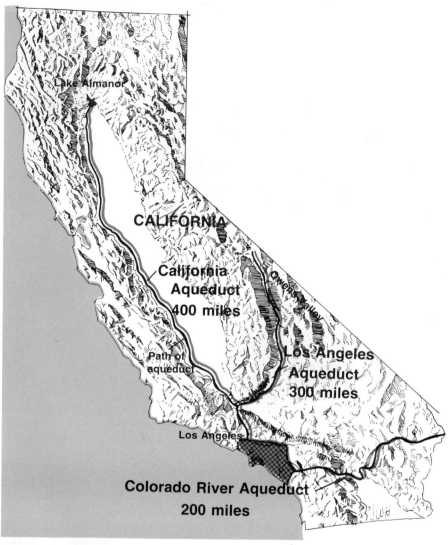

FIGURE 2a.4

Sources of water imported into the Los Angeles basin. Note the mountains surrounding it, across which water is transported.

and Mexico. Water would be stored in huge reservoirs at relatively high elevations (Fig. 2a.5). The climatic, hydrologic and ecologic results of such a scheme are virtually unknown, but some possible effects can be considered.

The numerous rivers, including the Yukon and the MacKenzie, which today flow into the Bering Sea and Arctic Ocean, carry relatively warm waters into these cold seas. Diversion of these rivers will result in a decrease in the temperatures of the polar waters, with unknown effects on climate and on the currents which circulate oceanic waters between arctic and tropical regions.

Certain effects on local water conditions in the donor basins also must be expected, although it is presently impossible, owing to lack of basic hydrologic data, to estimate their magnitude. Water table levels will decrease, with a drying up of shallow wells and a drop in flow of streams fed during the dry season by underground water (stream B in Figure 2a.1). Rates of ground water flow are likely to decrease. As will be seen in Chapter 16, this factor is of importance in determining the quality of the soil and hence the type of vegetation which it will support. It is possible that the northern forests, so well adapted to the relatively poor, water-logged soils of these regions, will be gradually replaced by other types of trees and grasses. Large amounts of stored water will never reach the intended site at all, but will evaporate from the storage reservoirs, short-circuiting the hydrologic cycle.

PROBLEMS OF DESERT IRRIGATION. Effects on the recipient basins, the desert areas, must also be considered, particularly if the water is intended for irrigation—for "making the desert bloom." Imported water contains some dissolved salt, and since a high proportion of irrigation water in the desert is lost by evapotranspiration, salts are left behind in the soil. Over a few years, the soil becomes too saline to grow crops. To prevent this, it is necessary to irrigate with a large excess of water to flush away the salts. Because the rate of groundwater movement in such areas is often very slow, whether the water table originally was high or low, there is all too often a rapid rise of groundwater levels under the irrigated land. The soil becomes water-logged, preventing adequate aeration of plant roots and hastening the rate of accumulation of salts at the surface. Hundreds of thousands of acres of agricultural lands throughout the world are lost each year in this way; the Indus Valley in West Pakistan (Fig. 2a.6) and parts of the southwestern United States are notable examples.

In some areas attempts to cure this problem have involved the use of tile drainage (Fig. 2a.7). In this system, excess irrigation water, containing high levels of dissolved salts, seeps into the joints between foot-long lengths of pipe made from clay tile or from concrete which are installed at depths of six or seven feet. The water table thus never rises to the level of plant roots, and the excess water is drained away in large concrete troughs at the edges of the fields.

Using such systems it may even become possible in coastal regions to irrigate certain crops with sea water. Although sea water has a high salt content, the roots of some plants are able to extract the water they need, leaving behind the excess salt. As long as this salt is not allowed to accumulate in the soil, but is washed away in the run-off water from the fields, the plants thrive. The main requirement is that the water table not be permitted to rise to a level where it will either interfere with the respiration of plant roots or allow water to reach the surface by capillary action and evaporate, leaving salt deposits on the surface. The hypersaline water returned to the sea will have little effect on the total salinity of the inshore waters.

FIGURE 2a.5

The proposed North American Water and Power Alliance scheme for interbasin transfer of water between Alaska, Canada, United States and Mexico.

FIGURE 2a.6

Salt-covered fields in the Indus Valley of West Pakistan contrast strongly with surrounding cultivated land. Accumulation of salts such as seen here occurs in irrigated soils with poor drainage. (From Water, by R. Revelle. Copyright © 1963 by Scientific American, Inc. All rights reserved.)

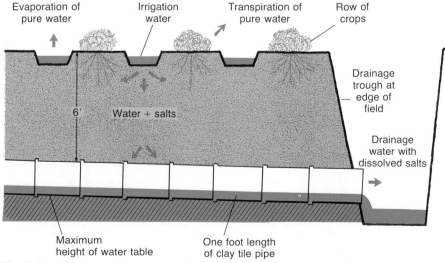

FIGURE 2a.7

Tile drainage system in an arid region. Excess quantities of irrigation water flush salts from the soil. The water seeps into joints of clay tile pipes, and flows to drainage trough at edge of field. Tile drainage prevents rise of water table.

Further experiments on the use of sea water for irrigation are obviously of immediate importance.

It is apparent that the greatly increased water losses from reservoirs, from surface evaporation and from transpiration, together with the need for excess water to flush away salts, must all be taken into account in calculating amounts of water required for irrigating arid regions. Unfortunately, one or more of these factors are all too often neglected.

THE QUALITY OF WATER

In the United States, about 50 per cent of all water consumed is used for industrial and domestic purposes. The quantities of water required for urban uses do not differ greatly between arid and moist regions. Evapotranspiration in arid regions is a major consideration in agriculture, but is much less significant in cities; thus, urban requirements may be considered irrespective of local climatology. For urban water requirements, however, quality becomes an important factor.

Aside from the amount of dissolved salts, we have so far been concerned only with the quantity of fresh water and not its quality. The problem of quality of water supplies is far more cultural than biological and is primarily associated with urban needs. On the other hand, both urban and agricultural uses can affect the quality of water discharged back into the biosphere. The dissolved and suspended wastes may be deleterious both to normal biological processes in fresh waters and to their esthetic and recreational values to man.

OUR CULTURAL ATTITUDE TOWARD WATER

In the sparsely populated world of primitive man, when per capita needs for fresh water were also small, the total wastes dumped into a stream by one village were readily removed before the water was used again by the next village downstream. The natural processes of dilution, filtration, and oxidation of organic wastes by bacteria were sufficient to purify the water. Men became accustomed to using the same water source for all purposes, and to discarding all wastes back into the stream.

Today, with enormously increased population in urban areas, and with per capita usage multiplied many times, we still cling culturally to our historical attitudes toward water. We expect to use the same water for all purposes: water of sufficient quality for drinking and bathing is also used for car washing, sewage disposal, lawn watering and frequently for industrial purposes. All "waste water," no matter to what degree it has been altered, is considered unfit for any further use and is discarded with little or no attempt at recycling.

It is all too obvious that the natural processes can no longer cope with the job of purifying the prodigious amounts of waste now poured into our water resource. Meanwhile, having contaminated our water supplies, we look afield for new sources rather than trying to recycle what we have. This is biologically unsound for two reasons. Extensive interbasin transfer, as we have seen, is likely to present several ecological problems whose magnitude cannot even be estimated because of our lack of knowledge about the hydrologic cycle. Also, the waters we now pollute are biologically altered in ways which are having an ever greater effect on their ability to support life.

TYPES OF POLLUTION

There is space here only to outline some of the major types of pollutants and their effects. Water pollutants may be broadly grouped into toxic substances which are poisonous to living organisms, nutrient wastes which upset the natural balance among the plants and animals inhabiting a lake or stream, and waste heat which kills sensitive organisms.

Toxic Pollutants. Toxic substances, such as heavy metals and long-lived pesticides, destroy life directly—although some organisms are more susceptible than others. Those released by industry could, for the most part, be relatively easily removed by chemical and physical treatment of water before its release. Internal recycling of industrial waste water would make such treatment more efficient. Technological innovations for cleaning industrial waste waters are being rapidly developed.

On the other hand, chlorinated hydrocarbons such as DDT and its relatives, heavy metals like mercury and arsenic, and other toxic substances used in agriculture are much more difficult to keep out of water resources. These chemicals are dispersed as sprays, dusts or in seed dressings over wide areas of fields and forests and eventually find their way into surface and ground water supplies. There is virtually no way to prevent the contamination of freshwater resources, or indeed of the global environment, once these long-lived chemicals are released. We have already noted in Chapter 1 how DDT has affected birds living far distant from its site of application. The best solution is to substitute alternative methods of carrying out the functions these chemicals now perform; some alternatives will be discussed in later chapters.

Nutrient Wastes. Pollution by nutrients is less directly dangerous to man and other organisms, but it is far more difficult to control than toxic pollutants. There are innumerable sources of excess nutrients—domestic sewage, household detergents, pulp mill effluents, food processing wastes, fertilizers and excreta from livestock are all major sources.

It is not immediately obvious why accumulation of excess nutrients should be deleterious. To understand the problem we must first consider the productivity of a body of fresh water—a lake, for example. Since the ultimate source of food for all animals is plants, it is the growth of plants, both microscopic phytoplankton or algae and macroscopic shore plants, by which the total productivity of a lake is measured. When plant nutrients are present in good supply, the lake is highly productive—it is called a **eutrophic** lake. (The word is derived from the Greek *eu*, "good," plus *trophé*, "nutrition.") When nutrients are in low supply, productivity is low and the lake is called an **oligotrophic** lake (*oligos* = "few").

Under natural conditions most lakes have a fairly short life span on the geological time scale; 10,000 or so years after its formation a medium-sized lake may become nearly filled with sediments. Eventually it completely disappears. Sooner or later during this aging process a lake passes from an oligotrophic condition, with crystal clear water, to a eutrophic state, full of algae and with a swampy odor; the process is known as **eutrophication.** The increase in nutrients results from the influx of natural sediments. As we saw in Chapter 2, the main requirements for plant growth, in addition to carbon dioxide and water, are two minerals, nitrate (NO_3^-) and phosphate ($PO_4^{\equiv}$). Since these minerals are usually the limiting nutrients for plant growth, the number of plants will increase if the minerals are increased. This is precisely what happens during the natural eutrophication of a lake. Phosphate is supplied by the sediment entering the lake, as a result of natural erosion, and nitrate forms from the gradual accumulation and degradation of dead organisms in the bottom of the lake.

MAN-MADE EUTROPHICATION. Waste matter added to fresh water by man is often rich in phosphate. This is especially true of many detergents. Chemical fertilizers are abundant sources of both phosphate and nitrate, not all of which are taken up by crops. Organic wastes, on the other hand, become sources of NO_3^- and $PO_4^{\equiv}$ for algae and other plants only after they have been broken down by bacteria in the water. This bacterial breakdown process requires oxygen; the more organic material broken down, the more oxygen used up. The amount of organic matter in waste waters is often expressed as the **biological** (bacterial) **oxygen demand**—known as B.O.D. The result is a depletion of the oxygen dissolved in the lake water due to bacterial action on organic wastes.

Meanwhile, algae and other plants flourish; the whole surface of the lake may become bright green with these microscopic plants. As the algae continue to multiply, however, some of them die, and they in turn are broken down by bacteria, recycling NO_3^- and $PO_4^{\equiv}$ and using ever more oxygen.

The net result is that eutrophied lakes can no longer support those animals which require high levels of dissolved oxygen. At first the trout and salmon disappear and are replaced by coarse fish such as carp (Fig. 2a.8). Eventually even these can no longer survive, and only a few worms and other animals able to live with very little oxygen are found. As decay processes increase, hydrogen sulfide and other odoriferous gases are produced, the water acquires a bad taste and becomes unfit for swimming or other recreation.

This man-hastened eutrophication process is common to many lakes in the world. The most famous (or infamous) case is Lake Erie, which is said to have aged 10,000 years in the past few decades. That is, without the additional nutrients introduced by man, Lake Erie would naturally have reached its present state in 10 millennia.

Not only lakes are affected by eutrophication; streams and rivers into which organic wastes are dumped also suffer the same fate. They no longer support fish, they become unpleasant and recreationally useless and they also provide a very poor quality water supply for cities downstream.

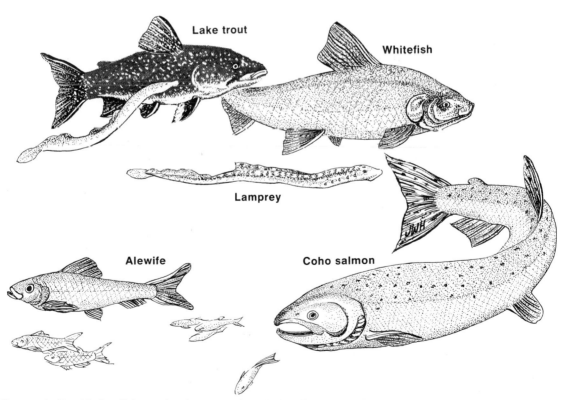

Changes in Great Lakes fish species due to man's activities. Fish originally common, now almost non-existent, include the lake trout (upper left) and the whitefish (upper right). Among the fish now common are the parasitic lamprey (center), the destructive alewife (lower left) and the coho salmon (lower right).

FIGURE 2a.8

It is not necessary for a lake to be bounded by large cities or fertile farmlands to undergo eutrophication. Lake Tahoe in the Sierra Nevada of California is rapidly eutrophying. The main villain in this case is phosphate released into the lake in huge quantities as a result of the use of bulldozers. Minerals, especially $PO_4^=$, are leached from the pulverized soil by rainwater and flow into the lake. The necessary nitrate is provided by sewage from resorts and subdivisions. This huge, deep lake, once blue, is rapidly acquiring a greenish tint around the edges from the thriving algae.

Fortunately, not all eutrophication is irreversible. The Schliersee and Tegernsee in Germany's Bavarian Alps, Lake Mendota near Madison, Wisconsin, and Lake Washington in the city of Seattle were all eutrophied from sewage wastes; in these cases steps were taken to alleviate the condition, and all the lakes are in the process of returning to their former state. In the case of Lake Washington, however, the problem was solved by diverting the sewage to Puget Sound, a solution which can hardly be regarded as an ultimate answer.

The problems in reversing or preventing eutrophication are varied and complex. As with toxic substances used in agriculture, the nutrients arising from fertilizers applied to farmlands are difficult to control because they come from such diffuse sources. Sewage and industrial wastes are more amenable to treatment; but both are, by our present economic standards, expensive to purify. Sewage treatment for suppression of human fecal bacteria has long been practiced as a public health measure, although this often amounts to no more than chlorination without removal of nutrients. Pilot projects undertaken by several communities have demonstrated the feasibility of complete treatment, where sewage water is eventually cleansed of nutrients by biological processes and used for recreational purposes and for watering public gardens.

ESTHETIC USES OF WATER

Although it is beyond the purpose of this book to delve into social and economic problems, a short discussion may prove useful.

Let us first consider the case of Lake Erie; it is theoretically possible that, were no more wastes to be dumped into it, in 20 to 30 years the water would have been sufficiently replaced and purified by natural processes to return the lake to its original condition. But the number of independent sources now emptying wastes into Lake Erie and the total population involved is so great that the concerted political and economic changes required are unlikely to take place. The losses of commercial fisheries and recreational uses are apparently of less immediate value than are the economic and political costs involved in stopping pollution.

As we go about obtaining more and purer water supplies, we shall be faced more and more not only with climatological and biological decisions but also with esthetic decisions. People will be required to make *value* judgments. How much land are we prepared to submerge under reservoirs? Are we willing to forego the shade of highly transpiring trees like sycamores, cottonwoods and willows, and uproot them in order to increase our groundwater supplies? How important are clear lakes and streams to our recreational needs?

As noted in Chapter 1, value judgments are presently founded solely on economics, and that which is economical is that for which people elect to pay money and from which others make a profit. Although people today are perfectly willing to buy vacuum cleaners to remove dust in their homes, for example, at the moment they are not willing to pay the costs of removing pollutants from used water and of recycling present water supplies. We prefer to cling to our historical assumption that water is infinitely available and wastes are infinitely disposable because our present values tell us this is more economical.

But if we value the esthetic losses we are now incurring it may become no longer economical to use water only once and to continue to contaminate our communal water resources. Furthermore, as we have tried to show in this chapter, ignoring the effects on the biosphere of our present trends in water usage goes far beyond merely the loss of esthetically pleasing lakes, streams and rivers. Eventually, economic costs will have to be paid to reverse or repair biological changes in the environment. Being more rational about the quality of immediate water supplies is one way of reducing both the present economic costs of pollution and the continuing need for increased supplies of fresh water. It will still be necessary to spend money, however, to purify and recycle our effluents. What is now only a matter of esthetic choice must eventually become a biological necessity.

SUMMARY

Although fresh water forms but a tiny fraction of all the world's water, there is probably enough for all man's uses in the foreseeable future. The major difficulties lie in the uneven distribution of fresh water relative to demand and in man's propensity for polluting it.

Building of dams, mining of ground water and interbasin transfer of water are all means of redistributing water supplies, and each poses environmental problems. Cloud-seeding is at present unsuccessful in this respect, but desalination holds promise for the future. Transfer of water to arid regions for agricultural purposes creates new problems which must be considered when such projects are being planned.

Changes in cultural attitudes about water quality and usage, now a matter of esthetic choice, will eventually become a biological necessity.

Gerard, R. D. and Worzel, J. L. 1967. Condensation of atmospheric moisture from tropical maritime air masses as a fresh water resource. Science 157:1300–1302. An imaginative scheme for obtaining fresh water on an oceanic island by using natural sources of energy.

Hasler, A. D. 1969. Cultural eutrophication is reversible. BioScience 19:425–431. An article describing the pollution of lakes in Bavaria and their return to a natural state after community action.

Leopold, L. B. and Tilson, S. 1966. The water resource. International Science and Technology, July 1966:24–34. A readable discussion of what is known and not known about fresh waters and man's effects upon them.

Powers, C. F. and Robinson, A. 1966. The aging Great Lakes. Scientific American Offprint No. 1056. A description of the formation of the Great Lakes and man's impact upon them.

Revelle, R. 1963. Water. Scientific American Offprint No. 878. An optimistic consideration of the potential for developing water resources.

Wagner, R. H. 1971. Environment and Man. W. W. Norton & Co., New York. Chapters 6 and 7 discuss our freshwater resource and man's effects upon it.

Wolman, M. G. 1971. The nation's rivers. Science 174:905–918. A detailed appraisal of water quality and the effects of man's activities.

READINGS AND REFERENCES

3

CELLS—THE BASIC UNITS OF LIFE

The tiny units called cells which make up the tissues and organs of plants and animals come in many shapes, but all share in common certain biochemical and structural features.

All living organisms, from bacteria to redwood trees, have a common unit of structure—the cell. A bacterium is but a single cell, while a redwood contains billions of cells, yet the cells of each share many common characteristics. Most individual cells are so small that they can only be observed at high magnifications under a microscope, but a few, such as the eggs of birds, are easily seen with the naked eye. To understand life itself, it is necessary to have an image of the living cell and how it functions.

In the course of its existence, a cell must take up nutrients from its environment, utilize these for energy, for repair or for growth, and dispose of its wastes. All of these processes must be regulated and coordinated. Reduced to its simplest terms, life is thus a series of complex chemical reactions which are carried out within the cells of an organism. It is our purpose in this chapter to become familiar both with the chemicals found in cells and with the main functions they perform in life processes—growth, maintenance and reproduction.

Cells contain a variety of chemicals, some of which, such as water and certain ions, are also found in the inanimate world as well. But there are other chemicals characteristic only of life which are known as *organic molecules.* (We shall exclude from our present discussion the numerous synthetic organic molecules made by man, such as DDT, which, although often similar to molecules found in nature, are not manufactured by living organisms.) As we shall see shortly, there are both small and large organic molecules and the latter, called *macromolecules,* are often component parts of structural entities within the cell, the cell *organelles.* A glance ahead to Figure 3.13 will give the reader an idea of the structural complexity of cells.

ELEMENTS OF LIFE

As noted in Chapter 2, only a few of the elements found on Earth are common in living systems. In order of importance, these are carbon (C), hydrogen (H), oxygen (O), nitrogen (N), phosphorus (P) and sulfur (S). All of these elements usually occur in the organic molecules of living cells. Other elements, especially minerals such as sodium (Na), potassium (K), calcium (Ca), magnesium (Mg) and chloride (Cl), are present in the form of ions. Water molecules (H_2O) are found in all cells. Again, for the reader unfamiliar with chemistry, a brief summary of the nature of chemical bonds, ions and molecules is given in Appendix I.

Within a cell, we can thus envision a mixture of ions and of large and small organic molecules, each associated to a greater or lesser extent with the universal solvent, water. Because, surprisingly enough, our understanding of the structure of liquid water is so incomplete, it is not yet possible to form a clear view of the chemical environment within a cell; we do know, however, that it is not at all like a mixture of the same chemicals in a test-tube.

Despite our general ignorance about water, we do understand some of its properties which are important to life. It is a **polar molecule** — that is to say, it has regions of positive and negative electric charge:

$$\text{negative region} \quad - \quad O\!\!\begin{array}{c}{}^{\nearrow H}\\[-2pt]{}_{\searrow H}\end{array} \quad + \quad \text{positive region}$$

For this reason, it is able to dissolve other chemicals which also have polar regions: for example, ions, such as Na^+ and Cl^-, and certain small organic molecules, like sugar. In the presence of water, a sugar cube or crystals of salt will dissolve (Fig. 3.1). Water is here the **solvent;** the molecules dissolved in it are called **solutes.**

ORGANIC MOLECULES, SMALL AND LARGE

Within cells there are four main types of organic molecules: fats, carbohydrates, proteins and nucleic acids. All are composed of C, H and O, and the proteins and nucleic acids regularly contain other elements as well. Each, because of its chemical nature, performs special functions in the living cell.

LIPIDS

Fats, or more fashionably, **lipids,** are largely water insoluble molecules. Most of them are made up only of C, H and O, with a great preponderance of C and H. The molecular structure of a typical fat, as pictured by a chemist, is shown in Figure 3.2A. Because this is a book about biology, not chemistry, we can simplify the structure of a fat, as in Figure 3.2B. Most fats consist of a small unit, **glycerol,** to which are attached three long-chain units, the **fatty acids,** composed mainly of carbon and hydrogen. Only the junction region is polar. A single fatty acid (Fig. 3.2C) has a small polar end and a long non-polar tail. We can thus conclude that fats are not easily soluble in water.

Fats serve three important functions in the organism. Because of their water insoluble nature they are useful as barriers between the watery solu-

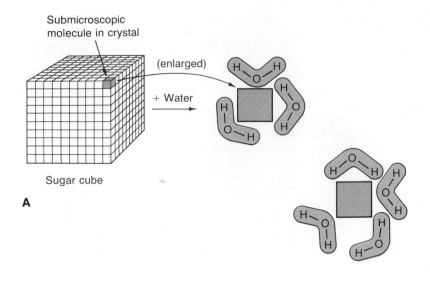

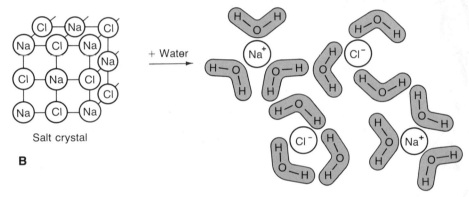

FIGURE 3.1

Solvent action of water.
  A. During solution of a sugar cube, individual sugar molecules, which have many polar regions, are surrounded by polar water molecules which are attracted to them.
  B. During solution of a salt, the ions separate, each being surrounded by a shell of oriented water molecules. Note that positively charged Na^+ attracts the negative O-region of water, and negatively charged Cl^- attracts the positive H-region.

tions outside and inside the cells and between various compartments within a cell—a point to which we shall return shortly. Fats also contain much chemical energy and thus serve for energy storage. When overweight persons are on a diet, their excess fat supplies them with needed energy and so gradually disappears. Fat also is an excellent thermal insulator; whale blubber, for example, prevents loss of body heat from a whale's internal organs to arctic waters.

CARBOHYDRATES

Like lipids, carbohydrates are composed of C, H and O, but the proportion of oxygen is much greater. The ratio of the three elements is 1:2:1, so carbohydrates are often conveniently abbreviated chemically as $[CH_2O]$.

The simplest carbohydrates are sugars, known more formally as **monosaccharides** (Greek *mono*, "one," plus *sakchari*, "sugar"). A common six-carbon sugar, **glucose**, is shown in the chemist's notation in Figure 3.3A. Note that each carbon has an oxygen atom attached to it. Glucose molecules more commonly take the form of the ring structure shown in Figure 3.3B. Again, to simplify our notation, we shall use the symbol shown in Figure 3.3C. Other sugars or sugar derivatives found in cells contain as few as two and as many as seven or more carbon atoms, but in their basic chemistry they are all similar; all are polar, owing to their oxygen atoms, and therefore readily dissolve in water.

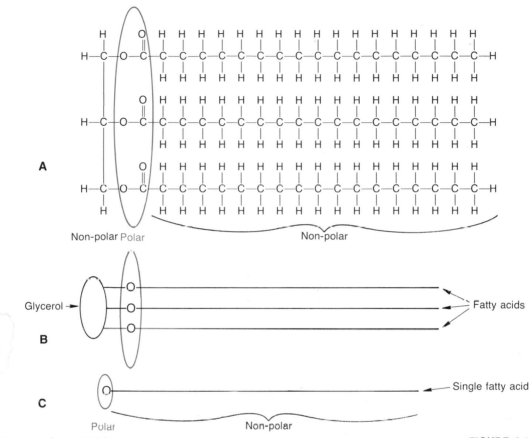

The structure of a typical fat.

A. The structure of a lipid shown in chemist's notation. Note the preponderance of C and H atoms. Only regions containing O are polar.

B. Our simplified notation. The small glycerol component is to the left, the three long-chain fatty acids to the right.

C. A single fatty acid, showing its small polar end and long, non-polar tail. The latter is not readily soluble in water.

FIGURE 3.2

Monosaccharides are often joined together in repeating units to produce macromolecules called **polysaccharides** (Greek *poly*, "many") (Fig. 3.4). Several thousand single sugars or **monomers** may be joined together to produce a polysaccharide or **polymer.** One can think of a polymer as a series of beads (the monomers), strung together to make a long chain. Despite their large number of polar groups, polysaccharides are often too large to dissolve easily in water.

Polysaccharides serve two main functions. One is to store energy, which can be used when food is in short supply. The common storage polysaccharide in animals is **glycogen**; in plants, it is **starch**, such as is found in potatoes or wheat seeds. The other function of polysaccharides is to provide structural supports. Some polysaccharides are so insoluble that they are especially suited to this function. In terrestrial plants, each cell is surrounded by a wall of **cellulose,** familiar to us as the major molecule in

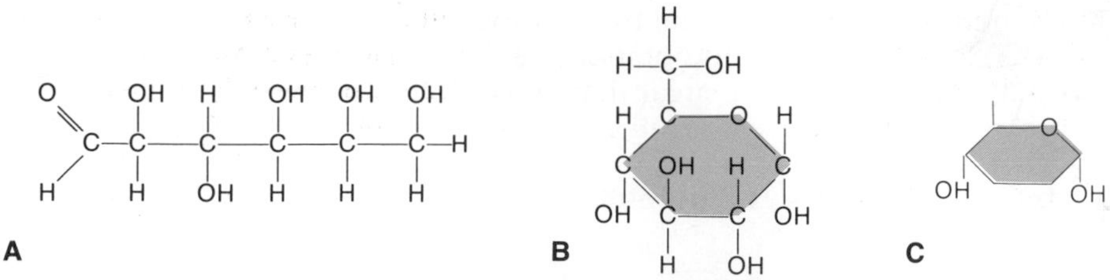

FIGURE 3.3

A

B

C

A simple six-carbon monosaccharide, glucose.

    A. Shown in linear form, in chemist's notation. Note that each C atom has on O atom associated with it; monosaccharides are thus polar and water soluble.

    B. The more commonly occurring ring-form of glucose.

    C. The shorthand notation we shall use for glucose, indicating only its most reactive OH groups.

trees, paper, cotton and linen. In animals, another polysaccharide, called **chitin,** forms the basic structure of crab and lobster shells and the cuticle or "shell" of insects. Bacteria also surround themselves with various substances composed of a special type of carbohydrate molecule; these are called **mucopolysaccharides.**

PROTEINS

    Whereas lipids and carbohydrates contain only C, H and O, **proteins** are characterized by, in addition to these three elements, a high content of nitrogen (N). Like polysaccharides, proteins are macromolecules, polymers made up of many smaller units. The monomers of proteins are the **amino acids,** curious molecules which have both an acidic region and a basic region. There are about 20 kinds of amino acids in the proteins of living organisms, each with its acidic and basic region, and a third region, often denoted by **R,** which differs from one amino acid to the next. A typical amino acid, valine, is shown in chemist's notation in Figure 3.5A; the simplified form we shall use for all amino acids is that shown in Figure 3.5B. The solubility in water of each amino acid depends to a large extent on the polar or non-polar nature of its R-region. In our example, valine, the R-region is non-polar and the molecule is only slightly soluble in water.

    There are many different proteins in nature, probably millions of them, all made from the same "alphabet" of 20 "letters," the 20 amino acids. Each protein is a "word" containing sometimes a thousand "letters" or more. The potential number of different protein "words" is almost unlimited!

    Proteins are formed by the chemical joining of the basic and acidic groups of amino acids in **peptide bonds.** In the resulting primary protein structure, a **polypeptide,** the variable R-regions of the amino acids stick out on each side (Fig. 3.6A). Certain regions of the polypeptide, in turn, tend to form weak bonds, causing it to wind into a helix or spiral; this results in the secondary structure of the protein (Fig. 3.6B). The coiled regions, in

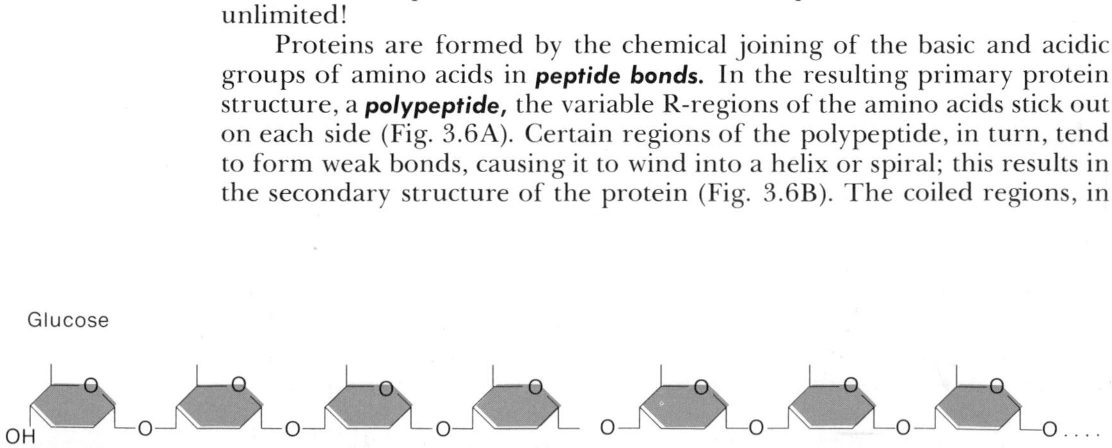

Glucose

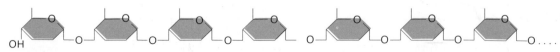

FIGURE 3.4   Diagram of part of the polysaccharide molecule starch, showing oxygen linkages between consecutive monomeric units of glucose.

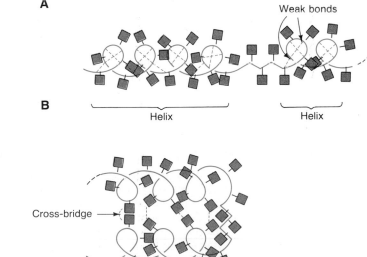

A

B

FIGURE 3.5

The structure of amino acids.

A. A typical amino acid, valine, shown in chemist's notation. The oxygens of the acidic region make it electronegative; the hydrogens of the basic region make it electropositive. The "variable" region is different for each amino acid and is often denoted by R. In this case R is a non-polar group.

B. Our shorthand notation for all amino acids. Solubility is mainly determined by the polar or non-polar nature of the R-group, which varies widely among the 20 different amino acids.

turn, may be attracted to each other to form various types of chemical cross-bridges between the protruding R-groups. This is called the tertiary structure of the protein (Fig. 3.6C). Some proteins have only a primary or secondary structure and are elongate **fibrous proteins.** Others have a tertiary, spherical form and are often called **globular proteins.**

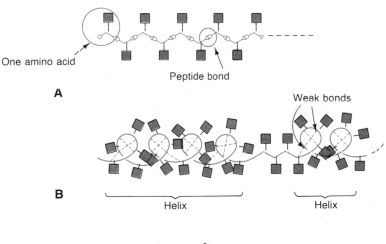

FIGURE 3.6

The structure of proteins.

A. Polypeptide primary structure; R-groups stick out on either side.

B. Helical regions, resulting from weak bonds along the polypeptide chain, produce secondary structure. (Peptide bonds have been omitted for simplicity).

C. Folding, with formation of cross-bridges, provides tertiary structure.

A and *B* are typical of fibrous proteins. *C* is the globular structure characteristic of enzymes.

The way in which a given protein coils and folds into a specific shape is thus determined by the sequential arrangement of the amino acids in the original polypeptide chain. Each type of protein has a unique shape, different from that of all other proteins. This **specificity** of protein shape is of enormous importance in biology, since shape determines function. Thousands of different jobs can be done in a cell, each by one of its thousands of different protein molecules.

Because the bonds resulting in the secondary and tertiary structure of globular proteins are usually not very strong chemically, they are easily ruptured, changing the shape of the protein and often destroying its function. This process, called **denaturation,** may be produced by heat—a property we make use of in cooking an egg or in sterilizing surgical instruments. In the latter case, heat destroys the proteins of bacteria, thus killing them. Even changing the salt concentration will denature a protein. A runny egg yolk will congeal if salt is sprinkled on it. The effect of excess salts on the function of protein is the main reason why man cannot drink sea water and survive. We cannot excrete the excess salts without losing more water than we drank; the accumulated salts meanwhile begin to denature our proteins.

Structural Proteins.   One job performed by proteins is to provide structural support in various tissues. Proteins of this group have only a primary or sometimes secondary structure and are thus elongate, fibrous molecules. Rigidity is attained by the formation of bonds between parallel molecules. One good example is **keratin,** the major protein of skin, hair, fingernails and feathers. In keratin, peptide chains lying side by side are attached to each other by cross-bridges between pairs of sulfur atoms (S–S). This provides rigidity and also explains the unpleasant sulfurous smell of burned hair and feathers. Chemicals used in permanents and hair-straighteners take advantage of the presence of S–S bonds in hair. The "permanent" or "straightener" solution dissolves the S–S bonds, and the "neutralizer" used subsequently allows the bonds to re-form between different molecules while the hair is held by curlers in a new position.

Contractile Proteins.   Another group of elongate proteins are the contractile proteins, found especially in muscle and familiar to us as the meat in our diet. Similar proteins are probably responsible for changes of shape in many other animal cells as well. These proteins have a very complex structure, which we shall consider later, in Chapter 12.

Enzymes.   A most important group of proteins are the globular **enzymes,** or **catalysts,** which speed up the rate of chemical reactions in the body without themselves being changed in the process. Living organisms must constantly process molecules to obtain energy for cell work, to build new tissues, and to make repairs. Each one of these processes requires many separate chemical reactions, so that literally thousands of different reactions are constantly taking place. Collectively these reactions constitute what is known as **metabolism.** Metabolic reactions would occur at a very slow rate indeed at ordinary temperatures were there no enzymes to speed them up.

NUCLEIC ACIDS

The fourth major group of molecules in living cells are also polymers, the **nucleic acids.** In addition to C, H, O and N, they contain another element, phosphorus (P). The building blocks of nucleic acids are the **nucleotides.** These are more complex monomers than amino acids, since they contain three sub-units, a nitrogen-containing base, a five-carbon sugar and a phosphate, joined together as shown in Figure 3.10.

There are two kinds of nucleic acids in cells. One kind, called **DNA** for deoxyribonucleic acid, has deoxyribose for its sugar; the other, called

The importance of enzymes to maintenance of life is so great that we shall have occasion to mention them many times in this text. It is worthwhile to spend a few paragraphs here to explain how they work.

Each reaction is catalyzed by a specific enzyme, which "recognizes" only one type of molecule in the cell — its **substrate.** During a reaction the enzyme temporarily binds with its substrate, bringing about a chemical change, as shown in Figure 3.7. The binding results from the complementarity of shapes between a specific part of the enzyme and its substrate, analogous to a lock and key arrangement. The tertiary folding of an enzyme determines its specific shape. If the shape of an enzyme is altered, by too much heat or by salts, for example, the fit will be less than perfect and the reaction will slow or stop.

HOW ENZYMES WORK

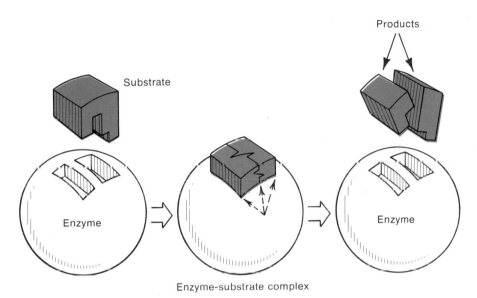

FIGURE 3.7

Diagram of an enzyme-catalyzed reaction.
Complementarity of shapes between enzyme and substrate leads to formation of a complex. Forces within the enzyme molecule, represented by arrows, cause chemical changes in the substrate — in this case, a splitting — which results in formation of new molecules, the products.

We can now understand in a general way how the metabolism of an organism may be controlled. Let us consider the case of a crab which has just eaten a full meal. This represents an "income," analogous to a paycheck, which may be "spent" in alternative ways (Fig. 3.8). What determines how much of this food will be converted to cash (energy) for immediate expenses, how much will be saved (as glycogen or fat) for future energy needs and how much will be invested for growth?

Each alternative pathway consists of a whole series of chemical reactions (shown by multiple arrows), each reaction being controlled by its specific enzyme. These pathways compete for the food. Its ultimate fate thus depends on how effective the enzymes in each pathway are. It is the regulation of enzyme activity, then, that determines which chemical processes go on in a living organism.

If our crab is short of immediate energy, for example, it can do several things to insure that its food is burned and energy is released. First, it can synthesize more enzymes in this pathway and less in the others. We shall consider this in more detail later in this chapter.

Second, it can provide enzyme **activators** for this pathway. We have seen that changes in the shape of an enzyme affect the reaction rate. Activators "improve" the shape of an enzyme and make the reaction go faster. An activator of some of the enzymes in the energy-yielding pathway in Figure 3.8 is magnesium ion (Mg^{++}).

A third control mechanism is the accumulation of **inhibitors.** These either change the shape of the enzyme so that it fits less well with the substrate or, because of their similarity in structure, compete with the substrate for its specific site, as shown in Figure 3.9A. During inhibition of its alternative pathways, our crab's energy-yielding pathway would become relatively more important. Often the reaction products themselves will inhibit the enzyme, providing a **negative feedback** control.

Immediate energy

Storage

Growth

FIGURE 3.8

Artist's representation of the alternative metabolic pathways open to an organism for utilizing its "income" —food. The multiple arrows indicate that many enzyme-catalyzed reactions take place in each pathway. The decision is determined by the internal state of the organism, and is carried out by an increase in activity of enzymes in one pathway and suppression of enzymes in other pathways.

Competitive inhibitors are commonly used in medicine, where they are known as **anti-metabolites.** As shown in Figure 3.9B, the drug sulfanilamide has a very similar structure to a substrate required by many bacteria, para-aminobenzoic acid. The antibiotic binds to the bacterial enzyme which metabolizes the substrate, and thus prevents the bacteria from growing.

The amount of each enzyme and of its activators and inhibitors present at a given time is controlled by a series of complex interactions between chemical mediators, producing a finely tuned balance among the hundreds of chemical reactions going on in a living cell. We shall have occasion to refer to some of these mediators and how they work later in the book.

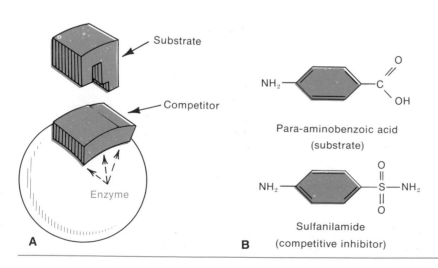

Substrate

Competitor

Enzyme

Para-aminobenzoic acid
(substrate)

Sulfanilamide
(competitive inhibitor)

A

B

FIGURE 3.9

Competitive inhibition of enzymes.

A. The enzyme is bound by the competitor which not only prevents the substrate from binding but, because of its shape, cannot be chemically changed by the enzyme.

B. Similarity in structure between the normal bacterial substrate para-aminobenzoic acid and the antibiotic drug sulfanilamide.

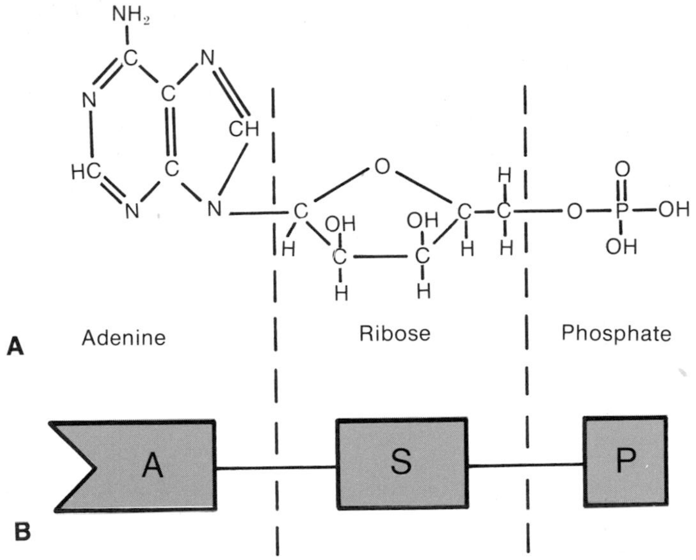

NH₂

A   Adenine   Ribose   Phosphate

B

A ── S ── P

FIGURE 3.10

The structure of the nucleotide adenylic acid, one of the monomers of nucleic acid.
 A. Chemist's notation. The nitrogenous base adenine is joined to a five-carbon sugar,
ribose, which is joined to a phosphate group.
 B. Our simplified notation for the same molecule. A, adenine; S, sugar; P, phosphate.

RNA for ribonucleic acid, has ribose, instead. There are also five main
kinds of nitrogen-containing bases—adenine, guanine, thymine, cytosine
and uracil—known most commonly by their abbreviations, **A, G, T, C** and **U**.
The nucleic acid in which the various nucleotides occur and the shorthand
symbol we shall use for each are shown in Table 3.1. Note that T occurs
only in DNA, and U occurs only in RNA.

The nucleotides join together between their sugar and phosphate
groups, as shown in Figure 3.11, to form very long polymers, often 100,000
or more monomer units in length! The sequence of nucleotides, as we shall
see later in the chapter, is of the greatest importance.

FIGURE 3.11

The structure of a nucleic acid polymer.
 Nucleotide monomers are joined together, as shown, between their sugar and
phosphate groups. One nucleic acid molecule may contain enormous numbers of
monomers linked together in this fashion.

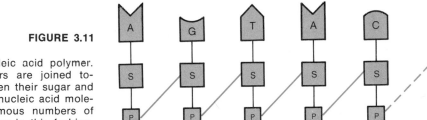

Nucleic acids have two very significant jobs to do. We have seen that
proteins, in their capacity as enzymes, control metabolism in organisms.
The nucleic acids, in turn, determine precisely which proteins are synthesized at any one time. They are thus the ultimate regulators of metabolism.
Shortly we shall consider how they carry out this function.

Nucleic acids not only regulate metabolism during the lifetime of an
individual organism, however; they also pass on genetic information from
one generation to the next. This function belongs exclusively to the mole-

cules of DNA. In these molecules there resides all the information needed for making an entire living organism—they are the blueprints of life!

This reproductive function of the DNA molecule requires that it be able to exactly copy itself so that no vital information is lost. For a long time it was a mystery as to how a molecule could precisely replicate itself. The secret was finally unraveled in 1953, when Watson and Crick published their analysis of the structure of DNA, which subsequently proved to fit all the experimental facts. This structure, as it turns out, is a **double helix**—two long threads of DNA spiraled about each other, as shown in Figure 3.12.

TABLE 3.1 The NITROGEN-CONTAINING BASES

Name	Our Notation	Nucleic Acid in which Found
Slightly larger molecules		
Adenine (A)		DNA and RNA
Guanine (G)		DNA and RNA
Slightly smaller molecules		
Thymine (T)		DNA only
Cytosine (C)		DNA and RNA
Uracil (U)		RNA only

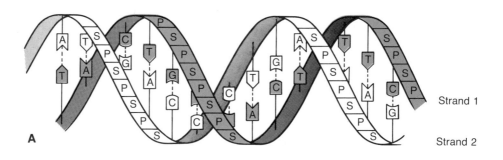

Strand 1

Strand 2

A

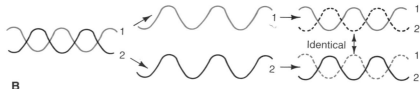

B

FIGURE 3.12

Double helix of DNA.

A. Complementarity of base pairs: A always pairs with T; C always pairs with G. The two strands, *1* (colored) and *2,* are "mirror images" held together by weak bonds shown as dashed lines.

B. When reproducing, the strands separate, strand *1* (colored) forming a new strand *2* complementary to itself while strand *2* produces a new strand *1* (colored). The new strands are shown as dashed lines. Therefore, each original produces its complementary half, resulting in two identical double helices.

Examination of the figure shows that the two spirals are joined in the center by weak bonds between their nitrogen-containing bases and also that the bases are always paired in a certain fashion: **A** pairs only with **T**; **C** pairs only with **G**. This base-pairing makes each half a "mirror image" of the other. To reproduce, the two strands simply unwind, and each one synthesizes a mirror image of itself. The end result is perfect self-duplication, as shown in Figure 3.12B.

THE GENERALIZED ANIMAL CELL

Although different cells in an organism have specific jobs to do, there are certain properties shared by virtually all cells. In describing these, we shall begin with a generalized animal cell, since they are somewhat less specialized than the cells of higher plants. A diagram of such a cell is shown in Figure 3.13.

Each structure in a cell has a characteristic chemical composition, and its structure and function are closely related. Until the development of the electron microscope about 25 years ago, only the larger elements in cells could be seen—the nucleus, nucleolus, mitochondria and large granules. The interior of cells was thought to be divided into only two compartments,

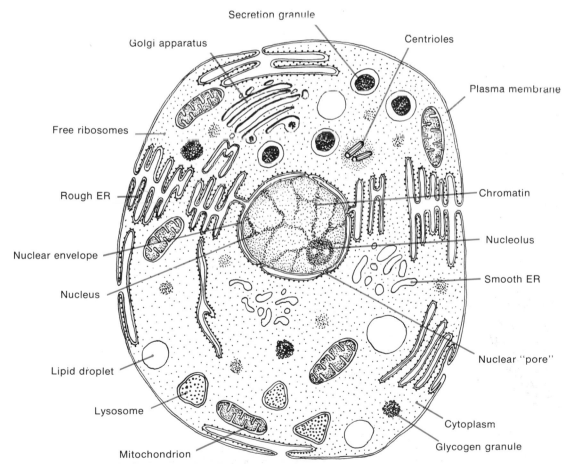

Diagram of a generalized animal cell. Although shown here as a round cell, animal cells may have many shapes. Note that ribosomes occur freely in the cytoplasm as well as attached to the endoplasmic reticulum.

FIGURE 3.13

{
Dark band
Light band
Dark band

A

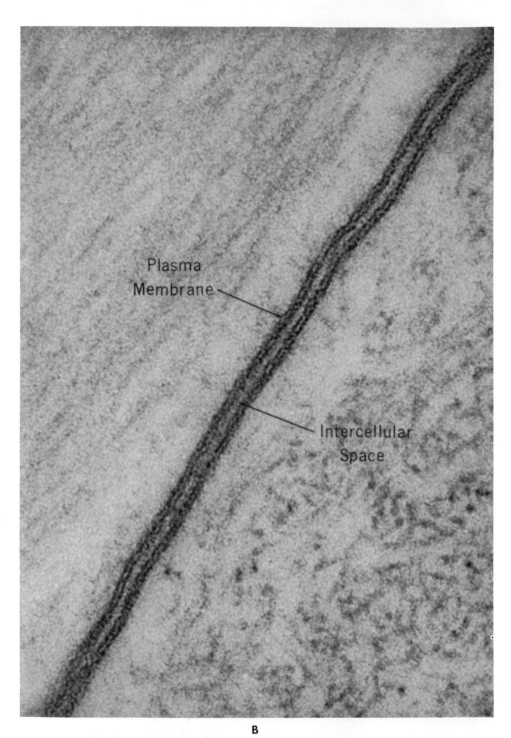

Plasma
Membrane

Intercellular
Space

B

FIGURE 3.14   The unit membrane.

 A. A sketch, showing the three-layered structure of membranes as seen in the electron microscope: two dark bands separated by a light band.

 B. Electronmicrograph of plasma membranes of two adjacent cells (magnification 260,000 ×). (From Fawcett, D. W. 1966. *The Cell: Its Organelles and inclusions.* W. B. Saunders Co., Philadelphia.)

the *nucleus* and the *cytoplasm.* The greater resolving power of the electron microscope, however, revealed many smaller structures, each with a complex substructure of its own. At about the same time, new techniques were developed for disrupting cells and isolating the individual organelles, thus permitting their specific functions to be identified.

MEMBRANES

There are many membranes in cells. The one around the outside of the cell is known as the *plasma membrane.* There are also membranes around organelles such as mitochondria, secretion granules and lysosomes; and the nucleus has a double set of membranes, the *nuclear envelope,* with thin regions at regular intervals, sometimes called "pores." Some membranes form long, hollow tubes, such as the *endoplasmic reticulum* and the *Golgi apparatus.* At very high magnifications in the electron microscope, all membranes look very much the same. They appear to have a three-layered structure composed of two dark bands separated by a light band of about the same thickness (Fig. 3.14). This arrangement has come to be known as a "unit membrane," and each membrane represented in Figure 3.13 has the same three-layered structure.

Chemically, membranes contain lipid and protein. Some of the proteins simply provide structure to the membrane, while others are enzymes with catalytic activity. Despite enormous amounts of research and a great

Let us pause to consider a physical process known as *diffusion.* If we place a sugar cube in the bottom of a glass of water and allow it to stand, the sugar eventually disappears (Fig. 3.15). At first, if we sip from the top with a straw, the water will taste barely sweet, but after several hours the sugar molecules diffuse to the top and the water will taste sweet at whatever depth we sample. The sugar molecules have moved along a *concentration gradient,* in this case from a high concentration at the bottom of the glass to a low concentration at the top, and have eventually become evenly distributed in the water. This tendency to uniformity in physical systems is characteristic throughout nature.

Next let us consider a selectively permeable membrane which is permeable to water but *not* to table sugar (Fig. 3.16). Such a membrane, initially separating two fluid compartments of equal height, one containing only water and the other con-

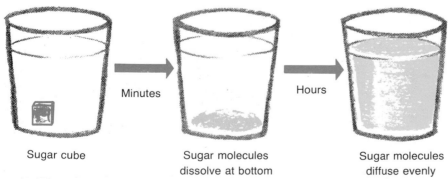

Sugar cube

Sugar molecules
dissolve at bottom

Sugar molecules
diffuse evenly

FIGURE 3.15

Concept of diffusion. In an undisturbed glass of water, at first the dissolved sugar molecules are clustered near the bottom, but eventually they are evenly distributed throughout the water.

taining water with sugar dissolved in it, will permit water to cross from the right side where the water has a high concentration, to the left side where the water is "diluted" by sugar molecules. After a few hours the liquid level on the right will decrease and that on the left will increase, producing a pressure difference between the two compartments. This is called the **osmotic pressure** of the sugar solution. The more concentrated the sugar solution, the greater will be its osmotic pressure. The actual movement of water across this selectively permeable membrane in response to the concentration difference is called **osmosis.**

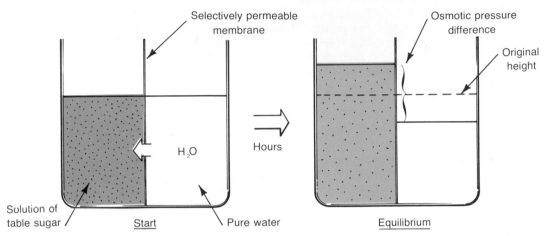

FIGURE 3.16   Osmosis across a selectively permeable membrane which passes only water and not table sugar molecules.

     Water diffuses into the sugar solution, leading to a pressure differential in the two compartments. Eventually this pressure difference, the osmotic pressure, becomes equal to the tendency for water molecules to move across the membrane, and an equilibrium state is reached.

     Because plasma membranes are more permeable to water molecules than to other molecules or ions, cells will swell when placed in dilute solutions and shrink in solutions more concentrated than themselves, as a result of osmotic movement of water. For example, the spherical egg of a sea urchin, normally shed into sea water, when placed in dilute sea water will swell and may eventually burst. On the other hand, when placed in sea water to which excess salt has been added, it will shrink (Fig. 3.17). Human red blood cells placed in solutions more dilute or more concentrated than blood do the same thing. Such solutions are said to be, respectively, **hypo-osmotic** and **hyperosmotic** to the cell. An **isosmotic** solution has the same osmotic pressure as the interior of the cell.

     Each dissolved molecule or ion, no matter what its size or charge, has the same effect on osmotic pressure. Thus, osmotically speaking, 1 protein molecule = 1 sugar

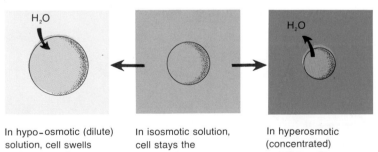

In hypo-osmotic (dilute) solution, cell swells and contents become more dilute

In isosmotic solution, cell stays the same size

In hyperosmotic (concentrated) solution, cell shrinks and contents become more concentrated

FIGURE 3.17

Changes in volume of a sea urchin egg due to osmotic movements of water.

     The cell's normal medium is sea water, which is isosmotic to the fluid inside the cell. In dilute sea water, the egg swells as water enters; in sea water to which extra salt has been added, the egg shrinks due to loss of water.

     Hyper = more: think of a *hyper*active person. *Hypo* = less: think of a *hypo*active person.

molecule = 1 sodium ion, even though the relative weights are 100,000 : 180 : 23. Although small molecules other than water can also cross cell membranes by diffusion, the rates are almost always much slower.

Diffusion, however, is not the only process by which molecules move across cell membranes. Some molecules are moved "uphill" against their concentration gradient. The cell membrane pushes some molecules from a dilute solution to a concentrated solution. In animal cells, for example, sodium ion (Na^+) is less concentrated inside cells than in the blood or lymph surrounding them. Na^+ diffuses inward in response to its concentration gradient but is quickly extruded from the cell. This process is known as **active transport**; the mechanism for removing Na^+, known as the "sodium pump," is situated on the cell membrane. Because it is an "uphill" process, active transport requires expenditure of energy by the cell. Exactly how this pumping process occurs is not clear, but a specific protein located in the cell membrane seems to be involved. The active extrusion of sodium, as we shall see later, is basic to the functioning of nerve and muscle cells in animals. Many other substances are actively transported into and out of cells by similar energy-requiring "pumps."

many theories, we still do not know exactly how the lipids and proteins are arranged in membranes.

The major function of membranes is to separate the internal liquid compartments of the cell, each of which has a different composition, and to separate the interior of the cell from the external medium around it. The lipids, because of their non-polar nature, make an excellent barrier to the movement of polar and charged molecules. But, like any barrier, be it the Great Wall of China or the United States Customs Service, cell membranes are not absolute. Water, ions, and certain small molecules such as glucose and amino acids, can pass through relatively easily. Except for water, however, the movements of small molecules are to a large extent controlled by the proteins in the membranes; only selected molecules are allowed to pass the boundary. Large molecules, such as proteins, usually cannot move across intact cell membranes at all. We thus say that cell membranes are **selectively permeable.**

MITOCHONDRIA

The word **mitochondrion** (plural, mitochondria) means "threadlike granule," a name given these organelles from their appearance in the light microscope. At the magnifications possible in the electron microscope, however, a complex structure of folded membranes and an internal matrix is revealed (Fig. 3.18).

Mitochondria are the "power plants" of the cell. In the presence of oxygen they burn glucose, releasing energy and producing carbon dioxide and water. If sugar is burned in a furnace, it will also be converted to CO_2 and H_2O, but only at very high temperatures. In the human mitochondria, however, this overall reaction takes place at body temperature, 37° C, owing to the catalytic action of enzymes. In fact, glucose oxidation occurs as a series of sequential reactions, each controlled by a single enzyme. It is believed that each enzyme is located at a specific site in a mitochondrion, either on one of the membranes or within the matrix. These all-important reactions are known collectively as **respiration** (not, however, to be confused with breathing); they are further discussed in Chapter 7. For the moment, the main thing is to realize that the mitochondria generate useful energy for the cell in a manner analogous to the generation of electrical energy by power plants when they burn coal, oil or gas.

ENDOPLASMIC RETICULUM

The **endoplasmic reticulum,** or **ER,** is an intricate network of hollow tubules found throughout the cytoplasm. Its unit membranes, like those of the plasma membrane, are composed of lipid and protein. In some areas of a cell, the endoplasmic reticulum is highly convoluted and its membranes have a smooth surface. This **smooth ER,** as it is called, has different functions in different cells, and in the next chapter we shall consider its role in the process of detoxification by liver cells.

Elsewhere, the ER tubules are expanded into flattened sheets—imagine very large tubules compressed sideways. In these regions, the outsides of the membranes of the endoplasmic reticulum are covered by small granules, the **ribosomes,** which give it a rough appearance—hence its name, **rough ER.** (The reader may wish to identify rough ER in Figure 3.22). A ribosome is a tiny sphere, with a slot in one side dividing it into a large and a small hemisphere. It is composed mainly of RNA, known as **r-RNA** (ribosomal RNA) to distinguish it from other types of RNA in the cell. In addition to the ribosomes located on the rough ER there are frequently small clusters of ribosomes scattered throughout the cytoplasm. Ribosomes are the site of protein synthesis.

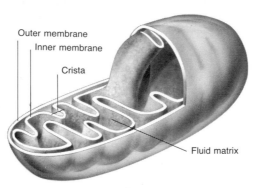

Outer membrane
Inner membrane
Crista
Fluid matrix

FIGURE 3.18

A

The structure of a mitochondrion.

A. Three-dimensional sketch, showing both the outer membrane and the inner membrane, which is folded into cristae (singular, crista).

B. Electronmicrograph of a mitochondrion (magnification 95,000 ×). (From Fawcett, D. W. 1966. *The Cell: Its Organelles and Inclusions.* W. B. Saunders Co., Philadelphia.)

Illustration continued on opposite page.

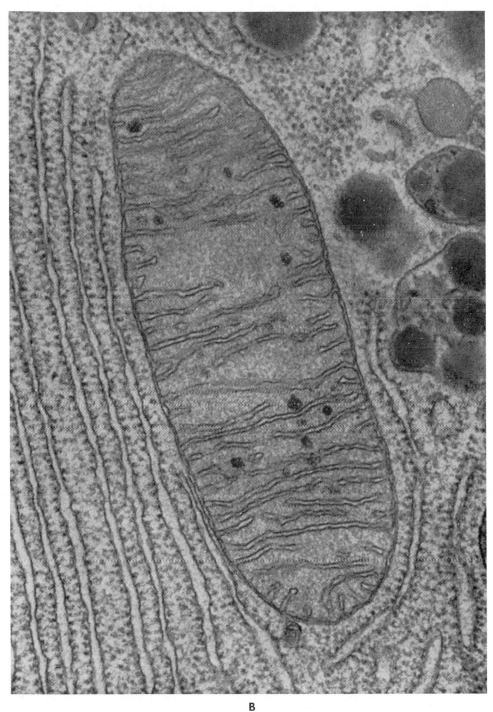

B

FIGURE 3–18 *Continued*

NUCLEUS

The **nucleus** of a cell is its control center, or "brain," and is required for its long-term survival. In Figure 3.19 are shown the results of an experiment on a single-celled animal, the amoeba, which demonstrate the essential role of the nucleus. An amoeba is cut in half with a glass needle. The half with the nucleus survives, regenerates its lost cytoplasm and eventually divides. The enucleated half lives but a few days and then dies. If a nucleus from another amoeba is transplanted into the enucleate half, however, it will live. It is apparent that the nucleus is essential for the continuance of life.

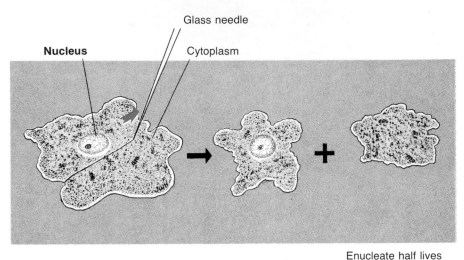

Glass needle

Nucleus   Cytoplasm

Amoeba

Nucleate half lives
and regenerates lost
cytoplasm

Enucleate half lives
a few days, then dies

FIGURE 3.19

Experiment demonstrating the essential role of the nucleus in cell survival.
An amoeba is cut in two with a glass needle. The half with the nucleus survives; the other half eventually dies. Transplantation of a nucleus into the enucleate half would allow it to survive.

From microscopic studies it has long been known that the nuclei of cells stain vividly with certain dyes. The small **nucleolus** usually stains intensely, while the staining of the rest of the nucleus is often diffuse and patchy; this diffusely stained material has been called **chromatin**. During cell division, however, the nucleolus disappears and the chromatin contracts into several discrete rod-shaped structures called **chromosomes.** Eventually it was learned that the staining of the nucleolus and chromatin is due to the presence of nucleic acids; the nucleolus contains RNA, while DNA is characteristically found in the chromatin of chromosomes. Between divisions the chromosomes unfold into very long, very thin filaments, too small to be seen clearly even in the electron microscope. The nucleolus, organized by certain regions of the extended chromosomes, is the site of synthesis of the RNA found in ribosomes.

Each type of organism has its own number of chromosomes, and all cells in a given organism, except its eggs and sperm, have the same number of chromosomes. In cells of all human beings, for example, there are 46 chromosomes. This constancy of chromosome number is what led to the hypothesis, now firmly established, that inheritance resides in the DNA of chromosomes. The two main functions of DNA, of chromosomes and of the nucleus, are thus identical: to transmit genetic information from one generation to another and to direct the activities of a cell during its lifetime. We have already noted how the double-helical nature of DNA permits self-duplication of two complementary strands. The mechanisms by which the DNA-containing chromosomes are exactly duplicated and precisely separated during cell division in order to transmit genetic information accurately from one generation to another will be considered in detail in Chapter 14. An outline of the means by which the nucleus directs protein synthesis and thus carries out its function as the "brain" of the cell is given in the section below, How Proteins Are Made.

HOW PROTEINS ARE MADE

As we have seen, proteins, particularly enzymes, are important regulators in the chemistry of the cell. They determine cell function and thus affect the well-being of the whole organism, allowing it to adapt to changing circumstances. The ultimate determinant of protein synthesis, however, is the DNA of the cell nucleus.

The means by which the DNA of chromosomes directs the synthesis of specific proteins and thus controls cell metabolism has been largely revealed by research in the last decade. It was known that DNA consists of four different nitrogen-containing bases, A, G, C and T, and also that the sequence of the 20 possible amino acids in a protein determines the way it folds into a unique shape and, therefore, its specific function. The problem was to convert from a language of four nucleotides to 20 amino acids. Without delving into the experimental details, which would require a whole book of their own to explain, we can briefly summarize how this translation of information occurs.

Earlier we saw that DNA exists as two long, complementary strands, wrapped around each other in the form of a double helix. In the unfolded chromosome of a non-dividing cell, there are associated with the DNA small proteins known as **histones.** It is believed that histones may be the molecules which prevent DNA from being continuously active, although other molecules, such as nuclear RNA, may also be involved. Along the length of the DNA in a chromosome there are many functional regions, each of which contains the information necessary for the synthesis of a specific type of protein. Such regions are called **cistrons** (Fig. 3.20A).

When a cell is stimulated to synthesize a particular protein, the DNA of the appropriate cistron is thought to lose its histone (or other control molecule) and to unwind temporarily from the double helix configuration. Precisely how this postulated unwinding occurs is not known, but it is believed that only one strand of the double helix participates in protein synthesis, as shown in Figure 3.20B.

FIGURE 3.20

Schematic diagram of part of a chromosome.

A. Several cistrons, each of which directs the synthesis of a single type of protein.

B. When a cistron is activated, it is hypothesized that certain histones (or other control molecules) are released from the DNA and it temporarily unwinds. Probably only one strand is active in RNA synthesis.

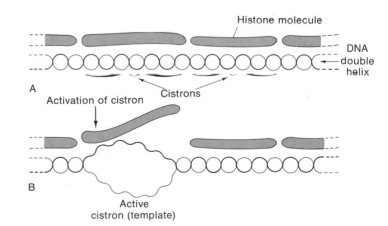

In the discussion which follows, the reader is referred to Figure 3.21, in which are diagramed the principal events in protein synthesis, and to Table 3.2, which gives the chemical rules for nucleic acid and protein synthesis. The first step in protein synthesis is the transfer of information from the DNA of the nucleus to the cytoplasm, where protein synthesis occurs. For this purpose, the strand of DNA on the activated cistron is used as a ***template*** for the synthesis of a long "messenger" molecule of complementary, mirror image RNA, called messenger RNA or ***m-RNA*** (see rules in Table 3.2). Messenger RNA passes out from the nucleus to the cytoplasm, perhaps through a "pore" in the nuclear membrane.

One end of the m-RNA from the nucleus, carrying the DNA message, attaches to the slot on a ribosome, either one attached to the rough ER or one free in the

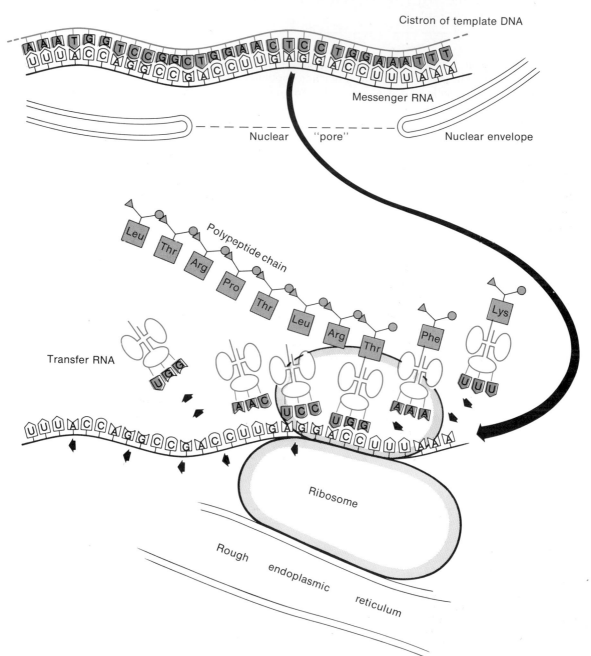

FIGURE 3.21   Summary of the steps in protein synthesis.

Arrows indicate movements of molecules. Messenger RNA, synthesized according to the rules for basepairing on the DNA template, passes to the cytoplasm where it moves over a ribosome on the endoplasmic reticulum. As each triplet of bases on messenger RNA passes over the ribosome, a transfer RNA molecule, with complementary bases at one end and an amino acid at the other, temporarily joins to the messenger RNA. The amino acid bound to its other end is simultaneously joined to the polypeptide chain, which will eventually fold into a protein.

cytoplasm. The m-RNA then moves over the ribosome, directing the synthesis of a protein molecule in the process. Although for the sake of clarity only one ribosome is shown in Figure 3.21, in fact a single m-RNA molecule moves over several ribosomes, synthesizing replicate copies of the same protein as it moves over each one.

At this point, molecules of a third type of RNA are necessary to convert nucleic acid language into protein language, thus "translating" the message. These small molecules of RNA, each folded into a clover-leaf shape, are known as transfer RNA or *t-RNA* since they are involved in transferring an amino acid to the site of protein synthesis. Transfer RNA is made by special regions of chromosomes, and there is a different type of t-RNA for each of the 20 amino acids occurring in the protein alphabet.

Transfer RNA molecules are double-headed. At one end they attach by complementary base-pairing to m-RNA, according to the rules in Table 3.2; at the other, they carry the appropriate amino acid, as shown in Figure 3.21. The rules which are obeyed in this translation process are known as the *genetic code.* In fact, it is a fairly simple code, although it took many experiments to unravel it. Each sequence of three bases in the DNA molecule, or their complementary bases in m-RNA, stands for one amino acid, as shown in Table 3.2.

The t-RNA molecules, as they attach to m-RNA while it is sliding over the ribosomes, bring the correct amino acids into contact with each other, and a specific polypeptide polymer is formed. As the polypeptide grows and becomes a protein, it folds into its unique shape, as determined by its sequence of amino acids; this shape, as we noted earlier, gives the protein a unique and specific function. Each m-RNA molecule is short-lived, producing only a few protein molecules before it is broken down. The cell thus retains its ability to produce different kinds of m-RNA, and hence of proteins, as new needs arise.

**TABLE 3.2
RULES FOR
BASE-PAIRING
AND THE
GENETIC CODE**

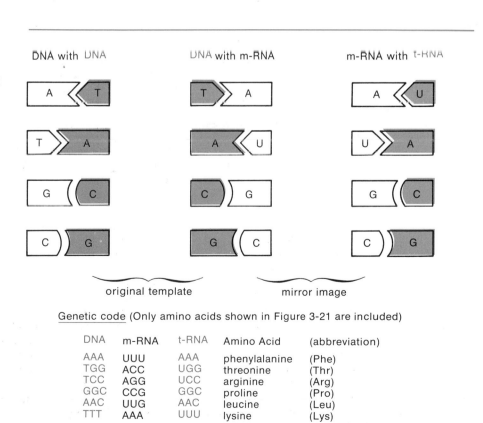

DNA with DNA   DNA with m-RNA   m-RNA with t-RNA

original template   mirror image

Genetic code (Only amino acids shown in Figure 3-21 are included)

DNA	m-RNA	t-RNA	Amino Acid	(abbreviation)
AAA	UUU	AAA	phenylalanine	(Phe)
TGG	ACC	UGG	threonine	(Thr)
TCC	AGG	UCC	arginine	(Arg)
GGC	CCG	GGC	proline	(Pro)
AAC	UUG	AAC	leucine	(Leu)
TTT	AAA	UUU	lysine	(Lys)

To summarize briefly, there are three kinds of RNA in cells: messenger RNA (m-RNA), which carries information from DNA to the cytoplasm; ribosomal RNA (r-RNA), which provides the framework on which protein synthesis occurs; and transfer RNA (t-RNA), which translates nucleic acid language into protein language. Complementary base-pairing insures correct transcription of information at each of the steps in which nucleic acid language is used; the code of a triplet of bases for each amino acid provides the translational step for converting to protein language. Biologists consider the "cracking of the genetic code" one of the great scientific landmarks of recent years.

GOLGI APPARATUS

In many cells, especially those which secrete much protein, such as cells of digestive glands, another tubular structure is found; this is called the **Golgi apparatus** (Fig. 3.22). It is named after an early Italian biologist who first described it in cells observed under the light microscope. Like rough ER, its three-dimensional structure is best envisioned as layers of highly flattened tubules. The primary function of the Golgi apparatus is believed to be the concentration and packaging of proteins which will eventually be secreted by the cell. Proteins synthesized by ribosomes are passed through the tubules of the rough ER to the Golgi apparatus and there are collected into membrane-bound secretion granules.

CENTRIOLES

The **centrioles** are paired structures composed of groups of short protein fibers. They are usually found near the nucleus and function during cell division to form focal centers for the mitotic spindle, a set of fibers which aids in the separation of chromosomes into the two "daughter" cells produced during cell division. It is of interest that plant cells, although they divide and have mitotic spindles, usually do not have visible centrioles. The details of cell division are discussed in Chapter 14.

ORGANELLES FOR FOOD STORAGE

Animal cells often contain glycogen granules and lipid droplets (Fig. 3.13), both of which serve for food storage. These are the cell's bank account, which can be converted to useful energy as the need arises.

LYSOSOMES

The prefix lys- indicates loosening or breakdown. **Lysosomes,** then, are membrane-bound organelles which contain numerous enzymes capable of breaking down or digesting a wide range of large molecules. These enzymes are similar to those that digest food in the intestines of animals. In some animals, particularly single-celled ones, food is taken inside the cell prior to digestion. In these cases, the lysosomes release their digestive enzymes into a membrane-bound cavity containing the food. But lysosomes also occur in many other cells, where they may act as "self-destruction" mechanisms in dying or damaged cells. When a cell becomes moribund, its lysosomes rupture, releasing enzymes which digest the cell contents; it can then be replaced by new, healthy cells. In fact, increased fragility of the membrane around lysosomes is characteristic of cell aging and may be hastened by environmental factors such as smog.

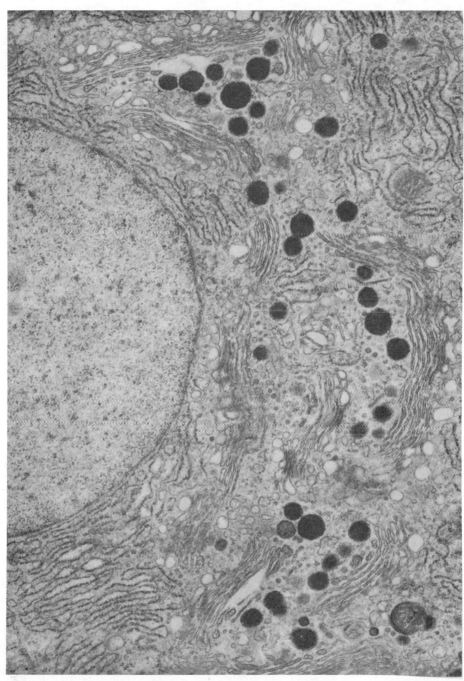

Electronmicrograph of part of an intestinal secretory cell (magnification 22,000 ×).   **FIGURE 3.22**
    All cell organelles involved in protein synthesis and processing are visible. The
nucleus is on the left. Rough endoplasmic reticulum is seen bordering the nucleus and
in the upper right corner. Several patches of Golgi apparatus (close-packed layers of
membranes) are packaging protein into dark secretion granules. (Photo by Dr. Daniel
Friend, from Fawcett, D. W. 1966. *The Cell: Its Organelles and Inclusions*. W. B. Saunders
Co., Philadelphia.)

THE GENERALIZED PLANT CELL

The plant cells we shall consider here are typical of the seed-forming green plants, the most familiar of all plants. Some of the variations found in other types of plant cells will be discussed in Chapter 5. In many ways, plant cells resemble animal cells, and there is no need to repeat our description of their common organelles. Both plant and animal cells have a plasma membrane, nucleus, endoplasmic reticulum, Golgi apparatus and mitochondria; the functions of these organelles are the same in both cell types. The cells of familiar plants, however, lack centrioles and glycogen granules but have several special, important features of their own. A cell typical of a higher plant is diagramed in Figure 3.23.

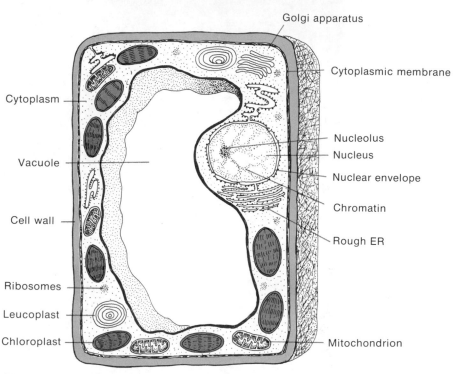

FIGURE 3.23

Diagram of a generalized cell from a higher plant. Organelles are described in text.

CELL WALLS

Plant cells have rigid **cell walls** external to the plasma membrane. These are usually composed of cellulose but are sometimes made of other complex polysaccharides which resist deformation. This means that even if plant cells are placed in solutions much more dilute than themselves (hypo-osmotic), they do not swell and rupture as would animal cells; the rigid cell walls maintain their shape and resist swelling. In fact, despite their relatively high internal osmotic pressure, terrestrial plants thrive with their roots in contact with the dilute salt solutions found in most soils. The hydrostatic pressure, resulting from the osmotic influx of water through the roots of terrestrial plants, provides **turgor** or stiffness to the leaves and stems. One has only to recall the difference between wilted and crisp lettuce to appreciate this. The turgor of terrestrial plants is essential for keeping their leaves stiff and able to adjust their alignment so as to receive maximum sunlight.

VACUOLES

Unlike animal cells, the cells of most higher plants possess one or more large central **vacuoles** which function as storage reservoirs for water, sugar, ions and pigments; the vacuoles also play an important role in the growth of plant cells, as we shall see in Chapter 14. Their main function, however, is to maintain cell turgor.

CHLOROPLASTS

The intracellular organelles known as **chloroplasts** (Fig. 3.24) are the most distinctive structures differentiating the cells of green plants from those of animals. Their membranous interior is reminiscent of that of mitochondria, but its greater complexity suggests more sophisticated chemical abilities. Mitochondria, found in both plant and animal cells, are capable of releasing energy for cell reactions from the oxidation of sugar during cellular respiration. Chloroplasts on the other hand, are capable of transforming the energy of sunlight into energy utilizable by cells for synthesizing complex organic molecules from simple molecules such as carbon dioxide and water. The functions of mitochondria and chloroplasts can be summarized as follows, the multiple arrows indicating that a series of reactions is involved in each process:

Mitochondria:

GLUCOSE + OXYGEN$\rightarrow\rightarrow\rightarrow$CARBON DIOXIDE + WATER + ENERGY for cell reactions

Chloroplasts:

SUNLIGHT + CARBON DIOXIDE + WATER$\rightarrow\rightarrow\rightarrow$OXYGEN + GLUCOSE for future energy needs

The details of these two fundamental biological processes are discussed in later chapters. For the moment it is sufficient to identify the reaction sequences in chloroplasts as **photosynthesis.** The detailed structure of a chloroplast, shown in Figure 3.24, reveals stacks of membranes, the **grana,** which are regions containing the light sensitive pigment **chlorophyll.** Chlorophyll transduces light energy into chemical energy which can then be used by the cell for work or for growth. Since green plants are the ultimate food source of virtually all other organisms, photosynthesis is a most important process; it is known to be adversely affected by several man-made pollutants, including DDT and smog. We shall consider these effects in Chapter 8a.

LEUCOPLASTS

Whereas animal cells often store their excess food in the form of glycogen, accumulated as glycogen granules, plants store their food as starch. Although small amounts of starch may be temporarily stored in the chloroplasts, the primary organelles in which starch is accumulated are the starch granules or **leucoplasts** (white bodies). Their function in plant cells is identical to that of glycogen granules in animal cells.

CYTOPLASMIC MOVEMENT

When living cells are viewed through the light microscope, the larger cell organelles, such as mitochondria, chloroplasts and storage granules, are sometimes seen to move about in the cytoplasm. In plant cells, this movement is called **cyclosis;** in animal cells, it is known as **cytoplasmic streaming.** The cause of the movement is unknown, as is its function, al-

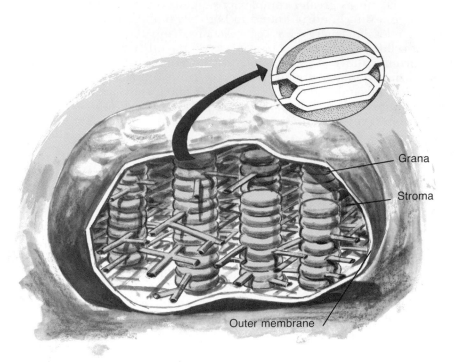

Grana

Stroma

Outer membrane

FIGURE 3.24

A

Structure of a chloroplast from a higher plant.

A. Diagram of chloroplast in three dimensions showing outer membrane, stacks of chlorophyll-containing grana and stroma. It is thought that many of the enzymes involved in energy capture during photosynthesis are organized in sequential fashion on the membranes of the grana. Other enzymes occur in the stroma.

B. An electronmicrograph of a chloroplast from a tobacco leaf (magnified 30,000 ×). (Courtesy of Dr. E. T. Weier, from Villee, C. A., and Dethier, V. G. 1971. *Biological Principles and Processes.* W. B. Saunders Co., Philadelphia.)

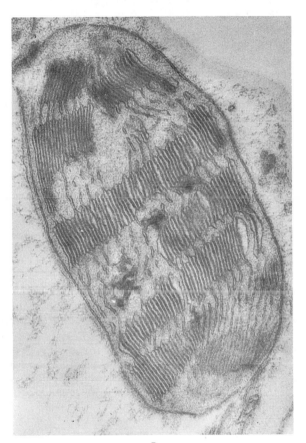

B

Figure 3.24 *Continued*

though one may suppose that it serves to mix molecules more rapidly than would occur by diffusion alone.

SUMMARY

All organisms share the fundamental life processes of growth, maintenance and excretion which are carried out in their cells by a multitude of enzyme-catalyzed reactions. To this extent, all cells also have common structural and functional characteristics and manufacture similar organic chemicals: chiefly, lipids, carbohydrates, proteins and nucleic acids. The macromolecules within cells are long polymers of similar repeating units; these units are such fundamental monomers as sugars, amino acids and nucleotides.

Cell membranes, composed of water insoluble lipids and proteins, provide selectively permeable barriers to the movement of ions and molecules into and out of the cell and between its various internal compartments. The cell nucleus, through the activity of the cistrons on its DNA molecules, directs the course of cellular reactions. Information sent to the cytoplasm in the form of complementary m-RNA is translated into protein language on the ribosomes by small t-RNA molecules; frequently the ribosomes are arranged along the intricate membranes of the rough endoplasmic reticulum. Proteins to be secreted from the cell are often packaged into secretory granules by the Golgi apparatus.

The unique shape of each protein determines its specific function. Regulation of cell metabolism can be effected either by synthesizing more enzymes in a particular pathway or by altering the shape of specific enzymes with activators or inhibitors. Energy for chemical reactions is released by the mitochondria during combustion of glucose. Excess nutrients are stored by a cell as lipid droplets, glycogen (in animal cells) or starch (in plant cells).

The DNA within the cell nucleus also provides a blueprint of information needed in cell reproduction. The complementary structure of the two strands of its double helix permits accurate self-duplication.

The cells of higher plants have additional structures. The rigid cell wall allows the cell to take up water osmotically without swelling, thus maintaining an internal pressure or turgor. The central vacuole helps maintain turgor and is a repository for ions, organic chemicals and pigments. Chloroplasts are the site of photosynthesis, in which sunlight is converted to chemical energy for cell maintenance and for synthesis of glucose from carbon dioxide and water.

READINGS AND REFERENCES

DeRobertis, E. D. P., Nowinski, W. W., and Saez, F. A. 1970. *Cell Biology.* 5th ed. W. B. Saunders Co., Philadelphia. A detailed account of the structure and function of various types of cells.

Fawcett, D. 1966. *The Cell.* W. B. Saunders Co., Philadelphia. Some of the best electron-micrographs in existence of the organelles of animal cells, together with brief descriptions of their function.

Kennedy, D. 1965. *The Living Cell; Readings from the Scientific American.* W. H. Freeman & Co., San Francisco. A well-illustrated collection of articles about cells and their properties, intended for the non-specialist.

McElroy, W. D. and Swanson, C. P. 1968. *Modern Cell Biology.* Prentice-Hall, Englewood Cliffs, New Jersey. Treats the material in this chapter, as well as some later ones, in greater detail but nevertheless in a readable style.

ALCOHOL: A DRUG AFFECTING CELL FUNCTION

Cells of the human body, like all cells, are finely tuned, complex machines, adaptable to limited changes in their environment such as the addition of moderate amounts of alcohol. Chronic large doses, however, produce changes in cell structure whose relation to altered cell function is just now becoming understood.

Alcohol has been known and used by human societies since prehistoric times, and its use and abuse have long affected cultural mores. Its use in moderate amounts is a socially accepted custom in many parts of the world. Used to excess, it becomes not only socially disruptive, but also a public health problem. For alcohol taken in large amounts is a drug, producing changes in body chemistry parallel to those caused by other depressants, such as barbiturates and tranquilizers. Unfortunately many individuals use alcohol in excess; there are approximately six to eight million persons addicted to alcohol in the United States, more than one in every 20 adults. Alcoholism is a greater medical and social problem than all other forms of drug abuse combined!

WHAT IS ALCOHOL?

The alcohol we drink is one of a family of chemicals, all called alcohols. Technically, any organic compound containing a hydroxyl group—an oxygen and hydrogen atom joined together—is an alcohol. Thus, all sugars are alcohols. The alcohol series to which the beverage belongs, however, are molecules composed of carbon and hydrogen which have a single hydroxyl (OH) group at one end. Three examples are shown in Figure 3a.1. The two-carbon ethyl alcohol, also known as *ethanol,* is the one found in alcoholic beverages. Other alcohols may be far more toxic; 15 to 30 grams (½ to 1 ounce) of methyl alcohol or wood alcohol may cause serious illness and even blindness. Larger amounts can be lethal.

THE PRODUCTION OF ALCOHOL

Alcoholic beverages are produced by the fermentation of sugar in fruit juices or other plant extracts by yeasts. In the process, carbon dioxide and alcohol are formed. The evolution of gaseous carbon dioxide by baker's yeast is used to make bread rise; it also causes the bubbles in beer and champagne, which are fermented in tightly closed containers. Starchy substances, such as barley, rye, corn and rice, cannot be attacked directly by yeasts. Germination of the grain must first take place, which causes release of an enzyme that splits the starch to sugar—a process known as malting.

Fermentation continues until the alcohol content reaches around 14 per cent, at which point the yeast cells are killed by the alcohol itself. At high concentrations, alcohol kills all living cells, and is often used as a disinfectant and as a preservative. Stronger spirits are obtained by distillation. Beer generally contains about 5 per cent alcohol, natural wine 8 to 14 per cent, fortified wine up to 20 per cent, and spirits or liquor, 40 per cent or more.

METABOLISM OF ALCOHOL

Although alcohol is often used externally—in skin bracers and face lotions, for rubbing alcohol, and in disinfectants—our concern is with its internal consumption. When swallowed, alcohol is completely absorbed by the stomach and intestines, passing directly into the blood (Fig. 3a.2). It circulates in the blood, diffusing into all organs of the body and having a marked effect on the central nervous system. Eventually 95 per cent of the alcohol taken in is broken down by oxidation in the liver, the remaining 5 per cent being excreted unchanged in the urine and breath. The latter fact forms the basis of "Breathalyzer" tests for drunken driving. Most alcoholic beverages contain little or no proteins or vitamins and have little nutritional value other than the calories contained in the alcohol itself or in the remaining unfermented carbohydrates.

The liver is the major organ which removes alcohol from the blood, converting it to other chemicals. The capacity of the liver to metabolize alcohol is limited, however; it can process only so much an hour, no matter how much has been drunk. This capacity varies considerably from one individual to another.

THE ROLE OF THE LIVER

Most ingested alcohol is initially metabolized in the liver by a two-step oxidation process in which hydrogen is removed from alcohol and oxygen is added to it (Fig. 3a.3). Two enzymes which are found in the cytoplasm of liver cells are involved. The resulting product, acetate, is a non-toxic, normally occurring chemical. Acetate leaves the liver via the blood and is even-

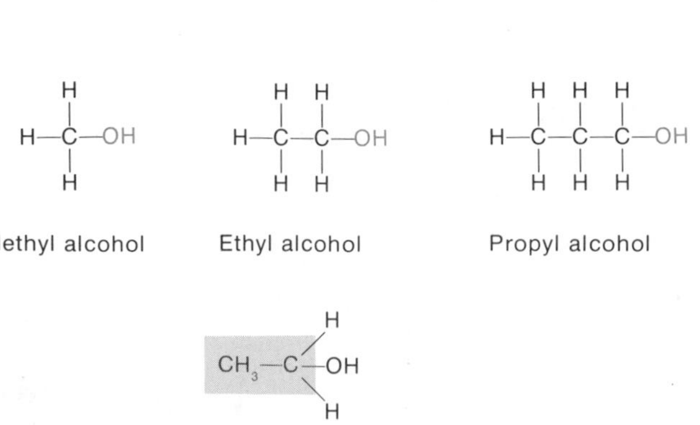

Methyl alcohol   Ethyl alcohol   Propyl alcohol

FIGURE 3a.1

Three simple alcohols, shown in chemist's notation. All have single OH— group at one end. Ethyl alcohol (ethanol) occurs in alcoholic beverages. A somewhat simplified formula of ethanol (which we shall use) is shown at bottom.

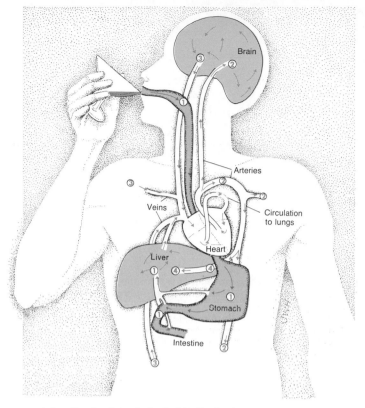

A schematized diagram of the distribution of alcohol in the body.

1. Ingested alcohol is virtually 100 per cent absorbed by stomach and small intestine, passing via blood vessels to the liver.

2. What the liver cannot immediately metabolize is pumped in the blood by the heart, first through the lungs and then via arteries to the peripheral tissues.

3. The alcohol is returned via the veins to the heart and recirculated.

4. At each recycling, a small fraction of blood enters the liver, which removes alcohol from it at a steady rate.

tually burned to carbon dioxide and water in other tissues of the body, providing them energy in the process.

During this preliminary oxidation of alcohol in the liver, a small hydrogen carrier molecule, known as **NAD** (for nicotine adenine dinucleotide), is needed to transfer the H atoms by a series of steps to the mitochondria. There the hydrogens are combined with oxygen to produce water, at the same time releasing energy for the liver cell.

PHYSIOLOGICAL EFFECTS OF ALCOHOL

The overt effects of alcohol consumption are familiar to almost everyone. Not all individuals react similarly — some become depressed or drowsy, while others are hyperactive, at least initially, probably due to suppression of inhibitions. Heart rate, brain waves, respiration, body temperature and other indicators of physiological function are affected. That judgment and

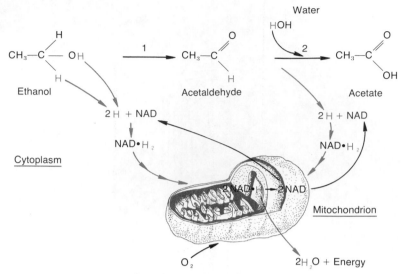

FIGURE 3a.3

Liver reactions which remove alcohol from the blood. The two enzyme-catalyzed oxidation steps, *1* and *2*, result in removal of hydrogen and addition of oxygen, producing
acetate which is non-toxic.
 The hydrogen atoms are transferred in several steps (multiple arrows) to mitochondria by the hydrogen transport molecule NAD. In the mitochondria, H atoms and O atoms
are combined to form water and releasing energy. The fate of H atoms is shown in color.

coordination are impaired is evident from the fact that alcohol is a factor in
more than half of the fatal automobile accidents in North America.
 Although, as we have seen, ingested alcohol is quickly carried via the
blood to all tissues of the body, it has its major effects on the nervous system, especially the brain. It acts as a depressant, reducing the ability of
nerve cells to conduct messages. It is still not certain how this effect is
brought about, but it seems to result from the action of alcohol on the functioning of the nerve cell membrane—the key structure in conduction of
nerve impulses, as we shall see in Chapter 12.
 Large quantities of alcohol drunk over a short period can be fatal.
Death usually results from severe depression of the breathing centers in
the base of the brain, but sometimes it is due to circulatory collapse.

EFFECTS OF CHRONIC ALCOHOLISM

 As long as the level of alcohol intake does not exceed the liver's capacity to metabolize it at a reasonable rate, there appears to be no permanent
damage to the body from its use. The equivalent of as much as five to six
ounces of straight alcohol per day, even if consumed regularly for years,
may do no harm. But prolonged consumption of nine or more ounces per
day may produce more or less permanent physiological changes. It is, of
course, necessary to remember that there are significant differences in the
way individuals metabolize alcohol; some are far more tolerant than others.
It has recently been found that many of the changes induced by chronic alcoholism, once attributed to the usually poor nutritional state of alcoholics,
can occur even if a balanced diet is being eaten.
 Heavy drinkers may suffer from disorders of the intestine, circulatory
system, lung, kidney, pancreas and nervous system. Sleep may be dis-

turbed; irreversible brain damage may occur, with permanent loss of memory; epilepsy, psychotic symptoms and chronic incoordination may develop. Sexual impotence is not uncommon after either acute or chronic usage of alcohol. The main organ affected by the chronic usage of alcohol, however, is the liver, and death may result from damage to this vital organ.

Before discussing the effects of alcohol on liver function, we might point out that many people who are not socially considered to be alcoholics—that is, their ability to function reasonably normally is not impaired—are still suffering to varying degrees from physiological damage. Thus, although they are not the derelict alcoholics of skid row, their bodies may be altered in significant ways which could prove dangerous, as we shall soon see. Likewise, a person does not need to be physically dependent on alcohol, nor to experience delirium tremens ("DT's") on abstention before exhibiting internal signs of alcoholism.

TOLERANCE TO ALCOHOL

Chronic drinkers of large amounts of alcohol acquire a tolerance to its effects. This appears to be the result of two adaptive mechanisms, one in the brain, one in the liver. At the same blood alcohol level, chronic alcoholics are less drunk than normal people. In some fashion, as yet not understood, their brain tissues compensate for the presence of alcohol. In addition, the liver cells of alcoholics undergo modifications of structure and function which are thought to increase the rate of removal of alcohol from the blood. Apparently the liver cannot greatly increase the amounts of the enzymes catalyzing the normal conversion of ethanol to acetate (Fig. 3a.3); instead it utilizes an alternative chemical pathway, which we shall discuss shortly.

CHANGES IN LIVER CELL STRUCTURE

It has long been known that the normal functioning of the liver is impaired by chronic alcoholism. After a few weeks of heavy drinking, fat tends to accumulate in the liver cells, producing so-called "fatty liver." With long-continued drinking, parts of the liver begin to degenerate and the normal cells are replaced by scar tissue, a condition known as **cirrhosis**. Passage of blood through the liver becomes impaired, causing its liquid components to be squeezed into the abdominal cavity; this condition is called **ascites**. As the ascites fluid accumulates, the abdomen becomes grossly distended. The ability of the liver to process foods is diminished, and the nutritional state of the patient deteriorates. Only in the past few years, however, have scientists begun to unravel the causes of these changes in terms of the structure and function of liver cells.

Normal Liver Cells. Liver cells perform a multitude of functions, and are metabolically the most versatile cells of the body. The liver stands as a sort of processing station between the gut and the rest of the body, converting raw materials of the diet into substances which are then transported by the blood to be used elsewhere.

A normal liver cell is very similar to the generalized animal cell discussed in Chapter 3 (refer to Figure 3.13). An electronmicrograph of a rat liver cell is shown in Figure 3a.4A. Note the rather extensive endoplasmic reticulum, most of which is covered by ribosomes, giving it a "rough" appearance. This is known as **rough endoplasmic reticulum** (rough-ER or RER); it functions in the production of a variety of enzymes and other proteins, chief of which are the plasma proteins of the blood (discussed further in Chapter 10). In a few regions of a normal liver cell, however, one finds **smooth endoplasmic reticulum** (smooth-ER or SER), whose vesicular membranes lack ribosomes.

Liver Cells in Alcoholism.   A change in the appearance of liver cells is observed, however, after chronic ingestion of various toxic substances; for instance, after prolonged ingestion of drugs such as barbiturates, of organic solvents like carbon tetrachloride, or of low levels of DDT, the amount of SER increases markedly. A similar result is found in animals or in human volunteers after chronic ingestion of alcohol. In humans, biopsy samples for study are taken through the rib cage with a small bore probe. Changes such as those seen in Figure 3a.4B are brought about after 10 to 14 days of heavy alcohol consumption and occur even if the subject is eating an otherwise normal diet. They are thus a direct result of the presence of alcohol. This leads us to ask: What is the function of the SER?

DETOXIFICATION FUNCTIONS OF SER

Not only does the liver process foodstuffs, it also is the first line of defense against potentially toxic chemicals that are swallowed and absorbed, among which is excess alcohol. In addition, the liver detoxifies some of the body's own natural waste products, converting them into forms which can be excreted.

Almost all detoxifying reactions involve non–energy-yielding oxidative steps. Either the substance is completely oxidized to carbon dioxide and water, as occurs with alcohol, or else it is partly oxidized by *hydroxylation*

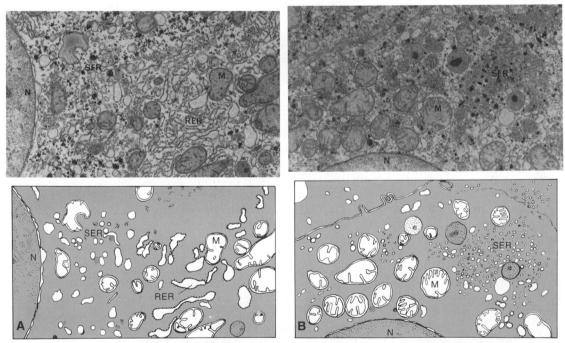

FIGURE 3a.4   Fine structure of liver cells, shown in electronmicrographs and matching drawings (magnification 11,000×).
A. Normal rat liver: Much rough-ER (*RER*) and little smooth-ER (*SER*). Dark granules are glycogen. *N,* nucleus; *M,* mitochondrion.
B. Liver of rat fed ethanol for 15 days: SER greatly increased; appears grey because tubules filled with electron-dense material. Plasma membranes of two adjacent cells run across top. (Electronmicrographs from Rubin, E., and Lieber, C. S. 1971. *Science* 172: 1098. Copyright © 1971 American Association for the Advancement of Science.)

and then is **conjugated.** In the latter step, some normal physiological compound, usually a derivative of an amino acid or sugar, is added to the molecule, making it relatively safe for blood transport and excretion by the kidneys. An outline summary of these reactions is shown in Figure 3a.5.

Although these reactions can occur in many tissues of the body, the major site of detoxification is the liver. Recently it has been shown that most of the oxidative steps involved take place on the smooth endoplasmic reticulum (SER), and to a lesser extent on the rough endoplasmic reticulum (RER) as well. As mentioned in Chapter 3, the mitochondria are the power plants of the cell; the oxidation of normal compounds by their enzymes releases energy for cell work. Thus, two different oxidative enzyme systems function in cells; one for energy metabolism, located in the mitochondria, and one for detoxification and excretion, sited mainly on the SER.

FIGURE 3a.5

Summary of detoxification reactions carried out by endoplasmic reticulum. A toxic compound or a normal body chemical which is to be excreted, such as a sex hormone, shown by *R*, is partially oxidized by hydroxylation. It may then be either oxidized completely or else conjugated with a normal compound, *X*, which renders it harmless.

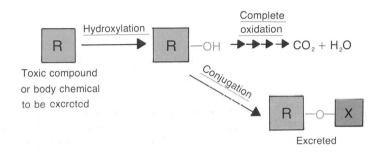

The enzymes of the SER may be separated from those of the mitochondria by **differential centrifugation,** as shown in Figure 3a.6. After breaking up the cells of a tissue such as liver in a blender, the investigator takes advantage of the different weights of the cell organelles in order to separate them. By repeatedly centrifuging (spinning) the resultant suspension at increasingly higher speeds, the various sub-cellular particles are brought to the bottom of the test tube. The nuclei are largest and centrifuge out at relatively low speeds. The remaining suspension is removed,

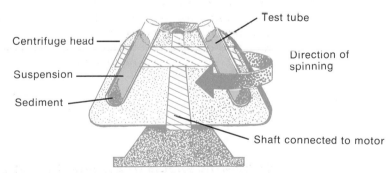

FIGURE 3a.6

Differential centrifugation of organelles from disrupted cells. At low speeds the nuclei are in the sediment; remaining organelles are in suspension. Recentrifuging the suspension at medium speeds removes mitochondria and lysosomes. Repeating the process at high speeds sediments microsomes.

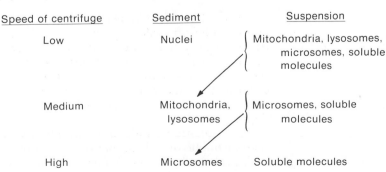

Speed of centrifuge	Sediment	Suspension
Low	Nuclei	Mitochondria, lysosomes, microsomes, soluble molecules
Medium	Mitochondria, lysosomes	Microsomes, soluble molecules
High	Microsomes	Soluble molecules

transferred to new tubes and centrifuged again at higher speed. The mitochondria and other structures such as lysosomes come down at intermediate speeds, and a fraction known as **microsomes** is centrifuged down at very high speeds. A final fraction, the **supernatant,** contains mainly soluble enzymes and small molecules.

The microsomes are small pieces of the fragile endoplasmic reticulum, which is disrupted during blending. The fragments still contain most of the enzymes originally bound to them in the living cell, however. On examining the chemistry of the microsomes, one finds that the oxidative enzymes involved in detoxification are located on them — a finding one would expect after observing an increase in SER upon chronic exposure to toxic compounds.

We can thus conclude that the SER and, to some extent, the RER are the site of detoxifying enzymes, and that these are increased in cells, especially liver cells, after ingestion of toxic compounds, including alcohol.

BIOSYNTHESIS OF SER ENZYMES

Although it is not known precisely how toxic substances cause an increase in the enzymes of the SER, it is probable that the normal pathways of protein synthesis are involved. Evidence for this is the following: Before a drug is detoxified, it must first bind onto the microsomes — specifically, onto a particular protein called **cytochrome P450.** It is known that liver cells use another protein, the enzyme δ-aminolevulinic acid synthetase (called ALAS for short), to produce part of the cytochrome P450 molecules. The amounts of ALAS are markedly increased soon after ingestion of drugs such as alcohol and barbiturates (Fig. 3a.7, A and B). It seems likely that this results in the synthesis of more cytochrome P450, thus providing more binding sites on the microsomes for toxic chemicals and hence causing more rapid detoxification. These latter speculations have yet to be demonstrated, however.

If inhibitors of protein synthesis are given to experimental animals prior to alcohol treatment, however, no increase in ALAS occurs. Two antibiotics which act as specific inhibitors at two different steps in enzyme synthesis are often used by biochemists. One of these, actinomycin, prevents transcription of genetic information from DNA to m-RNA. The other, puromycin, prevents translation of the RNA message into protein at the level of the ribosomes. Both these agents, as shown in Figure 3a.7, C and D, prevent the synthesis of increased amounts of ALAS after ingestion of alcohol by experimental animals. Alcohol is thus said to cause the **induction** of ALAS synthesis. It is not known whether alcohol induces synthesis of ALAS by acting directly on the DNA cistron or whether some intermediate chemical in the cell is involved.

SYNERGISTIC EFFECTS OF ALCOHOL AND DRUGS

It has long been known that not only do chronic consumers of alcohol become tolerant to alcohol, presumably because of adaptations in their central nervous system as well as the increased SER in their liver cells; they also develop tolerance to other drugs, such as barbiturates. A sober alcoholic requires more sedative than a non-drinker to produce the same effect. This suggests that the adaptations of both the brain and liver to chronic heavy usage of alcohol confer cross-tolerance which is shared by other types of depressants.

On the other hand, *anyone* who has been drinking heavily, whether he is an alcoholic or only an occasional drinker, is highly sensitive to barbiturates and similar drugs. A normal dose of sleeping pills easily becomes an

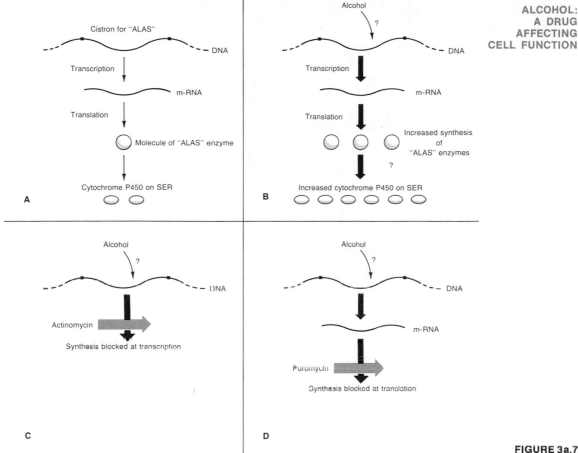

Evidence that alcohol stimulates synthesis of the enzyme "ALAS," which helps synthesize cytochrome P450, by acting on DNA.

A. The presumed pathway of "ALAS" synthesis is the same as that of any protein.

B. In the presence of alcohol, this pathway is thought to show increased activity (heavy black arrows). The question mark beside alcohol indicates that it is not known if the alcohol acts directly on DNA or via some intermediate. The question mark beside P450 synthesis indicates that this effect has not yet been demonstrated.

C. After prior treatment with the antibiotic, actinomycin, which blocks protein synthesis at the transcription step, no "ALAS" is formed in the presence of alcohol (colored arrow).

D. After treatment with puromycin, which acts at the transcription step, no "ALAS" is formed.

overdose after drinking, a fact which results in many tragic accidental deaths. This apparent paradox—resistance to drugs in sober alcoholics and sensitivity in anyone who is intoxicated—has long baffled physiologists and doctors and makes prescription of sedatives a tricky matter.

Explanation of the Paradox. The explanation of this apparent paradox has been slowly worked out. Studies on microsomes isolated by differential centrifugation from liver samples of animals or human volunteers who had continuously drunk large quantities of alcohol showed that not only were the enzymes necessary to oxidize alcohol increased but that *all* detoxifying enzymes increased. The results of one such experiment are shown in Figure 3a.8. It is clear that the induction of SER enzymes is a nonspecific effect. Any toxic substance, including alcohol, causes an increase in a whole battery of enzymes.

Anyone exposed to one drug or toxic substance for a long period of time will thus show an increased rate of detoxification of a second drug. For example, when human volunteers were fed a single dose of the tran-

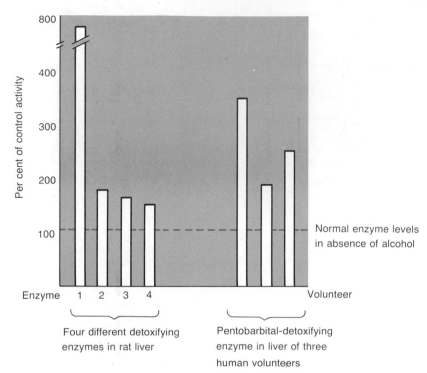

FIGURE 3a.8

Non-specific induction of detoxifying enzymes in liver microsomes of rats and man after chronic ingestion of alcohol. Levels in control animals are 100 per cent.

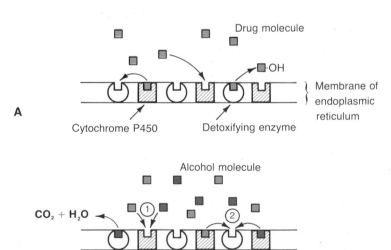

FIGURE 3a.9

Schematic diagram of detoxification of drug alone or in presence of alcohol.
   A. Drug molecules bind with cytochrome P450 sites, thus permitting detoxification by adjacent enzymes.
   B. Alcohol molecules may compete with drug either for P450 sites (*1*) or detoxifying enzymes (*2*).

quilizer meprobamate (Miltown), it required 16 hours for half the drug to disappear from their blood. After the same volunteers had consumed large quantities of alcohol daily for four weeks, the time taken for half an injected dose of meprobamate to disappear from their blood when they were sober was reduced to eight hours. Chronic alcohol consumption markedly *increased* the rate of detoxification of this drug. The same sort of non-specific induction of SER enzymes accounts for an increased tolerance to a variety of toxic substances in alcoholics.

On the other hand, the high degree of sensitivity of all inebriated persons to drugs and other toxic agents is more difficult to explain. It is probable that brain cells, although they exhibit some degree of cross-tolerance in alcoholics between alcohol and such other drugs as barbiturates, simply cannot withstand the additive effects of two depressants taken simultaneously. Another factor may be the competition between alcohol and a second drug for detoxification sites on the SER of liver cells. As noted above, cytochrome P450 is present in the SER, and there is evidence that any toxic substance must bind with it before being acted on by detoxifying enzymes. If this is true, then two toxic agents will compete either for cytochrome P450 or for detoxifying enzymes if they are present simultaneously. One can visualize this as in Figure 3a.9.

Evidence in support of this idea is given by the experiment shown in Figure 3a.10. A human volunteer was given an injection of the sedative Nembutal, a type of barbiturate; his blood was then sampled at regular intervals to measure the rate of removal of the drug by detoxification processes. Sixteen hours later, the subject was given a large dose of alcohol, followed by smaller amounts at regular intervals. The rate of disappearance of Nembutal from the blood slowed markedly soon after administration of alcohol.

Thus, although about 80 per cent of ingested alcohol is metabolized to acetate by the enzymes found in the cytoplasm of liver cells, the fact that

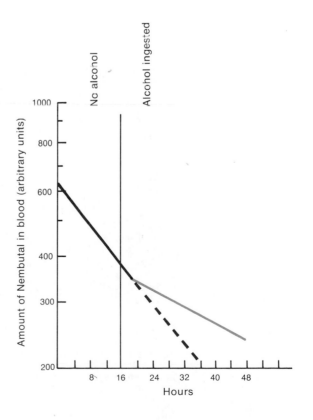

FIGURE 3a.10

Effect of acute alcohol ingestion on disappearance of drug from blood. Pentobarbitol (Nembutal) was ingested at 0 time and its concentration measured for 16 hours. Then approximately five ounces of alcohol were ingested, followed by one more ounce every two hours.

The rate of disappearance of the drug shows a sharp decline after alcohol ingestion. Dashed line indicates rate of drug metabolism had alcohol not been given.

some alcohol is oxidized by detoxifying enzymes of the SER means that excessive quantities of alcohol can seriously affect the ability of this system to remove other toxic substances from the blood. It is for this reason that doctors, on prescribing certain drugs, may prohibit the use of alcohol during the period of treatment. Since these same SER enzymes play a role in the metabolism and excretion of normal body chemicals, such as the sex hormones, it is possible that chronic alcoholism may also affect the physiological processes regulated by the these body chemicals.

OTHER EFFECTS OF ALCOHOL ON CELLS

Several other effects of alcohol on cell structure have been observed. In most cases, the significance of the observations is not yet clear, although one can surmise what functional changes may be taking place. We shall mention but two examples here: the effect of alcohol on the mitochondrial structure of liver cells, and its effect on the architecture of muscle cells.

CHANGES IN LIVER CELL MITOCHONDRIA

Earlier we noted that chronic ingestion of large amounts of alcohol results in fatty liver. This is indicated at the cellular level by the formation of an abnormal number of large lipid droplets in the cytoplasm. At the same time, many of the cell's mitochondria become associated with these droplets (Fig. 3a.11). The mitochondria also become abnormal in appearance, swelling greatly and losing their cristae; sometimes they accumulate large crystalline granules within their matrix. These changes suggest that the functioning of the mitochondria is no longer normal and that, in some way, they are associated with the chemical reactions taking place in the formation of the fat droplets. Since, as we shall see in Chapter 7, mitochondria play an important role in energy metabolism, we might suspect that the normal functions of these altered mitochondria are perhaps impaired.

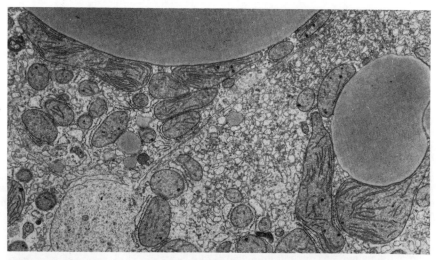

FIGURE 3a.11

Electronmicrograph of liver biopsy from human volunteer fed ethanol for 10 days (magnification 9,900×).

Many mitochondria are enlarged with abnormal cristae. SER is increased. Many fat droplets are present, which are surrounded by mitochondria. (From Rubin, E., and Lieber, C. S. 1967. Early fine structural changes in the human liver induced by alcohol. *Gastroenterology 52*:1–13. © 1967 The Williams & Wilkins Co., Baltimore.)

One of the symptoms of chronic alcoholism is muscular tenderness and weakness. The muscles become inflamed and show signs of degeneration. Recent studies were carried out on non-alcoholic human volunteers given large doses of alcohol for several weeks while eating an otherwise balanced diet. Electronmicrographs of muscle biopsies taken from these subjects indicated a decrease in the contractile proteins and an increase of both lipid droplets and endoplasmic reticulum, as shown in Figure 3a.12A. Six months after cessation of excessive alcohol intake, the muscles of the same volunteers had recovered a normal appearance (Fig. 3a.12B). Observations such as these offer an explanation for some of the symptoms of chronic alcoholism. How they come about still remains to be explained.

SUMMARY

Alcohol taken in large quantities becomes a drug which must be eliminated. Although moderate amounts are readily metabolized to acetate, a normal body chemical, by enzymes found primarily in the cytoplasm of liver cells, excessive quantities are eliminated by the aid of enzymes

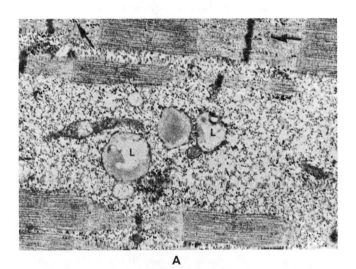

A

FIGURE 3a.12

Alcohol-induced changes in human muscle cells, as shown in electronmicrographs of biopsied tissue (magnification 13,500×).

A. After 28 days of alcohol ingestion, many contractile fibrils are replaced by lipid droplets, endoplasmic reticulum and other elements. *L,* lipid.

B. Six months after cessation of alcohol intake, muscle appears normal, being again composed predominantly of contractile fibrils. (From Song, S. K., and Rubin, E. 1972. Ethanol produces muscle damage in human volunteers. Science *175*:327–328. Copyright © 1972 American Association for the Advancement of Science.)

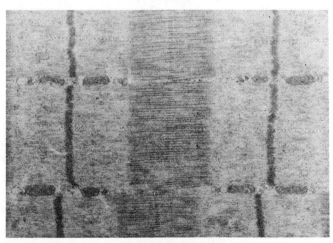

B

located on the endoplasmic reticulum of these cells, especially the SER. Prior to being metabolized by the SER, a molecule of alcohol must bind to a specific substance on its membrane, cytochrome P450. Both this binding site and the detoxifying enzymes of the SER and RER are shared by other toxic substances taken into the body as well as by certain natural compounds of the body normally destroyed by the liver. Alcohol thus competes with other chemicals for the body's defense mechanisms against toxic substances. This may partly explain the heightened sensitivity to drugs experienced by inebriated persons. Another cause lies in the additive effects of alcohol and other depressants on the brain. Accidental deaths are frequently encountered among inebriated persons who take overdoses of sleeping pills, tranquilizers or other depressants of the central nervous system.

Chronic alcohol consumption paradoxically leads to an opposite result. The sober alcoholic shows heightened tolerance to many depressant drugs, including barbiturates, tranquilizers and anesthetics. In part this is due to the as yet little understood cross-tolerance of the central nervous system to various drugs. In part it results from more rapid detoxification by non-specific liver enzymes originally induced by excessive alcohol consumption. The latter may also accelerate the metabolism of natural body chemicals, such as sex hormones, possibly upsetting normal body functions.

Other changes in cell structure resulting from alcoholism have been observed; their significance remains to be demonstrated.

READINGS AND REFERENCES

Israel, Y. and Mardones, J. (eds.) 1971. *Biological Basis of Alcoholism.* Wiley-Interscience, New York. A collection of articles by experts on the effects of alcohol on everything from cellular organelles to society.

Kissin, B. and Begleiter, H. 1971. *The Biology of Alcoholism.* Plenum Press. New York. Volume 1 (Biochemistry). Review articles by authorities in various fields on the effects of alcohol on cell structure and metabolism. Volume 2 (Physiology and Behavior) and Volume 3 (Clinical Pathology) are in press.

The Non-Medical Use of Drugs. 1970. Interim Report of the Canadian Government's Commission of Inquiry. First published by Information Canada; Penguin Edition, 1971. A useful paperback, written in layman's language, about physiological, social and legal aspects of drug use in North America.

Remmer, H. and Merker, H. J. 1965. Effect of drugs on the formation of smooth endoplasmic reticulum and drug-metabolizing enzymes. Annals of the New York Academy of Sciences *123*:79–96. A summary of the early evidence of non-specific action of drugs on liver cells.

Rubin, E. and Lieber, C. S. 1971. Alcoholism, alcohol and drugs. Science *172*:1097–1102. An excellent article summarizing many of the experiments discussed in this chapter.

Song, S. K. and Rubin, E. 1972. Ethanol produces muscle damage in human volunteers. Science *175*:327–328. Describes effects of alcohol on muscle cell structure.

Wallgren, H. and Barry, H., III. 1970. *Actions of Alcohol.* Elsevier, Amsterdam. Volume 1. Biochemical, Physiological and Psychological Aspects. Volume 2. Chronic and Clinical Aspects. An exhaustive and critical monograph of recent work relating to various aspects of the subject of alcohol.

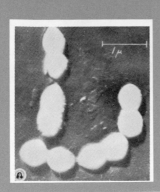

4

THE SIMPLEST FORMS OF LIFE

Bacteria, all too often regarded as categorically evil, perform many essential functions in the world, as do their relatives, the blue green algae. Only the enigmatic viruses can perhaps be regarded as altogether unfriendly to life, yet they, too, may prove useful to man.

It is not an easy matter to place all types of living organisms into distinct categories, such as plants or animals. Sponges, which are classed as animals, would hardly be thought of as such on casual observation, for they are firmly attached to their substrate, are virtually non-motile, and do not respond overtly to stimuli. The Venus flytrap, with all the characteristics of a higher plant, is yet able by means of relatively rapid movements to capture insects in its leaves. There is a group of single-celled, highly motile organisms called euglenoids which behave like animals but contain chlorophyll and are capable of synthesizing their own food; both botanists and zoologists have claimed these organisms over the years. Thus, any attempt to place a particular organism in a given group must to some extent be arbitrary. It is our purpose here, however, not to attempt a comprehensive classification of living organisms but to introduce the reader to the diversity of life forms and underline some of the functions of the major groups in the biosphere.

In this chapter we shall consider some of the simpler forms of life — the viruses, bacteria and blue-green algae. The two following chapters will survey, in turn, the plants and animals. A more complete classification of living organisms is presented in Appendix III.

BACTERIA

Having discussed in Chapter 3 the structure and function of typical animal and plant cells, we must now backtrack and consider those organisms which, although capable of performing the life functions of growth, maintenance and reproduction, exist as a single cell. Among the most familiar of these single-celled organisms are the bacteria, of which there is a great diversity throughout the world. Nevertheless, most bacteria share certain features in common.

BACTERIAL STRUCTURE

Bacterial cells are generally quite small, about 1/10,000 inch in length. A "typical" bacterium is shown in Figure 4.1. Note its cell wall, its single, circular chromosome not contained in a nucleus, and the free ribosomes; there are no endoplasmic reticulum, mitochondria or other distinctive organelles. The energy-yielding functions of the mitochondria are carried out by enzymes attached to the inner surface of the folded plasma membrane. Some bacteria are motile, bearing a long filamentous structure called a *flagellum*, which beats with a whip-like action. Not all bacteria are elongate; examples of the varied shapes and growth forms exhibited by different species are shown in Figure 4.3.

The bacterial cell wall is not composed of cellulose, like that of higher plants, but is made up of polysaccharides combined with proteins—substances known as *mucopolysaccharides;* lipids are also present in some species. These molecules are specific to each type of bacterium and, on contacting tissues of man or other higher animals, they exhibit *antigenic* properties; that is, they induce the formation of *antibodies* (a subject to be discussed further in Chapter 10). The antibiotic penicillin inhibits synthesis of new cell walls by bacteria and thus prevents their growth and division.

Despite their simple structure, bacterial cells are capable of complex chemical activities. Only a few bacteria contain a light sensitive pigment, bacteriochlorophyll, and are therefore capable of photosynthesis. Most species, like animals, require organic carbon as an energy source, although some can obtain energy from oxidation of inorganic molecules such as hydrogen sulfide (H_2S), sulfur (S), reduced iron (Fe^{++}) or ammonia (NH_3). Not all bacteria require oxygen. Those which do are called *aerobic* bacteria, while those which can live in the absence of oxygen are *anaerobic.* Among the latter are some of the bacteria most dangerous to man, especially those which produce potent toxins, such as the microbes that cause gas gangrene and botulism.

BACTERIAL GROWTH

Bacteria reproduce primarily by simple fission; the chromosome doubles and the cell divides in two. During such *asexual reproduction,* all offspring are genetically identical with the parent cell. Rarely, however, two cells come together and exchange parts of their chromosomes. On these occasions a recombination of genetic information occurs, such that the resulting cells are different from their parents; *sexual reproduction* has taken place.

Bacteria are capable of extremely rapid growth; under favorable conditions cells may divide every 20 minutes. This rapid rate of proliferation soon leads to an enormous population, as shown in Figure 4.2. Starting with a single bacterium, in six hours 18 divisions will have occurred, giving rise to 2^{18} offspring (or 250,000 bacteria). If no bacteria die and food is unlimited, as shown by the dotted line, in 24 hours 2^{72} bacteria—or about 4,000,000,000,000,000,000,000 —will have been produced! Such a population would weigh 200 tons. This type of growth pattern is known as a geometric increase; the absolute increase over each time interval is greater than that in the previous interval. Obviously bacterial populations never reach such proportions in any real situation. The solid line in Figure 4.2 indicates a leveling off in population size brought about by either a depletion of nutrients or an accumulation of toxic waste products.

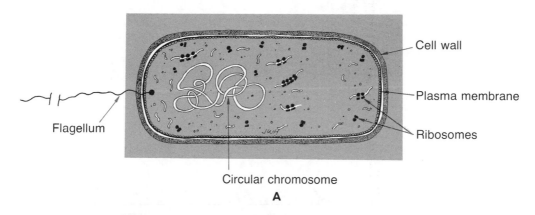

Cell wall

Plasma membrane

Flagellum

Ribosomes

Circular chromosome

A

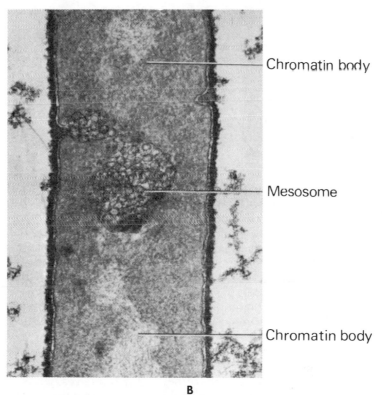

Chromatin body

Mesosome

Chromatin body

B

A typical bacterial cell.   **FIGURE 4.1**

  A. Diagrammatic sketch showing cell wall, circular chromosome, plasma membrane
and scattered clusters of ribosomes. Some but not all bacteria have one or more flagella
for locomotion.

  B. An electron photomicrograph showing part of the structure of a bacterial cell.
Note the presence of two chromosomes in preparation for cell division. (From van Iterson,
W. 1965. Bacterial cytoplasm. Bacteriological Reviews *29*(3):302, 303.)

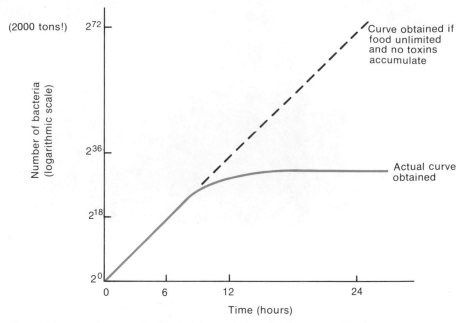

FIGURE 4.2

Geometric growth curve of a bacterial population dividing every 20 minutes.
In 24 hours, 2000 tons of bacteria would form if there were no limiting factors. (For explanation of exponentials, see Appendix II.)

FUNCTIONS OF BACTERIA

The roles played by bacteria in the biosphere are legion; they are to be found virtually everywhere. One indispensable function is the decomposition of dead organic matter into forms utilizable by plants and some animals. This recycling of nutrients occurs wherever there are dead organisms—in soils, at the bottom of lakes, in streams, in the oceans, in sewage plants and even in glaciers and snow banks, where "pink snow" often signifies the presence of bacterial populations. Another important job performed by bacteria is the conversion of nitrogen (N_2) in the air to nitrate (NO_3^-) which can be utilized by plants. **Nitrogen fixation,** as it is called, is carried out by special groups of bacteria. Some of these live in nodules on the roots of leguminous plants, such as peas, soybeans and clover; others are free-living forms existing in water or soil.

Bacteria also live in association with many animals, especially in the gut cavity, where they often perform special services for the host animal. Bacteria in our own digestive tracts, for example, produce and release several vitamins which we cannot synthesize for ourselves. Thus, when taking oral antibiotics it is wise to take vitamins at the same time, to make up for those no longer produced because one's normal complement of bacteria has been suppressed. Cows and other ruminants retain a host of intestinal bacteria capable of breaking down the cellulose of the grass they eat, thus permitting them to utilize food they could not otherwise digest.

It is probable that populations of many species of animals are limited by bacterial infection. In the past, disease-producing bacteria, as well as other infectious organisms, often took a heavy toll of human populations, a subject to be considered in relation to plague in Chapter 19. Some of these pathogenic bacteria are shown in Figure 4.3, together with two non-pathogenic species. The development within the past hundred years of public health measures, of aseptic clinical procedures and especially of antibiotics has largely controlled widespread bacterial epidemics in many parts of the world. In areas where population growth is exceeding the rate of increase of medical services, however, epidemics of infectious diseases are likely to become increasingly prevalent.

Man himself makes use of bacteria in many industrial processes. The production of cheese and yogurt are the result of bacterial activity; so is the manufacture of vinegar. A particular strain of bacteria provides the unique flavor of sourdough bread. Some antibiotics are also bacterial products, synthesized originally to protect one bacterial species from another. Dozens of other human uses of bacteria could be mentioned.

BLUE-GREEN ALGAE

Many simple photosynthetic organisms are called algae; in fact, the term covers a rather diverse group of organisms. The commonest algae,

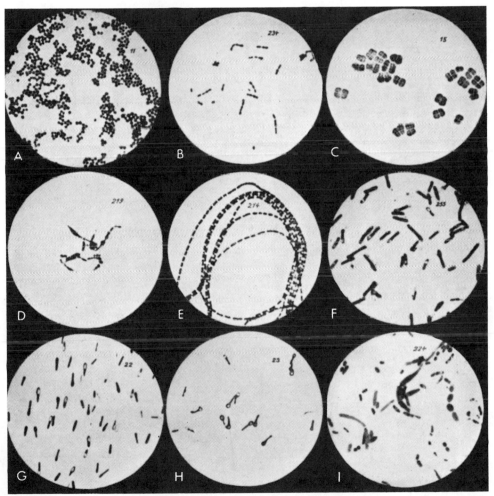

Various bacteria, some pathogenic, associated with human beings. Diversity of morphological types (all magnified 1000 ×) is shown.
    A. Normal coccus from human sputum
    B. Coccus causing bacterial pneumonia
    C. Normal coccus from stomach
    D. Bacillus causing diphtheria
    E. Anthrax bacillus
    F. Bacillus producing botulinum toxin
    G. Bacillus causing gas gangrene
    H. Bacillus causing tetanus
    I. Bacillus causing plague
  (From Dubos, R. J. (ed.) 1948. *Bacterial and Mycotic Infections of Man.* J. B. Lippincott Co., Philadelphia.)

FIGURE 4.3

the unicellular green algae, have a nucleus and a complex life cycle; the seaweeds or macroalgae are close relatives of theirs. They are all considered to be plants and will be discussed in the next chapter. The **blue-green algae,** on the other hand, are primitive, single-celled organisms somewhat similar to bacteria, as can be seen in Figure 4.4.

STRUCTURE OF BLUE-GREEN ALGAE

Like bacteria, blue-green algae lack such common organelles as mitochondria, endoplasmic reticulum and a Golgi apparatus. Their genetic material consists of short filaments of chromatin generally localized in one region of the cell. Their cell wall is composed mainly of cellulose; externally they often secrete a sticky, gelatinous sheath which protects the cell. The feature which most distinguishes them from bacteria, however, is their con-

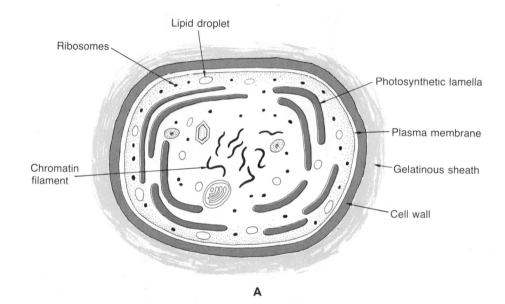

A

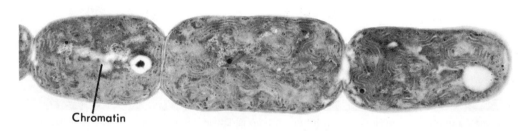

B

FIGURE 4.4   Structure of blue-green algae.
 A. Diagram of the major structures found in blue-green algae. Outside the cellulose-containing cell wall is a protective gelatinous sheath. Photosynthetic pigments are bound to simple membranous lamellae. Lipid droplets are common in the cytoplasm. Genetic material is located on chromatin filaments, generally restricted to one region of the cell.
 B. Electronmicrograph of a common filamentous blue-green alga, *Anabaena* (magnification 8000 ×). (From Carpenter, P. L. 1972. *Microbiology.* W. B. Saunders Co., Philadelphia.)

tent of photosynthetic pigments, located not in discrete chloroplasts but in long lamellae scattered throughout the cell. Characteristic pigments are the green pigment, **chlorophyll**, and a blue pigment, **phycocyanin**. Their collective name, **cyanophyte,** is derived from the name of the blue pigment. Some species additionally contain yellow and orange pigments and may be brightly colored. The blue-green algae are capable of photosynthesis and require carbon dioxide, water and light for their growth. They reproduce asexually by fission, and, although none of them has flagella, some species are capable of a curious oscillatory movement. Some exist as isolated cells, while others form filamentous threads of connected cells.

ROLE OF BLUE-GREEN ALGAE

Although, like bacteria, blue-green algae are among the most primitive of organisms, they perform several vital functions in the biosphere. Aquatic forms, found in both fresh and salt water, contribute to the overall productivity of their environment, providing an important source of food for animals. During natural or man-made eutrophication (enrichment) of lakes and lagoons, the blue-green algae increase enormously in number, especially during summer when light is plentiful. Together with other plants, they utilize the excess nutrients, particularly nitrate (NO_3^-) and phosphate ($PO_4\equiv$). The surface of a lake may turn bright green due to their presence. We shall discuss later some of the unpleasant results of these algal "blooms."

In nutrient-poor waters, by contrast, certain species of blue-green algae make a different sort of contribution to the overall productivity. Like nitrogen-fixing bacteria, they are capable of converting atmospheric nitrogen to nitrate, thus making this limiting nutrient available for the growth of other plants.

Although blue-green algae require moisture and cannot survive unprotected in dry terrestrial environments, a variety of species live in close association with certain fungi, giving rise to the familiar lichens, to be discussed further in Chapter 5.

VIRUSES

Viruses are a semantically perplexing subject—are they to be considered as living organisms or not? For a virus, by itself, has no living properties; it cannot grow, divide, metabolize nutrients or perform any other life functions. Only when these minute particles, far smaller than the average bacterium, are inside a living host cell, be it a bacterial cell, plant cell or human cell, do they become active and begin to multiply. Almost all species of organisms are attacked by viruses, although each virus is usually restricted to only one or a few types of host cells. In almost all instances, viruses produce diseases in the organisms they parasitize.

STRUCTURE OF VIRUSES

The structure of a typical virus is shown in Figure 4.5; it consists of an enlarged "head" and a "tail" piece. The tail and the coating of the head are protein, and the core of the head contains only nucleic acid, either RNA or DNA. Viruses have no mitochondria, ribosomes or other internal structures, and are totally unable to synthesize new molecules by themselves. The variety of sizes and shapes of viruses is to be seen in Figure 20.7.

Studies on viruses which infect bacteria—known as **phage,** which is

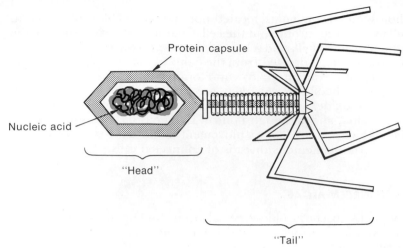

Protein capsule

Nucleic acid

"Head"

"Tail"

FIGURE 4.5   Elements of viral structure as exemplified by a bacteriophage, a virus which attacks bacterial cells. The "head" contains nucleic acid (in this case, DNA, although in some viruses it may be RNA). It is covered by protein which extends into the tail. (Not all viruses have a distinct tail, however; see Figure 20.7.)

short for bacteriophage or "bacteria-eating"—indicate that only the nucleic acid of the virus enters the bacterial cell (Fig. 4–6). Once inside, the viral nucleic acid usurps the role of the host cell DNA; it takes over the biochemical machinery of the host cell and directs the production of new molecules like itself. Eventually the whole bacterial cell becomes filled with new virus particles; it then ruptures, releasing them, and they in turn infect other cells. A similar sequence of events occurs in any type of host cell infected by a virus. As long as a virus particle does not dry out or get too hot it will remain infective indefinitely outside a host cell.

VIRAL DISEASES

Many plant and animal diseases are caused by viruses. In man, measles, mumps, chicken pox, poliomyelitis and numerous other diseases are all viral infections. Fortunately, each virus is usually specific for its own host organism, and often only attacks specific cell types within that host.

The mechanisms of resistance to viral infection are at present poorly understood. Some plant strains are virus-resistant, so genetic factors are probably important. In man, a non-specific protein, called **interferon,** is produced by infected cells and released into the blood soon after viral infection. This substance resists the initial infection, while specific antibodies, produced more slowly, result in longer-lasting immunity.

Currently attempts are being made to synthesize interferon in large quantities, but this is difficult since each type of interferon is effective only in the species which produces it. Viruses are known to cause some cancers in animals, but it has not yet been shown that they cause cancer in man. Because of their high host specificity, viruses may one day be successfully used to control pests. Since such viruses would attack only the target organisms, they would be far preferable to the non-specific, toxic insecticides and herbicides in use today.

SUMMARY

The microscopic bacteria and blue-green algae are structurally simple organisms which cannot be classified as either plants or animals. Despite their apparent lack of complexity, however, they are capable of carrying out important functions in the biosphere.

Most bacteria are non-photosynthetic, living on decaying organic matter, and are thus important in the recycling of nutrients. Growth is extremely rapid, being limited mainly by food supplies or accumulation of toxic wastes. Some species carry out nitrogen fixation; others perform digestive functions for host animals. Although some species cause disease, many types of bacteria are directly beneficial to man.

The blue-green algae contain chlorophyll and contribute to global photosynthesis. Some species are capable of nitrogen fixation, thus increasing productivity in oligotrophic waters. The only species found on land are those associated with fungi, with which they form lichens.

Viruses fall somewhere between the living and non-living world. These particles can reproduce only after injecting their nucleic acid into a host cell, where it takes over the cell's synthetic machinery and directs it to produce new viruses. Viruses are highly specific in the types of cells they infect. Resistance to viral infection in man results from the production of both interferon and antibodies. New uses for viruses in pest control may soon be developed.

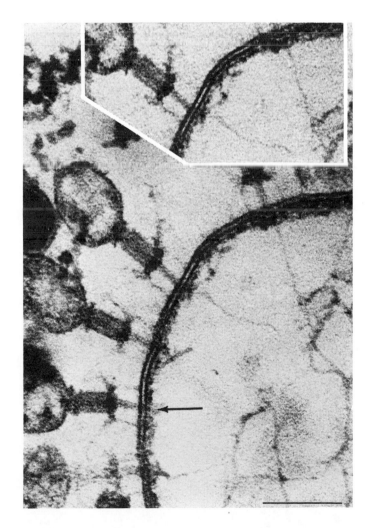

FIGURE 4.6

Phage particles attacking a bacterial cell. The tail attaches to the cell wall, and the DNA is injected through a tube (arrow) into the cell interior. Threads already visible within the cell are probably viral DNA. (Courtesy of the Institute for Cancer Research, Jeans Hospital, Philadelphia.)

READINGS AND
REFERENCES

Carpenter, P. L. 1972. *Microbiology*. 3rd ed. W. B. Saunders Co., Philadelphia. A useful, readily understandable introduction to microorganisms.

Curtis, H. 1966. *The Viruses*. Natural History Press, New York. Viruses in disease and research, written in layman's terms.

Dulbecco, R. 1967. The induction of cancer by viruses. Scientific American Offprint No. 1044. Describes laboratory experiments in which cancer has been induced in normal animal cells by viruses.

Echlin, P. 1966. The blue-green algae. Scientific American Offprint No. 1044. The nature and significance to man of these simple organisms is discussed.

Hilleman, M. R. and Tytell, A. A. 1971. The induction of interferon. Scientific American Offprint No. 1226. Discusses new research into ways of initiating interferon synthesis in humans as a means for protection against viruses.

Marples, M. J. 1969. Life on the human skin. Scientific American Offprint No. 1132. Considers the human skin as an environment for a great variety of microorganisms.

Sharon, N. 1969. The bacterial cell wall. Scientific American Offprint No. 1142. Describes recent studies on the structure and biosynthesis of bacterial cell walls.

Simon, H. J. 1963. *Microbes and Men*. Scholastic Book Services, Inc., New York. No. 5 in the series "Vistas of Science." Useful, compact book on both bacteria and viruses.

5

THE PLANT WORLD

What is a weed? A plant whose virtues have not yet been discovered.
RALPH WALDO EMERSON, 1803–1882

Everyone has an intuitive concept of what a plant is: it is generally green and has roots, stems and leaves. Yet there are organisms which have none of these characteristics but still are classified as plants: the molds, for example, and the mushrooms. In their structure and life form, however, these latter organisms more closely resemble green plants than they do animals. Although some taxonomists, the people who classify organisms into related groups, prefer to place such non-green plants into a separate category of their own, we shall consider both green and non-green plants together here.

In general, a plant falls within one of three categories, according to the degree of specialization of its cells into tissues serving distinct functions. These three major groups, in order of increasing structural complexity, are the **thallophytes,** the **bryophytes** and the **tracheophytes.**

THALLOPHYTES

The least complex of all plants are the **thallophytes.** Members of this group include both single-celled, microscopic plants and large, multi-cellular plants, such as seaweeds and mushrooms. The multicellular forms, however, have little tissue differentiation despite their size; their cells all tend to be similar in structure and function. Thallophytes may be conveniently subdivided into those with chlorophyll, the algae, and those without chlorophyll, the fungi.

ALGAE

The single-celled green algae, together with the simpler blue-green algae, are collectively known as **phytoplankton.** These tiny aquatic plants are abundant wherever there is light; they are found near the surface of the ocean and in freshwater lakes and streams. Unlike the blue-green algae, green algae have a distinct nucleus and intracellular organelles, similar to those described for "typical" plant cells in Chapter 3 (see Figure 5.1A). Many are motile. Some, such as the **diatoms,** have external shells impregnated with silica (Fig. 5.1B). When the cells die, the empty shells sink to the bottom, to be eventually compressed and solidified into sedimentary rocks. Another group of marine phytoplankton, the **dinoflagellates,** often contain other pigments in addition to chlorophyll; they become abundant at certain

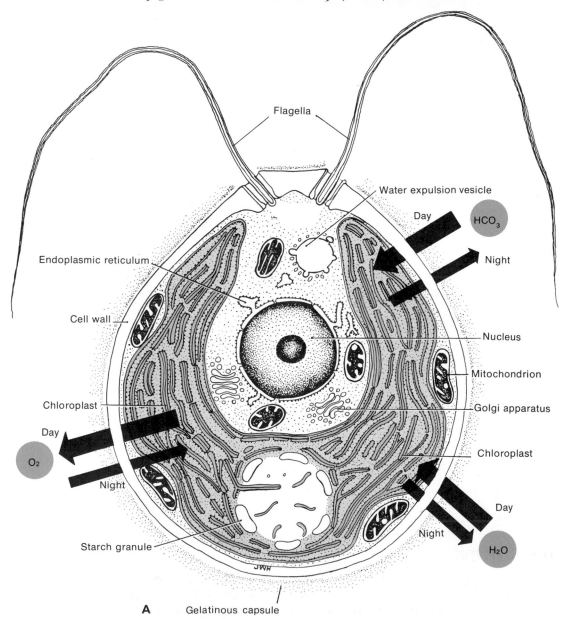

FIGURE 5.1 Single-celled green algae.
A. Internal structure of an unspecialized alga, *Chlamydomonas.*
B. A diatom, *Navicula crabo,* whose delicately sculptured shell is impregnated with silica. (Courtesy Macmillan Science Co., Inc., Chicago, Turtox/Cambosco.)
C. A dinoflagellate, *Gymnodinium,* covered with protective cellulose plates. (From Villee, C. A., and Dethier, V. G. 1971. *Biological Principles and Processes.* W. B. Saunders Co., Philadelphia.)

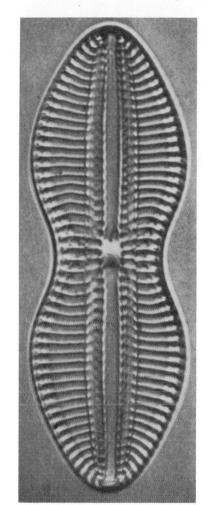

B

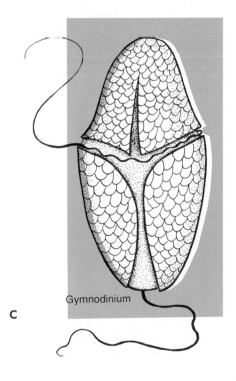

Gymnodinium

C

times of the year and are responsible for the "red tides." Some of these organisms produce toxins which are absorbed into the tissues of molluscs and other animals that feed on them, making the latter poisonous to fish and humans. Phytoplankton are the main food source for many aquatic animals. As a result of their photosynthetic activities, phytoplankton, especially diatoms and dinoflagellates, synthesize about half of all the organic matter produced in the world, and generate about 50 per cent of the atmospheric oxygen which is constantly being recycled. They are thus extremely important in the biosphere.

The multicellular algae include the seaweeds and some freshwater forms found in nutrient-rich lakes. In the oceans, seaweeds are restricted to shallow coastal waters, since the young plants, newly settled on the bottom, require sunlight to grow. Figure 5.2 shows one of the giant kelp, *Macrocystis*. Although the blades resemble leaves and the stipes, stems, these structures do not perform special functions as they do in higher plants. Nor is the holdfast a root; the plant derives no nutrients from the rock, absorbing all its needs directly from sea water. The gas-filled floats keep the stipes upright in the water.

In general, all the cells of multicellular algae have a similar appearance and function. In larger seaweeds such as kelp, more photosynthesis occurs in the upper blades, and excess organic molecules produced by them are transported, presumably from cell to cell, to the rest of the plant, but there are no specialized organs to serve this function, as occur in higher plants. The only cells in algae with a unique function are found in the reproductive organs, sporangia, located on the lower blades.

Kelp and other seaweeds provide habitats and food for a great number of animals, and are an important reason why fisheries along the coastlines are far richer than those in the open sea. Kelp is also harvested for its chemicals: the cell-wall polysaccharide, algin, for example, is used for thickening ice cream, candies and toothpaste. Kelp plants are also used for fertilizer.

ALTERNATION OF GENERATIONS

At this point it is useful to introduce the concept of alternation of generations, a characteristic feature of the life cycle of all plants. The simpler forms of life, discussed in Chapter 4, reproduce mainly by simple fission, but higher organisms, both plants and animals, regularly reproduce by sexual means. To accomplish this, certain events must take place, as outlined here:

1. All offspring produced by sexual reproduction obtain one-half of their genetic material, their chromosomes, from the female (♀) and one-half from the male (♂) parent. Therefore, all eggs and sperm must have one-half the adult number of chromosomes. For example:

   The human adult has 46 chromosomes in each cell.
   A human egg has 23 chromosomes.
   A human sperm has 23 chromosomes.

   We call 46 the diploid (2N) chromosome number and 23 the haploid (1N) chromosome number.
2. Therefore, at some stage in the reproductive cycle of a sexually reproducing organism, the number of chromosomes must be reduced from 2N to 1N. This occurs by a process known as **meiosis**, to be discussed in detail in Chapter 14.
3. In animals, the only haploid (1N) cells are the **gametes** themselves, the eggs and sperm. In plants, there are two distinct stages in the life cycle, one being haploid, the other diploid.

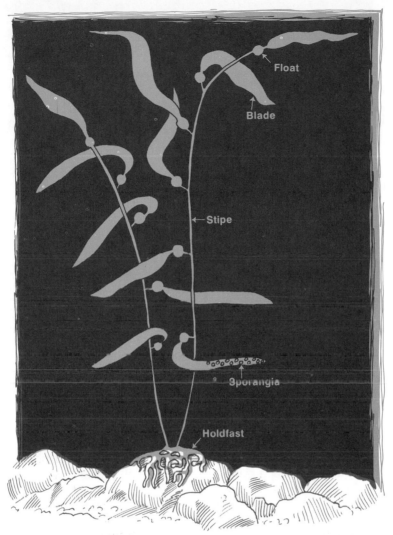

FIGURE 5.2

A stylized drawing of the giant kelp, *Macrocystis*. A full-grown plant bears several more stipes than shown here. Kelp forests are among the most productive communities of the sea.

1N stage = **gametophyte** or gamete-producing plant
2N stage = **sporophyte** or spore-producing plant

4. When male and female gametes unite during **fertilization**, a diploid (2N) individual is formed. By ordinary cell division this grows into the adult.
   Let us take the life cycle of the giant kelp, *Macrocystis*, as an example of the alternation of generations found in plants (Fig. 5.3). The large seaweed is the diploid, or 2N sporophyte stage. In the sporangia of its lower blades, meiosis takes place, producing haploid (1N) spores. The microscopic spores released from the sporangia are dispersed through the water, and eventually settle on the bottom of the ocean where they develop into tiny gametophyte plants. Although these plants look alike, they are sexually distinct. Each produces and releases haploid gametes, either male or female, which eventually unite to form a diploid (2N) **zygote**. The single-celled zygote begins to divide, becomes an **embryo,** and soon settles on the bottom to grow into an adult sporophyte plant, so completing the life cycle.
   There are thus two distinct stages in the life cycle of *Macrocystis*, the gametophyte generation and the sporophyte generation. This alternation of generations

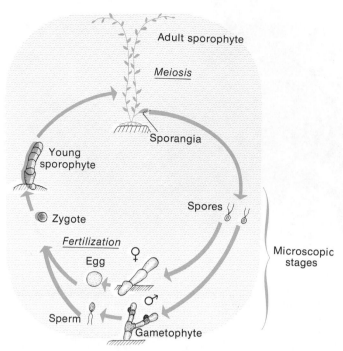

FIGURE 5.3
Alternation of generations in the kelp *Macrocystis*. Diploid (2N) sporophyte stage shown in color; haploid (1N) gametophyte stage in white. Meiosis produces 1N spores from 2N sporangium. Fertilization produces 2N zygote from 1N gametes. Both spores and gametes may be dispersed by currents.

is characteristic of all plants. In most lower plants, such as the majority of seashore algae, the fungi and the mosses, the gametophyte stage is the more prominent and the sporophyte stage is small or even microscopic. In higher plants, the situation is usually reversed: the sporophyte stage tends to be most prominent in ferns and seed plants, and the gametophyte stage is greatly reduced. The prominence of the sporophyte over the gametophyte stage cannot be used as an indicator of where a plant stands on the evolutionary scale, however — a point made by our example of the alga, *Macrocystis*.

It will be noted that there are two dispersive stages in the life cycle of *Macrocystis*, the flagellated spores and gametes; both are spread about, partly by their own motility but mainly by currents. This insures that as many sites as possible are colonized by adult plants. At least one and usually two such dispersive phases are found in the life cycles of all types of plants. Dispersion, which may occur as spores, gametes or embryos (seeds), makes certain that all possible favorable habitats are colonized by offspring. This is important to organisms which are non-motile, and is one of the biological functions served by the rather complex life cycle of plants.

FUNGI

The other important group of thallophytes is the fungi. Since they have no chlorophyll they cannot synthesize their own food and must feed on living or dead organisms. Like bacteria, they thus play an important role in decay processes and the recycling of nutrients.

Yeasts, Molds and Mildews. The yeasts, molds and mildews compose one group of fungi. The single-celled yeasts resemble the green alga *Chlamydomonas* (Fig. 5.1A), except that they have no chloroplast and no flagella. They are virtually ubiquitous in soils, where they help decompose organic molecules into inorganic nutrients useful to green plants. They also cause various diseases in plants and animals; one of the common forms of vaginal infection in women, for example, is due to the yeast *Candida albicans*.

The molds and mildews are multicellular organisms, with elongate cells strung end to end forming delicate white threads called **hyphae** within the substratum. The common bread mold, *Rhizopus*, is illustrated in Figure 5.4. In this particular group of molds, there are no cell walls between the hyphal cells, their nuclei being distributed throughout a continuous cytoplasm. Individual cell walls do occur in the hyphae of other types of molds and mildews, however. Every so often a cell will push upward and divide into a fruiting body, the sporangium; the sporangia are the fuzz often seen on moldy bread. Spores released from them are carried by wind or animals to new sites, where they germinate and grow into hyphae. Molds are also capable of sexual reproduction; this occurs in the form of a union of two hyphal cells to produce a diploid zygote, which in turn becomes a hypha giving rise to haploid spores.

Like yeasts, molds are common in soils, living mostly on dead organisms. The buried hyphae decay organic matter and return nutrients to the soil. Molds may cause diseases in plants, animals and man. Oiled sea birds, which are in a weakened state, are unusually susceptible to lung infections casued by a mold. It is this disease, rather than toxic effects of the oil itself, which kills many birds after an oil spill.

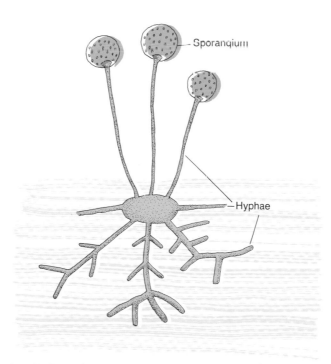

FIGURE 5.4

Structure of the black bread mold, *Rhizopus*. Hyphae in this species contain many nuclei (not shown) but no separate cell walls. The fruiting bodies or sporangia are the "fuzz" one usually sees; they are filled with minute spores.

Other molds are useful to man. The taste and color of blue cheeses, for example, are due to molds injected during processing. More importantly, penicillin and numerous other life-saving antibiotics are the products of molds.

LICHENS.   Some fungal species of this group, and a few related to mushrooms, live in close association with certain algal cells, forming the *lichens* (Fig. 5.5). Both blue-green and green algae may occur in these associations. The close relationship between the algae and the fungi is mutually beneficial. The algae, through their photosynthetic apparatus, provide organic molecules for the fungal cells, while the latter furnish protection and moisture for the water-dependent algae.

Frequently brightly colored due to orange and yellow algal pigments, lichens are often the first colonizers of bare rocks. Sand grains collect in the spreading lichens, and new soil begins to form in which seeds of grasses and other plants can eventually germinate. Lichens are thus of prime importance in the evolution of new terrestrial environments in such places as lava flows, talus slopes (rock falls) and scars produced by bulldozers.

Mushrooms and Their Relatives.   The second group of fungi are the multicellular puff balls, mushrooms and bracket fungi. Aside from being rather larger and more complex structurally, they do not differ greatly from molds. Their cells are all quite similar. Much of the plant, consisting of extensively branched hyphae, lies buried in decaying matter and all we see is the fruiting body bearing multitudes of spores, which are the main dispersive phase (Fig. 5.6). Like yeasts, molds and bacteria, the large fungi play an important role in the terrestrial environment, both through their contribution to nutrient cycles and by maintaining a light, aerated soil which is capable of retaining water and nutrients. The fruiting bodies of non-poisonous species are often used in gourmet cooking.

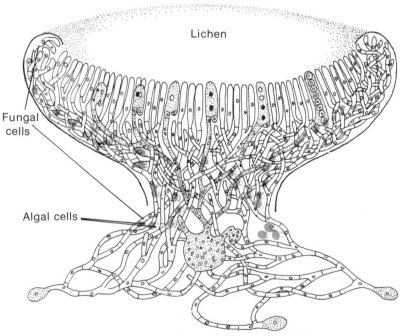

FIGURE 5.5

Structure of a lichen seen in cross-section. The fungal cells form a protective housing for the algae, shown in color. The algae provide food for the fungal cells.

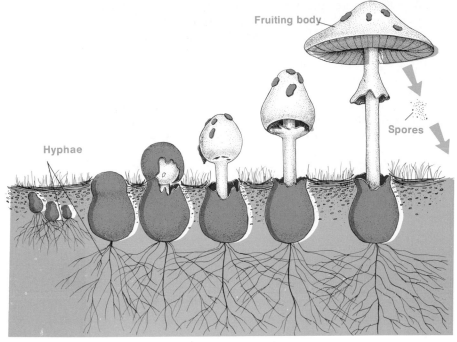

Fruiting body

Spores

Hyphae

FIGURE 5.6

Stages in the development of a mushroom. The fruiting body develops from the buried hyphae. On its undersurface sexual reproduction occurs, resulting in the formation of numerous sporangia. The latter release millions of spores. (From Villee, C. A., and Dethier, V. G. 1971. *Biological Principles and Processes*. W. B. Saunders Co., Philadelphia.)

BRYOPHYTES

Although fungi, as we have seen, have a terrestrial existence, they are restricted mostly to moist environments; the algae are almost all aquatic. The simplest forms of green plants found on land are thus the **bryophytes,** the liverworts and mosses. Adaptations to land among this relatively small group of plants include the differentiation of cells into primitive organs, each serving a specific function (Fig. 5.7). The haploid gametophyte stage is predominant in these plants, and consists of root-like **rhizoids** which penetrate the soil and of a green leaf-like structure in which photosynthesis takes place. In liverworts, the latter structure is generally a flattened **thallus** lying near the ground, while in mosses it is an upright stalk bearing "leaves." The latter are covered with a thin waxy covering to slow down water loss, and they contain numerous pores for the entry and exit of gases.

In contrast to most thallophytes, the sporophyte stage of the mosses and liverworts does not have an independent life of its own. Instead, it lives a sort of parasitic existence on the gametophyte plant from which it derives protection and nutrients. The sporophytes of mosses are sometimes several inches tall and may be mistaken for flowers. Male and female gametophyte plants are separate in liverworts, each producing their respective germ cells in specialized structures; the male sperm, released only after a heavy rain or heavy dew, swim to the female gametophyte and find their way to the egg which one of them fertilizes. The zygote, still attached to the female gametophyte, grows into a sporophyte; the latter produces a spore capsule in which, by meiosis, many spores are produced. A similar cycle occurs in mosses, except that eggs and sperm are produced by the same individual. This is the main dispersive phase of bryophytes.

Bryophytes generally lack an efficient internal vascular system, and cannot transport water and nutrients over long distances; thus, they are all relatively small plants. Because they are susceptible to water loss and have an absolute requirement for water during sexual reproduction, bryophytes are generally restricted to moist localities, which range, however, from the subarctic to the tropics. Certain species of mosses inhabiting swamps and bogs have given rise over the ages to vast peat beds; peat is still burned for fuel in many parts of the world and is also used in horticulture to improve soil quality. In nature, both liverworts and mosses break the force of rain and so prevent soil erosion. They serve as food for various animals, and also take part in the gradual conversion of ponds to bogs and, eventually, to dry land.

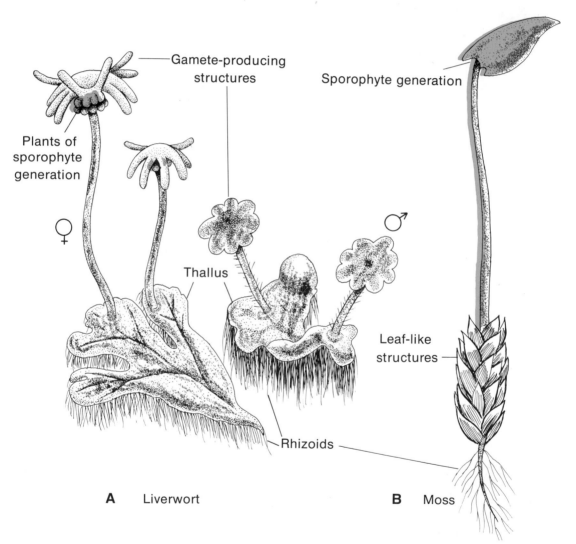

FIGURE 5.7   Two types of bryophytes, showing predominance of gametophyte (1N) generation and reduced sporophyte (2N) generation (color).

   A. A typical liverwort, *Marchantia*. Eggs and sperm are produced on umbrella-like structures growing out of the thallus. Sperm swim to female sporangium, from which the sporophyte plant grows.

   B. Mosses produce both eggs and sperm in sporangia (not shown) at the top of the gametophyte. The sporophyte grows up from the female sporangium on a long stalk ending in a spore capsule.

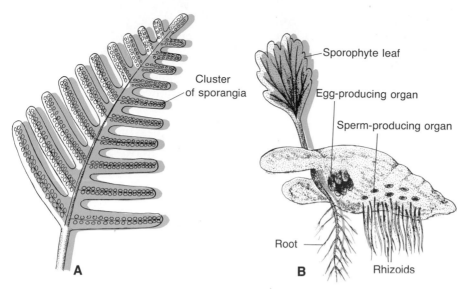

FIGURE 5.8

Sporophyte and gametophyte generations in a fern. Diploid stages are in color.
    A. Underside of sporophyte (2N) leaf, bearing rows of clustered sporangia.
    B. Underside of minute gametophyte stage, usually less than a quarter-inch long.
Sperm swim to egg-producing organs where fertilization occurs. The zygote develops
from this spot, germinating into a new sporophyte.

TRACHEOPHYTES

The true vascular plants or **tracheophytes** exhibit the highest degree of
organization of all plant types and are especially well adapted for life on
land. The sporophyte stage is always the predominant one among this
group. Basically, vascular plants are organized into a root system which
ramifies throughout the soil and serves to take up water and nutrients,
and a shoot system which bears lateral appendages, the leaves, in which
photosynthesis occurs. They also contain a well-developed vascular system
by which water and nutrients are transported throughout the plant. There
is great diversity among the higher plants, and we can recognize two fundamental groups: the non-seed plants and the seed plants.

NON-SEED PLANTS

This group includes the club mosses, horsetails and ferns, the last of
which, being most familiar, will be considered here. The fern plants which
we commonly encounter are the sporophyte generation which bear clusters
of sporangia on the backs of their leaves (Fig. 5.8A). The released spores,
poetically known as "fern seed," germinate during a moist period, since the
minute, heart-shaped gametophytes are susceptible to desiccation. From
their undersides, the gametophytes send out rhizoids which penetrate the
soil; also growing from this surface are the male and female reproductive
organs (Fig. 5.8B). As in mosses, fertilization occurs only when moisture is
present, allowing the sperm to swim to the egg contained within the female
sporangium. The young zygote begins to develop within this female
reproductive organ, from which it obtains moisture and nourishment until
its own leaves and roots form.

The dependence upon water of the entire gametophyte generation of
most non-seed tracheophytes restricts them to areas where moisture is
abundant for at least part of the year. A few species of ferns, however, are
adapted to drier climates. Ferns were far more abundant during the Carboniferous period of the Earth's history, some 300 million years ago, than

they are today. At that time, before seed-plants had evolved, the climate was moist and warm, and ferns the size of trees flourished. The great tree-ferns of modern tropical rain forests give some indication of what terrestrial vegetation was like then. The growth rate of plants during that period was so rapid that complete decay and recycling did not occur; much organic matter became incorporated into sediments, eventually to be converted to coal. Today's coal reserves thus represent stored chemical energy of photosynthesis, accumulated over some 50 million years—a thought to bear in mind if you should happen to be reading this by artificial light! This represents an energy bank account from the past—a non-renewable resource.

SEED PLANTS

The most diverse group of higher plants living today are the seed plants. Their success on land is due in large part to their ability to reproduce without the necessity of a water phase for transfer of sperm to eggs. The predominant phase is the sporophyte generation. The greatly reduced gametophyte stage is incapable of an independent existence and hence carries out its functions within the sporophyte plant. The seeds themselves, which contain the young sporophyte or embryo, represent a second important adaptation to a terrestrial existence. Not only does a seed possess a protective coat which prevents the embryo from drying out; it also contains a supply of nutrients to sustain the developing plant during its early growth at the time of germination. Within the seed, the embryo lies dormant until conditions are favorable for its germination. Seeds can withstand drought, cold and even passage through the gut of a bird, field mouse or human. They thus afford an excellent dispersive phase for the plants that produce them.

The details of the reproductive cycle in seed plants differ from one species to the next, and a multitude of names have been applied to the various structures involved. In order to simplify our description of the process we shall make use of the highly schematized diagram shown in Figure 5.9. This scheme should not be taken to represent any specific plant, however; it merely provides an outline of the steps in sexual reproduction in this group of plants.

Certain buds on the spermatophyte plant develop into tight clusters of modified leaves, the cones or flowers, in which the male sporangium and female sporangium are formed. In the conifers, these occur on separate cones; in the flowering plants, both are usually found in the same flower. The male sporangium, through meiotic division, produces spores; these divide once or twice to form the male gametophyte or **pollen** grain, which is generally covered by a protective coat and may bear wings to aid in its dispersal. Pollen is then transported, by wind, insects or birds, to the female sporangium, a process known as **pollination.** The female sporangium, meanwhile, also produces haploid spores by meiosis which divide to form female gametophytes. Note that these remain entirely within the female sporangium. When pollen falls on the receptive region of the female sporangium, it grows a hollow pollen tube which tunnels inwards to the female gametophyte. One or more of the haploid male cells enters, and one of them fertilizes the egg to form the diploid zygote.

The seed is formed from several tissues. The zygote itself begins to divide to form the embryo. The remaining gametophyte cells accumulate nutrients from the parent plant and become the nutritive **endosperm.** The former female sporangium forms the seed coat.

There are two major groups of seed plants, distinguished by whether the seed is bare—the **gymnosperms**—or whether the seed is surrounded by fruit—the **angiosperms.**

Gymnosperms. The bare-seeded plants include both the conifers, such as pines, firs, spruces and so forth, and the primitive cycads and ginkgos. In all species, the male and female sporangia are borne on different parts of the plant, usually as small and large cones respectively. The structure of a pine seed, typical of those of gymnosperms, is shown in Figure 5.10A. The nutrient endosperm is haploid. The embryo has already developed the rudiments of the first leaves; at its opposite end, the roots will develop.

Gymnosperms, particularly conifers, are of great importance in the biosphere, forming the major vegetation over large areas of the subarctic and temperate regions. They are, of course, of great commercial value as a source of wood for lumber, paper, pulp, plastics and other uses.

Angiosperms. The flowering plants have a great diversity of life forms. They comprise two major groups: the plants with parallel-veined leaves, such as grasses, palms, lilies, tulips, orchids and so forth; and the plants with net-veined leaves, such as most deciduous trees, perennial shrubs and the herbaceous plants like beans, columbines, sunflowers, violets and many others.

The male and female sporangia of angiosperms usually occur within the same flower, but a few species have separate male and female flowers, on either the same or different plants. Within the flower, the male sporangia develop as **anthers** on the tips of the **stamens;** within the anthers, the male gametophytes, or pollen, are formed. The female sporangia are the **ovules** at the base of the **pistil;** in the ovules the female gametophyte is formed (Fig. 5.11). During fertilization in angiosperms, two sperm enter the female gametophyte. One fertilizes the egg, the other fertilizes the

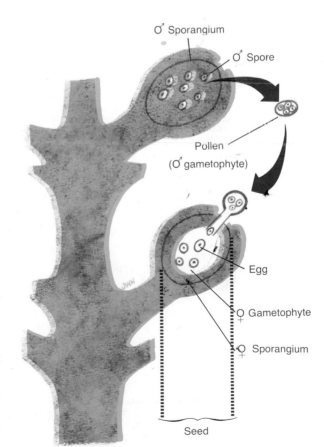

FIGURE 5.9

A highly schematized diagram of reproduction in seed plants. Sporophyte tissues (2N) are shown in color; gametophyte tissues (1N) in white.

Certain buds on the spermatophyte plant develop a ♂ sporangium and a ♀ sporangium. Each gives rise to a gametophyte. The ♂ gametophyte or pollen grain, on meeting the receptive female region, forms a pollen tube through which a sperm passes to fertilize the egg. The resulting zygote is enclosed by both ♀ gametophyte and sporangium to form the seed.

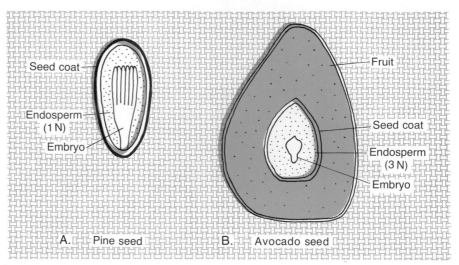

Seed coat

Endosperm
(1 N)

Embryo

Fruit

Seed coat

Endosperm
(3 N)

Embryo

A.   Pine seed   B.   Avocado seed

FIGURE 5.10

The seeds of plants. Tissues derived from parent sporophyte tissue are shown in color.
A. The bare seed of a pine tree, a gymnosperm.
B. The fruit-covered seed of an avocado, an angiosperm.

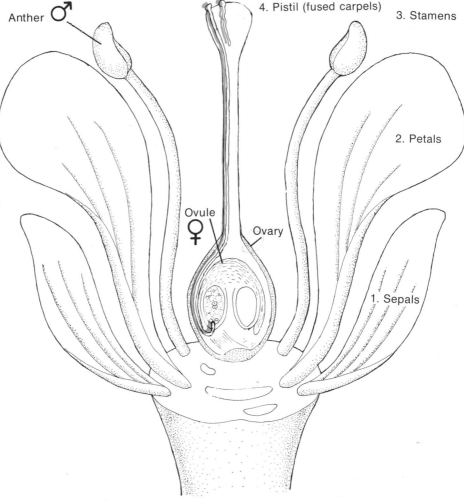

Anther ♂

4. Pistil (fused carpels)

3. Stamens

2. Petals

Ovule ♀

Ovary

1. Sepals

FIGURE 5.11

Essential parts of a flower.
The four modified leaf components contributing to a flower are identified on the right. The ♂ sporangia form in the anthers and produce ♂ gametophytes, the pollen. The ♀ sporangia form in the ovules, and produce the greatly reduced ♀ gametophytes. Pollen falling on the stigma grows a pollen tube down to the ovules, through which two sperm pass to fertilize the egg and to form endosperm.

fused nuclei of two other haploid female cells to produce a *triploid* endosperm. The tissues immediately surrounding the ovules provide the fruit which characteristically surrounds the seeds of angiosperms.

In some cases, such as peas and beans, the fleshy pod is the fruit and serves mainly a protective function, but more often the fruit also aids in dispersion of the seeds, since it is attractive to animals (Fig. 5.10B).

The functions of the angiosperms in the biosphere are legion. Like gymnosperms, they provide food and shelter for numerous animals and also help to retain nutrients within the soil and prevent its erosion. They also have many uses for man, providing him food, clothing, timber, fuel, drugs and dyes, to mention some of the more important.

THE UNITY OF THE PLANT ORGANISM

So far we have discussed only the life cycles and roles of the various plant types in the biosphere without consideration of their organization into functional organisms. Whereas almost all life processes are carried out by every cell of the less specialized thallophytes, in the higher forms of multicellular plants the various life functions are shared among discrete *tissues* and *organs,* each of which contributes to the survival of the whole plant. Since in subsequent chapters we shall be dealing with the specific roles of some of these organs, it is useful to describe here how the major parts of higher plants are interrelated within the organism as a whole. By way of example, let us use the most highly organized of all plants, an anglosperm.

A highly schematized diagram of a flowering plant, showing the interrelationships of its parts, is given in Figure 5.12. Basically, higher plants consist of a root system which extracts water and nutrients from the soil; an axial shoot system on which develop the leaves, where photosynthesis occurs, and the flowers, which serve a reproductive function. The growing points of the plant, where elongation occurs, are known as *meristems.* All parts of the plant are served by vascular tissue: the *xylem,* which conducts water and minerals upwards from the roots, and the *phloem,* which transports dissolved organic molecules in both directions, as indicated by arrows.

We have already considered the basic elements of flower structure and will discuss meristems later, in Chapter 14. Let us here examine the tissues of roots, stems and leaves, shown in diagrammatic cross-section in Figure 5.12.

Roots

Near its tip, the outer layer of cells on a root sends out numerous fine projections, the *root hairs,* which greatly increase the absorptive surface. Minerals and water pass inward to the xylem for transport to the rest of the plant. The root receives organic molecules, manufactured elsewhere in the plant, via the phloem.

Stems

Stems serve both to conduct sap and to support the shoot system. The inner bundles of xylem and outer bundles of phloem perform the former function. A central zone of *pith* serves for nutrient storage. Growth in diameter of both roots and stems is achieved by division of cells lying in a layer between the xylem and phloem, known as *cambium.*

The stems of annuals are generally supported by the cell walls and by turgor within the cells, but woody shrubs and trees require additional support. Each year, as the stem grows in diameter, new xylem and phloem are formed. The old, inner xylem dies, but its cell walls, which in addition to cellulose contain a second supportive substance called lignin, remain to provide support. A protective layer, the bark or cork, is also laid down on the outer surface of woody stems, and, like xylem and phloem, is added to each year by cell divisions in the cork cambium.

Leaves

A leaf in cross-section has an outer layer of cells which produce a protective waxy cuticle. (A cuticle also occurs on the stems of annual plants which have no cork.) Within the leaf are the cells which carry out most of its photosynthetic function. The underside of a leaf bears openings, the **stomata** (singular, stoma), through which oxygen, carbon dioxide and water vapor pass; the stomata, then, serve a respiratory function. Xylem and phloem are located in the veins.

Each organ of a higher plant thus has a distinct organization of specialized cells and tissues to perform particular functions.

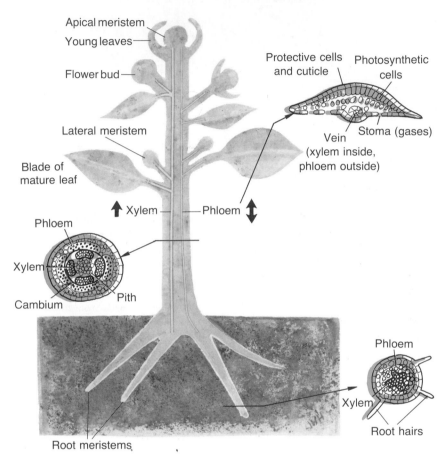

FIGURE 5.12

A highly schematized diagram of the main parts of a flowering plant. Xylem and phloem are represented on opposite sides for sake of clarity. Note the meristematic growing points. Cross-sections of a leaf, stem and root are shown, indicating important functional features. Further details are given in text.

SUMMARY

Plants range in structural complexity from the relatively undifferentiated algae and fungi to the seed plants with highly specialized tissues. All, however, exhibit life cycles characterized by an alternation of generations between haploid gametophytes and diploid sporophytes.

Among the thallophytes, the algae include unicellular phytoplankton, which account for half the world's photosynthetic activity, and multicellular seaweeds, which contribute to the productivity of coastal regions. The non-photosynthetic yeasts, molds and mushrooms play a role in decay processes and recycling of nutrients. Some are helpful to man, a few are harmful.

The bryophytes include the terrestrial mosses and liverworts, which have a limited distribution due to their dependence upon moisture. Some mosses produce peat and play a role in converting ponds to dry land.

The tracheophytes include the non-seed ferns, more prevalent during the coal-forming period than now, and the seed-forming gymnosperms and angiosperms. Both conifers and flowering plants are highly successful terrestrial plants which play important roles in the biosphere and are of great importance to man.

READINGS AND REFERENCES

Bell, P. R. and Woodcock, C. L. F. 1968. *The Diversity of Green Plants.* Addison-Wesley, Reading, Massachusetts. A well-illustrated, concise account, in paperback form, of structure and reproduction in groups of green plants.

Bold, H. C. 1968. *Morphology of Plants.* 2nd ed. Harper & Row, New York. A well-illustrated discussion of variations in plant structure.

Emerson, R. 1952. Molds and men. Scientific American Offprint No. 115. Considers the role molds have played in the course of human history.

Fuller, H. J. and Tippo, O. 1954. *College Botany.* Holt, New York. Still a valuable standard textbook on the structure and function of plant life.

Isaacs, J. D. 1969. The nature of oceanic life. Scientific American Offprint No. 884. Explains the importance of phytoplankton in the ocean.

Lamb, I. M. 1959. Lichens. Scientific American Offprint No. 111. Describes the structure of lichens and their functions in the biosphere.

Northen, H. T. 1968. *Introductory Plant Science.* 3rd ed. Ronald Press, New York. A clearly written introductory text on all aspects of plant life.

Went, F. 1963. *The Plants.* Time, Inc., New York. One of the *Life* (magazine) series on natural history, this beautifully illustrated book covers both structure and function of plants.

6

THE DIVERSITY OF ANIMALS

Throughout the Animal Kingdom there is a great diversity of body form, each type of animal being adapted to fulfill a specific role in the biosphere.

In the last two chapters we have surveyed the photosynthetic organisms, which are the producers of food, and the decomposers such as bacteria and fungi, which function in decay processes. In between stand the animals, the world's consumers, deriving nourishment either directly from plants or from other animals which feed on plants. The range of food types is great; bacteria, phytoplankton, fungi, herbaceous and woody plants, animals of all sizes and dead organic matter are all utilized by some species of animal.

Since food comes in many sizes and occurs in many places, it follows that the animals which feed upon it also have a great diversity of shapes and sizes which permit them to utilize a particular food source to maximum advantage. Some animals, including those most familiar to us, are actively motile, grazing on plants or pursuing moving prey. But many aquatic species are **sessile.** Being firmly attached during adult life to their substratum or living in permanent tubes within it, animals of this latter type must obtain food from the water that passes over them. Although some sessile species are able to capture large animals, the majority subsist on microscopic plankton, both plants and animals, which they filter from the water by a variety of ingenious devices. Such **filter feeders,** as they are known, play an important role in the economy of the sea and to some extent also in freshwater lakes and streams.

It is our purpose in this chapter, then, to describe some of the more significant groups of animals, pointing out how each is specialized to perform its specific function in the biosphere. Before beginning this somewhat awesome task, however, it may be useful to consider some of the organ systems which characteristically occur in animals.

Just as higher plants have cells organized into tissues and organs to perform specific functions, so do all but the most primitive animals have specialized organ systems. In order to discuss these meaningfully it is useful to consider a single type of animal. For this purpose, the earthworm, a most important animal in maintenance of soil quality, will serve admirably, since it exhibits in a relatively uncomplicated fashion all of the major organ systems found in higher animals. A side view of the front end of an earthworm is shown in Figure 6.1.

The first thing to note about the body plan of an earthworm is that it can be divided into three concentric layers—outer, middle and inner. The outer layer, or **epidermis,** is a relatively thin skin which secretes both a protective coating, the **cuticle,** and a slimy mucus to keep the body surface moist. The inner layer consists of the intestine, an elongate tube running the whole length of the animal and continuous with the exterior at the mouth and the **anus** (not shown). It functions in the processing and assimilation of food and is divided into distinct regions about which we shall have more to say in Chapter 9.

Within the middle layer, which is the most complex, are to be found the remaining organ systems. A second important feature of earthworms, and of many other animals, is that the middle layer contains a fluid-filled cavity, the **coelom,** which is subdivided by partitions or **septa.** Although appearing as thin lines in our diagram, in their third dimension the septa are virtually continuous circular sheets of tissue connecting the gut to the body wall, and thus subdividing the coelom into separate compartments. Such internal **segmentation,** with repetition of body parts, is characteristic of a great many, but not all, higher animals.

One of the most important features of animals is their ability to move, a function served by the muscular system. In earthworms, the major body musculature consists of two layers: an outer, circular layer, which constricts the body, and an inner, longitudinal layer, which shortens it. It will be noted that the earthworm has no hard skeleton against which the muscles can exert a force. Instead, they squeeze against the coelomic fluid, which, being incompressible, transmits the force to other parts of the body wall causing them to change shape. The coelom thus acts as a **hydrostatic skeleton.** With alternate contractions of its two muscle layers the earthworm moves by **peristaltic locomotion.**

In order to coordinate their locomotory movements, almost all animals have a nervous system. This system first receives information from the environment via sensory organs and integrates the information in central **ganglia,** which are clusters of

(Text continues on the following page.)

ORGAN
SYSTEMS IN
ANIMALS

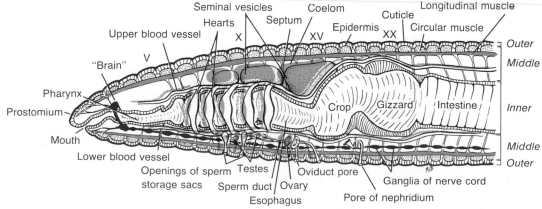

Side view of front end of an earthworm.

Note especially the three concentric layers of the body and the coelomic cavity within the middle layer, divided by septa. Individual organ systems are described in the text.

FIGURE 6.1

many interconnected nerve cells; then messages are sent to the appropriate muscles, which effect the response. The nervous system actually develops from the outer tissue layer, but in many animals, including the earthworm, it sinks inwards. The central nervous system of the earthworm is composed of a chain of connected ganglia; one slightly larger ganglion, the "brain," is located in the foremost part of the animal where most sensory information is received. Only a few of the peripheral nerves are shown in our diagram.

Animals greater than a fraction of a millimeter in thickness usually have a blood-containing circulatory system to distribute oxygen and nutrients to their tissues and to collect waste products of metabolism. The main features of the circulatory system of the earthworm are shown in light color in our diagram. Blood collected from various parts of the body passes forward in the upper blood vessel and is pumped by five pairs of contractile "hearts" to the lower vessel, whence it flows out to the peripheral regions. There is also a large blood sinus around the gut, and a pair of segmental vessels in each segment, neither of which is shown. Most large animals also have a specialized respiratory system, such as gills or lungs, where oxygen and carbon dioxide are exchanged between blood and surroundings. Earthworms, however, live continuously in moist soils and maintain a thin liquid film over the entire surface, which thus acts as a medium for gas exchange. They cannot survive for long in water-saturated soil, however, where the oxygen is quickly depleted, and hence often come to the surface after a heavy rain.

Larger animals usually also require an excretory system to process the body fluids, removing wastes and maintaining a suitable salt and water balance. The organs serving this function are called **nephridia** (from Greek *nephros*, "kidney").

FIGURE 6.2

Mating in earthworms.
    The undersides of the two worms are apposed for the exchange of sperm. (Courtesy of CCM: General Biological, Inc., Chicago.)

Those of the earthworm share the same basic structure as the kidneys of many other animals. Typically, there is a pair of nephridia in virtually every segment (we have shown but one, for the sake of simplicity). Body fluids pass into an open funnel and then through a long tube, where processing takes place; the urine is finally excreted to the exterior through a minute pore.

In some animals, the reproductive system is closely associated with the excretory system, often sharing the same ducts to the exterior. In the earthworm, the reproductive organs, shown in dark color, have become separate and are confined to but a few segments. Earthworms, and some other animals, are **hermaphroditic**—that is, both ovaries and testes occur in the same individual. Sex cells produced in the testes are stored in the **seminal vesicles,** where they mature into functional sperm. During mating, two individuals come together with their head ends overlapping (Fig. 6.2) and reciprocally exchange sperm, which pass from the pore of the sperm duct of one into the sperm storage sacs of the other. After the adults separate, the eggs and sperm are extruded into a mucus capsule where fertilization occurs and the young worms develop. Although sexual reproduction occurs in all animals, certain lower forms also reproduce asexually by budding. Both forms of reproduction may contribute to the dispersal of offspring of a particular species.

One further feature of the earthworm is its body symmetry; it has identical left and right halves, and is thus **bilaterally symmetrical.** Although the majority of animals have this same left-right symmetry, some animals appear to possess **radial symmetry,** where the body is similar in all directions about one central axis. In some cases, such radial symmetry is more apparent than real, for a study of the internal organization often reveals a hidden bilateral arrangement. Radial symmetry tends to occur among sessile or sluggish animals for whom direction of locomotion is inconsequential.

A COMMENT ON CLASSIFICATION

There are several possible ways to group animals together when discussing their various roles in the biosphere. One way would be to lump together all animals feeding on the same type of food; others would be to consider together all animals living in similar environments or all those sharing comparable body shapes. While each approach has some merit, we would find ourselves having to repeat many things over and over again about structures of individual animals. Instead we shall use the general scheme of classification devised by taxonomists in which animals are grouped together into **phyla.** All the animals within a given phylum share certain basic structural similarities, suggesting that they are related by a common evolutionary history. In each phylum, however, it is usual for there to be a secondary diversity of body form, as different groups evolved to fill different specific roles in the biosphere, a process known as **adaptive radiation.** Sometimes, as a result of this diversity within each phylum, we must expect to find similarities in body form, occasionally very striking, between quite unrelated animals. Forms that resemble one another, although differing phylogenetically, have become adapted to the same environmental circumstances; this phenomenon is known as **convergent evolution.**

On the whole we shall limit ourselves to the more important phyla, and the major subgroups within them, without being too concerned about levels of relatedness. A more formal classification of the animal kingdom is to be found in Appendix III. The sequence followed is approximately that believed to have occurred during the evolution of the various phyla.

PROTOZOA

The single-celled **protozoa** are in some respects similar to the unicellular green algae, although they have no cell wall and, except for the enigmatic euglenoids, contain no chlorophyll. Internally they bear the usual organelles of animal cells, and some forms are quite complex. Reproduction is mainly by simple fission, although sexual reproduction also occurs.

There are four main groups of protozoa, distinguished mainly by their mode of locomotion (Fig. 6.3). Changes in overall body shapes are characteristic of the amoebae and their relatives. Within this group are certain shelled forms, such as the **foraminifera,** which are a major component of the marine zooplankton. On the animal's death, its calcareous shell sinks to the bottom, eventually to become incorporated in sedimentary rock. Other species of amoeboid protozoa are parasitic, producing diseases such as amoebic dysentery, transmitted through fecally contaminated food or water.

Among the **flagellates,** which move by means of the whip-like action of their flagella, some forms are free-living, the most important of which are photosynthetic species such as the euglenoids and *Chlamydomonas* (see Figure 5.1A). Many flagellates, however, reside in the intestine or reproductive tract of various animals (Fig. 6.3B). The flagellates of the termite gut digest the cellulose of the wood eaten by the host, a function the termite cannot perform for itself. Some parasitic species find their way into the blood stream, usually through the bite of an insect, to cause severe diseases such as African sleeping sickness and leishmaniasis.

The **ciliates** characteristically are covered by numerous short cilia which beat in a synchronous fashion, driving the animal rapidly through the water. Most ciliates are carnivorous and free-living, being found both in marine and fresh water. The **sporozoans,** on the other hand, are all parasitic, and some cause serious diseases. The gregarines, some of which move by a curious gliding motion, infect mainly worms and insects, while forms known as coccidians include the protozoan *Plasmodium*, which causes malaria, one of the major killers of man. This parasite is transmitted in the salivary juice of several types of tropical mosquito. In temperate regions where these mosquitoes do not occur, transmission from an infected person by natural means is impossible. However, there have been reports of transmissions among drug users who presumably shared the same hypodermic needle.

FIGURE 6.3

Representatives of the four main groups of protozoans, as seen under the light microscope.

A. A foraminiferan, *Globigerina.* This shelled relative of the amoeba is abundant in the marine zooplankton. The thin extensions of cytoplasm are pseudopodia, used in food-gathering.

B. A flagellate, *Trichomonas vaginalis*, a common parasite of the human reproductive tract.

C. A freshwater ciliate, *Paramecium caudatum*, showing some internal structures.

D. The feeding stage of a gregarine sporozoan, *Coryella armata*, a parasite of water beetles.

(*A, B* and *D*, after Barnes, R. D. 1968. *Invertebrate Zoology.* W. B. Saunders Co., Philadelphia.)

See illustration on opposite page.

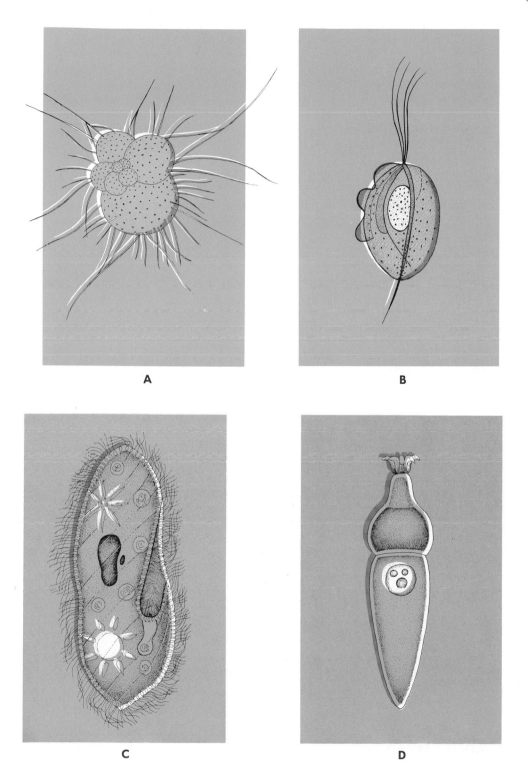

A

B

C

D

FIGURE 6.3

SPONGES

Sponges are extremely simple multicellular animals, often found encrusting rocks and reefs in the ocean; a few species live in fresh water. Their simple body plan (Fig. 6.4) indicates virtually no grouping of specialized cells into tissues. Being permanently attached, sponges are filter feeders. Even in small individuals, many liters of water are drawn daily through the lateral pores by the beating of the flagellated collar cells which line the internal cavity. Detritus and plankton are captured by the collar cells and digested either by them or by the motile amoeboid cells which transfer food throughout the body. Body rigidity is maintained either by spicules of calcium carbonate (shown here) or of silica, or by a meshwork of fibrous protein. There are no nerve or muscle cells, and sponges change shape extremely slowly, if at all. The lack of organization in these primitive animals can be demonstrated by squeezing pieces of sponge through a fine sieve; the resulting individual cells, if gently agitated, reaggregate into small groups, forming several tiny new sponges.

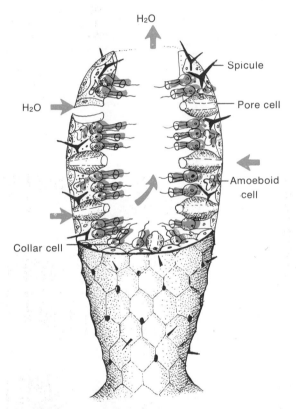

FIGURE 6.4

Part of a simple sponge cut away to show internal structure. Cells functioning in digestion are shown in color. (After Barnes, R. D. 1968. *Invertebrate Zoology*. W. B. Saunders Co., Philadelphia.)

COELENTERATES

A slightly higher order of organization is found among the **coelenterates** (from Greek *koilos*, "hollow," plus *enteron*, "intestine"). These organisms, which include jellyfish, corals and sea anemones, are thus hollow, radially symmetrical animals whose single cavity, the **gastrocoele,** functions as a gut. There are three main groups, whose similar body plans are schematically diagramed in Figure 6.5. There are two primary layers, an outer **ectoderm** and inner **endoderm,** with an intermediate gelatinous layer, the **mesoglea,** being prominent in jellyfish. The mouth, which is also the anus, is ringed with tentacles which bear numerous stinging capsules used in food capture by these carnivorous animals. The toxins they contain serve to immobilize the prey and are capable of causing extremely painful inflammations and even death in unwary swimmers who meet up with such forms as the Portuguese man-of-war or the Indo-Pacific sea wasp. Food captured by the tentacles is transferred to the gastrocoele where it is broken down and finally digested within the endodermal cells. Cells of both endoderm and ectoderm contain muscle fibers, which are innervated by a diffuse nerve net. Distribution of nutrients, gases and wastes occurs by simple diffusion.

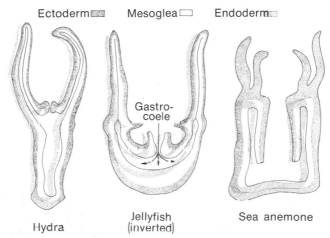

Ectoderm   Mesoglea   Endoderm

Gastro-coele

Hydra   Jellyfish (inverted)   Sea anemone

FIGURE 6.5

Schematic diagrams of the three main groups of coelenterates—hydrozoans, jellyfish and sea anemones. Each has basically two tissue layers, an inner endoderm and outer ectoderm. The jellyfish characteristically have a gelatinous intermediate layer, the mesoglea. (After Villee, C. A., and Dethier, V. G. 1971. *Biological Principles and Processes.* W. B. Saunders Co., Philadelphia.)

Most **hydrozoans** (Fig. 6.6A) are marine, colonial forms, with many individuals grouped together in a plant-like arrangement attached to pilings or to the blades of seaweeds. The Portuguese man-of-war, which belongs to this group, has a gas-filled float that allows it to drift with the currents (Fig. 6.6B). A few solitary hydrozoans, called hydras, are found in fresh water. The jellyfish are entirely marine, and most are unattached, pelagic animals, staying afloat by rhythmic contractions of the bell (Fig. 6.6C). The sea anemones, likewise, are marine, being prominent both in the intertidal zone and on reefs and rocky outcrops below the surface. Their near relatives, the corals, are the most important members of the entire group (Fig. 6.6D). These colonial forms secrete a calcareous skeleton around themselves as they grow. Coral reefs, which support a myriad of other forms of life, are formed by the continuing growth of these tiny animals, one generation forming on top of another. Such reefs are especially common along the continental margins in tropical regions and on the fringes of ancient volcanoes in the South Pacific and other areas, where, as the old crater gradually sinks beneath the surface, the corals continue to grow, forming a shallow lagoon. When the volcano finally disappears, all that is left is a coral atoll (Fig. 6.7).

A

B

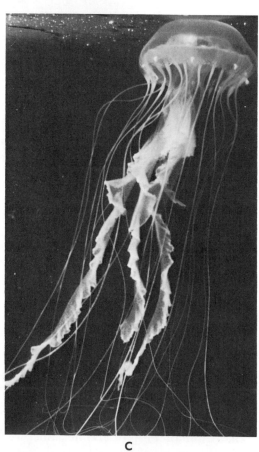

C

FIGURE 6.6   Representative coelenterates.

    A. A colonial hydrozoan, *Plumularia.* Individuals are connected by a stalked network. (From Barnes, R. D. 1968. *Invertebrate Zoology.* W. B. Saunders Co., Philadelphia.)

    B. A highly complex colonial hydrozoan, *Physalia,* the Portuguese man-of-war. Tentacles of the feeding individuals are covered with stinging capsules. (Courtesy New York Zoological Society.)

    C. A large jellyfish, the sea nettle, *Chrysaora.* (Courtesy of William H. Amos.)

Illustration continued on opposite page.

D

Figure 6.6 *Continued*
D. Various corals on the Great Barrier Reef off the northeast coast of Australia.
(Courtesy of the Australian News and Information Bureau.)

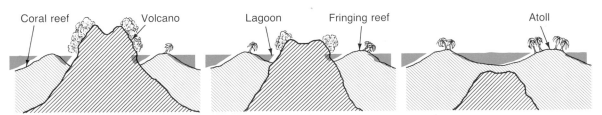

The origin of a tropical atoll. Coral begins to grow around the edges of a volcano, building
up as the volcano gradually sinks beneath the sea.

FIGURE 6.7

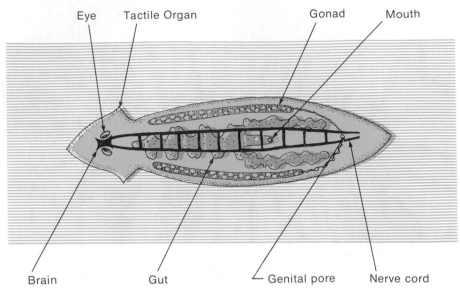

FIGURE 6.8

A schematized diagram of the major organ systems of a free-living flatworm. Musculature and excretory cells and ducts have been omitted for clarity.

FLATWORMS

The most primitive group of three-layered animals consists of the bilaterally symmetrical flatworms. These basically have a simple body plan as shown for the free-living form in Figure 6.8. The highly branched gut still has only one opening via the muscular, eversible pharynx which is used for capturing prey. Internally, the body is filled with a true middle layer, the **mesoderm,** containing muscle cells embedded in a semicellular matrix. These animals, being quite flat, require no circulatory system. They move over the substratum either by means of cilia on the lower surface or by waves of muscular contraction. Like earthworms, they are often hermaphroditic. The head end usually bears sensory organs, particularly the photosensitive "eyes," and the nervous system is primitively organized into a "brain" and a pair of interconnected nerve cords; these animals are capable of learning simple tasks, such as avoidance of light.

Members of the free-living group of flatworms are found in both marine (Fig. 6.9A) and fresh water. Other flatworms are parasitic, causing various diseases in animals and man. The **flukes** (Fig. 6.9B), which have a body form resembling that of the free-living species, have a complex life cycle involving at least two hosts, one of which is almost always a snail. We shall refer in more detail to this group in discussing the epidemiology of the disease schistosomiasis in Chapter 19a. The tapeworms, the third group of flatworms, have a body highly modified for a parasitic life in the intestines of their vertebrate hosts. The head end bears hooks and suckers by which the adult worm attaches to the host's gut wall (Fig. 6.9C). From it are proliferated a series of segments which are essentially factories for production of eggs and sperm (Fig. 6.9D). The human tapeworm may reach a length of 10 feet or more; its larval forms exist in the flesh of cattle. Tapeworms are protected from the host's digestive juices by a well-developed cuticle which is also specialized for absorbing nutrients predigested for it by the host—the only source of nourishment for these animals which have lost their own gut in the course of evolution.

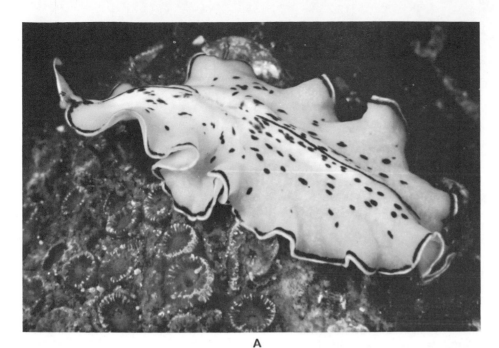

A

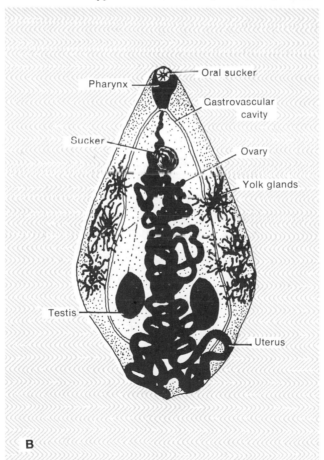

B

FIGURE 6.9

Three representative flatworms.
    A. A free-living marine flatworm. (From Gerking, S. D. 1974. *Biological Systems.* W. B. Saunders Co., Philadelphia. [In press.])
    B. A fluke, *Prosthogonimus macrorchis,* that parasitizes the oviduct of the domestic hen. Note sucker for attachment and feeding.

Illustration continued on the following page

ROUNDWORMS

The roundworms, or **nematodes**, are but one of several small worm-like groups, but ecologically and economically they are by far the most important. These animals have a complete digestive tube, longitudinal muscles and a thick yet flexible external cuticle; the sexes are separate.

Free-living nematodes occur virtually everywhere, and in enormous numbers; in a square meter of mud off Holland over 4 million were counted! Some are carnivorous, others herbivorous; some soil species, such as the potato-root eelworm, are highly destructive of crops. Parasitic roundworms cause diseases in virtually all groups of plants and animals. In man, nematodes cause **trichinosis** (Fig. 6.10) and **hookworm** disease, as well as **filariasis,** in which adult worms inhabiting the host's lymph nodes block the larger lymph vessels, causing the enlargement known as elephantiasis (Fig. 6.11).

ANNELIDS

The true segmented worms, among which the earthworms belong, are known collectively as **annelids.** All are elongate animals having a body subdivided internally into repeating segments, often separated by more or less complete septa, and all make use, to greater or lesser extent, of the coelom as a hydrostatic skeleton in locomotion. In general, the elongate body form is an adaptation to a burrowing existence, permitting these relatively large animals (some are several inches in length) to live within the substratum or in narrow crevices. The extended body length may also provide an increased reproductive potential in those species where gametes are formed in almost all segments. There are three main groups of annelids: the polychaetes, oligochaetes and leeches.

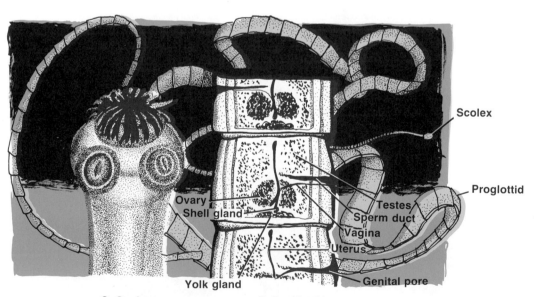

C **Scolex**   D **Proglottid**

Figure 6.9 *Continued*

C. The head end, or scolex, of the dog tapeworm *Taenia pisiformis.* Note hooks and suckers, for attachment only.

D. The reproductive segments or proglottids, of *Taenia pisiformis.* Genital pores are seen in each segment.

B, C and *D,* after Keeton, W. T. 1967. *Elements of Biological Sciences.* W. W. Norton, New York..

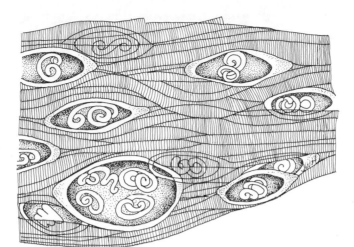

Trichinella larvae

FIGURE 6.10

Larvae of *Trichinella spiralis* within calcareous cysts in host muscle. The presence of such larvae causes the weakness and severe pain associated with trichinosis. (From Barnes, R. D. 1968. *Invertebrate Zoology*. W. B. Saunders Co., Philadelphia.)

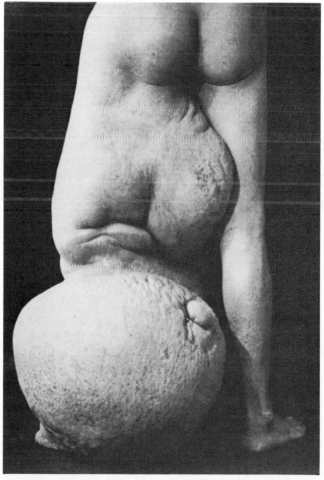

FIGURE 6.11

A severe case of elephantiasis resulting from chronic blockage of lymph nodes by the nematode worms which cause filariasis. (Courtesy of Mayo Clinic.)

A

FIGURE 6.12

B

Examples of the two main polychaete groups.

A. An errant polychaete, *Phyllodoce*, showing paired, bristle-bearing appendages on each segment. (From Barnes, R. D. 1968. *Invertebrate Zoology.* W. B. Saunders Co., Philadelphia.)

B. A sedentary polychaete, the fanworm, *Sabella.* All that is visible are the anterior tentacles, the "fan," used in filter feeding; the rest of the worm remains within its elongate tube made of sand-grains consolidated by an adhesive secretion. (Courtesy of D. P. Wilson.)

The **polychaetes** (Greek *poly,* "many," plus *chaitē,* "hair"), which are almost all marine or estuarine, are characterized by lateral segmental appendages bearing bristles. A great variety of body shapes is found among members of this group, which can be roughly divided into the errant or wandering forms (Fig. 6.12A) and the sedentary forms, primarily filter feeders, living in permanent tubes (Fig. 6.12B). Polychaetes are extremely numerous not only in the intertidal, where their tubes may be readily observed, but at greater depths, where in some places they are among the most numerous of mud-dwelling species. They thus form a major component of the marine environment as predators, scavengers and filter feeders, on the one hand, and as food for fish and shorebirds, on the other.

The second group of annelids are the **oligochaetes** (Greek *oligos*, "few," plus *chaitē*, "hair"), which have no lateral appendages and bear only short bristles on each segment. Included are the well-known earthworms and their small freshwater relatives, some of which are among the commonest animals within the oxygen-depleted muds of polluted lakes. The earthworms feed on dead organisms in the soil and are important both to its aeration and in recycling of nutrients.

The leeches, or **hirudineans,** are the third group of annelids. Without bristles or appendages, these rather short, flattened worms bear suckers at either end by which they move in inch-worm fashion over the substratum. Although found in marine and terrestrial habitats, they are most common in freshwater environments. Some leeches are predatory, feeding on other invertebrate animals, but most are blood-sucking **ectoparasites,** temporarily attaching themselves to the host while they feed. Medicinal leeches, now no longer used medically, are capable of storing two to five times their own weight of blood in a distensible crop, and require up to 200 days to digest a full meal. Some blood-sucking land leeches are serious pests of cattle.

MOLLUSCS

Living **molluscs,** including the familiar snails, clams and octopods, exhibit a diversity of body shape through adaptive radiation from what biologists believe was the primitive ancestor of this group. As shown in Figure 6.13, the major molluscan characteristics of this hypothetical early organism were (1) a muscular foot for creeping, (2) a protective shell, secreted by an underlying layer of **mantle** tissue and (3) a **mantle cavity**

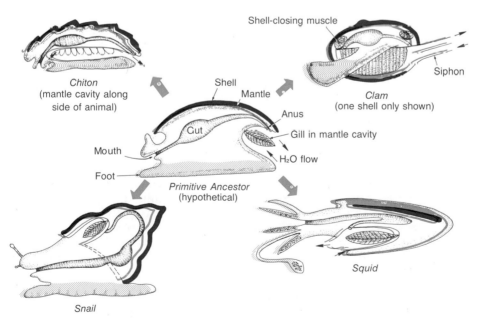

FIGURE 6.13

Adaptive radiation in the molluscs, showing changes from the basic body-plan of a hypothetical primitive ancestor to today's four major groups. Only the essentially molluscan features are shown.

within which was suspended a ciliated gill that created respiratory water currents. Note that the anus opens into this cavity. Such an animal was adapted to browsing on seaweed and microorganisms living on the substratum.

Two of today's four major groups of molluscs show relatively little change from this body form. The eight-shelled marine **chitons,** with a mantle cavity along either side of the body, are highly adapted for scraping tiny particles of food from the rocks they inhabit. The **gastropods,** which include snails, abalones, limpets, sea slugs and related forms, have undergone a torsion of the body such that the anus and mantle cavity are toward the front of the animal. Although most species are marine, the snails have colonized both fresh water and dry land. The majority of gastropods are browsers, but a number of marine forms are predators of other animals and certain tropical forms produce highly poisonous toxins. Abalone and the edible land snail, or "escargot," are important as food to many people.

The two-shelled molluscs, or **bivalves,** are sessile forms living within muds or attached to a solid substratum in marine, estuarine and freshwater environments. Food is obtained entirely by filter feeding, for which the ciliated gills are especially adapted. Fine food particles trapped in their mucus secretions are transferred to the mouth, which in these animals opens into the mantle cavity; extensions of the mantle cavity form the **siphons** for passage of water currents. The foot is used for locomotion in motile forms, but often the most prominent muscles are those which close the shells—the only part of a scallop which is usually eaten. Mussels, oysters, clams and other types of bivalves are commercially valuable, and all of this group, through their filter feeding activities, play an important role in aquatic environments.

The fourth molluscan group, the **cephalopods** (Greek *kephalē,* "head," plus *podos,* "foot"), are the highly motile, predatory marine octopus, squid, cuttlefish and chambered nautilus. In all but the latter the shell is greatly reduced or absent and the mantle entirely surrounds the "head." The foot is modified into several pairs of sucker-bearing tentacles used primarily for food capture; locomotion is brought about mainly by the enlarged mantle cavity which is surrounded by powerful muscles. Water is sucked into the cavity around its edges and then forcibly expelled through a funnel, producing a jet propulsion effect. In keeping with their active life and predatory habits, the cephalopods have a highly developed brain and sensory organs, including prominent camera-like eyes similar to those of vertebrates, an example of convergent evolution. Detailed studies of octopus behavior have greatly furthered our understanding of how nervous systems function. Species of this group have commercial value as food in various parts of the world.

ARTHROPODS

Among this group of highly successful and diverse joint-legged animals are the crustaceans, insects, centipedes, millipedes and arachnids, which include spiders and their relatives. The basic body plan of the **arthropods** (Greek *arthron,* "joint," plus *podos,* "foot") is similar to that of the annelid worms in consisting of repeating segments, each bearing a pair of appendages. It differs, however, in having a tough external covering, the **exoskeleton,** which is often calcified. To allow for movement the exoskeleton remains thin and flexible around the joints, but elsewhere it may be quite thick and hard. In order for the animal to grow, the exoskeleton must be shed at intervals, and after each such molt the animal is soft and defenseless until the new skeleton is hardened. Growth is thus discontinuous among the arthropods.

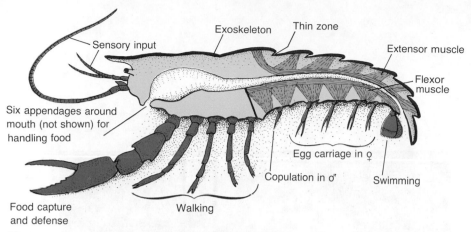

FIGURE 6.14

Diagram of a crayfish, showing general arthropod features. Note the thick and thin regions of the exoskeleton, the jointed appendages serving many functions and the antagonistic flexor and extensor muscles used in swimming.

Compared to those of annelids, the segments of arthropods tend to be less similar to one another; both the segments and the appendages they bear may be highly specialized to serve varying purposes. The diagram of the crayfish in Figure 6.14 exemplifies some of these general arthropod features. Note the exoskeleton, thinned in regions of bending, and the paired jointed appendages serving many different functions. The body is divided into two main regions, the fused head and thorax, and the jointed abdomen, or "tail," which functions in locomotion. Note the antagonistic muscles; the powerful **flexor** muscles pull the abdomen swiftly forward, forcing the animal backward, and the **extensor** muscles straighten it again. Many other crustaceans, as well as members of other arthropod groups, have a reduced number of appendages compared to the crayfish.

CRUSTACEANS

Mostly aquatic, the **crustaceans** show a broad diversity of body shape according to their habits. Most important in the economy of both the sea and freshwater systems are the various forms of small crustacea found near the surface water which comprise the majority of the **zooplankton** (Fig. 6.15). These animals feed on phytoplankton or each other and are a primary source of food for certain fish and whales.

Among the larger crustaceans are the familiar crabs, lobsters and shrimp, which are predators and scavengers, and the barnacles, which are important fouling organisms on ships and pilings. Being firmly attached to their substrata, barnacles have evolved a body form in which the upwardly directed appendages are used for filter feeding. Among the few crustaceans which have invaded land, the sow bugs are a familiar example.

INSECTS

Insects are primarily terrestrial, although some are secondarily aquatic. They are an extremely diverse and successful group, there being more types of insects than of all other animals put together! The segments of insects have become combined into three main body regions: the head, bearing usually one pair of sensory antennae, compound eyes and three pairs of appendages modified in various ways as mouth parts; a thorax with three pairs of legs and usually two pairs of wings; and an abdomen devoid of appendages in the adult. Neither the eyes nor the wings, however, are derived from segmental appendages, but have separate origins.

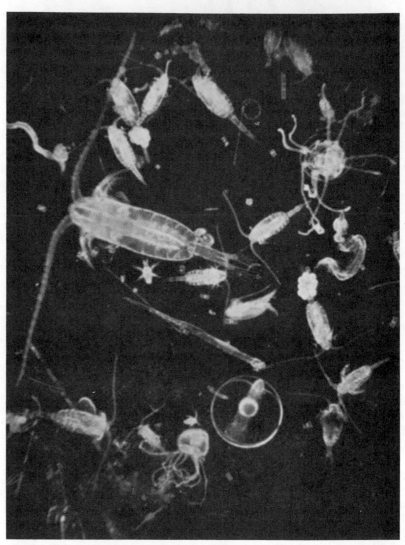

FIGURE 6.15

Living zooplankton. The shrimp-like animals in the picture are various small crustaceans. Two tiny jellyfish with long tentacles are also seen (magnification 16×). (Courtesy of D. P. Wilson.)

Specific adaptations to a terrestrial existence include not only the hard, chitinous cuticle, covered with a water-resistant waxy coating, but also the ramification throughout the body of a system of air ducts, the **tracheae,** which open to the exterior in each segment by means of a pair of tiny apertures, the **spiracles** (Fig. 6.16). Such an arrangement provides for direct exchange of gases between the individual cells and the tracheoles without need of a circulatory system, while at the same time preventing excessive loss of moisture.

The development of insects exemplifies the discontinuous growth of arthropods generally (Fig. 6.17). After the embryo hatches from the egg-shell it undergoes a series of molts producing successive larval stages called **instars.** During this period some insects, for example, grasshoppers, resemble miniature adults; in these forms the last larval instar molts directly into the adult. In other types, such as flies, beetles and moths, the larvae have a worm-like appearance. The last larval stage molts to produce a **pupa** which may become dormant and survive over winter. The adult form molts the pupal skin and emerges when conditions are favorable.

The functions of the insects in the biosphere are legion, and we shall have occasion to mention them frequently in later chapters.

MILLIPEDES AND CENTIPEDES

We here lump together two different but similar groups of arthropods which are primarily terrestrial, living in concealment under rocks and in decaying matter. Both are serially segmented and, like insects, have a water-resistant exoskeleton and a tracheolar system of tubules. The centipedes,

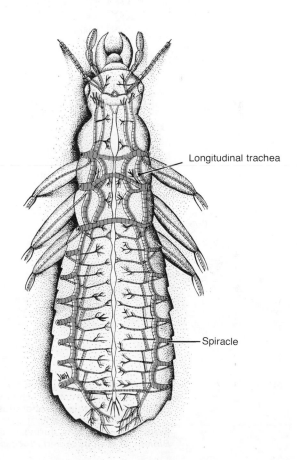

Longitudinal trachea

Spiracle

FIGURE 6.16

The tracheal system of an insect. Note the extensive ramifications throughout body and the spiracular openings in each segment. (After Barnes, R. D. 1968. *Invertebrate Zoology.* W. B. Saunders Co., Philadelphia.)

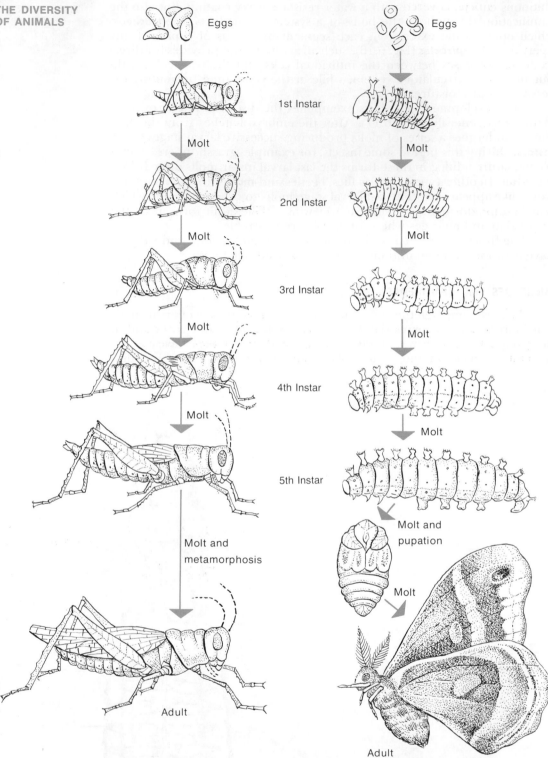

Eggs

1st Instar

Molt

2nd Instar

Molt

3rd Instar

Molt

4th Instar

Molt

5th Instar

Molt and
metamorphosis

Adult

Eggs

Molt

Molt

Molt

Molt

Molt and
pupation

Molt

Adult

FIGURE 6.17   Discontinuous growth in insects. Embryos hatch from eggs as larvae, and molt five times before becoming adults. Each larval stage is called an instar.
A. Direct development of grasshopper—larva resembles adult.
B. Indirect development of moth—larva differs greatly from adult and body must undergo great changes during pupal stage. (After Villee, C. A., and Dethier, V. G. 1971. *Biological Principles and Processes.* W. B. Saunders Co., Philadelphia.)

which are carnivorous animals, bear a single pair of appendages on each segment. The millipedes, or thousand-legged worms, however, have two pairs of legs in each adult segment as a result of the fusion of two sequential segments during development; these animals are herbivorous, feeding on both living and decaying plant matter, and thus contribute to soil recycling processes.

ARACHNIDS

Among the arachnids are the familiar terrestrial spiders, scorpions, ticks and mites; the marine horseshoe crabs and sea spiders are near relatives. The basic body plan of the predatory spiders and scorpions consists of two main regions, a fused head and thorax plus an abdomen. On the former are borne six pairs of appendages, of which the foremost two are used in feeling and the other four serve as legs for walking. In spiders producing poisons, these are secreted from the tips of the smaller pair of feeding appendages; in scorpions, poisons are produced by a gland at the tip of the abdomen. Species toxic to man are few, and deaths are usually restricted to young children. Spiders also have modified appendages, at the tip of the abdomen, called *spinnerets,* from which silk is extruded for building webs and cocoons. Alterations in the intricate web-building behavior of such forms as the orb-weaving spiders have furnished pharmacologists a means for distinguishing drug effects.

In the ticks and mites, the body segments are generally fused into a single unit. This group is extremely widespread. Many forms are ectoparasites of plants or animals, sucking nutrients from the host tissue and causing diseases such as mange and scabies. Other species are carriers or *vectors* of diseases such as Rocky Mountain spotted fever and Asian scrub typhus.

ECHINODERMS

Among the marine *echinoderms* (Greek *echinos,* "urchin," plus *derma,* "skin") are starfish, sea urchins, sand dollars, brittle stars, sea cucumbers and sea lilies, all of which possess a skin bearing calcified plates and spines. Members of this phylum occur from the intertidal to great depths and are very numerous in places. The body plan of these animals is radial, with five main axes of symmetry, as shown clearly in the starfish (Fig. 6.18). This

FIGURE 6.18

The underside of the starfish, *Astropecten irregularis.*
Note five-rayed symmetry and central mouth. The several rows of tube feet in each arm are faintly visible; the edges of each arm are guarded by heavy, movable spines. (Courtesy of D. P. Wilson.)

symmetry is a secondary adaptation to a sluggish mode of life, however; the young larvae are bilaterally symmetrical.

In addition to their spiny skins the echinoderms are characterized by multiple rows of hollow tube feet (see Figure 6.18) by which all but the permanently attached sea lilies achieve locomotion. These unusual organs arise as outgrowths of a special region of the coelom, the **water-vascular system.** Changes of shape of individual feet are brought about by contraction of muscles which forces fluid to move within them. The tips of the tube feet bear muscular suckers which grip the substratum or the food on which the animal is feeding.

Both herbivores and carnivores occur within this group. Sea urchins may be destructive of kelp beds in certain areas, and starfish are often troublesome predators upon oysters and other commercially important bivalves. The raw gonads of urchins are eaten as a delicacy by many, and dried sea cucumbers are a prized food commodity in the Orient.

CHORDATES

We now arrive at the last major group of animals, the chordates, among which are the animals with backbones, including man. The chordates are segmented animals possessing an internal supporting skeleton, the **endoskeleton,** whose major feature is a long, semi-rigid **notochord.** Other common features of the group are a hollow nerve cord running along the upper rather than lower body surface, and openings between the digestive tube and the exterior in the throat region, known as **gill slits.** These major characteristics are shown in the diagram of Figure 6.19.

The most primitive living members of the chordates are marine, and include the **tunicates,** or sea squirts, and the **lancelets.** Tunicate larvae superficially resemble frog tadpoles, and the sedentary adults, prevalent on pilings and rocks, utilize their greatly enlarged gill basket not only for respiration but also for filter feeding. The lancelets are elongate fish-like organisms living in sandy beaches of temperate and tropical regions and bear a close resemblance to the larvae of lampreys, which are vertebrates.

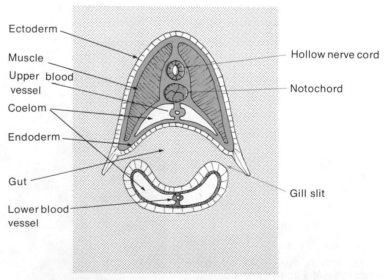

FIGURE 6.19

Schematic cross-section through throat region of a chordate, showing three characteristics of the group: hollow nerve cord; notochord beneath; gill slits connecting gut with exterior.

The major group of chordates, however, consists of the highly motile **vertebrates,** in which the endoskeleton is greatly developed as a support for the body and an anchor for the complex musculature. By contrast, all the other animals we have so far considered are lumped together as **invertebrates.** The vertebral column develops around and supersedes the function of the notochord and, being a series of separate but articulated segments, allows greater flexibility of the body. The ancestral vertebrates and today's primitive hagfish and lampreys have no limbs; but all the higher vertebrates possess two pairs of limbs supported on the bony limb girdles. The brain is enclosed within a protective bony case, the **cranium.**

Among the vertebrates, which include the various groups of fish, the amphibians, reptiles, birds and mammals, the same body plan has been adapted to a great variety of life styles during the course of evolution. The lobed fins of a primitive group of fish—once thought extinct but now known to still exist in deep waters off Africa—evolved into the articulated limbs of terrestrial amphibians and reptiles and were gradually modified to

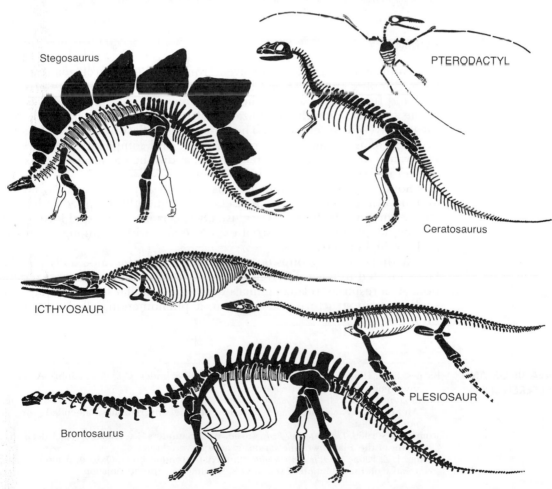

Adaptive radiation among fossil reptiles.

FIGURE 6.20

Stegosaurus, Ceratosaurus and *Brontosaurus* were all terrestrial or swamp-dwelling species, with limbs adapted to withstand their weight. The Pterodactyls had an elongated fifth finger to support the flight membrane. Both Ichthyosaurs and Plesiosaurs had elongate bodies and flap-like appendages for an aquatic existence. Birds and mammals evolved from two different groups of fossil forms, neither of which is shown here. (Redrawn from Storer, T. I. 1943. *General Zoology.* McGraw-Hill, New York.)

the wings of birds and the various types of legs found in mammals. The elongate body, so well adapted for the aquatic life of fish, became truncated, with loss of segments, among terrestrial forms. Today's marine mammals, especially the whales and dolphins, have secondarily acquired a fish-like body, an example of convergent evolution. The scales of fish became modified to the scutes of reptiles and, later, to the feathers and hair of warm-blooded vertebrates. The gills found in fish and amphibian tadpoles (and in some adult salamanders) appear in higher organisms only fleetingly during embryonic development; but derivatives of these structures are still present in such important organs as the tonsils, parathyroid glands and thymus. During the Age of Reptiles, between 200 and 100 million years ago, this group showed an enormous diversity of body form, all being adaptations of the same basic body plan to life on land, in the sea and in the air (Fig. 6.20). A similar degree of adaptive radiation exists among living mammals (see Figure 21.11).

The reptiles, birds and mammals, like the insects, exhibit special adaptations for life on land, including a water-impermeable skin, internal lungs and special provisions for development of the embryo, all features to be considered in later chapters. Likewise the roles played in the biosphere by various types of vertebrates are dealt with further on.

SUMMARY

Animals have gradually evolved from unicellular protozoa into ever more complex and organized forms exhibiting a great diversity of structure and function. First came the development of specialized cells, as seen in the sponges, then the organization of cells into tissues, exemplified by the two-layered coelenterates. The three-layered flatworms possess distinct organ systems. As animals increased in size and complexity, the need for respiratory and circulatory systems arose, as did the requirement for skeletal and muscular systems for locomotion.

Some, but not all, sessile animals have an apparent radial symmetry and are often filter feeders; browsers are frequently equipped with a protective shell, while predators are usually specialized for rapid locomotion. Special adaptations to terrestrial existence are evident among the insects and the higher vertebrates.

Within various groups one may observe how the same body plan has been adapted to meet different needs, a process known as adaptive radiation. As a result, unrelated animals which have similar life styles may show similarities in structure and function, a phenomenon called convergent evolution.

READINGS AND REFERENCES

Barnes, R. D. 1968. *Invertebrate Zoology*. 2nd ed. W. B. Saunders Co., Philadelphia. A detailed description of all the invertebrate phyla.

Buchsbaum, R. 1948. *Animals Without Backbones*. 2nd ed. University of Chicago Press, Chicago. An extremely well-illustrated account of the invertebrates, intended for the non-specialist.

Romer, A. S. 1969. *The Vertebrate Body*. 4th ed. W. B. Saunders Co., Philadelphia. A detailed account of the anatomy of the various groups of vertebrates.

Young, J. Z. 1963. *The Life of Vertebrates*. 2nd ed. Clarendon Press, Oxford. A most readable treatise on the evolution of vertebrates, relating structure to function.

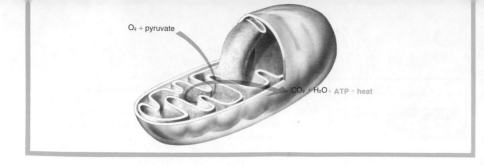

O₂ + pyruvate

CO₂ + H₂O + ATP + heat

7

ENERGY FROM CELLULAR RESPIRATION

All systems, including living cells, tend to become disorganized. To maintain their structural integrity and to grow and reproduce, cells oxidize reduced organic molecules, capturing a fraction of the energy released in high-energy phosphate bonds which can do work.

We are all aware that energy is necessary for life. We have also frequently heard that "energy is neither created nor destroyed." This statement, more formally known as the **First Law of Thermodynamics,** simply means that, in all physical processes, the total amount of energy remains constant, although the energy may be converted from one form to another. In a toaster, for example, electrical energy is converted to heat; in a hydro-electric generating plant, the potential energy of the falling water is converted into electrical energy.

This leads to an apparent discrepancy. If energy is never destroyed, why do we need a constant supply of it? Why does Earth require the continual input of energy from the sun to maintain life? If we have enough energy today, why do we need more tomorrow?

One reason is that the Earth loses much of the energy it receives from the sun. Energy is lost as radiant heat and reflected light, and this must be replaced. Another problem is that the sunlight which arrives in concentrated form at one point is soon dispersed, as heat, over a wide area. Heat, when uniformly distributed, cannot be used to do work, and work is essential to maintain life. Only *concentrated* forms of energy are useful to living organisms.

FREE ENERGY AND ENTROPY

Let us consider the total energy content of an isolated system. Part of this is concentrated into useful energy which can do work, and is called **free energy (G).** The rest is non-useful energy, or **entropy (TS)** (Fig. 7.1).

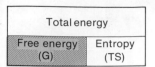

FIGURE 7.1

Energy distribution in a system. The total energy is distributed between two categories: G, the free energy, which can do work, and TS, the entropy, which increases with temperature. Entropy is unavailable energy and cannot be used to do work.

If we add external heat to a system, its total energy will increase, but most of the increase will be in the form of non-useful entropy. This unfamiliar term entropy is merely a way of expressing the fraction of energy that goes into molecular movement. These movements or vibrations make the system hotter. Its temperature increases, and the molecules mix more rapidly. Entropy is thus a measure of disorder or randomness. Conversely, free energy results from order, or non-randomness.

One or two simple examples may clarify these concepts. A file box full of carefully alphabetized reference cards is a system having a high degree of orderliness; it is useful. If the cards in the box are dumped on the floor, they become mixed and disorganized; they have lost part of their usefulness. The system has lost much of its order (free energy) and gained randomness (entropy). This change can be diagramed, as seen in Figure 7.2. Note that the change in free energy is expressed by ΔG. Another familiar example of a system containing a high degree of entropy is the mixture of useless refuse in one's garbage can. To return a disordered system to an ordered one requires input of work. The cards must be re-filed, or, if garbage is to be recycled and made useful, its constituent parts must be kept separate—the cans in one container, bottles in another, and paper in yet another. Anyone living in communities where recycling is practiced will no doubt agree that work is involved!

ENTROPY INCREASES IN THE NON-LIVING WORLD

In all systems, there is a natural tendency to increased randomness. Hence, maintaining order requires work. Let us take some examples from the non-living world. The masses of rocks which we recognize as mountains exist because cohesive forces are present which hold the rocks together. Otherwise the mountains would obey gravitational forces and collapse into a level heap of stones. During weathering processes, in fact, silt and rocks *are* carried downhill, and mountains slowly disappear. Instead of the organized mountains and valleys, a level plain results; the system is randomized and entropy has increased. New mountains are formed from geothermal forces generated inside the Earth. Heat and pressure cause stresses in the Earth's crust, resulting in the earthquakes and volcanoes which form new mountains.

Another example of an increase in entropy is found in the mixing of pure rainwater with salts leached from soils. The longer the ground water remains in the soil on its way to the ocean, the more salt it contains. To separate the salt from the water requires work. External energy, in the form of heat from the sun, causes water to evaporate from the ocean while the salts

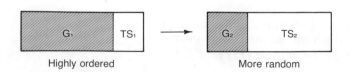

Highly ordered   More random

FIGURE 7.2

The redistribution of energy in a system resulting from randomization. Only ordered, non-random systems have a high free energy content and are capable of doing work. The difference between G_1 and G_2 is ΔG, the free energy change.

are left behind. To decrease randomness, work must be done, and the energy supply for doing the work must come from *outside* the system.

ENTROPY INCREASES IN LIVING ORGANISMS

The same laws apply to living systems. As we have seen, living cells are composed of large, complex molecules like DNA and proteins, which are organized into such complicated structures as chromosomes and membranes. But the tendency for randomization is always present. Moderate amounts of heat, for example, induce thermal motion of molecules, which tends to cause disruption of their organized structure and the breakdown of large molecules into smaller ones. Their physiological functions are lost, and they must constantly be replaced or repaired. If this tendency to randomization were not compensated, it would destroy the organization which characterizes life. Living systems therefore must constantly do work to prevent their own randomization and destruction.

The freshwater amoeba furnishes a good example. Water tends to diffuse into the cell due to the osmotic forces described in Chapter 3, and unless the cell can pump out the excess water it will soon rupture and lose its functional organization. To accomplish this, the amoeba has evolved a **water expulsion vesicle** (Fig. 7.3) which removes the excess water.

To summarize, all systems tend toward maximum entropy (randomness) and minimum free energy (order). The direction of this process can be reversed only by the addition of extra energy from outside the system—energy which is capable of doing work. This is in accord with the **Second Law of Thermodynamics**, which says that energy must be transferred from a concentrated, high energy region to a less concentrated, low energy region in order for work to be done.

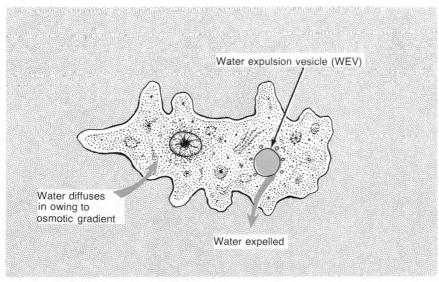

FIGURE 7.3

The water expulsion vesicle (WEV) of a freshwater amoeba.

As water moves into the amoeba as a result of the osmotic gradient, it is collected into small droplets which coalesce to form the water expulsion vesicle. The WEV collapses at regular intervals, expelling the excess water. The details of how this occurs are not yet clear. It is known, however, that the interior organization of the cell which causes the water to be expelled depends on the maintenance of an energy supply. Without energy, the vesicle would cease to function and the cell would rupture.

WORK REQUIRED TO RESTORE ORDER

In Figure 7.4, we can visualize two examples of this process. In part A, some of the heat energy transferred from the concentrated heat source, the furnace, to the cooler (less concentrated) atmosphere is used to drive the pistons of a steam engine. In part B, electrical energy from the highly concentrated source at the generator is transferred to a less concentrated region, the ground. In the course of this transfer, part of the energy is used to drive an electric motor. In both cases, there is a *net* randomization of energy as the high and low energy regions tend to equalize, resulting in an overall increase in entropy. Only a fraction of the energy transferred is able to do work—the rest must go into entropy in order for the process to proceed at all. Thus, increase in entropy is the ultimate driving force for all work done.

To remove heat from a cold body and transfer it to a hot one requires the input of free energy from another system. For example, a refrigerator

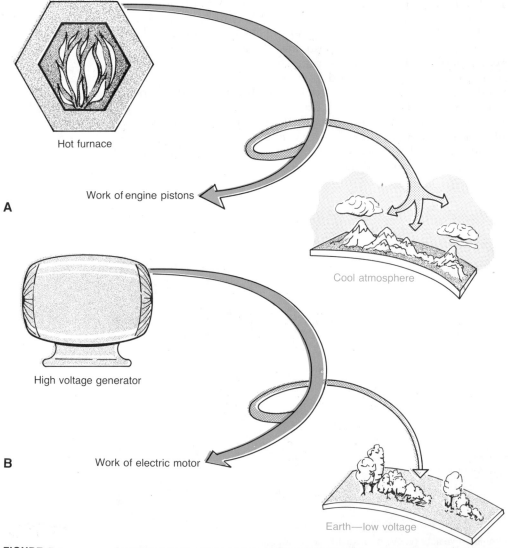

Hot furnace

Work of engine pistons

A

Cool atmosphere

High voltage generator

Work of electric motor

B

Earth—low voltage

FIGURE 7.4   According to the Second Law of Thermodynamics, work can be done only in the course of transfer of energy from a point of high concentration to a point of low concentration.
 A. Transfer of heat from a furnace to the atmosphere, during which part of the energy is used to drive a steam engine.
 B. Transfer of electrical energy from a generator to the ground, with part of the energy driving an electric motor.

removes heat from a cold body (the interior of the refrigerator) and transfers it to a hot body (the room). The source of energy for doing this work is the voltage drop which runs the refrigerator motor. The voltage supply, in turn, comes from power plant combustion, nuclear reactors or waterfalls, all of which are systems gaining entropy during power production. In Chapter 7a we shall consider in detail some of the consequences of entropy production—release of waste heat—by power plants.

NON-LIVING SOURCES OF ENERGY

So far we have considered more or less isolated systems. But the refrigerator, for example, interacts with the room it is in; the room interacts with the rest of the building; the building with the ground and air around it and so on. If we consider the Earth as a whole we can see that it is a semi-isolated system—a spaceship—whose free energy is constantly decreasing and whose entropy is increasing. Therefore Earth must have extra sources of energy to keep itself from running downhill. Where does this energy come from?

MAJOR SOURCES OF FREE ENERGY ON EARTH

The most important energy source is the sun's radiation. In the non-living world it is the heat energy rather than the light that does the most work. Evaporation of water, which drives the hydrologic cycle, is a function we have already discussed (Chapter 2a). Radiant heat also contributes to the weather, providing secondary forms of energy such as wind (often used by man) and electric discharges (not yet used by man). Without the sun's heat, the oceans would freeze, and another form of energy (tidal energy) would not be possible.

Tidal energy results from the combined effects of the Earth's angular momentum (rotation) and the gravitational pull of the moon and sun. These three forces result in the regular oscillation of oceanic waters known as tides. Recently power has been obtained by constructing tidal barriers across the mouths of wide rivers in areas with tidal ranges of 20 or 30 feet. The water trapped at high tide is used to drive generators as it flows back to the sea.

A third source of energy in the non-living world is geothermal energy. Nuclear reactions and great pressures generated at the center of the Earth result in a high temperature. The Earth's core is a highly concentrated heat source capable of doing work—an example is mountain building. The energy dissipated in hot springs and geysers could be used to do work, and there is considerable discussion of harnessing geothermal heat as a natural power supply.

In the living world, however, chemical energy is the main source of concentrated energy for doing work, and virtually all chemical energy required for life is a result of the photosynthesis carried out by green plants. Again, the ultimate energy source is the sun. We shall discuss the important reactions of photosynthesis in Chapter 8; our concern here is to describe the way chemical energy is utilized by organisms for doing work. How is the energy in a sugar cube or candy bar, for example, made available for running, thinking, perspiring and so on?

HOW CELLS OBTAIN ENERGY

The major energy-yielding reaction of all aerobic organisms, both plants and animals, is the oxidation of glucose to carbon dioxide and water, a process known as **cellular respiration** (not to be confused with breathing, or external respiration).

$$C_6H_{12}O_6 + 6\ O_2 \longrightarrow\longrightarrow\longrightarrow\longrightarrow 6\ CO_2 + 6\ H_2O + 686{,}000\ \text{calories}$$

(glucose) (oxygen)   (carbon   (water)   (free energy)
                             dioxide) ΔG

The reactants (glucose and oxygen) have more energy in their bonds than do the products (carbon dioxide and water). Such an energy-yielding reaction releases free energy (ΔG) which can do work. The multiple arrows indicate that there is a series of reactions involved.

In fact, the oxidation of any organic molecule results in a release of free energy. As shown in Figure 7.5, the sequential addition of oxygen to a simple two-carbon molecule like ethane allows a *stepwise* release of energy. Note that the more oxidized a compound is, the less free energy it contains. In the respiration reactions in the cell, the number of oxidation steps is much greater than the three shown for oxidation of ethane to oxalic acid in Figure 7.5, and the amounts of energy released at each step are smaller. The reason for this is that cells use standard amounts of energy, and only one standard unit can be made at any one reaction step. Any excess energy released by a reaction is wasted. Thus, the energy released in cellular reactions is never very much greater than that needed for making one standard unit of energy.

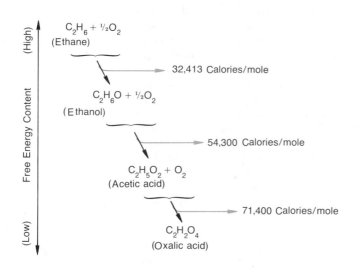

FIGURE 7.5

Free energy (ΔG) released in the stepwise oxidation of a two-carbon compound, ethane. Energy is expressed in calories. A mole is a standard amount of any substance; it contains 6×10^{23} molecules. Note that the more oxidized a compound is, the less free energy it contains.

ATP — THE STANDARD ENERGY UNIT

The standard energy unit of the cell is the molecule known as **ATP.** ATP stands for adenosine triphosphate, a molecule derived from adenylic acid, found also in RNA (refer to Figure 3.10). In fact, ATP is adenylic acid

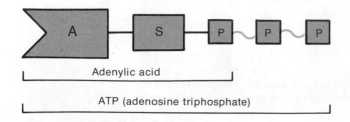

FIGURE 7.6

The standard energy unit of all cells, ATP, is composed of the nitrogen-containing base adenine *(A)* joined to the sugar ribose *(S)* and three phosphate groups *(P)*. The bonds between phosphate groups are high-energy bonds, designated by $\sim$. On being broken, they release free energy (ΔG) of about 8000 calories per mole.

with two more phosphate groups added onto the end, as shown in Figure 7.6. As before, the phosphate groups are denoted by **P**.

The sugar-to-phosphate bond is an ordinary, low energy bond, having a bond energy of about 3000 calories per mole—that is, if the bond is ruptured, that much energy would be released. The phosphate-to-phosphate bonds, however, are **high energy bonds,** with bond energies of about 8000 calories per mole. We shall denote them by the symbol ~**P**.

ATP is the energy currency for virtually all reactions requiring work in the cell. Although cells contain stored energy in the form of glycogen or starch, this cannot be used directly for doing work. We may regard it rather as energy stored in the bank which must be converted to cash in hand before being spent. ATP is like the dollar bill of the cell, which is formed by converting stored chemical energy, released during the oxidation of organic molecules, into a readily usable form. ATP thus takes part in two major types of reactions:

Energy-releasing

$$ATP \longrightarrow ADP + P_i + energy$$

Energy-requiring

$$ADP + P_i + energy \longrightarrow ATP$$

In energy-releasing reactions, ATP is broken down to ADP (adenosine diphosphate) plus inorganic phosphate (P_i). The energy released is used for all types of cellular work; osmotic work, muscle contraction, secretion of saliva and perspiration, synthesis of new tissue and so on. We shall discuss many of the types of work that cells perform in the following chapters.

It is the energy-requiring reaction, during which ATP is synthesized, that is our subject in this chapter. As we have seen, the laws of thermodynamics tell us, first, that a process requiring free energy can only occur if driven by a second energy-releasing process and, second, that part of the energy released must go toward an increase in entropy of the total system. ATP synthesis, therefore, must be *coupled* with one of the energy-yielding steps in the series of reactions known as respiration.

As we noted earlier, only one ATP can be formed in a single reaction. If the cell oxidized glucose to carbon dioxide and water in a single reaction step, only one ATP could be formed from ADP. Since the total energy released in the oxidation of a mole of glucose is 686,000 calories, and one ~P bond has an energy of approximately 8000 calories, we can calculate the energy yield from a one-step reaction:

$$\frac{8000}{686,000} \times 100 = 1.1\% \text{ useful energy yield}$$

Just over one per cent of the available energy would be trapped as ATP, the rest being dissipated as heat. In the cell, however, it is possible to obtain about 38 ~P bonds per molecule of glucose that is oxidized, by coupling ATP synthesis with several specific reactions in the cell.

$$\frac{38 \times 8000}{686,000} \times 100 = 44\% \text{ useful energy yield}$$

It is worth noting that most of today's power plants are only about 32 per cent efficient in capturing energy, so cellular efficiency is still better than that of the technologist! The remaining 57 per cent of the energy released by oxidation of glucose is dissipated as heat known as **metabolic heat.** Those living in or visiting snow-covered regions will have noted that snow tends to melt away from the base of a tree trunk. This phenomenon is partly due to radiant energy absorbed by the dark bark and transferred to the snow as heat and partly to the metabolic waste heat generated by metabolic reactions in the cells of the trunk.

OXIDATION-REDUCTION REACTIONS

Only a few of the many reactions occurring in the course of glucose oxidation are capable of releasing enough energy for synthesis of high energy bonds. These are the reactions which yield more than 8000 calories per mole, the amount of energy required to synthesize a mole of ATP from ADP and P_i. All of them are **oxidation-reduction reactions.**

If we return now to the overall oxidation of glucose, described previously, we note that its carbon atoms are oxidized to carbon dioxide, while its hydrogen atoms combine with oxygen to form water. In this latter reaction, we can say that the hydrogen atoms are being *oxidized* by oxygen. Or, conversely, we could say that oxygen is being *reduced* by addition of hydrogen atoms (actually by hydrogen ions and electrons, since hydrogen atoms do not exist freely in the cell). Every time an oxidation occurs, a reduction *must occur simultaneously.* Oxidation-reduction reactions are thus always coupled together. There are three kinds of oxidation-reduction reactions in cells:

1. Direct Addition or Removal of Oxygen. Addition of oxygen to a molecule is obviously an oxidation reaction; its removal would be a reduction reaction. In the cell, *molecular* oxygen (O_2) is never added directly to the glucose carbons at any stage of the sequence. As we shall see, the only reaction in which molecular oxygen takes part is the last one, where it combines with hydrogen ions and electrons to form water.

2. Removal or Addition of Hydrogen. The removal of two hydrogen atoms ($2 \, H^\bullet$) from a molecule is also an oxidation reaction, while the addition of $2 \, H^\bullet$ is a reduction reaction. Since hydrogen atoms cannot exist freely in any system, an oxidation and reduction reaction must be coupled together:

$$\text{coupled reactions} \quad \begin{cases} A + 2 \, H^\bullet \longrightarrow AH_2 \text{ (reduction of A)} \\ BH_2 \longrightarrow B + 2 \, H^\bullet \text{ (oxidation of B)} \end{cases}$$

net reaction (no free hydrogen atoms occur in the net reaction)

$$A + BH_2 \longrightarrow B + AH_2 \text{ (obtained by}$$

adding the two reactions together and cancelling terms appearing on both sides)

In living cells there are several special molecules which act as intermediate hydrogen acceptors and donors, transferring hydrogen atoms be-

Name	Oxidized Form	Reduced Form	Alternate Names
Nicotinamide adenine dinucleotide	NAD	NAD·H$_2$*	Coenzyme I; DPN
Nicotinamide adenine dinucleotide phosphate	NADP	NADP·H$_2$*	Coenzyme II; TPN
Flavin mononucleotide	FMN	FMN·H$_2$	
Flavin adenine dinucleotide	FAD	FAD·H$_2$	

Table 7.1
HYDROGEN
TRANSFER
MOLECULES

*For simplicity we have shown two H atoms bound to both NAD and NADP in their reduced form. The actual mechanism is rather more complex, but the net result is the same.

tween two molecules such as A and B in the example above. These are listed in Table 7.1. The ones with which we shall be most concerned are **NAD** and **NADP**. Note that the phosphate group is attached to NADP by a low energy bond. It cannot be used for doing work. Chemically, NAD and NADP contain part of the ATP molecule plus another sub-unit derived from the B-vitamin **niacin**. The details of their chemistry are unimportant here, but the role they play is very important. An example of how they function is the following:

coupled reactions $\quad \begin{cases} A + NAD \cdot H_2 \longrightarrow AH_2 + NAD \text{ (reduction of A)} \\ BH_2 + NAD \longrightarrow B + NAD \cdot H_2 \text{ (oxidation of B)} \end{cases}$
(need *not* be simultaneous)

net reaction $\qquad A + BH_2 \longrightarrow B + AH_2$

Thus NAD and NADP function in the transfer of two hydrogen atoms (2 H $\cdot$) from one molecule to another, even when these are at separate sites in the cell. In some reactions NAD is used; in others NADP. In either case, the NAD or NADP combines with the enzyme carrying out the reaction, and they are therefore known as **coenzymes.**

3. Removal or Addition of Electrons. The removal of a negatively charged electron (e^-) is called an oxidation, while the addition of an electron is called reduction. In the cell the addition and removal of electrons occur on a particular group of proteins, called **cytochromes,** which contain iron atoms. The iron easily switches between its oxidized (Fe^{+++}) and reduced (Fe^{++}) forms:

coupled reactions $\quad \begin{cases} Fe_1^{++} \longrightarrow Fe_1^{+++} + e^- & \text{(oxidation of } Fe_1^{++}) \\ Fe_2^{+++} + e^- \longrightarrow Fe_2^{++} & \text{(reduction of } Fe_2^{+++}) \end{cases}$
(must be simultaneous)

The iron atom attached to one cytochrome (Fe_1) is oxidized by losing an electron, which it passes to the iron atom of an adjacent cytochrome (Fe_2), which becomes reduced. (For a review of the nature of ions see Appendix I.)

The three types of oxidation-reduction reactions just discussed may be coupled in the cell with ATP synthesis in order to obtain energy:

coupled reactions $\quad \begin{cases} \text{oxidation-reduction reaction} \longrightarrow \text{energy} \\ ADP + P_i + \text{energy} \longrightarrow ATP \end{cases}$
(must be simultaneous)

This coupling is known as **oxidative phosphorylation,** and is the means by which the cell generates ATP currency, to be spent on whatever type of work is required.

RESPIRATION REACTIONS

The most important reactions involved in respiration are outlined in Figure 7.7. Although at first glance the system appears complex, it can be broken down into three subphases: **glycolysis,** the **Krebs cycle,** and the **electron transport chain.** Glycolysis can occur in the absence of oxygen and is therefore **anaerobic.** The reactions of the Krebs cycle and electron transport chain are closely linked to one another and require the presence of molecular oxygen. They are therefore **aerobic.** Let us consider these three subphases one at a time.

Glycolysis. During the first series of reactions, glycolysis, the original six-carbon glucose molecule (C–6) is split into two halves, each of which is oxidized by removal of 2 H $\cdot$ to form **pyruvic acid** (C–3) (Fig. 7.8). Two NAD $\cdot$ H$_2$ are formed in the process, and two ATP's are synthesized. In the

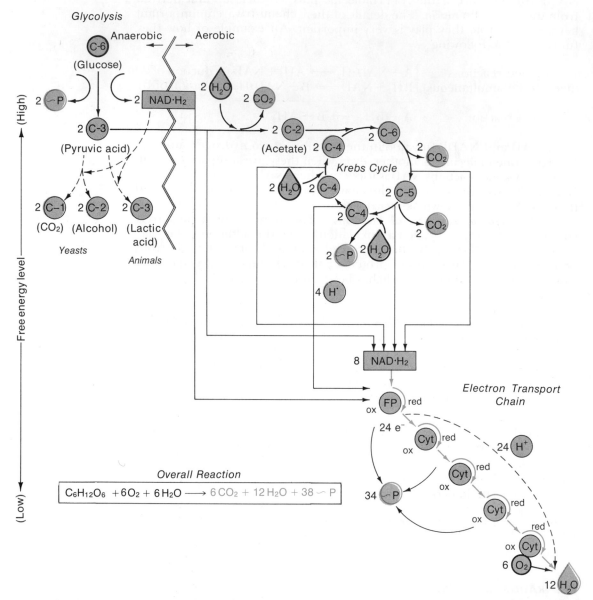

FIGURE 7.7   Outline of the reactions occurring during respiration. These may be subdivided into glycolysis, the Krebs cycle and the electron transport chain.

In the absence of oxygen, the final products of glycolysis are CO_2 and alcohol (in yeasts) or lactic acid (in animals). When oxygen is present, far more energy can be captured. The main products of the Krebs cycle are CO_2 and $NAD \cdot H_2$ plus H atoms. Hydrogen is then split by flavoprotein (*FP*) to hydrogen ions and electrons. The latter are transferred along the cytochromes (*Cyt*) of the electron transport chain (electron path shown in color) during which most of the ATP molecules generated by glucose oxidation are synthesized. Note gradual loss of free energy. Major reactants are outlined in black, major products are outlined in color.

absence of oxygen, this is as much energy as organisms can obtain from a molecule of glucose. The fraction of the total energy made available by anaerobic metabolism, then, is only two per cent:

$$\frac{2 \times 8000}{686,000} \times 100 = 2\% \text{ yield of total energy}$$

For yeasts the final products of anaerobic metabolism are alcohol and carbon dioxide. The NAD·H$_2$ gives up its 2 H$^•$ during formation of alcohol and is thus regenerated to NAD, which can react with further molecules of glucose as they pass through the glycolytic pathway. In animals metabolizing anaerobically, the hydrogen acceptor is pyruvic acid itself, and the final product is *lactic acid,* another C–3 molecule. During strenuous exercise, when insufficient oxygen reaches the muscles, lactic acid accumulates and gives rise to a sensation of fatigue.

The Krebs Cycle. If oxygen is present, however, a whole series of aerobic reactions is possible, and 36 more high energy phosphate bonds can be formed from the original molecule of glucose. This occurs by the further oxidation of pyruvic acid. First of all, carbon dioxide is formed in a series of reactions known as the Krebs cycle, named for the English biochemist, Hans Krebs, who discovered it (Fig. 7.9).

By the simple maneuver of adding water (H$_2$O) molecules at several stages and subsequently removing the hydrogen atoms, the cell is able to add the oxygen atoms necessary to form CO$_2$ without using molecular oxygen each time. An example of how this occurs in the case of the conver-

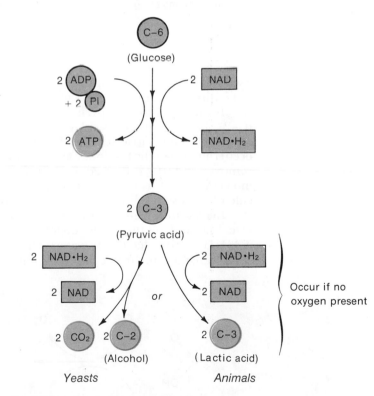

FIGURE 7.8

The reactions of glycolysis.

Glucose is split in half and, during the removal of hydrogen atoms by NAD, ATP is synthesized. In the absence of oxygen, NAD is regenerated by transferring H atoms to pyruvic acid to yield either carbon dioxide and alcohol (in yeasts) or lactic acid (in animals). If oxygen is present, pyruvic acid is further oxidized in the Krebs cycle.

Overall Reaction in Absence of Oxygen

Yeasts: $C_6H_{12}O_6 + 2 \text{ ADP} + 2 \text{ P}_i \longrightarrow 2 \text{ } C_2H_5OH + 2 \text{ } CO_2 + 2 \text{ ATP}$

Animals: $C_6H_{12}O_6 + 2 \text{ ADP} + 2 \text{ P}_i \longrightarrow 2 \text{ } C_3H_6O_3 + 2 \text{ ATP}$

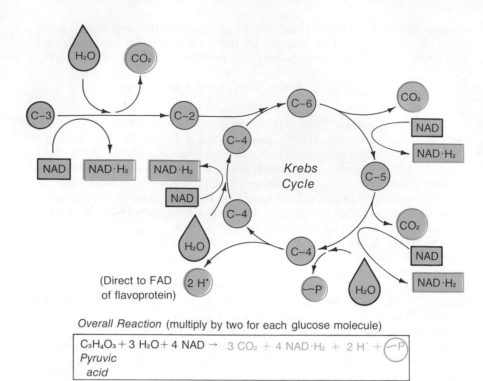

Overall Reaction (multiply by two for each glucose molecule)

$C_3H_4O_3 + 3\ H_2O + 4\ NAD \rightarrow 3\ CO_2 + 4\ NAD \cdot H_2 + 2\ H^\cdot + \sim P$
*Pyruvic
acid*

FIGURE 7.9

The Krebs cycle.
   In this series of reactions, pyruvic acid is oxidized to carbon dioxide, and reduced
NAD and H atoms are generated. Reactants are circled in black, products in color.

sion of pyruvic acid to acetic acid is shown in Figure 7.10. The extra
hydrogen atoms are transferred to the electron transport chain by means of
the hydrogen transport molecule, $NAD \cdot H_2$.
   In the Krebs cycle, pyruvic acid (C–3) is first converted to acetic acid
(C–2). The C–2 of acetic acid is then combined with a C–4 molecule to form
a C–6 molecule, citric acid (see Figure 7.9). In a further series of reactions
occurring in a cyclic fashion, water is added at two more points, hydrogen is
removed and CO_2 is formed. Eventually all that remains is the original C–4
molecule, which is thus regenerated in the cyclic process. For each molecule of glucose oxidized, two ~P bonds are synthesized in the Krebs cycle.
   The Electron Transport Chain.   The hydrogen atoms removed during
glycolysis and in the Krebs cycle are transferred, either directly or via
$NAD \cdot H_2$ to the electron transport chain (Fig. 7.11). The first step in this
chain is not a cytochrome, but a protein known as **flavoprotein**, to which is

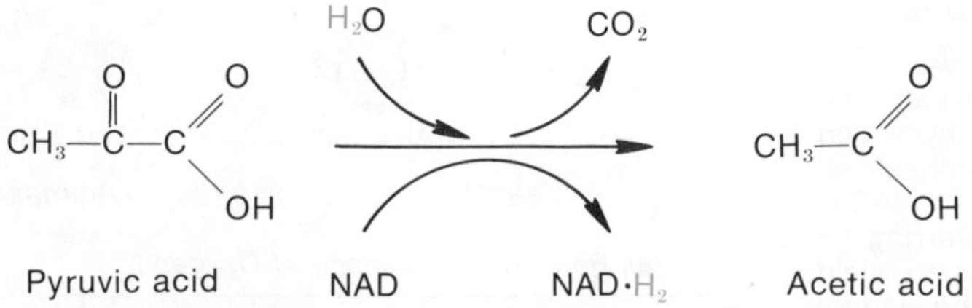

FIGURE 7.10   Pyruvic acid   NAD $NAD \cdot H_2$ Acetic acid

The net reaction in the conversion of pyruvic acid to acetic acid. By addition of a molecule
of water and the removal of its *hydrogen atoms* by NAD, the extra oxygen atom necessary
to form carbon dioxide is obtained.

bound the oxidation-reduction coenzyme, **FAD** (see Table 7.1). FAD accepts hydrogen atoms, either from $NAD \cdot H_2$ or from the Krebs cycle enzyme which releases $2\ H^\bullet$ directly, and in doing so becomes reduced. Reduced FAD splits these hydrogen atoms into two components—hydrogen ions and electrons.

$$2\ H^\bullet \longrightarrow 2\ H^+ + 2\ e^-$$

Each electron is transferred sequentially along the cytochromes of the electron transport chain, during which ATP is synthesized. For each molecule of glucose oxidized, the electron transport chain produces 34 molecules of ATP. Thus, the most important of the respiration reactions, so far as the generation of ATP is concerned, is the electron transport chain. Finally molecular oxygen binds with the terminal cytochrome, called cytochrome oxidase, which then combines electrons and hydrogen atoms with the oxygen molecule to form water.

RESUMÉ OF RESPIRATION

Although the details of the reactions we have been describing are quite complex, the overall picture, shown in Figure 7.7, is relatively easy to grasp. During glycolysis, glucose (C–6) is split in two, and by internal rearrangement of the resulting C–3 molecules a small quantity of energy is made available even in the absence of oxygen. In the presence of oxygen the Krebs cycle functions to produce carbon dioxide and to generate hydrogen

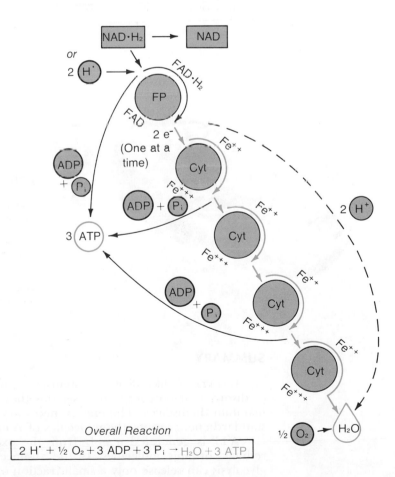

FIGURE 7.11

Generation of ATP by the electron transport chain.

Hydrogen atoms from $NAD \cdot H_2$ or direct from the Krebs cycle reduce FAD of flavoprotein *(FP)*. This then splits H atoms into hydrogen ions (H^+) and electrons. The latter are transferred down the cytochromes *(Cyt)* of the electron transport chain, reducing Fe^{+++} to Fe^{++} at each step. At three steps, oxidative phosphorylation occurs, with synthesis of ATP from ADP and P_i.

Overall Reaction

$$2\ H^\bullet + \tfrac{1}{2}\ O_2 + 3\ ADP + 3\ P_i \rightarrow H_2O + 3\ ATP$$

atoms for transfer to the electron transport chain, where almost half of the total energy of the glucose molecule is trapped as $\sim$P bonds. The oxidation-reduction reactions of the electron transport chain are thus the major site for ATP production.

The net reaction for the oxidation of glucose is as follows:

$$C_6H_{12}O_6 + 6\ O_2 + 6\ H_2O \longrightarrow 6\ CO_2 + 12\ H_2O + 38\ \sim P + heat$$

The six extra molecules of water used to produce CO_2 are regenerated at the end of the electron transport chain.

These reactions are common to all plant and animal cells and to some but not all bacteria. As noted in Chapter 3, most energy production occurs in the mitochondria of the cell. It is now possible to localize within the cell the reactions we have been describing, as shown in Figure 7.12. The glycolytic reactions occur in the cytoplasm of the cell, unassociated with membrane systems. The interior matrix of the mitochondrion, which is either fluid or, more probably a semi-fluid gel, contains the enzymes of the Krebs cycle, and the electron transport chain is located on the membranes of the mitochondrial cristae.

One last point seems pertinent. We have considered the reactions of respiration as if only sugars such as glucose were oxidized by the cell. This is of course not the case, since fats and proteins are also oxidized. Fats are first broken down to C–2 units (acetic acid) and enter the oxidative steps at that point (see Figure 7.7). Amino acids from proteins are generally converted to molecules identical with those found either in glycolysis or in the Krebs cycle. Thus all types of foods may be utilized for energy by the same pathway of respiratory metabolism.

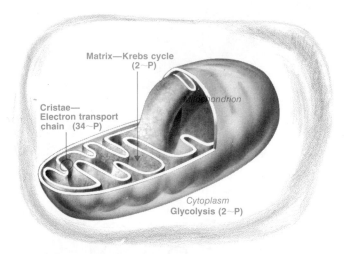

Matrix—Krebs cycle
(2~P)

Mitochondrion

Cristae—
Electron transport
chain (34~P)

Cytoplasm
Glycolysis (2~P)

FIGURE 7.12

The spatial distribution of respiratory enzymes in the cell.

The glycolytic enzymes are located in the cytoplasm, not associated firmly with any membranes. The Krebs cycle enzymes are located in the matrix of the mitochondrion, and the electron transport system is associated with the cristae of the inner membrane.

SUMMARY

Living cells, like all other systems, are constantly subjected to the forces of disorder or entropy. To offset this they must continually do work to maintain themselves. The energy necessary for this work is packaged in standard energy units, the molecules of ATP.

ATP is generated mainly from the oxidation of glucose and other reduced organic molecules. In the absence of oxygen, the reactions of glycolysis can release only a small fraction of the total energy available in glucose. If oxygen is present, however, almost half of the energy stored in glucose is captured in high energy phosphate bonds ($\sim$P).

In the Krebs cycle, the carbon atoms of glucose are oxidized to carbon dioxide. In the closely linked electron transport chain, hydrogen atoms are split into hydrogen ions and electrons. The latter, during passage through the cytochrome chain, give up their energy in coupled oxidative-phosphorylation reactions, during which ATP is synthesized.

Besides glucose, other compounds in the cell, such as proteins and fats, can be oxidized to yield the ATP currency needed by the cell for its maintenance and growth. The energy not captured as $\sim$P is dissipated as metabolic heat.

READINGS AND REFERENCES

Dawkins, M. J. R. and Hull, D. 1965. The production of heat by fat. Scientific American Offprint No. 1018. Describes how body heat is maintained by metabolism of special "brown" fat reserves.

Lehninger, A. L. 1960. Energy transformation in the cell. Scientific American Offprint No. 69. An account intended for the layman of the reactions involved in cellular respiration.

Lehninger, A. L. 1965. *Bioenergetics: The Molecular Basis of Biological Energy Transformation.* W. A. Benjamin, New York. A brief, readable account of the way cells obtain energy.

McGilvery, R. W. 1970. *Biochemistry.* W. B. Saunders Co., Philadelphia. An excellent introduction to the details of biochemical reactions within the cell.

Scientific American. September, 1971. (Offprint Nos. 661–671.) Entire issue devoted to energy in the biosphere. The ideas appearing in the first half of this chapter and some of the practical aspects discussed in Chapter 7a are considered in the various articles in this issue.

7a

ENERGY, ENTROPY AND POLLUTION

At some point, man must set a limit to the amount of extra heat he imposes on the Earth's heat budget. Even if this limit is far from being reached, local environmental effects of excessive heat may soon force him to curtail his ever-increasing demands for energy.

The three subjects of this chapter are inextricably woven together. In Chapter 7 we considered the way in which the total energy in any system is divided between its *free energy,* which is capable of doing work, and its non-useful energy, called *entropy.* We defined entropy as a measure of the randomness of a system, or of its uniformity. We also saw that work can only be done in the course of transferring energy from a concentrated energy reservoir to a reservoir containing less energy. This statement is a result of the Second Law of Thermodynamics. We also noted that any system, left to itself, tends to become randomized—that is, it loses free energy and gains entropy; it tends to become uniformly mixed.

This chapter deals with the energy content of the Earth's surface and with the effects of man's present and future activities on our planet's heat budget, both locally and as a whole. It will be useful to keep in mind the fact that uniformly dispersed energy, no matter how much of it there is, cannot do work. Work can only be done in the course of a downhill transfer of energy from a concentrated state to a dispersed state, and only part of the energy transferred can do work.

ENERGY BUDGET OF THE EARTH'S SURFACE

Let us first consider Earth as a part of the solar system. The sun is the main source of concentrated energy in this system, and over the course of time the sun is gradually losing this energy by radiation to the rest of the solar system—to space and to the planets, including Earth. The sun's energy is gradually being spread evenly throughout the whole system, the entropy of which is thus increasing. Virtually all of the natural energy now available for doing work on Earth results from the transfer of heat and light from the hot sun to the cool Earth. At that far, far distant point in time when the sun and Earth reach the same temperature, no further net energy transfer will occur, and the Earth's main source of energy will have disappeared.

Although only a small fraction of the sun's total energy falls on Earth, the input of energy from the sun represents more than 99.999% of the annual energy input at the Earth's surface. For the surface of the Earth to remain at an average constant temperature, the Earth must either re-emit all of this radiant energy back into space or else store part of the energy in some form which does not heat its surface. Stored energy is a form of **potential energy,** and can eventually be used for doing work.

A familiar example of a potential energy store on Earth is the snow on mountains. The sun's heat causes water to evaporate from the oceans and some of it is stored as snow at high elevations. When it melts and flows downhill, it is capable of doing work such as driving water wheels or turbines. Another example of stored energy is fossil fuel—coal, oil and gas. In this case the energy is stored as chemical energy which can be released later by combustion.

Most of the solar radiation which falls on Earth, however, is eventually returned to outer space as heat or reflected light. A large fraction is lost immediately, by reflection from clouds or non-absorptive surfaces such as deserts. A smaller quantity is converted to other forms of energy—winds, ocean currents, the potential energy of snow and the chemical energy trapped by plants in the course of photosynthesis. In the natural cycle of events, however, almost all this energy in the past has eventually been reconverted to heat and lost again to space. Winds, for example, generate frictional heat in the air, which is lost to space. Likewise, most dead organisms are metabolized by bacteria, during which most of their chemical energy is once again converted to heat. Only a small fraction of the sun's energy has remained stored on Earth, mostly as fossil fuels. The rest has all been lost to space. The average temperature of the Earth's surface has thus remained nearly constant over long periods of time.

Solar radiation, and those forms of energy which it secondarily produces, such as winds, falling water and chemical energy trapped by photosynthesis, will all eventually be dissipated to space as heat. Therefore, man's use of these as sources of energy cannot affect the natural heat balance between the total energy received by Earth and the amount re-emitted back into space. Neither sailing a boat, nor generating electricity from waterfalls or windmills, nor burning wood nor consuming plants and animals affects the Earth's energy budget. The heat generated from these natural energy sources would be produced whether or not man retrieved some of the energy as useful work in the process. On the other hand, burning large quantities of fossil fuels, long removed from the Earth's energy balance sheet, or generating electricity from "unnatural" sources such as nuclear or thermonuclear reactors, imposes a thermal stress on the Earth's heat budget.

Almost all of the energy we now utilize in the United States, however, is supplied by sources not included in the Earth's natural heat budget; it comes in two forms. About one-third is electrical energy, only about 20 per cent of which is produced by non-polluting hydroelectric plants, the remainder being generated by power plants burning either fossil or nuclear fuels. The other two-thirds of our energy may be called "portable" energy—mainly gas, coal and oil—which is burned directly for power or heating. But no matter in which way energy is supplied, nor how efficiently it is used for doing work, *eventually all of this energy is converted to heat which is added to the Earth's total heat budget* (Fig. 7a.1).

PROJECTED ENERGY DEMANDS

Our first question, then, is, What are the likely energy demands of the future, and how will they affect the average temperature of the Earth's surface? This is a difficult question to answer, partly because we know so little about the Earth's present heat budget and partly because predictions about future demands can only be based on present trends. In attempting to answer this question, we are therefore faced with some of the problems about predictions discussed in Chapter 1a.

In Table 7a.1 are listed the per capita energy requirements predicted for the United States through the year 2000. Note that total energy demands are expected to be about twice what they are now in just 30 years' time and also that much of the increase will come from generation of electrical power. The *average* per capita consumption throughout the world today is only one-seventh what it is in this country; therefore, if the underdeveloped countries do industrialize, they will contribute proportionately greater increases to the world's total energy utilization in the future.

A great range of predictions now exists about what will happen to the

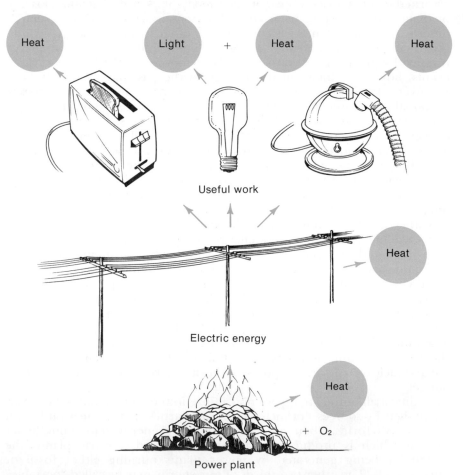

FIGURE 7a.1

An example of the way in which all energy is eventually dissipated as heat and added to the Earth's heat budget.

Only part of the energy from fuel is converted to electrical energy at the power plant. More heat is dispersed during transmission, through frictional losses. In doing work, such as in the appliances shown, the remaining electrical energy is converted to heat. (Light, as we shall see in Chapter 15a, is but another form of heat energy.)

Earth's surface temperature if present trends of increasing energy utilization continue. We shall consider here the two extremes, which serve to indicate how little we now know about what might happen and how much we need to know to make more accurate predictions.

A highly pessimistic estimate has been made by ecologist LaMont C. Cole, which may in fact be based on rather more rapid increases in energy production than currently exist; Dr. Cole says that these amount to 7 per cent per year. Although man now adds only an additional 0.002 per cent to the energy received on Earth from the sun, according to Cole's projected increases, in just 90 years we shall be adding enough extra heat to increase the *average* temperature of the Earth by 1° C. This would have a marked effect on the geographical distribution of plants and the animals which depend on them. In only 108 years, the temperature would increase by 3° C, which would melt the ice caps, raising sea level by some 400 feet and inundating many major cities and coastal agricultural lands. In 130 years, the average temperature of the Earth would be the same as today's hottest spot at Massawa in Ethiopia—the Earth would become uninhabitable. Even though Dr. Cole's time scale may be slightly accelerated, if his fundamental assumptions are correct, then we must begin immediately to cut down on the rate of growth of our energy demands.

By contrast, a highly optimistic prediction has been made by two physicists, A. M. Weinberg and P. R. Hammond. They estimate that even if the world population increases to about three times what it is now and at the same time everyone uses twice the energy now being used by the average American, the effects on the Earth's heat budget will be negligible, increasing the surface temperature by not more than 0.2° C! The sources of the enormous discrepancies between these two views obviously must be resolved if we are to have reasonably accurate predictions to guide us in determining how much extra heat load the Earth can safely tolerate.

Problems about which there is much less disagreement, however, are the local heating and polluting effects of electricity-generating plants, which are constantly increasing both in size and number and will provide an ever greater proportion of our energy supplies in the future.

PROBLEMS GENERATED BY POWER PLANTS

"Conventional" power plants are fired by fossil fuels, especially coal and oil. Energy for generating electricity is derived from chemical energy during combustion, and at the same time, heat, water, carbon dioxide, ash and a variety of toxic gases are emitted into the environment. In nuclear power plants however, no combustion takes place, and only heat and a

TABLE 7a.1 PAST, PRESENT AND PREDICTED COMPARISON OF PER CAPITA ENERGY UTILIZATION IN UNITED STATES, 1950–2000	1950	1970	1980	2000
Electricity generated* (kwhr)	2165	6500	11,650	28,250
Energy input into power plants** (kwhr)	9930	23,650	30,250	66,400
Total energy utilized† (kwhr)	65,900	79,600	104,700	145,300

Data from U.S. Bureau of Mines statistics, including Information Circular 8384.

*Useful energy produced by power plants.
**Total energy consumed by power plants.
†Total energy consumed for all purposes, both electric and "portable."

small amount of radioactivity are produced as unwanted by-products. We can consider these three factors—atmospheric additives and pollutants, waste heat and radioactivity—in turn.

ATMOSPHERIC SIDE EFFECTS

The most apparent and annoying side effects of "conventional" power plants are smog and smoke. Although up to two-thirds of the smog in larger industrialized cities is produced by cars, much of the rest comes from power plants. The effects of smog fall into two categories, local climatological effects and toxic effects. The latter are discussed in detail in Chapters 8a and 9a, but it is appropriate to deal with the consequences to local climate here.

Firstly, the dust particles released into the air, known as fly ash, behave as would a cloud cover. They reflect incoming solar radiation, preventing it from reaching the Earth's surface, and thus tend to produce a cooling effect. This, however, is generally more than offset by a second, opposite effect, known as the **greenhouse effect,** which requires a brief explanation.

The Greenhouse Effect. The main products of fossil fuel combustion are carbon dioxide and water; the reaction is similar to that for metabolism of glucose by cells.

$$\text{fossil fuel} + \text{oxygen} \longrightarrow \text{carbon dioxide} + \text{water}$$

If either CO_2 or H_2O is increased in the atmosphere a warming effect on the surface of the Earth results. The reason is shown in Figure 7a.2.

When energy from the sun falls on Earth it arrives mostly as light, with a smaller quantity of heat. Some of the light energy is reflected unchanged—from clouds, oceans and deserts. Some of it, however, is absorbed by gases in the atmosphere and by soils, water and plants, which reemit it almost entirely as heat. It is this heat which determines the surface temperature of the Earth. If the amount of heat being absorbed is increased for some reason, then the temperature of the Earth's surface will also increase.

Carbon dioxide and water vapor both transmit sunlight, and thus their presence in the atmosphere does not significantly affect the amount of energy reaching the Earth's surface. But the energy re-radiated from the Earth as heat is readily absorbed by both CO_2 and H_2O. Instead of being

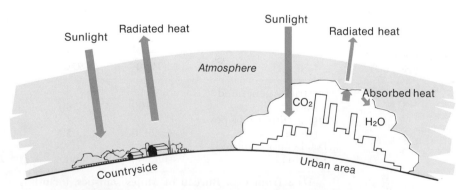

FIGURE 7a.2

The "greenhouse effect."
In the countryside, the sun's energy falling on Earth is mostly converted to heat, which is reradiated to space, and so the average surface temperature remains constant. Over urban areas, where CO_2 and H_2O accumulate, they absorb some of the reflected heat and warm the atmosphere, thus raising the temperature of the local surface to a new level.

radiated into space, this heat is trapped in the atmosphere, warming the Earth's surface. This is known as the greenhouse effect, since exactly the same thing occurs in the humid conditions in a greenhouse, where the water vapor and glass transmit light and absorb heat, preventing the re-radiation of heat to the air outside. As a result, the local climate of most industrialized cities is indeed like a greenhouse; the average temperature is several degrees warmer than that of the surrounding countryside.

Is There a Global Greenhouse Effect?   The average content of carbon dioxide in the world's atmosphere has been increasing in recent decades as a result of the accelerated combustion of fossil fuels (see Figure 18.7), and it has been claimed that this is having an effect on the global temperature. It is almost impossible to monitor small trends in world temperature, however, and no conclusion on this point is yet possible.

Combustion and Oxygen Supplies.   It is all too often said that burning fossil fuel is using up the world's oxygen supply. As will be shown in Chapter 8a, this is simply not possible. Only a minute fraction of the fossil organisms which helped produce today's oxygen supplies are in concentrated enough form to be used as fuel. Fortunately, running out of oxygen is the least of our worries!

WASTE HEAT

Our most important concern today arising from power production is the large quantity of waste heat which is released into the local environment around power plants, known as **thermal pollution.** Although this waste heat may one day prove useful in aquaculture and desalination, so far most of its environmental effects are detrimental, as we shall see below. But first let us discover the source of this waste heat and the ways it gets into the environment.

Both fossil fuel- and nuclear-fired power plants work on the same principle. Steam generated at high pressure in a boiler or reactor flows through a turbine which is connected to a generator. As we saw in the last chapter, only part of the total energy available in such a concentrated energy source can be employed in useful work—the generation of electricity. The rest is lost as entropy, in the form of heat. In cellular metabolism, we saw that the efficiency of useful energy obtained from food is somewhat less than 50 per cent. Power plants are rather less efficient. In electricity-generating plants fired by fossil fuels, the fraction of energy captured for useful work is only about 32 per cent, and it is decreasing because of environmental constraints on power companies. The remaining 68 per cent is dissipated as heat. In today's nuclear power plants, the efficiency of useful energy production is even less—only 25 per cent.

Dissipating Waste Heat.   In the construction of a power plant it is necessary to build in a mechanism for removing the waste heat so that the plant itself does not become overheated. In most contemporary plants, water is used as coolant to prevent overheating; the water is in turn cooled in one of three ways, each of which involves eventual transfer of heat to the atmosphere. Cooling is accomplished either by discarding the heat directly to the atmosphere through **cooling towers** (Fig. 7a.3), or by running large quantities of water through the plant and returning it to a lake or stream.

In so-called "wet towers" the heated water is run over the surface of screen-like baffles within the huge, hyperbolically shaped cooling towers which are ventilated by an air inlet at the bottom. Evaporation of part of the water causes cooling of the rest, which is then recirculated to the plant (Fig. 7a.4A). In the "dry tower" method (Fig. 7a.4B), the hot water remains in pipes which are cooled by forced ventilation of air driven through the tower by a fan; about 3 per cent of the plant's power capacity is required to run the fan. Cooling towers, especially dry towers, are expensive to build

FIGURE 7a.3

Three wet cooling towers under construction in 1968 for the TVA power plant being built on the Green River in Kentucky. Each tower is 437 feet high and 320 feet in diameter—almost large enough to hold a football field! (Courtesy of Tennessee Valley Authority.)

and operate and are used only where water supplies are inadequate for more direct cooling methods. It is far cheaper to discard the heated water directly back into the original water supply—a lake, stream, or artificially constructed cooling pond. From there, the heat is eventually lost to the atmosphere.

Each method has its disadvantages, and the ultimate result in each case is a local heating of the air in the vicinity of the power plant. In the presence of condensation nuclei, such as dust or smoke particles, the super-saturated warm air from wet towers, cooling ponds or lakes readily forms mists and fogs when it contacts the cold air above. The fogs that frequently cause poor visibility at urban airports, for example, are often partially due to the local power plants. Since somewhat more waste heat is formed by nuclear power plants, the problem is even greater with them than with fossil fuel-fired plants.

RADIOACTIVITY

One of the great advantages of nuclear power plants is that they do not produce smog and fly ash, although they do emit heat and small quantities of radioactivity into the environment, about which there is a great deal of public concern.

The design of one of today's fission-type reactors is shown in Figure 7a.5. The nuclear chain reactions which take place in the uranium-containing fuel elements heat the surrounding water, generating steam which turns the turbines connected to the generators. This primary water is then condensed by a second source of cool water drawn from the ocean, a river

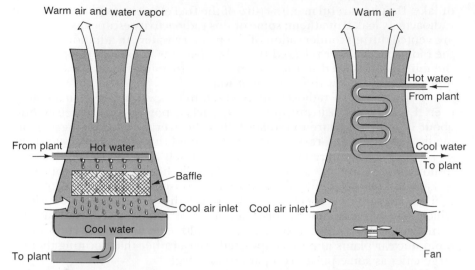

Figure 7a.4

Two types of cooling towers used to dissipate waste heat from steam-electric power plants. Such towers are often larger than a football field at their base and rise to 400 feet or more.

A. Wet tower. Part of the heated water evaporates from the baffle, cooling the rest of the water, which is then returned to the plant. The warm air contains much water vapor. Extra water must be supplied to the returning cool water to make up its volume.

B. Dry tower. The heated water remains in the pipes and is cooled by conduction of heat to moving air currents. The air suction from the hyperbolic towers usually needs to be augmented by a fan, which uses a significant fraction of the plant's energy production. No water is lost in this system, however. Dry towers have not been built at any large plant in the United States, since they add substantially to both investment and running costs, by current economic standards.

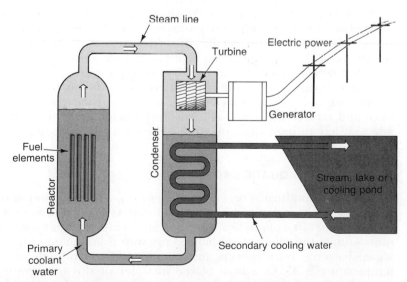

Figure 7a.5

A schematic diagram of a contemporary nuclear power plant. Uranium in the fuel elements undergoes fission, generating heat which causes the primary coolant water to boil, generating steam which turns the turbines. The primary water is cooled by a second source of cooling water, usually drawn from a stream, lake or artificial cooling pond.

or lake. Despite careful manufacture of the fuel elements, small amounts of radioactivity leak from them; some of this radioactivity escapes as gases that are vented during condensation of the primary water. In addition, some of the radioactive matter released from the elements bombards the pipes and the minerals dissolved in the coolant water, forming new radioactive isotopes which escape with the effluent water.

The amounts of radioactivity released in this way are very small, however. Even a person living quite near such a power plant receives only about 1/50 the exposure to radiation that the average American gets from medical and dental X-rays each year! Provided that the present standards set by the Atomic Energy Commission are closely adhered to and that continuous monitoring is carried out, the public health dangers from such leakage are minimal and will be more than offset by a decreased combustion of air-polluting fossil fuels. Although the probability of accidents which would release uncontrolled amounts of radioactivity from nuclear plants is remote, it is not zero. Hence, it would seem prudent to continue to build nuclear plants in underpopulated areas, rather than locating them in large cities as some power companies now urge.

The major problem of radioactivity from nuclear plants, now and in the near future, is the disposal of the spent fuel of the reactor. Today's technology for reprocessing fuel elements produces large quantities of highly radioactive liquids—some millions of gallons a year—which are at present being stored in huge underground tanks. Unfortunately, these wastes will remain highly radioactive for centuries, while the lifetime of the tanks can be measured in decades. Sooner or later, reduction of the bulky liquids to solids and their incarceration in old salt mines will be necessary. Solutions to this long-term problem are discussed in the article by Weinberg and Hammond listed at the end of the chapter.

BIOLOGICAL SIDE EFFECTS OF POWER PLANTS

In addition to the local climatic changes brought about, at least in part, by various types of power plants, the smog produced by fossil fuel-fired plants has an adverse effect on living organisms. We shall consider this problem in Chapters 8a and 9a. We shall also, in Chapter 15a, examine the biological effects of radiation from various sources, including nuclear power plants. Thus, our main concern here is with the effects on the biosphere of the entropic waste heat emitted by power plants.

In power plants where wet or dry cooling towers are used, the heat is immediately dissipated to the atmosphere. As we noted, because of the cost of building and operating such towers, it is more usual to use the water only once and return the heated effluent to the original surface water supply, where it has an immediate impact on the local environment.

HEAT DEATH OF AQUATIC ORGANISMS

The temperature of oceans and lakes ranges from a minimum of $-2°$ C in the polar seas to a maximum of about $+30°$ C in the Red Sea (or perhaps slightly higher in tropical swamps). At water temperatures exceeding $35°$ C, only a limited number of specialized organisms are able to survive—a few algae, bacteria, roundworms and protozoa. (Even though human body temperature is $37°$ C, a man placed in water of this temperature would soon die from overheating since he could not rid himself of his excess metabolic heat through perspiration.) Heated power plant effluents which raise water temperatures to these levels will destroy almost all life. In most states, legislation exists barring such calamitous thermal pollution in natural bodies of water, but even more moderate temperature increases can result in significant biological alterations.

Each species of organism is able to survive over a definite temperature range, and some species have a wider temperature tolerance than others. Thus, even though not all life in a pond or stream receiving heated effluents may be killed, the types of organisms able to survive in such waters will be quite different from the organisms found under natural conditions. For example, in waters heated above 30° C, the growth of green algae is suppressed and the blue-green algae take over. Animals are even more sensitive to high temperatures, and few live in water over 30° C. But much lower temperatures, although not lethal to adults, often prevent successful reproduction. The eggs of the carp, for instance, do not develop at temperatures greater than 24° C. The small crustacean *Gammarus*, an important member of the freshwater zooplankton, produces only female offspring when water temperature exceeds 8° C. In coolant waters maintained continuously above this temperature, sexual reproduction in this species would soon come to a halt. Dozens of other examples could be cited.

Another problem arising from the discharge of heated waters to streams, lakes and ponds is that the amount of heat discarded can fluctuate, causing wide temperature variations, even within a single day. Many organisms can adapt to a wide range of temperatures, spanning 10° C or more, but such adaptation takes time—about two days for warmer environments and up to several weeks for colder ones. The reason is that metabolic processes must be altered, and many new enzymes must be synthesized. All this must be accomplished slowly without distrupting the normal functions of the organism. Sudden changes in temperature are usually lethal. Figure 7a.6 shows daily fluctuations of as much as 9° C downstream from a power plant in a river in England; adjustment to this is virtually impossible for most animals. Even though such wide fluctuations are unusual, power plant shutdowns are not; every time this happens to a plant that dumps heated water into a stream, massive kills of fish and other organisms will occur.

EFFECTS ON OXYGEN SUPPLY

The other main effect of heat on the quality of water is the decrease in oxygen supply. As the temperature of water increases, less and less oxygen dissolves in it. Trout, salmon and other desirable fish have high oxygen requirements and can live only in cold, highly oxygenated waters. As the temperature rises, the rate of metabolic reactions increases, causing a higher demand for oxygen at the same time that there is less oxygen available. Although many adult animals, given time to adapt, will reduce their metabolic rates as the temperature increases, developing larvae, algae and bacteria do not. Frog eggs, for example, develop six times faster at 20° C than at 10° C, and survive less well in oxygen-poor warm waters.

FIGURE 7a.6

Daily fluctuations in river temperature downstream from a power plant on the River Lea in England. Fluctuations of as much as 9°C in a few hours prevent thermal acclimation by plants and animals.

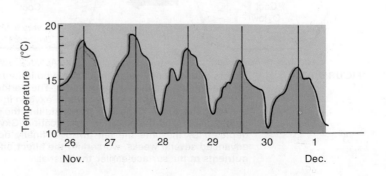

Another factor causing a fall in oxygen content of warm water is the increased growth of algae. As we saw in Chapter 2, the resulting dead algal cells increase the **biological oxygen demand** (B.O.D.) of the water. The subsequent bacterial decay processes use up even more oxygen, further worsening the oxygen deficiency already present due to high temperatures.

One further comment regarding the use of lakes as sources of cooling water for power plants should be made. As we saw in Chapter 2, temperate lakes are thermally stratified during the summer—the warm epilimnion floats on the cold hypolimnion, and the oxygen content of the latter is gradually depleted. Not until the fall turnover does the entire lake become fully oxygenated again. Some power plants, such as that originally proposed for Lake Cayuga in New York, draw cold water from the hypolimnion and discharge it to the already warm epilimnion (Fig. 7a.7). The reasons for doing this are to get the coldest water possible for cooling and to dissipate the added heat as rapidly as possible from the surface of the lake to the atmosphere. As was pointed out by many ecologists, this technique would have effectively increased the degree of thermal stratification of Lake Cayuga and extended the period of stratification by several weeks. The organisms currently supported by this lake are highly adapted to the present seasonal cycles. Both the further depletion of oxygen before delayed mixing in the fall and the hastening of spring mixing, bringing nutrients to the surface and thus initiating algal growth earlier in the year, would have upset the balance between the various species of plants and animals. Algae would have increased and oxygen-dependent organisms in the hypolimnion would have decreased, adversely affecting the biological health of the lake. Although public pressure finally forced the power company to build cooling towers in the case of Lake Cayuga, other lakes could be threatened in the future.

POWER GENERATION IN THE FUTURE

Decisions about how much power is to be produced, by what means and where must be based on a great number of considerations. As we have already seen, we are not yet in a position to estimate how much of an increase in energy utilization will bring about changes in global temperature. Meanwhile, other factors—biological, esthetic, economic and political—must be taken into account in making decisions about future power supplies.

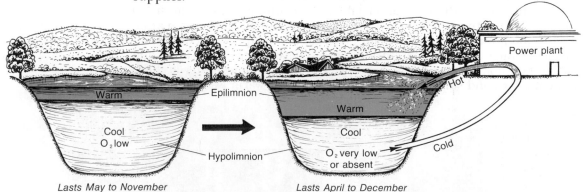

Lasts May to November   *Lasts April to December*

FIGURE 7.7   Effects of using cold waters from the hypolimnion of a lake for power plant cooling.
On the left is the lake beforehand. The period of summer thermal stratification lasts from May to November, by which time oxygen in the hypolimnion is low. On the right are shown the changes in the lake after installing the power plant. The epilimnion increases in depth, prolonging the period of stratification. Oxygen in the hypolimnion becomes highly depleted by the time the fall turnover finally occurs, in December. Spring turnover is advanced several weeks, worsening the effect on oxygen depletion and also introducing nutrients at the surface earlier than usual.

The supplies of coal and oil, especially of the low sulfur variety now required to be used in most urban power plants to cut down on noxious sulfur dioxide fumes, are limited. In the future, more and more of our total power will be generated by nuclear power plants. Although there are also distinct limits to the amount of high-grade uranium ores available for the fuel elements of today's fission-type nuclear plants, it is probable that in a few years breeder reactors will become feasible. In this type of nuclear reactor more fuel is generated than is consumed, and only low-grade granitic ores, of which there is an almost boundless supply, are needed. But the problem of dissipating waste heat remains unchanged and will set an eventual limit on how much power can be generated.

As we have seen, the cheapest way to dissipate heat is to use river, lake or ocean water, but this often leads to undesirable biological changes. To limit the rise in temperature of a river to 3° C or less, a 4000 megawatt nuclear plant (about four times larger than our biggest plants today, but of a size projected for the future) would require a water flow of about 20,000 cubic feet per second. Few rivers in the United States could provide this requirement.

Attempts to limit by statute the amount of heat added to natural bodies of water have not been altogether satisfactory. The Water Quality Act of 1965 provides that standards of water quality for interstate streams and coastal waters should be set by individual states and approved by the Department of the Interior. In several cases, the temperature criteria proposed by individual states have not met with Federal approval. Much needs to be done also to define the terms and conditions for applying water temperature standards once they are approved, for monitoring temperature changes and for certifying that standards are indeed being met. Present regulations function in a very haphazard way.

Where natural water supplies are limited, artificial cooling ponds are often constructed to provide a surface for transferring heat to the atmosphere. The water is continuously recycled between plant and pond. Such ponds must be large, however—about one to two acres per megawatt of electricity. Thus a 4000 megawatt plant of the future would require 4000 to 8000 acres (6.5 to 13 square miles) of cooling ponds! Many suggestions for using such heated ponds for recreation and increased food production (aquaculture) are of great merit. On the negative side, evaporative water loss, as we have seen, will increase local fog conditions and will also short-circuit the hydrologic cycle (see Chapter 2a). The management of such ponds is a complex and difficult task, since sudden temperature fluctuations due to climatic changes or changes in level of plant operation could cause massive kills of the thermophilic ("heat-loving") species which would be cultivated in them.

The more expensive cooling towers are a way of avoiding thermal addition to water resources, but, as we have seen, they have problems of their own. Wet towers lose large quantities of precious water through evaporation, which also has local climatic effects, while dry towers are costly to build and operate.

ALTERNATIVE SOURCES OF MAN-MADE POWER

Three new types of man-made energy sources are under study at the moment: magneto-hydrodynamics (MHD), fuel cells and nuclear fusion. In MHD, hot gas is forced at high velocity through a pipe lying between the poles of a powerful electromagnet. As the gas molecules cut the lines of magnetic force they become charged; these charges are removed by electrodes embedded in the pipe and become electric current. MHD is only about 50 per cent efficient, however, and also requires a source of heat to warm the gas initially. It also is not altogether free from side effects of local thermal pollution.

Fuel cells consist of a positive and a negative electrode immersed in a conducting solution or electrolyte. Spontaneous chemical reactions occur between the solution and the electrodes, generating electric energy from chemical energy. Such fuel cells could supply power to individual buildings, doing away with power lines. Presumably trucks would deliver chemicals to replenish the electrolyte at regular intervals.

Production of power by nuclear fusion reactions is further into the future, since many technical problems remain to be solved. In principle, energy is generated in the same way as in the sun or in a hydrogen bomb. Deuterium, an isotope of hydrogen which is plentiful in the ocean, is heated to 50 or 100 million degrees centigrade. Under these extreme conditions, the electrons and nuclei of the individual deuterium atoms become dissociated from each other, forming a gaseous, highly ionized "plasma." When free nuclei collide, as they frequently do at such high temperatures, they join together to form a helium nucleus, simultaneously releasing enormous quantities of energy.

The only way to control the plasma is to contain it in a huge magnetic field—no material could possibly withstand the enormous temperatures. The released fusion energy is trapped in the magnetic field, setting up in it an electric current which is tapped directly. In this direct transfer of energy, virtually no waste heat would be formed. It is believed that with well-designed fusion plants, radioactive leakage would be minimal—and certainly far less than that of present day fission reactors and tomorrow's breeder reactors.

In all three cases, of course, there will still be an increased thermal stress on the environment. This will occur during conversion of electrical energy into work—be it in electric motors, heating appliances, or light bulbs. There are still limits to the total amount of extra energy man can introduce onto the surface of the Earth.

NATURAL POWER RESOURCES

The trapping of solar and geothermal energy and putting them to useful work, on the other hand, can in no way increase the heat load of our global environment. The reason, of course, is that these energy sources are already part of the Earth's heat budget. No matter what route is taken between the time they reach the Earth's surface and the time they are eventually lost to space, the net result over time is the same. This is true whether the energy does work along the way or not.

Green plants already make use of solar energy in photosynthesis, the subject of our next chapter. Only a small fraction of the total sun's energy reaching them is converted by plant cells into chemical energy, however. For man's use the difficulty with solar energy lies in its low density. It would require huge trapping devices to capture enough sunlight, for example, to equal the power output of a very modest power plant. On the other hand, ingenious small-scale uses of direct solar energy, and also of its secondary energy forms—winds, tides and waterfalls—are constantly being invented, and all help to reduce the demands for man-made energy. Perhaps in the future we will return to windmills and watermills as supplemental energy sources.

Another source of virtually non-polluting energy is animal work. For warm-blooded animals such as man, horse and ox, about 80 per cent of the food burned is utilized to maintain body temperature. The calories required for muscular work are only a small fraction of the total. And since those calories are derived from thermally non-polluting photosynthesis, any increased requirements from utilizing animal work would not affect the global heat budget. As far as man is concerned, additional output of physical labor might have great beneficial effects on health. Many of the diseases of old age, especially breakdowns of "the heart-lung machine," are directly attributable to sedentary life styles now prevalent in industrialized countries. Home exercise machines, for instance, might easily be attached to battery generators to supply some of the domestic power now being produced by power plants.

LIMITING ENERGY DEMANDS

It is evident that man simply cannot go on expanding his energy requirements forever. Sooner or later a steady state must be achieved. In determining how much energy will be consumed and for what purposes, account will have to be taken of one other need for energy—the energy needed for recycling. At present, with but few exceptions, we treat our material resources as non-returnable. Raw materials are taken from concentrated sources (forests, mines and so forth), are used once and are then discarded in a dispersed form. In the process, there is an increase in entropy and a decrease in free energy of the world. Metals, plastic, glass and garbage are all mixed together in refuse dumps—their randomness has increased. To restore order requires work—the input of outside energy.

Recycling must inevitably take place. Our initial supplies of material resources are already dwindling while the environment suffers from being used as a garbage can. Sooner or later total recycling will be both an economic and biological necessity. In any long-range calculations about energy requirements, a sufficient part of the total energy consumed must be budgeted for recycling material resources.

SUMMARY

The ultimate limit to man's utilization of energy is set by the energy budget of the Earth and its ability to radiate heat into space. At present we have no reliable estimate as to what that limit is. Meanwhile, our relatively inefficient power-generating plants create local environmental problems. As nuclear plants gradually replace the conventional fossil fuel-fired plants, the smog problem will decrease, but the problem of waste heat will remain. Besides the effects on local climate, the utilization of surface waters for cooling purposes upsets the ecological balance of lakes, streams and coastal waters. Construction of large cooling ponds will require many acres of land, and will affect the local atmosphere as well. Nuclear fusion reactors, being theoretically more efficient than current power plants, would be preferable, but great technological difficulties remain to be solved before they become practical. Meanwhile, widespread, small-scale utilization of natural energy sources—solar radiation, winds, tides, geothermal energy and animal power—could alleviate some of the demands for man-made energy. In any case, we must begin budgeting a fraction of our energy supplies for recycling of material resources—a need so far ignored in our calculations about energy requirements.

Clark, J. R. 1969. Thermal pollution and aquatic life. Scientific American Offprint No. 1135. Discusses the various effects of heated effluents on aquatic organisms in a readable fashion.

Cole, LaMont C. 1969. Thermal pollution. BioScience 19:989–992. An ecologist's calculations about the effects of man's projected energy utilization on the Earth's heat budget.

Gough, W. G. and Eastlund, B. J. 1971. The prospects of fusion power. Scientific American Offprint No. 340. Describes the nature of fusion reactors and the technical problems which must be surmounted to make them feasible.

Lowry, W. P. 1967. The climate of cities. Scientific American Offprint No. 1215. Man creates his own urban environmental climate through his industry, automobiles, buildings and pavements.

Luce, C. F. 1971. Energy: Economics of the environment. *In* Helfrich, H. W. Jr. *Agenda for Survival: The Environmental Crisis—2.* Yale University Press, New Haven. A very accurate account of the realities which now face us in the realm of power production, written by the Chairman of the Board of the Consolidated Edison Company of New York.

Ninety-first Congress—Joint Committee on Atomic Energy. August, 1969. Selected materials on environmental effects of producing electric power. U. S. Government Printing Office, Washington, D.C. A complete and revealing account of the hearings held before a committee of Congress on the various environmental problems associated with power production.

Wagner, R. H. 1971. *Environment and Man.* W. W. Norton & Co., New York. Chapters 8 and 9 give clear and concise explanations of power plants, thermal pollution and radiation emanating from nuclear plants and their operation.

Weinberg, A. M. and Hammond, P. R. 1970. Limits to useful energy. American Scientist 58:412–418. A highly optimistic projection by two physicists of the outlook for power generation if breeder reactors become feasible.

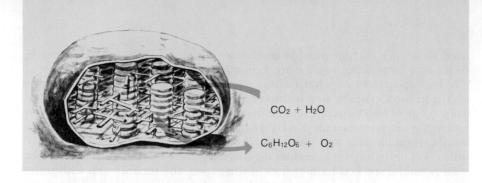

CO$_2$ + H$_2$O

C$_6$H$_{12}$O$_6$ + O$_2$

8

PHOTOSYNTHESIS

The reactions by which glucose is synthesized from carbon dioxide and water in the chloroplasts of illuminated green plants are the basic support of all other forms of life.

We have seen in the past two chapters how the chemical energy present in reduced carbon molecules—such as the carbohydrates, fats and proteins of cells—can be released by combustion processes. In living cells, both plant and animal, the sum of all reactions in this process is known as respiration; part of the chemical energy is converted to high-energy phosphate bonds in ATP, which in turn is used to perform work in the cell. In fossil fuel-fired power plants, the burning of reduced carbon molecules, produced years ago by green plants, releases heat; part of this is used to generate electrical energy, which in turn is capable of doing work. The net equation for both these processes is:

reduced carbon molecules + oxygen $\longrightarrow$ carbon dioxide + water + energy

The subject of this chapter, **photosynthesis,** is concerned with the reverse reaction—the formation of energy-rich, reduced carbon molecules from carbon dioxide, water and an outside source of energy, sunlight. The reactions of photosynthesis are essential for the continuance of life on Earth and are therefore worth considering in detail.

The primary site of photosynthesis is the pigment-containing chloroplast, in which light energy is converted into chemical energy. The overall reaction may be written as:

$$6 \ CO_2 + 12 \ H_2O + light \longrightarrow C_6H_{12}O_6 + 6 \ H_2O + 6 \ O_2$$

In this scheme we have indicated the fate of the various atoms involved. Note that 12 water molecules are split during photosynthesis, forming both oxygen and six new water molecules. The remaining hydrogen atoms are used to reduce carbon dioxide to glucose which, as we saw in Chapter 7, contains a large amount of stored chemical energy.

In anticipation of what follows we can state here that photosynthesis occurs in two steps: the first is the conversion of light energy into the chemical energy of two familiar intermediate compounds, ATP and the reduced hydrogen carrier molecule $NADP \cdot H_2$; the second is the synthesis of glucose from carbon dioxide and water, during which the ATP and $NADP \cdot H_2$ are utilized. This is known as **carbon dioxide fixation.** The first step takes place mainly on the stacked membranes of the grana, the second step within the stroma of the chloroplast (refer to Figure 3.24). We shall consider these steps one at a time.

THE HILL REACTION

Much of our information about photosynthesis comes from studies on suspensions of carefully isolated chloroplasts which retain the ability to synthesize glucose. Some years ago it was noted that, in the absence of carbon dioxide, no glucose was synthesized by illuminated chloroplasts, but oxygen was still released if an electron acceptor molecule was present. This indicated that water was being split during photosynthesis, and that this required the presence of light:

$$\text{light} + H_2O \longrightarrow \tfrac{1}{2} O_2 + \underbrace{2 H^+ + 2 e^-}_{\substack{\text{combined with electron} \\ \text{acceptor molecule}}}$$

It was subsequently determined that a naturally occurring molecule in the cell, the hydrogen carrier NADP, could act as the electron acceptor molecule, being reduced by the hydrogen ions and electrons of water to $NADP \cdot H_2$. This entire process has been called the Hill reaction for the Englishman who first observed it. The equation could next be written:

$$\text{light} + H_2O + NADP \longrightarrow \tfrac{1}{2} O_2 + NADP \bullet H_2$$

It was further found that isolated chloroplasts, still in the absence of CO_2, but with added ADP and inorganic phosphate (P_i) are capable of synthesizing ATP. Thus, both chemical energy and hydrogen atoms are products of illuminated chloroplasts. The final equation is thus:

$$\text{light} + H_2O + NADP + ADP + P_i \longrightarrow \tfrac{1}{2} O_2 + \mathbf{NADP \bullet H_2} + \mathbf{ATP}$$

The formation of ATP and $NADP \bullet H_2$ constitutes the first step in photosynthesis; both molecules, as we shall see shortly, are subsequently utilized in the synthesis of glucose.

CAPTURE OF LIGHT ENERGY

It was long known that the pigments located in the chloroplasts of green plants, especially the green pigment, **chlorophyll,** were essential for photosynthesis. The colorless roots of plants, or leaves bleached of their pigments by being kept in the dark, are incapable of photosynthesis. Chlorophyll is a moderately complex molecule, although it is not a protein. It contains within its structure an atom of magnesium. Several other pigments are also present in varying amounts in chloroplasts, of which the most common are the **carotenoids,** ranging from yellow to orange-red in color. A blue pigment, **phycocyanin,** and a red pigment, **phycoerythrin,** occur in various algae. All of these are photosensitive—that is, they are capable of being activated by light.

EVIDENCE FOR THE ROLE OF PIGMENTS

Let us consider the case of chlorophyll first. If we extract chlorophyll from the leaves of a plant and shine visible light through a solution of it in a test tube we observe that only a fraction of the light is transmitted, the rest being absorbed by the chlorophyll. If we now use a prism to form a spectrum in which the light is split into its component colors—just as raindrops split sunlight to form a rainbow—and shine each color in turn through the test tube, we can determine the **absorption spectrum** for chlorophyll. The results of such an experiment are shown in Figure 8.1. Note that since each color of light vibrates at a given frequency, it can be defined by its **wavelength,** which is expressed in nanometers (or 10^{-9} meters). Chlorophyll absorbs blue and red light, transmitting green and yellow wavelengths.

If, however, we repeat this experiment with the leafy part of a plant such as a frond of "sea lettuce" from the green alga, *Ulva,* we obtain a somewhat different absorption spectrum because of the presence of accessory pigments, as shown by the solid line of Figure 8.2. If we next expose the same plant to different colors of light, and this time measure the amount of photosynthesis that occurs at each wavelength, we obtain the dashed curve of Figure 8.2, called the **action spectrum.** Slightly different action and absorption curves are obtained for various kinds of plants, but for any given plant there is a close correlation between the two. It is apparent that the ability of a plant to carry out photosynthesis corresponds closely to the light absorbing capacity of its combined pigments.

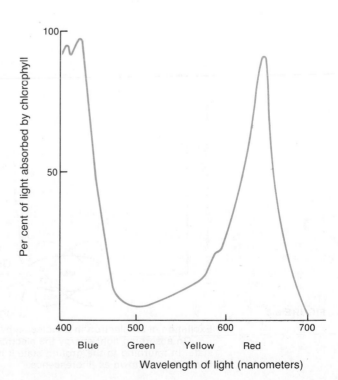

FIGURE 8.1

The absorption spectrum of a chlorophyll solution. Most of the absorption occurs in the blue and red parts of the spectrum, green and yellow light being transmitted.

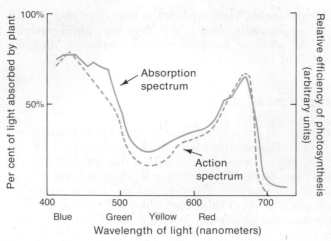

FIGURE 8.2

The absorption and action spectra of a green alga, the seaweed *Ulva*. Note close correspondence between light absorbed and photosynthesis carried out at each wavelength.

PHOTOACTIVATION

Any colored substance is capable of absorbing some of the light energy which falls upon it; in fact, this is why it is colored. We can thus reasonably ask: What is special about the pigments found in chloroplasts?

Under ordinary conditions, the electrons of a molecule are said to be in a **ground state,** energywise. By this we mean that all the electrons are at their lowest energy levels. If the molecule receives extra energy, however, this is sometimes transmitted to one of its electrons, which achieves an **excited state.** Light energy, which comes in discrete energy units called **photons,** is often capable of exciting electrons in colored molecules when the light is of the appropriate wavelength to be absorbed (Fig. 8.3). A molecule with an electron in an excited state is unstable, and one of several things

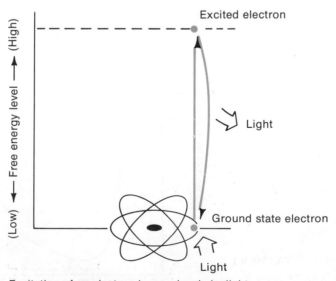

FIGURE 8.3

Excitation of an electron in a molecule by light.

On absorbing light energy the electron is raised from the ground state to an excited state. In returning to the ground state it may give off its extra energy as light, as shown here. This is known as fluorescence.

may result. The electron may lose its energy as heat as it returns to its ground state; or it may re-emit the energy as light, called **fluorescence;** or the whole molecule may split, absorbing the excess energy in the process.

Most molecules give off absorbed light energy as heat, but chlorophyll extracted from plants and placed in solution in a test tube fluoresces when illuminated by red or blue light. Obviously, in the chloroplast, fluorescence would achieve nothing; incoming light energy would simply be re-emitted as light of a different color. Instead, as we shall see shortly, the excited electron is transferred to nearby electron acceptor molecules in the chloroplast membranes, temporarily endowing them with its excess energy.

THE ENERGY TRAP SYSTEM

But although chlorophyll is the major pigment of photosynthesis, it is, as we have seen, by no means the only pigment involved. Moreover, only one in every several hundred chlorophyll molecules is capable of transferring its excited electron to the electron acceptor molecules in the chloroplast. It is now believed that most of the chlorophyll as well as the accessory photosensitive pigments in chloroplasts act as energy gathering devices, funneling the energy of their excited electrons into two energy traps, which we shall call Trap I and Trap II (Fig. 8.4). Only from the special chlorophyll molecules in these two traps are energized electrons transferred to other molecules. Thus both the pigments and their arrangement within the chloroplast are important for the capture of light energy.

THE LIGHT REACTIONS

Our next step is to discover how the energy of the excited electrons funneled to the chlorophyll molecules of Trap I and Trap II is converted into the chemical energy of ATP and NADP·H_2 — that is, how the overall Hill reaction is brought about. At the present moment it is possible to present only a tentative picture of how this happens, as shown in Figure 8.5. Basically we can think of it as a parallel scheme to that found in the electron transport system of mitochondria. The excited electrons, raised to a high energy level, as indicated by the left-hand scale of the figure, are transferred from one molecule to another within the chloroplast membranes, in a series of oxidation-reduction reactions. At each downhill transfer of an electron, some of its energy is lost. Finally the electron loses all its energy and returns to a ground state. Like the cytochromes of mitochondria, the molecules involved in this electron flow in the chloroplasts

FIGURE 8.4

The energy trap system of chloroplasts.
   A. An artist's conception of how the photosensitive pigments may be arranged on grana membranes. The central chlorophyll molecule receives energy from surrounding pigment molecules.
   B. A schematic diagram of how the energy trap system works. Only the special chlorophyll molecule of the trap can emit an energized electron.

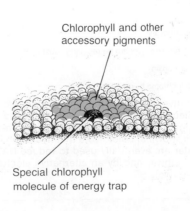

Chlorophyll and other accessory pigments

Special chlorophyll molecule of energy trap

A

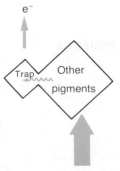

e^-

Trap

Other pigments

B

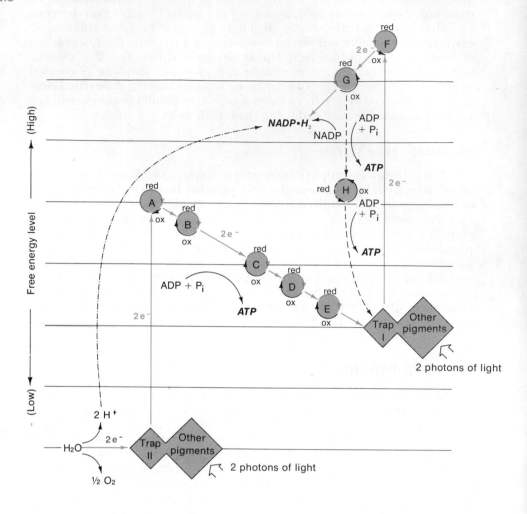

Photosystem II   Photosystem I

Identity of electron acceptor molecules

A—Substance Q

B—Plastoquinone

C—Cytochrome b 559

D—Cytochrome f

E—Plastocyanin

F—Ferredoxin-cytochrome reducing substance

G—Ferredoxin

FIGURE 8.5   H—Cytochrome b_6 (or 553)

A tentative scheme of the light reactions of photosynthesis.

Electrons from chlorophyll are raised to high energy levels by light and then lose the energy stepwise in a series of oxidation-reduction reactions. The main path of electron flow is shown in color. Note that hydrogen ions and electrons produced from the splitting of water are eventually used to reduce NADP to $NADP \cdot H_2$. ATP is produced both in the main pathway and by an alternative electron flow pathway in Photosystem I (shown by dashed line).

are each capable of accepting and giving up electrons. In fact, some of the molecules in chloroplasts are cytochromes.

Several things may be noted about this diagram. First, there are two photosystems, I and II. Light energy striking the pigments of Photosystem II is funneled to the chlorophyll of Trap II, which sends an energized electron to molecule A. As the electron loses energy during passage from one electron acceptor molecule to the next, a molecule of ATP may be synthesized from ADP and P_i. The electron eventually finds its way to the chlorophyll of Trap I, which must meantime have lost one of its own electrons to make space for the new one. In fact, the electron released by photoactivation of Trap I is even more highly energized, and reduces molecule F, which passes it along to G, which in turn transfers it to NADP. Finally, to replace the electron originally lost by Trap II, water is split by an as yet unknown process which requires manganese ions (Mn^{++}) and chloride ions (Cl^-). The other products of this dissociation of water are molecular oxygen (O_2) and hydrogen ions (H^+).

Since the entire cycle must go around twice to provide the two electrons needed to reduce NADP, we have doubled the number of photons and electrons in our diagram.

$$NADP + 2\,H^+ + 2\,e^- \longrightarrow NADP \cdot H_2$$

The position of $NADP \cdot H_2$ on our energy scale indicates that it contains considerable chemical energy in its reduced state. The net results of this entire process are the formation of a molecule of $NADP \cdot H_2$ by electrons and hydrogen ions obtained originally from water, and the formation of about one ATP molecule per two electrons.

There is, in fact, considerable uncertainty about the number of ATP molecules synthesized for each two electrons passing through the two photosystems, but it is generally agreed that they are too few to supply all the energy necessary for carbon dioxide fixation in the next phase of photosynthesis. To make up the deficit, electrons energized by Photosystem I may take an alternate route from molecule G, not going to NADP but returning via one or two more electron acceptor molecules to the chlorophyll of Trap I. In the process it is believed that two molecules of ATP are formed. The synthesis of ATP by Photosystems I and II is known as **photophosphorylation,** since energy for the addition of inorganic phosphate to ADP comes from light.

The reactions we have discussed so far all require light and hence are collectively known as the **light reactions,** which in their overall result are equivalent to the Hill reaction.

THE DARK REACTIONS

In contrast to the light reactions, a leaf or a chloroplast suspension which has acquired a moderate store of $NADP \cdot H_2$ and ATP in the light is capable of synthesizing glucose for a time after the light is switched off. This is the second major step in photosynthesis, during which carbon dioxide is synthesized into glucose, utilizing ATP and $NADP \cdot H_2$ in the process. Collectively these reactions are known as the **dark reactions,** which overall may be written:

$$2\,NADP \cdot H_2 + 3\,ATP + CO_2 \longrightarrow\longrightarrow\longrightarrow [CH_2O] + 2\,NADP + 3\,ADP + 3\,P_i$$

Again, the multiple arrows indicate that several reactions are involved during the incorporation of one molecule of carbon dioxide into carbohydrate.

Note that we have used our standard formula [CH₂O] to denote the basic carbohydrate unit (see Chapter 3, page 60).

The dark reactions, which, as we have noted, occur in the stroma of the chloroplast, are made up mainly of 14 or 15 separate steps; these collectively are known as the **Benson-Calvin cycle** for the two scientists who were primarily responsible for working out its complex details. A somewhat simplified diagram of the cycle is shown in Figure 8.6. It is not necessary for you to commit this cycle to memory, however, in order to grasp the essentials of how glucose is formed.

In the Benson-Calvin cycle, carbon dioxide (CO_2) molecules are added one at a time to five-carbon sugar molecules already present in the cell, forming an unstable six-carbon sugar which quickly splits into two three-carbon sugars. There then ensues a complex breaking and re-forming of compounds of different chain lengths. For every six turns of the cycle, one new molecule of glucose ($C_6H_{12}O_6$) is formed. The intermediate sugar compounds all have a low energy phosphate group bound to them, which is not shown in the diagram.

The most important thing to note, however, is that the cycle depends upon the presence of both ATP and NADP·H₂ molecules. Recent studies indicate that alternative pathways for the incorporation of carbon dioxide into organic molecules, in addition to the Benson-Calvin cycle, exist in plant cells, although it is not yet possible to estimate their relative contribution to total carbon dioxide fixation.

A summary of all the reactions of photosynthesis is given in Figure 8.7, to emphasize the interrelationships of the light and dark reactions in the formation of reduced organic molecules. For energy-requiring processes taking place elsewhere in the plant cell, ATP is synthesized as needed by the oxidation of glucose in its mitochondria, just as occurs in the cells of all other organisms.

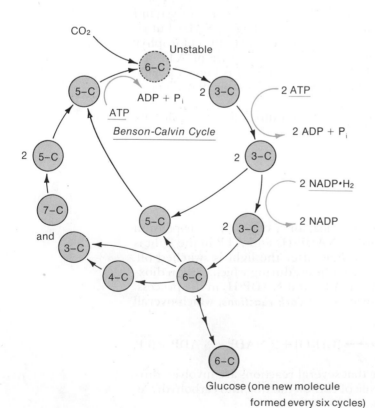

FIGURE 8.6

The dark reactions of the Benson-Calvin cycle. The number of carbons in each sugar is designated by the symbol 3–C, 5–C and so forth. All of these sugars also have low energy phosphate groups bound to them, which are not shown. The complex shuffling between compounds of different chain lengths enables one CO₂ molecule to enter the cycle at a time. One new 6–C chain, glucose, is generated for each six turns of the cycle. Note that 18 ATP and 12 NADP·H₂ molecules, generated in the light reactions, are used in the synthesis of one glucose molecule.

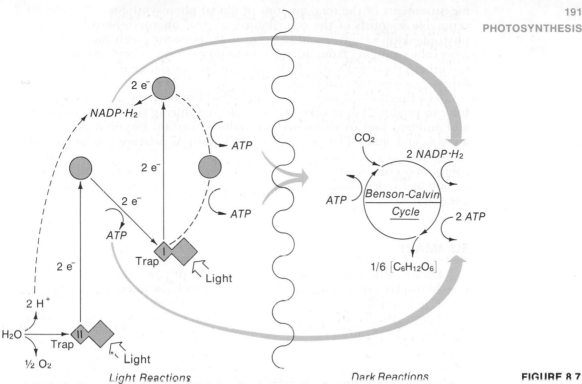

FIGURE 8.7

A simplified summary of all the photosynthetic reactions.
    The light reactions supply both ATP and NADP·H₂ for carbon dioxide fixation during the dark reactions.

THE SYNTHESIS OF OTHER ORGANIC MOLECULES

Although only sugars are synthesized by photosynthesis, cells must have other types of molecules—the proteins, fats and nucleic acids described in Chapter 3. It was already mentioned in our discussion of respiration in Chapter 7 that such molecules may be oxidized in the cell after being converted to one of the carbohydrate molecules of the respiratory pathway. Providing the necessary ATP is present to supply the required energy, these interconversion steps are reversible. Thus, it is possible for components of the Benson-Calvin and Krebs cycles to be changed into the building blocks of protein, fat or nucleic acids. The sum total of all these interconversions involves perhaps a hundred or so individual enzymes, each catalyzing a separate reaction. Together this whole complex is known as *intermediary metabolism,* the details of which belong to the subject of biochemistry. A simplified schematic representation is shown in Figure 8.8 to give the reader a picture of how these interconversions take place.

THE SIGNIFICANCE OF PHOTOSYNTHESIS

The world's supply of reduced organic carbon molecules is produced almost entirely by photosynthesis in green plants. A few bacteria are also able to synthesize organic molecules from carbon dioxide, but their contribution is negligible. Perhaps half of the world's photosynthesis is carried out by phytoplankton and macro-algae in the sea. It is a measure of our ignorance that we do not know exactly what proportion of the total photosynthesis occurs in the oceans; nor do we even have any accurate measurements of the total amount of global photosynthesis. Estimates by reputable scientists of the contribution to total photosynthesis by marine phytoplankton vary from 30 per cent to 90 per cent—with the most common values ranging from 40 per cent to 60 per cent. The remaining photosynthesis is carried out by freshwater and terrestrial green plants—algae, lichens, mosses, ferns and spermatophytes.

The functions of photosynthesis in the biosphere are essentially threefold: to produce O_2; to take up CO_2 released during respiration of plants and animals; and to synthesize the reduced carbon required as food by animals and man. The last of these is known as **primary production**. Although the first two functions are of importance in maintaining a constant composition of the atmosphere, it is the primary production of reduced carbon molecules which is of greatest importance to man, as we shall see in later chapters.

SUMMARY

Photosynthesis comprises a complex series of reactions taking place in the chloroplasts of green plant cells during which the energy of sunlight is converted into chemical energy. It occurs in two steps.

During the light reactions, the absorption of light by the pigments of the chloroplasts causes the excitation of an electron in each of the special chlorophyll molecules of Traps I and II; the energy of the electron is gradually released in a sequence of oxidation-reduction reactions, during which ATP and $NADP \cdot H_2$ are synthesized. Hydrogen ions and electrons for reduction of NADP are generated initially by the splitting of water, during which oxygen is released as a by-product.

In the dark reactions, during which carbon dioxide fixation occurs, ATP and $NADP \cdot H_2$ are used up. In a complex series of reactions known as the Benson-Calvin cycle, carbon dioxide molecules are added one at a time to phosphorylated sugar molecules in the chloroplast. Six turns of the cycle produce one glucose molecule. Other types of organic molecules—proteins, fats and nucleic acids—are made from various sugars and acids of the

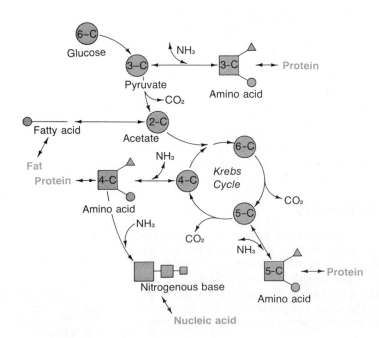

FIGURE 8.8

The fundamental reactions of intermediary metabolism. Note how small carbohydrate molecules of different chain lengths can be used to synthesize fatty acids, amino acids (with addition of ammonia, NH_3) and nitrogenous bases. Thus all of the organic molecules necessary in a cell can be constructed from modifications of the carbon-chain framework of glucose, initially formed in photosynthesis.

Benson-Calvin and Krebs cycles in a series of reactions known as intermediary metabolism.

Photosynthesis by green plants is responsible for the production of virtually all the world's reduced organic carbon molecules.

Bassham, J. A. 1962. The path of carbon in photosynthesis. Scientific American Offprint No. 122. Describes the experiments carried out by Benson, Calvin and others in elucidating the Benson-Calvin cycle and pathways of intermediary metabolism in plants.

Bishop, N. I. 1971. Photosynthesis: the electron transport system of green plants. Annual Review of Biochemistry 40:197–226. A detailed account of the various electron acceptor molecules involved in photosynthesis and of the problems in determining the exact path of electron flow.

Haxo, F. T. and Blinks, L. R. 1950. Photosynthetic action spectra of marine algae. Journal of General Physiology 33:389–422. A comprehensive survey of the relation between the pigments contained in various algae and their photosynthetic action spectra.

Rabinowitch, E. I. and Govindjee. 1965. The role of chlorophyll in photosynthesis. Scientific American Offprint No. 1016. Explains the oxidation-reduction processes initiated by the capture of light energy in chloroplasts.

Rabinowitch, E. I. and Govindjee. 1969. Photosynthesis. 2nd ed. John Wiley and Sons, New York. A comprehensive treatise on all aspects of photosynthesis.

Salisbury, F. B. and Ross, C. 1969. Plant Physiology. Wadsworth Publishing Co., Belmont, California. Contains an excellent description of photosynthesis as it occurs in the laboratory and in nature.

8a

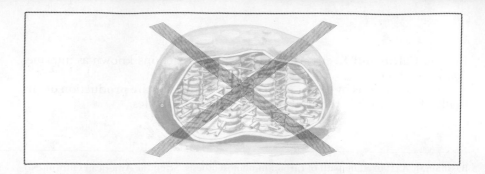

MAN'S IMPACT ON GLOBAL PHOTOSYNTHESIS

Man now possesses the potential for interfering with photosynthesis on a worldwide basis. How this may be occurring and its implications for the future are of major importance.

Photosynthesis is essential for the continuance of life on Earth as we know it. Virtually all organisms—man, other animals, most bacteria, fungi, and even green plants themselves—depend upon the synthesis of reduced carbon compounds and on the stabilization of atmospheric carbon dioxide and oxygen, which result from the photosynthetic equation:

$$6\ CO_2 + 6\ H_2O \longrightarrow C_6H_{12}O_6 + 6\ O_2$$

Various recent reports from the scientific world indicate that man's activities may be affecting photosynthesis on a global scale. It is the purpose of this chapter to review these studies and attempt to evaluate them. As we proceed, the reader will become aware of the limitations of our present knowledge about how much photosynthesis actually occurs in the world, and to what extent the activities of man affect it.

As noted earlier, perhaps half of the world's total photosynthesis is carried out by unicellular green algae in the sea. On the other hand, man himself is more directly dependent on terrestrial plants. Although a small fraction of man's diet is obtained from the sea, and may be somewhat increased by extension of mariculture, by far the preponderance of our food is produced on land. Terrestrial crops also supply fibers for clothing, paper pulp, organic chemicals, solvents and innumerable other products useful to man. Thus, the most immediate effects to man of his interference with photosynthesis are those occurring in the terrestrial environment.

DEPLETION OF TERRESTRIAL PLANTS

While man has been able to increase greatly the productivity of certain areas through irrigation, application of fertilizers, mechanized farming and the introduction of improved strains of crops, in other places his thoughtlessness or ignorance has seriously diminished the productivity of the land. We do not at present possess the data necessary to draw up an exact balance sheet of man's total impact over the centuries on the productivity of terrestrial regions. But since the world population is still increasing extremely rapidly, we need all the productive land available, and therefore a brief look at past errors is of much value.

AGRICULTURAL FACTORS

Large climatic changes have no doubt affected once fertile areas of the world, and these changes have often been abetted by man. Many scientists believe that overgrazing and cutting of timber resulted in denudation of once productive lands of the southern and eastern shores of the Mediterranean, extending the margins of the great natural deserts in this region (Fig. 8a.1).

Other, more recent examples of man's carelessness and lack of understanding are to be found in the United States, where inept agricultural practices in northern Texas and Oklahoma, particularly the extensive sowing of single crops, produced the Dust Bowl wastelands in the 1930's. Continued drought caused repeated crop failures, and the topsoil of the denuded land—over a hundred million acres—was soon eroded away. As many as 40 years are required even for useful grazing lands to become reestablished in such areas.

Overgrazing has often been blamed for the deterioration during the past 60 years or so of grasslands in the semiarid regions of the western states, such as shown in Figure 8a.2. As grazing becomes heavier, the wild grasses are gradually replaced by shrubs unpalatable to cattle, such as mesquite and burroweed. It seems, however, that the effect of grazing, at least in the more southern part of the region, is an indirect one. The cattle remove much of the inflammable dry grasses, reducing the frequency of prairie fires. It is thought that such fires, by killing the young woody shrubs, permit the grasses, whose seeds and roots are more fire-resistant, to

Overgrazing by sheep in this Middle Eastern landscape has left the soil unprotected by vegetation. Note the desert-like appearance of the background. (From Wagner, R. 1971. *Environment and Man.* W. W. Norton & Co., New York.)

FIGURE 8a.1

A

B

FIGURE 8a.2

Grassland subjected to varying intensities of grazing in southern Idaho.
    A. Light grazing, with excellent grass cover.
    B. Moderate grazing; a few mesquite are visible.

Illustration and Legend continued on opposite page

D

FIGURE 8a.2 *Continued*
   C. Heavier grazing; many mesquite and less grass.
   D. Very heavy grazing; virtually no grass, only mesquite and herbaceous "weeds."
   (From Graham, E. H. The re-creative power of plant communities. *In* Thomas, W. L. *Man's Role in Changing the Face of the Earth.* Copyright 1956, University of Chicago Press, Chicago.)

remain dominant. This is an instance where fire control may produce undesirable side effects.

Other man-made losses of useful agricultural or forest lands may be mentioned briefly. Overuse of pesticides has been known to adversely affect agricultural land use. In California's Central Valley, for example, an insecticide called Azodrin, belonging to the **organophosphate** group (see Chapter 12a), was extensively used to control bollworms in cotton. But the pesticide proved more effective in killing predators of the pest than in killing the pest itself. Bollworm populations actually increased, resulting in less production of cotton rather than more. The panicked farmers, responding to the advice of salesmen, used even more Azodrin in successive years with ever more disastrous results. The final imbalance of insect populations and loss of soil organisms has caused large areas to become unusable for several years.

Attempts to apply temperate zone farming practices in the tropics also result in loss of farmland. Tropical soils are often of poor quality, and once large areas of forest are removed, as in plantation farming, the thin topsoil is quickly eroded, exposing a layer of soil containing iron and aluminum to oxidation. In a few years, the metal oxides are baked by the tropical sun into a permanent rock-like crust, called **laterite,** on which neither crops nor forest will grow. Native practices of mixed crop farming on small, temporary clearings used for only a few years give far lower annual yields than plantations, but they do not permanently destroy the land for agricultural use.

LAND USE

The rate of urban expansion in western societies, particularly into once prime agricultural land, is enormous, amounting in the United States alone to about one million acres per year over several decades. Farms are replaced by pavement—freeways, streets, parking lots and gas stations. It is estimated that one-third of Los Angeles County, once a rich agricultural area, is now covered with buildings, asphalt and concrete. If present trends remain unchecked, today's semi-rural interurban areas will disappear along both our coasts, and over much of the Midwest.

Another result of urbanization, as we noted in Chapter 2a, is the frequent need to import water. In some cases, this has led to a loss of agricultural lands. Diversion of water from the Owens Valley of California to the thirsty Los Angeles basin converted thousands of acres of semi-arid grassland, once supporting many cattle, into a desert.

Forest lands are also being destroyed as a result of man's activities. Clearing for agriculture sometimes results in erosion and permanent loss of land to all vegetation. Logging usually produces only temporary changes, but without good forest management, permanent erosion can occur, especially along old logging roads. Conscientious replanting is necessary to avert permanent damage. Widespread use of defoliants, both for military purposes and for brush-control, can produce undesirable effects. One-seventh of South Vietnam, a country about half the size of California, has been adversely affected by the heavy application of herbicides (see Figures 14a.3 and 14a.4). On much of the mature forest land that has been sprayed the natural wet forests have been replaced largely by undesirable bamboo.

ATMOSPHERIC POLLUTION

Air pollution is another factor affecting forests in widespread areas near large cities—several hundred thousand acres of forest are threatened around Los Angeles alone. In this area the trees affected are primarily ponderosa pine, which not only provide timber and recreational amenities, but also stabilize the watershed, a factor of great importance to an area subject to flash flooding.

Although we shall discuss the way in which smog is formed and how it enters the leaves of plants in detail in Chapter 9a, a few remarks here on its effects on photosynthesis are appropriate. One of the first changes seen in affected trees is a yellow mottling of the needles (Fig. 8a.3), due to the de-

FIGURE 8a.3

Needle mottling in a pine tree.

Note that the older needles, lower on the stem, are affected more than the younger needles at the tip. (From Richards, B. L., *et al.* 1968. Ozone needle mottle of pine in southern California. Journal of the Air Pollution Control Association *18*:73–77.)

struction of chlorophyll within the leaf, while leaving the yellow carotenoids untouched. This effect is caused by a particularly toxic compound in smog called **ozone.** In experiments in which pine trees in greenhouses were exposed either to pure air or air containing various amounts of ozone, it was found that ozone decreases both the rate of carbon dioxide uptake by the trees and the amount of stored carbohydrate within the needles (Table 8a.1). The concentrations of ozone used in these experiments were similar

Table 8a.1
EFFECTS OF
OZONE ON
PHOTO-
SYNTHESIS IN
PINE TREES

	Atmosphere to which trees exposed for 30 days		
	Air (control)	Ozone 0.15 ppm*	0.30 ppm*
Per cent of initial rate of CO_2 uptake:	120%	52%	33%
Per cent of initial stored carbohydrate:	124%	not measured	63%

*ppm = parts per million of ozone in air. Ozone was added only nine hours per day, simulating natural daily fluctuations. (Based on data from Miller, P. R., Parmenter, J. R., Jr., Flick, J. H., and Martinez, C. W. 1969. Ozone dosage response of ponderosa pine seedlings. Journal of the Air Pollution Control Association *19*:435–438.)

to those frequently found in air near Los Angeles and other large cities. Whereas the control trees in pure air substantially increased their rate of photosynthesis due to growth during the 30 day period, the ozone-treated trees showed marked inhibition of photosynthetic capacity.

Toxic fumes from industries are also responsible for denuding lands of their original vegetation. Great areas east of Butte, Montana and in the Copper Basin at Copperhill, Tennessee had their vegetation killed years ago by fumes from copper smelters. Even though fumes are no longer released by more modern methods of smelting, the eroded lands have remained barren (Fig. 8a.4).

FIGURE 8a.4

Part of Copper Basin at Copperhill, Tennessee.
The once luxuriant forest was killed by smelter fumes and, despite the absence of fumes today, the eroded land remains barren. (From Odum, E. P., 1971. *Fundamentals of Ecology.* W. B. Saunders Co., Philadelphia.)

EFFECTS OF DDT ON PHOTOSYNTHESIS

The discovery in 1939 by the Swiss entomologist, Paul Müller, that DDT is highly toxic to insects and relatively non-toxic for humans, quickly led to its widespread use as an insecticide for control of insect-spread diseases and crop pests. Millions of lives were saved, especially in malaria-infested parts of the world, and Dr. Müller subsequently received the Nobel prize. Not until Rachael Carson's book, *Silent Spring*, was published in 1962 were some of the dangers of DDT brought forcefully to the public attention.

WHAT IS DDT?

DDT is but one of a family of chlorine-containing chemicals, known as **chlorinated hydrocarbons,** synthesized by the chemical industry as pesticides. Another group of chlorinated hydrocarbons, which have toxic ef-

FIGURE 8a.5

Chemical structures of two chlorinated hydrocarbons. The benzene rings make these compounds far more soluble in fats than in water.
A. DDT stands for para, para'-*di*chloro*di*phenyltrichloroethane (or more formally, 1,1,1-trichloro-2,2-bis (p-chlorophenylethane).
B. PCB's come in many forms, depending on how many chlorine atoms are attached to the possible X positions of the parent molecule shown above.

Where X may be either an
H-atom or a Cl -atom

A. DDT

B. PCB

fects similar to DDT, are the polychlorinated biphenyls or **PCB's.** These are manufactured in greater amounts than DDT for use in plastics, paints and other manufactured products, and, like DDT, are now widely dispersed throughout the world. The formulas for DDT and PCB's are shown in Figure 8a.5.

All these compounds have two important properties in common. They are highly resistant to natural processes of degradation, and are thus fairly stable in the biosphere for many years after initial release. They also are far more soluble in fats than in water, and thus readily accumulate in the tissues of plants and animals. An animal eating plants containing DDT therefore excretes little in its aqueous urine, but instead stores the pesticide in its own tissues, especially in its fats. If such an animal is eaten by a second animal, and it, in turn, by a third, a gradual build-up in DDT concentration occurs at each step. This process has led to the accumulation of high levels of chlorinated hydrocarbons in animals at the top of the food chain, a process known as **biological magnification.** It is also possible that large animals accumulate higher levels simply because they live longer. As we have already mentioned, many species are threatened with extinction as a result, and many others, especially fish in certain localities, have accumulated so much DDT as to be unfit for human food.

DDT ASSOCIATED WITH PHYTOPLANKTON

In addition to the toxic effects on animals and possibly man from consuming DDT-containing foods, there is a possibility that photosynthesis itself, the very basis of life, may be affected by chlorinated hydrocarbons. Although these chemicals do not readily find their way into the leaves of terrestrial plants, which are protected by a thick waxy cuticle, there is some concern that aquatic algae, which have no such protection, are being affected. When samples of phytoplankton collected off the coast of California were analyzd for DDT, for example, a threefold increase—from 0.2 ppm to 0.6 ppm—was found between samples collected in 1955 and those collected in 1969. Some 1969 samples contained nearly 1 ppm (ppm = parts per million). It is not clear, however, how much of the pesticide was inside the cells and how much only adsorbed to the outer surface.

Our next question is, Does DDT affect photosynthesis, and if so, are present levels of environmental DDT having an effect? The first part of this question can be answered with a firm "yes"—DDT can markedly decrease photosynthesis. This was demonstrated by a series of experiments on four species of living phytoplankton.

LABORATORY EXPERIMENTS

In order to determine the ability of the algae to photosynthetically fix carbon dioxide in these experiments, radioactive carbon atoms, provided in the form of bicarbonate ion, were supplied to the algae. Such radioactive bicarbonate is indicated by the formula $H^{14}CO_3^-$. After a period of time, the amount of carbon incorporated into organic molecules within the cells could be determined by measuring their radioactivity.

Cell suspensions in culture flasks were exposed to DDT and radioactive bicarbonate ($H^{14}CO_3^-$) for 24 hours, during which time they were illuminated to permit photosynthesis (Fig. 8a.6). Then the cell suspension was filtered and washed to remove any $H^{14}CO_3^-$ adhering to the outside of the cells, and the amount of radioactive carbon incorporated by the cells into reduced organic molecules was measured, using an instrument which detects radioactivity.

In another series of experiments, the effects of DDT at a concentration of 100 ppb (ppb = parts per billion) on the growth of the same four algal

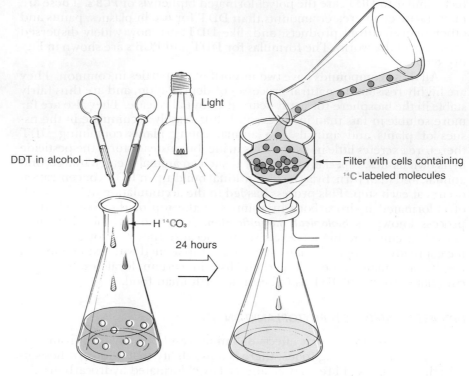

Light

DDT in alcohol →

← Filter with cells containing
^{14}C-labeled molecules

H^{14}CO$_3$

24 hours

FIGURE 8a.6

Experiments testing effects of DDT on photosynthesis by marine algae.
The algal suspension, to which DDT and H^{14}CO$_3^-$ have been added, is allowed to photosynthesize for 24 hours in the presence of light. The cells are collected on a filter. The amount of photosynthesis which has taken place is estimated by counting the radio-activity in the cells.
Control cultures received only alcohol, without added DDT.

species was measured over a week. The effects of DDT both on the amount of radioactive carbon incorporated into the algal cells and on their growth are shown in Figure 8a.7. Note that the DDT concentration scales in the photosynthesis experiments are logarithmic rather than linear. Interestingly enough, the four species of algae do not react in the same way to DDT. One species, *Dunaliella*, is quite resistant, while two others, *Skeletonema* and *Cyclotella*, show a significant suppression of both photosynthesis and growth due to DDT.

Therefore, in answer to the first part of our question, DDT clearly can affect photosynthesis in certain species of phytoplankton if present at high enough concentrations. Let us next consider the other half of our question, Are present levels of environmental DDT having an effect? Unfortunately, in the experiments described above, the final concentration of DDT in the cells was not measured. It is possible that much of the DDT added to the culture flasks never reached the cells at all. For one thing, the concentrations used far exceeded the solubility of DDT in water. (DDT is highly nonpolar and only dissolves easily in alcohol, ether and other organic solvents.) Therefore, some of the DDT probably precipitated out. Some DDT may also have evaporated together with water from the surface of the flask, a phenomenon known as codistillation. Much probably adsorbed onto the surface of the flask, since DDT has a high affinity for glass. To obtain a more precise answer to our question about the effects of environmental DDT on photosynthesis we must turn to two other studies.

Experiments on the passive absorption of the pesticide by unicellular algae from sea water containing only .015 ppb DDT have shown that the amount of DDT associated with cells very quickly reaches 1 to 2 ppm, simi-

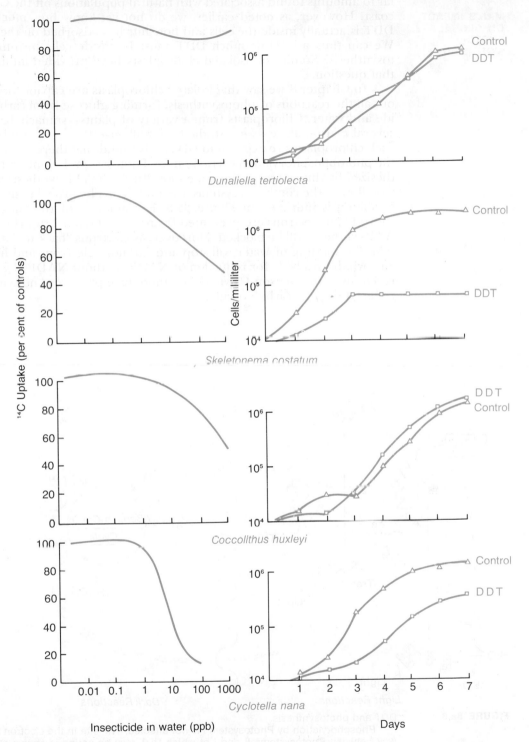

Effect of DDT on photosynthesis and growth of four species of phytoplankton.

Left. Photosynthesis, measured as uptake of radioactive carbon, ^{14}C, in presence of various DDT concentrations. (Note that DDT scale is logarithmic.) 100% equals photosynthetic rate of cells with no DDT added.

Right. Effect of 100 ppb of DDT on growth, expressed as cells per milliliter (cells/ml), compared to control cultures without DDT. △ control; □ DDT.

(After Menzel, D. W., *et al.* 1971. Marine phytoplankton vary in their response to chlorinated hydrocarbons. Science *167*:1724–1726.)

FIGURE 8a.7

lar to amounts found associated with natural populations off the California coast! However, as noted earlier, we do not yet know how much of this DDT is actually inside the cells and how much is adsorbed on the outside. We can thus ask, How much DDT must be inside cells to inhibit photosynthesis? Studies on isolated chloroplasts have provided an answer to that question.

In Chapter 8 we saw that isolated chloroplasts are capable of carrying out all the reactions of photosynthesis, forming glucose from carbon dioxide and water. Chloroplasts from a variety of plants—spinach, barley and several marine algae—were studied, and all gave the same results. When such chloroplasts are exposed to DDT, it is found that there is no effect on the photophosphorylation carried out by Photosystem I alone; ATP is synthesized by this system at the same rate. But both ATP synthesis and electron flow in the combined systems are inhibited. The probable site of DDT inhibition is indicated in Figure 8a.8. Examination of this diagram will show that if electrons are prevented from passing from Trap II to Trap I, ATP synthesis will be blocked. Moreover, as electrons "back up" in the system, the splitting of water will stop and no more electrons and hydrogen ions will be available for reduction of NADP. Without NADP·H$_2$, the dark reactions of the Benson-Calvin cycle cannot take place, and no synthesis of reduced organic carbon occurs.

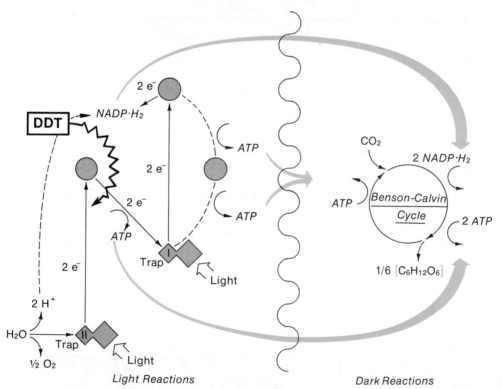

FIGURE 8a.8   DDT and photosynthesis.
      Phosphorylation by Photosystem I is not affected. The site in the electron flow pathway between Photosystems II and I at which DDT may be acting is shown by a zigzag arrow. This causes inhibition of synthesis of both ATP and NADP·H$_2$ by the combined systems

When the experimental DDT concentration is 1.5 ppm, electron flow in Photosystems I and II of isolated chloroplasts is inhibited about 20 per cent; at 15 ppm, it is inhibited 75 per cent. As mentioned above, levels of 1 to 2 ppm DDT are known to be associated with phytoplankton in nature.

Nevertheless, photosynthetic capacity of some species of algae is unaffected by such concentrations of DDT, while that of other species is significantly lowered. The various data we have been discussing are summarized in Table 8a.2.

Table 8a.2 SUMMARY OF DATA CONCERNING DDT AND PHOTO-SYNTHESIS		DDT concentration
	DDT in oceans (after 30 years' use)	.015 ppb*
	DDT found in natural phytoplankton populations (1969)	0.6–0.8 ppm*
	DDT taken up experimentally by phytoplankton exposed to 0.015 ppb	1–2 ppm
	DDT causing 20% inhibition of photosynthesis in isolated chloroplasts	1.5 ppm
	DDT causing 75% inhibition of photosynthesis in isolated chloroplasts	15 ppm

*ppb = parts per billion, or 1 part in 10^9.
**ppm = parts per million, or 1 part in 10^6.

EVALUATION OF DDT DATA

Our information so far leads to the hypothesis that even though DDT from the environment is associated with unicellular algae, much of it may only be adsorbed on the outside or stored at sites in the cells (in lipid droplets, perhaps) where it does not affect the functioning of chloroplasts. Since all isolated chloroplasts are equally sensitive to DDT, it is likely that DDT penetrates to the chloroplasts of non-resistant species of algae more readily than it does in resistant forms, and that it cannot penetrate at all through the waxy cuticle of the leaves of terrestrial plants. This is only a guess, however, and most of the experiments have lasted only a few days. We cannot assume from them that even today's relatively low levels of DDT in the oceans are having no effect on photosynthesis. It is sobering to reflect that the DDT concentrations now found in the ocean accumulated in less than 30 years. Even though the use of this pesticide in the United States has been greatly restricted, it is still being manufactured here for use in many other parts of the world and therefore will continue to increase in the oceans. In addition, the manufacture and use of PCB's, which may produce effects on photosynthesis similar to DDT, still continue.

The algal species found to be most affected by DDT belong to the most important groups of phytoplankton in the oceans. If their growth is affected, they may be replaced by other, more resistant species, but we know that many planktonic animals are capable of feeding on only one sort of alga. It is thus possible that the entire food chain in the ocean could be significantly altered by the presence of chlorinated hydrocarbons.

EFFECTS OF LIGHT AND TEMPERATURE

Little consideration has been paid to the effects man may be having on the amount of sunlight reaching the Earth's surface and the impact of this on worldwide photosynthesis. More consideration has been given to climatic changes, which would affect the rate of plant metabolism. Both subjects—man-made changes in light energy and climate—are highly disputed by scientists, and no very firm conclusions can be drawn about either. We shall simply list briefly some of the points which are presently being argued about man's impact on these factors.

FACTORS LEADING TO A DECREASE IN PHOTOSYNTHESIS

Fine particles of dust and moisture in the air, known as **aerosols,** are all too evident over large cities, and the possibility that they may be accumulating globally has been suggested. Both positive and negative evidence has been cited, although most meteorologists conclude that no definite trends are yet apparent. Reflection and refraction of sunlight from such aerosols could decrease photosynthesis directly (less light energy falling on plants) and indirectly (cooling of Earth's surface causing reduction in the rate of plant metabolism).

Another contribution by man leading to a decrease in photosynthesis is an increased cloudiness, or **turbidity,** of coastal waters. The continental coastlines are by far the most productive regions of the oceans, the open sea being a virtual desert, biologically speaking. It is into these coastal waters that many large cities now dump sewage containing large quantities of suspended particles. In addition, intensive agriculture along river valleys leads to accumulation of fine particles of soil in rivers flowing out to sea. Both contribute to an increasing turbidity of coastal waters, which in some areas seriously interferes with light penetration. As a result, the distribution of photosynthetic organisms is restricted to the uppermost surface layers and shallower waters, and productivity is decreased. Coastal waters also receive, of course, toxic substances — heavy metals, pesticides, detergents used to clean oil spills and so forth — which also are deleterious to populations of plants (and animals, as well).

FACTORS LEADING TO AN INCREASE IN PHOTOSYNTHESIS

It is often suggested that the combustion of fossil fuels is increasing the carbon dioxide level of the atmosphere and that this will produce a heating or "greenhouse" effect (see Chapter 7a). An increased temperature would speed up plant metabolism, and, providing soil moisture was not decreased, an increase in photosynthesis would result. There would also be a direct effect on plant growth. Since CO_2 is frequently a limiting factor in photosynthesis, higher atmospheric levels should increase growth rates.

Actual information about the relative contributions of these two opposing effects of man's activities is insufficient to draw any conclusions at present; the difficulties of measuring small but important changes on a global scale are prodigious. Since either an increase or decrease in photosynthesis would obviously be of great importance, all these factors should be carefully monitored.

CONSEQUENCES OF A REDUCTION IN PHOTOSYNTHESIS

Although there is at present no accurate estimate of the total impact of man's activities on global rates of photosynthesis, we have seen that the potential for producing great changes exists. Environmental pollutants such as DDT and smog, poor agricultural practices and urban sprawl all tend to decrease photosynthesis. In contrast, expansion of agriculture and mariculture and extended irrigation increase it.

In concluding this chapter, let us consider some consequences of a reduction in world-wide photosynthesis. If the rate of the photosynthetic reaction

$$6 \ CO_2 + 6 \ H_2O \longrightarrow C_6H_{12}O_6 + 6 \ O_2$$

were to decrease, the following qualitative changes would occur:

1. An increase in global H_2O;
2. An increase in atmospheric CO_2;
3. A decrease in atmospheric O_2;
4. A decrease in reduced organic carbon compounds.

We may consider the significance of each of these in turn.

1. Increases in global water would go unnoticed, since the per cent change would be infinitesimal.

2. Build-up of carbon dioxide, on the other hand, could result from respiration by plants and animals and from man's combustion of fossil fuels. The present small levels of 0.03 per cent CO_2 in the atmosphere may be increasing slightly (some estimate by 12 per cent since 1800), but much of the extra CO_2 dissolves in the oceans. Even if a greenhouse tendency results, increases in atmospheric aerosols are likely to offset it. Also, an excess of CO_2 would tend to increase photosynthesis in remaining plants, helping maintain a balance.

3. A loss in atmospheric oxygen is often heralded by some as a threat to man and animals. While it is true that plants play a major role in maintaining O_2 levels in the atmosphere, oxygen is present in such vast quantities (21 per cent of the Earth's atmosphere) that enormous changes would have to occur before much effect would be seen. Recollect that there is only four-fifths as much oxygen at Denver as at sea level, yet plants and animals there do not suffer from oxygen deficiency! The use of oxygen breathing machines in large cities like Tokyo is necessary *not* because pollution has removed oxygen from the air, but because toxic pollutants seriously interfere with the efficiency of breathing. Even if all photosynthesis in the world were to stop altogether, and animals and bacteria oxidized all plants and organic debris now present, less than 1 per cent of atmospheric oxygen would be used up. Therefore, changes in the rate of photosynthesis cannot noticeably affect man's oxygen supply.

4. The real loss, however, would be the disappearance of reduced organic molecules. Long before oxygen supplies ran out, sources of edible foods would disappear. Landscapes and oceans would lose their fauna as more and more animals lost out in the competition for less and less food. Even if man could raise sufficient food for himself (a possibility in real doubt) by selection of productive, pollution-resistant crops, a global decrease in photosynthesis would impoverish his whole environment — the scenery, the weather and many resources on which he now depends.

SUMMARY

The photosynthetic reaction carried out by green plants is the mainstay for all living organisms, and there are indications that man's activities may be interfering with it on a regional if not global scale. On land, several factors are at work. Large regions of the Earth's surface have been denuded by poor agricultural and lumbering practices; among these are one-crop agriculture, overgrazing, overlogging, plantation farming in the tropics, and overuse of pesticides. Urbanization is also reducing photosynthetic activity over ever larger areas. Both herbicides and urban air pollutants are significantly decreasing photosynthetic activities in many regions. Among air pollutants, ozone in smog is responsible for decreases in both carbon dioxide fixation and carbohydrate storage in susceptible plants, especially some species of pine trees.

In the oceans, the primary threat to photosynthesis may be DDT and other synthetic chlorinated hydrocarbons which persist in the environment. The association of DDT with phytoplankton has been demonstrated. Even small concentrations applied to isolated chloroplasts significantly reduce electron flow through Photosystems I and II.

Man's effects on other environmental factors, such as light and temperature, may also be affecting photosynthesis on a global scale. No balance sheet can yet be drawn for the total impact of the various activities of man on photosynthesis, however.

The major danger to the biosphere from a reduction in photosynthesis is not a change in atmospheric composition but the loss of food resources for all organisms.

READINGS AND REFERENCES

Bowes, G. W. 1972. Uptake and metabolism of 2,2-bis (p-chlorophenyl)-1,1,1,-trichloroethane (DDT) by marine phytoplankton and its effect on growth and chloroplast electron transport. Plant Physiology *49*:172–176.

Bowes, G. W. and Gee, R. W. 1971. Inhibition of photosynthetic electron transport by DDT and DDE. Bioenergetics *2*:47–60. This paper and the preceding listing describe some of the original experiments discussed in this chapter.

Broecker, W. S. 1970. Man's oxygen reserves. Science *168*:1537–1538. Argues that reported threats to oxygen supplies are groundless.

Cox, J. L. 1970. Low ambient level uptake of ^{14}C-DDT by three species of marine phytoplankton. Bulletin Environ. Contam. and Toxicology *5*:218–221.

Cox, J. L. 1970. DDT residues in marine phytoplankton: Increase from 1955 to 1969. Science *170*:71–73. The two papers by Dr. Cox include some of the data discussed in this chapter.

Humphrey, R. R. and Mehrhoff, L. A. 1958. Vegetation changes on a southern Arizona grassland. Ecology *39*:720–726. Observations over 50 years indicate replacement of grazed grasslands by shrubs is mainly due to decreased prairie fires.

Menzel, D. W., Anderson, J. and Randtke, A. 1970. Marine phytoplankton vary in their response to chlorinated hydrocarbons. Science *167*:1724–1726.

Miller, P. R., Parmenter, J. R., Jr., Flick, B. H. and Martinez, C. W. 1969. Ozone dosage response of ponderosa pine seedlings. Journal of the Air Pollution Control Association *19*:435–438. Original studies showing effects of ozone on growth, photosynthesis and starch formation in pine trees.

Richards, B. L. Sr., Taylor, O. C. and Edmunds, G. F. 1968. Ozone needle mottle of pine in Southern California. Journal of the Air Pollution Control Association *18*:73–77. Experiments and observations demonstrating ozone as the major cause of needle-mottling.

Rasool, S. I. and Schneider, S. H. 1971. Atmospheric carbon dioxide and aerosols: effects of large increases on global climate. Science *173*:138–141. A careful analysis of the potential effects of CO_2 and dust on future climate.

Thomas, W. L., Jr. (ed.) 1956. *Man's Role in Changing the Face of the Earth.* University of Chicago Press, Chicago. (With the collaboration of C. O. Sauer, M. Bates and L. Mumford). Report of an international symposium in Stockholm, 1955, covering subjects discussed in this chapter and much else. An excellent source book, one of the first and most complete on the subject.

van den Bosch, R. 1969. The toxicity problem—comments by an applied insect ecologist. *In* Morton, M. W. and Berg, G. G. (eds.) *Chemical Fallout.* Charles C Thomas, Springfield, Illinois. The Azodrin story is told by the scientist whose own research predicted the events which took place.

NUTRITION AND GAS EXCHANGE

9

Does a man dine well because he ingests the requisite number of calories?
— WALTER LIPPMANN, *American journalist*

Thus far we have considered the physical environment of plants and animals, the organic molecules they contain in their cells and the means by which energy is obtained both for synthesizing these molecules and for other life processes. The purpose of this chapter is to discuss the types and amounts of nutrients which various organisms must obtain from their environment to carry out their life functions. We shall include here not only the organic molecules and minerals but also the gases, oxygen and carbon dioxide, which various organisms require.

There are essentially two classes of organisms with respect to food requirements. Those organisms which synthesize their own food are known as **autotrophs** (Greek *autos,* "self," plus *trophos,* "feeder"). By far the most important group of autotrophs are the chlorophyll-containing green plants. As we noted in Chapter 4, there are also a few photosynthetic bacteria which utilize bacteriochlorophyll to trap the energy of sunlight. In addition, there are several types of autotrophic bacteria which use chemical sources of energy, such as Fe^{++}, instead of light for synthesizing reduced organic compounds. Since neither group of bacteria is of major significance in the biosphere, we will restrict our consideration of autotrophs to green plants.

The other main group of organisms, from a nutritional standpoint, are the **heterotrophs** *(heteros,* "other," plus *trophos,* "feeder")—those organisms which feed upon reduced organic molecules. Included are most bacteria, the fungi and other non-green plants and all animals. Heterotrophs depend upon photosynthetic organisms for their existence. The nutritional requirements of autotrophs and heterotrophs are therefore quite different.

NUTRITION OF GREEN PLANTS

We have already considered in Chapter 2 some of the nutritional requirements of green plants, and what follows will serve as a review of our earlier discussion. The environments from which plants obtain their nutrients fall into two major categories—aquatic and terrestrial.

AQUATIC PLANTS

Those plants living in an aqueous environment must obtain all their nutritional requirements from the water itself—all nutrients are in solution, even the gases. These nutrients may be conveniently grouped according to the amounts of each required.

The major nutrients are those required for photosynthesis—namely, carbon dioxide and water. Since all nutrients must be dissolved, carbon dioxide is obtained by aquatic plants in the form of bicarbonate ion (HCO_3^-).

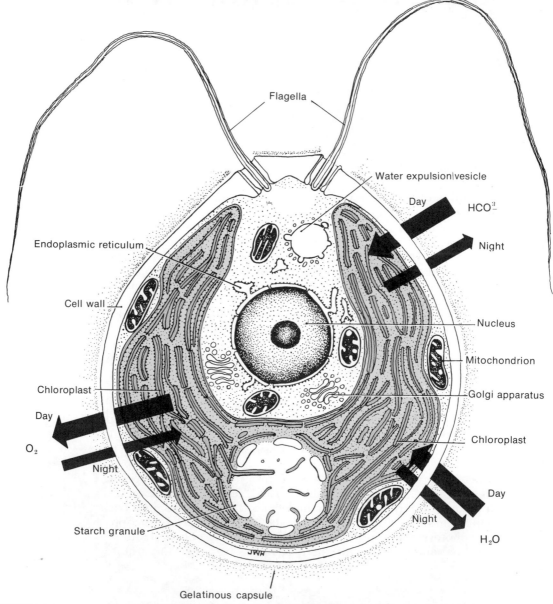

FIGURE 9.1   The exchange of major nutrients by an aquatic plant cell. Pictured is the unicellular green alga *Chlamydomonas,* which has two flagella and a diffuse chloroplast. During photosynthesis, in the daytime, HCO_3^- and H_2O are taken up and oxygen is released. At night, owing to respiration, the overall processes reverse; there is a net uptake of oxygen and release of HCO_3^- and H_2O. The size of the heavy arrows shows the relative importance of day and night processes.

$$CO_2 \quad + \quad H_2O \longrightarrow H_2CO_3 \longrightarrow H^+ \quad + \quad HCO_3^-$$

<table>
<tr><td>carbon
dioxide
(gas)</td><td>water</td><td>carbonic
acid</td><td>hydrogen
ion</td><td>bicarbonate
ion
(solution)</td></tr>
</table>

Green plants also require oxygen for respiration during the hours of darkness. A diagram of the exchange of major nutrients by the cell of an aquatic plant, such as a unicellular alga, is shown in Figure 9.1. Note that the exchanges occurring during photosynthesis in the daytime are greater than those resulting from respiration at night. There is thus a net synthesis of new tissue, and the plant grows.

In addition to the water and carbon dioxide used in photosynthesis, plants also require a number of minerals for synthesis of various organic molecules and as intracellular ions. Some of these are required in relatively large amounts, and are known as **macronutrients.** A list of these nutrients, and the role they fulfill in the cell, is shown in Table 9.1.

Table 9.1
MACRO-
NUTRIENTS
REQUIRED BY
GREEN PLANTS*

Element	Form Obtained	Role in Cell
Potassium	K^+	intracellular ion
Calcium	Ca^{++}	intracellular ion
Nitrogen	NO_3^-	protein; nucleic acid
Phosphorus	$PO_4^=$	nucleic acids; ATP
Sulfur	SO_4	protein; polysaccharide
Iron	Fe^{++}	cytochromes
Magnesium	Mg^{++}	chlorophyll

*Listed in order of abundance in plants, within each functional category.

Other nutrients, supplied as ions, are required in very small (trace) amounts. Among these are copper (Cu^{++}), manganese (Mn^{++}), zinc (Zn^{++}), boron (in the form of borate ion, $B_4O_7^-$) and molybdenum (in the form of molybdate ion, $MoO_4^=$). All the functions of these **micronutrients** are not known, although some are required for activation of certain enzymes. Manganese ion, as noted in Chapter 8, is essential for splitting of water during photosynthesis. Although sodium ion (Na^+) is usually found in plants, it is not required by most of them.

TERRESTRIAL PLANTS

Plants living on land have similar nutritional requirements to those living in water, but their nutrients are obtained in somewhat different ways. Photosynthesis takes place only in those parts of the plant exposed to sunlight, mainly in the leaves. The water required for photosynthesis reaches the leaves mostly from the roots, while carbon dioxide and oxygen are exchanged directly with the air. Figure 9.2 recalls our earlier description of the anatomy of a leaf from a higher plant.

The **stomata,** or openings into a leaf, permit passage of gases between the interior of the leaf and the air. All cells within the leaf are thus in direct contact with air, with which they directly exchange oxygen, carbon dioxide and water. The air within a leaf is maintained at or near 100 per cent humidity at all times. This is accomplished by the **guard cells** surrounding the stomata, which change shape and control the size of the openings. When water is scarce, they lose turgor, and the stomata close. When water is available, glucose and ions accumulate in the guard cells, causing an osmotic influx of water. As the guard cells swell, the stoma opens (Fig. 9.3). Similar

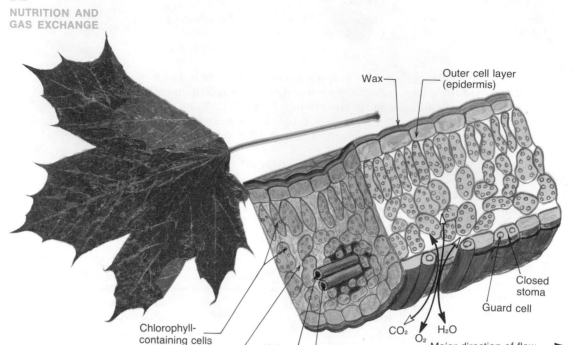

FIGURE 9.2   A whole leaf (left) and section through the middle of a leaf (right).
 The xylem and phloem run together, forming the central vein. The outer layer of cells (epidermis) protects the leaf and is covered by wax which is impermeable to water or air. Stomata on the bottom of the leaf are protected by guard cells which can change the size of the opening by changing shape. Gases exchange through the stomata between the outside and inside of the leaf.

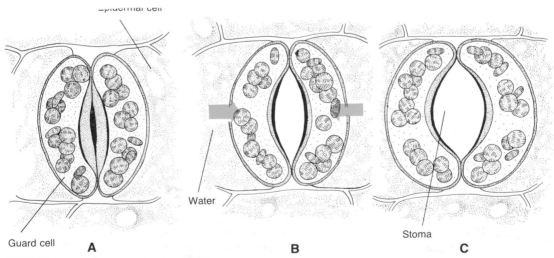

FIGURE 9.3   How guard cells control stomatal size.
 A. In the absence of turgor pressure, the guard cells are collapsed and the stoma is closed.
 B. When water is plentiful, the guard cells accumulate glucose and ions which cause an osmotic influx of water.
 C. Turgor pressure causes the guard cells to expand laterally, opening the stoma.
 (After Villee, C. A., and Dethier, V. G. 1971. *Biological Principles and Processes.* W. B. Saunders Co., Philadelphia.)

openings are also found on the stems of terrestrial plants, which therefore have no need for a gas-transporting circulatory system such as occurs in most animals.

The roots of terrestrial plants are in contact with capillary water in the soil, with which they exchange gases during their respiration. Since light does not reach the roots, they have no chloroplasts, and no photosynthesis occurs in them. Roots take up mineral nutrients from the soil, as we noted earlier, while obtaining the organic molecules necessary for their growth from the photosynthetic regions of the plant above ground.

NUTRIENT REQUIREMENTS OF HETEROTROPHS

All fungi and other non-green plants, most bacteria, and all animals depend ultimately upon green plants as a major source of food. The types and amounts of food required form the subject matter of courses on nutrition, a science which came into its own in the first quarter of this century with the discovery of the nature and role of vitamins.

TYPES OF FOOD REQUIRED

The nutritional requirements of heterotrophs fall into four convenient categories: energy sources, organic precursors, vitamins, and minerals.

Energy Sources. As we saw in Chapter 7, all reduced organic molecules—carbohydrates, fats and proteins—can be oxidized via respiratory pathways, resulting in the production of ATP, the energy currency molecule. The amount of energy available in various foods is usually expressed in terms of calories per gram of food. The calorie listed in most diet books is the so-called "large" calorie, which is 1000 times the standard calorie. It should more properly be called a *kilocalorie*. Table 9.2 gives the approximate caloric content of the three types of food, together with some foods which fall mainly within a given category. Weight-watchers will already be well aware that cream, avocados and nuts are fatty foods containing forbidden calories, but these also contain smaller amounts of protein and carbohydrate. The energy requirements of most animals are met by eating food equal to about one-tenth of one per cent (0.1%) of their body weight daily. Warm-blooded animals, however, require considerably more.

Organic Precursors. Aside from pure energy requirements, most heterotrophs require particular reduced organic molecules as building blocks for new tissue. Consider proteins, for example. These macromolecules are formed from about 20 different amino acids, some but not all of which an animal can synthesize for itself from other foods in its diet. Those amino acids which an animal cannot make must therefore be taken in as

Table 9.2 CALORIC CONTENT OF MAIN TYPES OF FOOD		
Carbohydrate (Sugar, flour, rice, potatoes)		4.2 kilocalories per gram*
Fat (Butter, oils, nuts, cream)		9.5 kilocalories per gram*
Protein (Meat, eggs, fish, cheese)		4.2 kilocalories per gram*

*28 grams = 1 ounce.

part of its diet, and are known as **essential amino acids.** Different animals have different requirements. For man, 10 are required, and not all proteins are capable of supplying all of them. Zein, the protein of maize, lacks several amino acids essential for man—it is thus an "incomplete" protein. Casein, or milk protein, on the other hand, contains all the essential amino acids needed by man.

Another well-advertised requirement is for polyunsaturated fats, which play a role in cell membranes. Man is incapable of synthesizing certain of these and must obtain them from vegetable fats in his diet. The amounts of each of these organic precursors that are required are relatively small, however—a few millionths of the body weight per day.

Vitamins. A few organic molecules are required in even smaller quantities—on the order of a few billionths of the body weight per day. These substances have come to be known as **vitamins,** and many of them, particularly the B-vitamins, play an important role as co-factors for enzymes in intermediary metabolism. **Niacin,** as we have seen, is a component part of the important hydrogen carrier molecules NAD and NADP, and **riboflavin** is a component of FAD (see Chapter 7).

Diets deficient in vitamins lead to a variety of deficiency diseases. Lack of vitamin C, for example, results in **scurvy**—a disease formerly prevalent among sailors deprived of fresh foods for long periods. A deficiency of vitamin D prevents the normal absorption and deposition of calcium and phosphate in bones, leading in children to **rickets,** which is characterized especially by deformed extremities (Fig. 9.4). Before vitamin D can be effective, however, it must be activated by ultraviolet light falling on the skin. Too little niacin (a B-vitamin) produces **pellagra**, once common in the southern United States among the poor whose diet was mainly cornmeal and small amounts of pork. **Beri-beri,** a disease resulting from a deficiency of **thiamin** (vitamin B_1), is common in parts of Asia where polished rice forms the staple food. This vitamin plays a role in those reactions of cellular respiration in which carbon dioxide is removed. Thiamin occurs only in the seed coat or bran of rice and other grains, which is lost in the polishing process. A diet consisting mainly of whole grain cereals, whole milk, fresh fruits and vegetables and meat or fish contains more than enough vitamins for the average person. Deficiency diseases are still prevalent among malnourished populations, however. Unfortunately, so-called "modern" food processing often discards or destroys vitamins; hence the need for fortified foods and vitamin pills even in an affluent society.

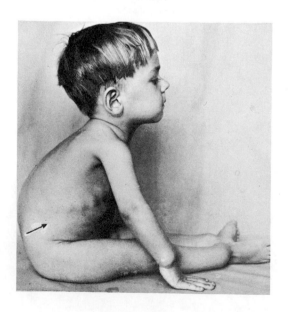

FIGURE 9.4

Rickets in a small child.
    Note the malformation of the ribs (arrow) and of the wrists and ankles. The deficiency of vitamin D, by decreasing the body's ability to use calcium and phosphorus, has resulted in soft, malformed bones. (Photo courtesy of Dr. Niilo Hallman, from Villee, C. A., and Dethier, V. G. 1971. *Biological Principles and Processes.* W. B. Saunders Co., Philadelphia.)

Nutritional requirements for vitamins vary from one species of animal to another. Rats, for example, are able to synthesize certain vitamins that man cannot make. It is therefore not always possible to apply information obtained from studies on laboratory animals directly to man. Most bacteria are capable of making the vitamins they require, and human intestinal bacteria provide man with a secondary source of several vitamins. Hence it may be wise to supplement one's diet with vitamins while taking oral antibiotics, which kill the normal intestinal bacteria as well as those causing disease.

Minerals. Although minerals are an important part of one's diet, they are usually available in plentiful amounts. Such ions as sodium (Na^+), potassium (K^+) and chloride (Cl^-) are found in all foods as well as in most water supplies. Other elements, however, although required in smaller amounts, may be deficient in some diets. Iron (Fe) is required for synthesis not only of the cytochromes of the electron transport system but also of hemoglobin, and deficiency of this element results in anemia.

The mineral iodine forms part of **thyroxin,** a hormone produced by the thyroid gland, which regulates body metabolism. People living far from the iodine-rich sea often have iodine-deficient diets, which result in **goiter,** or swelling of the thyroid. Severe deficiency leads to dwarfing (Fig. 9.5) and mental retardation—the "village idiot" was a common figure throughout central Europe until quite recently. The introduction of iodized salt and of refrigerated transport of seafood has largely eliminated this disease in many countries.

FIGURE 9.5

A group of seven dwarfs in Switzerland. Iodine deficiency in the area resulted in insufficient thyroxin production and growth retardation of these people. Compare their size and appearance with that of normal individual in background. (From Grollman, A. 1947. *Essentials of Endocrinology.* J. B. Lippincott, Philadelphia.)

In summary (Fig. 9.6), food is used both for maintenance of the organism and for growth. Maintenance includes such things as muscular work, osmotic work, excretion of wastes and so on. For all these processes ATP, which is obtained from oxidation of foods, is required. Food not required for maintenance can be used for growth—for synthesis of macromolecules. Energy from ATP is required here also.

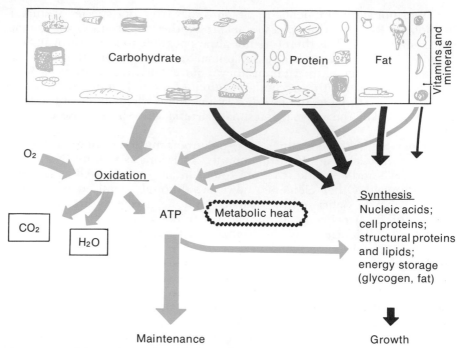

FIGURE 9.6

Maintenance

Growth

A summary of the pathways followed by various types of food.

The sizes of the arrows are intended to approximate the relative importance of the various routes. The sizes of the compartments show the approximate proportions of each food type in the human diet.

BODY TEMPERATURE REGULATION

We saw in Chapter 7 that the capture of energy during cellular respiration is remarkably efficient. Nevertheless, about 50 per cent of the energy is lost as waste heat (Fig. 9.6). Plants and cold-blooded animals lose much of this heat readily to the environment, and their body temperature thus tends to follow that of their surroundings. Because the rates of chemical reactions are highly temperature-dependent, the activity of such organisms is affected by environmental temperature. On a cold morning, insects and lizards may be observed to move sluggishly, but after basking in the sun they become quite active. Note that basking is a behavioral adaptation by which an animal can increase its body temperature above that of its surroundings.

Thermal Insulation. Birds and mammals, on the other hand, maintain an almost constant body temperature, and are equally active over a wide range of environmental temperatures. By insulating the body surface, heat loss is controlled (Fig. 9.7). If the surface temperature drops, hair or feathers are raised by reflex contraction of erector muscles, increasing the volume of the insulating layer of air trapped beneath them. Even though our body hair is vestigial (no longer useful), this reflex still produces "goose bumps" when we are cold. Fat layers under the skin add further insulation. Heat loss is also controlled by constriction and dilation of blood vessels supplying the skin. In hot climates, heat loss may be increased by evaporation of perspiration in man and some other mammals. Other animals—birds and dogs, for example—evaporate water from the respiratory tract by panting.

Metabolic Regulation. Increasing the amount of food oxidized is another means of increasing body temperature, since more waste heat is produced. In warm-blooded animals, as much as 80 per cent of the food

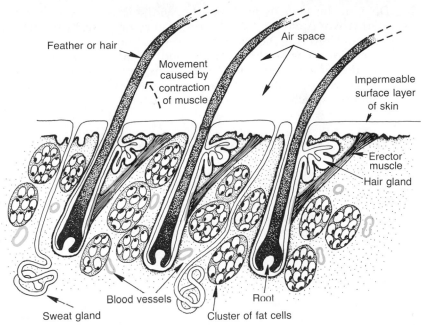

FIGURE 9.7

A section of skin through a hypothetical warm-blooded animal showing the possible ways the skin can be used for temperature regulation. Cooling is achieved by dilation of blood vessels and by sweat glands (not present in birds or all mammals). Heat retention is achieved by erection of hair or feathers, by increasing the fat layer under the skin and by constriction of blood vessels

eaten is used to maintain body temperature. The thyroid gland has long been known to control rates of oxygen consumption in mammals by secretion of the hormone **thyroxin.** Several suggestions as to how thyroxin may act on cells have been made, and two are shown in Figure 9.8. By stimulat-

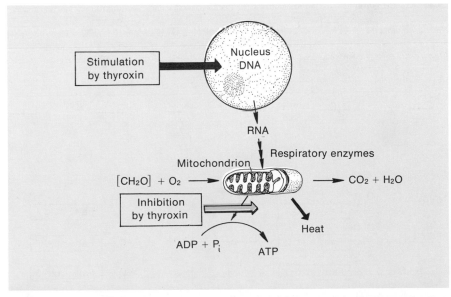

FIGURE 9.8

Hypotheses proposed for the action of thyroxin on respiratory rate and heat production. Stimulation of DNA to direct synthesis of more respiratory enzymes would increase oxygen consumption by burning more food. Decreased efficiency of ATP synthesis would generate more waste heat; overall respiration would increase to supply needed ATP.

ing DNA in the nucleus to direct the synthesis of more respiratory enzymes, thyroxin could increase the respiratory rate. Or, by inhibiting ATP synthesis and lowering the efficiency of respiration, thyroxin could increase the generation of waste heat. More food would need to be oxidized to produce the same amount of ATP. Perhaps both hypotheses are partially correct. Thyroxin, it should be added, has many other important functions, some of which we shall come across later.

DIGESTION AND ABSORPTION OF FOOD

Whether food is burned for energy or used for building new tissue, the large molecules of carbohydrate, lipid and protein first must be broken down into their constituent parts before they are taken into the body and utilized—they cannot pass directly across the cell membranes. The polymers must be digested to monomers. For this purpose all heterotrophs have evolved a series of digestive enzymes.

DIGESTIVE ENZYMES

The polymers of all large molecules are split into monomers by the addition of water molecules—a type of reaction known as *hydrolysis* (Fig. 9.9). This can be brought about in a test tube by addition of strong acid or alkali and heating to high temperatures. Such conditions, of course, are lethal to cells, which instead use specific enzymes for the purpose.

There are three major groups of digestive enzymes: the **carbohydrases,** which split carbohydrates; the **proteases,** which split proteins; and the **lipases,** which split lipids. (Note the suffix **-ase,** denoting "enzyme.") As each large molecule is split into smaller subunits, a new enzyme is required. A list of the major digestive enzymes is given in Table 9.3.

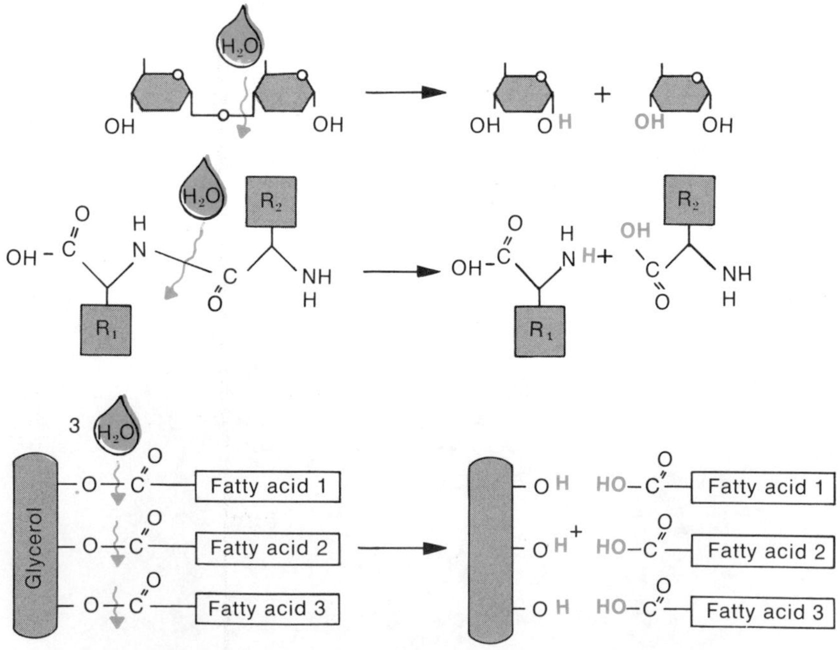

FIGURE 9.9

Examples of hydrolytic reactions, in which large molecules are split by water.
  A. The dimer (two-unit) molecule of maltose is split to two glucose monomers.
  B. A peptide dimer is split into two amino acid monomers.
  C. A lipid is split into its subunits, glycerol and three fatty acids.

Although the hydrolytic reactions of digestion are energy-yielding reactions, they do not release sufficient energy for ATP synthesis, and so this energy is not utilized by organisms. A fraction of the energy present in food, therefore, is lost during the initial stages of digestion.

Table 9.3
IMPORTANT
DIGESTIVE
ENZYMES OF
HETEROTROPHS

Enzyme	Acts Upon	Produces
Carbohdrases		
Amylase	starch (polymer)	maltose (dimer)
Cellulase*	cellulose (polymer)	cellobiose (dimer)
Maltase	maltose (dimer)	glucose (monomer)
Sucrase	sucrose (dimer)	glucose and fructose (monomers)
Cellobiase*	cellobiose (dimer)	glucose (monomer)
Proteases		
Pepsin	protein (large polymer)	polypeptide (small polymer)
Trypsin	protein (large polymer)	polypeptide (small polymer)
Peptidases	polypeptide (small polymer)	amino acids (monomers)
Lipases	lipids (tri-esters)	fatty acids and glycerol

*These enzymes are not produced by most animals, despite the abundance of cellulose in the diet. (One exception is the shipworm, *Teredo*, which bores into pilings and submerged timbers.) Herbivores subsisting mainly on cellulose—termites, snails, cows—harbor symbiotic bacteria or protozoa in their guts which produce these enzymes and digest foods for their hosts.

SITE OF DIGESTION

So far we have considered only the chemistry of digestion and have ignored where it occurs. Since only small molecules can pass into a cell, all digestion must take place outside. However, in the case of a single-celled organism such as an amoeba, living in an aqueous environment, the release of enzymes outside the cell into the surrounding water would benefit the animal little. Currents would wash the enzyme away before food digestion could occur. Therefore, many single-celled organisms have devised a mechanism for walling off part of the environment, as shown in Figure 9.10.

Food captured by pseudopodia becomes surrounded by cell membrane, forming a **food vacuole.** This is really a miniature, temporary digestive cavity, and the food remains external to the rest of the cell. Enzymes from secretory granules (perhaps lysosomes) inside the cell are released into the food vacuole. The precise mechanism by which large protein enzymes pass across cell membranes is not understood, although it is possible that the membranes of the secretory granule and food vacuole join together to form an opening between them. When food has been digested to the level of monomers, it is absorbed into the cell.

This type of digestion has come to be known as **intracellular digestion,** which is rather a misnomer, since the food vacuole is never a part of the cell proper. Intracellular digestion is also used by sponges, coelenterates, flatworms and some molluscs.

Bacteria and fungi living in dry places or in regions where water movement is restricted, on the other hand, secrete digestive enzymes directly onto their food, a process known as **extracellular digestion.** Bread molds, for example, secrete digestive enzymes directly onto the bread, digesting starch to glucose and proteins to amino acids. Likewise, bacteria usually thrive in soils or lake sediments, where their externally secreted enzymes remain in close contact with the organic matter which forms their food. In neither of these cases is the manner of enzyme release from the cells understood.

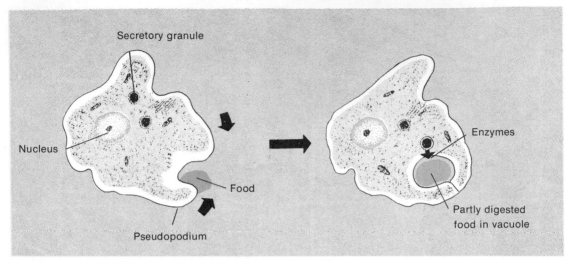

FIGURE 9.10   Capture of food and formation of food vacuole by *Amoeba*. Enzymes are released into the vacuole from secretory granules (perhaps lysosomes).

Most multicellular animals, however, permanently wall off part of the exterior environment within themselves to form a digestive tract. Within this restricted space, high concentrations of digestive enzymes can be maintained. As noted in Chapter 6, coelenterates and flatworms have only a single opening to the digestive tract, and practice intracellular digestion, but in more advanced animals the gut is a long tube with an opening at each end, the mouth and the anus. In its simplest form, such a gut has a uniform appearance along its whole length, and both digestion and absorption are carried out by the endodermal cells lining its cavity (Fig. 9.11).

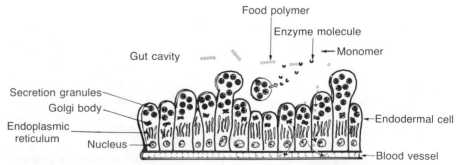

FIGURE 9.11   Schematic diagram of one side of a simple gut showing stages in digestive processes. The gut cavity, containing food, is at the top. Enzymes synthesized by the endoplasmic reticulum of endodermal cells are formed into secretion granules by the Golgi apparatus. Eventually the tip of the cell pinches off into the cavity, the secretion granules rupture and enzymes are released. The polymers of food (shown in color) are hydrolyzed to monomers which are absorbed by gut cells and transferred to the blood.

Animals with specialized diets, however, have the gut subdivided into distinct regions adapted for various functions.

The food of heterotrophs comes in several sizes. Small molecules, such as dissolved amino acids and sugars, are but a minor part of the diet of most animals and require no special processing prior to absorption. The remaining food sources can be roughly divided between fine particles in suspension and large particles which require mechanical breakdown. Animals feeding on fine particles capture their food by specialized filtering devices and hence are known as *filter feeders.* Several types of filter feeders were mentioned in Chapter 6. The best-studied group are the bivalves—clams, oysters and mussels—which trap minute food particles in mucous secretions that are transported by cilia to the mouth. Since the food is already very small, consisting mostly of single-celled algae and detritus, it is easily attacked and digested by enzymes. Animals feeding on large foods, such as macroscopic plants or animals, must mechanically break down the food into small pieces before it can be digested.

FUNCTIONS OF COMPLEX DIGESTIVE SYSTEMS

Processing of large food particles prior to its final absorption often requires several steps. A simple gut, such as shown in Figure 9.11, is found in relatively few animals. In most species, various regions of the gut are modified to perform distinct functions appropriate to the animal's diet (Fig. 9.12).

As food is received into the mouth and **pharynx,** preliminary maceration by jaws, beaks or teeth often occurs. Associated **salivary glands** may be present which add lubricants (mucus) and enzymes to initiate digestion. A storage region may follow behind the mouth; this is especially useful to animals feeding only infrequently, whenever food chances to be present. The crops of birds, leeches and bedbugs are good examples. In man, food simply passes directly from the mouth via a long tube, the **esophagus,** to the stomach.

The region known as the **stomach** may serve two functions: further grinding of food, and preliminary digestion. In various animals, such as birds, some insects and some snails, there are two separate regions. The highly muscular **gizzard,** often lined with horny "teeth" and sometimes

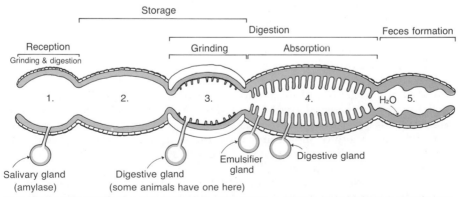

FIGURE 9.12

Subdivisions found in the digestive tracts of animals. *1,* mouth and pharynx. *2,* esophagus or crop. *3,* stomach or gizzard. *4,* small intestine or midgut. *5,* large intestine or hindgut.

Not all animals show such distinct subdivisions. Variations are often related to diet: Carnivores generally have short guts, while herbivores, whose food is harder to digest, have long ones; filter feeders require no pharynx nor storage regions; aquatic animals do not need to resorb water.

containing gravel or sand grains, pulverizes the food. In the secretory region of the stomach, proteases and (in vertebrates) large quantities of hydrochloric acid, are released into the stomach cavity, where preliminary protein digestion occurs. The stomach lining is normally protected from self-digestion by a thick layer of mucus and by a barrier of tightly fused epithelial cells, although prolonged mechanical irritation or hyperacidity can cause ulcer formation.

The small intestine, often called the midgut in many animals, is the region in which digestion is completed and the important process of absorption occurs, which we shall describe in detail shortly. This region often has glands associated with it which produce not only digestive enzymes but also **emulsifiers** (detergents) for dispersing fats. In man, the **pancreas** secretes several digestive enzymes into the small intestine; others are produced by the gut cells themselves. Bile produced by the liver contains **bile salts**, which also have a detergent effect.

The last part of the gut, called the large intestine or, more generally, the hindgut, serves for feces formation. The feces contain indigestible food, mucus secreted by the gut to facilitate food passage, bacteria which normally live in the intestinal cavity of most animals, parts of dead intestinal cells and whatever secretions have been added by the various glands that empty into the gut. In vertebrates, for example, the bile contains, besides the emulsifiers just mentioned, several waste products of metabolism. In terrestrial animals, such as man and insects, water resorption occurs here to prevent excess water loss with the feces.

Food is moved along the intestine either by cilia on the surface of the endodermal cells or, more commonly, by peristaltic contractions of circular and longitudinal muscles surrounding the gut. These are analogous to the peristaltic contractions used for locomotion by earthworms, described in Chapter 6.

ABSORPTIVE PROCESSES

The cells of the small intestine usually have a greatly increased surface area to facilitate the absorption of monomers released by digestion. The outer surface of the cells is often covered with numerous finger-like projections, called **microvilli** (Greek *mikros*, "small," plus *villus*, "hair"), which are so small they can be seen clearly only by using an electron microscope (Fig. 9.13A). These should not be confused with cilia, however, which may also be present. Microvilli are true extensions of the cell membrane and contain cytoplasm. In addition, the whole surface of the intestine in this region may be highly folded to increase the surface area available for absorption (Fig. 9.13B).

Once food is digested, how is it finally absorbed into the body? Do molecules simply diffuse down a concentration gradient, or are they actively taken up? In almost every case which has been studied carefully, it turns out that amino acids and sugars do not simply diffuse into the endodermal cell, but are transported across its membrane. It is thought that specific **carrier molecules,** presumably proteins, are located in the membrane and facilitate the movement of molecules across the boundary. Although we do not yet know how such carriers actually function, a useful model for visualizing the process is shown in Figure 9.14.

In some cases the concentration of a particular substance inside the cell is greater than that outside, and so transport inward must occur against a concentration gradient. Such a process is called **active transport** and requires expenditure of energy derived from the splitting of ATP. Although it is not known how ATP energy is coupled with active transport, ATP-splitting enzymes (ATPases) have been found on cell membranes. When these are inhibited, or when ATP production is prevented by

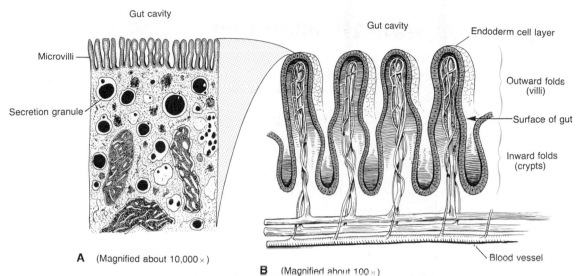

A (Magnified about 10,000×)

B (Magnified about 100×)

Adaptations for increasing the absorptive surface of the intestine.

A. The outer surfaces of individual cells often have up to 3000 microvilli; these enormously increase surface area. (Only outer part of a single cell is shown.)

B. The entire intestinal wall may be thrown into deep folds, producing villi which protrude into the gut cavity and crypts which sink into the wall itself.

FIGURE 9.13

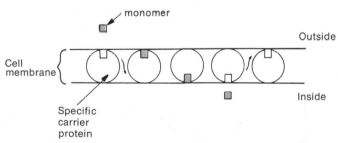

FIGURE 9.14

Highly schematized model of a carrier molecule in a cell membrane, transporting a monomer into the cell. The steps, from left to right, represent the sequence of events in time. The monomer binds with a specific site on the carrier which then transports it inside. After releasing the monomer, the carrier returns to its original position. We do not yet know the shape of these hypothetical carrier molecules nor the changes they undergo during transport. (After Pardee, A. B. 1968. Science *162*:632–637.)

poisoning the electron transport system with cyanide, active transport ceases.

Substances known to be actively transported across the endodermal cells of the intestine from the gut cavity on one side to the blood on the other include amino acids, sugars and some ions. The circulatory system then distributes these to other cells—liver, muscle, nerve, kidney and so on.

GAS EXCHANGE IN HETEROTROPHS

Having obtained energy-yielding molecules, heterotrophs must also obtain one other "nutrient"—oxygen, for respiration—which brings us to the final subject of this chapter. We should first clearly distinguish, however, the two meanings that the word "respiration" has in biology. Cellular respiration, the subject of Chapter 7, is understood to mean those oxidative reactions involving the Krebs cycle and electron transport systems, wherein reduced organic molecules and oxygen are consumed and carbon dioxide, water and ATP are produced. External respiration, on the other hand, refers to the exchange of oxygen and carbon dioxide between an animal and its environment—what the layman commonly considers respiration. It is the latter with which we are now concerned.

TYPES OF RESPIRATORY ORGANS

Very small heterotrophs, including oxygen-utilizing bacteria and fungi, single-celled protozoa and the very thin flatworms, obtain sufficient oxygen by direct diffusion to supply their needs. But animals thicker than about one millimeter (1/25 inch) generally require both a circulatory system to transport oxygen to their internal cells and also a specialized region of the animal where gases are exchanged between the blood and the environment—namely, a respiratory organ. The functioning of circulatory systems in gas transport is discussed in Chapter 10.

Gases pass into and out of animals purely by diffusion. However, they must first dissolve in a thin film of water on the outside of the cells of the respiratory organ before entering an animal. In those animals whose whole body surface is continuously moist, the entire surface may function as a respiratory organ, exchanging gases between environment and blood. A common example is the earthworm; another is the frog. Although frogs have lungs and can breathe air when on land, their respiratory needs while under water are met by gas exchange across the skin. In both animals the skin is well supplied with blood vessels.

Most animals, however, have special regions which serve as respiratory organs, and these can be lumped broadly into external structures, or gills, and internal structures, such as lungs. Various types of respiratory organs are shown in Figure 9.15. Gills are mostly limited to aquatic animals, since their moist surfaces would soon dry out if they were exposed to air. The anatomy of gills varies from relatively simple finger-like structures on some marine worms to the highly complex gills of bivalve molluscs and bony fish. In all cases, however, the gill is abundantly provided with capillaries, and the distance between blood and environment is usually measured in hundredths of an inch or less.

Internal respiratory organs are found in both aquatic and terrestrial animals. Some of the ocean-dwelling sea cucumbers, members of the group called echinoderms, pump water in and out of the anus, to which are attached internal branched tubes called "respiratory trees." But by far the most well-known internal respiratory organs are the lungs of terrestrial

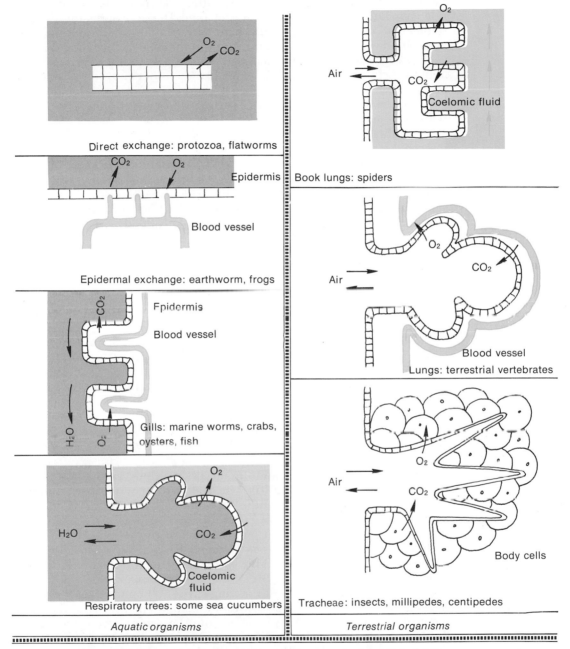

Direct exchange: protozoa, flatworms

Epidermal exchange: earthworm, frogs

Gills: marine worms, crabs, oysters, fish

Respiratory trees: some sea cucumbers

Book lungs: spiders

Lungs: terrestrial vertebrates

Tracheae: insects, millipedes, centipedes

Aquatic organisms　　　　*Terrestrial organisms*

The major types of gas exchange systems in animals. Black arrows indicate ventilatory currents of water or air; blue arrows indicate blood or body fluid circulation. Note absence of circulatory system for gas transport in insects, where tracheae penetrate into tissue cells. All terrestrial organisms have narrow openings to exterior to reduce water loss. (After Keeton, W. T. 1967. *Elements of Biological Science.* W. W. Norton & Co., New York.)

FIGURE 9.15

animals, where the problem is to exchange gases with the air without undue loss of precious water. Spiders have a sort of combined invagination-evagination type of lung known as a "book lung." The lungs of terrestrial vertebrates are outgrowths of the digestive tract. In frogs, these are little more than a pair of hollow sacs, rather like balloons, which provide relatively little surface area for gas exchange. In the progression from amphibians to reptiles, to birds and mammals, the lungs become more and more subdivided and branched, ending in millions of tiny air-sacs, the **alveoli.** The greatly increased exchange surface permits these groups to lead a highly active life. In all cases, lungs, like the other respiratory structures so far considered, are well supplied with blood capillaries. In Chapter 9a we shall describe the human lung further and examine the effects of air pollutants on it.

Another important group of terrestrial animals, the insects, have an internal respiratory system which is not associated with a circulatory system. Instead, they have a highly branched system of **tracheae,** or minute tubes, which carry oxygen to every cell. The tracheae open to the exterior by means of tiny openings on the body surface, the **spiracles,** which possess valves to prevent water loss (see Figure 6.16). Such an arrangement can function only in organisms with a relatively small body size.

VENTILATORY MECHANISMS

Although for some animals it is sufficient merely to expose the respiratory surface to the environment to obtain sufficient gas exchange, for more active forms, some sort of ventilatory movements are required. Animals living in aquatic environments where oxygen occurs at low concentrations (about six to nine milliliters per liter — air has 200 milliliters per liter) must either move the gills through the water or pump a current of water over them. Fish, for example, take water into their mouths while the opercular flaps (the hard covering over the gills) are shut. The mouth is then closed, and the muscles of the throat contract, forcing the water out past the gills. A few fish, such as herrings, do not make such ventilatory movements at all. Instead, they swim continuously with the mouth open, so that water constantly flows over the gills.

As an example of ventilation in an air-breathing animal we can consider breathing in man (Fig. 9.16). The coelom of man is divided into three compartments: the lung cavity, the abdominal cavity and the heart cavity. The first two of these play a role in breathing. The lungs are elastic organs: they stretch easily to receive air and contract readily to expel it. In certain pathological conditions, such as **emphysema,** elasticity is partially lost, making breathing more difficult. This is discussed in the next chapter. The primary muscle involved in breathing is the **diaphragm,** which, when relaxed, arches upward against the lung cavity. When the diaphragm contracts, it flattens, increasing pressure in the abdominal cavity and decreasing it in the lung cavity. Air pressure from the atmosphere forces air into the lungs, expanding them. When the diaphragm relaxes, the pressure changes are reversed, and air is expelled from the lungs, aided by their elasticity.

Only the diaphragm is used for breathing while one is at rest. (Lying in bed, one can watch the abdomen rise and fall.) When more air is required, as during exercise, or when breathing is made difficult, as in emphysema or an asthma attack, the muscles of the rib cage and of the shoulder are employed to further increase the size of the lung cavity during inspiration and to decrease it during expiration.

The rates of ventilation of almost all animals are affected by the levels

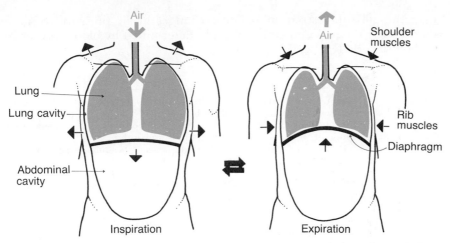

FIGURE 9.16

Ventilatory movements in man. Contraction and relaxation of the diaphragm and of certain rib and shoulder muscles cause a decreased pressure in the lung cavity; this results in air flowing into the lungs. The lungs expand. Relaxation of the diaphragm and contraction of different rib and shoulder muscles squeeze the lung cavity, expelling air.

of oxygen and carbon dioxide in the blood. In aquatic animals, it is mainly the oxygen availability which determines ventilatory rate, whereas in terrestrial animals, including man, a rise in blood carbon dioxide level stimulates the breathing centers at the base of the brain. This causes the heavy breathing associated with strenuous exercise. Sensitive regions in the brain detect a build-up of carbon dioxide and send stimuli to the ventilatory musculature. No doctor would ever give a patient carbon dioxide-free oxygen to breathe, since this would inhibit the normal breathing reflexes; small quantities of CO_2 are always added to therapeutic oxygen supplies.

SUMMARY

Photosynthetic autotrophs require carbon dioxide and water as major nutrients, together with numerous elements in the form of dissolved ions. From these all the chemicals required by the plant are synthesized.

Heterotrophs (bacteria, fungi, animals) ultimately utilize reduced organic molecules produced by autotrophs. Most food eaten by heterotrophs is used for energy-requiring maintenance work, any excess being channeled to growth. Specific requirements of heterotrophs include organic precursor molecules, vitamins and several minerals, each of which serves a special function.

Food is digested by hydrolysis, which is catalyzed by specific digestive enzymes. This usually occurs either in food vacuoles within cells or in the cavity of the gut. The gut may also serve to macerate food, to store it and to form feces. Microvilli on cells and folding of the gut wall enormously increase surface area available for absorption. Most monomers are taken up by carrier molecules, often against concentration gradients.

Uptake of oxygen and release of carbon dioxide, which occurs by diffusion, are frequently associated with special respiratory organs in larger animals. These contain numerous blood vessels, and the distance between blood and environment is always small. Aquatic animals often pump water over the gills; terrestrial animals use changes in air pressure to fill and empty their lungs. Ventilatory rates are determined by blood oxygen and carbon dioxide levels.

**READINGS AND
REFERENCES**

Carlson, A. J., Johnson, V. and Cavert, H. M. 1961. *The Machinery of the Body.* 5th ed. University of Chicago Press, Chicago. An excellent beginning text of human physiology.

Comroe, J. H. 1966. The lung. Scientific American Offprint No. 1034. A readable and informative description of the structure and function of human lungs.

Davenport, H. W. 1972. Why the stomach does not digest itself. Scientific American *226*:86–93. Explains the safety mechanisms that maintain the protective barrier against gastric acid.

Guyton, A. C. 1969. *Function of the Human Body.* 3rd ed. W. B. Saunders Co., Philadelphia. A clearly written introductory textbook of human physiology.

Hock, R. J. 1970. The physiology of high altitude. Scientific American Offprint No. 1168. Human physiological adaptations to low oxygen pressures are discussed.

Neurath, H. 1964. Protein-digesting enzymes. Scientific American Offprint No. 198. Describes the structure of proteases and explains how this is related to catalytic function.

Pardee, A. B. 1968. Membrane transport proteins. Science *162*:632–637. A review of research into the specific molecules which transport monomers and ions across cell membranes.

Prosser, C. L. and Brown, F. A. 1961. *Comparative Animal Physiology.* 2nd ed. W. B. Saunders Co., Philadelphia. A comprehensive text on the subject, with detailed data and references. Chapters 5 and 6 are pertinent to material discussed in this chapter.

"SMOG": SOME EFFECTS ON PLANTS AND MAN

As photochemical smog becomes more prevalent, its most toxic component, ozone, can be expected to damage the chloroplasts of non-resistant plants and the lungs of man and other animals.

People living in large cities in many countries are all too aware of the presence of air contaminants generated by the activities of man (Fig. 9a.1).

New York City smog. (Courtesy of Planned Parenthood—World Population.)

FIGURE 9a.1

9a

A blue or brown haze in the air, which often causes both irritation of mucus membranes and respiratory complaints, has become known as "smog." The word was coined from the combination of "smoke" and "fog." Smog is mainly the result of combustion products released from a variety of sources into the atmosphere, where, under favorable meteorological conditions, they accumulate and react with the constituents of normal air. The combustion of gasoline by vehicles contributes about one-half of the total 164 million tons of pollutants released into the atmosphere over the United States each year. Burning of fossil fuels for space heating and in electricity generating plants is the second major contributor, and the mixed pollutants released from industrial smokestacks are the third.

TYPES OF SMOG

The myriad pollutants produced in various locations make the smog of each area unique. Basically, however, two types of smog can be distinguished. When the predominant smog source is the combustion of fossil fuels with a high sulfur content, particularly low-grade coal, the characteristic pollutants are oxides of sulfur, especially sulfur dioxide (SO_2). In the presence of oxygen, SO_2 is converted to SO_3. Thus SO_2 has a reducing effect; such smogs are known as **reducing smogs.** When SO_3 combines with water droplets in the air, sulfuric acid is produced which is highly damaging to buildings, plants and the lung tissues of man and other animals. Other pollutants resulting from combustion of fossil fuels include soot, ash and a variety of partially oxidized chemical substances. This type of smog formerly was prevalent in the winter months in London and other large English cities and was responsible for many deaths, especially among the aged. Chronic bronchitis, a condition associated with this type of air pollution, is known throughout Europe as "the English disease." A law passed in the past decade prohibiting the burning of high-sulfur fuels has virtually eliminated this type of smog from much of England, although it still occurs in other places.

The other main type of smog does not usually contain large amounts of sulfur, and it requires sunlight to catalyze its formation. For this reason it is called **photochemical smog;** it is very common in cities where automobiles are numerous, the most infamous case being Los Angeles. To understand

Gas	ppm**
Nitrogen (N_2)	780,900
Oxygen (O_2)	209,400
Argon (A)	9300
Carbon dioxide (CO_2)	315
Neon (Ne)	18
Helium (He)	5.2
Methane (CH_4)	1.0–1.2
Krypton (Kr)	1
Nitrous oxide (N_2O)	0.5
Hydrogen (H_2)	0.5
Xenon (Xe)	0.08
Nitrogen dioxide (NO_2)	0.02
Ozone (O_3)	0.01–0.04

TABLE 9a.1
GASES IN
NORMAL DRY*
AIR

*Amounts of water vapor, of course, vary from place to place.

**ppm = parts per million.

(From Stern, A. C. (ed.) 1968. *Air Pollution.* 2nd ed. Volume 1, p. 27. Academic Press, New York.)

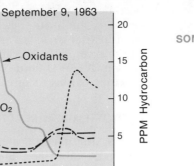

FIGURE 9a.2

Hour of day (e.s.t.)
Air pollutants during a smog day in Cincinnati.

its formation, we must consider the gases of normal, unpolluted air, which has a nearly constant composition throughout the world. Table 9a.1 gives a more extensive list of the components of normal air than was presented in Chapter 2. Of the less abundant gases present, note especially that oxides of nitrogen (N_2O and NO_2) and ozone (O_3) occur only at very low levels in normal air. Note also that nitric oxide (NO), another oxide of nitrogen, and carbon monoxide (CO) are both absent.

If the air over a large city is monitored, however, one detects significant increases in the normally occurring oxides of nitrogen, especially on a sunny day. Moreover, new substances not found in normal air appear: nitric oxide (NO), oxidants and hydrocarbons (Fig. 9a.2). These increases begin about the time of the morning traffic rush. Note that the NO (nitric oxide) level increases *before* that of NO_2 (nitrogen dioxide). The brown tint to the sky characteristic of photochemical smog is due to NO_2. About noon, there is a peak of oxidant, which is mostly **ozone,** together with several other oxidizing substances. There is a second peak of oxides of nitrogen following the evening rush hour, but no oxidants are formed at night. Hydrocarbons, products of incomplete combustion of gasoline and other fuels, also peak at the rush hours. Photochemical smog is thus produced during daylight hours, reaching a maximum around midday or early afternoon.

FORMATION OF PHOTOCHEMICAL SMOG

The basis for photochemical smog formation was first elucidated by A. J. Haagen-Smit in the 1950's. Even today, the details of all the chemical reactions are not completely understood. The reactions are extremely complex, but it is only necessary to understand a few important features to grasp how smog is formed. Point number one is that NO_2 (nitrogen dioxide) is not the main product of combustion processes. As seen in Figure 9a.2, NO (nitric oxide) is the first gas to appear in the atmosphere. It is released in auto exhausts and is the primary oxide of nitrogen produced by any combustion process. Fossil fuels—coal, gasoline, oil and natural gas—are mainly hydrocarbons. They are composed mostly of carbon and hydrogen, with a little sulfur and oxygen. When these are burned with air, the main products are carbon dioxide (CO_2) and water (H_2O). Carbon monoxide (CO) is also formed, but in much smaller quantity. However, CO levels along busy freeways may rise to dangerous levels, causing drivers to become drowsy and unalert. The production of nitrogen oxides is the result of combustion at high temperatures, such as occurs in power plants, in industrial furnaces and in the internal combustion engine of cars. Air is the source of the oxygen needed for combustion, but it also contains nitrogen. When O_2 and N_2 are mixed at high temperature and pressure they produce oxides of nitrogen, mostly NO.

PHOTOCHEMICAL REACTIONS

A summary of the reactions which produce photochemical smog is shown in Table 9a.2. Step I is the conversion of NO to NO_2. This happens in two ways. NO may be directly oxidized by oxygen or ozone. The reaction with oxygen is slow, however, and early in the day there is little ozone present to generate NO_2. An alternative for NO_2 formation is by a series of reactions involving CO, NO and O_2.

Once NO_2 is formed, the stage is set for the absorption of sunlight and the further generation of the characteristic products of smog (Step II). NO_2 is a brown gas capable of absorbing ultraviolet light energy; in doing so, it splits into NO and atomic oxygen, $O^\bullet$, which contains excess energy and is therefore very reactive. (All such reactive molecules are underlined in the outline.) Atomic oxygen quickly combines with oxygen to produce ozone (O_3). This can only happen, however, if energy-absorbing molecules are available. As luck would have it, the incompletely burned hydrocarbons (HC) from exhaust are present to facilitate the reaction. Some of these hydrocarbons are also capable of absorbing energy from sunlight and become highly reactive **free radicals** ($\underline{HC^\bullet}$).

In the next stages of smog formation (Step III), ozone and oxygen interact with hydrocarbons and free radicals in a complex series of chain reactions, only a few of which are indicated. The final products of these reactions are a variety of highly oxidized hydrocarbon molecules, far more toxic and irritating than those originally emitted in the exhausts from automobiles (Step IV). The most infamous of these are a group of com-

---

Step I: Generation of NO_2

$$NO \quad + \quad oxidant \longrightarrow NO_2$$
(from exhaust) $\quad$ (O_2 slow; O_3 fast)

$$CO \quad + \quad NO \quad + \quad O_2 \rightarrow\rightarrow\rightarrow \quad NO_2 \quad + \quad CO_2$$
(from exhaust) $\quad$ (from exhaust)

Step II: Photochemical Reactions

$$\begin{cases} NO_2 & + \text{ light} \longrightarrow NO + O^\bullet \text{(free oxygen atom)} \\ O^\bullet & + O_2 \longrightarrow O_3 \text{ (reaction facilitated by hydrocarbon)} \end{cases}$$

$$HC \quad + \text{ light} \longrightarrow HC^\bullet$$
(hydrocarbon from exhaust) $\quad$ (free radicals)

Step III: Chain Reactions

$$O_3 \; + \; HC \longrightarrow HCO^\bullet$$
$$HC \; + \; HC^\bullet \longrightarrow HC^\bullet + HC^\bullet$$
$$HC^\bullet + O_2 \longrightarrow HCOO^\bullet$$

Step IV: Final Products

PAN; aldehydes; acroleins

pounds called peroxyacylnitrates—or **PAN**s, for short. They contribute, along with ozone, to the total oxidants measured in smog.

In addition to the effects of these oxidants on plants and man, which we shall consider shortly, they also cause millions of dollars worth of damage annually by oxidizing rubber, paint and other vulnerable materials.

To summarize briefly, the primary ingredients for production of photochemical smog are oxides of nitrogen, especially NO, partially burned hydrocarbons and sunlight. The end products are ozone and irritating substances such as PAN and aldehydes. Note that ozone is not released from car exhausts, but is produced in the atmosphere. On a heavy smog day in a city such as Los Angeles, up to 90 per cent of the ultraviolet light may be absorbed by smog, thus making it difficult to acquire a suntan or to activate vitamin D, the factor which prevents rickets.

INFLUENCE OF WEATHER ON SMOG FORMATION

Meteorological conditions are an important factor in formation of both reducing and photochemical smogs. It is obvious that accumulation of pollutants in the atmosphere and reactions between them can only take place if the molecules are kept close together. Mixing with unpolluted air disperses the smog-producing gases, and reactions occur only slowly. Smog is thus most likely to accumulate when wind speeds are low. It also forms more readily when air is trapped over a city by a **temperature inversion** (Fig. 9a.3). Normally the air near gound level is warmer than the air at higher altitudes and tends to rise and be dispersed. But frequently a layer of even warmer air settles several hundred feet over a city and prevents the ground-level air from rising. This is known as a temperature inversion, and it is quite common in some parts of the world. Such an inversion may last for several days during settled conditions. The smog-producing molecules are confined to a small volume of air over the city and accumulate rapidly.

EFFECTS OF SMOG ON PLANTS

Both types of smog are capable of extensively damaging both crops and natural vegetation. Despite the enormous economic losses involved we know relatively little about how damage is produced, although the symptoms have been well documented. In reducing smog, sulfur dioxide is the most toxic substance for plants. Both broad-leaved plants and conifers are susceptible. In the United States, power plants and smelters are the main sources of SO_2, and vegetation may be severely damaged or totally destroyed over areas extending several miles. Even though the plant producing the pollution may have ceased operating, recovery is often slow, requiring decades (see Chapter 8a).

THE ROLE OF OZONE

Of the various components of photochemical smog, PAN and especially ozone are the most toxic for plants. Again, both broad-leaved plants and conifers are affected. Of all the various pollutants in photochemical smog which affect plants, ozone is most prevalent and has been most studied. For many years horticulturists and foresters observed a yellow speckling or mottling of the leaves of plants growing near large cities afflicted with smog. At first, plant pathologists suspected viruses and insects as causes, but finally the disease was traced to the smog itself—in particular, to ozone. By growing plants in greenhouses in which the quality of the air could be carefully controlled, it was possible to show that some of the ef-

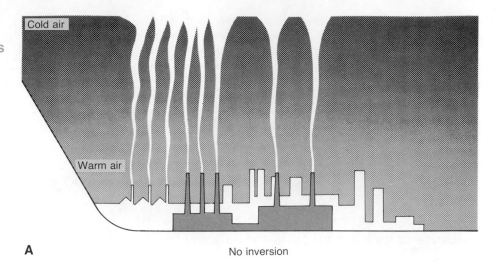

A   No inversion

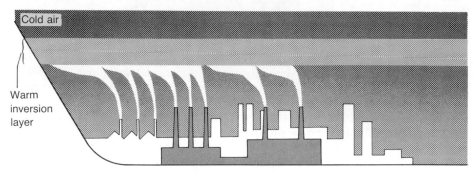

FIGURE 9a.3   **B**   Inversion

Weather effects on smog formation.
   A. Normally warm air from a city rises, carrying pollutants with it.
   B. When temperature inversions occur, a layer of warm air traps pollutants beneath.

fects observed on plants growing in smoggy environments could be exactly duplicated by adding small quantities of ozone to the pure air inside.

Damage to Crops.   Studies on broad-leaved plants, such as tobacco and bean plants, were carried out in such experimental greenhouses. It was found that at high humidities, the effects of ozone were much greater than when the air was dry — much more of the leaf was mottled. The reader will recall that the normal gas exchange in plant leaves occurs through the stomata. Oxygen is expelled during daylight hours and carbon dioxide is taken in. Water vapor also passes through the stomatal openings as the result of transpiration. The guard cells of the stomata are particularly sensitive to humidity. During dry weather they remain closed, preventing excessive water loss from the plant, whereas in humid weather they are open, permitting free entry of carbon dioxide into the leaf. Plants therefore photosynthesize at a higher rate in humid than in dry conditions.

The correlation between the size of stomatal openings and degree of ozone damage strongly suggests that ozone must enter the interior of a leaf before producing its deleterious effects. Thus conditions which would ordinarily be most favorable to growth are the most damaging when photochemical smog is present. The ozone attacks the chloroplasts of the leaf and so destroys the very means by which the plant produces energy and nutrients for its growth (see Chapter 8a).

The susceptibility of broad-leaved plants varies considerably, but many crops once grown in areas such as the Los Angeles basin—spinach and grapes, for example—can no longer be grown there. Even many of the famous citrus orchards of southern California are now severely damaged by smog—up to 50 per cent of the crop in some areas is lost owing to air pollution, and many growers find it more economical to sell their land for subdivisions, which inevitably bring more cars, more freeways and more smog!

Damage to Forests. Conifers are also affected by smog, and once again ozone has been shown to be the main culprit, although PAN is also toxic. The effects of ozone on pine trees have been particularly well documented. As with broad-leaved plants, experiments under controlled conditions in greenhouses provided the final proof necessary to pinpoint smog as the ultimate agent. Young, healthy trees exposed to concentrations of ozone similar to those occurring in forests bordering large cities show the same pathological changes. Older needles are affected first. Mottling appears especially in the more mature cells at the tip of the needle. (Pine needles grow from their base; hence the tip is the oldest part.) As the effects spread to the rest of the needle, it becomes yellow, then brown, and finally drops off.

In southern California the major cities are ringed by steep mountains covered predominantly with ponderosa pine. These forests serve several important functions for man: they provide recreational areas for the nearby urbanites; they are a source of timber; and they act as a watershed, holding rain water in the soil and preventing flash floods and erosion. Unfortunately, ponderosa is one of the pine species most susceptible to smog. In some areas more than half the pines are seriously damaged—and they will never recover. It is a common sight to see trees which have lost all but the preceding season's growth of needles. Such trees are at a serious disadvantage, for they can no longer photosynthesize at rates sufficient to meet their needs, nor can they transpire at rates sufficient to keep the sap flowing normally. They become easy targets for invasion by bark beetles, which deliver the coup de grâce which finally kills the tree. Although the ponderosa pine is the most sensitive conifer, many other species, including other pines, fir and cedar, are blighted. Ornamental pine trees grown in parks and gardens are also being destroyed.

If smog levels are not reduced, in several years the pine forests of southern California will be affected over wide areas. Unless resistant trees are planted, natural forms of vegetation less suited to the local climate and soil will replace the dead pines. As a result, the total ecology of the area will be altered—there will be fewer timber trees, more floods and a less pleasant recreational area. Nor are these changes limited to southern California; it happens to be the area where smog effects have been most studied.

EFFECTS OF OZONE ON ANIMALS AND MAN

As we have seen, smog has far-reaching effects on plants. It also affects animals, including man. City-dwellers are well acquainted with the irritating symptoms produced by PAN, aldehydes and other components of photochemical smog. The oxides of nitrogen are also a health hazard and have been shown to affect lungs adversely. But the real villain is ozone. It is about 10 times more toxic to humans than oxides of nitrogen or PAN, and its main target is the lung.

Most of our information regarding the physiological effects of ozone comes from studies on mice. Scientists distinguish two major types of effects — the immediate or acute effects resulting from breathing high levels of ozone for a short time; and long-term or chronic effects which occur after prolonged exposure to lower levels of ozone.

ACUTE EFFECTS

The initial effects of exposure to ozone are primarily on the lung, which presents an enormous surface of relatively unprotected cells. When animals or people are exposed to ozone for the first time, their rate of oxygen uptake is decreased. This is the result of two effects.

First, there is an acute inflammation of the lungs together with **edema** of the air spaces (Fig. 9a.4). Edema means an increase in water between cells. In the case of the lungs, this includes a partial filling of the tiny air sacs, or **alveoli.** The water in the alveoli decreases the efficiency of the lung by considerably increasing the distance across which gases must diffuse. The membranes of the alveolar cells appear to be weakened by ozone and permit water from the tissue to leak into them. A person with such inflamed and edematous lungs is literally "half-drowned."

The other immediate effect of ozone is an inhibition of the normal breathing reflex. Normally there is a complex feedback pathway between the muscles involved in breathing and the brain, which automatically regulates the rate and depth of breathing to meet the demands of the body. After inhaling ozone this reflex pathway is partly inhibited. Breathing becomes shallow, and the normal rate of oxygen consumption is decreased. Animals exposed to ozone are thus undergoing stress similar to that experienced at high altitudes — a decrease in the ability to obtain oxygen.

Cellular Changes. How these acute effects of ozone are brought about is only beginning to be understood. The reason for the inhibition of the breathing reflex is totally unknown, but recent investigations on chemical changes in lung tissue of ozone-exposed mice suggest a possible explanation for edema formation. In this experiment two groups of mice were studied: one group was given air to breathe; the other, air containing 0.4 to 0.7 ppm ozone. After four hours the mice were killed and their lungs removed. The tissue was extracted with a fat solvent, which removed all of the lipids from the cells, including those in the cell membranes. The lipids were then studied in a spectrophotometer, an instrument which measures the amount of light of different colors absorbed by a chemical. Each substance has its own **absorption spectrum** which permits the chemist to identify it (see Chapter 8). An absorption spectrum is rather like a chemical "fingerprint" — with the added advantage that certain changes in absorption indicate specific changes in the composition of the substance being studied.

When the researchers compared the absorption spectra of the lipids extracted from lungs of control and ozone-exposed mice they found a quite distinct difference (Fig. 9a.5A). There was an unusual hump in the spectrum of the lipids from the ozone-exposed lungs. This hump exactly resembled that found by other workers who had, for quite different reasons, exposed pure samples of **polyunsaturated fats** to highly oxidizing conditions (Fig. 9a.5B). Only polyunsaturated fats produced this peculiar spectrum after such treatment.

The scientists working with the mice put forward the following hypothesis: Polyunsaturated fats are located primarily in cell membranes; they are essential for maintaining normal membrane permeability, and thus play a role in separating fluid spaces within and between cells (see Chapter 3). (This is one important reason for obtaining enough polyunsaturated fats in one's diet.) Exposure of these lipids to ozone results in a chemical change known as "peroxidation" — extra oxygen atoms combine

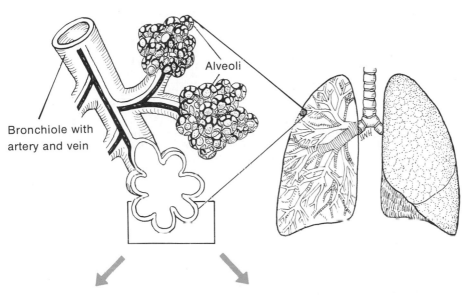

Alveoli

Bronchiole with
artery and vein

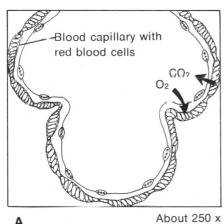

Blood capillary with
red blood cells

CO₂

O₂

A   About 250 x

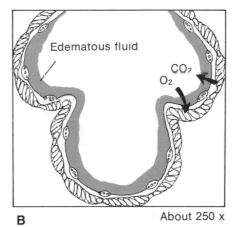

Edematous fluid

CO₂

O₂

B   About 250 x

FIGURE 9a.4

Diagram of human lung, showing circulation of air and blood to alveoli.
    A. Normal lung—gases exchange readily between air space and blood, passing
across extremely thin cells lining alveoli.
    B. Edematous lung—alveoli are partially filled with fluid, which greatly increases the
distance across which gases must diffuse, slowing down the rate of exchange.

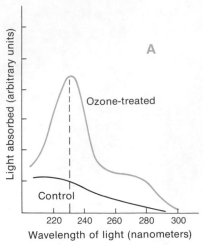

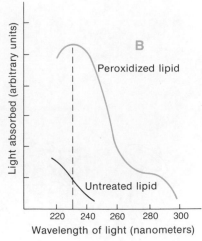

FIGURE 9a.5

Comparison of ultraviolet absorption spectra from two different experiments.

A. Absorption spectra of lipids extracted from lungs of control and ozone-treated mice. (After Goldstein, B. D., *et al.* 1969. *Archives of Environmental Health 18*:631–635.)

B. Absorption spectra of untreated and peroxidized polyunsaturated lipids. (After Bolland, J. L., and Koch, H. P. 1945. *Journal of the Chemical Society 1945*:445–447.)

At a wavelength of 230 nanometers (indicated by dashed line) the treated lipids in both graphs absorb much more light than do the untreated lipids.

with the lipid molecule because of the highly oxidizing activity of ozone, and the absorption spectrum is changed. During the peroxidation reaction, highly reactive free radicals are formed, and it is thought that these damage the cell membrane, making it much more permeable to water. In the case of lung tissue, this results in fluid passing into the alveoli, the condition known as edema.

Although the data are not yet sufficient to prove the hypothesis beyond a reasonable doubt, it is the most probable explanation for the edematous response proposed so far. It is a sad fact that this effect of ozone is the only one about which we have any very direct information as to how the effect is produced, and even that information is far from complete.

Factors Affecting Ozone Toxicity. There are three factors which intensify the acute effects of ozone. Temperature has a marked effect. An increase in temperature of 8° C (15° F) will double the toxic effect of ozone. Age is also a factor. Young, growing animals are much more susceptible than healthy adults. Exercise also has an effect. In experiments with mice, it was found that these animals could survive one part per million (1 ppm) ozone if they were allowed to remain at rest, but if forced to exercise 15 minutes per hour, the mice died. The immediate parallel with the dangers to young children is all too apparent—smog accumulates on warm, still, sunny days; the very fact of their youth makes children susceptible; children exercise spontaneously. The city of Los Angeles, where smog is common, realizing this danger, calls "school smog alerts" on days when oxidant levels are expected to reach 0.35 ppm (Fig. 9a.6). On these days, recess activities are limited to quiet play. This level does not carry a very great safety margin, however, (a factor of about 3) compared to that found lethal for exercising mice. Fortunately, high ozone levels lead to unconscious suppression of activity.

Tolerance to Acute Exposure. Tolerance to the acute effects of ozone is also known to occur. It has been found in mice that previous exposure to sublethal doses of ozone permits animals to survive concentrations which would otherwise be lethal. The physiological mechanisms underlying ozone tolerance are not understood, although several theories have been advanced. *Tolerance does not, however, prevent the development of chronic symptoms of ozone exposure.*

CHRONIC EFFECTS

Long-term or chronic pathological changes due to smog affect not only the lung but the whole body. The effect on the lung is an irreversible decrease in respiratory capacity known as **emphysema.** Emphysema results from several types of chronic insult to the lung, including asthma and heavy cigarette smoking, as well as exposure to toxic air pollutants, especially sulfur dioxide and ozone. The walls of the respiratory tubes or **bronchioles** become fibrous and constricted, so there is more resistance to the movement of air. It becomes harder to breathe. At the same time, the walls between the alveoli break down, causing a decrease in the surface available for gas exchange (Fig. 9a.7). Each breath is less effective.

Other effects of breathing ozone which have been observed in animals are changes in hair texture, stiffening of the joints and decreased visual acuity. All these symptoms are characteristic of normal aging. In short, chronic exposure to ozone seems to accelerate the normal rate of aging in experimental animals. Precisely how these effects are brought about at the cellular level is not known.

Measuring the chronic effects of ozone on humans is extremely difficult, and numerous surveys have been and are being conducted to estimate the risks to the general population. There are two enormous difficulties of the sort discussed in Chapter 1a. First, the actual degree of exposure of individuals is almost impossible to measure. Unless a person has lived for years in one place, and that place happens to be very close to an air monitoring station, no exact record of his exposure is available. The second complication is that many other factors, such as asthma, type of occupation, smoking habits, past respiratory illness and so on produce an enormous variability in the population. It becomes very difficult to assess which effects are due to ozone and which to other variables.

AIR POLLUTION AND PUBLIC HEALTH

Some scientists engaged in epidemiological research on air pollution believe that, because so many other factors affect respiratory capacity and aging processes, the slight additional effect of chronic exposure to ozone is of little consequence to the health of the population as a whole. Others disagree and believe that photochemical smog increases the incidence of respiratory illness in people who would otherwise remain healthy. Most physicians in large cities agree that air pollution is highly dangerous to those patients already suffering from respiratory incapacity from other causes. Such patients are advised to live elsewhere.

Although it has not been shown conclusively that ozone and other air pollutants have a damaging effect on the average citizen, placing some known facts side-by-side, as in Table 9a.3, gives rise to sobering thought.

The data speak for themselves. Even though it is not possible to quantify the effects due to oxidants on the health of the general public, the safety margin is obviously dangerously low. Air pollution monitoring is quite extensive in California, hence much of our data derives from that state, but cities elsewhere also suffer from high levels of either oxidants or sulfur dioxide, which have rather similar long-term effects on health.

Synergism with Other Pollutants. In this chapter we have scarcely mentioned the other toxic substances in smog—carbon monoxide, oxides

LOS ANGELES COUNTY

AIR POLLUTION CONTROL DISTRICT

434 South San Pedro Street, Los Angeles, California 90013/629-4711

LOUIS J. FULLER
Air Pollution Control Officer
ROBERT L. CHASS
Chief Deputy

To All Schools in the Los Angeles Basin

Gentlemen:

On March 3, 1969, the Los Angeles County Medical Association unanimously adopted the following resolution:

> "Because SMOG is an increasing health hazard which
> may seriously affect the lungs of young people, the
> Committee on Environmental Health of LACMA strongly
> recommends that when the forecast concentration of
> OZONE (oxidants) in the atmosphere reaches 0.35
> part per million, Los Angeles County students through
> high school, in any identified air monitoring zone,
> should be excused from strenuous indoor and outdoor
> activities. The OZONE (oxidants) information will be
> transmitted by the Air Pollution Control Officer to the
> Board of Education in the specific zones affected by
> the air pollution."

At its meeting on April 1, 1969, the Los Angeles County Board of Supervisors adopted an order directing the Air Pollution Control Officer to implement the above resolution.

The news media are the most practical method of transmitting forecast information on ozone levels. Most of the County's radio, television, and press will carry this information as a "School Smog Warning" from the time the forecast is released (afternoon before the school day), and, in addition, the following radio stations, among others, will carry the warning on their news broadcasts between the hours of 7 a.m. and 9 a.m.: KNX, KFWB, KFI, KLAC, KGFJ, KPOL, KABC, KGIL, and KHJ.

School Smog Warnings will be carried by the news media only when the Air Pollution Control District forecasts ozone levels to reach or exceed the level contained in the resolution.

FIGURE 9a.6   Copy of a letter from the Air Pollution Control Officer to all schools in the Los Angeles Basin, April, 1969.

Illustration continued on opposite page

To All Schools in the Los Angeles Basin   -2-

Any questions concerning health effects should be directed to the Los Angeles County Medical Association. Information regarding forecasts may be obtained from the District's Public Information and Education Division, 629-4711, extension 66101.

Very truly yours,

Louis J. Fuller
Air Pollution Control Officer

LJF:sw

Attachments

FIGURE 9a.6 *Continued*

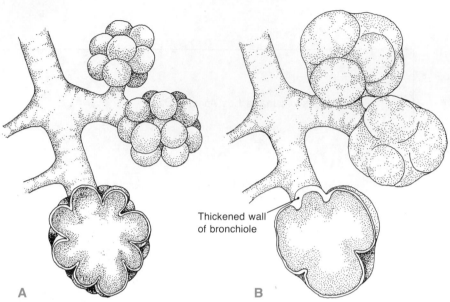

FIGURE 9a.7 A B

Diagram of human lung showing breakdown of alveolar walls during development of emphysema.
    A. Normal lung.
    B. Lung with emphysema—alveolar walls have disappeared, decreasing the surface available for gas exchange. The walls of the bronchioles have become thick and less elastic and are more resistant to the flow of air.

Lethal to mice forced to exercise 15 minutes per hour: 1 ppm
Changes in mouse lung chemistry after 4 hours: 0.4–0.7 ppm
Oxidant level for school smog alert in Los Angeles: 0.35 ppm

Maximum hourly average oxidant:*

Area	Station	May 1970	Sept. 1970
San Francisco Bay	Downtown San Francisco	0.13	0.14
	San Leandro	0.26	0.17
	Burlingame	0.24	0.29
South Coast Basin	Downtown Los Angeles	0.27	0.25
	Pasadena	0.46	0.45
	Riverside	0.35	0.51
San Diego Basin	Downtown San Diego	0.24	0.15
	El Cajon	0.26	0.21
	Oceanside	0.40	0.18

**TABLE 9a.3
SUMMARY OF
INFORMATION
ABOUT OZONE
EFFECTS AND
CONCENTRATION
LEVELS IN AIR**

*The maximum oxidant during the month recorded for a full hour (value given is average during the hour). Data from California Air Quality Data, Vol. II, 3, and Vol. II, 5. Published quarterly by the California Air Resources Board, 2180 Milvia St., Berkeley, California 94704.

of nitrogen, sulfur dioxide, PAN, lead. All are present and are acting simultaneously on the health of people. Virtually no experimental studies exist to show whether the effects of two or more toxic agents together have only additive effects or whether they act synergistically, producing an effect greater than the sum of the two independent effects.

<table>
<tr><td align="center">Additive Effect</td><td align="center">Synergistic Effect</td></tr>
<tr><td align="center">Agent A + Agent B</td><td align="center">Agent A + Agent B</td></tr>
<tr><td align="center">↘ ↙</td><td align="center">↘ ↙</td></tr>
<tr><td align="center">Effect = A + B</td><td align="center">Effect = C
[C > (A + B)]</td></tr>
</table>

Synergistic effects are common in medicine; persons ill from one disease are more susceptible to another type of infection than are healthy people. Much research on the total health effect of all air contaminants remains to be done.

SUMMARY

Two major types of air pollution are recognized. The main noxious compounds in reducing smog are sulfur dioxide and sulfuric acid, produced from combustion of high-sulfur fossil fuels. In photochemical smog, hydrocarbons and nitrogen oxides released from combustion at high temperatures and pressures react in sunlight to form toxic oxidants and irritants such as PAN. Temperature inversions hasten rates of smog accumulation.

The most toxic substance of photochemical smog, ozone, enters leaves of plants via the stomata, causing mottling. The destruction of chloroplasts leads to early aging and leaf-drop. Weakened plants are highly susceptible to other diseases. In many areas near large cities extensive destruction of crops and forests by both types of smog is occurring. Only resistant species will survive.

Smog, especially ozone, is also injurious to human health, affecting mainly the gas exchange capacity of the lungs. Breakdown of cell membranes from lipid peroxidation may be the cause of edema after acute exposure. Tolerance to such exposure gradually develops, but tolerance does not prevent such permanent changes as emphysema and early aging. The degree to which smog represents a public health hazard is disputed by scientists, but present levels of ozone in some large cities leave only a small margin of safety.

READINGS AND REFERENCES

Bolland, J. L. and Koch, H. P. 1945. Course of autoxidation reactions in polyisoprenes and related compounds: IX. The primary thermal oxidation product of ethyl linoleate. Journal of the Chemical Society (London) 1945, pp. 445–447. Studies on peroxidation of polyunsaturated fats.

Goldstein, B. D., Lodi, C., Collinson, C. and Balchum, O. J. 1969. Ozone and lipid peroxidation. Archives of Environmental Health (Chicago) 18:631–635. Investigation of effects of ozone on lipids of mouse lung tissue.

Hepting, G. H. 1964. Damage to forests from air pollution. Journal of Forestry 62:630–634. Reviews types of air pollutants damaging trees in the United States and elsewhere.

Miller, P. R. and Millecan, A. A. 1971. Extent of oxidant air pollution damage to some pines and other conifers in California. Plant Disease Reporter 55:555–559. A timely report on forest areas already damaged and those in danger from air pollution.

Miller, P. R., Parmeter, J. R., Jr., Flick, B. H. and Martinez, C. W. 1969. Ozone dosage response of ponderosa pine seedlings. Journal of the Air Pollution Control Association 19:435–438. Experiments demonstrating effect of ozone on photosynthesis and growth of pine trees.

Otto, H. W. and Daines, R. H. 1969. Plant injury by air pollutants: influence of humidity on stomatal apertures and plant response to ozone. Science *163*:1209–1210. Experiments showing that ozone must enter leaf via stomata to cause damage.

Richards, B. L., Sr., Taylor, O. C., and Edmunds, G. F., Jr. 1968. Ozone needle mottle of pine in southern California. Journal of the Air Pollution Control Association *18*:73–77. Greenhouse experiments demonstrating that ozone is the main component of photochemical smog damaging pine trees.

Stern, A. C. (ed.) 1968. *Air Pollution*. 2nd ed. Academic Press, New York. Volume 1: *Air Pollution and Its Effects*. Volume 2: *Analysis, Monitoring and Surveying*. Volume 3: *Sources of Air Pollution and Their Control*. An excellent source of information on all aspects of air pollution. The following chapters of Volume 1 are especially pertinent to this chapter:

Goldsmith, J. R. Effects of air pollution on human health.

Haagen-Smit, A. J. and Wayne, L. L. G. Atmospheric reactions and scavenging processes.

Robinson, E. Effect on the physical properties of the atmosphere.

Stokinger, H. E. and Coffin, D. L. Biologic effects of air pollutants.

Tebbens, B. D. Gaseous pollutants in the air.

Symposium on trends in air pollution damage to plants. 1968. Phytopathology *58*:1075–1113. Six articles dealing with all aspects of pollution damage to vegetation.

Westberg, K., Cohen, N. and Wilson, K. W. 1971. Carbon monoxide: Its role in photochemical smog formation. Science *171*:1013–1015. Explains how carbon monoxide accelerates ozone production.

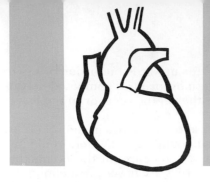

CIRCULATORY SYSTEMS

10

The constancy of the internal environment of larger organisms is provided by the circulation of body fluids which are processed by various organs.

In small organisms, or those with body tissues arranged so that all cells are near the surface, the nutrients and oxygen needed for maintenance and growth reach the individual cells by diffusion, and waste products are removed in the same manner. In order for certain organisms to obtain the advantages of larger size, it has been necessary for them to evolve internal circulatory systems to supply the needs of their innermost cells, thereby supplementing diffusion processes.

CIRCULATION IN HIGHER PLANTS

Primitive plants, even the giant seaweeds, transfer substances directly from cell to cell, but the tracheophytes—ferns, conifers and flowering plants—have specialized conducting tissues through which water and solutes are distributed to the various organs. There are two types of conducting pathways: the *xylem,* which transports water and mineral nutrients upward from the roots to the rest of the plant, and the *phloem,* whose function is to conduct the products of photosynthesis from the leaves to regions where it is needed, such as buds, flowers, fruit and roots (see Figure 5.10).

THE ASCENT OF SAP

The inner ring of conducting tissue in a plant stem, the xylem, has long been known to conduct water and minerals upward from the roots, but the driving force, until recently, remained a mystery. It is estimated that a pressure about 30 times greater than that of air pressure at sea level is required to push water through a narrow xylem tube to a height of 100 meters—that is, from the ground to the top of a giant redwood tree. Studies on the structure and function of the cells forming the xylem tubes showed that these cells need not be alive for upward conduction of sap to occur and, in fact, mature xylem tissue in most plants is composed only of non-living cell walls. Therefore, forces either in the roots or in the leaves must drive the sap upward.

One might suppose that the concentration of solutes in the root cells, being generally greater than that in the surrounding soil water, would pull water inward through osmotic diffusion. To some extent this occurs, but the osmotic pressures so developed in root cells are never great enough to push water a great distance up the stem, and may function only minimally in the roots of plants, such as mangroves, which are immersed in solute rich sea water.

The most likely explanation for the rise of sap lies in the powerful attractive forces between the water molecules themselves, which has given rise to the **cohesion theory.** According to this theory, the driving force for upward movement of water in the xylem arises from **transpiration** in the leaves (Fig. 10.1). As water passes out through the stomata, vapor pressure inside the leaf falls slightly and more water molecules evaporate from the cell surfaces. Evaporation of water molecules occurs mainly during the day, when the stomata are open and the sun's heat increases evaporation rates.

It is thought that the water molecules in the cells of the leaf, in the xylem tubes and in the cells of the root form a continuous column of liquid water. Therefore, as water molecules evaporate in the leaves, cohesive tension is developed in the entire water column, pulling the remaining water molecules together. This powerful cohesive force pulls water from the roots upward to the leaves. More water enters the root hairs osmotically from the soil. Measurements of the cohesive forces between water molecules show that they are, in fact, great enough to drive water to the top of even the tallest tree.

If gas spaces were to occur in the system, breaking the water column, then the ascent of sap would stop. Tiny bubbles do form occasionally, but the structure of xylem is such that they do not ordinarily break the water column. The cell walls of xylem contain minute pits, connecting them with parallel tubes. When bubbles form, they are trapped above the pits, leaving the water column intact. The pits thus act as check valves.

Whether transpiration also plays a direct role in transport of nutrients from soil to leaves is still debated by plant physiologists. Some think that the role of transpiration is only secondary and that nutrient transport can occur, at least to some extent, in the absence of transpiration. Nevertheless, transpiration is essential for growth in most plants, including many major crops. The reason is that the stomata of the leaves must remain open for gas exchange to occur. Water lost by evaporation through the open stomata must be replaced by transpiration, or else the guard cells will close and photosynthesis will cease (see Chapter 9).

TRANSLOCATION OF ORGANIC MOLECULES

The **translocation** to other parts of the plant of organic molecules synthesized in the leaves occurs mainly via the tubular cells of the phloem. Unlike xylem, the conducting cells of phloem tissue remain alive, but the nucleus and vacuole disappear. It is possible that the phloem tube cell is maintained by its smaller companion cells (Fig. 10.2). The end walls of the conducting cells become modified into **sieve plates**, which connect with the cell above and below.

The problem of how fluid moves through phloem tissue is almost as puzzling as that for xylem. The most likely theory is that bulk water flow occurs, driven by osmotic pressure differences. Bulk water flow implies a current of water, much like that in a stream, in which dissolved molecules are carried at the same rate as the water molecules. According to this

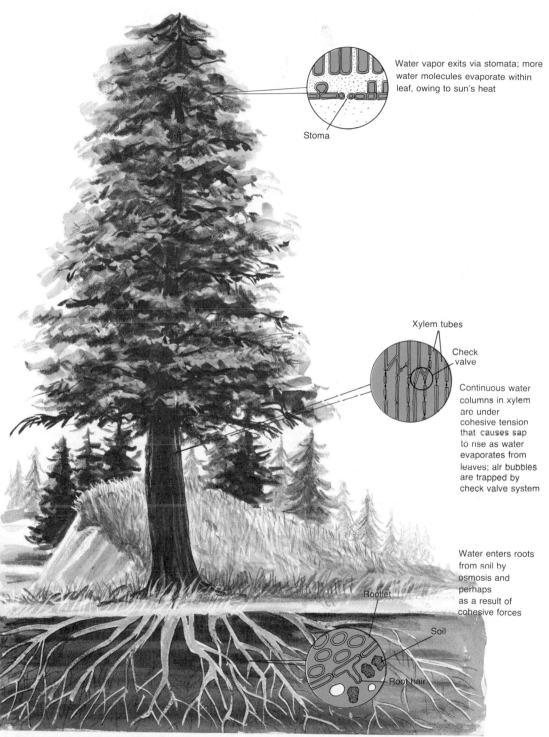

Water vapor exits via stomata; more water molecules evaporate within leaf, owing to sun's heat

Stoma

Xylem tubes

Check valve

Continuous water columns in xylem are under cohesive tension that causes sap to rise as water evaporates from leaves; air bubbles are trapped by check valve system

Water enters roots from soil by osmosis and perhaps as a result of cohesive forces

Rootlet

Soil

Root hair

The cohesion theory of the ascent of sap. Water is indicated by color; solid color—liquid state; dots—vapor state. Note that there is a continuous liquid water column from roots to leaves.

FIGURE 10.1

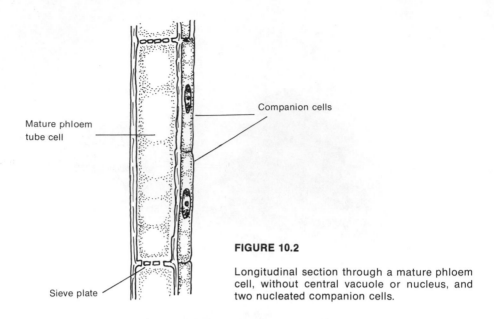

Companion cells

Mature phloem
tube cell

Sieve plate

FIGURE 10.2

Longitudinal section through a mature phloem
cell, without central vacuole or nucleus, and
two nucleated companion cells.

theory, osmotic pressure in the "producer" cells of leaves is high due to accumulation of sugars, amino acids and other products of photosynthesis.
By contrast, osmotic pressure is relatively low in the "consumer" cells of
buds, fruit or roots, where small organic molecules are either oxidized during respiration or synthesized into polymers which contribute little to osmotic pressure. Water thus enters the producer cells osmotically from the
xylem. The resulting hydrostatic pressure within these cells pushes water,
together with its dissolved molecules, into the phloem. It then travels to any
region where osmotic pressure and hence hydrostatic pressure is lower
(Fig. 10.3). Note that fluid in the phloem contains far more dissolved particles than does that in the xylem.

In addition to distributing water and nutrients, the circulatory system
of plants transports chemical messengers or **hormones** from one part of the
plant to another; these serve to integrate the growth and metabolism of its

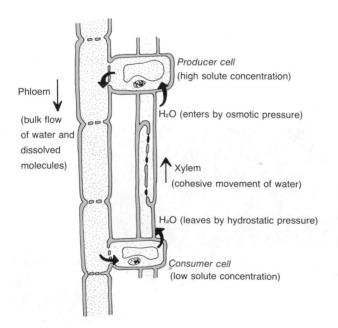

Producer cell
(high solute concentration)

Phloem

(bulk flow
of water and
dissolved
molecules)

H₂O (enters by osmotic pressure)

Xylem
(cohesive movement of water)

H₂O (leaves by hydrostatic pressure)

Consumer cell
(low solute concentration)

FIGURE 10.3

Bulk flow theory of transport in phloem.
Producer cells photosynthesize many
small molecules, causing osmotic entry of
water from xylem. Their increased hydrostatic
pressure squeezes water and solute into
phloem, through which it travels to consumer
cells where small molecules are removed,
reducing their osmotic pressure. Water returns to xylem, which has a low hydrostatic
pressure owing to internal cohesive forces.

various organs. We shall have more to say about plant hormones in Chapter 14.

CIRCULATION IN ANIMALS

The main function of circulatory systems in animals is the maintenance of a constant internal environment for the body cells, supplying them with oxygen and nutrients and removing their wastes. To perform this function, the blood itself is processed by many organs—the gut, liver, lungs, kidneys, to mention but a few. To integrate the metabolic activities of the various organs, the blood also transports hormones, whose functions we shall consider shortly. In many animals, the blood also acts as a first line of defense against foreign organisms such as bacteria and other parasitic organisms. Before discussing these various functions, a few words about the anatomy of circulatory systems and the composition of blood will be useful.

ANATOMY OF CIRCULATORY SYSTEMS

Circulatory systems fall into two anatomical categories, as shown schematically in Figure 10.4. Each consists of a muscular pumping organ, the heart, of arteries which distribute blood to the tissues, and of veins which collect it and return it to the heart. In an **open circulatory system,** blood flows in large sinuses which bathe the tissue cells directly. Such systems are found in most arthropods, in many molluscs and in a few other animals. Because the pressure generated during contraction of the heart is quickly dissipated in such an open system, the rate of circulation is rather slow. In more active animals possessing such a system—crabs and lobsters, for example—contraction of the locomotory muscles assists in the circulation of the blood.

In **closed circulatory systems,** characteristic of segmented worms, cephalopods and vertebrates, blood remains confined within minute, thin-walled capillaries as it passes through the tissues. The enormous surface area made available by the numerous small capillaries is essential for the efficient exchange of molecules between blood and cells.

The contraction of the heart is capable of generating high hydrostatic

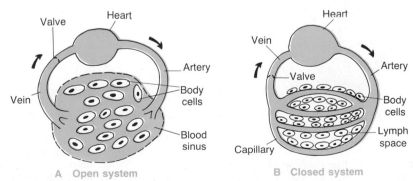

A Open system   B Closed system   **FIGURE 10.4**

Schematized diagrams of the two main types of circulatory systems in animals.

A. Open system. The heart pumps blood via arteries into a large blood sinus which bathes all cells. Return is via veins bearing valves. The veins are often very short or absent, the blood draining directly from the sinus to the heart.

B. Closed system. The blood remains in minute, thin-walled vessels, the capillaries, and never comes into direct contact with body cells. Lymph occurs between the cells.

pressures in closed systems, which have therefore evolved thick-walled arteries to withstand the force. Arteries are also elastic, however, acting to dampen the initial pressure after each heartbeat and to sustain the pressure over a longer period. In arterial hardening, this elasticity is lost, and dangerously high peripheral blood pressures may result. The veins, by contrast, are thin-walled and, in both open and closed systems, characteristically contain valves to prevent backflow of blood.

The heart is located so that it immediately receives blood from or sends it to the respiratory organ. The next organs in the circulatory pattern are those whose tissues require highly oxygenated blood, such as the nervous system. Thus the overall anatomy of the circulatory system insures that the highest concentrations of oxygen are received by the tissues which need it most. Among the more active vertebrates, which have a high oxygen requirement, the heart acts as a double pump, circulating blood in two separate circuits, one through the lung and one through the rest of the body. In birds and mammals, the heart is divided internally into two completely separate halves, thus insuring that the two circulations do not mix and that blood is fully oxygenated as it circulates to the body tissues (Fig. 10.5).

The hearts of many animals are interesting in their ability to initiate

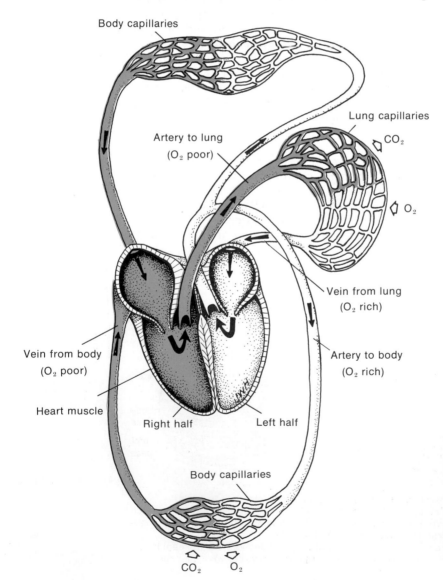

Body capillaries

Lung capillaries

CO_2

Artery to lung
(O_2 poor)

O_2

Vein from lung
(O_2 rich)

Vein from body
(O_2 poor)

Artery to body
(O_2 rich)

Heart muscle

Right half   Left half

Body capillaries

CO_2 O_2

FIGURE 10.5

Simplified diagram of human circulation. The heart (shown in front view) is divided into two parallel halves. The right half pumps oxygen poor blood to the lungs; the left half pumps oxygen rich blood to the body. (After Kogan, B. A. 1970. Health: Man in a Changing Environment. Harcourt Brace Jovanovich, New York.)

their own contraction without benefit of a nervous stimulus. Such **myogenic contractions** may continue for hours in a frog heart after its removal from the body. Although nerves to the heart regulate the rate and force of its contraction, the ability to initiate its own beat is of fundamental importance in the hearts of all vertebrates, including man. When this function fails, as occurs in some human heart diseases, an electronic device can be worn which artificially stimulates the heart. In emergency, heart massage may be employed to induce the return of normal myogenic contractions.

COMPONENTS OF BLOOD

Blood is a complex mixture of many substances, and its composition varies from one animal to another. In most animals, specialized cells circulate within the blood, in which case the non-cellular, fluid fraction is called **plasma.** In the plasma are dissolved a great variety of ions and molecules, and after a meal it may also contain a suspension of tiny lipid droplets. The major components of human blood and the primary functions they perform are listed in Table 10.1.

The space between the tissue cells of animals with closed circulatory systems is filled with a fluid known as **lymph** (see Figure 10.4). In vertebrates, its composition is very similar to that of blood plasma, except that it contains much less protein. We shall have more to say about lymph circulation shortly.

MAINTENANCE OF BLOOD VOLUME

Any excessive decrease in blood volume will lead to a drastic fall in blood pressure and a consequent loss of effective circulation. Hemorrhaging from damaged blood vessels is a not infrequent occurrence in animals, and two types of clotting have been evolved by animals to prevent excessive fluid loss in these circumstances.

Blood Clotting. The most primitive type of clotting, common among invertebrates, is cellular clotting. Since many invertebrates depend upon their coelomic fluid to act as a hydrostatic skeleton for locomotion, maintenance of its volume is of importance to their survival; clotting of both blood and coelomic fluid occurs in these animals. Cells resembling small amoebae, and therefore known as **amoebocytes,** normally occur singly in the blood and coelomic fluid (Fig. 10.6A). When wounding occurs, the surfaces of these cells undergo a remarkable change. They become "sticky" and adhere both to each other and to the edges of the wound, forming a clot (Fig. 10.6B).

The clotting of vertebrate blood, on the other hand, takes place in the plasma. The clotting of mammalian blood has been studied most. During

TABLE 10.1 MAJOR COMPONENTS OF HUMAN BLOOD	Component	Function(s) Performed
	Cells and cell fragments	
	Erythrocytes	Oxygen transport
	Leucocytes	Defense against infection
	Platelets	Blood clotting
	Plasma	
	Water and salts (mainly Na^+ and Cl^-)	Ion and water balance
	Small organic molecules absorbed from gut (glucose, amino acids, lipids)	Cell nutrition
	Nitrogenous waste products (ammonia, urea, uric acid)	Waste disposal
	Proteins	{ Maintenance of blood volume Clotting Immunity

clotting, a series of changes takes place which ultimately results in the conversion of a globular protein, **fibrinogen,** into an extended protein, **fibrin.** Cross-linkages formed between fibrin molecules produce the clot. An outline of the reaction sequence is shown in Figure 10.7. The initial stimulus, or trigger, is a wound-induced breakdown either of tissue cells or **blood platelets,** which are tiny non-nucleated cell fragments circulating in the blood. These damaged cells release an enzyme, **thrombokinase,** which catalyzes the conversion of a plasma protein, **prothrombin,** into **thrombin.** Like thrombokinase, thrombin also has enzymatic activity, converting fibrinogen to fibrin. Both of these enzymatic reactions require calcium ion (Ca^{++}), which is therefore essential for clotting.

The clotting of mammalian blood is really far more complex than we have shown it. A variety of factors is required for the process, and many of them are not yet chemically identified. The absence of any one of several such factors leads to a form of the inherited disease, **hemophilia,** characterized by excessive bleeding (see Chapter 15).

In some people, spontaneous clots, called **thromboses,** may form if platelets are accidentally damaged by rough regions on the walls of blood vessels. This frequently occurs in diseases such as **arteriosclerosis** (hardening of the arteries). If such clots block circulation through a blood vessel supplying an important organ, such as the heart or brain, severe impairment of function may result. Thromboses can thus be a cause of heart attacks and strokes.

Blood clotting serves several important functions: it stops bleeding; it closes a wound and prevents bacterial entry into the exposed tissues; also forms a temporary framework into which cells can migrate to heal the wound.

The Role of Plasma Proteins.   Aside from fibrinogen, the major plasma proteins in man are albumins and globulins, which together constitute between seven and eight per cent by weight of the plasma. Their contribution to the total osmotic pressure of blood is small but, as we shall see, very important. During the normal circulation of blood through tissue capillaries, the relatively high blood pressure forces water and *small* dissolved mole-

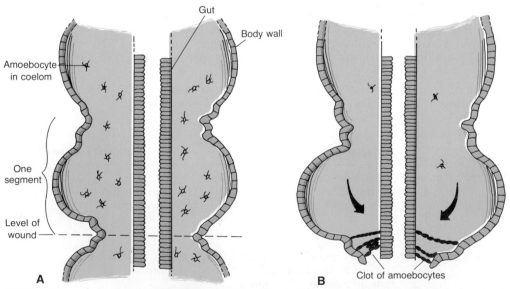

FIGURE 10.6   Diagram of cellular clotting in a segmented marine worm, drawn in longitudinal view.
A. Intact animal: amoebocytes are isolated in coelom.
B. After the worm loses its rear segments (perhaps to a predatory sea-bird) the amoebocytes migrate to the wound, adhering to each other and to the surrounding tissues. Undue loss of coelomic fluid (color) is thus prevented.

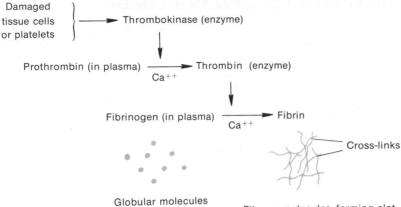

FIGURE 10-7

Sequence of events during clot formation in mammalian blood. Once damaged cells release thrombokinase, all further reactions occur in the plasma. Cross-linkages between elongate fibrin molecules result in clot formation.

cules, but *not* the larger proteins, through the capillary walls into the lymph space between the cells. The fluid so lost from the blood is returned to it in two ways.

Lymphatic fluid from the tissue spaces eventually collects within the lymphatic vessels, which structurally resemble veins. These vessels, from various parts of the body, join together to form the lymphatic duct, which empties into the venous blood system just before it enters the heart. The flow of lymph is relatively slow, however, and by itself cannot compensate for the loss of blood volume in the capillaries.

A large fraction of the fluid squeezed from the blood in the capillaries is returned to it by the osmotic effect of the blood proteins, as shown in Figure 10.8. At the arterial end of the capillary, blood pressure exceeds the osmotic pressure difference between blood and lymph, and fluid is lost. As the blood loses fluid, however, its protein concentration rises, increasing the osmotic pressure of the blood. At the same time the hydrostatic pressure falls. Toward the venous end of the capillaries the total pressure differential is reversed and fluid re-enters the blood. If the liver, which produces the plasma proteins, is diseased, if the kidneys are damaged and plasma proteins are lost in the urine or if blood pressure is abnormally high, then the osmotic effect of the proteins is insufficient to restore blood volume. The net effect is a reduced circulatory efficiency and an increase in tissue fluids, known as *edema.*

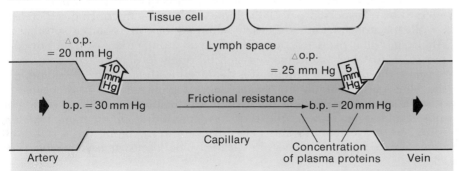

FIGURE 10.8

Diagram of effects of blood pressure and osmotic pressure on water movements across the walls of capillaries. (Relative osmotic pressures are shown in color.)

Blood normally has a slightly higher osmotic pressure than surrounding tissue fluids (Δ o.p. = 20 mm Hg) tending to draw water into the capillaries. But at the arterial end of the capillary, blood pressure tending to force fluid outwards (b.p. = 30 mm Hg) more than compensates for the inward osmotic pressure and the net effect is movement of fluid outward.

Proteins remain behind, however, causing an increase in the Δ o.p. to 25 mm Hg. Meantime blood pressure falls to 20 mm Hg due to frictional resistance in the capillary. The net pressure difference is now reversed and fluid flows inward.

THE CONSTANT INTERNAL ENVIRONMENT

The internal environment of most animals is generally maintained within fairly narrow physicochemical limits which may differ greatly from that of the external surroundings. These limits are not the same for all animals, and each species tends to regulate the composition of its blood and other body fluids to meet the conditions for which its cells are best adapted. This characteristic maintenance of a constant internal set of conditions is known as **homeostasis** (Greek *homoios*, "the same," plus *stasis*, "standing"). In order to achieve this constant internal environment, the blood must be acted upon by various organs. In the discussion which follows we shall be mainly concerned with human blood, noting, however, important variations found in other animals.

SALT AND WATER BALANCE

The major constituents of blood plasma are water and ions, especially sodium (Na^+) and chloride (Cl^-) ions. Almost the entire osmotic pressure of the blood and body fluids is, in fact, due to ions. Although their internal composition may be quite different from that of blood, the total osmotic pressure within animal cells is almost always the same as that of the blood. Hence there is little tendency under normal conditions for water to enter or leave animal cells, in strong contrast to the situation in plants. If the proportion of salt relative to water in the blood increases, however, as occurs in man when he becomes dehydrated, then the body cells tend to shrink. On the other hand, the cells of a marine animal placed in fresh water tend to swell unless the animal is able to remove the extra water from its blood and other body fluids. We shall consider the means by which animals regulate the salt and water balance of their blood in Chapter 11.

GAS TRANSPORT

Supplying oxygen to cells and removing carbon dioxide is an important function of the blood of most animals. Let us consider oxygen transport first. The amount of oxygen which will dissolve in water or blood is limited to about 0.6 volumes per 100 volumes of fluid. Even in rapidly circulating blood, such a low concentration is sufficient to meet the metabolic needs of only the most sluggish, inactive animals. The cells of active animals require far more oxygen, and for this purpose animals have evolved **respiratory pigments** to facilitate the transport of oxygen. The respiratory pigment of man, for example, permits about 20 volumes of oxygen to be transported per 100 volumes of blood—an enormous increase!

Respiratory pigments, as their name implies, are colored substances. All are globular proteins containing a metal atom, either iron (Fe) or copper (Cu), to which molecular oxygen becomes loosely attached. In some animals, the respiratory pigment is dissolved in the plasma. An example is the blue, copper-containing **hemocyanin** of molluscs and crustaceans. In other animals, the pigment is enclosed within specialized blood cells. The most familiar example is the red, iron-containing pigment **hemoglobin**, found in the vertebrates and some invertebrates. Among the vertebrates, hemoglobin is contained within the red blood cells or **erythrocytes** (Fig. 10.9A).

In hemoglobin, iron is not directly attached to protein. Instead, it is associated with a special group within the molecule, known as **heme.** Iron-containing groups similar to the heme of hemoglobin also are found in the cytochromes of electron transport chains. A schematic representation of part of a vertebrate hemoglobin molecule is shown in Figure 10.9B. A complete vertebrate hemoglobin molecule is made up of four such units, and thus has four iron atoms capable of binding oxygen. When oxygen binds with the metal atom of a respiratory pigment, however, there is no oxida-tion-reduction reaction as occurs in cytochromes; no energy is released in the process.

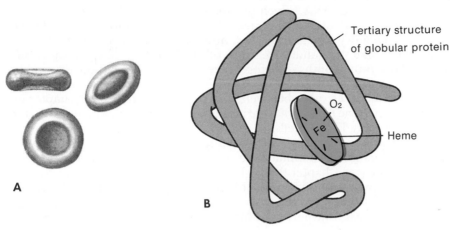

FIGURE 10.9

Oxygen transporting system in mammals.
   A. Mature circulating mammalian erythrocytes have no nuclei. They are flattened, slightly concave cells. Their shape thus provides a large surface area for gas exchange. (From McCauley, W. J. 1971. *Vertebrate Physiology.* W. B. Saunders Co., Philadelphia.)
   B. One of the four subunits of a mammalian hemoglobin molecule. The special heme group, containing an Fe atom, lies near the center. It loosely binds molecular oxygen. Note the specific tertiary folding of the protein part of the hemoglobin.

In lungs or gills, where the oxygen concentration is high, O_2 molecules readily associate (combine) with the respiratory pigment. At the tissues, however, the oxygen concentration is low as a result of cellular respiration, and the pigment readily dissociates (gives up) its oxygen (Fig. 10.10).
   Respiratory pigments frequently have different colors depending on whether or not they are combined with oxygen. Hemoglobin, for example, is red when oxygenated, and purplish-blue when unoxygenated; arterial blood is thus red, and venous blood bluish in color. People suffering from blood diseases in which hemoglobin is insufficiently oxygenated may show a bluish tinge to their skin. "Blue babies" are blue due to imperfect separa-tion of the two sides of the heart (see Figure 10.5). Only part of their blood

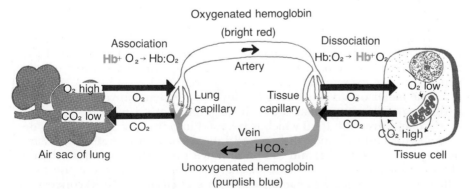

FIGURE 10.10

The transport of blood gases in man.
   Association of oxygen and hemoglobin occurs in the lungs, where oxygen concen-tration is high (Hb:O_2). Arterial blood is bright red.
   Dissociation occurs at the cells, where oxygen concentration is low. Venous blood is purplish blue due to the presence of unoxygenated hemoglobin (Hb). Carbon dioxide released from cells is transported mainly as dissolved bicarbonate ion (HCO_3^-).

is circulated through the lungs; the rest, still unoxygenated, crosses to the left side of the heart and is sent to the body without oxygen.

In man, hemoglobin is synthesized within the erythrocytes, of which there are about five million per cubic millimeter (1 millimeter = 1/25 inch). These cells are produced in bone marrow. Initially each red cell has a nucleus, containing DNA, which produces messenger RNA, which in turn directs the synthesis of hemoglobin in the cytoplasm. Before being released into the circulatory system, however, human red blood cells extrude their nuclei and, like the enucleate half of an amoeba, have a short life span—about four months—during which they cannot synthesize new proteins. Circulating erythrocytes are essentially a bag of hemoglobin, serving mainly to transport oxygen between lung and tissue. In the next chapter we shall consider some of the ways various environmental pollutants affect the functioning of red blood cells.

There are two ways in which the amount of oxygen reaching the tissues can be increased. An immediate change can be produced by increasing breathing rate (see Chapter 9) and heart rate. If demand for extra oxygen is prolonged over several days, as at high altitude, more red blood cells are formed in the bone marrow.

The other gas transported by blood is carbon dioxide (CO_2), a product of cellular respiration. After diffusing from the cells to the blood, it is mostly transported to the lungs as bicarbonate ion:

$$CO_2 + H_2O \longrightarrow H_2CO_3 \longrightarrow H^+ + HCO_3$$

| carbon dioxide | water | carbonic acid | hydrogen ion | bicarbonate ion |

The acidic hydrogen ions are neutralized by negatively charged regions of plasma proteins. This neutralizing effect of the blood is called its **buffering capacity;** the acidity of the blood is thus kept constant. In the lungs the reactions just shown are reversed, releasing free CO_2 which is exhaled. A small fraction of blood CO_2 is bound directly to hemoglobin and is carried by it from the tissue cells to the lungs.

NUTRIENT AND WASTE TRANSPORT

The blood may be likened to the trucks on a freeway which transport food from farms to hungry people and remove their garbage to a convenient dump. Nutrient molecules—sugars, amino acids and lipids—absorbed by the gut are dispersed to cells via the circulatory system. After eating, the levels of these in the blood, especially of lipids, may rise temporarily, but, as we shall see shortly, the concentrations tend to be closely regulated by homeostatic mechanisms. In a similar way, the waste molecules of metabolism—carbon dioxide and certain nitrogen-containing compounds—are removed from cells and transported to the lung and kidney respectively. The nature of these nitrogenous wastes and the means by which they are excreted are dealt with in Chapter 11.

The major organ controlling the level of nutrients in vertebrate blood is the liver. This important organ in fact has many functions. The liver forms **bile,** a greenish fluid which is released into the gut. Bile contains detergents for emulsification of ingested fats, detoxified waste products and poisons (see Chapter 3a) and the breakdown products of worn out hemoglobin. Synthesis of most of the plasma proteins also occurs in the liver. The liver is the major site of **intermediary metabolism**—the interconversion of fats, amino acids and proteins described in Chapter 8. It also functions to convert ammonia removed from amino acids into less toxic waste products, such as urea. The liver is a major site of storage of carbohydrate in the form of glycogen, and thus plays an important role in maintaining the level of one of the most important nutrients of the blood, glucose.

Each organ modifies its level of activity in response to changes in those blood constituents which it processes. In addition, the function of particular organs can be affected by chemical regulators circulating in the blood—the **hormones** (Greek *hormōn*, "that which excites"). The regulation of blood sugar levels provides an excellent example of the interaction of several such homeostatic regulators.

The nõrmal concentration range of blood sugar in man lies between 70 and 100 milligrams of glucose per 100 milliliters of blood. The level may rise briefly after a meal and may decrease briefly after hard exercise. As blood sugar is used up between meals by cellular respiration, the level is continuously replenished from the two main carbohydrate storage centers of the body, liver and muscle. Liver glycogen is mainly responsible for maintaining blood sugar, while muscle glycogen serves as a reserve for ATP synthesis during muscular work. The blood sugar level in man is under the control of no less than four different **endocrine glands,** (Greek *endon*, "within," plus *krinein*, "to separate"; hence "internally secreting"). Each gland produces its own hormone(s). The major endocrine organs in man are shown in Figure 10.11. All but the parathyroid, which regulates calcium and phosphate metabolism, and the gonads are known to produce hormones affecting blood sugar level.

THE HOMEOSTATIC REGULATION OF BLOOD SUGAR LEVELS

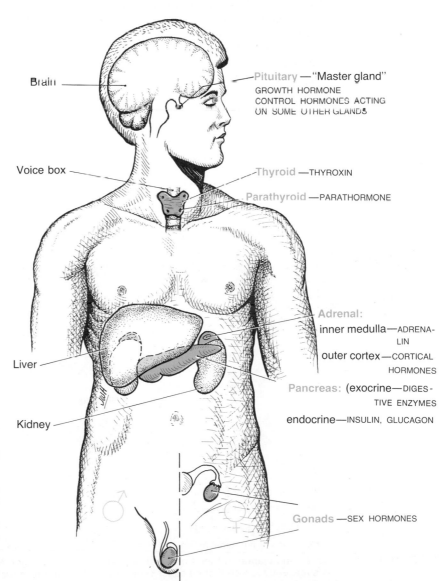

Brain

Pituitary —"Master gland"
GROWTH HORMONE
CONTROL HORMONES ACTING
ON SOME OTHER GLANDS

Voice box

Thyroid —THYROXIN

Parathyroid —PARATHORMONE

Liver

Adrenal:
inner medulla—ADRENA-
LIN
outer cortex—CORTICAL
HORMONES
Pancreas: (exocrine—DIGES-
TIVE ENZYMES
endocrine—INSULIN, GLUCAGON

Kidney

FIGURE 10.11

The most important endocrine organs in man are indicated on the right side of the diagram, together with the hormones they produce. Note that the pancreas produces both digestive enzymes and hormones. Organs indicated on the left are added for reference orientation.

Gonads —SEX HORMONES

Several hormones act to increase blood sugar levels. The most important of these is **adrenalin**, produced by the **adrenal medulla,** which causes both liver and muscle to release glucose following glycogen breakdown. Adrenalin is released when an animal is alarmed—when it is angry or frightened—and is thus often called the "fight-or-flight" hormone. In addition to increasing blood sugar, it causes increases in pulse rate, breathing rate and blood pressure, all necessary for unusual exertion.

Some of the several hormones produced by the **pituitary gland** also increase blood sugar. One of these, growth hormone, acts directly on the liver, while two others act indirectly by stimulating intermediate endocrine glands, the **thyroid** and the **adrenal cortex.** The hormones of these latter two glands, **thyroxin** and **cortical hormones,** respectively, cause fat and protein breakdown in various parts of the body. The resulting fatty acids and amino acids are then converted to glucose in the liver.

The **pancreas** is the only endocrine gland which produces a hormone decreasing blood glucose, namely **insulin.** Insulin increases the rate of glucose uptake by both liver and muscle cells, where it is converted into glycogen. When too little insulin is produced by the pancreas, the blood sugar rises and the disease **diabetes mellitus** (meaning excess sweet urine) results. Normally the kidney does not permit glucose to be excreted in the urine, but the extremely high blood sugar levels associated with diabetes exceed the capacity of the kidney to retain it, and some sugar leaks out into the urine. Diabetes mellitus is usually an inherited disease; prior to the discovery of insulin and the introduction of replacement therapy, it was often fatal, although the course of the disease was generally slow.

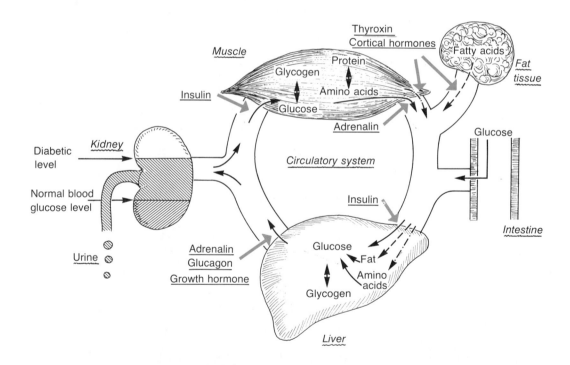

FIGURE 10.12   Schematic diagram of the hormonal regulation of blood sugar. Normally blood glucose is sufficiently low so that no sugar overflows into the urine. In the absence of insulin, the levels increase, resulting in diabetes mellitus.

To complete the picture, a second hormone produced by the pancreas should also be mentioned: **glucagon.** This hormone has an effect opposite to that of insulin, causing a release of glucose into the blood from the liver. A summary of all these complex interactions is shown in Figure 10.12.

The control of blood sugar level, then, is an extremely complicated and delicately balanced process. Which hormones are secreted at a given time depends not only upon the blood sugar level at the moment, but upon a variety of other factors, such as the nutritional state, the degree of stress, the age of the organism and so on. Such complexity is characteristic of all the homeostatic mechanisms acting on components of blood. Many feedback mechanisms between the various endocrine glands and other body organs exist to maintain the constant internal conditions on which the cells of organisms depend.

DEFENSE FUNCTIONS OF BLOOD

Bacteria, viruses, yeasts, molds and parasitic animals all require a source of organic food, and the blood and tissues of animals provide ideal sources. Any wound in the gut, lungs or skin provides ready access for these organisms and, once inside, their rapid growth at the expense of the host tissues would soon lead to death of the host were it not checked. The blood serves as the main line of defense against disease-causing organisms.

PHAGOCYTES

The immediate defense against such invaders is the phagocytic activity of blood cells. Although not all blood cells are phagocytic, most animals contain at least some cells of this type in their body fluids, and similar cells are often scattered throughout other tissues of the body. These are the **amoebocytes** of invertebrates and the white blood cells, or **leucocytes,** of vertebrates. They have the ability to recognize a foreign object in the body as well as dead cells of the animal itself. Because these cells ingest invading organisms in much the same way as an amoeba engulfs its food (see Figure 9.10), they are known as **phagocytes.**

Phagocytes tend to collect near a wound (Fig. 10.6), forming a barrier against invasion of the body by infectious agents. In man, the alveoli of the lungs are well supplied with phagocytes called **dust cells.** These remove dust, bacteria, particles from cigarette smoke and other foreign substances not already trapped in the nose or on the mucus lining of the air ducts leading into the lungs. The walls of the mouth, nose and digestive tract are also well supplied with phagocytic cells which prevent invasion of the body by infectious organisms. The engulfed bacteria or other foreign material is digested by the phagocytes to simple, harmless compounds which may then be excreted. In some cases too much material must be processed and an abscess occurs. Many overworked phagocytes die and accumulate as pus, which is sometimes extruded from the body surface.

IMMUNITY

A second line of defense against foreign substances, the immune response, depends on the formation of specific proteins, called **antibodies.**

These globular proteins circulate in the blood and immobilize the infectious agent, facilitating its removal. How antibodies are formed is one of the most elusive questions in modern biology, despite an enormous amount of research. We can, however, outline some of the known steps in the process.

Antibody Formation. All foreign proteins and certain other molecules foreign to the body have the property of inducing antibody formation and are therefore called *antigens.* Examples are the polysaccharides of bacterial cell walls, the proteins of viruses, the enzymes added to detergents and the proteins of a kidney or heart transplanted from one person to another. Only identical twins have identical proteins, the proteins of other people being sufficiently different to cause an immune response.

Each antigen induces the synthesis of a *specific* antibody in two of the tissues where white blood cells are produced, the *spleen* and the *lymph nodes.* These organs are associated with the lymphatic system. The spleen lies near the stomach, and lymph nodes are scattered throughout the body. At least two different groups of white blood cells are involved: one which picks up the antigen in the first place, and a second which makes the antibody. How these two cell types interact is still a mystery.

Antibody Function. Each antibody is highly specific for a particular antigen, just as an enzyme is for its substrate. This suggests that the antibody has a specific region of recognition where it binds with the antigen, forming an antigen-antibody complex. In fact, it is now believed that most antibodies bear two such specific sites (Fig. 10.13A).

When antibodies bind with the antigenic regions of a foreign organism one of several things may happen. Sometimes **agglutination** takes place, as shown for the bacterial cells in Figure 10.13B. In other cases, the antibodies cause the foreign cells to disintegrate or *lyse.* Antigen-antibody complexes also often facilitate phagocytosis of the invader by white blood cells.

Antibodies are formed during the first exposure to an infectious agent, but it may take several days before the concentrations of antibody are high enough to control the disease. Upon a second exposure, the body may "remember" its past contact, and within a day or so, before the disease can spread, high levels of antibodies are found in the blood. One has built up an *immunity.* After once having chicken pox, measles or whooping cough, one does not get the disease again. Immunity to other agents, such as cold viruses, is often short-lived, however, lasting but a few weeks. Precisely which cells retain this memory for making a specific antibody is not yet known.

Lower animals, including fish, reptiles and some invertebrates, exhibit immune-type responses, although the presence of specific antibodies has been reported only in one or two cases. However, accelerated rejection of second grafts of foreign tissue has been demonstrated in animals as primitive as earthworms, indicating that they are capable of a true immune response.

Suppression of the Immune Response. Although the immune response is an essential defense mechanism against infection, it occasionally causes medical problems. As we have noted, skin grafts and organ transplants between different people result in antibody formation and eventual graft rejection. Treatment with the drug **cortisone,** which suppresses the function of the lymphatic system where antibodies are formed, slows down antibody formation and graft rejection, but at the same time leaves the patient without defenses against infections.

Occasionally the body makes antibodies against its own proteins, causing inflammation and tissue breakdown. Diseases such as rheumatic fever and certain types of hemolytic anemia are caused by antibodies formed against one's own heart muscle and red blood cells, respectively. These **autoimmune diseases** may sometimes be successfully controlled with cortisone.

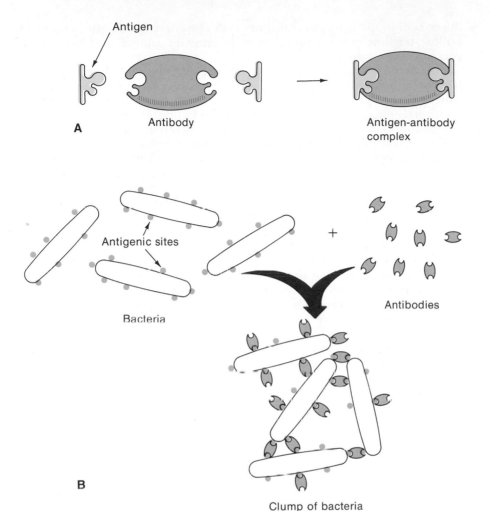

A

Antigen

Antibody

Antigen-antibody
complex

Antigenic sites

Bacteria

+

Antibodies

B

Clump of bacteria

FIGURE 10.13

Antigen-antibody reaction.
A. Diagram showing interaction of antigen molecules with specific antibody. Note similarity to enzyme-substrate complex formation.
B. Agglutination of bacteria. Antibodies attach to several antigen sites on bacteria, resulting in clumping.

Hypersensitivity. *Allergies* are the result of hyperactive antibody formation. A small amount of foreign substance, such as dust or pollen, induces an enormous immune response at the site of antibody contact. In the course of the immune reaction, potent chemicals called **histamines** are released into the tissues from certain white blood cells, causing swelling and inflammation. Different people are hypersensitive in different organs. When eyes or nose are affected, one has hay fever; when the bronchial tubes leading to the lungs are affected, the disease is called asthma; if skin rash develops, the allergy is called eczema or "hives." The rashes developing from poison oak and poison ivy are also allergic-type reactions.

SUMMARY

There are two circulatory systems in higher plants. The xylem serves to move water and mineral nutrients upward, the forces being generated by transpiration from the leaves coupled with the cohesive tension of water in the xylem tubes. Translocation of nutrients via the phloem from producer cells in the leaves to consumer cells elsewhere is thought to occur by bulk water flow brought about by differences in osmotic and hydrostatic pressure within the plant.

Animal circulatory systems, which may be open or closed, typically have a pumping organ and arteries and veins. Capillaries are present in closed systems. Loss of blood after wounding is prevented either by cellular (in invertebrates) or plasma (in vertebrates) clotting. In animals with closed circulations, plasma proteins also act osmotically to maintain blood volume in the capillaries and thus play a role in normal lymphatic circulation.

Blood and other body fluids provide a constant environment for the tissues. Salt and water balance is maintained by kidney action on the blood. Gas exchange is effected by the respiratory organ and by respiratory pigments, which loosely bind oxygen and also help transport carbon dioxide. Food from the gut and wastes from the cells are also transported in blood. Maintenance of constant blood glucose levels in vertebrates depends directly upon at least six hormones, acting on muscle, liver, fat and other tissues. Such complex homeostatic mechanisms are necessary for keeping blood components within the narrow limits required by body cells.

The blood also defends the body against invading organisms. Phagocytic blood cells—amoebocytes or leucocytes—act first by surrounding and engulfing foreign matter. Later, cells of the lymphatic system produce specific antibodies against the antigenic substances of the invader. These immobilize foreign cells and speed their removal. This immune response can be suppressed with cortisone to prevent rejection of organ transplants or to control autoimmune diseases.

READINGS AND REFERENCES

Adolph, E. F. 1967. The heart's pacemaker. Scientific American Offprint No. 1067. Describes the important role of myogenic contraction in maintenance of heartbeat.

Clegg, P. C. and Clegg, A. C. 1969. *Hormones, Cells and Organisms: The Role of Hormones in Mammals.* Stanford University Press, Stanford. A brief account of mammalian hormones, with emphasis on homeostatic mechanisms.

Mayerson, H. S. 1963. The lymphatic system. Scientific American Offprint No. 158. Extends our brief account of the physiological role of this second circulatory system in man.

Nossal, G. J. V. 1969. *Antibodies and Immunity.* Basic Books, New York. Describes recent research in this important field in layman's terms.

Perutz, M. F. 1964. The hemoglobin molecule. Scientific American Offprint No. 196. A readable account of the structure and function of vertebrate hemoglobin written by one of the men who contributed much of the original research.

Prosser, C. L. 1973. *Comparative Animal Physiology.* 3rd ed. W. B. Saunders Co., Philadelphia. (In press.) Chapters 8 and 11 provide a broad, comparative coverage of the circulatory systems of animals.

Salisbury, F. B. and Ross, C. 1969. *Plant Physiology.* Wadsworth Publishing Co., Belmont, California. See especially Chapters 6, 7, 8 and 9 for a comprehensive coverage of water movements and circulation in plants.

Wood, J. E. 1968. The venous system. Scientific American Offprint No. 1093. Explains the active role that veins play in maintaining normal circulation.

Zimmerman, M. H. 1963. How sap moves in trees. Scientific American Offprint No. 154. A non-technical, well-illustrated account of the circulation of fluids in trees.

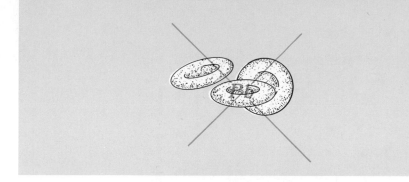

ENVIRONMENTAL LEAD – A HEALTH PROBLEM

10a

Present levels of lead in the environment are reflected in the blood of city dwellers. Even without overt clinical symptoms, lead may be adversely affecting general public health, and it is definitely a cause of anemia and neurological disease among ghetto children.

Lead is one of the elements known as heavy metals. Others of concern are mercury, zinc, copper and cadmium. All are relatively dense sub stances, and all are toxic to living organisms. Lead has long been used by man; the word plumbing derives from the Latin word for lead, *plumbum* – hence its chemical symbol, Pb. Until quite recently, domestic water supplies were regularly supplied in lead pipes. Lead has always been present at very low levels in the soils and waters of the Earth, but recent studies show the amounts have increased many times in the past few decades. The source of this lead and its potential effects on the health of man are the main subjects of this chapter.

SOURCES OF INCREASED ENVIRONMENTAL LEAD

The first indications of an increase in environmental lead were obtained about a decade ago. Geochemists studying lead in the ocean found far higher concentrations at the surface than in deep waters, whereas one would expect a fairly uniform concentration at all depths. This suggested an increased atmospheric fallout in recent years, since it takes several centuries for lead to sink from the surface to the depths of the ocean. To test whether there had indeed been a recent global increase in atmospheric lead, the geochemists took core samples from the centuries-deep snow-packs of Greenland and Antarctica. The annual layers, like growth rings of a tree, could be counted and dated. Analysis of the different layers indeed showed an enormous increase of lead washed from the atmosphere in rain and snow in the past few decades (Fig. 10a.1). About 800 B.C. there was very little lead in the global atmosphere; lead smelting during the industrial revolution brought about a gradual increase; the introduction of lead compounds to gasoline in 1924 coincided with a sharp upward trend.

263

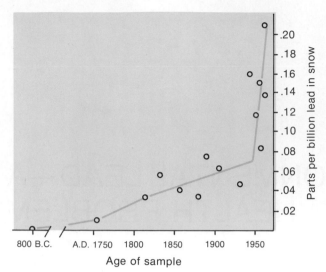

FIGURE 10a.1

Age of sample

History of atmospheric lead pollution as revealed by Greenland snow strata. Note gradual increase with Industrial Revolution and sharp increase after addition of lead to gasoline in 1924.

NON-UNIFORM DISTRIBUTION OF LEAD

The next step was to examine air and soil near large cities and compare their lead content with samples taken far from heavy concentrations of automobiles. Data for San Diego, California, a city of nearly one million inhabitants, are summarized in Table 10a.1.

As can be seen, air in the urban area is far more contaminated than elsewhere. Even air around a pier jutting into the ocean and exposed to almost constant sea breezes has 50 times more lead than that at the top of a 6000 foot mountain located 35 miles inland, and 500 times more than air in the open ocean. Although the quantity of lead in urban air is only micrograms per cubic meter (μg/m³), a concentration that may seem negligibly small to the layman, it is, as we shall see shortly, sufficiently high to affect the health of people continuously exposed to it. Its real significance can be appreciated by expressing lead as a fraction of the total dust in the air. Dust settling on the roof of San Diego's downtown library, for example, averages two per cent lead and has reached as high as seven per cent. These concentrations of lead are as high as those in the ores taken from some commercial lead mines! This continual rain of atmospheric lead onto cities constitutes a threat not only to people breathing it but more especially to young children who may happen to eat lead-contaminated dirt.

The Environmental Protection Agency considers atmospheric lead concentrations in excess of 2 μg/m³ to be "associated with a sufficient risk of adverse physiologic effects to constitute endangerment of public health."

Air Source	Lead Content (μg/m³)*
Open ocean (Pacific)	0.001
Scripp's pier	0.5
Roof of downtown library	2.1
Mount Laguna (6000 ft: above inversion layer)	0.01

TABLE 10a.1 AVERAGE VALUE OF ATMOSPHERIC LEAD IN SAN DIEGO, 1970

*Micrograms per cubic meter; a microgram equals 0.000000035 ounce; a meter is about one yard. (From Chow, T. J., and Earl, J. L. 1970. Lead aerosols in the atmosphere; increasing concentrations. Science *169*:577.)

The present standard set by the California State Air Resources Board in 1971 says that no air in California shall have a lead concentration exceeding an average value of 1.5 µg per cubic meter for more than 30 days. In 1970, however, San Diego's air *averaged* 2.1 µg for the whole year! During a 12 month period in 1968–1969, Los Angeles air *averaged* 3.6 µg lead per cubic meter. Air near freeways may have far higher values. A concentration of 8.9 µg per cubic meter was found near a freeway in Detroit in 1961, although the data were only made public in 1969. Short-term values as high as 71 µg/m³ have been recorded! It is clear that statutory limits are not now being met.

When samples of soil near urban areas were compared with those taken at remote places, once again the lead content in cities was strikingly higher. As can be seen in Table 10a.2, the values in each area are approxi-

TABLE 10a.2 LEAD IN SOILS FROM VARIOUS PLACES*

Source		Lead Content (ppm)
Antarctica		
South Victoria Land (a "clean" control area)		10–15
Mexico		
Mexico City		179
United States — urban areas		
Washington, D.C.	Ellipse Park	243
New York	Central Park	834
Los Angeles	MacArthur Park	3357
Honolulu	Irwin Park	1087
United States — rural areas		
New Hampshire		18
California		8

*Based on unpublished data from Chow, T. J. Scripps Institution of Oceanography, La Jolla, California.

mately proportional to the number of cars in use. The levels at remote areas are between 8 and 15 ppm, about the same as has always been found in soil. Lead falling out from the atmosphere in these regions does not seem to be accumulating in soils; instead it washes out to sea. But the quantities falling out near cities are far greater, and so lead is accumulating in urban soils.

A profile of lead in samples of soil and vegetation taken along U.S. Highway 1 in Maryland is shown in Figure 10a.2. The soil was sampled at

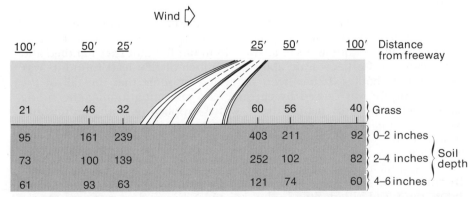

FIGURE 10a.2

Lead fallout beside U.S. Highway 1 in Maryland. Values are given in parts per million (ppm). Three depths of soil, as well as grass on the surface, were sampled at different distances from the road.

several depths and at various distances from the road. There is a clear correlation between lead content and distance from the freeway, with lead fallout being distinctly higher on the leeward side. Plants do not take up lead through their roots. Thus, the lead associated with the vegetation in Figure 10a.2 is adsorbed onto the surface of the leaves, but it can be washed off only with great difficulty.

IDENTIFYING ENVIRONMENTAL LEAD

Despite the circumstantial evidence that environmental lead levels today are enormously increased over natural levels and that automobile fuel is the source of this lead, gasoline manufacturers, and particularly the companies manufacturing lead additives, have claimed other sources contribute as much or more to lead pollution. Coal, for example, also contains considerable lead which reaches the atmosphere on combustion. Burning old car batteries has also been blamed.

Recently, however, a unique fingerprinting technique has been developed which allows precise identification of a source of lead. There are four *isotopes* of ordinary lead, which vary only in their weight: ^{204}Pb, ^{206}Pb, ^{207}Pb and ^{208}Pb. Each ore deposit has a slightly different ratio of these isotopes, and can thus be distinguished from all others. Since oil companies obtain lead from different sources, the fallout due to a particular gasoline can be identified. When isotope ratios of lead in gasoline and soil samples of several American cities were analyzed by Dr. T. J. Chow and his colleagues at Scripps Institution of Oceanography, they found that the two types of samples from each city always showed virtually identical isotope ratios—gasoline and soil lead originally came from the same ore! The mines from which the leads were obtained are located in Mexico, Peru and Canada. High-grade United States ores are used for storage batteries and other purposes, and their isotope fingerprints are not found in the environment. Nor are the lead isotope ratios of coal found in soil samples.

There seems no question that the major source of environmental lead contamination is the antiknock compound, tetraethyl lead. About four grams of lead are added per gallon to "ethyl" grades of gas and two grams to "regular" grades. In the entire United States this currently amounts to an annual release of 500 million pounds of lead into the atmosphere!

LEAD IN LIVING ORGANISMS

LEAD LEVELS IN MAN

With such increased amounts of lead in the environment, one may well ask whether it finds its way into people and, if so, whether it hurts them. These are difficult questions to answer. Man has for centuries handled and used lead, but we know little about how much of that lead found its way into people's bodies. It has been suggested by some that the fall of the Roman empire was caused by chronic lead poisoning among the ruling classes. These wealthy people drank water supplied in lead conduits and wine mulled in lead-lined casks. The theory runs that a high incidence of stillbirths due to lead toxicity caused the dying off of this powerful aristocracy.

The total amount of lead in the body is known as the **body burden.** Because of its similarity to the element calcium, a major component of bone, most lead is deposited in the skeleton, where it does no harm. On the other hand, lead circulating in blood reaches other tissues, such as kidney and brain, where it can cause damage. Thus, blood lead levels are of importance in estimating potential danger to health from lead in the body.

Blood lead levels among various groups in the United States today are shown in Table 10a.3. There appears to be a definite correlation between blood lead concentration and exposure to environmental lead. The higher values in smokers result from the lead content of cigarettes. Note that for most people in the United States, blood lead concentrations are at least one-quarter that producing clinical symptoms of lead poisoning—0.8 ppm. So far isotope fingerprints have not been performed on lead in human blood; the samples required are too large. We have only circumstantial evidence that present human lead burdens are mainly due to leaded gasoline.

TABLE 10a.3 BLOOD LEAD CONCENTRATIONS OF VARIOUS GROUPS IN THE UNITED STATES*		
Suburban non-smokers	0.11	ppm
Suburban smokers	0.15	
Urban non-smokers	0.22	
Urban smokers	0.25	
Traffic policemen	0.31	
Parking lot attendants	0.34	
Garage mechanics	0.38	
Persons with lead poisoning	0.80	

*Data from HEW Environmental Health Services. 1965. *Air Pollution*: Survey of lead in the atmosphere of three urban communities. Publication No. 999–AP–12.

Is the amount of lead in people today significantly higher than that in primitive man? According to some authors, the answer is an emphatic "Yes—at least one hundred times higher." Others point to blood lead levels in remote tribes in New Guinea which are as high as those found among American urban populations. Although the question cannot be answered with certainty, it seems probable that levels in primitive man were generally considerably lower than in today's urbanite. But the real question is, Are present-day blood lead concentrations a danger to health? Before approaching this question, however, let us briefly consider some of the ways lead enters the human body.

ROUTES OF ENTRY

Under normal conditions most lead finds its way into the body by way of the digestive tract. Food and water in the daily diet contain about 300 micrograms of lead. Of this, however, only about five per cent (15 μg) is absorbed as lead ion, the remainder being lost harmlessly in the feces. In the air we breathe, however, there is also considerable lead. Assuming an average value of 1.5 μg per cubic meter for urban areas (a conservative estimate, as we have seen), this amounts to about 30 μg lead breathed in per day. The fraction of lead absorbed from the lungs, however, is much greater than from the gut, amounting to 40 per cent (12 μg per day). Thus, food and air contribute about equally to the human body burden of lead in the average urbanite.

Other sources of lead intake should also be mentioned. Industrial exposure is now quite carefully monitored, although occasional overdoses occur. Until about 15 years ago most interior paints contained large quantities of lead. Even today's "lead-free" interior paints still contain significant amounts. Outdoor paints contain large amounts of lead, and so do some types of putty. Both are a particular problem in ghetto areas, where leaded paints and putty peeling off dilapidated buildings are often eaten by young children. **Pica,** the habit of putting non-food items in the mouth and chewing and swallowing them, is prevalent in a proportion of all children from six months to two years of age, whatever their nutritional status. Its cause is unknown, although some pediatricians believe it may be related to iron deficiency. It is estimated that a quarter-million children in the United States suffer from chronic lead poisoning from eating chipped paint and putty. Most of them live in slums. It is probable that such children are also exposed to highly polluted air as well. We shall discuss the effects of lead pollution on these children later.

FIGURE 10a.3

The little girl in the picture, Kelly, has been brought to St. Louis Children's Hospital by her mother. She is nineteen months old, and is suffering from chronic lead poisoning. (From Craig, P. P. and Berlin, E. 1971. The air of poverty. Environment *13*(5):2–10.)

Another common source of lead is the glaze found on certain types of inexpensive pottery. Drinking acidic or carbonated beverages from lead glazed vessels is a particularly effective way to get lead poisoning. The carbonic acid dissolves the glaze. We are indeed exposed to a great deal of lead in our man-made environment.

LEAD IN OTHER ANIMALS

Before considering the toxic effects of lead in man it is worth underlining the degree to which the environment is polluted by discussing the incidence of lead poisoning in animals unwittingly exposed—namely, animals in zoos. Zoos are almost always situated in the midst of urban environments. Since they contain a wide spectrum of animals, some of which are likely to be more sensitive than others to a given pollutant, it is not surprising that these animals often serve as "guinea pigs" for their human admirers.

In 1971, chronic lead poisoning was reported among various animals at Staten Island Zoo in New York City. Several leopards fell ill and one died. Tissue analysis showed the cause to be lead poisoning. Snakes were also dying after losing muscular coordination, and they, too, contained high amounts of lead. The blood of many other animals, both sick and well, was tested and found to be contaminated with lead, often in amounts far higher than those considered toxic for man.

In seeking the source of this lead, investigators found the food, water and bedding free of heavy metals. The paint in the cages, however, was found to contain as much as three per cent lead, although the paint cans were labeled "lead-free indoor paint!" More than that, plants and soil on the zoo grounds contained amounts of lead equal to or higher than those found near freeways. Animals living in outdoor cages without paint had the highest lead levels in their bodies. Even dead mice found around the buildings were loaded with lead. Although the exact source of the lead found in these animals has yet to be identified, circumstantial evidence strongly points to atmospheric fallout.

TOXIC EFFECTS OF LEAD

Lead poisoning can produce a variety of symptoms. Acute poisoning results in weakness, dizziness, nausea, muscular spasms and pains and other unpleasant effects. Far more insidious is chronic poisoning, which may remain undetected for a long period of time. Vital organs such as the kidney, liver and brain may be irreversibly damaged. Anemia is a common symptom. Our concern in the following discussion is primarily with the degree to which the health of the average American is being affected by today's lead pollution.

EFFECTS ON THE CIRCULATORY SYSTEM

One of the earliest effects of chronic lead poisoning is on the oxygen-carrying capacity of blood. In ways not yet understood, lead interferes with at least two steps in the synthesis of hemoglobin by red blood cells. The reader will recall that hemoglobin consists of a globular protein and an iron-containing unit called **heme** (Chapter 10). Heme is synthesized in a series of reactions which convert the precursor molecule, δ-aminolevulinic acid or **ALA**, to porphyrin; iron (Fe) is then added to make heme. The heme in turn is combined with a protein known as globulin to form hemoglobin (Fig. 10a.4). Two steps in this process are affected by lead: ALA-dehydrase, the enzyme which converts ALA to porphyrin, is inhibited, and so too little porphyrin is synthesized; the step which combines heme with globulin is also inhibited, the unused porphyrin being excreted in the urine as coproporphyrin. The result is a hemoglobin deficiency in the red blood cells—**anemia.**

All the above reactions take place in newly forming red blood cells in the bone marrow. Lead can also affect mature, circulating red blood cells. Normally, hemoglobin gradually ages, becoming converted to **methemoglobin,** which cannot bind with oxygen. Toward the end of its four month lifetime, a red blood cell carries less oxygen than previously. Lead increases the conversion rate of hemoglobin to methemoglobin, causing the already anemic cells to function less well in carrying oxygen. Lead also hastens the rate of destruction of red blood cells by spleen and liver. All these combined effects of lead produce severe chronic anemia in patients with lead poisoning. Too little hemoglobin is synthesized, and what is formed has a shortened period of usefulness in transporting oxygen.

Combined Effects of Other Pollutants.   Lead is not the only pollutant which affects the oxygen-carrying capacity of blood. At least three other widespread pollutants interfere, as shown in Figure 10a.5. Carbon monoxide (CO), released in large quantities from automobiles, combines 200 times more readily with hemoglobin than does oxygen. The carboxyhemoglobin thus formed can no longer participate in binding oxygen. Ozone (O_3), a major component of photochemical smog, slows the release of oxygen from hemoglobin in tissue capillaries. Nitrate (NO_3^-) is a component of artificial fertilizers which frequently seeps into well water. When such water is drunk by babies and young children, their intestinal bacteria convert nitrate to nitrite (NO_2^-). Nitrite is absorbed into the blood and oxidizes the iron of hemoglobin from Fe^{++} to Fe^{+++}. This oxidized form of hemoglobin, methemoglobin, is unable to combine with oxygen. (The conversion of NO_3^- to NO_2^- is insignificant in adults, and they are therefore little affected by this pollutant.) The reader may be dismayed to learn that there have so far been no studies to determine what synergistic effects (interactions) can result from combinations of these several common pollutants in one individual.

Lead and Resistance to Infection.   Another important function of the blood—defense against bacterial infection—is also impaired by low doses of lead. In recent experiments, mice were injected daily with very small amounts of lead, insufficient to produce clinical symptoms. The mice were then inoculated with bacteria which cause typhoid fever. It was found that far fewer bacteria (one-fifth to one-tenth) were required to cause death in the lead-treated mice as compared to control mice unexposed to lead. Exactly how lead interferes with resistance to bacterial infection is not yet known. It may either inhibit phagocytic activity of white blood cells or interfere with the synthesis or function of antibodies.

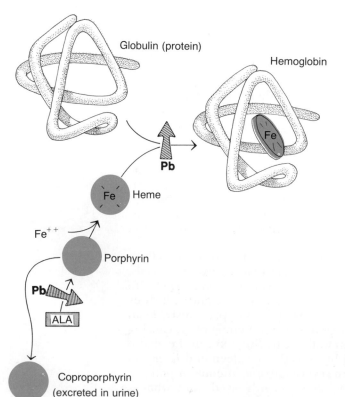

FIGURE 10a.4

Effects of lead on hemoglobin synthesis by red blood cells.
    Inhibition of the enzyme ALA-dehydrase interferes with porphyrin synthesis. Inhibition of the combination of heme and globulin causes a deficiency in hemoglobin formation. Unused porphyrin is excreted in the urine as coproporphyrin.

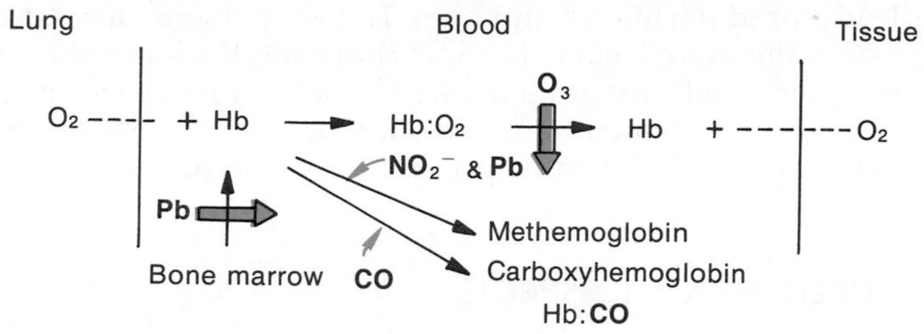

FIGURE 10a.5

Effects of several environmental pollutants on oxygen-carrying capacity of human blood. Lead (Pb) interferes with hemoglobin (Hb) synthesis. Ozone (O_3) interferes with oxygen release to the tissues. Carbon monoxide (CO) competes with oxygen for the binding site on hemoglobin. Nitrite (NO_2^-) and Pb both hasten the conversion of hemoglobin to methemoglobin, which cannot combine with oxygen.

EFFECTS ON THE CENTRAL NERVOUS SYSTEM

People with chronic lead poisoning often exhibit neurological symptoms. This is an especially severe problem among the ghetto children mentioned earlier. Behavior problems, poor discipline, inadequate interpersonal relationships and inability to comprehend the abstract may all be manifested during early stages of lead poisoning. When the condition remains unrecognized and untreated (as all too often occurs, since the symptoms are vague) the damage may become permanent.

A common result of severe chronic lead toxicity is encephalopathy, or swelling and distortion of the brain. In addition to recurrent convulsive seizures, paralysis may occur. Twenty-five per cent of children so affected die. Even more distressing is the fact that permanent brain damage results in almost all children who survive. The average I.Q. of one group of children with lead-induced encephalopathy was 80, compared to a norm of 100. The extent to which undiagnosed, asymptomatic lead poisoning is responsible for underachievement of children in many areas is probably far higher than is now recognized.

How lead acts in the brain to produce the observed symptoms is not known. Virtually no experimental work on animals has been carried out in this area, primarily because the public health aspect of lead poisoning in large segments of the population has so long been ignored.

EFFECTS ON OTHER ORGANS

Lead probably has harmful effects on all tissues. The liver is frequently damaged in severe lead poisoning, with jaundice (yellowing of the skin) resulting from destruction of its bile passages. The most common effect on the kidney is a deterioration of the arteries, which seriously affects its function. Reproductive cells in the gonads are injured, and severe exposure of women to lead decreases fertility and affects the development of the fetus. Again, in no case is it known how lead produces its effect.

TREATMENT OF LEAD POISONING

Both chronic and acute forms of lead poisoning are treated with *chelating agents,* organic molecules which have the capacity to bind lead. Once bound, the chelated lead can be excreted by the kidney. The treatment itself, however, may lead to complications. If there is still lead in the stomach, the orally-given chelator will hasten its absorption and may produce a higher degree of poisoning. Likewise, chelators tend to release lead stored harmlessly in bones. In fact, persons may become ill from lead poisoning many years after incorporating lead into their bodies. Common diseases which are accompanied by a high fever can cause lead stored in bones to be released to the blood. Obviously, treatment for lead poisoning is less preferable than its prevention in the first place.

PUBLIC HEALTH ASPECTS

Several studies in large cities show that at least one in 15 ghetto children is affected by serious lead poisoning, with blood levels greater than 0.5 ppm. Children appear to be more susceptible to the effects of lead than adults, perhaps partly due to the habit of pica. This represents an enormous public health problem in terms of diagnostic screening and treatment, not to mention removal of the environmental sources, the dilapidated slum housing.

But what of the rest of the people living in urban areas, especially those heavily exposed to lead—the parking lot attendants, car mechanics and policemen? What of people living near crowded freeways? Until just a few years ago it was assumed that there were no physiological effects until blood lead levels in adults reached 0.8 ppm—the point at which clinical symptoms appear. That this is not true is shown by a study of the amount of the enzyme ALA-dehydrase in the red blood cells of people exposed to different amounts of environmental lead. This key enzyme in hemoglobin synthesis is still present in erythrocytes even after they are released from the bone marrow. The results are shown in Figure 10a.6.

If lead were without any effect until it reached 0.8 ppm, then the dashed line should have been obtained. In reality, there is a close negative correlation between blood lead and the amount of ALA-dehydrase. We can conclude that lead at *any* concentration has a physiological effect. It is thus likely that the other known effects of lead also occur in proportion to the blood or tissue concentration, and do not suddenly appear at some threshold value.

It is apparent that every effort should be made to decrease all sources of environmental lead. Removal of lead from gasoline would have a significant effect both on that breathed in via the lungs and that eaten on food contaminated by fallout. Such measures are of especial importance to those people in the population who are particularly susceptible to the effects of lead. These include not only children but also all those people suf-

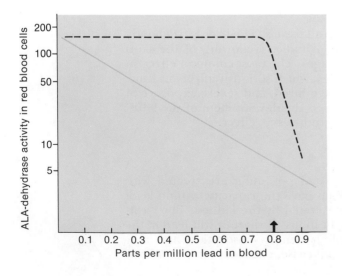

FIGURE 10a.6

Correlation between blood lead concentration and activity of the hemoglobin synthesizing enzyme, ALA-dehydrase, in red blood cells.

The solid line shows recent data. (Note the logarithmic scale on the vertical axis.) It refutes earlier ideas (dashed line) that lead has no effect until it reaches some threshold level at which clinical symptoms appear (arrow).

fering from anemia. Many forms of anemia are inherited. Sickle-cell anemia, prevalent among blacks, is characterized by the presence of abnormal hemoglobin; so are some forms of hemolytic anemia. Persons with kidney disease are also at special risk from lead poisoning. Several million Americans suffer from these diseases, and exposure to lead is far more dangerous for them than for others.

Early in 1972 the Environmental Protection Agency announced a new set of regulations concerning lead additives in gasoline. By 1974 all service stations must provide one grade of "lead-free" gas which contains less than 0.05 grams of lead per gallon. By January, 1977, no grade of gasoline in the United States will be permitted to exceed 1.25 grams of lead per gallon. Providing the amounts of gasoline consumed do not increase markedly, such regulations should go far toward reducing, although not altogether eliminating, the major source of environmental lead.

OTHER HEAVY METALS

Although the day may be fast approaching when lead will no longer be a major pollutant, it has served as an example of the problem of heavy metals in general. Mercury is another heavy metal often in the news. Both inorganic mercury ion and organic mercury compounds are found in the environment. The latter are formed from inorganic mercury, mainly by bacterial action. Mercury differs from lead in being rapidly concentrated in food-chains—a phenomenon known as *biological magnification.* Thus fish are frequently found to have high levels of mercury in their tissues, and ingestion of them has caused human fatalities, especially in Japan. In that country, large amounts of waste mercury were dumped by industries into bays from which fish and shellfish for human consumption were taken.

Although the best evidence we have indicates that the activities of man have only doubled the natural background levels of environmental mercury, the fact that any excess is quickly taken up by plants and animals means that man is continuously exposed to it in his food. Only 18 per cent of the mercury produced for industrial use is recycled; the rest is dispersed into the environment. Manufacturing industries using large amounts of mercury are chemical plants making chlorine and caustic soda, and electronics plants. Organic (methyl) mercury is also used as a fungicide (killer of molds). It is applied to seeds to protect them from mildew during germination and is also used in pulp mills. Treated seeds eaten by wild game birds have often caused the accumulation of toxic levels of mercury in their tissues. Hunters in parts of Canada and California have been warned not to eat the fowl they shoot! Another, perhaps the largest, source of environmental mercury is the burning of fossil fuels.

The symptoms of mercury poisoning indicate that its main toxic effects are on the nervous system. Drowsiness and headache are early symptoms, followed by loss of coordination, weakness, paralysis and death. In chronic or severe poisoning, persons who survive are usually permanently retarded. Mercury at concentrations far below those causing symptoms in adults can produce fetal damage, and at even lower levels it is known to cause chromosome breakages in human cells. The consequences of such genetic damage will be considered in Chapter 15a. The full public health aspects of mercury remain to be assessed, however.

Another heavy metal much in the news is cadmium, which is released into the environment by many industrial processes. Cigarette smoke also contains significant quantities of this element. At present, little is known of the environmental distribution of cadmium and of other metals which may be toxic, such as copper, zinc and vanadium. Although all are known to inhibit enzymes at high concentrations, their potential effects on man and

other living organisms at the concentrations now found in them has yet to
be investigated.

SUMMARY

A widespread environmental contamination by lead released into the
atmosphere from combustion of leaded gasolines has been clearly demonstrated by isotope fingerprinting techniques. Concentrations of lead in
urban environments, especially near freeways, are many times greater than
elsewhere. This lead has adverse effects upon animals and probably on
humans living in urban areas.

Lead in the blood of the human population is correlated with its environmental distribution. In urban America, air and food contribute about
equally to the total body burden of lead. Children in ghettos, where environmental concentrations are particularly high, may be subject to permanent mental retardation from lead poisoning. It is also likely that people
with subclinical blood levels of lead may suffer adverse effects on their
health. This is especially true for persons with anemia due to other causes.
Lead interferes with hemoglobin synthesis and shortens the life span of red
blood cells. Interactions between lead and other pollutants affecting the
oxygen-carrying capacity of the blood have not yet been investigated.

In addition to lead, other heavy metals give cause for concern. Several
effects of mercury, for example, although potentially dangerous, are not
easily detected.

**READINGS AND
REFERENCES**

Bazell, R. J. 1971. Lead poisoning: Zoo animals may be the first victims. Science *173*:130–131.
Describes deaths among animals in the New York City zoos from lead poisoning.
Chow, T. J. 1970. Lead accumulation in roadside soil and grass. Nature *225*:295–296. Data
about lead pollution near freeways.
Craig, P. P. and Berlin, E. 1971. The air of poverty. Environment *13*(5):2–9. A popular ac-
count of lead pollution in ghettos.
Gilfillan, S. C. 1965. Lead poisoning and the fall of Rome. Journal of Occupational Medicine
7:53–60. Proposes a theory about lead toxicity among classical Roman aristocracy.
Grant, N. 1971. Mercury in man. Environment *13*(4):2–15. Discusses major sources of environmental mercury and some of its known effects on man.
Hemphill, F. E., Kaeberle, M. L. and Buck, W. B. 1971. Science *172*:1031–1032. Lead
suppression of mouse resistance to *Salmonella typhimurium*. Experiments demonstrating a
decreased resistance to bacterial infection after chronic lead poisoning are described in
this original research paper.
Hernberg, S., Nikkanen, J., Mellin, G. and Lilius, H. 1970. δ-amino levulinic acid dehydrase as
a measure of lead exposure. Archives of Environmental Health *21*:140–145. Studies
correlating decrease in a key hemoglobin synthesizing enzyme with exposure to lead.
Murozumi, M., Chow, T. J. and Patterson, C. C. 1969. Chemical concentrations of pollutant
lead aerosols, terrestrial dusts and sea salts in Greenland and Antarctic snow strata.
Geochimica et Cosmochimica Acta *33*:1247–1294. Original data showing increasing
global atmospheric lead fallout in recent decades.
Skerfving, S., Hansson, K. and Lindsten, J. 1970. Chromosome breakage in humans exposed
to methyl mercury through fish consumption. Archives of Environmental Health *21*:133–
139. Recent studies on the effects of mercury poisoning on human chromosomes.
Smith, H. D. 1964. Pediatric lead poisoning. Archives of Environmental Health *8*:68–73. A
pediatrician's description of the symptoms, diagnosis and outcome of lead poisoning in
ghetto children.

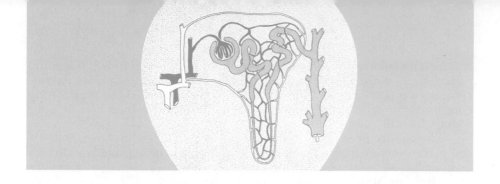

OSMOREGULATION AND EXCRETION

11

The homeostatic functions performed by osmoregulatory and excretory organs permit animals to live in a great diversity of environments.

In considering the physiology of whole organisms we have so far been mainly concerned with the means by which plants and animals acquire nutrients and oxygen and distribute them to their cells. We have also noted that circulatory systems provide a constant internal environment and that this phenomenon of **homeostasis** depends upon the processing of the body fluids by other organs. In this chapter we shall consider two of these homeostatic mechanisms—namely, **osmoregulation** and **excretion.**

Osmoregulation is concerned with maintaining a constant osmotic pressure and thus primarily deals with problems of water and ion balance. Excretion, on the other hand, is concerned with ridding the body of waste products which become toxic if allowed to accumulate in high concentrations. Although these are really two distinct physiological problems, they are generally considered together for a very good reason. Waste substances, the by-products of normal metabolism, are usually soluble substances and therefore must be excreted dissolved in water. This means that excretion is closely tied to water balance, and in higher animals the same organ, the kidney, frequently serves both functions. We shall therefore first consider the water balance of organisms and then, the mechanisms for excretion of the waste products of metabolism.

OSMOREGULATION: THE PROBLEM

We may begin by considering the osmotic problem faced by a cell bounded only by its own thin plasma membrane. In Chapter 3 we examined the effects of placing such a cell either in a solution more concentrated than itself (hyperosmotic medium), in which the cell shrinks because of loss of water, or in a solution more dilute than the cell (hypo-osmotic medium), in which it swells as a result of influx of water (Fig. 3.17). In both cases the water moves down a concentration gradient as it tends to equalize its concentration on both sides of the cell membrane. It is helpful to think of the concentration of water as being *less* in a salt solution than it is in pure water — the salt "dilutes" the water.

The picture presented in Chapter 3 is slightly oversimplified. When a cell is transferred from an isosmotic solution, in which it neither shrinks nor swells, to a more concentrated or more dilute solution, not only does water movement occur across the membrane: ions also will diffuse inwards or outwards as they tend to equalize their own concentrations. This reflects the general tendency of all substances to mix uniformly, to increase entropy or randomness, which we discussed in Chapter 7. The important thing to remember is that cell membranes are far more permeable to water than to such other substances as ions. The final result of placing a cell in hyperosmotic or hypo-osmotic salt solutions is shown in Figure 11.1. At first, the cell rapidly changes volume, as water movement occurs. Then, as salts and water move slowly in the opposite direction, the original volume is nearly restored. The volume changes shown are greatly exaggerated for the sake of clarity.

The extent to which cells can regulate their volume when placed in concentrated or dilute solutions is largely dependent on the ability of the cell membrane to actively assist in the movement of salts (slow phase shown in Figure 11.1). If the new medium has the same composition as the normal medium of the cell and differs from it only in concentration of solutes, then the cell membrane is often able to **actively transport** salts and other solutes in order to restore its volume. This function is considered later in the chapter.

Not all cells are capable of volume regulation, however — many depend rather upon the maintenance of a constant internal osmotic concentration in the body fluids. The ability of various organisms to provide a constant internal environment in the face of osmotic changes in their surroundings varies greatly. Let us consider some of the osmotic problems faced by various organisms.

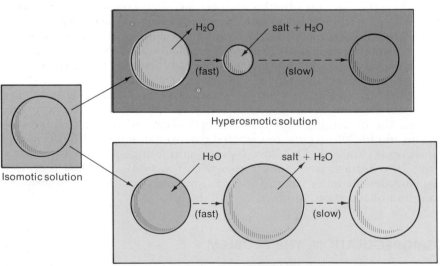

FIGURE 11.1

Sequence of osmotic volume changes in a cell bounded only by a plasma membrane. When a cell is transferred from an isosmotic solution to a hyperosmotic or hypo-osmotic solution, rapid water movements occur which cause shrinkage or swelling, respectively. Slower salt and water movements result in partial reversal of these volume changes. The density of color indicates the relative concentrations of cells and solutions.

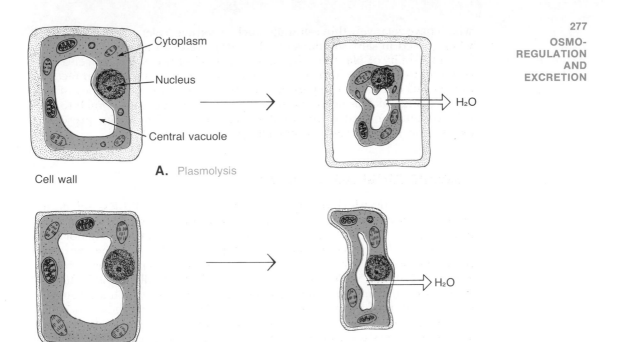

A. Plasmolysis

Cell wall

B. Wilting

FIGURE 11.2

Plasmolysis and wilting in plant cells.

A. In plants with a thick cell wall (trees and shrubs), osmotic loss of cell water causes shrinkage of cytoplasm and central vacuole. The pulling away of cytoplasm from the cell wall is called plasmolysis.

B. In plants with a thin cell wall (lettuce and celery) loss of cell water also causes shrinkage of the cell and vacuole. In this case the fall in turgor pressure causes collapse of the cell wall, or wilting.

OSMOREGULATION IN PLANTS

As we noted in Chapter 10, the solute concentration inside the cells of higher plants is maintained at a level greater than that in the circulatory fluids, in strong contrast to the situation in animals. Water thus osmotically enters plant cells from the sap, but because of their rigid walls, the cells do not swell like the cell shown in Figure 11.1. The resulting internal hydrostatic pressure results in **turgor.** Since plant cells are permeable to salts as well as water, although to a lesser degree, they would gradually lose salts to the more dilute sap if there were no mechanism for retaining them. The internal osmotic pressure of plant cells is maintained by the active pumping of ions from the sap into the cytoplasm and central vacuole.

As we have already seen, the osmotic gradients in plants play a role both in the absorption of water from soil by roots and in the bulk flow of sap through the phloem. Plants whose roots are immersed in water with a high salt concentration, such as tropical mangroves, appear to take up water by cohesion rather than osmotically. The salts entering the rootlet with the water are then actively extruded by the root cells, so that the solute concentration in the xylem is extremely low, as it is in most other plants.

When terrestrial plants are deprived of water they undergo water stress. As water is lost from the leaves, the solute concentration in the sap rises until eventually the osmotic differential is reversed and favors water movement *out* of the cells. The cells shrink, pulling away from their walls (Fig. 11.2). This phenomenon is called **plasmolysis.** In some plants, such as woody bushes and trees, the cell walls are so thick that one observes no immediate change in the organism after plasmolysis has occurred. If water stress is prolonged, the leaves eventually die and drop off. But those plants which have relatively thin cell walls, such as lettuce, celery and so forth, wilt when placed in salty solutions or when deprived of water.

The ability of plants to withstand changes in the osmotic concentration of their environment is far greater than that of most animals. In fact, the deleterious effects of a hyperosmotic environment on some types of salt-hardy plants, such as beets and tomatoes, may be due not so much to total salt concentrations as to the toxic effects of a particular ion, present in excess, which interferes with normal growth and metabolism.

OSMOTIC PROBLEMS IN ANIMALS

Animal cells, which do not have rigid cell walls, are far less able to tolerate fluctuations in osmotic pressure, and therefore require a controlled internal environment. In Chapter 2 we saw that there are four major environments with respect to salt and water availability—marine, brackish, freshwater and terrestrial—and that each poses its own problems for the animals inhabiting it. Before discussing these individually, it will be helpful to consider the range of internal solute concentrations which are found in the body fluids of the main groups of animals. These are shown in Figure 11.3. It should be noted that our presentation here is oversimplified. There are animals, such as the brine shrimp, *Artemia*, which live at higher salinities than sea water, and there are also animals whose body fluids are more dilute than any we have shown. Such special cases are beyond the scope of our present survey, however.

Figure 11.3 contains the essentials for understanding the osmotic problems faced by various animals and therefore requires detailed explanation. It is immediately apparent that not all animals normally have the same amount of solutes in their body fluids. By and large, animals fall into two groups—those with internal osmotic concentrations similar to sea water, and those with osmotic concentrations equal to about one-third that of sea water. Let us look at the five categories illustrated.

Marine invertebrates and also the hagfish, which are primitive vertebrates whose ancestors always lived in the sea, are characterized by a solute concentration in their body fluids virtually equivalent to that of sea water. These animals experience no problems with water balance, since their body fluids are isosmotic with the environment. Small adjustments in the individual salt ions are necessary, but on the whole, their blood is very similar to sea water, both in total concentration and in ionic composition.

Animals living in brackish waters, where salinity fluctuations are common, also tend to have an internal solute concentration similar to that of their environment. However, as shown in Figure 11.3, the range of internal osmotic pressure in these animals tends to be somewhat less than that of the external medium. We shall return to the adaptations made by these animals shortly.

Both invertebrate and vertebrate animals may be found in fresh water, and one is struck by the fact that their body fluids are always hyperosmotic to the environment. Salt is at a premium, and water is all too abundant. In most freshwater animals, the body fluids have a concentration about equal to one-third that of sea water, and it appears that this is near the *minimum* osmotic pressure which can be tolerated by the cells of most animals, even though they are living in far more dilute media.

The next group of animals is the cartilaginous fishes—the sharks and rays. At this juncture we must point out that all fish, except for the primitive hagfish mentioned above, have evolved from freshwater ancestors. The predecessors of sharks and rays lived in fresh waters, and hence developed cells which came to require an internal environment equal in osmotic pressure to that of other freshwater animals—namely, about one-third that

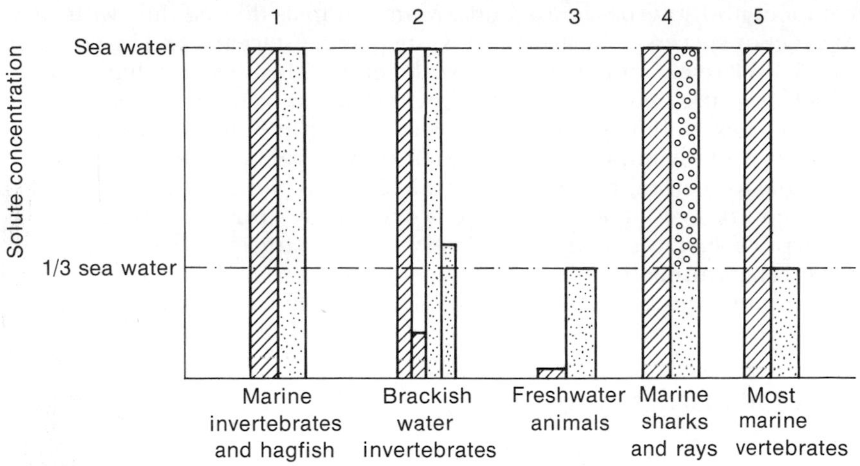

The osmotic relations of aquatic animals. The environmental osmotic pressure (solute concentration) is shown on the left in each case (striped histograms) and the osmotic pressure of body fluids on the right (stippled histograms). The larger circles indicate urea in the body fluids of cartilaginous fish (group *4*).

For brackish water organisms, the two sets of histograms indicate the approximate limits of the osmotic ranges for both environment and body fluids.

FIGURE 11.3

of sea water. On becoming sea-dwellers, however, the sharks and rays evolved a means for obtaining osmotic equilibrium with the environment without actually changing their internal salt concentrations. This alternative solution is the retention of a substance called **urea** in their blood and tissues. Urea is a normal waste product of protein metabolism in many animals, and we shall speak more of its nature below. But whereas most animals excrete urea in their urine as fast as it is formed, the cartilaginous fish retain it. Thus they can live in the sea without experiencing any osmotic water loss from their tissues or blood.

Most marine-dwelling vertebrates, however, including bony fish, turtles, sea birds, whales and porpoises, still retain the internal salt concentrations of their freshwater ancestors. Living as they do in the ocean, they are faced with a surfeit of salt and a deficiency of pure water.

It will be noted in Figure 11.3 that terrestrial animals have been omitted. These animals have internal salt concentrations similar to those of the inhabitants of fresh water, and the water they drink is also generally low in salts. The difference arises in the availability of water. For many semi-arid and desert species, the actual problem is too little water rather than too much. Thus, in terms of their osmotic balance, such animals are more nearly in the position of the marine vertebrates, having an overabundance of salt relative to the amount of water available.

We thus see that, except for the marine invertebrates and the hagfish whose internal osmotic pressure is the same as their environment, most animals face either one of two situations: they tend to become desiccated owing to an insufficient supply or excessive loss of water, or they tend to become flooded with water. A few animals, such as earthworms, may face both extremes at different seasons of the year.

SOLUTIONS TO OSMOTIC STRESS IN ANIMALS

Animals have evolved numerous mechanisms for maintaining constant internal salt and water concentrations under the most adverse circumstances. One solution is to envelop the body surface in an impermeable covering which slows down or prevents unwanted exchanges between the environment and internal body fluids. Many animals do just this, with varying degrees of success. The arthropods, such as crustaceans and insects, have a tough, relatively impermeable exoskeleton. Most worms have a cuticle which is thick enough to slow down water movements, and most vertebrates have a horny skin or scales to serve the same purpose. But these are never totally impermeable. Even if they were, however, it would still be essential for certain regions of an animal's body to be in direct contact with the medium — recall both the respiratory surfaces for gas exchange and the absorptive surfaces of the gut. Every animal (and plant, for that matter) must retain intimate contact with its environment if it is to acquire the nutrients and gases it needs, and these areas of contact are also readily permeable to water and ions.

ACTIVE TRANSPORT OF IONS

Whenever an animal is living in a situation of osmotic stress, it must perform osmotic work to maintain constant internal conditions. Theoretically, animals living in an aqueous environment could pump either water or salts in the appropriate direction. Thus a frog, living in a freshwater pond, could theoretically pump salts inward or water outward to keep its blood hyperosmotic to the environment. Most animals effectively do both, but the movement of water across cell membranes seems always to be brought about by local movements of ions. The water follows by passive diffusion. Thus, osmoregulation is generally achieved by means of the *active transport* of ions. The membrane mechanisms involved are known as *ion pumps*. One of the most important ions which is so transported is sodium ion (Na^+), and its transport system is commonly called the *sodium pump*.

The Sodium Pump. Both animal and plant cells are capable of pumping ions, especially sodium (Na^+) and potassium (K^+), across their membranes. At the moment, we are concerned with the role of ion pumps in animals. Although cells and body fluids of animals contain the same total osmotic pressure, their ion composition is quite different. Body fluids have a preponderance of sodium ion, whereas cells contain mostly potassium ion (Fig. 11.4A).

This disparity in ion composition inside and outside cells would quickly disappear because of diffusion if it were not maintained by an active pump, and, as one might expect for such an uphill process, energy obtained from the splitting of ATP is required. It has repeatedly been found that poisons which interfere with the reactions of cellular respiration by which ATP is synthesized cause a loss of such active transport. Since the maintenance of these sodium and potassium gradients is essential for the reactivity of cells to stimuli — especially those of nerves and muscles, as we shall see in Chapter 12 — the functioning of sodium pumps is vital to organisms.

In recent years, a specific enzyme has been identified in cell membranes which catalyzes the splitting of ATP, releasing its energy for active transport. Although there are many ATP-splitting enzymes in cells (called **ATPases**), linked with various energy-requiring processes, the one found in membranes is distinct in requiring sodium and potassium ions to activate it. This [Na^+-K^+] activated ATPase performs its catalytic function only when potassium concentrations outside the cell and sodium concentrations inside the cell are higher than normal. Thus the enzyme is active only when pumping is required (Fig. 11.4B).

The reader will recall, from our discussion in Chapter 9 of molecular transport across the gut during absorption of food monomers, that specific carrier molecules have been hypothesized for active transport across cell membranes. Whether the [Na^+-K^+] activated ATPase itself is the transport

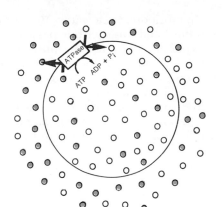

FIGURE 11.4

$\bullet = Na^+$

$\circ = K^+$

A

B

Diagram of the distribution of positive ions in cells and body fluids.

A. Potassium ions (K^+) are about 10 times more concentrated than sodium ions (Na^+) inside cells and about 10 times less concentrated outside. K^+ tends to diffuse out and Na^+ tends to diffuse in (arrows).

B. The ion pump, associated with [$Na^+ - K^+$] activated ATPase in the membrane, reverses the effects of diffusion, moving ions uphill against their concentration gradients. ATP is split in this energy-requiring process, producing ADP and P_i (phosphate). ATP is regenerated by energy-yielding reactions in the cell.

molecule involved in ion transport is not yet clear. What is of importance at the moment is that very high concentrations of this enzyme are found in all those tissues where active transport of ions is known to occur—the gut, kidneys and regions of the body surface where osmoregulation is carried out.

So far we have described ion pumps only in relation to moving ions into or out of cells, but not across cells. Our knowledge about how ions, or other molecules, are moved between the blood and environment, across a layer of intact cells, is still rather rudimentary. One can hypothesize that, by having an ion pump on the appropriate membrane of those cells which function in osmoregulation, an animal can transport ions from its blood to the environment or vice versa. A tentative model of how such an osmoregulatory cell layer could function is shown in Figure 11.5.

The main positive ion involved in osmoregulation is sodium. Although potassium ions are known to be important to the pumping process, the exact role of potassium is not known, and it has hence been omitted from the diagram. The low intracellular concentration of sodium allows its diffusion into the cell from both the medium and the environment, and the sodium is then pumped outward or inward, as necessary, to maintain the osmotic pressure of the blood. The main negative ion, chloride (Cl^-), is thought to diffuse passively in the same direction as the active transport of sodium ion. This passive diffusion occurs as a result of the electrostatic attraction between positive (Na^+) and negative (Cl^-) charges, which always tend to remain equal. Some animals, however, have chloride pumps as well.

OSMOREGULATION IN VARIOUS ENVIRONMENTS

Having considered the general problem of osmoregulation and the means by which it may be achieved, let us turn to actual animals living in different situations and discover which of the several options various animals utilize to maintain an internal salt and water balance. We shall follow the sequence shown in Figure 11.3.

Brackish Water Animals. Marine invertebrates, as we have seen, do not need to osmoregulate, since their body fluids have virtually the same composition as sea water. Invertebrates living in bays and estuaries, on the other hand, must either passively conform to the salinity fluctuations they experience, or else they must osmoregulate by actively pumping ions. Both mechanisms—conforming and osmoregulation—are found among brackish water animals, as shown in Figure 11.6.

The flatworm, *Procerodes*, is exposed daily in the estuaries in which it lives to sea water when the tide is in and to fresh water when the tide is out. This animal does not regulate and is an **osmoconformer.** (Compare the volume and concentration changes of this animal to the cell in Figure 11.1.) In dilute media, salts never leave the body in sufficient amounts to allow the body volume to return to normal after influx of water, and the animal remains swollen. The cells of *Procerodes* therefore must tolerate both osmotic dilution and change in volume. Such a range of passive tolerance in estuarine animals is uncommon, however. The marine spider crab, *Maja*, is able to survive a small amount of dilution. In doing so, it first takes up water and swells slightly; but soon internal salts are lost and the animal shrinks to its original size, although the internal medium is now dilute. Hence *Maja* is also an osmoconformer, but it is able to regulate its volume. Unlike *Procerodes*, the cells of *Maja* can withstand only a small degree of dilution, and the animal dies at salt concentrations less than 75 per cent that of sea water.

The Chinese mitten crab, *Eriocheir*, lives part of its life in freshwater streams and part of it in the sea. It has a very low permeability to both salts and water, and is able to maintain both its volume and its internal osmotic

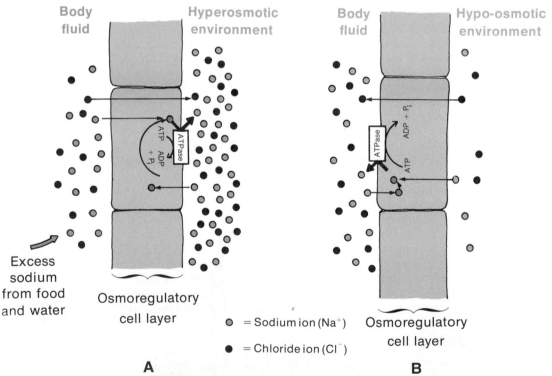

FIGURE 11.5

A model for the regulation of sodium and chloride ions in (*A*) hyperosmotic and (*B*) hypo-osmotic environments. The intracellular sodium ion concentration is thought to be kept at a low level by the ion pump, so that sodium diffuses into the cell from both directions. It is then pumped either outward to the environment (*A*) or inward to the body fluid (*B*). Chloride ions (Cl⁻) follow passively.

Plain arrows indicate passive diffusion processes; heavy solid arrows indicate active transport.

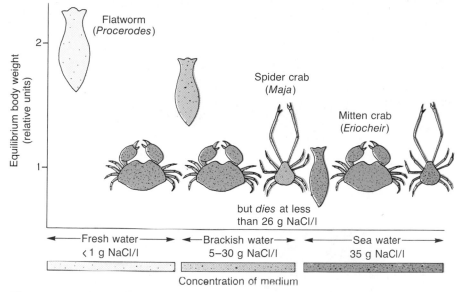

FIGURE 11.6

Three types of adjustment to dilution of the environmental medium. The color of the animals indicates internal osmotic pressure and should be compared with that of the various media below. (Animals not shown to scale.)

Procerodes is an osmotic conformer with minimal volume regulation but tolerates a wide range of salinities.

Maja is an osmotic conformer over a narrow salinity range but adjusts its volume.

Eriocheir is an osmotic regulator at all salinities, maintaining a constant osmotic pressure and volume.

pressure at all environmental salinities. *Eriocheir* is therefore one of the most perfect **osmoregulators** known. Most brackish water animals show only partial regulation, coupled with partial dilution. Osmoregulation in *Eriocheir* is carried out mainly by the gills, which actively transport ions inwards when the animal is in dilute water. The kidney excretes the excess water but does not reabsorb salts, and so the urine is isosmotic: it has the same concentration as the body fluid.

Freshwater Animals. Animals living constantly in fresh water, like *Eriocheir* during its residence in a freshwater stream, must both acquire extra salt and get rid of excess water. An excellent example is the freshwater bony fish shown in Figure 11.7A. Water diffuses inward across the body wall, while salts leak out. Fish do three things to compensate for this tendency to osmotic dilution: they drink as little water as possible, swallowing only food; they actively absorb salts via the gills; and they actively reabsorb salts in the kidney, producing large volumes of hypo-osmotic urine. Both gills and kidney have active sodium pumps.

Marine Vertebrates. The marine bony fish face the reverse set of problems, tending to lose water to the ocean (Fig. 11.7B). To compensate for this they drink sea water, absorbing both salt and water in the gut, and then excrete the excess salt back to the environment, using a sodium pump located on the gills. Their kidneys are unable to actively pump the major salts from blood to urine; the urine is thus isosmotic with the body fluids, and the kidneys do not perform an osmotic function. By minimizing the amount of urine formed, however, marine fish conserve considerable water.

Other vertebrates living in the ocean have evolved different means for excreting excess salt. Certain marine turtles have tear glands capable of secreting salt, and sea birds such as ducks, gulls, petrels and albatross have **nasal salt glands** which produce a highly saline secretion often seen as salt

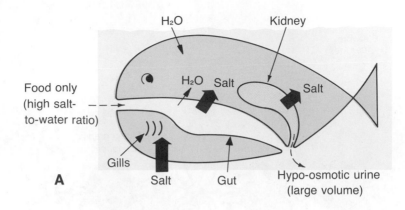

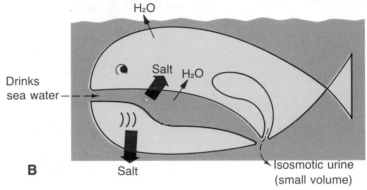

FIGURE 11.7

Osmoregulation in bony fish.
  A. Adaptations to hypo-osmotic fresh water.
  B. Adaptations to hyperosmotic sea water.
The thin arrows indicate passive diffusion; the heavy arrows indicate active transport.

crystals on the upper side of the beak. In both types of glands, as well as in the gills of fish, the amounts of [Na^+-K^+] activated ATPase enzyme are very high. Another means for conserving water found in reptiles and birds is the production of only a small volume of urine. Since this phenomenon is related to the type of nitrogen excretion they utilize, we shall consider it further below.

The means by which marine mammals conserve water is not well understood. All mammals have kidneys capable of excreting excess salts, thus producing hyperosmotic urine. Also, many of these animals avoid drinking sea water and feed mainly on bony fish whose osmotic pressure is the same as their own. The osmotic work has already been done by the fish! Such a solution is not possible for the great baleen whales, however, that feed mainly on plankton, whose salt content is the same as sea water. How they osmoregulate is not known. No external salt secreting glands comparable to nasal salt glands have been discovered in mammals so far.

Terrestrial Animals.   Terrestrial animals face very similar problems to those of marine vertebrates—too much salt relative to the amount of water available. Reptiles, birds and mammals, especially those living in arid regions, conserve water both by seeking shade or humid burrows in the soil and by means of special physiological mechanisms. In birds and reptiles, the isosmotic urine passes into the hind part of the gut, called the **cloaca,** where water from both urine and feces is reabsorbed before the two wastes are excreted together in a semi-solid form. In mammals, water reabsorption from the feces occurs in the large intestine, helping to conserve body water. In insects the kidney is an outgrowth of the gut. It forms an isosmotic urine which passes into the hindgut, where water reabsorption takes place.

Precisely how water reabsorption occurs in terrestrial animals is not well understood. Detailed studies of the anatomy and physiology of the insect hindgut, however, suggest that the epithelial cells involved in water reabsorption are capable of producing very high *local* concentrations of salts by active transport processes. Water diffuses from the gut into these regions, and then flows through a complex system of minute tubules back into the body fluids. The effective result is the *active transport of water* from a region of high solute concentration (feces) to one of low solute concentration (body fluids).

STRUCTURE AND FUNCTION OF THE MAMMALIAN KIDNEY

The thoughtful reader will perhaps have been wondering how it is possible to regulate osmotic pressure by pumping ions if the rates of water movements are as great as has been stated. Why does the water not simply follow the ions and cancel the gains made by active ion transport? There are two reasons why this does not happen. One is that the circulatory system usually permits the body fluids to remain in contact with osmoregulatory cells for only a short period of time, thus limiting the amount of water which diffuses after the ions. The other reason is that the cells of osmoregulatory organs often are far less permeable to water than are other cells of the body. Both these mechanisms are found in the human kidney, and hence it serves as an excellent model of how excretory organs function.

The human kidney is composed of about one million independent subunits called **nephrons,** whose upper regions lie in the outer zone of the kidney, the **cortex** (Fig. 11.8). These have elongate tubular loops which dip

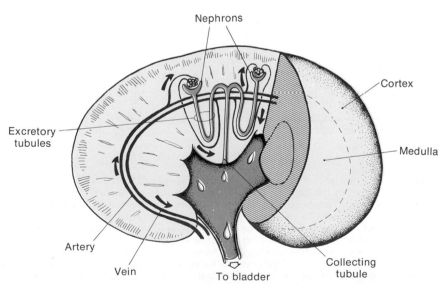

FIGURE 11.8

Schematic diagram of whole human kidney cut in half, with two nephrons (much enlarged) and circulatory system. Outer region (cortex) contains capsules; inner region (medulla) contains tubules. Direction of blood circulation indicated by heavy arrows.

into the central region or **medulla** of the kidney. Each nephron begins with a cup-shaped capsule within which a knot of capillaries nestles. Blood pressure forces water and the small soluble blood components into the capsule, leaving proteins and cells behind. The fluid in the capsule is thus called a **blood filtrate.**

The filtrate passes along the lengthy **excretory tubule** lined with cells capable of active transport of ions and small molecules. These cells actively reabsorb useful substances, such as glucose, from the filtrate. The reabsorbed molecules then diffuse into nearby blood capillaries. These cells also

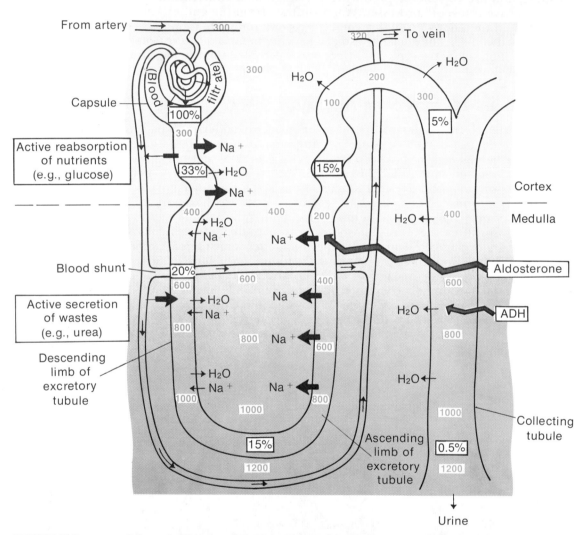

FIGURE 11.9   Schematic diagram of single nephron with associated circulation.
    Blood filtrate in the capsule flows through the loop of the excretory tubule to the collecting tubule. Osmotic pressure in kidney tissue indicated by intensity of color. Colored numbers refer to milliosmoles of solutes dissolved per liter of tissue water. Note that solute concentration in urine shown here is four times that of blood. Per cents in squares are fraction of original filtrate volume remaining in tubules.
    Aldosterone stimulates sodium-pump activity.
    Antidiuretic hormone (ADH) increases water permeability in the collecting tubules. If ADH is absent, urine that is more dilute than blood is produced, since water cannot diffuse so readily back into the kidney tissue.

secrete waste substances, like urea, effectively transporting them from the capillaries to the tubular fluid (Fig. 11.9). The extensive capillary network associated with the kidney tubules thus permits a continuous exchange of substances between tubular fluid and blood at all stages. Numerous shunts, or by-passes, in this network allow regulation of blood flow through the medulla. The tubular fluid finally flows into **collecting tubules,** where reabsorption of water occurs.

Osmoregulation in the human kidney is carried out in the following way. Sodium ions are actively reabsorbed from the filtrate into the kidney tissue by cells both near the capsule and in the ascending limb of the secretory tubule. Sodium pumps occur at these sites. This produces an osmotic gradient in the medulla of the kidney, indicated by the color in Figure 11.9. Water tends to leave the filtrate by diffusion, but this happens at only two places—only the descending limb and the collecting tubule are permeable to water. The ascending limb is impermeable. Because of the high concentration of sodium ion maintained in the medulla of the kidney, more than 99 per cent of the water in the blood filtrate is returned to the blood! Thus large volumes of blood can be processed while only a small volume of urine is formed. About 120 times more filtrate is formed than is actually excreted in the urine. This allows all the body fluids to be processed by the kidneys about sixteen times each day, insuring continuous maintenance of homeostatic conditions.

Control of Urine Concentration. In man and most other mammals, the urine may be either hyperosmotic or hypo-osmotic to the blood, depending on the body's water balance. The control of urine concentration is performed by three independent processes, two of which are hormonal. The level of activity of the sodium pumps is controlled by a hormone from the adrenal cortex known as **aldosterone.** A disease in which the adrenal glands fail to produce enough of this hormone (named "Addison's disease" after its discoverer) used to be invariably fatal due to the kidney's inability to retain salt. Today, however, treatment with aldosterone is possible.

A second hormone, **antidiuretic hormone** (ADH), is produced by specialized cells in the brain and released into the blood from a region of the pituitary gland. ADH is capable of increasing the water permeability of the collecting tubule, allowing water to flow osmotically from the urine into the hyperosmotic medullary region of the kidney. ADH is released into the circulation in high concentrations after eating a salty meal (such as a ham dinner garnished with dill pickles and olives!) and results in formation of a concentrated urine. In the absence of ADH, a dilute urine is produced.

Alteration of blood flow in the kidney also affects urine concentration. As we have seen, the salt gradient in the tissues of the medulla causes water to enter this region from the tubules. The resulting high hydrostatic pressure forces water back into the blood. Obviously, salt will also tend to diffuse into the blood, especially from the more concentrated regions of the medulla. Therefore, when water but not salt retention is required, the upper shunt of the blood capillary network (Fig. 11.9) is used. Blood flow to the deeper regions of the medulla is thereby minimized, diminishing the amount of salt which diffuses back into the blood.

The functioning of the kidney is severely affected by a wide variety of diseases, including many infections, metabolic diseases (such as Addison's disease) and heart disease. In addition, many toxic substances are especially damaging to the kidney, among which the common heavy-metal environmental pollutants, lead and mercury, damage both blood vessels and tubules. In acute cases, the tubules may become totally non-functional, all filtrate simply diffusing back to the blood and no urine being produced. Death may result in a few hours. Mercury is used clinically as a diuretic because it blocks water reabsorption, but after prolonged exposure it can produce permanent damage to the tubules. Lead may cause chronic nephritis (kidney disease) and, eventually, kidney failure.

NITROGENOUS WASTES

Nitrogen is a waste product mainly of animal metabolism. Plants, being the original synthesizers of amino acids, are much better equipped with the necessary enzymes to re-use their discarded nitrogen. In the course of metabolism in all types of organisms, molecules are constantly being destroyed and replaced. Some molecules are unstable and break down spontaneously; others are destroyed because their functions are no longer required. As cells adapt to new situations, for example, they require different enzymes. Although the carbon framework of these molecules is easily oxidized or converted to another use, the nitrogen that is associated with proteins and nucleic acids cannot usually be re-used by animals.

TYPES OF NITROGENOUS WASTE PRODUCTS

The initial nitrogen compound formed from protein breakdown is **ammonia** (NH_3). This is a gas which readily dissolves in water to form ammonium hydroxide (NH_4OH), the same substance that one buys in the market under the name "household ammonia." It is also a component in the solutions sold for hair permanents. Anyone having experience with either of these products is well aware that ammonia is a highly irritating substance, and would be extremely toxic to the body if it were allowed to accumulate.

Other nitrogenous waste products are **urea** and **uric acid,** the formulas for which are shown in Figure 11.10. Urea is highly water soluble and is relatively non-toxic. Uric acid is not at all toxic and only sparingly soluble in water; it tends to precipitate out into crystals.

NITROGENOUS WASTES IN RELATION TO WATER AVAILABILITY

From the above discussion it is apparent that only animals living in aquatic environments where water is abundant can afford to excrete all their nitrogen in the form of ammonia. Thus, marine and brackish water invertebrates and freshwater animals excrete ammonia as their main waste product (groups 1, 2 and 3 in Figure 11.3). Those animals with rather less water available, either in the environment itself or internally for urine formation, excrete mainly urea. Included in this group are the marine fishes and also the mammals. Urea, because it is a water soluble substance, requires a moderate volume of urine for its excretion.

There are three groups of animals, however, for whom water is at a great premium during at least part of their life cycle, and which therefore excrete most of their nitrogenous wastes as insoluble uric acid. Insects, especially desert forms, face continuous water shortage, and all but the most aquatic insects produce a semi-solid urine containing mainly uric acid. Both reptiles and birds lay their eggs on dry land, and the only water avail-

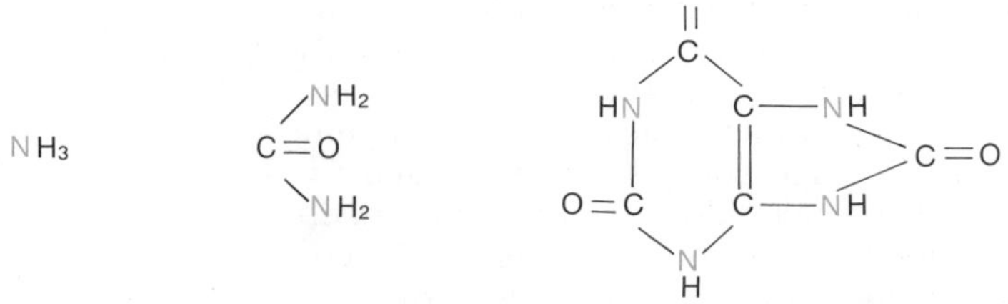

FIGURE 11.10

Chemical structures of the three major nitrogenous wastes. Ammonia and urea are both very soluble in water; uric acid is only sparingly soluble. Ammonia is highly toxic; urea is non-toxic at low concentrations; uric acid is non-toxic.

Ammonia   Urea   Uric acid

able to the developing embryo is that contained in the egg. This water has to suffice both for body water for the hatchling and for diluting the metabolic wastes which accumulate during embryonic development. Not even urea would be non-toxic in such minimal amounts of water. Hence, reptiles and birds begin, as embryos, to excrete mainly uric acid. In all these animals, the uric acid crystallizes out, forming a smooth white paste in the urine. One advantage of this precipitation of uric acid is that it allows the osmotic reabsorption of water in both kidney and cloaca, since a solid substance exerts no osmotic pressure.

The primary sites of ammonia formation in vertebrates are the liver and kidney. Urea is produced mainly in the liver. Small amounts of uric acid are formed and excreted by all animals as a result of nucleic acid breakdown. Uric acid is primarily synthesized in the liver and transported in the circulation to the kidney, where excretion takes place. In man, if uric acid elimination by the kidney is defective, the extremely painful disease known as gout develops, in which uric acid crystals precipitate out, especially in the joints.

SUMMARY

Osmoregulation—the ability to regulate internal water and salt levels at concentrations different from those found in the environment—is a problem faced by many animals. Plants, on the other hand, are generally not faced with this problem, although water deficiency causes wilting, and some salts are toxic at high concentrations. Animals may face desiccation, due to too much salt or too little water in their surroundings, or flooding, due to too little salt relative to the amount of water. Most marine vertebrates and terrestrial animals have the former problem, whereas freshwater animals have the latter. Organisms living in brackish waters are regularly exposed to changing conditions, to which they either passively conform or actively regulate.

The main way in which animals maintain a constant internal salt and water balance is by pumping ions, primarily sodium. Ion pumps are probably located on the membranes of all cells, but they are especially active in osmoregulatory organs such as kidneys, gills and nasal salt glands. The pumping of salts requires the expenditure of energy. A membrane-sited enzyme, [Na$^+$-K$^+$] activated ATPase, provides this energy by splitting ATP. Negatively charged chloride ions are thought to move passively, following positively charged sodium ions, thus maintaining electrical neutrality.

In animals, water transport against a concentration gradient appears to occur by means of local osmosis. Water permeability can be reduced by water-resistant body coverings. Vertebrates also produce hormones which regulate the water-permeability of their osmoregulatory organs, allowing them to produce concentrated or dilute urine as required.

The major nitrogenous waste products of protein and nucleic acid metabolism are ammonia, urea and uric acid. Ammonia is highly toxic and is produced primarily by aquatic animals. Less toxic urea is excreted by animals with only moderate desiccation problems, whereas uric acid is produced by organisms which have only limited water supplies.

Pitts, R. F. 1968. *Physiology of the Kidney and Body Fluids.* Year Book Medical Publishers, Inc., Chicago. A comprehensive treatise of the human kidney, written with extreme clarity.
Royal Society of London. 1971. A discussion on active transport of salts and water in living tissues. Philosophical Transactions B 262:83–342 (no. 842). Organized by R. D. Keynes.

Up-to-date accounts of various aspects of secretion and osmoregulation written in a clear style by major researchers in the subject.

Schmidt-Nielsen, K. 1959. Salt glands. Scientific American Offprint No. 1118. A well-illustrated account of the structure and function of nasal salt glands in marine birds and reptiles.

Smith, H. W. 1953. The kidney. Scientific American Offprint No. 37. An account for the layman about the homeostatic functioning of the kidney, written by the man who contributed most to our understanding of that organ.

Smith, H. W. 1961. *From Fish to Philosopher.* Doubleday & Co., Garden City. An easily read explanation of the role of the kidney in vertebrate evolution from water to land.

Society for Experimental Biology. 1965. The state and movement of water in living organisms. (Symposium No. 19.) A series of papers covering many aspects of water transport. See especially the excellent discussion by J. M. Diamond on transport of water and solutes across membranes.

12

HOW NERVES AND MUSCLES FUNCTION

The ability to respond rapidly to environmental stimuli is a characteristic of animals. It depends upon the coordinated functioning of receptors, nerve cells and effectors.

Although plants are capable of a certain degree of movement in response to environmental stimuli—for instance, houseplants kept near a window orient their leaves toward the light—such movements are slow. They are brought about by changes in turgor within the cells, which may result in cell elongation or growth. In contrast, the responses of animals are rapid, and are usually the result of reversible changes in cell shape—in particular, the contraction of muscles. How responses in animals are brought about forms the subject matter of this chapter.

Even single-celled organisms are capable of responding to environmental stimuli. The direction of cytoplasmic streaming within an amoeba, for example, is altered as a result of the nearby presence of food or of a noxious substance, either of which causes an appropriate change in direction of movement. Changes in the direction of ciliates are brought about by alterations in beating patterns of cilia. More complex animals use muscles and glands to effect responses to stimuli. Such responses involve three components: a **receptor** which receives the stimulus; a **central nervous system** which transmits and analyzes the information received; and an **effector** which brings about the response. Examples are shown in Figure 12.1.

HOW NERVES FUNCTION

Cells of all animals possess the property of **irritability:** they are able to respond to changes in their immediate environment. Since irritability is best developed in the nervous system, we shall use nerve cells to describe its basis.

Nerve cells, or **neurons,** come in many shapes but share certain features in common. The vertebrate motor (muscle innervating) neuron, shown in Figure 12.2, will serve as an example. It has a large nucleus, surrounded by cytoplasm; these together constitute the central part of the cell, the "cell body." The cytoplasm bears extensions or processes by which the cell receives information and transmits it to other cells. The numerous processes which receive information from receptors or from other nerve cells are called **dendrites** and are usually short; these constitute the region of input into the cell. The single process which transmits the information to other neurons or to effectors is the **axon;** this is the output region of the cell. In large animals some axons may be several feet in length. As shown in Figure 12.2, there are numerous sources of information coming into a nerve cell, and the axon transmits to several other cells as well. At these junctions, where the axon terminals often appear as swollen knobs, the membranes of the two cells are not touching, but are separated by a tiny space, the **synapse.** We shall investigate how synapses work shortly.

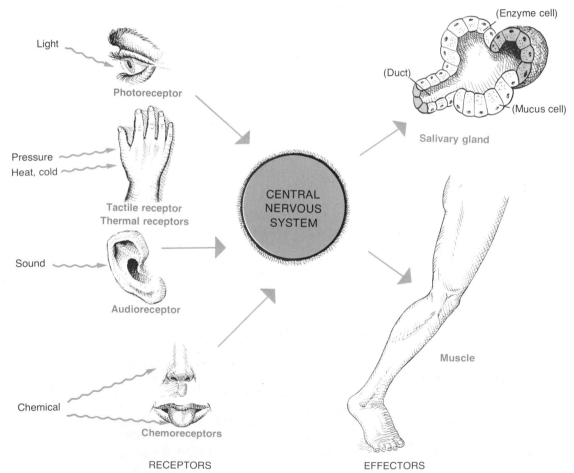

FIGURE 12.1   Components of response by animals to external stimuli.

Receptor organs receive information, which is then transmitted to the central nervous system. After analysis, the nervous system sends information to effectors which perform the appropriate response. Effectors include glands, such as the salivary gland and certain endocrine glands, and also muscles. Organs shown are those of man. Analogous organs are found in many other animals.

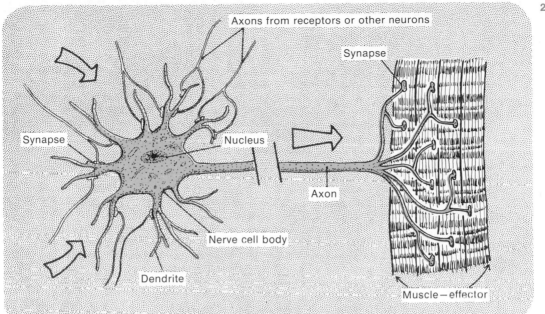

Anatomy of a motor neuron.

FIGURE 12.2

Messages from either receptor cells or other neurons are received by dendrites. They are then passed on by the axon either to other nerve cells or, as shown here, to an effector. Junctions between cells are known as synapses. Large arrows indicate direction of information flow.

THE BASIS OF IRRITABILITY

Many theories of irritability have been proposed, but the one most widely accepted is based on electrical potentials which exist across plasma membranes, known as **membrane potentials.** As we noted in Chapter 11, the distribution of sodium (Na^+) and potassium (K^+) ions on the inside and outside of cells is not the same. In nerve cells, the concentration of K^+ inside is about 20 times higher than that outside the cell, and the concentration of Na^+ outside is about 10 times higher than that inside. This condition, as we shall discover shortly, is essential for conduction of nerve impulses. The unequal distribution of sodium and potassium ions is maintained by the sodium pump, which actively extrudes Na^+ from the cell and replaces it with K^+. The pump, of course, requires energy from the splitting of ATP in order to function. If Na–K exchange were the only process involved in ion distribution, no membrane potential would result, since the equal exchange of Na^+ and K^+ ions does not alter the proportion of positive and negative charges across the membrane. Another factor must be involved.

This factor is the presence of more protein inside the cell than outside. Most proteins have an excess number of negatively charged groups. It is the proteins within the cell that are mainly responsible for balancing the positive charges of the potassium ions. These proteins are large and cannot pass across the cell membrane; they thus represent immobile or "fixed" charges. Outside the cell, however, the positive charges of Na^+ are mainly balanced by chloride ions (Cl^-). Chloride ions are small and can diffuse across the cell membrane. Because there is little Cl^- within the cell, chloride ions tend to diffuse inward to "equalize" their concentration on both sides of the cell membrane. As they do so, they upset the charge balance; remember that the negatively charged proteins must remain inside. The inside of the membrane thus becomes negative relative to the outside, establishing the membrane potential (Figure 12.3).

When the inside charge becomes great enough, it repels the further entry of Cl⁻ into the cell, since like charges repel one another. At the point when the tendency for Cl⁻ to diffuse inward, due to its concentration difference, is exactly balanced by the repulsion of further Cl⁻ entry due to excess internal negative charges, an equilibrium potential is established. For nerve cells, this generally amounts to about −70 millivolts (a millivolt = 1/1000 volt, and is abbreviated mV). In other words, if we arbitrarily set the potential outside the cell to zero, then the inside potential is −70 mV. This

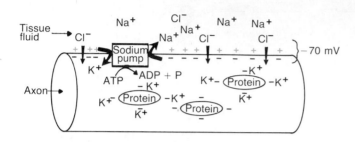

FIGURE 12.3

Schematic diagram of formation of resting membrane potential in nerve axon.

High K⁺ concentration inside and high Na⁺ concentration outside are maintained by the sodium pump. Cl⁻ tends to diffuse to equalize its concentration on both sides. As a result, the inside of the membrane becomes negative relative to the outside by −70 millivolts. The resting membrane behaves as if it were impermeable to Na⁺.

is called the **resting potential**, since it exists in the inactive or resting nerve. Our explanation of the factors involved in establishing the resting potential is somewhat oversimplified, since we have ignored the tendency of Na⁺ and K⁺ to migrate in response to electrical and chemical gradients. In fact, the resting membrane behaves as if it were impermeable to Na⁺, a condition which no longer applies when it is conducting an impulse.

ACTION POTENTIALS

The function of the nervous system is to transmit information from one part of the body to another, which it performs by transfer of electrical energy—the nerve impulse. The conditions for generating such an impulse depend upon the presence of the resting potential described above, especially on the unequal distribution of + and − charges *and* on the unequal distribution of Na⁺ and K⁺ across the cell membrane.

An impulse is initiated by partial **depolarization** of a small region of the cell membrane; some of the charge difference is lost, and the membrane potential moves toward zero. Depolarization occurs either upon receiving a stimulus from another nerve cell or by experimental application of a nega-

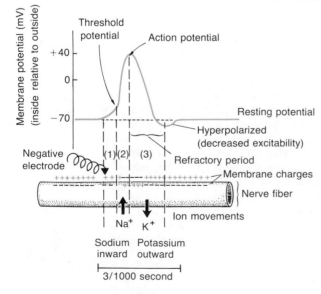

FIGURE 12.4

Changes in local membrane potential after partial depolarization by a negative electrode.

Once the threshold potential is reached, sodium ions flow in rapidly, producing a reversal in membrane charge, the action potential. This is followed by a rapid outflow of potassium ions, restoring the resting potential. Excess loss of K⁺ often causes temporary hyperpolarization.

tive electrode to the outside of a nerve axon, as shown in Figure 12.4. A series of rapid changes occur in the depolarized region of the membrane:

1) The initial stimulus (from the negative electrode), causes partial depolarization of the membrane. The potential moves toward zero.
2) If the stimulus is strong enough, a **threshold potential** is reached, at which point membrane permeability to sodium ion abruptly increases and Na^+ rushes into the cell, down its concentration gradient. This results in an immediate local reversal in membrane polarization, known as the **action potential.** The outside becomes negative relative to the inside.
3) Recovery occurs through a second change in permeability of the membrane, this time to potassium ion: K^+ rushes out, restoring the resting potential. Often excess K^+ leaves the cell, causing temporary **hyperpolarization;** the inside is even more negative than usual. During this recovery period, the nerve is unresponsive or **refractory** to further stimuli; it also has decreased excitability while hyperpolarized — a larger stimulus is needed to depolarize it to the level of the threshold potential.

It is important to realize that these changes occur very rapidly — in a few thousandths of a second. Only a small number of Na^+ and K^+ ions flow across the membrane with each impulse, and one impulse has little effect on the total ion concentrations inside and outside the cell.

It is the action potential which results in propagation of the impulse (Fig. 12.5). The instantaneous charge differences cause a flow of charges along the inner and outer surfaces of the membrane, depolarizing adjacent regions. When threshold potentials are reached in these neighboring regions, rapid Na^+ movements occur, resulting in action potentials in them also. These in turn stimulate inactive adjacent areas, and so on. Thus, the action potential moves along the nerve fiber in a sort of chain reaction.

As shown in Figure 12.5, an axon is capable of conducting an impulse in *both* directions and does so when artificially depolarized by an electrode. During normal functioning, however, most neurons conduct only in one direction, away from the nerve cell body. This happens because the nerve cell receives stimuli only at the dendrite-bearing end.

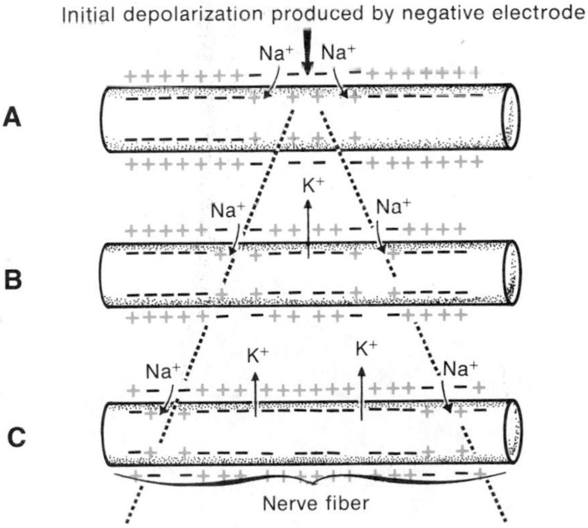

FIGURE 12.5

Propagation of a nerve impulse.
  A, B and *C* are time sequences in the propagation of an impulse after stimulation by a negative electrode, showing how an impulse travels along a nerve fiber. The action potential generated at the site of initial depolarization causes charge flow in neighboring regions of the membrane. When these, in turn, reach the threshold potential they are further depolarized by inflow of sodium ions generating the action potential anew.

When a nerve is stimulated, it either conducts or not, according to whether the threshold potential is reached, leading to an action potential. All responses are the same; there is no gradation of response. This is called the **all-or-none principle** of nerve conduction. Conduction rates are extremely rapid, reaching speeds of 60 feet per second or more in active animals. Certain nerves in mammals conduct up to 300 feet per second!

Although a single impulse has very little effect on the ion distribution across the membrane of a nerve fiber, many nerve cells, especially in the brain, conduct repeatedly; some fire off impulses several hundred times a second. After a few thousand impulses, the Na^+ and K^+ concentration gradients would be badly depleted were there no means of restoring them. The membrane-bound sodium pump restores the ion gradients after a series of impulses, thus maintaining the gradients necessary for generating an impulse. The pump depends on a continuous supply of ATP, and poisoning the respiratory reactions by which ATP is generated quickly stops the functioning of a nerve cell. In man and many other animals, the nervous system is unable to obtain sufficient amounts of ATP from anaerobic glycolysis and requires a constant supply of both oxygen and blood sugar to meet its energy needs. Insufficiency of either causes unconsciousness or death.

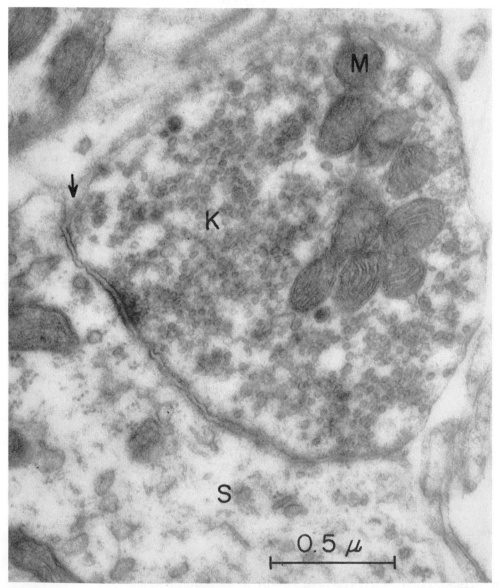

FIGURE 12.6   Synapse on a motor neuron.

 A. Electronmicrograph of synaptic junction (magnified about 65,000 ×). (From Palay, S. 1958. Experimental Cell Research (Suppl.) 5:275–293.) *M,* mitochondrion; *K,* axonal knob; *S,* postsynaptic dendrite. Arrow indicates synaptic space.

Illustration continued on opposite page.

As we have just seen, a nerve cell, once stimulated, always transmits the same amount of information in accordance with the all-or-none principle. It remains, then, to explain how the same stimulus can elicit different responses on different occasions or in different individuals. We can ask, How is information processed in the nervous system? To a large extent, we still do not understand how processing occurs. If we did, many aspects of human behavior—from temper tantrums of children to assassinations of presidents—might be explicable. It is possible, however, to explain the existence of alternative pathways in relatively simple systems. These pathways are largely determined by the functioning of synapses.

Except in very primitive animals, a synapse conducts in one direction only, accounting for the one-way conduction of impulses by neurons. A synaptic junction as it appears at very high magnification in the electron microscope is shown in Figure 12.6A. The electrical impulse arriving at the end of the axon must be transmitted across the synaptic space and act on the receiving cell on the other side. The Nobel Prize in Biochemistry and Medicine for 1970 was awarded to three scientists who spent many years elucidating the way in which a synapse works; a diagrammatic interpretation is shown in Figure 12.6B.

The swollen tip of the axon of a transmitting cell contains many tiny, membrane-bound vesicles filled with a special chemical known as a **neurotransmitter.** When the impulse arrives at the synapse, it causes a fixed number of vesicles near it to break down and release their contents into the synaptic space. The neurotransmitter then diffuses across the space—a distance of only a few molecular diameters—and binds to specific sites on the thick, **postsynaptic membrane** of the receiving cell. This results in a change

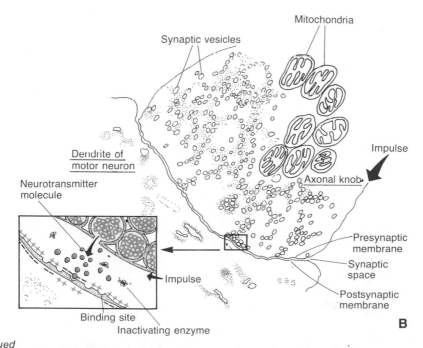

Continued

B. Drawing of *A*, identifying the important structures. Inset shows in schematic fashion how neurotransmitter molecules, released from a synaptic vesicle by an impulse, cross the space. They bind to specific binding sites on the postsynaptic membrane, causing either depolarization (shown here) or hyperpolarization. Inactivating enzymes in the synaptic space quickly destroy neurotransmitter molecules.

TABLE 12.1 COMMON NEUROTRANSMITTERS*

Name	Formula	Main Functions in Man	Inactivating Enzymes
Acetylcholine		Excitation of locomotory muscles Inhibition of heart	cholinesterase
Noradrenalin (related to adrenalin)		Constriction of blood vessels, maintenance of blood pressure, etc.	monoamine oxidase (MAO) and catechol-O-methyltransferase (COMT)
5-hydroxytryptamine		Stimulates visceral muscles "Tranquilizer" in brain	monoamine oxidase (MAO)

*Amine groups are circled.

of polarization of the postsynaptic membrane. Some neurotransmitters cause a hyperpolarization of the postsynaptic membrane; others cause a partial depolarization. The former are called **inhibitory synapses** since a greater stimulus is needed to reach the threshold potential. The latter are called **excitatory synapses,** since they tend to produce an action potential. We shall consider the role of each shortly.

Several types of neurotransmitters have been identified; the commonest of them are listed in Table 12.1. All are amines, which are irritating substances if applied to exposed tissues. Amines are closely related to amino acids, and many of them are synthesized from these protein building blocks.

If neurotransmitters remained in the synaptic space after their release, they would continue to exert their powerful effects on the postsynaptic membrane, and rapid changes in the nervous system's responses would not be possible. The knee-jerk reflex, for example, would become a continuous extension of the leg instead of a brief kick! However, certain enzymes released into the synaptic space rapidly destroy the neurotransmitters. These inactivating enzymes, which are listed in Table 12.1, must act very quickly to permit continued functioning of the nervous system. One molecule of the enzyme **cholinesterase** is capable of destroying about 20 million acetylcholine molecules per minute.

The Summation of Input. As we have noted, some neurotransmitters hyperpolarize the postsynaptic membrane, whereas others depolarize it. Each nerve cell in the central nervous system receives information from many sources (see Figure 12.2). Some of these are excitatory, releasing neurotransmitters which depolarize the membrane; others are inhibitory, producing neurotransmitters which hyperpolarize the membrane. Thus, the response of a particular nerve cell at a given moment will depend on the *sum* of the excitatory and inhibitory stimuli it receives at that instant. If the membrane around the nerve cell body receives a preponderance of excitatory stimuli, it will become sufficiently depolarized to reach the threshold potential and generate an action potential in its axon. If too many inhibitory stimuli are received, then the hyperpolarized membrane will be refractory to excitation, and no action potential will result.

Synaptic Modifiers. Many drugs and poisons affect synaptic junctions. These can act in several ways, of which a few examples may be given here. (We shall discuss the effects of such substances in further detail in Chapter 12a.) Many drugs mimic the action of neurotransmitters. For instance, cocaine and amphetamines ("speed") act as stimulants, resembling noradrenalin. Other substances inactivate binding sites for neurotransmitters on the postsynaptic membrane. The poison strychnine prevents binding of the inhibitory transmitter at motor synapses in the spinal cord of vertebrates. In the presence of strychnine, only excitatory impulses reach the muscles; if touched lightly, the animal goes into tetanus (continuous contraction of all locomotory muscles). Botulin, a toxin which is the product of certain bacteria that grow in improperly sterilized canned foods, prevents release of the motor excitatory transmitter acetylcholine and thus causes paralysis.

FACILITATION OF NERVOUS PATHWAYS

If a particular nerve cell receives excitatory stimuli regularly, it gradually requires fewer simultaneous stimuli to generate an action potential. This phenomenon, known as **facilitation,** permits the modification of the response of the nerve cell according to previous information received by it. Facilitation of this sort may last from minutes to hours. Other, more permanent modifications of nervous pathways are involved in memory, which is discussed in Chapter 13.

MUSCLE EFFECTORS

The nervous system communicates information to both glands and muscles, which then produce the appropriate responses to the original external or internal stimulus. Glands associated with the digestive tract, as well as certain gland-like cells in part of the brain, are innervated by nerves which induce the release of their secretions. We shall consider the gland-like cells in the brain as they relate to human reproduction in Chapter 22.

Muscles, however, are the most important means by which animals respond to environmental stimuli. Although heart muscle and a few other muscles contract spontaneously (see Chapter 11), most muscles require stimulation from the central nervous system in order to contract. Like nerves, muscles have a highly polarized cell membrane. In some animals, the muscles are innervated directly by both excitatory and inhibitory nerve fibers. In the vertebrates, though, the voluntary muscles have only excitatory fibers; inhibition of the excitatory nerve occurs centrally in the nerve cord. A **neuromuscular** (nerve-muscle) **junction** is functionally very similar to the synapse pictured in Figure 12.6.

There are far more muscle cells than motor neurons in the human body. Thus the hundreds of individual cells of a single muscle—for example, the biceps—are grouped together in **motor units,** each unit being innervated by the same set of excitatory nerve fibers. All cells in a unit contract at once. Since there are many such units in a whole muscle, a graded series of responses is possible, even though an individual motor unit obeys the all-or-none principle each time it contracts. Far fewer units contract in lifting a five pound sack of potatoes than in lifting a 50 pound barbell.

There are two types of muscle activity: that which maintains **tonus,** or continuous tension, and that which brings about movement. Tonus is essential for maintenance of posture when an animal is awake. Various motor units within a postural muscle, such as those of the back, take turns developing tension, but the whole muscle does not change in length. Movement, on the other hand, requires a shortening of the whole muscle. During this type of contraction, the muscle changes shape—it becomes shorter and fatter—although its volume remains virtually the same.

MECHANISMS OF CONTRACTION

It has long been known that muscle cells contain very elongate, fibrous proteins which, even when extracted from the cell, are capable of changing shape under the right conditions. During the past two decades or so, studies both of muscle cell structure using the electron microscope and of muscle chemistry have produced a theory of contraction which so far fits the experimental data. Figure 12.7 summarizes this theory.

Part of a very elongate muscle cell is shown in Figure 12.7A. Note that it has many nuclei located around the perimeter of the cell. The cell also contains many **myofibrils,** which are composed of the elongate contractile proteins, and between them lies the ordinary cytoplasm of the cell, especially rich in endoplasmic reticulum and mitochondria (Fig. 12.7B). At regular intervals on the cell surface, the cell membrane forms "pores" that dip down into the cell and are closely associated with the endoplasmic reticulum. Between these regions lie the contractile units.

There are two types of contractile proteins in the contractile units, which are arranged in a specific relationship with one another (Fig. 12.7C): thin filaments, containing molecules of **actin;** and thick filaments, containing molecules of **myosin.** The myosin molecules are believed to bear little extensions along their length which are able to form cross-bridges with the actin-containing filaments. Myosin also has enzymatic activity and is capable of splitting ATP. When a relaxed muscle cell is stretched, the actin and myosin filaments barely overlap.

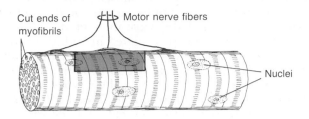

Part of a single muscle cell

A

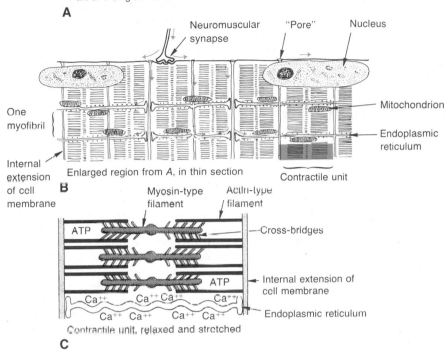

Enlarged region from *A*, in thin section

B

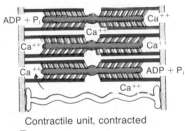

Contractile unit, relaxed and stretched

C

Contractile unit, contracted

D

FIGURE 12.7

Diagram of a muscle cell.

    A. Part of a single muscle cell, showing peripheral nuclei, central myofibrils and innervating nerve.

    B. Enlarged view of part of *A* showing relations between neuromuscular synapse, "pores" of muscle cell membrane, endoplasmic reticulum, mitochondria and contractile units, which have dark and light regions. Path of nerve impulse within cell indicated by colored arrows.

    C. A contractile unit in relaxed, stretched state has little overlap between thin actin filaments and thick myosin filaments. ATP is present.

    D. A contractile unit in contracted state. When an impulse depolarizes extension of cell membrane between the contractile units, calcium ion (Ca^{++}) is released from the endoplasmic reticulum. This causes splitting of ATP by myosin. Simultaneously, cross-bridges between actin and myosin are rapidly formed, broken and re-formed, resulting in contraction.

The currently most accepted theory of muscle contraction is the following: When an excitatory nerve impulse arrives at the muscle, acetylcholine is released, causing local depolarization of the muscle membrane. This wave of depolarization is propagated in both directions along the membrane and reaches the endoplasmic reticulum internally. The membranes of the endoplasmic reticulum are thereby induced to release calcium ion (Ca^{++}), which is tightly bound to them when the muscle is relaxed. In a way not yet clearly understood, the free Ca^{++} causes cross-bridges between actin and myosin to form, break and re-form, over and over again. In this way, the actin and myosin molecules slide over one another, resulting in a high degree of overlap (Fig. 12.7D). This is known as the "sliding filament" theory of muscle contraction. During the process, myosin splits ATP, providing the source of energy for contraction. When the membranes of the cell become repolarized, the free Ca^{++} is once again bound to them, ATP is no longer split and the muscle relaxes. There are still many details of this process which remain to be explained, however.

ENERGY FOR CONTRACTION

ATP is the immediate source of energy for muscle contractions, as we have just seen. It may be analogized to "cash on hand." However, the amount of ATP in a cell is sufficient to sustain only a few contractions and must be replenished if continued contraction is to occur. Despite the presence of mitochondria, stored glycogen and blood capillaries bringing oxygen, the muscle cells of most active animals are not able to synthesize ATP fast enough from respiratory reactions to meet the needs of sudden spurts of muscular activity. For this purpose most animals have evolved a special type of high energy storage molecule, known as **phosphagen.** We can think of this as a type of energy "savings account." Different animals have slightly different types of phosphagens. The specific chemistry of these compounds is not important; they are all nitrogen-containing bases but are different from those in nucleic acids. What is of importance is that they are reservoirs of high energy phosphate bonds, which can replenish ATP after it is split during contraction.

$$\text{ATP} \xrightarrow{\text{myosin ATPase}} \text{ADP} + \text{P}_i$$

$$\boxed{\text{Nitrogenous-base}} \sim \text{P} + \text{ADP} \longrightarrow \text{ATP} + \boxed{\text{Nitrogenous-base}}$$

(phosphagen)

Even the phosphagens, however, are not present in unlimited supply, and so a constantly contracting muscle—such as the leg muscle of a long-distance runner—must obtain energy from glycolysis and respiration, burning stored glycogen, the "fixed assets" of the cell's energy resources. Despite the heavy breathing and increased heart rate of the long-distance runner, his circulatory system is unable to meet the oxygen and glucose demands of his leg muscles, and the only remaining source of energy is the glycogen stored in the muscles themselves (see Chapter 10). In the absence of oxygen, this glycogen can yield only relatively small quantities of ATP by means of glycolysis. The end product of this reaction pathway is lactic acid (see Chapter 8), some of which passes into the blood, the rest remaining in the muscle cells. It is probable that sensations of fatigue during prolonged exercise are due to accumulation of lactic acid in the body.

THE OXYGEN DEBT

Extended muscular exertion leads to a so-called "oxygen debt." The muscles have gone on contracting in the absence of sufficient oxygen to maintain their normal ATP and phosphagen levels, and lactic acid has accumulated. To restore the *status quo* after exercise, an animal must respire extra oxygen, reconverting the lactic acid first to glucose and then to glycogen. These reactions require ATP, synthesized during the oxidation-reduction reactions of the electron transport chain. Thus, the long-distance runner continues to breathe heavily and his heart continues to beat faster than usual for several minutes after the end of a race, while he "repays" his oxygen debt.

RECEPTORS

Receptors are specialized cells or organs which are capable of transducing or converting an environmental stimulus into the action potential of an impulse. Environmental stimuli come in many forms; light, sound, pressure, heat, cold, gravitational force and chemicals are all perceived by various animals. It should be noted, however, that not all possible forms of environmental stimuli are perceived by all animals. For example, the human eye responds only to the so-called visible wavelengths of light; we do not "see" ultraviolet nor infrared, nor can our eyes detect the extremely short wavelengths of X-rays (which are a form of energy related to light). Other animals are able to see colors of light invisible to us, and some insects are even able to respond to X-rays. Although electrical stimuli are perceived by all animals, only a few animals are known to sense the forces of a magnetic field. Some interesting experiments with birds suggest that the Earth's magnetic field may be detected by them during long-distance migrations.

Each type of receptor is specialized to receive a particular kind of stimulus. Receptors contain modified nerve cells, called *sensory cells,* which are capable of absorbing one type of energy—light, heat, or pressure, for example. The absorption of this energy causes a depolarization of the membrane of the sensory cell; it fires an impulse down its axon which leads to the central nervous system. Since space does not permit us to consider details of all the various types of receptors known in animals, we shall restrict ourselves to a few examples.

PHOTORECEPTORS

Photoreceptors range all the way from simple, cup-like structures capable only of detecting the intensity of a light source, to the complex eyes of insects, octopus and man, which have lenses and are capable of image formation (Fig. 12.8). In all cases, there is a group of sensory cells which contain a photosensitive pigment, usually a derivative of vitamin A combined with a protein. This pigment breaks down on absorption of light energy, releasing chemical energy which in some way affects the cell membrane and causes a change in its polarization. Resynthesis of the photosensitive pigment requires ATP, and such cells therefore have numerous mitochondria. Behind the sensory cells there is usually a layer of cells which contain black, absorbing pigments to prevent scattering of light, which would confuse the image.

Color vision, once thought to be rare among animals, may be quite widespread. Observations on behavioral preferences of animals suggest that many vertebrates and insects can distinguish between different colors. Color vision requires the presence of several additional photosensitive pigments, each sensitive to a limited range of wavelengths (colors) of light.

The eyes of octopus and man function much like a camera, the cornea and lens focusing the image on the sensory cells of the retina, which are analogous to the grains of silver salts on photographic film. Impulses from the retina are transmitted by way of the optic nerve to the brain, where

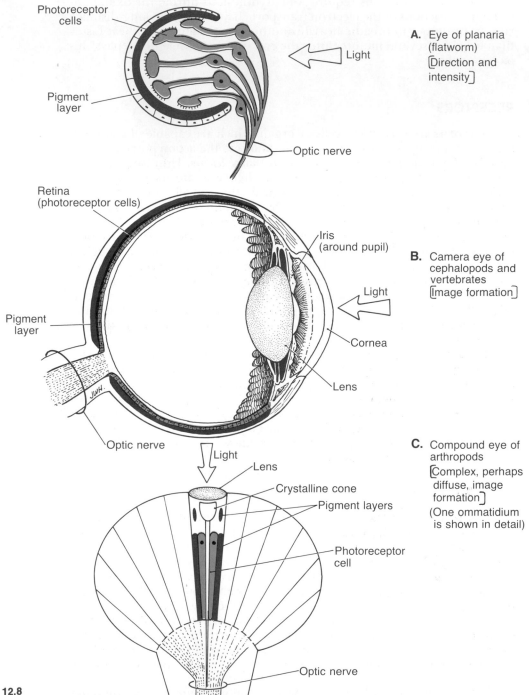

Photoreceptor
cells

Pigment
layer

Optic nerve

Light

A. Eye of planaria
(flatworm)
[Direction and
intensity]

Retina
(photoreceptor cells)

Pigment
layer

Optic nerve

Iris
(around pupil)

Light

Cornea

Lens

B. Camera eye of
cephalopods and
vertebrates
[Image formation]

Light

Lens

Crystalline cone

Pigment layers

Photoreceptor
cell

Optic nerve

C. Compound eye of
arthropods
[Complex, perhaps
diffuse, image
formation]
(One ommatidium
is shown in detail)

FIGURE 12.8

Representative types of photoreceptors (eyes).
The conversion of light energy into an action potential occurs in photoreceptor cells, which contain a photosensitive pigment. Photoreceptors are usually enveloped by a layer of pigment which absorbs excess light, preventing scattering and formation of a hazy image. Eyes capable of forming images have a lens (*B* and *C*) that focuses light on photo-receptor cells. Camera eyes (*B*) have only one lens and form sharp images of limited scope. Compound eyes (*C*) have only a few photoreceptors in each unit (ommatidium) and probably form many overlapping hazy images. What is actually perceived by animals with compound eyes is not known.

there is a spatial correspondence with specific areas of the retina. A sharp blow on the head which causes some of the neurons in the brain to discharge impulses thus gives one the impression of "seeing stars."

The eyes of insects and some crustaceans, on the other hand, consist of many small, independent photoreceptor units, called **ommatidia**, each with its own lens, cluster of photoreceptor cells and surrounding pigment. Such eyes are known as **compound eyes.** Each ommatidium gathers light from only a small part of the entire visual field and records its mean intensity in the corresponding area of the brain. The total image is thought to be a mosaic picture made up of dots of different intensities, much as a newspaper photograph looks under a magnifying glass. Although the image itself is hazy, the wide visual angle and short image retention time of some insect eyes make them especially well suited for vision during rapid flight. Whereas man cannot distinguish light flickers of more than about 50 per second, some flies can detect flicker frequencies of 265 per second. Thus, a fly in a movie theater does not see the apparently continuous motion on the screen that we do; instead it sees a rapid series of stills.

MECHANORECEPTORS

Mechanoreception, the detection of changes in pressure, is widespread among animals. Distortion of the sensory cell membrane or of a minute, hair-like process extending from the receptor cell induces membrane depolarization which results in an impulse. Cells of this type are involved in many types of perception. Tactile receptors in the skin detect external pressure due to changes in their shape. Sand grains or, in man, calcium carbonate granules, bend hairs on sensory cells in the balancing organ of the inner ear; auditory receptors in another part of the inner ear of man possess fine hairs which are stimulated in response to certain frequencies of vibration (Fig. 12.9A). The distortion of stretch receptors in muscles (Fig. 12.9B), causes them to send impulses to the brain indicating the state of contraction of all voluntary muscles—information which is necessary for maintenance of posture and balance and for coordinated movement.

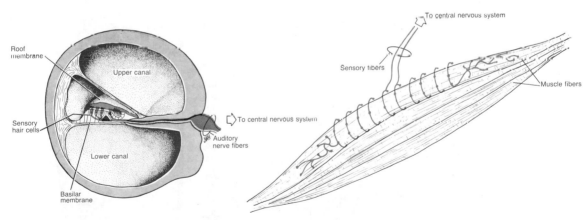

A Enlarged view of the auditory part of the inner ear, seen in cross-section

B Stretch receptors in muscle

Examples of mechanoreceptors.

A. Sound receptors in the human ear. Vibrations reaching the fluid-filled canals of the inner ear produce movements of the basilar membrane; these cause the hair cells to touch the roof membrane. Action potentials generated by distortion of the hairs are transmitted to the brain via the auditory nerve. The basilar membrane varies in length in different parts of the inner ear, each length being resonant with a given pitch or frequency of vibration.

B. Stretch receptor in a voluntary muscle. If the muscle is passively stretched, the distortion stimulates the receptor; it "tells" the central nervous system to send a contraction impulse. (Similar receptors in tendons at the ends of muscles are stretched when a muscle actively contracts; they send impulses to the central nervous system which result in relaxation of the muscle.)

FIGURE 12.9

CHEMORECEPTORS

Chemoreception, the ability to smell and taste, is the least well understood of all sensory mechanisms, despite its widespread occurrence and enormous importance to animals. We still do not know how the nose discriminates differences in perfumes or how dogs distinguish one another by scent. Although man's sense of smell is rather poorly developed, many animals detect odors at great distances. Chemical communication between animals is extremely important. Chemoreception is involved not only in finding food and eliciting feeding responses but also in release of sex attractants and mating behavior, in establishing and defending territories and in detecting predators. DDT and other environmental contaminants, although causing no overt damage to an animal, may interfere with its chemoreceptors and disrupt its normal behavior patterns. Experiments with bullhead catfish indicate that courtship behavior necessary for mating may fail in the presence of relatively low levels of DDT.

THE ORGANIZATION OF NERVOUS SYSTEMS

In the course of evolution, there has been a tendency both to increase the number of neurons in the nervous system and to organize them into clusters of nerve cells known as **ganglia** (singular, ganglion). Parasitic roundworms, primitive animals which infect the guts of many animals, have only 162 nerve cells and are capable only of simple types of behavior. A highly developed animal such as an octopus, which is able to distinguish and remember complex information and to make a variety of complex responses, has about one billion nerve cells. Man has about 20 billion.

In the simplest nervous systems, nerve cells are scattered in a **nerve-net**

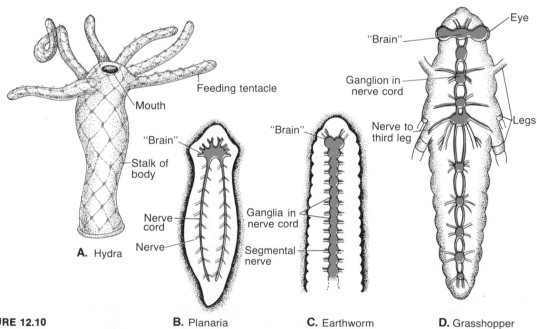

FIGURE 12.10

A. Hydra   **B.** Planaria   **C.** Earthworm   **D.** Grasshopper

Increasing complexity of nervous systems.

A. The simple nerve-net of the coelenterate, *Hydra,* has no ganglia. Synapses conduct in both directions.

B, C and *D.* Central nervous systems of the flatworm (*Planaria*), the earthworm and the grasshopper show increasing organization into ganglia. Cephalization is apparent in the increasing size and importance of the brain. The ganglia contain many association neurons. Synapses conduct in one direction only.

throughout the body. Such is the case in coelenterates—jellyfish, hydroids and sea anemones (Fig. 12.10). Synpases conduct in both directions, and a stimulus at one point often elicits a general reaction from the whole animal. In these animals, stimulus strength and frequency appear to play a major role in determining whether the whole body or only the local region near the site of stimulation responds.

As nervous systems increase in size and complexity on the evolutionary scale, they become more highly organized (Fig. 12.10, parts B to D). The additional neurons are grouped into centrally located ganglia, which constitute the **central nervous system** or CNS. Sensory fibers leading to and motor fibers going away from the CNS form the **peripheral nervous system.** Most of the centrally located neurons are small and have as many as 200 dendritic (input) synapses. They are known as **association neurons,** and function both in analyzing simultaneous information and in modifying it in relation to past experience before passing it on to effectors. During evolution, there has also been a tendency for one or more ganglia in the head to become more and more important, forming the "brain." This process is known as **cephalization.** The most highly organized brain among all animals is that of man.

The functions of the nervous system are (1) to receive and relay information; (2) to filter, and so select, information, reinforcing or inhibiting it as necessary; and (3) to store information so it can be used for comparison at a later time. In animals with highly organized nervous systems, the latter two functions are carried out mainly in the central nervous system.

THE REFLEX ARC

The transmission of information from sense organs to effectors by way of central pathways constitutes a **reflex arc.** The pathway may consist of only one or a few synapses, or it may involve a complex passage via numerous synapses, at each of which processing and modulation of the information may occur. The differences, however, are more of degree than of kind.

In a simple reflex arc, such as the familiar knee-jerk reflex of man, only three cell types are normally involved: (1) the stretch receptors in the muscle, which send fibers to the spinal cord; (2) the motor neurons which activate the muscles in the front part of the thigh; and (3) the thigh muscles themselves (Fig. 12.11). Even such a simple arc, however, is susceptible to behavioral modification. By consciously tensing the opposing muscles, one

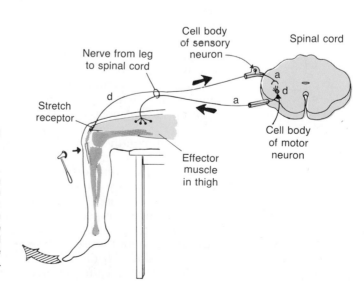

FIGURE 12.11

The knee-jerk reflex in man.
Only three cell types are involved. The hammer, by stretching the tendon and also the thigh muscle, stimulates an impulse in the muscle stretch receptor. This impulse activates a motor neuron in the spinal cord which depolarizes the thigh muscle, causing it to contract and extend the lower leg. Nerve pathways are shown by heavy arrows. *a*, axon; *d*, dendrite.

can suppress the knee-jerk reflex. Similarly, one can train oneself not to drop a hot object, even if it is burning one's fingers. Obviously, information from the brain can modify even the simplest response.

Since, as we have seen, single nerve cells and muscle motor units respond in an all-or-none fashion, an animal would be capable of only a single response to a given stimulus if modification of the response were not possible. We have already noted that facilitation of a pathway may occur after continued use. It is also possible for a response to be inhibited if a stimulus is repeatedly received.

A simple example exists in the inhibition of motor neurons in man by a negative feedback system (Fig. 12.12). As excitatory stimuli are received from the sensory neuron, they are transmitted to the motor neuron in the spinal cord, but the axon of the latter divides in two. One part carries the excitatory impulse to the muscle, while the other innervates a small cell, called a **Renshaw cell,** which sends an *inhibitory* impulse back to the motor neuron. The more excitatory stimuli received, the more inhibitory stimuli are fed back to the motor cell. In other circuits, positive feedback occurs, where the recurrent fiber is excitatory rather than inhibitory.

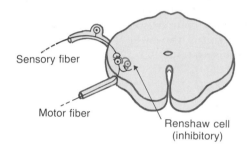

Sensory fiber

Motor fiber

Renshaw cell
(inhibitory)

FIGURE 12.12

Central modulation by negative feedback.
  The excitatory impulse reaching the motor neuron stimulates not only the muscle but also the Renshaw cell, which releases an inhibitory neurotransmitter to the motor neuron.

From these simple circuits it is possible to see how highly complex feedback loops, involving many interacting neurons in the spinal cord and the brain, can create a high degree of modulation in the central nervous system. To a large extent, then, the behavior of which an animal is capable depends upon the anatomy of its nervous system and the degree to which synaptic transmission can be modified.

SUMMARY

The responsiveness of organisms to environmental stimuli depends on the property of irritability. This is due to an imbalance of positive and negative ions on the two sides of a cell membrane, resulting in a membrane potential. When the membrane is depolarized, sodium flows in, initiating an action potential which passes along the membrane as an impulse. Subsequent outflow of potassium restores the original potential.

The nervous system is the most highly irritable tissue in multicellular organisms and conducts impulses at high speed. It functions both to transmit and to analyze information. Each nerve cell forms many chemically mediated synapses with other nerve cells, or with receptors or effectors, thus providing many alternative circuits in the central nervous system. Both excitatory and inhibitory synapses occur on the same cell, mediated by different neurotransmitters. Temporary modification of pathways occurs by facilitation.

Both glands and muscles are effectors, producing responses to information transmitted to them by the nervous system. Muscles contain the contractile proteins, actin and myosin, which slide over one another after calcium is released from the endoplasmic reticulum in response to membrane depolarization. This process requires energy from breakdown of ATP, which is replenished initially from stored phosphagens and ultimately from oxidative metabolism of glycogen.

Receptors contain sensory cells which transduce energy received from the environment into the action potential of an impulse. Light, mechanical vibrations, heat and chemicals can be detected by various animals. Internal receptors are also present to signal the state of contraction of muscles and other internal information.

In the course of evolution, nervous systems have tended to increase in both size and complexity of organization. Basically, responses may be regarded as reflex arcs, mediated by a few to many neurons. Anatomical and physiological modifications of the simple reflex arc permit the central modulation of information, making complex behavior possible.

Baker, P. F. 1966. The nerve axon. Scientific American Offprint No. 1038. A readable account of experimental studies on giant nerve axons from the squid.

Eccles, Sir John. 1965. The synapse. Scientific American Offprint No. 1001. Explains how neurotransmitters perform their functions.

Guyton, A. C. 1969. *Function of the Human Body*. W. B. Saunders Co., Philadelphia. Section six deals extensively with the human nervous system.

Heimer, L. 1971. Pathways in the brain. Scientific American Offprint No. 1227. New techniques provide insight into brain circuitry.

Hoyle, G. 1970. How is muscle turned on and off? Scientific American Offprint No. 1175 Describes recent studies on the sequence of reactions occurring during muscular contraction and relaxation.

Huxley, H. E. 1969. The mechanism of muscular contractions. Science *164*:1356–1366. An account of the sliding filament theory of contraction by one of its originators.

Katz, B. 1966. *Nerve, Muscle and Synapse*. McGraw-Hill, New York. A highly readable paperback explaining the basic fundamentals of the functions of nerves and muscles.

**READINGS AND
REFERENCES**

12a

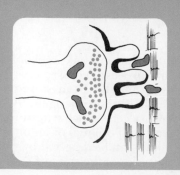

DRUGS, TOXINS AND THE NEUROMUSCULAR JUNCTION

The mechanism of the all-important synapses between motor nerve axons and muscles serves as an excellent and easily studied model of synaptic function in general. Many of the chemicals which affect the neuromuscular junction are also of medical or environmental significance.

The synaptic junction between nerve and muscle is readily accessible to certain drugs and to other chemicals which may find their way into the body. The effects wrought by some of these substances, such as botulinum toxin and curare, have been known for a long time, even though it was not understood how they acted. Other compounds, such as the organophosphorus pesticides synthesized by the chemical industry, are recent comers to the environmental scene. All have intrinsic interest since all are potentially lethal. A study of the effects of these chemicals serves the additional purpose of aiding our understanding of how a synapse works.

THE NEUROMUSCULAR JUNCTION

The **neuromuscular junction,** or synapse between nerves and muscle, is also known as the nerve-muscle junction or the myoneural junction. It is similar in overall structure to the generalized synapse of Figure 12.6. The neuromuscular junction is somewhat more complex than other synapses, probably because the impulse must affect not just a small surface area of another nerve cell, but the relatively large surface of a muscle fiber. The branched tip of the nerve axon lies in a trough-like depression on the surface of the muscle cell (Fig. 12a.1), and in this trough the membrane of the muscle cell repeatedly folds inward, greatly increasing the surface area available for neurotransmitter receptor sites (Fig. 12a.2).

In this chapter we shall limit ourselves to a study of those neuromuscular junctions which are associated with voluntary muscles; the neurotransmitter at these synapses is **acetylcholine.** A brief review of the steps in transmission of excitation from nerve to muscle is given in Figure 12a.3, upon which our further discussion will be based.

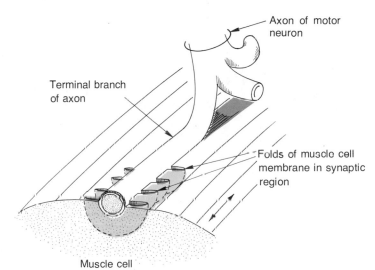

Axon of motor neuron

Terminal branch of axon

Folds of muscle cell membrane in synaptic region

Muscle cell

FIGURE 12a.1

Schematic diagram of nerve ending on a muscle cell.

Note the branched end of the axon buried in a trough of the muscle cell membrane. The latter has many folds in the synaptic region, greatly increasing the area available for synaptic binding sites. (After Birks, Huxley and Katz. 1960. Journal of Physiology *150*:134–144.)

Note that three steps are involved: (1) release of acetylcholine from synaptic vesicles into the synaptic space; (2) binding of acetylcholine in a lock-and-key fashion to the specific receptor sites on the postsynaptic membrane of the muscle, leading to its depolarization; (3) destruction of acetylcholine by hydrolysis, catalyzed by the specific enzyme, **cholinesterase.** Interference with any of these three steps has marked effects on the response of the muscle cell to a nerve stimulus.

INHIBITION OF ACETYLCHOLINE RELEASE

An example of inactivation of the mechanism which releases acetylcholine is the disease known as **botulism.** Botulism, which is fortunately a rare occurrence, is caused by the most potent toxin known in the world; only 60 billionths of a gram will kill a man. The toxin is a fairly large protein molecule, produced by certain microorganisms found in soil, known as *Clostridium botulinum.*

Although worldwide in distribution, these bacteria do not grow and produce toxin to any extent in the soil itself; for this they require both the absence of oxygen and an enriched medium. In soil, only the harmless spores are present; these will germinate and multiply only under favorable conditions. Ideal surroundings for growth of the bacteria are cooked, smoked, spiced or canned foods. Only prolonged high temperatures and pressures such as those obtained by food packing industries or home pressure cookers will kill the heat-resisting spores during food preparation. If insufficiently sterilized foods are allowed to stand for a time without refrigeration, the bacteria multiply wherever oxygen is absent, and the toxin then accumulates. No cases of poisoning are known from eating freshly

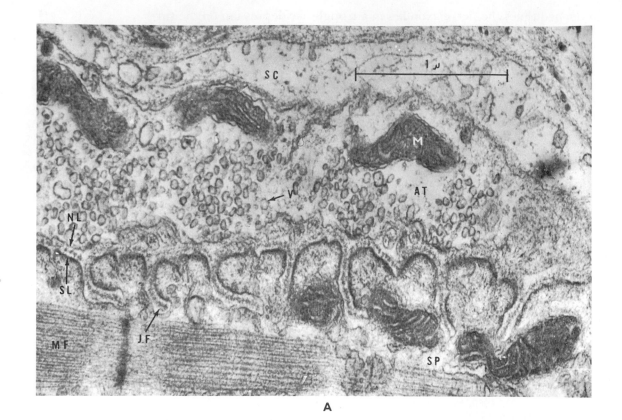

A

B

FIGURE 12a.2   Details of the neuromuscular junction.

    A. An electronmicrograph of a neuromuscular junction. Axon (*AT*), bounded by supporting cell (*SC*) above, muscle cytoplasm (*SP*) below. Note the presence of mitochondria (*M*) in both cell types; the vesicles (*V*) containing acetylcholine; the folds (*JF*) of the muscle cell membrane; and the muscle filaments (*MF*). (From Birks, Huxley and Katz. 1960. Journal of Physiology *150*:134–144.)

    B. A drawing of *A*, to clarify details of relationships. Acetylcholine must cross synaptic space and bind with specific sites on thickened regions of muscle cell membranes.

313

DRUGS, TOXINS
AND THE
NEURO-
MUSCULAR
JUNCTION

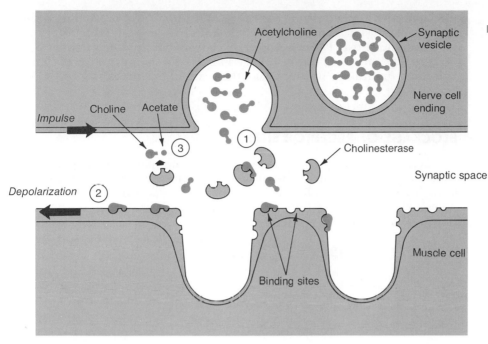

The steps in transmission of an impulse at a neuromuscular junction, shown in schematic diagram.
  1. The nerve impulse releases acetylcholine from the synaptic vesicle into the synaptic space.
  2. Acetylcholine binds in a lock-and-key fashion with binding sites on the muscle cell membrane, causing depolarization.
  3. Acetylcholine is finally destroyed by the enzyme cholinesterase. This enzyme either acts in the synaptic space, as shown here, or else may be attached to the muscle cell membrane. In either case, the end result is destruction of the neurotransmitter.

prepared food, either raw or cooked; nor will the spores germinate and grow in the digestive tract.

Usually, but not always, infected food looks and smells bad; cans may be swollen with accumulated gas, a by-product of the metabolism of these bacteria. The botulinum toxin itself is readily destroyed by heat; boiling for five minutes insures its total destruction. The toxin, unlike most other proteins, is highly resistant to the digestive enzymes of the gut and, in a way not clearly understood, is able to pass intact across the gut wall and into the body. About one to three days must usually elapse between eating poisoned food and the onset of symptoms, during which time the molecules of toxin find their way to the nervous system, particularly to the neuromuscular junctions. The signs of botulism are vomiting, constipation, and paralysis of eye, throat and respiratory muscles.

Botulinum toxin acts upon the presynaptic nerve ending of the neuromuscular junction, interfering with the release of acetylcholine (Step 1 in Figure 12a.3). Synthesis of the neurotransmitter is not impaired, nor is the ability of the muscle membrane to respond to electrical depolarization. It is the synapse itself which is affected; no matter how many nerve impulses pass down the axon, the synaptic vesicles do not release acetylcholine into the synaptic space. The muscle membrane is never depolarized and paralysis results. Precisely how the toxin prevents acetylcholine release is not yet known.

Although botulism is a rare affliction, those people who are poisoned face an uncertain recovery. **Antitoxin** is available. It is produced by injecting animals with partially inactivated toxin, against which they then form a specific antibody, the antitoxin (see Chapter 10). When serum of these immunized animals is injected into poisoned patients, the antitoxin is some-

314

DRUGS, TOXINS
AND THE
NEURO-
MUSCULAR
JUNCTION

times effective in neutralizing the effects of the toxin and preventing death. Failure may result from using too little antitoxin or from administering it too late to be of use. Patients who recover from botulism may exhibit signs of partial paralysis of the voluntary muscles for months after the initial poisoning, indicating how resistant the toxin is to breakdown within the body.

BLOCKAGE OF RECEPTOR SITES

Curare is a neurotoxin obtained from a vine native to parts of South America, and for centuries the Indians living there applied it to arrow tips to paralyze their prey. Like botulinum toxin, it acts at the neuromuscular junction by blocking transmission of the excitatory nerve impulse. Curare, however, acts at a different site from that of the bacterial toxin, and its action is due to a certain similarity between its chemical structure and that of acetylcholine (Fig. 12a.4).

FIGURE 12a.4   Chemical structures of the neurotransmitter acetylcholine and the postsynaptic blocking agent curare.
Regions of similar structure which may compete for specific binding sites on the muscle cell membrane are indicated by color. Underneath are shown simplified models.

Because of this similarity to acetylcholine, curare molecules are able to bind to the receptor sites of the muscle cell membrane, thus competing with acetylcholine molecules and blocking their action. Curare itself, however, is not capable of depolarizing the postsynaptic membrane, and so the muscle remains relaxed or paralyzed (Fig. 12a.5).

Curare has been used medically for many years—at low doses to prevent muscular spasms and at higher doses as a paralytic agent during certain types of operations. More recently, curare has been employed in animal experiments in order to further our knowledge of chemical transmission at the neuromuscular junction.

EXPERIMENTS WITH RADIOACTIVE CURARE

It is now possible to prepare curare which is labeled with radioactive carbon, ^{14}C. The radioactivity permits the experimenter, by means of a special technique which counts radioactive molecules, to measure how many molecules of curare are actually bound to each neuromuscular junction.

315

DRUGS, TOXINS
AND THE
NEURO-
MUSCULAR
JUNCTION

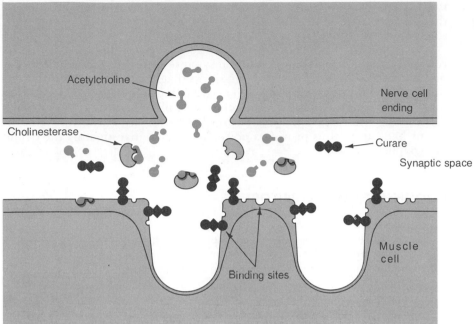

A schematic diagram illustrating how curare competes with the neurotransmitter acetyl-choline for binding sites on the postsynaptic membrane.

Since curare does not depolarize the muscle membrane, contraction cannot occur and the muscle is paralyzed. Note that curare does *not* bind with the enzyme cholinesterase, which continues to destroy acetylcholine.

(Only a few of the 2.6 million binding sites are shown.)

Experiments in which mice were injected with ^{14}C-curare have shown that a *maximum* of 2.6 million molecules of the drug are bound at each neuromuscular synapse. (Although this may seem a very large number of molecules to those who are unfamiliar with chemistry, it is in fact an almost infinitesimally small quantity.) Under these conditions, paralysis is complete, and all of the animals die of respiratory failure (Fig. 12a.6). Such experiments have been interpreted to mean that there are approximately 2.6 million binding sites on the postsynaptic membrane of each synapse.

FIGURE 12a.6

Graph of number of curare molecules bound per synapse after injection of different amounts of radioactive curare (^{14}C-curare) into mice. Only when virtually all receptor sites are filled is paralysis complete, causing death. (After Waser, P. G. Relation between enzymes and cholinergic receptors. *In* Mongar, J. L., and de Reuck, A. V. S. 1962. (Ciba Foundation Symposium) *Enzymes and Drug Action.* Little, Brown and Co., Boston.)

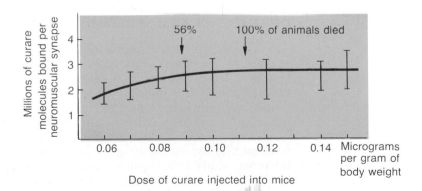

When lower doses of curare are injected, fewer molecules are bound at each neuromuscular synapse, and some of the animals are able to continue breathing movements and so survive. In the survivors, enough receptor sites remain free to react with acetylcholine and thus to cause an action potential. The conclusion reached from these experiments is that only a fraction, perhaps less than a third, of the total binding sites at each synapse

316

DRUGS, TOXINS
AND THE
NEURO-
MUSCULAR
JUNCTION

need be available to acetylcholine in order for sufficient depolarization of the membrane to occur to cause an action potential with subsequent contraction of the muscle.

It is known that about six million acetylcholine molecules must be released from a nerve ending into the synapse in order to produce an action potential. The above experiment shows that less than a million of them need to cross the synaptic space and arrive at binding sites in order for the muscle to contract. Apparently the other five million molecules never reach the binding sites, being inactivated by cholinesterase on the way. Neurotoxins like curare thus provide a powerful tool for investigating the details of synaptic function.

CHOLINESTERASE INHIBITORS

In contrast to curare, which blocks acetylcholine action and produces paralysis, there is another group of chemicals which have the opposite effect: they potentiate the transmission of an impulse across the neuromuscular junction by inhibiting the enzyme cholinesterase.

CLINICAL DRUGS

Two important chemicals used medically as cholinesterase inhibitors are **physostigmine,** obtained from the seed of an African bean-like plant, and a similar compound, **neostigmine.** These two drugs are used in cases of an overdose of curare. By inhibiting the enzyme, they increase the number of acetylcholine molecules available in the synapse. The neurotransmitter molecules are then able to compete with and displace curare molecules at the binding sites on the muscle cell. If the concentration of acetylcholine molecules becomes sufficiently great, they displace enough of the curare molecules to bring about an action potential in the muscle cell membrane.

In the rare disease known as myasthenia gravis, there is a malfunctioning of the receptor sites on the postsynaptic muscle membrane; the normal number of acetylcholine molecules released by an excitatory impulse fails to cause an action potential. This disease can sometimes be treated by careful administration of the cholinesterase inhibitors physostigmine and neostigmine; they effectively increase the concentration of neurotransmitter in the synaptic space, and thus help to overcome the deficient functioning of the binding sites.

OTHER USES OF CHOLINESTERASE INHIBITORS

A variety of cholinesterase inhibitors has been developed for chemical warfare—the modern nerve gases, which are not gases at all but very fine powders. These include the chemicals known as Tabun, Sarin and Soman. Of these, Sarin kills fastest; only one milligram (about 30 millionths of an ounce) is fatal when applied to the skin. These poisons cause continuous muscular contraction which results in convulsions and death.

The potential of such substances for use as insecticides was quickly realized, and similar compounds, somewhat less hazardous than the potent nerve gases, are being widely used today to replace DDT. Although they have the advantage of being broken down in the environment far more rapidly than DDT, they are much more toxic to man and have caused significant illness from poisoning among farm workers exposed to large doses of them.

The two main groups of cholinesterase inhibitors now in widespread use as pesticides are the organophosphorus compounds and the carbamates. Some common trade names of these chemicals are Amiton, Diazinon, Malathion, Parathion, Carbaryl and Sevin, most of which are

317
DRUGS, TOXINS
AND THE
NEURO-
MUSCULAR
JUNCTION

components of insecticides sold for use in home and garden. Neostigmine is also a carbamate type of compound.

Since all of these compounds produce their effects in a similar way, we shall limit our discussion to the action of organophosphorus compounds. In the following discussion we shall attempt not only to explain the actions of these neurotoxins but also to further our understanding of the mechanisms of enzyme action as exemplified by the enzyme cholinesterase.

We have already noted that this enzyme destroys the neurotransmitter acetylcholine by breaking it down in the synaptic space. The reaction involved is the following:

$$CH_3\text{-}N^+\text{-}CH_2\text{-}CH_2\text{-}O\text{-}C\text{-}CH_3 + H_2O \xrightarrow{\text{Cholinesterase}} CH_3\text{-}N^+\text{-}CH_2\text{-}CH_2\text{-}OH + HO\text{-}C\text{-}CH_3$$

Acetylcholine · · · · · Choline · · Acetate

Inspection of this reaction reveals that a molecule of water is inserted between the two halves of the acetylcholine molecule, a phenomenon known as **hydrolysis.** (Hydrolytic or water-cleaved reactions, where a molecule of water splits a bond in the substrate, are also characteristic of digestive reactions, as described in Chapter 9.) Like all enzymes, the enzyme cholinesterase binds with its substrate, acetylcholine, to form an **enzyme-substrate complex** which, in a way not completely understood, greatly facilitates the splitting of acetylcholine.

Let us next look at a model—a scientist's visualization—of the **active site** of the cholinesterase molecule, the region which binds acetylcholine and causes it to be split by water. In Figure 12a.7 we have shown a schematized diagram of this active site. Both the acetate and choline parts of acetylcholine are believed to bind to specific regions of the enzyme (Fig. 12a.7A).

The first step in the enzyme-catalyzed cleavage of acetylcholine is the formation of an enzyme-substrate complex. Complex formation appears to result from the attraction of non-polar (uncharged, hydrophobic) regions of the substrate to similar regions at the active site of the enzyme; they fit together in a lock-and-key fashion. Note also attraction between oppositely charged + and − regions. The second step is the splitting of acetylcholine by transfer of a hydrogen atom from the enzyme to the bond between choline and acetate—the choline then falls away from the enzyme. The third step is the splitting of the acetate-enzyme complex by a molecule of water, yielding free acetate and the original enzyme.

Organophosphorus compounds (and also carbamates) have structures resembling that of acetylcholine (Fig. 12a.7B). The insecticides combine almost as well as acetylcholine with the active site of the enzyme cholinesterase—thus "competing" with the natural substrate. But if competition for the active site of the enzyme were all that happened when these pesticides react with cholinesterase, far higher concentrations of them would be needed than are actually found necessary to produce the desired result—hyperactivity, convulsions and death of an insect. As shown in Figure 12a.7B, the organophosphorus compound Amiton, like acetylcholine, is split by the enzyme, but step 3 does *not* occur. The enzyme rids itself of the phosphorus end of the molecule only with the greatest difficulty and thus remains almost permanently blocked. One molecule of cholinesterase can

318

DRUGS, TOXINS
AND THE
NEURO-
MUSCULAR
JUNCTION

split 300,000 molecules of acetylcholine per second, but requires several minutes to hydrolyze and release a single molecule of insecticide! A comparison of the normal, uninhibited enzyme with the inhibited form is shown in Figure 12a.8. In the presence of pesticides, acetylcholine thus

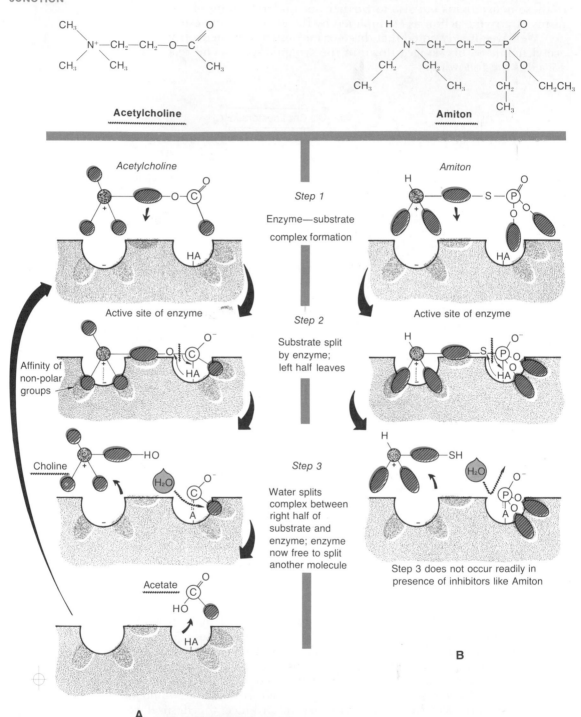

FIGURE 12a.7   The binding and splitting of the natural substrate acetylcholine by the active site on the enzyme cholinesterase (A), compared with the binding and splitting of the organophosphorus inhibitor Amiton (B).

    Note structural similarity of acetylcholine and Amiton (above). (Amiton contains a phosphorus atom, P—hence it is an organophosphorus molecule.) Note also lock-and-key fit both between non-polar regions of enzyme and substrate and between charged regions, + and −. (Further details in text.)

319

DRUGS, TOXINS
AND THE
NEURO-
MUSCULAR
JUNCTION

FIGURE 12a.8

Comparison of uninhibited and inhibited cholinesterase.
   A. Uninhibited enzyme quickly splits its normal substrate, acetylcholine, acting on 300,000 molecules per second.
   B. Enzyme inhibited by P-containing half of organophosphorus pesticide cannot bind acetylcholine, which accumulates in synapse, causing convulsions.

remains intact and continues to cause depolarization of the muscle cell membrane; convulsions result.

CONTRAST BETWEEN CURARE AND CHOLINESTERASE INHIBITORS

The thoughtful reader may be wondering why curare and pesticides, which both have strong chemical similarities to acetylcholine, should act in opposite ways within the synapse. Why do both compounds not have the same effect? The answer lies in the differences between the **specificity** of the active site of the enzyme, on the one hand, and of the receptor sites on the muscle cell membrane, on the other. Each responds in a different way to the acetylcholine molecule. They are two different "locks" opened by the same "master key," acetylcholine. Curare and the pesticides are like "sub-master keys," fitting only one or the other lock, but not both. As we have seen, accidental overdoses of curare given clinically are treated by the cholinesterase inhibitor neostigmine. Curare and neostigmine both resemble acetylcholine, but each acts at a different site in the synapse.

THE ENIGMA OF DDT

It is of interest to note that, despite its wide usage and much experimentation, we still do not know how the insecticide DDT produces its effects on the nervous system. The end results are the same as for organophosphorus compounds—hyperactivity, convulsions and death. Although at high concentrations, DDT exhibits some anti-cholinesterase activity, this is not believed to be its primary action. The high solubility of DDT in fats causes it to accumulate not only in adipose (fat storage) tissue, but also in nervous tissues which contain much fatty material called **myelin.** Myelin is formed from the membranes of supporting cells which are tightly wrapped in spirals around the axons of nerve cells (Fig. 12a.9). These myelin sheaths insulate the axons from one another and also speed the rate of impulse conduction.

It has recently been found that DDT binds tightly with a type of fat which is characteristic of cell membranes and particularly of myelin. If this bound DDT is affecting the properties of the membranes, it may explain how this insecticide acts in the nervous system, whose normal functioning depends to such a large extent on its membranes. Only further experiments, however, will clarify the biological action of DDT.

320

**DRUGS, TOXINS
AND THE
NEURO-
MUSCULAR
JUNCTION**

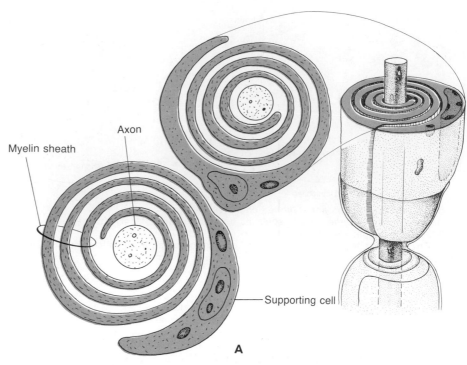

Myelin sheath

Axon

Supporting cell

FIGURE 12a.9

A

The myelin sheath.

A. Schematic diagram of two axons, cut in cross-section, showing their myelin sheaths. These are composed of the spirally wrapped membranes of supporting cells surrounding the axons.

B. An electronmicrograph of a myelin sheath (193,000 ×). Certain fats in myelin and in the axon membranes bind DDT tightly. We have yet to learn how this bound DDT may affect nerve function. (Micrograph by Dr. J. D. Robertson, from Fawcett, D. W. 1966. *The Cell: Its Organelles and Inclusions.* W. B. Saunders Co., Philadelphia.)

Illustration continued on opposite page.

SUMMARY

The neuromuscular junction is susceptible to the action of various drugs and chemicals which can greatly affect the normal conduction of impulses between motor nerves and the muscles they innervate. Examples of the different kinds of chemical effects which can be produced serve to illustrate further the mechanisms involved in synaptic transmission.

Botulinum toxin acts, by means as yet unknown, to prevent the release of the neurotransmitter acetylcholine. It thus produces muscular paralysis, since excitatory impulses cannot be transmitted at the neuromuscular junction. Curare has the same end result but acts at a different site in the synapse. Its structural similarity to acetylcholine allows curare to compete for binding sites on the postsynaptic muscle membrane, but it does not cause depolarization of the membrane, and so muscle contraction is prevented. If more acetylcholine is made available, as occurs when the enzyme cholinesterase is inhibited, the paralytic effects of curare can be reversed.

Inhibitors of cholinesterase include not only medically useful drugs such as physostigmine, but also war gases and certain pesticides, the organophosphorus compounds and carbamates, now commonly used as replacements for DDT. These chemicals kill by interfering with acetylcholine hydrolysis; the excess neurotransmitter causes convulsions and eventual exhaustion of the affected organism. The pesticides are nonspecific, acting on many types of animals, including insect pests, beneficial insects and man. Inhibition of the enzyme cholinesterase is due to the almost irreversible binding of part of the pesticide molecule with the enzyme.

321
DRUGS, TOXINS
AND THE
NEURO-
MUSCULAR
JUNCTION

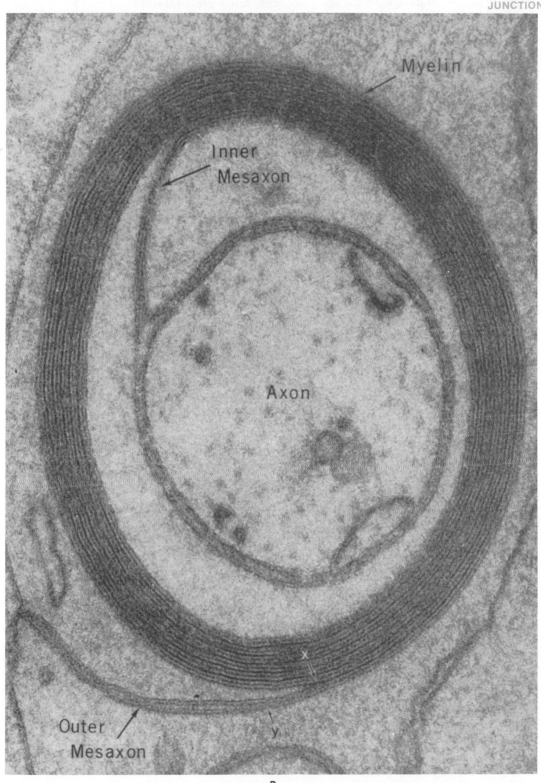

B

FIGURE 12a.9B

322

DRUGS, TOXINS
AND THE
NEURO-
MUSCULAR
JUNCTION

The enzyme is no longer free to react with its normal substrate, ace-tylcholine.

Despite its many years of use, it is still not known precisely how DDT causes death in insects. It may result from an effect of DDT on membrane fats, for which the insecticide has a great affinity.

READINGS AND REFERENCES

de Reuck, A. V. S. (ed.) 1962. *Curare and Curare-like Agents.* Ciba Foundation Study Group No. 12. Little, Brown & Co., Boston. An interesting collection of experimental studies on the action of curare.

Ingram, M. and Roberts, T. A. 1967. *Botulism. 1966.* Chapman & Hall, Ltd., London. A study of the worldwide distribution of the spores and the disease.

Mongar, J. L. and de Reuck, A. V. S. (eds.) 1962. *Enzymes and Drug Action.* Ciba Foundation Symposium. Little, Brown & Co., Boston. Includes papers on drugs affecting enzymes at synapses in the nervous system.

O'Brien, R. D. 1967. *Insecticides, Action and Metabolism.* Academic Press, New York. Further reading for the student who is curious about other types of insecticides in addition to cholinesterase inhibitors.

O'Brien, R. D. 1969. Phosphorylation and carbamylation of cholinesterase. Annals of New York Academy of Sciences. *160*:204–214. A theory on how these pesticides exert their inhibition of cholinesterase and so cause death.

Thomas, K. B. 1964. *Curare: Its History and Usage.* Pitman Medical Publishing Co., Ltd., London. A readable book of curare's uses, past and present, with details of its effect on the myoneural junction.

Tinsky, I. J., Hague, R. and Schmedding, D. 1971. Binding of DDT to lecithin. Science *174*:145–147. Evidence for a possible site of toxic action of DDT—on a membrane-located fat.

Aggression versus Appeasement

BEHAVIOR

13

The way in which an animal responds to stimuli is of primary importance to its survival. The study of animal behavior is thus of great interest to biologists generally as well as to those interested in human behavior.

In Chapter 12 we discussed the functioning of the organ systems necessary for the behavior of animals—the sense organs, the central nervous system and the muscle and gland effectors. Our subject in this chapter is the use to which these organ systems are put by various animals in carrying out their behavior. Under this broad heading come all those phenomena which are regulated by the central nervous system or, in the case of a single-celled amoeba, by its general irritability. In addition to the obvious forms of behavior such as eating, mating or fleeing from danger, we also include remaining still (as certain animals do when a predator is near) and even sleeping. Such "involuntary" responses as heart rate and blood pressure are greatly affected by environmental stimuli and, in man, can even be consciously controlled (see Chapter 13a).

For many years, the behavior of animals was studied mainly for what it could reveal about human behavior. Psychologists studied animals under laboratory conditions, subjecting them to controlled experimental conditions. In almost all cases, the animals were responding to situations unfamiliar to them, and often little account was taken either of the unusual stresses placed on the animal or of its ability to perceive the strange stimuli presented to it. The experiments were too frequently designed and interpreted in terms of human experience.

Several decades ago, however, a new approach to the study of behavior developed in Europe, initiated by the Austrian scientist Konrad Lorenz. Lorenz and his students turned their attention to the systematic observation and testing of animal behavior under natural conditions, a science known as *ethology*. This approach has now become widespread, and although there is an enormous amount still to be learned about animal behavior, ethologists have already demonstrated that much behavior has great significance for survival of the individual. In other words, under natural conditions, behavior has adaptive value (a concept pursued further in Chapter 21).

THE NATURE OF BEHAVIOR

Before beginning our discussion of the details of behavior, during which we shall use examples from many different types of animals, it is necessary to emphasize two points. First, the behavior of a particular *type* of animal will be limited by its genetically inherited organ systems: by the nature of the stimuli its sense organs can detect; by the size and organizational complexity of its nervous system (see Chapter 12); and by the capability of its effectors. Thus naturally blind cave-dwelling fish cannot respond to visual stimuli, jellyfish with diffuse nerve nets cannot be trained to perform complex tasks and wingless insects cannot fly. Second, the behavior of an *individual* animal will alter from one occasion to the next, since each past experience to some degree or other modifies its later response to a given stimulus.

Inherited behavioral potential, also known as instinctive or **innate** behavior, was long thought to be synonymous with the relatively fixed responses characteristic of much animal behavior, whereas modification of behavior, or **learning,** was thought to be possible only in limited types of behavior patterns. Recent experiments have shown, however, that certain apparently rigid behavioral responses are learned rather than innate—for ex-

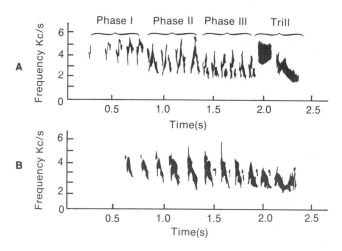

FIGURE 13.1

Sound spectrograph recordings of the chaffinch song.

A. The normal adult chaffinch song lasts about 2.0 seconds and has three main phases, each of successively lower frequency, followed by a terminal trill.

B. The song of a bird raised in isolation is very simple, contains no separate phases and no terminal trill.

(From Hinde, R. A. 1970. *Animal Behavior.* McGraw-Hill, New York.)

ample, the characteristic songs of chaffinches and other song birds develop to their fullest extent only if the young bird hears other members of its species singing; the song does not develop automatically in birds hand-reared in isolation (Fig. 13.1).

FIGURE 13.2

A

B

Prey capture by the cuttlefish.

A. The attention of the slowly swimming cuttlefish is caught by movements of its prey, a shrimp-like crustacean.

B. When at the correct distance, the cuttlefish rapidly shoots forth its long tentacles.

(After Tinbergen, N. 1965. *Animal Behavior.* Time, Inc., New York.)

On the other hand, many stereotyped behavior patterns characteristic of a species are modified by experience. The cuttlefish, *Sepia*, for example, catches its food by orienting toward it and then striking at it with its long tentacles (Fig. 13.2). A successful catch depends upon being at the correct distance from the prey. Newly hatched squid are nearly always successful on their first attempt—thus exhibiting an ability to assess distance without prior experience. Yet as the animal grows, the correct distance between itself and its prey also increases, and so the behavior must be gradually modified to accommodate this change. There is thus no sharp distinction between "innate" and "learned" behavior.

COMPONENTS OF BEHAVIOR

As we noted above, the ultimate function of behavior is survival of the individual (or in some instances, the species). But the immediate cause of a particular behavior pattern, even in man, is seldom the result of conscious awareness of one's own survival. You wave and smile at a passing friend not because you are thinking that social behavior is essential for the survival of man but because you recognize the person and wish to acknowledge this. This is a culturally learned response which is performed so often as to be almost an unconscious reflex. Although we cannot know how much animals are aware of the ultimate benefits to themselves of their actions, it seems likely that most animal behavior, whether genetically inherited or learned from past experience, is made in response to an immediate signal or **stimulus.** Upon reaching the central nervous system, the stimulus is modified not only by other incoming stimuli, including those signalling the internal physiological state of the body, but also by past experience. Finally, after processing the information, the nervous system relays signals to the body's effectors, producing a response. Details of this central processing are considered further in relation to the human brain in the next chapter. These steps have been outlined in Chapter 12, and may be summarized here as follows:

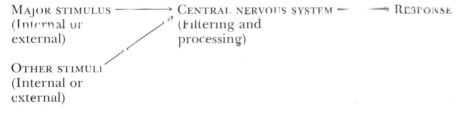

Responses involving movement can be grouped into two types: the specific response of a particular set of muscles to a stimulus, and the general response of various sets of muscles, leading to a complex pattern of behavior. An example of the first is the rapid extension of the lower leg in a knee-jerk reflex (see Figure 12.11). An example of the second is the food-searching behavior of a hungry animal in response to internal stimuli such as a low blood sugar level. Obviously, the nature of the stimulus plays a role in determining the type of response elicited. In general, we shall be most concerned with complex responses, involving a large number of muscles.

THE EFFECTIVE STIMULUS

In the course of observing animal behavior, ethologists have noted several things about the sorts of stimuli that elicit certain behavioral responses in animals. For one thing, an animal may sometimes respond to a stimulus and sometimes not. It is as if the animal were "tuned-out" on one occasion and "tuned-in" on another. Apparently the central nervous systems of many animals act as filters, letting through only those signals which are significant at a given instant or for a given state of the animal. Two examples will help clarify this point.

If recording electrodes are implanted in the auditory regions of a cat's brain, it can be shown on a meter receiving signals from the electrodes that the brain neurons are firing in synchrony with a metronome while the cat is sitting quietly. The cat is consciously hearing the metronome (Fig. 13.3A). But if a mouse runs in front of the cat, its attention to the metronome ceases and it no longer hears the ticking sound, as shown by the absence of synchronized signals (Fig. 13.3B). Thus, only those stimuli which actually reach the higher centers of the brain are capable of eliciting a behavioral response.

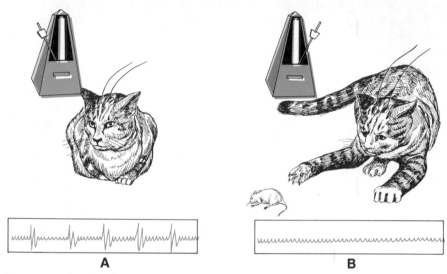

FIGURE 13.3

A   B

Awareness of a stimulus.
   A. The quiescent cat "hears" the ticking metronome, as shown by synchronized signals picked up by electrodes implanted in auditory regions of its brain.
   B. Cat diverted by mouse no longer "hears" ticking sounds, as shown by absence of synchronized signals in recording from electrodes.
   (After Tinbergen, N. 1965. *Animal Behavior.* Time, Inc., New York.)

A second example is furnished by mating behavior. In many animals, unlike man, mating occurs only during specific seasons of the year. During the early summer, when its breeding season begins, the male white-crowned sparrow, for instance, exhibits territorial behavior (see Chapter 19). It flies about its territory, delimiting the boundary by singing its characteristic song, and reacts with specific aggressive behavior toward other males that approach too closely. In the fall and winter, these same males readily congregate in large flocks with other members of their species, both male and female; only during the breeding season do other males elicit aggression.

This behavioral change is related to the amount of sex hormone being produced by the gonads. As day length increases in the spring, impulses reaching the brain from the eyes are thought to stimulate certain gland-like cells of the brain, which in turn act upon the pituitary gland; the latter then releases hormones stimulating the growth of the gonads and the amount of sex hormone they secrete (Fig. 13.4). This type of brain-pituitary-gonad relationship is discussed in further detail in Chapter 22.

The hormonal triggering of sexual behavior is but one of many examples of the internal stimuli which initiate or modify animal behavior.

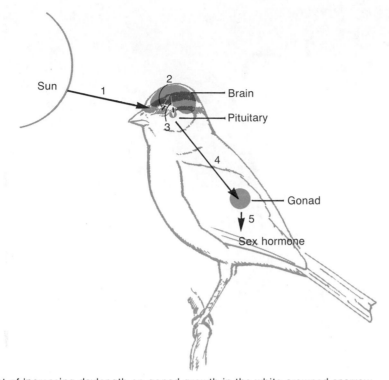

FIGURE 13.4

Effect of increasing daylength on gonad growth in the white-crowned sparrow,
Light striking the eye (*1*) sends signals to the brain (*2*). Certain secretory cells there produce a substance which activates the pituitary (*3*), which in turn releases gonad-stimulating hormones (*4*). The gonad increases its secretion of sex hormones (*5*), triggering stereotyped mating behavior.

Other internal "drives" affecting behavior are hunger, thirst, parental care of the young and so forth.

Limits Imposed by Sense Organs. In addition to being responsive to a stimulus, either because of its internal state or because the stimulus supersedes other, less "important" stimuli in the central filtering mechanism, an animal must of course be able to detect a stimulus by means of its sense organs. Although briefly touched upon in Chapter 12, a few specific examples here will underline this point.

It has long been known that many animals have a more acute sense of smell than does man and also that certain other animals have better visual discrimination. The insect-eating falcon, for example, can distinguish a dragonfly at a distance of half a mile, whereas man cannot see one further off than 100 yards.

An animal's ability to sense signals is limited not only by its perceptual acuity, however, but also by the range of each type of energy it can perceive. A bat, for instance, emits sound signals at what we call "ultra-high frequencies." The most acute human ear hears air vibrations of but 20,000 per second, whereas bats produce and hear sounds of over 100,000 vibrations per second. The sounds emitted by bats are used as a type of echolocation or sonar—they bounce back from tree limbs and other obstacles into which the bat might fly in the dark and also from moths and other insects upon which it feeds (Fig. 13.5). Were our ears capable of detecting these sounds, the quiet tranquility of an evening beside a mountain lake would be impossible.

For many insects the visible world is quite different from the world we

FIGURE 13.5

Photograph of a brown bat about to catch a falling meal worm.

Even in the dark, a bat is capable of flying continuously with no collisions. The animal's constantly emitted high-frequency sounds are reflected back to it from solid surfaces and are detected by its large, acute ears. Other animals using echolocation are marine mammals such as dolphins and whales. (Courtesy of Mr. Frederic Webster.)

see. The light spectrum to which they are sensitive is shifted toward slightly shorter wavelengths than those visible to man (refer to Figures 8.2 and 15a.9). Despite the large number of red flowers, most insects cannot see the longer red wavelengths. However, since these red-appearing pigments usually absorb blue light also, they do in fact appear colored to insects. At the shorter end of the spectrum, insects see colors which we cannot see—the so-called ultraviolet colors. Many flowers produce pigments which absorb ultraviolet light, and several of these, as they appear to us and to insects, are shown in Figure 13.6.

FIGURE 13.6

A   B

A group of daisy-type flowers from Florida, photographed in visible light (*A*) and ultraviolet light (*B*).

To us, the flowers appear yellow with some having dark centers, whereas to insects which can detect shorter, ultraviolet wavelengths the flowers have a much different appearance and are more distinguishable from one another. (From Eisner, T., *et al.* 1969. Science *166*:1172. Copyright 1969 by the American Association for the Advancement of Science.)

The Sign Stimulus.   Having observed that a particular stimulus elicits a predictable response in an animal, ethologists began to ask some questions: To what aspects of the total stimulus is an animal responding? Is it the total shape, movement and color of a potential mate that induces courtship behavior? Is it the sight, sound or movement of a hungry infant that causes its parents to feed it? In other words, ethologists began to ask, What are the specific signals to which an animal is reacting?

A great many experiments on a wide variety of animals responding to various stimuli have been carried out, and in almost every case it has been found that only certain specific elements of the total stimulus are necessary to induce the appropriate response.

One of the favorite animals of ethologists is the three-spined stickleback, a small fish inhabiting both coastal marine waters and freshwater lakes and streams. During the non-breeding season, males and females share a neutral grayish-green coloration which serves as camouflage against predators. As the breeding season approaches, the fish migrate to shallow freshwater spawning grounds. The males then develop a red belly,

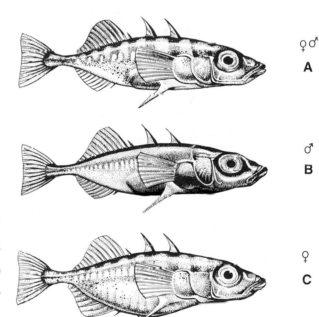

FIGURE 13.7

Changes in appearance of sticklebacks at breeding.
    A. Non-breeding animals have similar shape and neutral coloration.
    B. Breeding male develops a red belly (shown here in blue).
    C. Breeding female, swollen with eggs.

and the lower surface of the females becomes swollen as eggs accumulate in their abdomens (Fig. 13.7). Each male fish soon establishes a territory which it defends from other males. Whenever a strange male enters its territory the male assumes a threatening posture and exposes its red underside to the intruder, which generally suffices to scare off the stranger.

Experiments were next conducted on breeding fish kept in tanks to determine what triggered the male fish to perform its threatening display. A variety of models were used (Fig. 13.8), and it was found that the upper model of a non-breeding stickleback was relatively ineffective, whereas any of the four lower models with a red belly elicited aggressive behavior, despite their otherwise crude shapes. Models with the red patch on the upper side were less effective. Many other examples discovered from work on various animals have shown that only a particular part of the total stimulus is usually necessary to elicit a specific behavioral reaction.

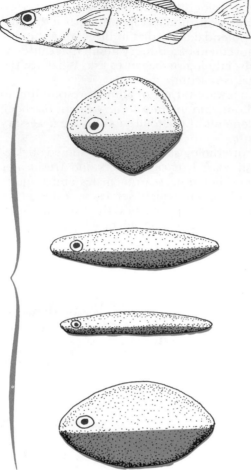

FIGURE 13.8

Models used to identify stimulus characteristics eliciting aggressive behavior in sticklebacks.

Upper model of non-breeding stickleback produced no response, whereas all of the cruder models with red bellies (in color) did. (After Tinbergen, N. 1951. *The Study of Instinct.* Clarendon Press, Oxford.)

FIXED ACTION PATTERNS

In the course of observing the behavior of an animal, an experimenter often observes that a particular response is performed more often in response to one stimulus than to another. Thus, a red belly usually induces aggressive behavior in a stickleback, whereas the swollen belly of the female usually induces a quite different behavior, a zig-zag courtship dance (Fig. 13.9), which in turn leads to the head-up acceptance signal of the female. The male then leads her to the nest which he has previously built of plant material while establishing his territory. Once the female enters the nest the male performs a shivering movement which stimulates the female to lay her eggs. The male then enters the nest and spawns, and the eggs are fertilized. During the week required for development of the eggs the male regularly fans oxygenated water through the nest, without which the embryos would die.

In the course of the above events, each behavior by one partner acts as a sign stimulus eliciting the next behavioral pattern in the other partner. However, events seldom progress as smoothly as shown in our diagram. After the zig-zag dance, for instance, the female may fail to follow the male to the nest. In addition, not infrequently the male may behave in a way inappropriate to mating, such as attacking the female or ignoring her and returning to the nest. Eventually, however, one or other partner repeats one of the stages in the behavior pattern. Only after long hours of observation, during which cues and responses have been statistically correlated, can the scientist begin to unravel the overall behavior of these fish (Fig. 13.10). When he finally achieves a meaningful sequence, the observer then can assign to each response something called a **fixed action pattern** — a response which, more often than not, follows after a particular stimulus. Thus, even the most stereotyped forms of animal behavior are subject to vagaries of response which the observer cannot fully explain.

The Nature of Drives. At any given moment an animal receives a whole host of stimuli, both external and internal, any of which may affect its behavior. Even the most stereotyped of responses to external stimuli can be made only if there are no overriding internal stimuli to interfere and, by the same token, only if prior internal conditions are met. As we have seen, mating behavior occurs only in animals whose internal hormonal levels have provided a responsive state.

These internal stimuli, or "drives," are often difficult to estimate quantitatively. Even a relatively simple internal signal such as thirst can be measured experimentally only by knowing the amounts of salt, water and food taken in by an animal and its salt and water losses from feces, urine, perspiration and respiration over a given period. Measuring all these

1. Female appears, gives head-up display.

2. Male swims zigzag to female.

3. Female swims, head-up, toward male.

4. Male swims toward nest.

5. Female follows.

6. Male shows the nest.

7. Female enters nest.

8. Male tremble-thrusts.

9. Female spawns.

10. Female leaves.

11. Male enters and fertilizes.

12. Male ventilates eggs in nest.

FIGURE 13.9

A diagram summarizing the various events that occur during mating behavior of the stickleback. (Note that the belly of the male, shown in blue color here, is actually red; further details in text.)

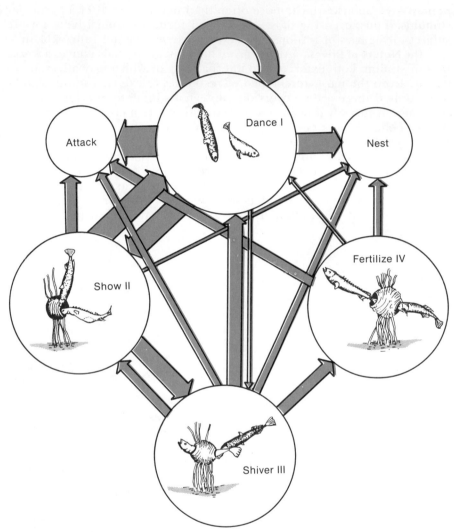

FIGURE 13.10

The frequency of various sequences of behavioral activities during mating in the ten-spined stickleback.

Directions of arrows indicate the sequence of behavior patterns, and thicknesses of arrows the frequency with which each sequence occurred. Note the complexity of the actual behavior. (After Marler, P., and Hamilton, W. J. III. 1966. *Mechanisms of Animal Behavior.* John Wiley and Sons, New York.)

parameters is a difficult task. It is far more difficult, however, to assess such drives as sex or parental behavior. In such cases, we know little about the internal signals, hormonal or otherwise, which affect an animal's behavior. By a detailed analysis of its blood, we can determine the levels of circulating hormones, but other factors, such as the receptivity of its tissues to the hormones, its past experience and its other internal physiological signals, are far more difficult to measure accurately.

CONFLICT IN BEHAVIOR

Much active behavior of animals involves orientation with respect to an object. In food searching, aggression and mating, the animal moves toward an object, whereas in threatening situations it moves away. Not uncommonly, approach and withdrawal come into conflict. This often occurs during aggressive behavior in birds, where two adversaries cannot decide whether to attack or flee one another. Sometimes the internal conflict is resolved by a combination of both types of behavior, neither of which is completed. This results in an aimless mosaic movement (Fig. 13.11A). Sometimes the aggressive pecking which would have been directed at an adversary is redirected to some other object, such as a twig or a leaf (Fig. 13.11B). And sometimes an animal performs an irrelevant activity, such as preening, a phenomenon known as *displacement activity* (Fig. 13.11C). In this latter case it is thought that the two conflicting responses, although both take precedence over the displacement activity in the hierarchy of the animal's behavioral repertoire, cancel one another; the less important activity is released from its prior inhibition.

Conflict situations in human behavior are often resolved in similar ways; an angry man, inhibited by cultural restrictions from expressing aggression directly, may pace the floor, strike at inanimate objects or scratch his head (Fig. 13.12).

FIGURE 13.11

Behavior resulting from conflict situations.
A. An angry gull shows a threatening head and wing posture but remains unsure and frozen in position.
B. An aroused blackbird attacks a leaf rather than the rival bird which stimulated its aggressive behavior.
C. A starling faced with a rival cannot decide whether to attack or flee, so preens itself instead.
(After Tinbergen, N. 1965. *Animal Behavior.* Time, Inc., New York.)

Mosaic movement

A

Redirected response

B

Displacement activity

C

Mosaic
movement

Redirected
response

Displacement
activity

FIGURE 13.12

Behavioral conflict in man may be resolved in ways similar to those used by other animals. (After Tinbergen, N. 1965. *Animal Behavior.* Time, Inc., New York.)

BEHAVIORAL MODIFICATION

We are all aware that we are constantly modifying our behavioral responses to stimuli—we are constantly learning new patterns. Such behavioral modification occurs to some extent in all animals, even single-celled protozoa.

WHAT IS LEARNING?

Attempts to analyze the processes involved in learning have preoccupied behaviorists for many years. At least in animals with discrete nervous systems, it is thought that modification of the ease with which impulses cross synaptic junctions plays an important role. However, it is still not known what sorts of modification occur at the cellular level to bring this about.

The Role of Memory. The major factor in learning is memory of a prior event. Thus, response to a stimulus will continue to be different only as long as the subject remembers what it learned previously. Memory depends upon storage of information in the nervous system, and this occurs only if the subject is aware of the information at the time it is received. The first requirement for learning, then, is that the subject be attentive.

There appear to be two types of information storage. **Short-term memory storage,** where the memory lasts from a few minutes to a few hours, is commonly experienced in man. We have all "memorized" a telephone number long enough to dial it correctly, but cannot recall it a few hours later. A similar phenomenon has been demonstrated in animals.

The second step in remembering is the transfer of information to **long-term memory storage,** which requires constant repetition of the learning experience or rehearsal. Thus to remember a telephone number permanently, one must look it up and dial it many times or else repeat it over and over to oneself. Once this process has taken place, at least in man, it is thought that the memory trace or **engram** lasts throughout life, even though it may only be recalled with great difficulty. Thus the third component in the total learning process is being able to recall information. These steps may be summarized as follows:

(1) TEMPORARY LEARNING
Perception $\xrightarrow[\text{(requires attention)}]{}$ short-term storage $\longrightarrow$ recall

(2) PERMANENT LEARNING
Short-term storage $\xrightarrow[\text{(requires rehearsal)}]{}$ long-term storage $\longrightarrow$ recall

Further details of these processes are dealt with in Chapter 13a in relation to learning in man.

The Role of Reinforcement. As we have seen above, attention to a stimulus is necessary for learning to occur. For a stimulus to be worthy of attention, it must usually carry an expectation of something to follow; it must be associated with a reinforcement. Only when such a correlation is established can the stimulus bring about a modification of behavior. Let us examine this concept in relation to several types of learning.

HABITUATION. Many sedentary animals, such as marine worms living in tubes, must expose themselves in order to feed on the small organisms which form their diet. When a predator's shadow falls on a worm, it responds by a quick reflex withdrawal into its tube. Its eyes are unable to distinguish, however, between the innocuous shadows of seaweed washing overhead and those produced by a real danger, such as a flatfish. If a worm were to duck into its tube at every shadow, it would lose valuable feeding time, so it makes a "compromise"—it stops responding to a repeated shadow, as would be cast by seaweed drifting back and forth. Without the reinforcement of danger the withdrawal response wanes. Such suppression of reflex behavior is called **habituation.** If no shadows occur for a period of time, the original reflex withdrawal returns; habituation is only a temporary form of learning, at least in simple animals.

CONDITIONED LEARNING. Many animals learn to respond to a stimulus through repeated experience. A young mammal, soon after birth, does not produce saliva until food is placed inside its mouth. But after weaning the young animal learns to salivate at the sight or smell of food, in anticipation of eating. We have all experienced the same thing on smelling Thanksgiving dinner cooking or perhaps even while watching Julia Child preparing some mouth-watering dish on television. A stimulus other than the taste of the food elicits the response—we become **conditioned** to a substitute stimulus.

The Russian scientist Ivan Pavlov discovered around the turn of the century that he could train dogs to salivate in response to a completely neutral stimulus—the ringing of a bell. This occurred if the bell was rung just prior to each presentation of food. Gradually, the bell alone became a sufficient stimulus to elicit the salivation response. The dog became conditioned to the sound of the bell. However, if Pavlov continued to ring the bell, time after time, without presenting food, gradually the salivation response ceased. Without occasional reinforcement, the sound of the bell lost its meaning.

Pavlov's experiments, however, do not tell us much about how animals learn in their natural environment. Far greater insights have been gained by studying the effects of reinforcement on behavior itself, or what is known as **operant conditioning.** It is the behavior which is reinforced rather than a neutral stimulus.

Let us imagine an untrained rat placed in a cage with a lever and a light (Fig. 13.13). At first the rat simply explores the cage and perhaps pushes the lever. When it does so, food appears. The rat quickly learns that pressing the lever produces food, and as long as it is hungry it will continue to push the lever. The sight of the lever is a stimulus for producing food much as Pavlov's bell was a stimulus for salivation. But in this case the response made by the animal produces a result—food. There is **positive reinforcement** for pushing the lever.

Now let us suppose that the experimenter changes the situation so that only when the light is on is food obtained when the lever is pressed. At first a hungry rat pushes the lever all the time, but it soon associates reward with the light. It has thus learned that its actions are successful only if it pushes the lever when the light is on. It has since been found that rats can learn a long series of correct maneuvers necessary to obtain a positive reward such as food (Fig. 13.14).

In much the same way a rat or other animal can be trained to avoid a punishment or **negative reinforcement.** It will climb a pole or escape to a second chamber in the cage after learning that a light or some other stimulus will soon be followed by an electric shock delivered to the floor of the compartment it is in.

TRIAL-AND-ERROR LEARNING. To some extent the operant conditioning experiments we have described above are only possible because the animal learns by making mistakes or errors. Thus, **trial-and-error learning,** which is an extension of operant conditioning, depends to a large degree upon reinforcement. A rat quickly learns to run a maze from beginning to end, making the correct right and left choices, if there is a reward to be had or a punishment for mistakes.

But there is some evidence that a rat that receives no reinforcements in such an environment may nevertheless be learning his way about. During the period when neither food nor electric shock is forthcoming he may wander aimlessly about. Yet as soon as reinforcement is added, such a rat runs the maze far more quickly than does a rat that has never before experienced it. Such experiments suggest that animals may register information about their environment even in the absence of a reinforcing experience. Much remains to be discovered about such non-reinforced learning, espe-

FIGURE 13.13

Rat in a test box, normally completely enclosed, finds that by pushing the lever it can obtain food. The experimenter may complicate the problem by providing food only when the light is switched on. (After Tinbergen, N. 1965. *Animal Behavior.* Time, Inc., New York.)

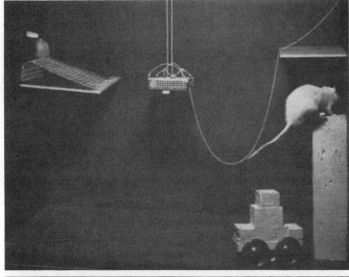

FIGURE 13.14

A rat learns to obtain a food reward placed on the wire screen at upper left by performing a complex sequence of maneuvers. (Photos by Frank Lotz Miller, reproduced by permission of Black Star Publishing Co.)

cially in animals with relatively simple nervous systems and behavior patterns.

INSIGHT LEARNING. Among higher animals, particularly mammals, the solving of problems such as those solved by the rat in obtaining food in Figure 13.14 can often be shown to involve remembering earlier, somewhat different situations. During such *insight learning,* the animal relates the usefulness of responses learned on a previous occasion to somewhat different stimuli, correlating parallels in the two situations. Much controversy surrounds the meaning of insight learning. Some hold that because the two sets of stimuli are different, "abstract ideas" are necessary to relate them; others see it merely as a substitution of one set of stimuli for another related set.

Whatever the disagreements among scientists, the differences in behavior among animals are striking. The raccoon in Figure 13.15, for ex-

FIGURE 13.15

Lack of insight behavior in the non-primate raccoon. Only by trial and error will the animal learn to increase its available rope length by going around the second stake. An adult primate would solve the problem quickly by observation alone. (After Keeton, W. T. 1967. *Elements of Biological Science.* W. W. Norton & Co., New York.)

ample, is frustrated because it does not perceive that by going around the second stake it can reach its goal, the bowl of food. A chimpanzee or man would soon recognize the problem and solve it without resort to trial-and-error.

THE DEVELOPMENT OF BEHAVIOR

As we have already noted, behavior, even in its most stereotyped form, is modified by learning during an individual's lifetime, although the extent and duration of modification vary greatly from one type of animal to the next. Let us examine some of the ways in which an animal's early environment can affect its later behavior.

IMPRINTING

Although originally described in birds, this type of behavioral modification appears to be common in many vertebrates whose young are mobile at birth but still require parental care. Shortly after hatching, a young duckling, for instance, goes through a critical period of several hours during which it learns to follow any suitable object which moves or emits a sound, a process known as *imprinting.* Normally, of course, this object is the solicitous parent who remains nearby, and the offspring thus learns to identify with its own species. Once this identification is made it remains fixed or imprinted and can only be reversed with the greatest difficulty.

If the parent is removed during this critical period, by accident or by experimental design, the youngster imprints on the nearest suitable object, be it a matchbox, a mechanical toy, a strange animal or even the experimenter (Fig. 13.16). It is thought that many animals born in captivity, such as great condors and giant pandas kept in zoos, may not mate because they are imprinted on their keepers rather than on their own species. Once imprinting has occurred it plays a major role in the subsequent behavior of an animal.

SENSORY STIMULATION

The nervous systems of many animals, especially birds and mammals, are not fully formed at birth. Although the nerve cells do not undergo any further divisions, they continue to enlarge and establish synaptic connections. There is a gathering body of evidence indicating that maximum development of the nervous system and of learning ability depends upon sensory stimulation, especially during development. Both rats and chimpanzees raised in the dark, for example, later perform far less well on tests requiring visual perception than do control animals raised under standard conditions.

Studies on the influence of early environment on the anatomy and chemistry of the brains of young rats indicate that the more stimulating the environment, the bigger is the brain, especially the cerebral cortex, the region where most of the association centers involved in learning are located. (The functions of this region in the human brain are discussed further in Chapter 13a.) Rats were raised either in groups in large cages provided with a variety of objects which were changed each day (Fig. 13.17) or in isolation in small wire cages without added objects. In addition to a larger cortex, the brains of rats in the more stimulating environment had larger nerve cells with fewer but larger synaptic connections, more RNA per nerve cell, and more of the enzyme acetylcholinesterase, which plays a role in synaptic transmission in the brain. The rats from the enriched environment also showed enhanced learning ability in some but not all test situations. It is not yet possible, however, to utilize studies such as these to explain the role of environment on the development and learning capacity of the human brain.

Studies on the development of behavior in young primates are of special interest. Unlike other animals which are born with relatively rigid behavioral patterns, which may later be modified, young primates are poorly equipped in this respect and depend to a great extent upon interactions with their parents or with other young for development of normal behavior patterns. Infant rhesus monkeys, for instance, require security engendered by physical contact with the mother (Fig. 13.18A) before they will explore a strange environment (Fig. 13.18B). Infants deprived of this security often withdraw and become autistic: they do not respond normally to playthings or to other young, instead retreating in fright (Fig. 13.18C). Playing with other young is also an important factor in normal socialization.

Although much remains to be learned, it is evident that there is an important link between early environment and behavioral development that plays its greatest role in higher vertebrates, especially primates.

FIGURE 13.16

A young duckling automatically follows a mechanical toy duck on which its behavior has become imprinted. (From "Imprinting" in Animals, by E. H. Hess. Copyright © March, 1958 by Scientific American, Inc. All rights reserved.)

FIGURE 13.17   Young rats in a highly complex environment.
    Such rats show increases in brain size, especially of the cortex, and have larger neurons and synapses and more of certain chemicals in their brains than do litter-mates reared in simple wire cages without added objects. (After Case, J. F., and Stiers, V. E. 1971. *Biology: Observation and Concept.* Macmillan, New York.)

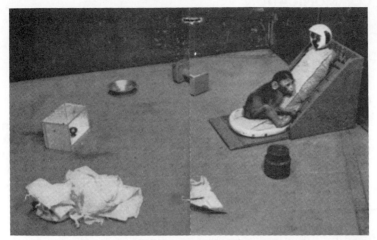

A

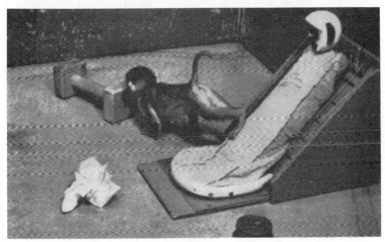

B

C

FIGURE 13.18

Behavior of a young rhesus monkey upon being placed in a strange environment.
A. A cloth surrogate mother provides reassuring contact.
B. The infant then explores its new cage and plays with objects.
C. An infant without a cloth surrogate mother crouches fearfully in the corner.
(From Love in Infant Monkeys, by H. F. Harlow. Copyright © June, 1959 by Scientific American, Inc. All rights reserved.)

COMMUNICATION

Although many of the behavioral situations already considered involve communication between animals, let us turn to a few specific examples to indicate some of the various mechanisms used by animals to exchange information.

COMMUNICATION THROUGH MOVEMENT

Many animals make postural signals indicating their intent toward others—we can all distinguish between the meaning of the raised ruff and bared teeth of a dog (even if it does not growl) and that of its prancing and tail-wagging behavior. Animals also use bodily movements to signal other kinds of information, and sometimes such behavior patterns are difficult for man to interpret. As in the case of the stickleback, only patient observation reveals the significance of apparently random behavior.

One of the most interesting examples of this type of behavior is that used by foraging honey bees to communicate the location of a source of pollen or nectar to other workers. This information is communicated by a complex "dance" performed within the hive; it contains two pieces of information: distance (flying time) and direction. The dance takes the shape of a figure "8" repeated over and over again (Fig. 13.19). Distance is indicated by the speed at which the dance is performed and by the number of waggles of the abdomen executed in the middle of the figure eight. With increasing distance, the speed is reduced, but the number of waggles is increased. Direction of the food source is communicated relative to the direction of the sun. But since the inside of the hive is dark, a substitute sun must be indicated. For this, gravitational force is used, and the dance is performed on a vertical surface within the hive. Waggling directed upward indicates the food source is toward the sun as a bee leaves the hive; waggling downward indicates that food is located away from the sun. Waggling in oblique directions indicates food located at a given angle to the sun. As the dance is repeated, other worker bees copy it several times and then leave the hive in search of the food.

SOUND COMMUNICATION

Although we are all familiar with the sounds of many animals, their functions as signals are not always so apparent. We have noted that the songs of birds play a role in establishing territories and attracting mates. The songs of male grasshoppers, produced by rubbing together modified wings on their backs, are specific signals helping females to locate them, as are also the croakings of male frogs. Among both frogs and grasshoppers, the quality of the sounds emitted is specific to each type, or species, and attracts only females of the same species. Hybrid frogs emit songs halfway between those of the two parents, emphasizing the genetic component to the quality of the sound they produce.

CHEMICAL COMMUNICATION

Chemical communication is perhaps the most primitive and widespread type of communication used by animals, since an anatomically complex organ is not required to either produce or detect a chemical. Because man has such a poor sense of smell and has yet to devise instruments capable of detecting only a few molecules of a specific chemical—as can the chemical receptors of many animals—chemical communication is usually discovered only through careful observations of behavior.

The most studied group of animals in this respect are the insects, which produce scented secretions for many purposes—to discourage pre-

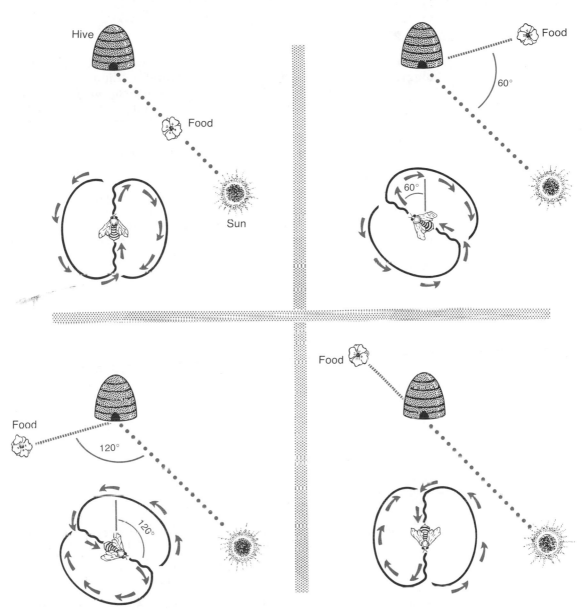

Communication of direction of food source by bees.

The vertical plane of the hive indicates direction of the sun. The angle of waggling movement away from the vertical indicates angle of food source in relation to sun and hive. The degree of waggling indicates flight time required to reach food (not shown here). (After Villee, C. A., Walker, W. F., Jr., and Smith, F. E. 1968. *General Zoology*. W. B. Saunders Co., Philadelphia.)

FIGURE 13.19

dators (certain moths), to mark a trail to food (ants), to help foragers return to the colony (bees) and to attract mates (virtually all insects). When scents are used for communication within a species they are known as **pheromones;** we shall have occasion later in the book to mention these substances again.

SUMMARY

Behavior results when a stimulus elicits a response in an animal. Ethology, the observation of behavior under natural conditions, has greatly increased our understanding of the adaptiveness of behavior to survival.

Effective stimuli depend upon the ability of sense organs to detect them and upon the attentiveness of the animal, which may be affected by other external and internal stimuli received simultaneously. Frequently only a part of the stimulus object—the sign stimulus—is necessary to elicit a response. Responses are often stereotyped, being known as fixed action patterns, but there is no sharp distinction between learned and innate behavior, since the latter can be modified by experience. A sequence of fixed action patterns between two or more individuals may lead to integrated behavior patterns, such as commonly occurs during mating. When two stimuli leading to opposite responses are presented, the conflict is often resolved by alternative behavior such as displacement activity.

Modification of behavior, or learning, depends upon permanent modification of the nervous system, the memory engram, although little is known about how information is stored and retrieved. Reinforcement of a response appears essential at all levels of behavioral modification, although awareness of new stimuli may occur without such reinforcement.

During development of the nervous system, environmental stimuli play an important role in determining later behavior; one striking example of this is imprinting. Sensory stimulation has been shown to affect cortical development and brain chemistry, and in primates it is essential for development of normal social behavior.

Many types of communication are utilized by animals for exchange of information, including visual signs, postural movements, sounds and chemical signals, the latter being perhaps the most primitive and widespread. Pheromones used for communication between members of the same species are becoming of ever greater interest to man.

READINGS AND REFERENCES

Contemporary Psychology. 1971. Readings from Scientific American. W. H. Freeman, San Francisco. A collection of articles slanted primarily toward human behavior but including several dealing with experimental studies on learning and memory.

Dethier, V. G., and Stellar, E. 1970. *Animal Behavior.* 3rd ed. Prentice-Hall, Englewood Cliffs, New Jersey. An excellent paperback introducing the reader to modern ethology and neuroanatomy.

Frontiers of Psychological Research. 1966. Readings from Scientific American. W. H. Freeman, San Francisco. Includes among its articles several on ethology, perception, development of behavior and learning.

Hinde, R. A. 1970. *Animal Behavior: A Synthesis of Ethology and Comparative Psychology.* 2nd ed. McGraw Hill, New York. A comprehensive analysis of current ideas about behavior; difficult reading but essential for those interested in definitions about behavior.

Marler, P., and Hamilton, W. J., III. 1966. *Mechanisms of Animal Behavior.* John Wiley and Sons, New York. A broad and readable coverage of animal behavior from the ethologist's viewpoint.

Rosenzweig, M. R., Bennett, E. L., and Diamond, M. C. 1972. Brain changes in response to experience. Scientific American, February, 1972, pp. 22–29. An account of the experiments demonstrating that early environment can affect the way the brain develops.

Tinbergen, N. 1965. *Animal Behavior.* Time, Inc., New York. One of the Life Nature Library series, this beautifully illustrated book gives a fascinating account of animal behavior by one of the world's best known ethologists.

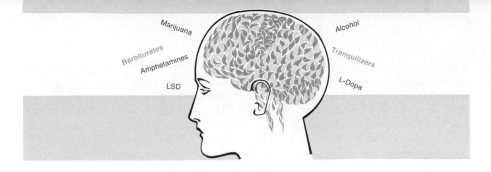

DRUGS AND THE HUMAN BRAIN

The organ which most distinguishes man from all other animals is his brain. Despite enormous amounts of study, only recently have we begun to understand how some of the functional pathways within this complex organ are reflected in human behavior and how they can be modified by drugs.

The unraveling of the anatomical complexities of the human brain, with its billions of nerve cells, has only begun. Tracing out the almost innumerable functional connections and relating them to human behavior is perhaps the most formidable task in all biology. There is still an enormous gap between what the psychologist or psychiatrist observes as behavior and what the anatomist or physiologist describes as going on in the brain. In the past decade or so, the advent of powerful new techniques for studying the brain, including the experimental study of a variety of new drugs, has deepened our understanding of the general functioning of the human brain. The use and abuse of such drugs also have great medical and social significance.

In this chapter we shall attempt the difficult task of explaining some of the more important facts now known about the workings of the human brain. Then, by means of a few selected examples, we shall try to correlate the effects of drugs on behavior with what is known about the chemistry of various pathways in the brain.

THE HUMAN BRAIN

The simple reflex arc, such as the knee-jerk described in Chapter 12, is never as simple as shown in higher animals, especially man. For example, a conscious patient is quite aware that the doctor has tapped his knee with a hammer and that his leg has automatically responded; moreover, by consciously tensing the antagonistic muscles, he can suppress the response. The stimulus initiated by the hammer not only goes to the spinal column but is also transmitted to the brain, which can modify the response. The brain thus serves as a central receiver for a variety of sensory information which it processes and then may either ignore, store for future use or act upon immediately by sending impulses to appropriate muscles and glands.

The whole nervous system may be likened to an enormous and complex system of reflex arcs, where sensory information passes upward in the spinal column to the brain, is processed by association neurons and is finally transmitted downward again to effectors. Along the ascending pathway between a sensory receptor and the conscious levels of the brain there are numerous synapses which modulate incoming sensory signals. Some of these signals are filtered out and suppressed—a person thus "hears" at a crowded party, for instance, only the voices of those he is conversing with, unless someone across the room mentions his name or some other word that has special significance. Moreover, the responses one makes after receiving a stimulus are continuously monitored, so that one is to some degree aware of what one is saying or doing at a given instant. There is a constant exchange of information between input and response at all levels, conscious and unconscious.

In general, the human brain is organized so that the less conscious activities, such as breathing and swallowing, are regulated by the lower, evolutionarily older regions of the brain, whereas conversation, thinking, awareness of a rock concert or a Beethoven symphony, and execution of delicate movements by a surgeon wielding a scalpel, are all coordinated by the higher centers. However, a person who has mumps may suddenly find himself all too aware that his swallowing is impaired. In other words, under certain circumstances, awareness and even control of functions usually carried out unconsciously are quite possible (Fig. 13a.1). These higher

FIGURE 13a.1

The discovery that yogis can consciously regulate their heart rate, blood pressure and the amount of oxygen they consume has led to attempts to apply the same principles in medicine. (*Magnum Photo,* by Marilyn Silverstone.)

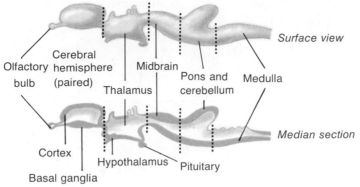

Surface view

Olfactory bulb · Cerebral hemisphere (paired) · Midbrain · Pons and cerebellum · Medulla · Thalamus · Cortex · Hypothalamus · Pituitary · Basal ganglia

Median section

The generalized vertebrate brain, showing its five major regions. (After Romer, A. S. 1970. *The Vertebrate Body*. W. B. Saunders Co., Philadelphia.)

centers of awareness are associated with the evolutionarily most recent parts of the nervous system, located in the cerebral cortex. In man, this region of the brain is developed as in no other animal.

The human brain is thus organized into several structural levels, each with its ascending sensory pathways and descending motor pathways. Each level is connected with the one above and the one below.

MAJOR SUBDIVISIONS OF THE BRAIN

The regions of the human brain may perhaps best be studied in the order of their evolutionary sequence from the **medulla,** which connects the brain to the spinal cord, forward to the **cerebral cortex,** the most advanced and complex region. The generalized vertebrate brain shown in Figure 13a.2 may help to orient the reader. In the course of evolution, as new functions were added and old ones expanded, the front regions of the brain became increasingly enlarged, taking over the ultimate control of functions once performed entirely by older, more primitive regions. Thus, in the human brain the cerebral cortex has become greatly enlarged, overshadowing all other regions in size and complexity of structure (Fig. 13a.3).

The Medulla. The brain stem or medulla is essentially an extension of the spinal cord, controlling basic bodily functions such as heart rate, blood pressure and breathing reflexes. Even though these functions can be modified by higher centers—a frightening experience causes increases in heart rate, blood pressure and respiratory rate—they continue to work even when one is deeply unconscious, a feature which makes the techniques of modern anesthesiology possible.

The Cerebellum and Pons. The outgrowth from the roof of the brain in front of the medulla, the **cerebellum,** is the center for coordination of movements and for maintenance of posture, muscle tone and balance. Impulses are exchanged between this region and the cerebral cortex, where voluntary movements are initiated; the two regions thus act together to bring about normal behavioral movements. Damage to the cerebellum often causes a patient to experience difficulty in making discrete movements, such as reaching out to grasp an object. The arm wavers in its course rather than following a direct path (Fig. 13a.4). Severe damage to the cerebellum results in loss of ability to stand or walk.

The **pons,** or "bridge," is a continuation of the main fiber tracts running between the medulla and higher regions of the brain. The medulla, pons, midbrain floor and thalamus together constitute the brainstem.

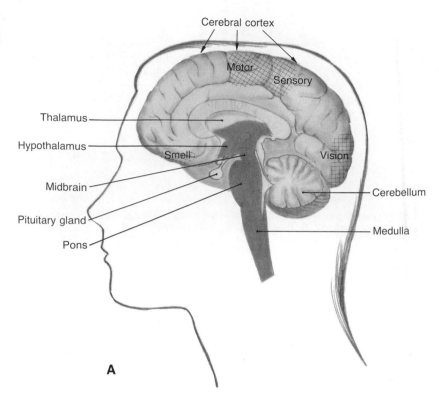

Cerebral cortex

Motor

Sensory

Thalamus

Hypothalamus

Smell

Vision

Midbrain

Cerebellum

Pituitary gland

Medulla

Pons

A

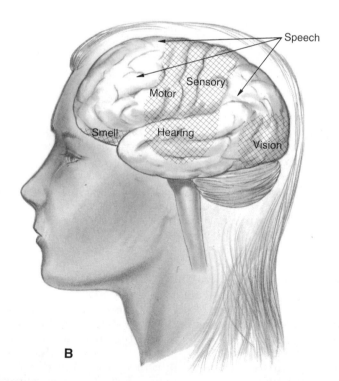

Speech

Sensory

Motor

Smell   Hearing

Vision

B

FIGURE 13a.3

Major regions of the human brain.

A. Median section. The components of the brainstem are shaded more darkly. Specific sensory and motor regions of the cerebral cortex are indicated by color crosshatching.

B. Outer view. Note how the cerebral cortex has overgrown the rest of the brain. Specific sensory and motor regions of the cortex are shown in color crosshatching; areas associated with speech are indicated by arrows.

The Midbrain. In front of the cerebellum, the brainstem continues into a region known as the *midbrain.* In lower animals, the roof of this region is the major coordinating center for optic and auditory stimuli, but in man these functions reside primarily in the cerebral cortex, and the midbrain roof consists of but two small bumps. Of greatest interest in the human midbrain are the important ascending (sensory) and descending (motor) axonal tracts that pass through its floor, branching and synapsing with other fibers along the way. We shall return to them shortly.

The Thalamus and Hypothalamus. The *thalamus* is the terminus of the ascending sensory tracts and operates as a relay center between them and the cerebrum. It also has various coordinative functions. The *hypothalamus,* although small in size, plays a central role in moderating many behavioral drives related to the internal physiological state of the body, such as thirst, hunger, body temperature and sexual appetite. Stimulating electrodes placed in these specific physiological control centers of an experimental animal will call forth drinking, eating, temperature adjustment or sexual behavior. Through its relations with the pituitary gland, the hypothalamus also controls water balance and regulates the secretion of various hormones affecting metabolism. In Chapter 22 we shall explore its control over human reproductive physiology.

The hypothalamus has also been called "the seat of the emotions." When electrodes are implanted into certain parts of the hypothalamus of an animal and mild shocks are delivered, the animal exhibits those behavioral patterns which we call emotions—rage, in which it bares teeth and claws and may attack the experimenter, or fear, when its pupils dilate, its eyes dart from side to side and it searches for an escape route. Electrodes implanted into yet other regions of the hypothalamus and nearby centers apparently deliver a pleasurable response, and rats may continuously press a lever to "reward" themselves with shocks to these areas, even ignoring food in the process (Fig. 13a.5).

The hypothalamus does not, however, act in isolation from other parts of the brain in its role of controlling behavior related to the survival of the organism. If the cerebral cortex is removed from a cat, for example, and the animal is lightly stroked, instead of purring as usual the cat flies into an apparent rage. The cerebral cortex, therefore, normally exerts a modifying influence upon the hypothalamus. Moreover, the hypothalamus in man is necessary only to *exhibit* emotion; patients with lesions of the hypothalamus cannot display their feelings, but still report having them. These feelings may arise in part from the more ancient parts of the cerebral cortex.

The Cerebrum. The *cerebrum* is a paired structure developing as two lateral outgrowths from the front of the brain. In the most primitive vertebrates it consists of a basal region, containing the *basal ganglia,* which

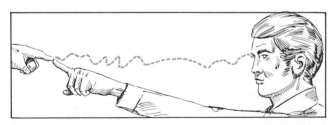

FIGURE 13a.4

Tremulous movement in patient with cerebellar damage.
As the finger moves from the nose to touch the finger of the examiner, it traces a wavering course. The antagonistic muscles of the arm are not being synchronized, and so large overshoots and overcompensated corrections occur. (After Ruch, T. C., *in* Stevens, S. S. (ed.) 1951. *Handbook of Experimental Psychology.* John Wiley and Sons, New York.)

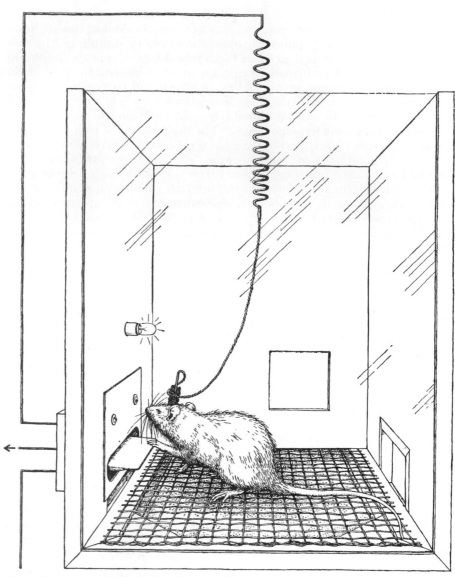

FIGURE 13a.5

Rat with electrode implanted in "pleasure" center of brain quickly learns to stimulate itself by pressing lever, which rewards it with a stimulus. (From Pleasure Centers in the Brain, by J. Olds. Copyright © 1956 by Scientific American, Inc. All rights reserved.)

serve for motor coordination, and a small region of cortex associated with the sense of smell (Fig. 13a.6). In mammals and especially in man, this "old cortex," as it is called, loses much of its importance and is overgrown by the new cortex, or *neocortex,* which becomes deeply folded in order to fit within the cranial space. New connections are formed between the basal ganglia and the neocortex, and the latter now takes over direction of the most complex motor functions.

As shown in Figure 13a.3, only small areas of the human cerebral cortex are specifically related to sensory and motor functions. The rest consists of *association cells* which process incoming information and correlate it with past experience, thus carrying out the intellectual functions of the human brain and making it possible for you to read and understand this book. Note the large amount of association area devoted to language and speech in the human cortex.

Although the brain is often conveniently described by the five subdivisions we have listed above—medulla, cerebellum and pons, midbrain, thalamus and hypothalamus, cerebrum—none of these regions functions

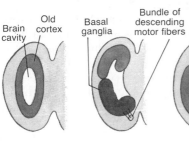
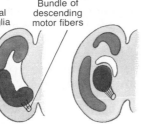
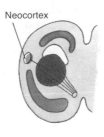
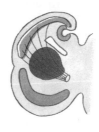
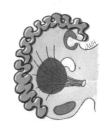

| Early embryonic stages | Primitive reptile | Advanced reptile | Primitive mammal | Advanced mammal |

The evolution of the cerebral hemispheres (left hemisphere, shown in cross-section). **FIGURE 13a.6**

Early in development, the old cortex is above and the basal ganglia are below. In reptiles, the major cortex is the old cortex, and all locomotion is controlled only by basal ganglia. In higher animals, motor control is taken over by the neocortex, which largely displaces the old cortex. (After Romer, A. S. 1970. *The Vertebrate Body*. W. B. Saunders Co., Philadelphia.)

in isolation from the others: there are numerous interconnections and feedback loops between the regions, so that responses to stimuli are integrated throughout the brain. Although this is a great inconvenience for the neurologist studying brain function, without the complexity of his own brain such studies would be impossible for him; it is an essential prerequisite for the sophisticated behavior patterns characteristic of man.

By placing minute recording electrodes in various parts of the cortex and monitoring the electrical activity evoked in them after stimulating

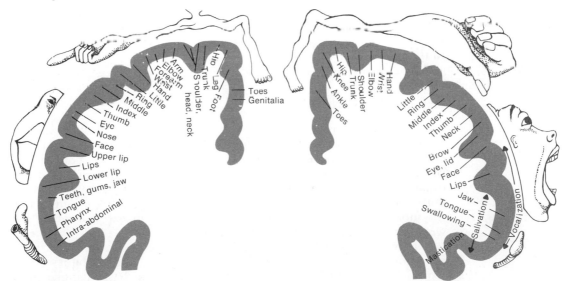

Sensory area   *Motor area*

Maps indicating the importance of various parts of the cerebral cortex (color) in relation to organs controlled by it. Sensory, left; motor, right. Note prominence of areas devoted to hands, mouth and tongue. (After Keeton, W. T. 1969. *Elements of Biological Science*. W. W. Norton & Co., New York.)

FIGURE 13a.7

parts of the body surface, it is possible to map specific cortical regions responding directly to peripheral stimuli. These are localized mainly in the "sensory" region shown in Figure 13a.3. The areas of this sensory region corresponding to different parts of the body are indicated in Figure 13a.7A. Note the extreme importance of the mouth, tongue and hands. Auditory, visual and olfactory stimuli also produce electrical potentials in the specific cortical regions identified in Figure 13a.3.

In a similar way, electrical stimulation of other parts of the cortex produces discrete movements of certain muscles. When a map of the muscles affected is superimposed on the cortex, it is again found that only a very small region is directly involved in muscular movement (the "motor" region of Figure 13a.3). Here, too, special emphasis is put on the muscles controlling mouth, tongue and hands, as shown in Figure 13a.7B. These motor pathways appear to pass directly from the cerebral cortex to the motor neurons of the spinal column. Nevertheless, their ultimate effect may be modified by other impulses reaching the motor neurons from lower levels of the brain, as we shall see shortly.

It is a matter of common experience that the same stimulus repeated on different occasions does not produce an identical response. A mother may drop a pot with a hot handle on one occasion but may let it burn her hand on another if dropping the pot could cause her baby underfoot to be scalded by its contents. Obviously the spinal reflex pathways have been modulated by other parts of the brain. Even in less extreme situations, our responses are continuously modified in terms of past experience and present circumstances. This modulation seems to occur not only by direct interconnections between one part of the brain and another, but also by diffuse pathways which affect the levels of activity of various parts of the brain.

In fact, the diffuse control systems are of great interest to psychologists, since they play an important role in regulating human behavior. They also are significantly affected by drugs. In the following discussion we shall consider three of these diffuse control systems: the basal ganglia of the motor system; the reticular activating system, which plays a role in wakefulness, learning and memory; and the limbic system, which has a generally inhibitory effect on other parts of the central nervous system. To some degree all these systems overlap one another, and all are characterized by *reverberating loops*—that is, they cycle information continuously, in circular pathways, between the outer cerebral cortex and lower control centers in the thalamus and brainstem.

THE BASAL GANGLIA

The direct transfer of impulses from the motor region of the cortex to motor neurons in the spinal column occurs via a special tract of axons known as the *pyramidal pathway* (Fig. 13a.8). In addition, modifying impulses reach the spinal motor neurons not only from the cerebellar coordinating system (not illustrated), but also from elsewhere in the cortex; both help to integrate muscular activity. These modifying impulses reach the motor neurons in the spinal cord by way of the *extrapyramidal pathway,* in which several synapses may occur.

Extrapyramidal fibers orginating in the basal ganglia exert mostly inhibitory impulses on the spinal motor neurons. This inhibitory effect is *increased* by release of *acetylcholine* (ACh) in the basal ganglia and is *suppressed* by *dopamine* (D), a chemical related to noradrenalin. Dopamine is synthesized by neurons in the *substantia nigra* (black substance) of the midbrain, whose axons end in the basal ganglia.

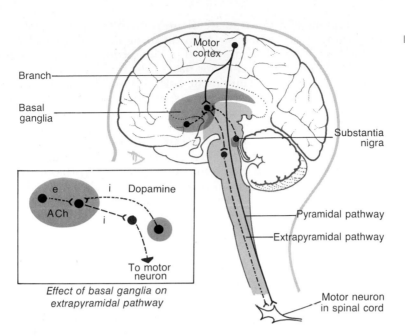

Effect of basal ganglia on
extrapyramidal pathway

Cortical motor control and the basal ganglia.

    Two sets of motor impulses are sent to spinal motor neurons. The pyramidal pathways directly connect the motor cortex to the spinal neuron.

    Extrapyramidal pathways from the cerebellum (pathways not shown) and the basal ganglia also affect spinal motor neurons. Impulses from the basal ganglia have an inhibiting function, which is augmented by ACh and suppressed by dopamine.

FIGURE 13a.8

PARKINSON'S DISEASE

    Normally the balance between acetylcholine and dopamine in the basal ganglia is such that coordination of muscular activity is almost perfect and movements are carried out smoothly. In Parkinson's disease, however, the cells of the substantia nigra are unable to manufacture the neurotransmitter dopamine. The resulting imbalance of acetylcholine causes an excessive inhibition of spinal motor neurons, thus interfering with normal muscle coordination. The disease, which may either be inherited or caused by a rare virus infection, is characterized by partial paralysis, muscular rigidity, a mask-like facial expression, uncontrollable muscular tremor and a shuffling gait.

    Until quite recently, treatment of Parkinson's disease relied upon suppression of the excess acetylcholine in the basal ganglia by administration of atropine and other drugs which block ACh receptor sites in the brain. Although somewhat successful, these drugs had unwanted side effects, since they also inhibited ACh-mediated arousal mechanisms located elsewhere in the brain; these will be discussed shortly. In the past few years much success has been obtained with some patients by treatments with a compound known as **L-DOPA.** In patients with Parkinson's disease, the nerve cells of the substantia nigra are unable to make the initial conversion of the amino acid tyrosine into L-DOPA, a precursor of their neurotransmitter, dopamine.

$$\text{tyrosine} \xrightarrow{\text{enzyme 1}} \text{L-DOPA} \xrightarrow{\text{enzyme 2}} \text{dopamine}$$

(defective in
Parkinson's disease)

Dopamine, if given to a patient, is unable to pass from the blood into the brain, but L-DOPA is. Upon diffusing into the cells of the substantia nigra, L-DOPA is then converted by the second enzyme into dopamine. A single biochemical defect in one part of the brain is thus responsible for this disease. The large quantities of L-DOPA needed to control Parkinson's disease produce certain side effects, including marked loss of appetite and an excessively increased sexual drive, but these are greatly compensated for by the fact that some patients who once were in wheelchairs are now candidates for broken legs from skiing accidents!

Interestingly enough, certain drugs used as tranquilizers tend to produce symptoms of Parkinsonism if given over prolonged periods. The calming effect of these drugs results from their ability to reduce the brain's content of monoamine-type neurotransmitters, of which dopamine is one. Let us turn next to the diffuse control system that regulates the normal sleep-awake cycles and may be involved in those anxiety states for which tranquilizers are prescribed.

THE RETICULAR ACTIVATING SYSTEM

As we have already noted, all sensory impulses entering the brain pass through the brainstem—the medulla, pons, midbrain and thalamus—before being relayed to the cortex. During passage through this region, the incoming fibers bifurcate, one branch going straight through, via the thalamic relay, to the cortex, the other branch entering a diffuse network of fibers and neurons in the brainstem, the *reticular formation.* Part of the reticular formation is taken up with the extrapyramidal motor pathways which we have already mentioned; part of it is concerned with relaying information from the cerebellum; and yet another part acts as a sensory filter, suppressing background "noise" and selecting only that information which is significant for relay to the cortex (Fig. 13a.9).

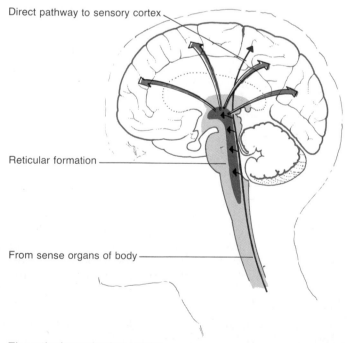

Direct pathway to sensory cortex

Reticular formation

From sense organs of body

FIGURE 13a.9

The reticular activating system.
Branches of the axons carrying sensory input to the brain enter the reticular formation, which filters out background noise and alerts the cortex to significant information.

This *reticular activating system* (RAS), as it is called, plays an important role in arousal and conscious awareness in the cerebral cortex, a state necessary for learning and memory to take place. It thus sets the stage, so to speak, for intellectual activity. The RAS is also the control center for sleep-awake cycles. If a particular part of the reticular system is destroyed in experimental animals, they fall into a permanent state of deep sleep or coma. On the other hand, electrical stimulation of the reticular system of an intact sleeping animal causes it to awaken.

HOW THE RAS WORKS

It is not yet certain precisely how the RAS system carries out its job of filtering information and causing arousal, but a simplified "working hypothesis" may prove useful, even though it is only tentative.

There seem to be at least three neurotransmitters involved in the action of the RAS on the cortex. Two of these are monoamines, produced mainly by neurons in the midbrain part of the RAS which send fibers both to the thalamus and directly to the cortex. The two monoamines serve opposite functions: 5-HT (5-hydroxytryptamine) causes drowsiness and sleep, while noradrenalin causes behavioral alertness. The third neurotransmitter involved is acetylcholine, which is believed to affect the level of electrical activity of cortical neurons.

Acetylcholine functions in the brain quite differently from the way it does at muscle endings, where it produces its effect and is then destroyed in a few thousandths of a second. In the brain, the effects of ACh may persist for many seconds. In order for one to become consciously aware of a sensory input to the cortex, the cortical cells must continue firing for at least half a second after stimulation, a relatively long time compared to the time required for transmission of an impulse at a neuromuscular junction. In deep sleep or during anesthesia, cortical cells fire only a brief volley of impulses and awareness of stimuli does not occur; but during arousal, they continue to fire and so one becomes aware of incoming sensory information. It is thought that ACh diminishes the resting potential of cortical cells, allowing them to reach the threshold potential more easily; they then send impulses in rapid succession during a period of stimulation.

SLEEP-AWAKE CYCLES AND THE EEG

Whether one is asleep or awake, the cortical neurons fire off spontaneous impulses at regular intervals, and these can be detected by placing recording electrodes on the surface of the scalp. Voltages received from the brain are amplified and recorded as an *electroencephalogram,* or *EEG.*

As one is falling asleep, certain 5-HT centers of the reticular system send slowly spaced, synchronized impulses throughout the cortex which result in a series of fairly regular, slow "brain waves"; at the same time, the body muscles are relaxed (Fig. 13a.10A). These slow waves characterize most of one's sleeping hours. When one awakens, the impulses in the cortex become much more rapid and are desynchronized: the electrodes, which record the *average* electrical activity, now show a higher frequency of waves and a lower amplitude—nerve cells are firing out of synchrony, and so the net voltage at any instant is smaller than during slow-wave sleep. Muscle tone is restored during arousal, as shown by recordings from the postural muscles of the neck (Fig. 13a.10C).

At intervals while one is asleep, an unusual situation prevails; the brain waves are desynchronized, as though one were awake, but the body muscles are even more flaccid than during regular sleep, although the limbs may twitch occasionally (Fig. 13a.10B). (This condition may often be observed in a sleeping cat or dog.) Little sensory information is filtered through to

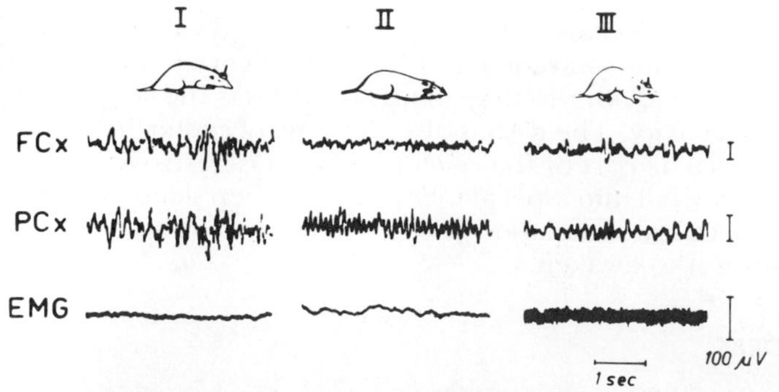

I   II   III

FCx

PCx

EMG

1 sec   100 μV

FIGURE 13a.10

EEG and muscle activity in the rat during periods of sleep and wakefulness. (FCx = EEG of front of cortex; PCx = EEG from back of cortex; EMG = recording from postural muscles of neck).

I. Normal sleep. Slow, high-amplitude brain waves and slight tonus in neck muscles. (If EMG were spread out sideways, recording would look like a regularly zigzagged line.)

II. REM sleep. EGG waves are flatter and faster. Muscles completely relaxed as shown by thin, wavy line.

III. Awake: EGG waves show alert pattern as in REM sleep, but increased muscle tonus gives an active EMG.

(From *An Introduction to Psychopharmacology.* 1971. Edited by Richard H. Rech, Ph.D., and Kenneth E. Moore, Ph.D. Raven Press, New York, © 1971.)

the cortex, and arousal as a result of sensory input is more difficult during this period of sleep. This is the time when dreaming is thought to occur. It is as if the brain were living a life of its own, completely shut off from the outside world. About one quarter of one's total sleeping time is occupied in this paradoxical condition, which is characterized by rapid, involuntary movements of the eyeballs, and hence is known as **REM**, or **rapid-eye-movement, sleep.** Its function remains obscure, although persons or animals chronically deprived of REM sleep through excessive background noise or drugs often are tired and irritable.

During REM sleep, behavioral arousal is at a minimum, whereas the EEG activity of the brain is at a maximum. The onset of REM sleep is thought to be triggered by a specific noradrenalin-producing center in the midbrain, the **locus coeruleus,** which not only stimulates the cortex, perhaps via the cholinergic arousal system, but also suppresses incoming sensory information and outgoing motor impulses.

STIMULANTS AND TRANQUILIZERS

As might be expected from our preceding discussion, drugs which block 5-HT synthesis tend to produce insomnia, and drugs which block noradrenalin synthesis tend to cause drowsiness while simultaneously diminishing REM sleep. In contrast, those drugs which mimic the action of 5-HT tend to cause sleep and those which mimic noradrenalin produce wakefulness.

Amphetamines. One of the commonest group of noradrenalin-mimics are the **amphetamines** (Fig. 13a.11A). Among these are amphetamine sulfate (benzedrine or "bennies"), dextro-amphetamine (dexedrine or "dexies") and methamphetamine ("crank," "crystal" or "meth"). It is thought that the amphetamines, although chemically resembling noradrenalin, do not act at the binding sites on the postsynaptic membrane but instead are able somehow to facilitate release of stored noradrenalin from the presynaptic axon (Fig. 13a.11B).

The amphetamines are used clinically to elevate mood, induce euphoria, increase alertness and reduce fatigue. They thus have an anti-depressant effect on the brain. In addition to their general effects, they also act at specific noradrenergic synapses in the hypothalamus, suppressing those concerned with food drive and exciting those which cause a rise in body temperature. In the peripheral regions of the body the amphetamines increase respiratory and heart rates, mimicking the action of adrenalin, the "fight-or-flight" hormone.

The unsupervised, non-clinical use of amphetamines often leads to abuse. A chronic consumer of high doses may go for days without sleep owing to continued cortical arousal. Eventually the subject experiences hallucinations similar to those occurring in undrugged volunteers who are kept constantly awake by external stimuli. It thus appears that hallucinations are due to lack of sleep and not directly to the drug itself. In severe overdoses of amphetamines, the subject may develop a schizophrenic-type state in which he is no longer able to distinguish reality from imagination and often exhibits paranoia (exaggerated suspicions of others and delusions of personal grandeur).

Other well-known stimulants of the central nervous system are co-

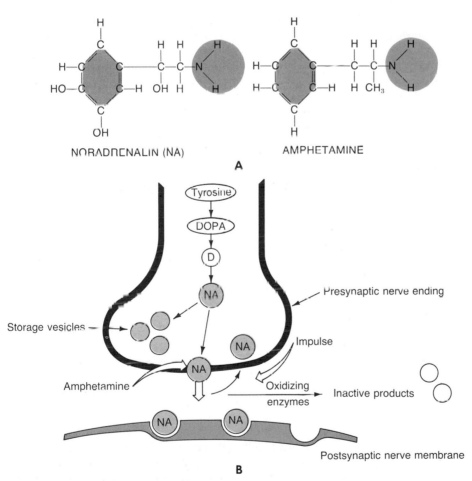

NORADRENALIN (NA)   AMPHETAMINE

A

B

FIGURE 13a.11

Amphetamine at the noradrenergic synapse.
    A. Similarity in chemical structures of noradrenalin and amphetamine.
    B. Some details of a noradrenergic synapse. Following synthesis, noradrenalin (NA) may either be stored in vesicles or remain free in the cytoplasm. When an impulse reaches the end of the axon, NA is released. This release is thought to be enhanced by amphetamines, thus increasing the chance that a molecule of NA will reach a binding site on the postsynaptic membrane before it is either oxidized in the synapse or taken up again into the original axon. Among the destructive enzymes is monoamine oxidase. (After Rech, R. H., and Moore, K. E. 1971. *An Introduction to Psychopharmacology.* Raven Press, New York.)

caine, caffeine and nicotine, whose mechanisms of action, however, are less clearly understood. Certain drugs which prevent destruction of noradrenalin, the inhibitors of monoamine oxidase (see legend, Figure 13a.11B), are also used to treat depression.

Tranquilizers. The earliest drug used by clinical psychologists to control hyperactive psychotic states such as the manic phase of manic-depression was *reserpine.* This drug acts by severely reducing the brain stores of *all* the monoamine neurotransmitters—noradrenalin, dopamine and 5-HT. Its calming effect upon mood is probably due to reduction of noradrenalin. But because it also depletes dopamine reserves in the substantia nigra, reserpine may induce symptoms of Parkinson's disease. Nowadays it is seldom used clinically, being supplanted by other tranquilizers with fewer side effects.

Among the most commonly administered tranquilizers are the phenothiazines; one of these is chlorpromazine. Despite their wide clinical usage, however, little is yet known about how these drugs produce their effects. More recent drugs, such as imipramine, combine the anti-depressant effects of a stimulant with a tranquilizing action, gaining for the patient "the best of both worlds." When the sites of action of these compounds are finally discovered we shall have come a long way in our understanding of the chemistry and anatomy of those parts of the brain controlling our moods.

Other drugs used as central nervous system suppressors include the *barbiturates* (sleeping pills) and alcohol. Both appear to act by generally depressing all nervous function, acting first upon the cortex and inducing a sleep-like EEG. The minor tranquilizers, such as meprobamate (Miltown), seem not to act upon the RAS-cortical arousal pathway but upon other modulating regions of the brain, including the limbic system, to be discussed later. Non-prescription sleeping compounds such as *Sominex* and *Compoz* contain anticholinergic drugs (that is, chemicals antagonizing ACh) and presumably act by interfering with the diffuse acetylcholine pathways of the cortical arousal system.

DRUGS, MEMORY AND LEARNING

As we saw in the preceding chapter, the mechanism by which memory—the storage and retrieval of information—occurs is still poorly understood. There are several steps in the memory process:

1. the perception of information;
2. the short-term storage of information (up to three hours);
3. the long-term storage of information (indefinite retention);
4. the retrieval of information.

In general, perception and short-term storage depend upon a state of arousal or awareness. A person who is asleep or unconscious or distracted does not pay attention to the stimuli presented and cannot recollect their existence even immediately afterward.

Transfer to long-term storage depends upon two things: conscious rehearsal or repetition of the information, which also requires arousal; and synthesis of protein by the brain cells. (Precisely which proteins are synthesized and what function they serve remains obscure.) Inability to transfer information from short-term to long-term memory results in *amnesia* or loss of memory for recent events.

The mechanisms for retrieval also are not well understood. We have all experienced inability to recall a name when occasion demands, only to awake in the middle of the night with it on the tip of the tongue. It is not always clear in studies of memory whether it is the recording of the memory itself which has failed to take place or the inability of the subject to retrieve the information.

Protein Synthesis and Long-term Memory. Studies on learning and memory in mice have provided interesting information on the way drugs affect the memory process. The antibiotic acetoxycycloheximide, which, like puromycin (Chapter 3a), specifically interferes with protein synthesis, can effectively prevent long-term memory if it is given just before or during the period of learning (Fig. 13a.12). Mice so treated remember what they have learned three hours later but have forgotten it after six hours. But if the protein-synthesis inhibitor is not given until 30 minutes *after* the learning period (colored arrow in the figure), transfer to long-term memory storage has already occurred and virtually no effect on long-term memory is observed. Thus, transfer of information from short-term to long-term memory storage normally occurs very soon after initial learning takes place.

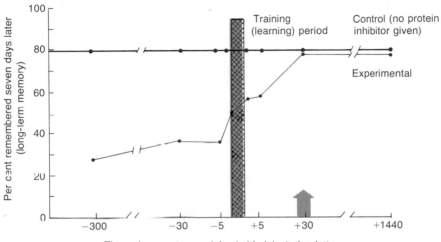

FIGURE 13a.12

Effects of an inhibitor of protein synthesis, acetoxycycloheximide, on long-term memory in mice.

Mice were trained to escape an electric shock when a light was switched on (conditioned learning). Their ability to remember seven days later was severely impaired if protein synthesis was inhibited before, during or shortly after training, but not 30 minutes or more afterward (colored arrow). (After Barondes, S. H., and Cohen, H. D. 1968. Memory impairment after subcutaneous injection of acetoxycycloheximide. Science *160*:556–557.)

Arousal and Long-term Memory. The same experimenters who studied the effects of acetoxycycloheximide on learning next asked a very important question: What happens if protein synthesis is allowed to take place again *before* the short-term memory has disappeared? Can transfer still take place? To test this, they gave very small doses of acetoxycycloheximide to mice during the training period, which wore off before the critical three hours were up; protein synthesis could again occur. If the mice were left undisturbed during this time, total amnesia resulted without transfer of information to the long-term memory store. But if the arousal state of the mice was increased by injection of amphetamines or certain other drugs, or by giving the mice a severe electric shock, permanent storage of information occurred (Fig. 13a.13). In some way, increased arousal causes rehearsal of the temporarily stored information which then results in its permanent retention. Thus, amphetamines and other stimulants improve long-term memory, at least in experimental animals.

However, before the anxious student rushes out to buy "speed" before an exam he should recall that such drugs also produce other effects—increases in heart rate, blood pressure and respiratory rate, together with general irritability, insomnia, palpitations, blurred vision, anxiety, chest pains, bowel dysfunction and a whole host of other unpleasant side effects. Drugs therefore are not recommended as a substitute for more mature approaches to coping with academic pressures.

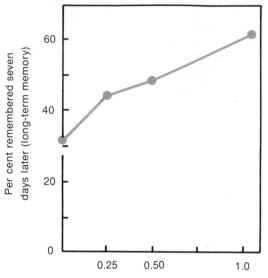

Per cent remembered seven days later (long-term memory)

Amount of methamphetamine injected into mice
(milligrams/kilogram of body weight)

FIGURE 13a.13

Effects of methamphetamine on long-term memory storage.
    The ability of mice to remember what they learned seven days earlier is increased with increasing amounts of amphetamines given during the time of transfer from short-term to long-term memory. These mice show less recall than the untreated control mice in Figure 13a.12 because they were given low doses of acetoxycycloheximide wore off. Arousal, which can be enhanced by amphetamines, is necessary for this transfer. (After Barondes, S. H. 1970. Cerebral protein synthesis inhibitors block long-term memory. International Review of Neurobiology *12*:177–205.)

Drugs Inhibiting Memory.   Most of us have experienced the effects of overindulgence in alcohol and are mildly embarrassed by our inability to recollect detailed events of a particularly convivial occasion. We suffer from a mild form of amnesia. It is perhaps not surprising that marijuana and LSD also have been found to affect memory. Recent studies on the ability of human subjects to recall information presented to them while they are under the effects of marijuana suggest that it is the inability to concentrate which affects retention. In particular, the transfer of information from short-term to long-term memory is impaired because rehearsal does not take place. Neither initial learning nor subsequent retrieval of information is affected, however. Marijuana, LSD and alcohol all seem to interfere with information transfer from short-term to long-term memory, but precisely how they have this effect is not yet known. They may act directly upon some point in the protein synthetic pathway or may have an indirect effect on the state of arousal which seems also to be necessary. Perhaps these two processes — arousal and protein synthesis by neurons — are intimately linked biochemically.

HALLUCINOGENIC DRUGS

Most of the 5-HT-containing cells of the brain are located in the midbrain reticular formation and send axons to the thalamus and hypothalamus, to the basal ganglia and to parts of the old cortex. Aside from the 5-HT center initiating sleep, little is known about the specific role of the remaining 5-HT neurons. It is thought that they have a generally inhibitory action on higher centers, and may modulate the strength or impact of incoming stimuli.

The hallucinogenic drugs — LSD (or ergot), mescaline (from the peyote cactus) and psilocin (from certain mushrooms) — all bear a chemical resemblance to 5-HT, and all probably act by antagonizing the action of this neurotransmitter (Fig. 13a.14). Although much research has been done on the chemical means by which these drugs may compete with 5-HT in the brain, no clues as to how they produce their effects on subjective experience have been forthcoming. It is likely that the action of the hallucinogens is a diffuse one, affecting many parts of the brain rather than just one region.

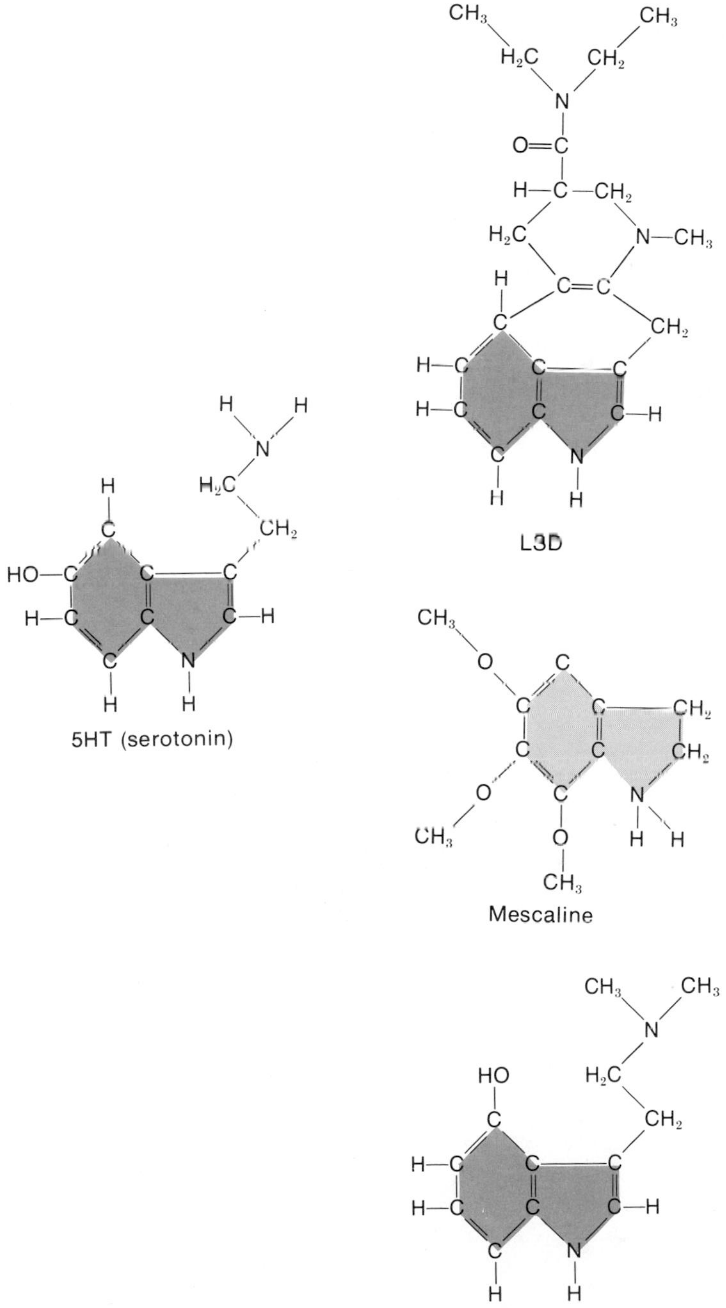

5HT (serotonin)

L3D

Mescaline

Psilocin

FIGURE 13a.14

Comparison of chemical structure between the natural neurotransmitter, 5HT (serotonin) (on the left) and hallucinogens (on the right).

THE LIMBIC SYSTEM

The third of the diffuse control systems we shall consider is the *limbic system.* As shown in Figure 13a.6, the "old" cortex of the primitive vertebrate brain becomes greatly reduced in man. Nevertheless, it retains its important connections with the hypothalamus and the midbrain; together these structures constitute an inner border or "limbus" of the neocortex (Fig. 13a.15), with which they have many connections. The limbic system to some extent overlaps with the basal ganglia (the amygdala is sometimes included in one system, sometimes in the other) and with the reticular formation, with which it has many reciprocal connections.

For a time it was thought that the limbic system was primarily concerned with emotions, since accidental destruction of one area (the amygdala) resulted in extreme passiveness and hypersexuality, while destruction of another region (the septum) produced uncontrolled aggressiveness; but this idea has been somewhat modified recently. In terms of behavior, the limbic system is apparently the region where internal and external sensations are fused, thus giving one a subjective sense of individuality and a conviction of the reality of oneself and one's surroundings.

The limbic system exerts its behavioral effects by means of a generalized inhibitory control over other parts of the brain. The amygdala and septum are mainly concerned with controlling the emotional centers of the hypothalamus, whereas the hippocampus acts on the cortex and may function in learning. Damage to this region in man impairs the transfer of information from short-term to long-term memory storage. It is possible that the hippocampus plays a role in arousal. On the other hand, direct electrical stimulation of the hippocampus in unanesthetized patients (a quite painless procedure) may evoke detailed memories of events from the dis-

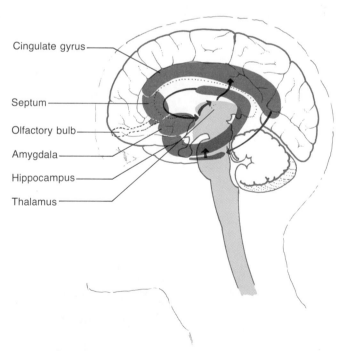

Cingulate gyrus

Septum

Olfactory bulb

Amygdala

Hippocampus

Thalamus

FIGURE 13a.15

The limbic system.
This system, shown in color, lies on each side of the brain; it has connections with the neocortex and also with the thalamus, hypothalamus and reticular formation. Note the existence of an internal limbic-thalamic circuit (black arrows) linking the several regions of the limbic system with one another.

tant past which were believed to be forgotten. Some scientists suggest that the hippocampus may therefore be the physical site of memory storage but it seems more likely that it takes part in initiating recall. The close association of the limbic system with the endings of the olfactory tracts (Fig. 13a.15) may explain the unusually strong power of odors to elicit recall of people and events from out of the past.

It is obvious that learning, perhaps the most important aspect of human behavior, is a highly complex process involving many parts of the brain. As psychologist R. N. Leaton has pointed out:

> The total pattern of learning involves the interaction of many functional systems in the brain. Among these are sensory, motor, motivational, attentional, and general processes of excitation and inhibition. An impairment in any one of these systems will reflect on the animal's capacity for learning and memory.*

One particularly interesting facet of the limbic system is the functional interconnection of its various parts, as shown in Figure 13a.15. Impulses generated in one region move in a reverberating circuit throughout the entire system. Thus, in a normal person, no one part of the controlling action of the limbic system over the rest of the brain is ever completely isolated from its overall effect. The limbic system is yet another example of a diffuse control center helping to modulate and integrate the functioning of the more direct, reflex-type pathway in the brain.

CONCLUDING COMMENTS

The student of human behavior trying to explain specific abnormalities in terms of discrete control centers and pathways in the brain is faced with an extremely complex task. For one thing, most areas of the brain have diffuse connections with many other areas. For another, the neurotransmitters which mediate various responses are not localized to one particular region of the brain. Thus, as we have seen, acetylcholine is involved not only in augmenting inhibitory impulses from the basal ganglia to the extrapyramidal pathways of the motor system but also in mediating cortical arousal and behavioral awareness. Noradrenalin not only causes alertness when one is awake, through a diffuse innervation of the cortex, but also interferes with sensory input during REM sleep. Each neurotransmitter serves a variety of functions which often are conflicting in their end result on overt behavior.

It is not surprising that drugs which interfere with or facilitate the action of a particular neurotransmitter often have ambiguous effects upon human behavior. It is extremely difficult to distinguish the primary target of a drug from its side effects on other centers of the brain. Even small changes in the chemical makeup of psychotropic drugs can greatly affect their overall action, presumably because their relative effects on different sites within the brain are altered.

The effects of drugs may also depend greatly upon the previous experience and present surroundings of the subject, both of which may determine the chemical state of his brain at a given moment. The naïve user of LSD, for example, may experience an extremely bad "trip," yet this same drug has proved extremely useful, when given under supervised clinical conditions, for alleviating the pain and anxiety of patients suffering from incurable cancer. It is likely that many years will elapse before we can fully explain the actions of most of the psychopharmacological drugs now in current medical or social use.

*Leaton, R. N. 1971. The limbic system and its pharmacological aspects. Chapter 4, *in* Rech, R. H., and Moore, K. E. (eds.) *An Introduction to Psychopharmacology*. Raven Press, New York.

SUMMARY

Although general functions can be assigned to the five major anatomical regions of the human brain, it is clear that these are applicable mainly to direct neuronal pathways. There also exist diffuse control centers which modify these direct, reflex-type pathways, bringing about an integration of behavioral responses.

The motor pathways leaving the cortex are modulated by coordinating centers in the cerebellum and by inhibitory impulses arising in the basal ganglia of the cerebral hemispheres. In Parkinson's disease there exists an excess of inhibition from the basal ganglia, which may be treated by anticholinergic drugs or, more effectively, by L-DOPA, which helps restore the normal balance between dopamine and acetylcholine in the basal ganglia.

The reticular activating system is another diffuse control center whose function is to modulate sleep-awake cycles. Arousal depends upon a noradrenalin release system, which in turn excites the diffuse acetylcholine arousal network in the cerebral cortex. Amphetamines and other drugs facilitating the release of noradrenalin cause arousal, whereas tranquilizers seem to act by depleting brain stores of this neurotransmitter. The arousal system plays a role in memory and learning, functions which also require protein synthesis by cortical neurons. Drugs which suppress either arousal or protein synthesis can impair long-term memory and permanent learning.

The limbic system may act as a general inhibitor of other regions of the brain, including specific parts of the hypothalamus. It also seems to play a role in learning and memory.

Since each of the various neurotransmitters is not localized to one part of the brain but fulfills many roles, drugs which mimic or inhibit the action of a particular transmitter may produce a complex effect. Much research remains to be done before we can explain the normal functions of the human brain and the effects of various drugs.

READINGS AND REFERENCES

Abel, E. L. 1971. Marijuana and memory: Acquisition or retrieval? Science *173*:1038–1040. Research on humans suggests the drug decreases the ability to concentrate necessary for rehearsal prior to storage of information.

Barondes, S. H. 1970. Cerebral protein synthesis inhibitors block long-term memory. International Review of Neurobiology *12*:177–206.

Barondes, S. H. and Cohen, H. D. 1968. Memory impairment after subcutaneous injection of acetoxycycloheximide. Science *160*:556–557. This paper and the preceding one are reports of the original experiments described in this chapter.

Hicks, R. E. and Fink, P. J. (eds.) 1969. *Psychedelic Drugs.* Grune & Stratton, New York. A collection of reports dealing mainly with clinical uses of drugs to treat depression, anxiety and other psychic complaints. See especially the use of LSD for terminal cancer patients.

Klawans, H. L., Jr. 1968. The pharmacology of Parkinsonism. Diseases of the Nervous System *29*:805–816. Explains the cause of this disease and its treatment with various drugs.

Krenjevic, K. 1967. Chemical transmission and cortical arousal. Anesthesiology *28*:100–105. Suggests a mechanism for the role of acetylcholine in arousal.

Malitz, S. 1971. L-*DOPA and Behavior.* Raven Press, New York. Use is now being made of the "side effects" of L-DOPA to treat depression and impotence.

McGeer, P. L. 1971. The chemistry of mind. American Scientist *59*:221–229. A readable though diffuse account of how various drugs are believed to act in the brain.

Physiological Psychology. 1972. W. H. Freeman, San Francisco. A book of readings from Scientific American, with articles covering many subjects discussed in this chapter.

Rech, R. H. and Moore, K. E. 1971. *An Introduction to Psychopharmacology.* Raven Press, New York. An excellent starting place for the non-specialist reader wishing further information about where and how drugs affect the brain.

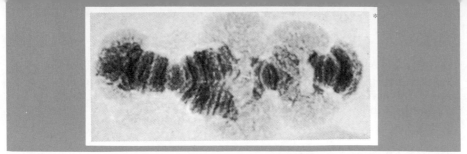

W. H. Benjamin, New York

REPRODUCTION AND DEVELOPMENT

The cellular division mechanisms of mitosis and meiosis insure that the all-important genetic information in the chromosomes is transmitted in viable form to the next generation. Subsequent embryonic development depends upon the decoding of this information in an orderly sequence of steps, mediated mainly by chemical messengers.

Every individual organism, be it a bacterium, orchid or man, has a finite life span. The continuation of life thus depends upon the reproduction and development of new individuals. Parents must produce offspring if a species is to survive.

There are two types of reproduction: **asexual reproduction,** in which a new individual is budded off from the parent, and **sexual reproduction,** in which a new individual develops from a single cell, the product of the union of an egg and a sperm. Asexual reproduction is most prevalent among bacteria, yeasts, protozoa and some plants and primitive animals. Most gardeners are well aware that many of their favorite plants can be propagated from cuttings or by division of roots, bulbs and so forth. But it is probable that almost all organisms, even those which commonly utilize asexual processes, are capable of sexual reproduction. The story of how the resulting single-celled offspring grows and develops into a multicellular individual is the subject of this chapter.

The process of embryonic development has two components: an increase in size, due mainly to an increase in cell numbers; and an increase in complexity. From one single cell many different cell types are formed in the embryo—muscle, nerve, glands, skin and so forth. Each cell type comes to possess distinct enzymes and other proteins necessary for its particular function. This process, known as **cellular differentiation,** results from the activation of specific regions of the genetic blueprints encoded in the DNA of the chromosomes. The messenger RNA molecules produced by these activated regions, or **cistrons,** then direct the formation of specific enzymes appropriate to each particular cell type, as described in the special section of Chapter 3 entitled "How Proteins Are Made." This specific activation

*From Watson, J. D. 1970. *Molecular Biology of the Gene.*

process is one of the most fascinating subjects of biology, and we shall consider it later in the chapter. But first we must discover the mechanisms by which the necessary genetic information is transmitted intact from one cell to another and from one generation to the next.

CELLULAR REPRODUCTION

Genetic information resides almost entirely in the chromosomes of the cell nucleus, and it is with the events in the nucleus that we shall be mainly concerned in considering cellular reproduction. As noted in our discussion of the life cycle of plants in Chapter 5, most sexually reproducing organisms have an even number of chromosomes, known as the **diploid** or **2N** chromosome number. This varies from one species to the next. In man, there are 46 chromosomes in every cell of the body (except for the gametes). Since half of the chromosomes of a diploid cell are derived from each parent, the **gametes,** the eggs and sperm, must have a **haploid** or **1N** number of chromosomes (in man, this is 23).

$$\underbrace{1N + 1N}_{} \quad = \quad 2N$$

haploid gametes   diploid zygote
(egg and sperm)   (fertilized egg)

The two haploid sets of chromosomes tend to be identical in appearance. For each chromosome derived from the male parent, there is one of similar size and shape from the female parent. These matched pairs are known as **homologous chromosomes.**

CHROMOSOME DUPLICATION

Before a cell can divide, its genetic material must be exactly copied if each of the two daughter cells is to receive a full complement of genetic information, a condition essential to the survival of most cells. This entails the replication of the long threads of DNA. The two complementary strands of the DNA double helix unwind, each synthesizing a mirror image of itself as shown in the highly schematic diagram of Figure 14.1A. The chromosome now consists of two identical halves, each called a **chromatid.** These remain joined together at a constricted region, the **centromere,** which plays an important role in separation of the chromatids during cell division (Fig. 14.1B).

All of the chromosomes in a cell are duplicated more or less simultaneously, an event which may occur hours or even days before a cell divides. It is still not known exactly what triggers chromosome duplication, thus setting into motion the events which eventually lead to cell division. It is a subject of great interest, however, especially for our understanding and control of cancer. As we shall see later in the chapter, certain hormones may play an important role in the regulation of cell division.

Once the chromosomes have duplicated, the stage is set for cell division to occur. We should logically consider first the divisions leading to gamete formation, known as **meiosis,** since these come first in the development of a new individual. But the process of ordinary cell division, called **mitosis,** is somewhat easier to understand and we shall begin with it.

MITOSIS

The complex events which take place during mitosis insure that one chromatid of *each* doubled chromosome goes to each of the two daughter cells. The stages of mitosis are outlined in Figure 14.2, in which we have illustrated a cell with four chromosomes, or two homologous pairs. The names of the various phases are given merely as a matter of convenience, since the events flow in a continuous series from one to the next. Note, however, that a nondividing cell is said to be in **interphase.**

The originally elongate chromosomes, not even visible as discrete entities in electronmicrographs, become highly contracted and are easily seen in stained dividing cells under the light microscope. The nuclear membrane disappears and the centrioles move to opposite sides of the cell.

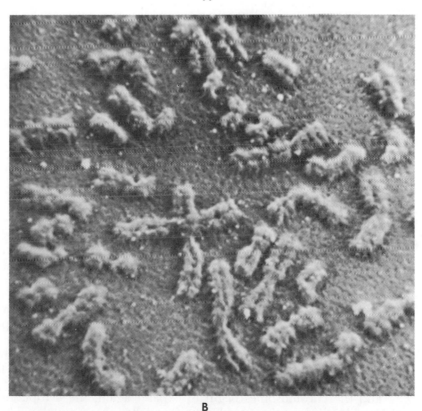

FIGURE 14.1

Doubling of chromosomes prior to division.
   A. Schematic diagram of DNA duplication. The complementary threads (——— and — — —) of the double helix unwind, each synthesizing the mirror image of itself (color). The daughter double helices, called chromatids, remain joined at their centromeres.
   B. High power scanning electronmicrograph of doubled human chromosomes. Chromatids and centromere regions are clearly seen on some of the chromosomes. (Electronmicrograph from Markert, C. L., and Ursprung, H. 1971. *Developmental Genetics.* Prentice-Hall, Inc., Englewood Cliffs, N.J.)

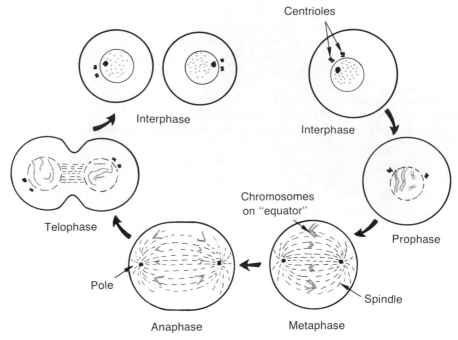

FIGURE 14.2

Mitosis in an animal cell containing four chromosomes.

Interphase. At some time prior to cell division, each chromosome is doubled to consist of two chromatids, the future daughter chromosomes.

Prophase. The doubled chromosomes, their chromatids attached at the centromeres, begin to contract; the nuclear membrane breaks down; and the centrioles migrate to opposite sides of the cell.

Metaphase. The mitotic spindle forms; fully contracted chromosomes migrate to the equator of the spindle; and some of the spindle fibers attach to the centromere of each chromosome, connecting it with the poles of the spindle.

Anaphase. The spindle fibers attached to the chromosomes contract, separating the daughter chromatids — the chromosomes of the daughter cells; the cell begins to elongate.

Telophase. Cytoplasmic constriction occurs; the chromosomes begin to expand again; a nuclear membrane begins to form; the centrioles divide.

Interphase. The two daughter cells, separated, now each contain identical sets of four chromosomes (expanded and no longer visible).

There they form two poles, from which radiate fibers composed of protein derived from the cytoplasm; this structure is known as the **mitotic spindle**. (A mitotic spindle also forms in plant cells, even though they lack centrioles.) As the chromosomes continue to contract, they migrate to the center of the cell, on the "equator" of the spindle. Some of the spindle fibers attach to the centromeres of the doubled chromosomes in such a way that the two chromatids are each connected to the opposite poles of the spindle.

The spindle fibers attached to the chromosomes then contract, rapidly separating the daughter chromosomes and pulling them to the spindle poles. Meantime, the cell as a whole elongates. As the spindle disappears and the chromosomes begin to expand again, the cytoplasm constricts, usually dividing the cytoplasmic organelles about equally. The centrioles also divide at this time, in preparation for a future cell division. Finally, the daughter cells separate and become two new interphase cells, each with a distinct nuclear membrane and four chromosomes. The chromosome sets are identical in the two daughter cells. The entire process of mitosis requires from 20 minutes to several hours, depending on the cell type involved.

MEIOSIS

We next come to the steps by which the haploid gametes, the eggs and sperm, are produced from the original diploid germ cells of the reproductive organs or gonads. At first it might seem simplest for these cells merely to divide by mitosis, but without first doubling their chromosomes. This raises an obvious problem: During mitotic division, there would be no way of insuring that each daughter cell would receive one of *each* pair of homologous chromosomes. An essential part of meiosis, then, is that the two homologous chromosomes join together prior to division, making it certain that each gamete receives a complete chromosome set. Let us follow through the stages of meiosis and then summarize what is accomplished.

The first step in meiosis, as in mitosis, is the doubling of DNA in each chromosome of the diploid germ cell. At this stage, there are 4N chromatids present in the nucleus, and therefore *two* divisions will be required to reduce this to the required 1N number. During meiosis, the homologous pairs of chromosomes do not behave independently, as they do in mitosis; instead they join together in exact alignment along their entire length, forming **tetrads** (Fig. 14.3). Each tetrad thus consists of four chromatids, two from each of the homologous chromosomes. While the chromatids are lying side by side in a tetrad, it is possible for them to exchange corresponding parts of their DNA with one another. This process, known as **crossing-over,** has important consequences for genetics and will be referred to again in Chapter 15.

The divisions in meiosis are similar to those in mitosis with respect to spindle formation and cytoplasmic division. During the first meiotic division the tetrads line up on the equator of the spindle; two chromatid strands of each tetrad are pulled to one pole of the spindle and two to the other, producing two 2N cells (Fig. 14.4). Unlike the case for daughter cells after mitosis, the DNA complement of these cells is not the same, since the identical chromatids of a tetrad both go to one or other cell.

During the second meiotic division, the centromeres divide and the remaining two chromatids of each chromosome are separated, giving rise to two 1N gametes. Each original germ cell thus produces four daughter cells.

In our example, four genetically different types of gametes are possi-

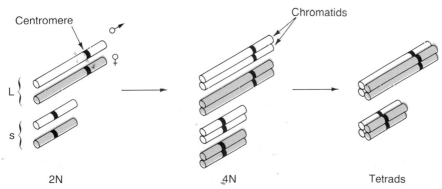

| 2N | 4N | Tetrads | **FIGURE 14.3** |

Tetrad formation in germ cells during meiosis.

In our example, there are two pairs of homologous chromosomes, one large *(L)* and one small *(s)*; those derived from the female (♀) parent are shown in color, those from the male (♂) are uncolored. Early in meiosis, each chromosome doubles, forming two chromatids (as in mitosis). There is now twice the diploid number of chromatids present, or 4N. Next, the homologous chromosomes pair together, forming tetrads, each composed of four chromatids.

ble, depending on the way the chromosomes are distributed: the two we have shown in Figure 14.4, and two others, in which the gametes possess both chromosomes from the same parent.

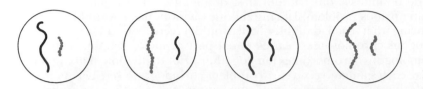

All four types occur in equal proportions during gamete production in a gonad, since they result from the chance alignment of the individual tetrads on the spindle equator during the first metaphase. Had crossing-over occurred in either chromosome, gametes with still other genetic constitutions would have been produced. In contrast to mitosis, where daughter cells are identical, *meiosis produces unlike cells.*

The consequences of meiosis are the following:

1. Each gamete receives a haploid (1N) number of chromosomes, one from each homologous pair.
2. The germ cells of an organism produce, through segregation of chromosomes, several genetically different types of gametes, insuring genetic variability among the offspring.
3. Tetrad formation permits exchange of DNA between homologous chromosomes (crossing-over), further increasing genetic differences between gametes and hence between offspring.

The genetic variability inherent in sexual reproduction is extremely important for the evolution of new types of organisms, as we shall see in later chapters.

Before leaving the subject of cell reproduction, we should make one

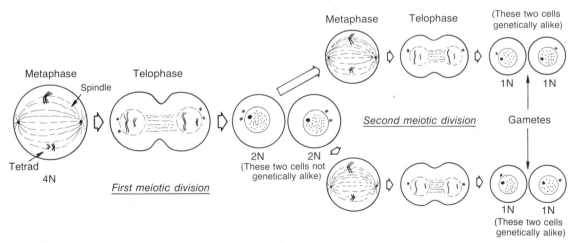

FIGURE 14.4   Important stages in meiosis.

There are two divisions. During the first meiotic division, the tetrads line up on the equator of the spindle during metaphase. At telophase, two chromatids of each tetrad go to each daughter cell. We have shown the female set (colored blue) of one tetrad going to the same cell as the male set (uncolored) of the other tetrad. An alternative would be for both colored sets to go together to one cell. In either case, the daughter cells are not identical with respect to their DNA.

During the second meiotic division, the remaining chromatids separate, producing identical haploid gametes.

further point. The student should not assume that the processes of mitosis and meiosis always proceed without a flaw. Occasionally two chromatids may fail to separate properly. This **nondisjunction**, as it is called, results in one extra chromosome in one daughter cell and a deficiency in the other. Human ova with an extra chromosome are sometimes fertilized and give rise to inherited abnormalities. The most familiar example is Mongolism, in which the affected child is mentally retarded and has a characteristic facial appearance.

DIFFERENTIATION OF GAMETES

In most animals, male **spermatozoa** (sperm) are distinctly different from female **ova** (eggs). In the male **testis,** each germ cell produces four sperm. Following the meiotic divisions, which take place as we have described, most of the remaining cytoplasm around each haploid nucleus is lost, and the male gamete consists only of densely packed DNA in the "head" region, to which is attached a flagellum ("tail") which provides motility for the sperm. Energy for beating of the flagellum is generated by mitochondria in the "middle piece" (Fig. 14.5).

The **ovary** of the female, on the other hand, produces diploid germ cells which enlarge from about 10 microns in diameter to 100 microns or more prior to meiosis; thus their volume increases 1000 times or, if it is the egg of a bird or reptile, many thousands of times (Fig. 14.5). The cytoplasm contains large quantities of stored protein, glycogen and fat, forming the **yolk** that will nourish the embryo before it begins to feed for itself or to receive nourishment from the mother. Around the time the female germ cell is released from the ovary, the two meiotic divisions occur. These divisions do not divide the cytoplasm equally, however. Instead of four gametes, only one is formed, the rest of the DNA being discarded in two tiny cells called polar bodies. Despite their great difference in size, however, eggs and sperm carry virtually equivalent amounts of genetic information.

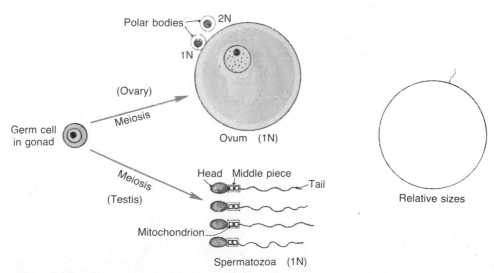

Comparison between eggs and sperm.

    Large eggs contain cytoplasm full of stored food for embryo. Unused chromatin is discarded in polar bodies during meiosis. Small sperm contain DNA in head and a few mitochondria in middle piece for supplying energy to flagellum (tail).

    Despite great difference in size, ovum and sperm contain equal amounts of genetic information.

FIGURE 14.5

FERTILIZATION

Although a few organisms, such as aphids and some plants, exhibit *parthenogenesis* (the development of an unfertilized, haploid egg into a new individual), in most organisms an embryo develops only after union of male and female gametes. A single sperm fertilizes each egg, usually only the head and middle piece penetrating into the interior (Fig. 14.6). A fertilization membrane may lift from the egg surface, preventing entry of further sperm. The male and female nuclei then unite, forming the diploid nucleus of the **zygote.** Such fertilization of the egg by a sperm may occur external to the female's body, or it may occur internally.

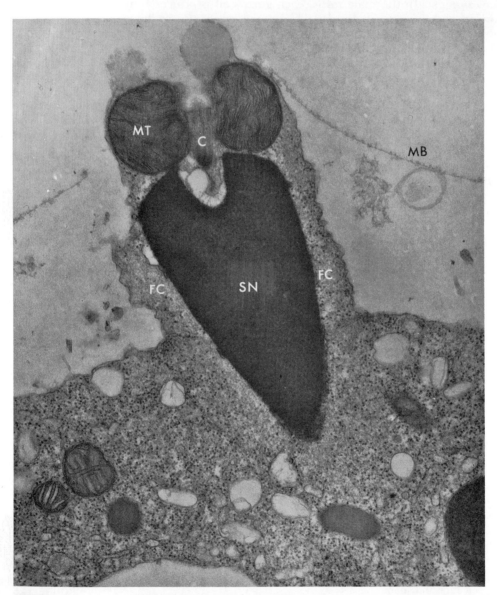

FIGURE 14.6   Fertilization of the sea urchin egg.
   Only a small part of the egg is shown in this electronmicrograph. The head *(SN)* and middle piece of the sperm are being drawn into the egg cytoplasm by the fertilization cone *(FC)*. Note the mitochondria *(MT)* and the fertilization membrane *(MB)* lifted above the egg surface. (Courtesy of Professor E. Anderson.)

EXTERNAL FERTILIZATION

In external fertilization, which is limited almost entirely to aquatic organisms, eggs and sperm are shed into water in great numbers—an event known as **spawning.** Only a fraction of the eggs are fertilized by chance meetings with sperm, and the embryos develop independently of the parents. Thus plants and animals utilizing external fertilization must release vast numbers of eggs and sperm into the water to insure the continuation of the species; there is much "genetic wastage."

Timing of spawning is important to successful reproduction. Eggs and sperm must be released nearly simultaneously to insure fertilization and at the right season to insure optimum environmental conditions for development of the embryos. In some species, temperature, light and other environmental factors trigger synchronous spawning; in others, chemical and behavioral communication between male and female are also necessary.

INTERNAL FERTILIZATION

When the embryo is to develop within the protective environment of the female's body or within a protective shell which is produced by the female after fertilization takes place, fertilization must be internal. This type of fertilization is especially common among terrestrial animals, where the dry environment would be unsuitable for the development of an unprotected embryo. Complex patterns of **courtship behavior** are usually necessary to bring the male and female together (refer to Figure 13.9) and permit the male to **copulate** with the female, injecting sperm into her reproductive tract. Partly by their own locomotory activity but often helped by contractions of the female's reproductive tract, the sperm find their way to the ova and fertilize them. Then, either the female lays fertilized eggs (as occurs with land snails, insects, birds) or development occurs internally and live offspring are born subsequently (as with some fish, some reptiles, almost all mammals). Further details of internal development in humans are covered in Chapter 22.

DEVELOPMENT OF THE EMBRYO

Once an egg is fertilized and becomes a diploid zygote it is competent to develop into a new individual. As we noted above, development consists of two processes: (1) growth by mitosis and cell enlargement and (2) differentiation of cells into specialized tissues and organs. At the moment, details of the control mechanisms governing both these processes are poorly understood. As more information is gathered, however, it appears that the regulation of both the growth of parts and their differentiation is the result of chemical agents; as one group of cells begins to develop it releases substances which activate or suppress neighboring cells, and these in turn affect still other cells. The consequence of this orderly sequence of interactions in time and space is the development of an integrated organism.

Since we cannot discuss here the many types of embryonic development that occur among living organisms, we shall select two examples: the development of a flowering plant and the development of higher animals. Flowering plants, in fact, have provided us with considerable information about the mechanisms controlling cell division and cell enlargement, while studies on higher animals have been particularly useful in elucidating some of the mechanisms involved in cellular differentiation.

DEVELOPMENT OF A SEED PLANT

After fertilization, but while the seed is still in the female ovary, the zygote within it has undergone several cell divisions to produce an embryo

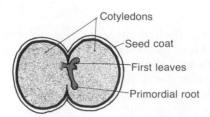

A   Enlarged seed cut open

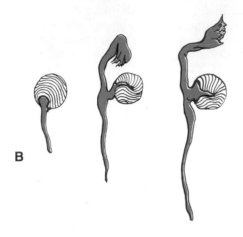

B

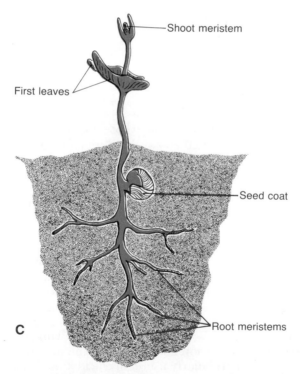

C

FIGURE 14.7

Development of pea seedling.

   A. Even before germination, the young embryo has formed the primordia of its first leaves and root and has incorporated virtually all of the endosperm into the cotyledons.

   B. During germination, first the root, then the shoot appears.

   C. In a few days the roots are well developed and the first leaves are opening. Mitotic regions are shown in black stipple.

(Fig. 14.7A). The largest part of the embryo at this stage consists of the enlarged **cotyledons**. These cellular structures have already transferred virtually all of the endosperm nutrients to themselves (refer to Chapter 5 for review of seed structure). Above the attachment of the cotyledon are the first leaves, and below, the primordial root. The embryo at this stage may lie dormant within the seed for months, or even years, without losing its ability to germinate. Seeds of the arctic lupine, estimated to be 10,000 years old, when transferred from the frozen soil of the tundra and planted in a favorable spot germinated in a few days!

During germination the seed takes up water, and soon the growing root of the embryo ruptures the seed coat. In a short time, the leaf-bearing shoot also emerges and continues to grow (Fig. 14.7B). The cells of the young plant receive nourishment from the cotyledons until the root system is sufficiently well established to take over the job (Fig. 14.7C).

Cell division occurs only at particular parts of the plant—namely, the tips of the root system, the tip of the growing shoot and sometimes also in lateral shoots located at the base of each leaf. These growing points are indicated in Figure 14.7C, and are seen more clearly in the longitudinal section of the growing tip of the water plant *Elodea*, in Figure 14.8. These **meristems**, as they are called, produce new cells by mitosis; after each division the daughter cell lying farthest away from the meristem then elongates and differentiates into a distinct tissue type—leaf, xylem, phloem or such. In general, the tip meristem tends to be more active than the lateral meristems.

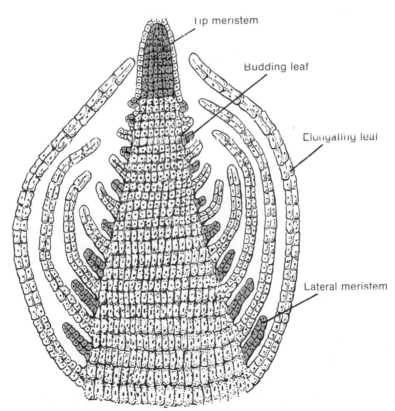

Tip meristem

Budding leaf

Elongating leaf

Lateral meristem

FIGURE 14.8

Section through the growing tip of the water plant *Elodea*.
New cells are produced by mitoses in the tip meristem (colored region at top). Budding leaves already contain all the cells needed to reach their final size. Further growth occurs by elongation of existing cells, which later become specialized (differentiated). Lateral meristems (remaining colored regions) may begin dividing later if lateral stems are needed.

The earliest indications that the pattern of growth in developing plants is at least partly controlled by chemical messengers, or **hormones**, traveling from one part of the plant to another came from studies of **phototropic** responses—the bending of plants toward light. If a potted plant is placed near a window so it is illuminated only on one side, it responds by bending toward the light source (Fig. 14.9).

Careful experiments conducted in 1926 on oat seedlings by a young Dutchman, Frits Went, helped to explain an observation made by earlier investigators—namely, that plants illuminated from the side bend only if the tip is intact. Went removed the tip from a seedling and placed it on a block of agar, a gel-like

HORMONAL CONTROL OF PLANT GROWTH

FIGURE 14.9   Phototropism in a plant.
Experiments show that bending is due to excess auxin production on the shaded side of the growing tip, shown in color.

substance. After several hours, he placed the agar on the cut end of another seedling, which was then kept in the dark. If he placed the agar to one side rather than dead center, the seedling bent as it would in the light (Fig. 14.10). Some substance was clearly passing from the agar to the stem, and when it was more concentrated on one side, this caused an excess elongation of cells on that side of the stem, resulting in bending. Went named this substance **auxin**, from the Greek aux-ein, "to increase." The most common naturally occurring auxin, now isolated from several plants, is **indolacetic acid**, or IAA.

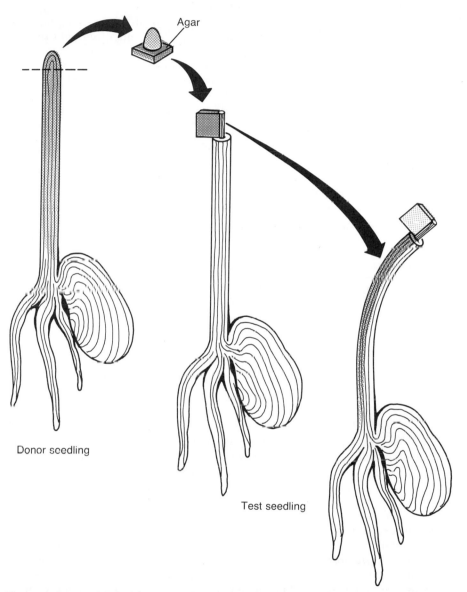

Agar

Donor seedling

Test seedling

FIGURE 14.10

The experiment of Frits Went demonstrating the production of auxin, a cell-elongating hormone, by the growing tip of oat seedlings.

It soon became apparent that auxins also play an important role in the elongation of stems during the normal development of plants. Auxin, or one of its metabolic products in the cell, acts in some way on the plant cell wall to bring about breaking of its chemical bonds. The wall is thus softened and able to expand owing to the cell's turgor pressure. It is thought that in this way much of a plant's growth is regulated.

Auxins not only affect stem growth, however; they also have effects on other parts of the plant that influence its ultimate development, and not all of these effects are due to cell elongation. Some of them are the following:

1. Initiation of root formation (auxins are used in the rooting of cuttings);
2. Suppression of root elongation;
3. Suppression of lateral bud growth in some species (hence the practice of removing the central, auxin-producing shoot in order to increase lateral shoots in chrysanthemums, stocks, and other garden plants);
4. Stimulation of mitosis in stems and trunks of woody plants;
5. Initiation of flowering in some plants;
6. Prevention of premature abscission (dropping) of leaves and fruit (fruit growers spray auxins onto trees so that they will retain fruit until ripe for picking).

Auxins do not affect all parts of the plant equally and thus greatly influence the relative proportions of its parts. Roots are more sensitive than leaf buds, and leaf buds more sensitive than stems (Fig. 14.11). At high concentrations, inhibition of growth sometimes occurs, perhaps due to toxic effects of the hormones.

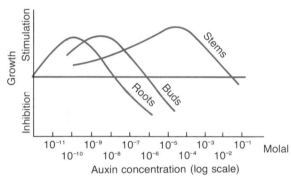

FIGURE 14.11

Differential effect of auxin on different tissues. Roots are far more sensitive than stems, and are killed by too much auxin. (Note log scale of auxin concentration.) (After Gorbman, A., and Bern, H. A. 1962. *A Textbook of Comparative Endocrinology.* John Wiley and Sons, New York.)

Other Plant Hormones

Another group of hormones with extraordinary growth-stimulating properties are the **gibberellins.** Produced by growing tips, these hormones stimulate both cell division and cell elongation (Fig. 14.12). Gibberellins are also present in seeds, where they play an important role in initiating digestion of stored nutrients at the time of germination.

Several plant substances related chemically to the nitrogen-containing bases of nucleic acids, as well as a few other quite different plant chemicals, have the ability to stimulate mitosis in cultures of plant cells. This capacity has earned them the name of **cytokinins** (since cytokinesis is another term for cell division). The cytokinins appear to play a role in the early development of the embryo following germination and also to stimulate the development of lateral buds, in direct antagonism to auxins. The "bushiness" of a plant may thus depend on a delicate balance between these two types of hormones.

Several other plant growth substances have been described, but the auxins, gibberellins and cytokinins are the best studied so far. It has become more and more evident in the past few decades that these hormonal growth regulators affect both cell division and cell growth at selected sites in the plant and thus play a major role in its development from embryo to adult. Why plant hormones affect one cell type more than another, and how they bring about their effects on cell division and cell growth, are presently under investigation.

FIGURE 14.12

Effect of gibberellin on cabbage plant.

Untreated plant, at left, has virtually no stem between leaf whorls. Treated plant, at right, demonstrates extraordinary stem elongation together with flower formation. (Photo courtesy of S. H. Wittwer.)

DEVELOPMENT OF HIGHER ANIMALS

The early development of animals varies in detail from one group to the next, but the basic pattern is similar, at least for the higher animals, those which have a body cavity or **coelom.** The problem is to get from a single-celled zygote to a three-layered embryo with a digestive tract and a coelom. The stages in early animal development shown in Figure 14.13 do not represent a particular animal, although with varying degrees of modification they could be adapted to a great many different species; rather, they are a generalized scheme to give an overview of animal development.

After numerous divisions, the embryo develops into a hollow ball of cells, the **blastula,** its fluid-filled cavity being known as the blastocoele. As a result of cell movements, the blastula becomes indented on one side to form a two-layered **gastrula.** The inner cell layer represents the earliest stage of the digestive tract. From this inner layer, pouches may form, giving rise to a third layer of embryonic tissue, which contains a central cavity, the coelom. (The third layer and its cavity may form in alternative ways, but the end result is essentially the same.) Finally, to complete the initial stages of development, the embryo elongates and a second opening is formed at the other end of the gut. In some animals, this becomes the mouth, in others, the anus, but the distinction is not important here.

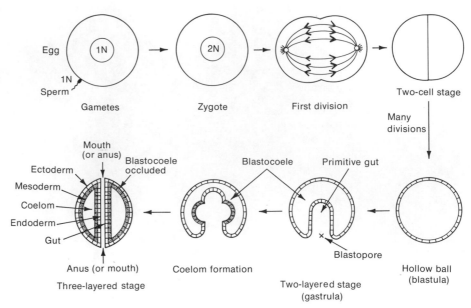

FIGURE 14.13

Early embryonic development of animals.
 The zygote divides many times to form a hollow ball, the blastula. One end invaginates, forming a two-layered gastrula. Coelomic pouches then form from the primitive gut, giving rise to a three-layered animal. The gut becomes open at both ends. (Note that no growth occurs, since the embryo cannot feed until later stages of development. The cells become smaller at each division, but all contain identical diploid [2N] nuclei.)

The result is a three-layered embryo. As development continues, each layer subsequently gives rise to specific organs characteristic of the particular species. The outer layer, or **ectoderm,** produces the skin and structures associated with it, such as skin glands, hair and similar structures; it also gives rise to the central nervous system. The middle layer, or **mesoderm,** produces muscle, skeleton, blood vessels, and excretory and reproductive organs. And finally the inner layer, or **endoderm,** becomes the lining of the gut, from which the digestive glands are formed.

It will be noted that until the three-layer stage has been reached the embryo does not increase in size. The cells have in fact become smaller at each subsequent division. Only later, when the embryo is able to feed for itself or begins to receive nourishment from its mother, does it increase in size. Even at this early stage, however, it is apparent that some cellular differentiation has occurred.

The Question of Cellular Differentiation.   One question long raised by embryologists was the following: Does the nucleus of a cell in a partially developed embryo still retain the potential for producing a complete individual, or is it so specialized, perhaps through loss or permanent suppression of part of its DNA, that it is forever limited in its capabilities? In other words, do all cell nuclei of the embryo retain the original genetic potential of the zygote from which they developed? Experiments by R. Briggs and T. J. King in the United States showed clearly that *any* cell nucleus of the embryo retains all of the capabilities of the original zygote. After many technical failures, they finally demonstrated that a nucleus from a frog gastrula transplanted to an enucleated frog egg would allow the egg to develop into a normal frog embryo (Fig. 14.14). Later J. B. Gurdon did the same thing, only using nuclei from tadpoles and even from adult frogs. These experiments, which have been repeated many times, support the hypothesis that no genetic information is lost during mitosis and cellular differentiation, and that all cells of the body contain equal amounts of functionally competent DNA.

These experiments face us squarely with another question. If the DNA complement is identical in all cells of the body, how is it that there can be so many cell types? In other words, how does cellular differentiation come about? Although we have no direct answer to this question in animals (or in plants, for that matter), information from bacteria suggests a possible mechanism. In these microorganisms, studies have shown that certain enzymes are not present all of the time, but only under special conditions. Only at these times are the cistrons responsible for directing the synthesis of these particular enzymes activated; most of the time these regions of the DNA are inhibited by a *repressor*. When the repressor is itself repressed or put out of action, then the cistron begins to produce its specific messenger-RNA, which in turn leads to synthesis of the special enzyme molecules.

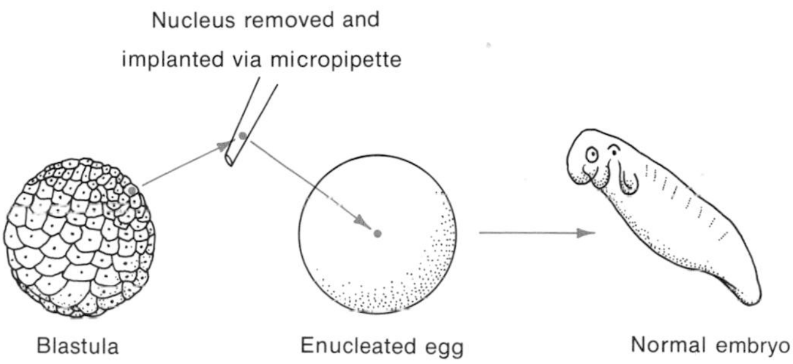

FIGURE 14.14

Nucleus removed and implanted via micropipette

The transplantation of the nucleus from a partially differentiated frog gastrula cell into an enucleated frog egg. With its new nucleus, the egg goes on to develop into a normal embryo. (After Villee, C. A., and Dethier, V. G. 1971. *Biological Principles and Processes.* W. D. Saunders Co., Philadelphia.)

Blastula   Enucleated egg   Normal embryo

We do not know if this model fits higher organisms or, if so, what substances act as repressors of DNA in their cell nuclei. Some experiments have indicated that the *histones* attached to the chromosomes may perform this function (see Chapter 3), but more recently it has been suggested that other molecules associated with chromosomes, such as special types of RNA, are the key repressor agents.

Whatever the actual DNA repressors are which allow only part of the total genetic information to be active in each specific cell type, there is some interesting circumstantial evidence that hormones may play a role in the de-repression or activation of DNA during various stages of animal development. Most of this information has come from studies of hormones and chromosomes in insects.

Insects, like other arthropods, have a hard exoskeleton and thus must undergo discontinuous growth; their development is marked by a series of larval molts. At each molt, the larva not only increases in size but differentiates toward a more adult-like individual (Fig. 14.15).

HORMONES AND CHROMOSOMES IN INSECT DEVELOPMENT

The Role of Hormones

The regulation of both growth and differentiation is under hormonal control. In insects two hormones control larval development. One of these, called *juvenile hormone,* when released into the body fluids causes cell division and growth, but has no effect on further cell differentiation. Under experimental conditions where only juvenile hormone is present in the animal, the larva simply increases in size at each molt but does not progress toward adulthood.

The other hormone, *ecdysone,* is responsible for the differentiation of the embryo towards the adult form. In animals where only ecdysone is present, the larva often becomes a precocious, miniature adult at the next molt. It is believed that a balance exists between these two hormones which brings about the normal development of an insect larva, a balance which swings gradually in favor of ecdysone and adult characteristics at each successive molt.

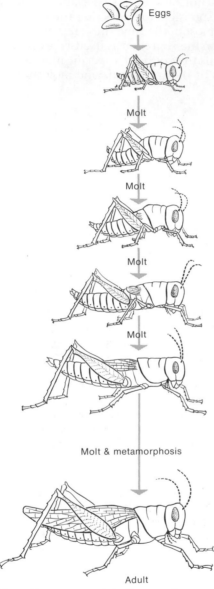

Eggs

Molt

Molt

Molt

Molt

Molt & metamorphosis

Adult

FIGURE 14.15

Discontinuous growth in the grasshopper.

At each larval molt the young grasshopper not only increases in size, but also be-comes more like the adult. Note especially the changes in relative wing size. See text for details of hormone control. (From Villee, C. A., and Dethier, V. G. 1971. *Biological Princi-ples and Processes.* W. B. Saunders Co., Philadelphia.)

The Role of Chromosomes

Some insects — namely, the two-winged flies — have giant chromosomes in the cells of their salivary glands. These chromosomes are composed of several hundred identical chromatid strands lying side by side, and the two homologous chromosomes often remain synapsed together. Of particular interest to geneticists is the presence of distinctive bands along the length of these chromosomes, which in many cases have been identified as regions where specific genetic traits are located.

It has long been known that these bands sometimes swell — a property known as *puffing.* In recent years it has been shown that the puffs are regions where the DNA of the chromosomes has become unfolded and is actively synthesizing RNA (Fig. 14.16). In one specific case, it was shown that certain protein secretions of the salivary gland were produced only when a particular band on one of the chromosomes was present in an active, puffed state (Fig. 14.17). This was the first

A

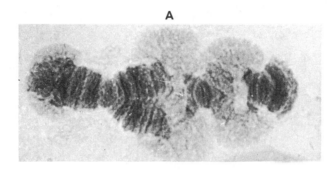

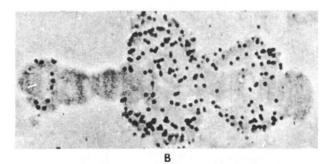

B

FIGURE 14.16

Evidence for RNA synthesis at puffed regions of giant salivary gland chromosomes.

The living cells are exposed to a radioactive precursor of RNA, which is rapidly combined into RNA molecules at sites of active RNA synthesis. Radioactivity is then detected by overlaying the killed chromosomes with photographic film. The radioactive disintegrations from labeled RNA cause silver grains to be formed in the same region of the film, thus identifying the most active regions of RNA synthesis.

A. Stained chromosome showing bands and three puffed regions.

B. The same chromosome covered with photographic film. Note that silver grains occur almost only over puffed regions.

(W. Beermann, from Watson, J. D. 1970. *Molecular Biology of the Gene.* W. A. Benjamin, New York.)

direct evidence that a known region of DNA on a chromosome was active in directing the synthesis of a specific protein in a cell. Thus:

$$DNA \longrightarrow m\text{-}RNA \longrightarrow PROTEIN$$

Careful studies of the chromosomal puffing patterns in different tissues at different times during development of fly larvae have shown that

1. Different regions of DNA are active at different stages of development (Fig. 14.18);
2. Different regions of DNA are active in different tissue types.

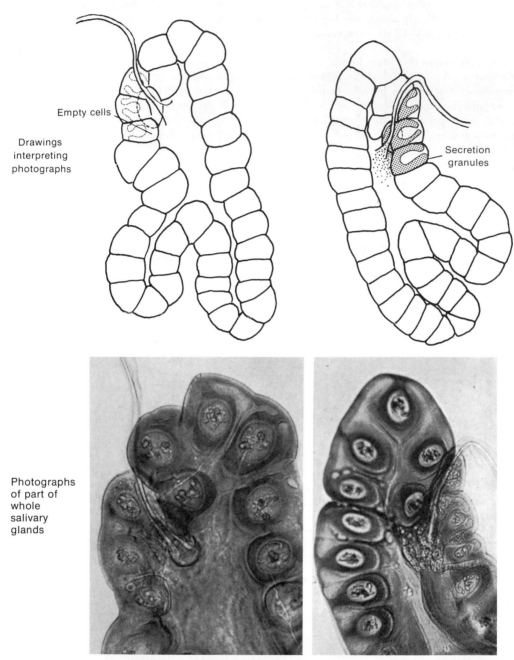

Empty cells

Drawings
interpreting
photographs

Secretion
granules

Photographs
of part of
whole
salivary
glands

FIGURE 14.17 Evidence for correlation between puffing at a particular chromosomal band and synthesis of a special protein.
Salivary gland on left comes from a fly larva deficient in puffing of a particular band. No specific protein granules are formed. Cells of the gland on the right come from a normal larva, exhibiting puffing of the necessary band. These gland cells are able to synthesize their specific product. (Photograph from Beermann, W.: Ein Baldiani-Ring als Locus einer Speicheldrüsen-mutation. Chromosoma (Berl.) *12*:1–25 [1961]. Berlin-Göttingen-Heidelberg: Springer-Verlag.)

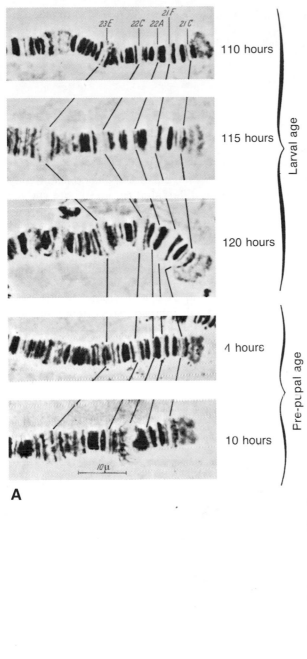

A

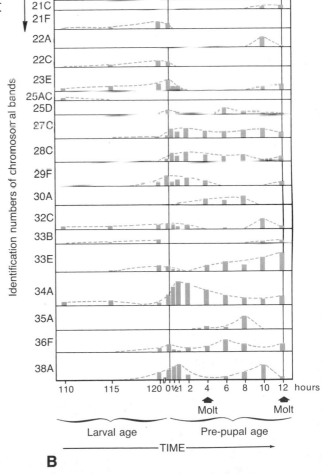

B

FIGURE 14.18

Chromosome puffing and developmental stages in a
fly larva.

A. Photographs of the same region of a chromosome at sequential stages in development. Corresponding bands are traced by lines. Note changes in
puffing patterns.

B. Histogram of degree of puffing relative to time.
The most puffing occurs about four hours before
each molt.

(From Ashburner, M. Patterns of puffing activity
in the salivary gland chromosomes of *Drosophila*.
I. Autosomal puffing in a laboratory stock of *Drosophila
melanogaskr,* Chromosoma (Biol.) *21*:398–428, 1967.
Berlin–Heidelberg–New York: Springer.)

One cannot escape the conclusion that chromosomal puffs are visible indicators of specific DNA activity during the differentiation of cells in developing insects, and that they are directing the synthesis of the specific proteins needed to carry out the functions of the cell.

Do Hormones Activate DNA?

The next question—Do insect hormones activate the specific bands of DNA and cause them to puff and begin RNA synthesis?—is still without a precise answer. Although it is clear that hormones, not only in insects but also in other animals, play an important role in development and cellular differentiation, we still are unclear as to their precise mechanism of action. It is true that injection of ecdysone into an insect will cause one or two puffs to form. And in vertebrates it has been shown that steroid hormones, which are chemically similar to ecdysone, are capable of binding to chromatin, as might be predicted for a DNA-activating agent. It is tempting to think that these hormones have a direct action on chromosomal DNA, but they could just as well be acting at an earlier stage—activating an activator, so to speak (or perhaps neutralizing a repressor)—or at a later stage, affecting the processes of transcription or translation of genetic information, or even later still, influencing the way the proteins carry out their specific function. Some of these possibilities are outlined in Figure 14.19.

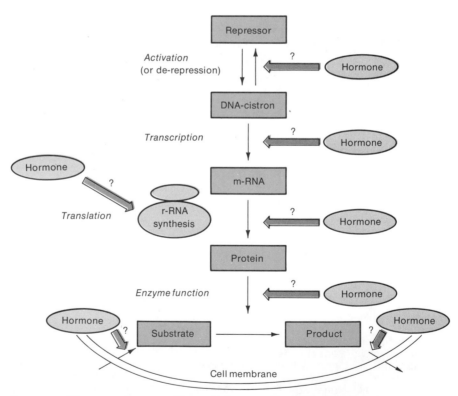

FIGURE 14.19

Some possible ways a hormone like ecdysone may regulate cell function.
Many other possibilities, not shown here, have also been suggested. For example, many feedback loops are believed to exist, and a hormone might act at one of them rather than at some site on the direct pathway as suggested here.

SUMMARY

Reproduction of organisms depends ultimately on reproduction of cells, or cell division. During asexual reproduction, new cells are formed by mitosis. The primary purpose of mitosis is to reproduce exact copies of the DNA in the chromosomes of the parent cell so that each daughter cell receives a complete set of genetic information. This is accomplished by separation of the doubled chromosomes on the mitotic spindle.

Sexual reproduction requires a reduction in number of chromosomes during gamete formation. To achieve this, the DNA is first doubled in the germ cells (4N); then homologous chromosomes join together, forming tetrads, a step which insures that each gamete will receive one member of each pair of homologous chromosomes. Two meiotic divisions take place, producing four haploid (1N) gametes, each containing one of each pair of homologous chromosomes. Sexual reproduction results in greater genetic variability among offspring than is possible during asexual reproduction.

External fertilization requires production of many gametes and occurs almost exclusively in aquatic organisms. Internal fertilization is common among terrestrial species, where protection of the developing embryo is essential. When a sperm enters an egg, the nuclei join to form a zygote, which then divides repeatedly, growing and differentiating into the cells and organs of the adult. Differentiation of each cell type is believed to result from activation of only a selected part of the total DNA, the rest remaining suppressed.

The partially developed embryo of a seed plant may lie dormant for years, but upon germination, gibberellins and perhaps other hormones are released, stimulating renewed growth. Mitoses in the developing plant occur only at meristematic regions, further growth resulting from cell enlargement. The latter is under control of auxins and gibberellins produced by the tip meristem. Auxins also appear to regulate cell division and cell growth elsewhere in the plant; their suppression of lateral shoot development is antagonized by yet a third group of hormones, the cytokinins. Normal plant development thus appears to be modulated by a variety of hormonal controls.

The zygotes of higher animals develop by a sequence of stages involving cell division and cell movement into a basically three-layered embryo with a coelom and digestive tube. Each layer then gives rise to specialized cells and organs characteristic of the particular species.

The mechanisms of cellular differentiation have been well studied in insects, where growth in size and differentiation toward the adult form are controlled by two hormones, juvenile hormone and ecdysone. In certain flies, it has been shown that puffing patterns on giant chromosomes, indicating sites of DNA activity, are correlated with the stage of development. It is not yet known how hormones and chromosomes interact during normal insect development.

Ashburner, M. 1967. Patterns of puffing activity in the salivary gland chromosomes of *Drosophila*. Chromosoma 21:398–428. Evidence for the functioning of different regions of DNA at different stages in development.

Gurdon, J. B. 1968. Nucleic acid synthesis in embryos and its bearing on cell differentiation. Essays in Biochemistry 4:25–68. Summarizes experiments on timing of DNA and RNA synthesis in relation to observable events in frog development.

Gurdon, J. B. 1968. Transplanted nuclei and cellular differentiation. Scientific American Offprint No. 1128. Describes original transplantation experiments.

Markert, C. L. and Ursprung, H. 1971. *Developmental Genetics.* Prentice-Hall, Englewood Cliffs, N. J. Outlines our present knowledge of molecular mechanisms involved in cellular differentiation.

Mazia, D. 1961. How cells divide. Scientific American Offprint No. 93. A review of the significance of mitosis by the man who first showed that the mitotic spindle was a real entity.

van Overbeck, J. 1968. The control of plant growth. Scientific American Offprint No. 1111. Describes how plant growth is regulated by hormones.

Waddington, C. H. 1962. *How Animals Develop.* Harper & Row, New York. A lifelong embryologist produces a readable book on his subject.

Watson, J. D. 1970. *Molecular Biology of the Gene.* 2nd ed. W. A. Benjamin, New York. An extremely readable account of almost everything we know about chromosomes and how they perform their functions.

MAN'S USE OF HORMONES IN THE ENVIRONMENT

The potent chemicals which regulate the natural development of plants and animals have been isolated by man, who now uses them to manipulate organisms in the environment for his own ends.

As we saw in the preceding chapter, chemical messengers called hormones play an important role in the development of both plants and animals. In this chapter we will examine some of the present and proposed applications of these potent agents to specific goals of man. Of the many examples available, we shall consider two in some detail: plant auxins and insect hormones. Both have important applications to agriculture and thus to increasing the production of food for the growing world population.

WEED-KILLERS

Natural plant auxins, of which the only one so far chemically identified is *indolacetic acid* or IAA, have differential effects on the growth of the stems, buds and roots of plants. At high concentrations, which maximally stimulate the growth of stems, the growth of roots is inhibited (refer to Figure 14.11). At these concentrations the plant tissues tend to increase their entire metabolic rates; more enzymes are produced, and turnover of cell chemicals is speeded up. It is believed that excess auxin upsets the natural checks and balances, with the result that metabolism becomes uncontrolled and causes death of the plant.

The significance of this fact as a potential tool for military defoliation of enemy crops was quickly recognized soon after auxins were discovered. Natural auxin, however, would need to be applied at prohibitively high concentrations, since tissue enzymes are present in plants which quickly destroy IAA. During World War II, a search was begun for synthetic chemicals which could be mass-produced and would mimic the effects of natural auxins without being destroyed by the plant's enzymes. Although the search was not successful in time for use in that war, in the late 1940's the potent chemical 2,4-D, a product of that effort, was first marketed for a more peaceful purpose, as a weed-killer. Subsequently, other similar substances, such as 2,4,5-T and 2,4,5-TP (Silvex), were also developed. These

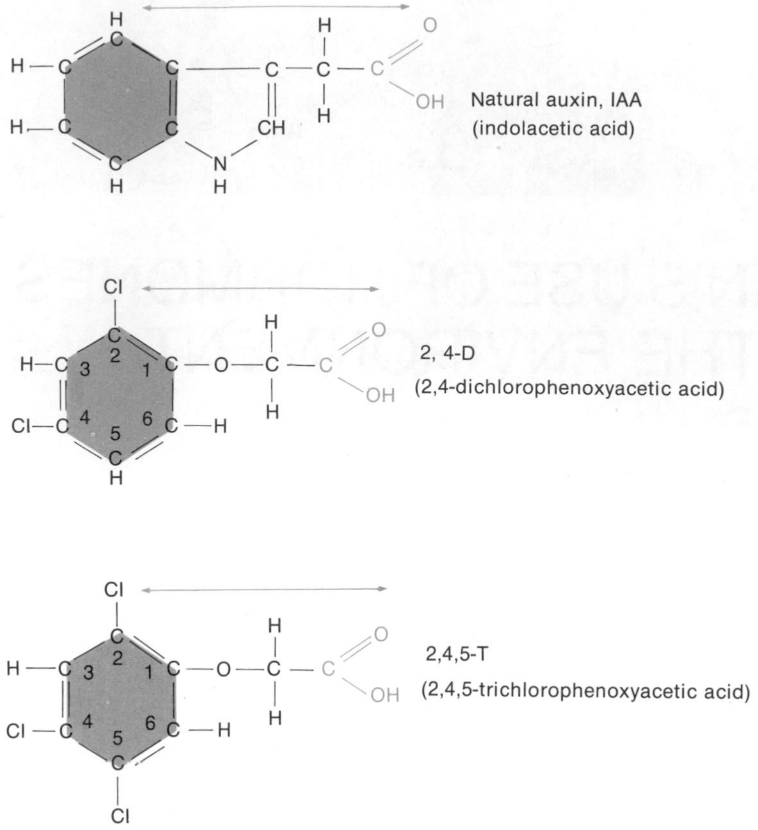

FIGURE 14a.1

Similarity, emphasized by color, between chemical structures of natural auxin (indolacetic acid) and synthetic weed-killers, 2,4-D and 2,4,5-T. (Numbers refer to carbon atoms in ring part of compound, and are used to identify carbons to which chlorine atoms are attached.)

have close structural similarities to the natural auxin (Fig. 14a.1) and produce similar effects.

USES IN AGRICULTURE

In a natural environment, there is a great diversity in types of plants and animals which coexist in a system of checks and balances. In such a situation, each type of organism interacts with many others, and a stable situation is maintained which changes only slowly with time. When man developed agriculture, however, he began selectively growing only those plants of benefit to himself and eliminating species which originally participated in a stable system. Because this balance is upset in a field sown with a single crop, plant species which would not become abundant in a complex environment are able to thrive and often overgrow the crop itself. "Weeds," then, are a natural outcome of agriculture.

Until recently, weeds were controlled mainly by repeated plowing prior to planting, by laboriously removing tares (weed seeds) from grain before sowing and by hoeing or mechanical cultivation of crops after germination. Despite these efforts, weeds still claimed significant proportions of the total crop, and in the last decade or so the use of synthetic auxins has been introduced in an attempt to further control weeds, improve crop yields and decrease labor costs.

Weed-killers are particularly useful for protecting grass-type crops such as corn, rice and wheat, since the broad-leaved weeds (dandelion, dock and so forth) are many times more sensitive than are the grasses. This also explains the popularity of weed-killers with home-owners who have dandelion-infested lawns. There has been much speculation as to why these auxin-like weed-killers are so much more effective against the dicotyledonous plants (those with broad leaves) than the monocotyledonous species (the narrow-leaved plants, such as grasses). Even yet there is no clear-cut explanation, although it is likely that the more sensitive species either take up and distribute 2,4-D throughout the plant faster or metabolize it more slowly than do the resistant forms.

Weed-killers, which also go by the name of **herbicides,** are in addition used to facilitate the harvest of certain broad-leaved crops. In particular, the picking of cotton is greatly speeded if the leaves are removed beforehand by spraying the plants with 2,4-D. In addition to its use as a weed-killer, 2,4-D is often sprayed on fruit trees, such as citrus, to prevent **abscission,** or premature falling of the fruit. This is another function of the natural auxin in plants that is mimicked by the synthetic auxins. In this instance, of course, much lower concentrations of 2,4-D are used which do not damage the tree.

USES AS DEFOLIANTS

The use of herbicides as weed-killers has been directed mainly at annuals—plants that live one season only. But herbicides have also been applied against perennials such as shrubs and trees. For this purpose, 2,4-D is less effective than its sister compounds, 2,4,5-T and Silvex. The effect of herbicides on woody plants is mainly on the leaves, which, shortly after application, wither and die. The rest of the plant is not harmed directly, and providing it has sufficient stored nutrients to carry it over, it may produce another crop of leaves. Repeated sprayings, however, result in eventual death, since most of the organic molecules needed by a plant for growth and survival are synthesized in its leaves.

FIGURE 14a.2

Control of white pine blister rust by spraying intermediate host, wild currants, with 2,4,5-T. (U.S. Department of Agriculture.)

At first, herbicides were used mainly to clear rights of way along power lines and beside railroads and highways, but their use soon spread to other jobs. A 1300 mile long swath has been cleared by spraying herbicides in the coniferous forests to permanently mark part of the Canada–United States boundary. The U. S. Forest Service also uses 2,4,5-T to control white pine blister rust. This disease is caused by a parasitic fungus which has a complex life cycle—part of the time the fungus lives on wild currant and gooseberry bushes on the forest floor; the rest of the time it lives beneath the bark of white pines, where it causes considerable damage, killing many trees each year. Formerly, control of the disease was possible only by digging out the intermediate host plants—the currant and gooseberry bushes; selective spraying with herbicides now does the job quicker and with less work (Fig. 14a.2).

Another projected use of 2,4,5-T and Silvex is for brush control in regions susceptible to brush fires. In particular, several of the large, sprawling cities in the southwestern United States have expanded into highly flammable chaparral regions (chaparral being a type of brush cover characteristic of semi-arid regions). The Forest Service is considering the use of herbicides to clear the brush from extensive areas around these inhabited regions.

The most publicized of all uses of herbicides has been their military application in Vietnam to defoliate forests. Approximately one-tenth of the country has been sprayed from the air, using doses of herbicide about 10 times greater than are used in the United States. The results of such spraying may be seen in Figure 14a.3.

SIDE EFFECTS OF HERBICIDES

The release of these potent agents into the environment has raised considerable criticism with respect to both the environment and public

FIGURE 14a.3   A rubber plantation in Southeast Asia defoliated as a result of aerial spraying with herbicides. The defoliant was carried by the wind many miles from the intended jungle target area.

health. Suspicions that 2,4,5-T may be dangerous to humans first arose from its use in Vietnam. Shortsightedly, however, no attempt was made between the years 1962 and 1970, when herbicides were being used there, to accurately measure their effects on humans; consequently, no definite conclusions can be drawn. On the other hand, evidence for their effects on the Vietnamese landscape are far clearer.

ECOLOGICAL EFFECTS OF HERBICIDES

Repeated spraying of large areas of Vietnam has resulted in widespread destruction of tropical forests composed mainly of broad-leaved

FIGURE 14a.4

Aerial photograph at top shows an unsprayed mangrove forest in South Vietnam. Aerial photograph at bottom shows a mangrove forest which was subject to herbicide spraying in 1965. One spraying kills essentially all trees. The larger trunks appear to have been removed by the local population for firewood. (From Herbicides in Vietnam: AAAS study finds widespread devastation. Science 171:43. Copyright © 1971 by the American Association for the Advancement of Science.)

trees. In some areas, these have been largely replaced by bamboo—a grass-type plant. The situation is analogous, then, to the effects of herbicides on weed-infested fields of grain or corn, except that in Vietnam bamboo is considered a weed, whereas the forests are more useful to man. In other areas, where bamboo has not replaced the forest cover, the absence of vegetation has resulted in flooding and erosion (Fig. 14a.4).

The environmental effects of herbicides used in the United States have not yet been accurately assessed. It is to be expected, however, that any widespread, non-selective application, such as that projected for brush control, will have a secondary effect on the associated animals. Herbivores (animals feeding on plants) are highly selective in their choice of food. This is particularly true of insects, many species feeding only upon a single plant. Our knowledge about the food preferences and limitations of most animal species is as yet too sparse to allow exact predictions as to the effects on animals of a change in food supply through use of herbicides. It is certain, however, that major changes in animal populations will result if these compounds are applied over extended areas.

Two other factors may be mentioned briefly. Changes in vegetation, besides affecting food supply, also result in altered habitats, so that nesting sites normally used by certain animal species are no longer available; the number of birds in sprayed areas of Vietnam has been greatly reduced. Herbicides may also have direct toxic effects on other forms of life, especially soil microbes and aquatic organisms affected by run-off. Fish are known to be particularly susceptible; the concentrations of herbicides used in Vietnam were high enough to have been toxic to fish, although no studies of the overall effects on fish populations have yet been made there. The total environmental consequences of herbicide application clearly require further investigation. Many ecologists believe that herbicides, together with non-specific insecticides such as DDT, are ecological "drugs" whose use should be legally limited to licensed professionals in the same way as are drugs used to treat the human body.

PUBLIC HEALTH EFFECTS OF HERBICIDES

The herbicides 2,4-D and 2,4,5-T are virtually non-toxic to adult human beings, and for many years it was believed that their use presented no hazard to human health. Unfortunately, a number of other synthetic chemicals, developed for various purposes, were also believed to be innocuous and were widely used before their potential dangers were discovered. One such compound, the artificial sweetener cyclamate, was recently shown to induce bladder cancer in rats and was subsequently removed from diet drinks, even though no direct effects on humans had been demonstrated. A far more tragic case occurred with the tranquilizer thalidomide, which was withdrawn from use only after thousands of deformed babies were born to mothers who had taken the drug. Many intelligent but severely crippled children are alive today in Europe because of insufficient testing. Fortunately, thalidomide was not released by the Food and Drug Administration for use in the United States before its side effects were discovered.

It is argued by some scientists that herbicides may be capable of producing birth defects similar to those caused by thalidomide. During early stages of human embryonic development, various groups of cells are differentiated to form a specific organ or structure. If such a group of cells is inhibited during this critical time, that organ does not develop normally. In the case of thalidomide, cells destined to form arms and legs were severely inhibited; affected babies were often born without arms, so that their hands were attached directly to their shoulders. Others were born with fingers missing, or without legs. Similar effects have been observed in chick embryos treated with various pesticides (Fig. 14a.5). This type of birth defect, in which a particular organ system is deformed, is called a **teratogenic defect** (from Greek *teratos*, "monster," plus *genés*, "born").

Assessing Safety. The reader may well wonder why it is not possible to determine the potential dangers of a new chemical, prior to its widespread use, by conducting experiments on animals. Although such experiments are indeed carried out, two problems exist. One has to do with sample size—the number of animals studied. If a defect can be expected only once in 10,000 human births, then the usual test study on 10 or 20 litters of animals (each consisting of 8 to 10 offspring) is obviously inconclusive; only 200 or so animals have been examined. All too often, screening tests involve such small numbers of animals that low-level effects which might occur in the human population are not detected. Experiments involving more animals are often precluded by lack of funding to the agencies charged with carrying out the tests.

There is a second problem which is even more difficult to overcome, however: man and experimental animals do not necessarily respond identically to toxic chemicals. The case of thalidomide is a tragic example. The drug was extensively tested in animals prior to being placed on the market, and it did not cause any birth defects in them when given at the dose levels proposed for use in man. Not until later was it found that only at relatively high doses does thalidomide cause defects in experimental animals; pregnant mice require 30 milligrams per kilogram body weight per day; rats, 50; dogs, 100; and hamsters, 350. Tragically, the dose needed to produce birth defects in the embryos of pregnant women was only one-half (0.5) milligram per kilogram body weight per day. Man is thus 60 times more sensitive than a mouse, the most sensitive animal tested! From this example we can see that it is important to add a large safety factor to the dosages determined as "safe" for experimental animals when applying them to man.

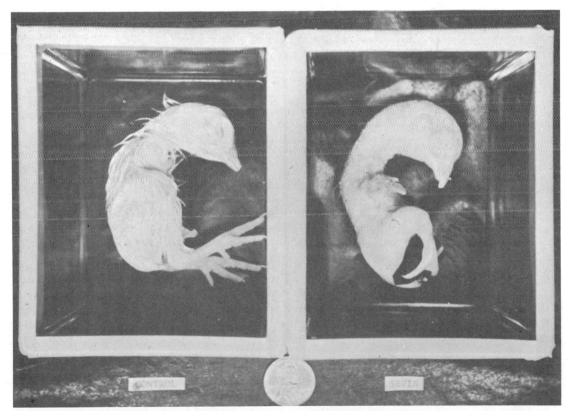

Teratogenic effects produced in a chick by injecting a toxic substance (in this case, the pesticide carbaryl) into the egg. Normal chick on left, just before hatching. Deformed chick on right, of same age. Note deformed limbs and beak and absence of feathers.

Similar effects are seen in animals treated with other teratogenic agents, such as thalidomide and herbicides.

FIGURE 14a.5

Another problem in assessing the safety of potentially toxic substances, such as herbicides, is the detection of functional birth defects. An experimental animal or human infant may appear perfectly normal at birth, yet suffer from dysfunction of some vital organ. Kidney, lung or brain malfunction, for example, even though present at birth, may not become apparent for months or, in the case of man, years. Few tests on animals are designed to detect such latent defects.

Results with Herbicides. The possible teratogenic effects of auxin-like herbicides have become a matter of much dispute. Early studies on animals showed that severe birth defects were produced by these chemicals, but later it became clear that the main culprit was a contaminant called *dioxin.* Dioxin is a by-product formed during herbicide synthesis, especially that of 2,4,5-T, and may be present in commercial samples at levels ranging from 0.5 to 30 ppm. The early tests for teratogenicity were done with herbicide samples containing high levels of dioxin. Since then, the manufacturers of the herbicides in question claim to have reduced dioxin levels in their products to negligible levels. They also state that they have removed other contaminants previously present which are converted to dioxin when plants sprayed with 2,4,5-T are burned.

The above information on the putative safety of auxin-like herbicides produced by new manufacturing methods was known to the director of the Environmental Protection Agency, William Ruckelshaus, in August, 1971; despite this information, he still felt there was sufficient potential risk from the use of 2,4,5-T itself to continue an earlier ban on its use on food crops. Decisions such as this one, as to whether a possibly dangerous (yet beneficial) chemical should be used before lengthy and exhaustive tests have shown it to be perfectly safe, come up ever more often. What is a risk? What is a benefit? How are they to be weighed, one against the other? These are questions of increasing importance.

INSECT CONTROL

Like weeds, insects are a major cause of agricultural crop losses. Until the introduction of chemical poisons around the turn of the century, farmers were relatively helpless to prevent crop destruction by insects. Songs such as "The Ballad of the Boll Weevil" attest to the frustration of the southern sharecropper in the United States:

> De first time I seen de boll weevil
> He was settin' on de square.
> De next time I seen de boll weevil,
> He had all of his fam'ly dere,
> Jus' a-lookin' for a home, jus' a-lookin' for a home.

Embodied in this one stanza is the ecological fact that monocrop culture (in this case cotton) invites rapid multiplication of pests, unchecked by natural predators. In Salt Lake City, there stands a monument to the sea gulls that migrated to the lake just in time to devour a plague of crickets which threatened the early settlers with starvation. For much of human existence, prayer and acceptance were man's primary means of coping with crop-destroying insects, although the use of small fields separated by hedgerows, of crop rotation and of the burning of unused parts of plants after harvesting all helped to control severe infestations. The advent of chemical poisons was a great boon to farmers.

Who are these insect enemies, against whom so much effort has been expended? Some 800,000 insect species are known, and it is estimated that there are about three million insect species altogether in the world—far more than all other known animal and plant species combined. Of these, more than 99 per cent are either innocuous or beneficial to man. Many in-

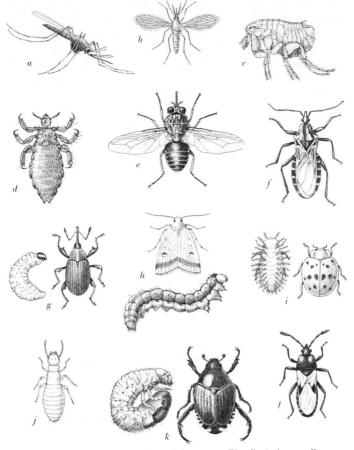

FIGURE 14a.6

Twelve of the approximately 4000 insect species harmful to man. The first six are disease carriers, the second six are crop pests. (Not all drawn to same scale.)
  a. the *Anopheles* mosquito (malaria)
  b. the sand fly (leishmaniasis)
  c. the rat flea (bubonic plague)
  d. the body louse (typhus)
  e. the tsetse fly (sleeping sickness)
  f. the kissing bug (Chagas' disease)
  g. the boll weevil
  h. the corn earworm
  i. the Mexican bean beetle
  j. the termite
  k. the Japanese beetle
  l. the chinch bug

(From Third Generation Pesticides, by C. M. Williams. Copyright © 1967 by Scientific American, Inc. All rights reserved.)

sects are indispensable, such as the insects which pollinate various crops (e.g., bees) or feed on harmful insects (certain beetles, including ladybugs). Only about 0.5 per cent of known insect species are harmful, destroying crops or carrying disease (Fig. 14.6). However, these 4000 or so species represent a major problem to mankind, and the trick is to control the pest insects without either harming beneficial organisms, destroying the environment or poisoning ourselves in the process!

NON-SPECIFIC POISONS

The first chemicals used extensively for insect control were of two sorts—natural poisons derived from plants, such as rotenone, nicotine and pyrethrum; and heavy metals, particularly lead arsenate, which are highly toxic to man and other animals. Both have disadvantages in their use. The natural pesticides are rapidly destroyed in the environment and must be repeatedly applied to be effective, whereas lead arsenate accumulates in soil, destroying soil organisms.

The development in the 1940's of DDT and other long-lasting chlorinated hydrocarbons with a high toxicity for insects and virtually no effects on man after brief exposure seemed the ultimate answer. We have already seen, however, that the spread and accumulation of chlorinated hydrocarbons throughout the world is having a boomerang effect — pelicans and eagles are dying off; Coho salmon in Lake Michigan are unfit to eat. Continued use of these compounds will further increase their concentration in the oceans, with the possibility of significant effects on photosynthesis by some species of marine phytoplankton (see Chapter 8a).

Pesticide Resistance. Another factor which has greatly limited the usefulness of chlorinated hydrocarbons, as well as earlier insecticides, is the ability of insects to become resistant to them. The great reproductive capacity of insects provides a large number of genetically varying offspring, and occasionally an individual is produced which is capable of detoxifying a poison such as DDT. Such an individual has an enormous advantage over its fellows, and its offspring survive and multiply further, creating a new generation of insecticide-resistant individuals.

Insects utilize detoxifying enzymes, very much like those in human liver (see Chapter 3a), to destroy toxic substances taken into their bodies. Some species of insect pests not only are resistant to DDT — they even thrive on it! To combat insect resistance, manufacturers have begun to add other chemicals which block the insects' detoxifying enzymes, thus hoping to maintain the potency of their pesticides. No doubt insects resistant to the added chemicals will also shortly appear.

Even the short-lived carbamates and organophosphorus compounds, synthesized to replace DDT for use against resistant species, are meeting the same fate; insects are becoming resistant to them also. Moreover, as potent inhibitors of the enzyme cholinesterase, they are more acutely toxic to man than their chlorinated hydrocarbon predecessors (see Chapter 12a).

We thus can visualize a growing, hungry world population depending more and more on greater agricultural yields, which in turn depend on rescuing crops from insects, which in turn are becoming resistant to insecticides almost as fast as the chemical industry can synthesize them. Meantime, the accumulating insecticides are destroying the environment and proving dangerous to man. What are the alternatives?

USE OF INSECT HORMONES

What is obviously needed is a substance which is toxic only to insects and to which they cannot become resistant. Such a substance may be at hand in the form of insect hormones. As we saw in the last chapter, two hormones are required for normal insect development: **ecdysone**, which causes differentiation toward the adult stage of the life cycle, and **juvenile hormone**, which causes growth but not maturation. In the normal course of larval development, the balance between these two hormones gradually alters in favor of maturation to the adult.

Experiments carried out on developing insects during the past 20 years have shown that imbalance of the normal levels of these two hormones interferes with normal development. In particular, either the absence of ecdysone or an excess of juvenile hormone during the later stages of development prevents the insect from becoming an adult. Although it continues to molt, it becomes a giant larva, incapable of reproduction (Fig. 14a.7). Such giant larvae often die through inability to molt again successfully. Juvenile hormone also prevents normal development of the eggs of

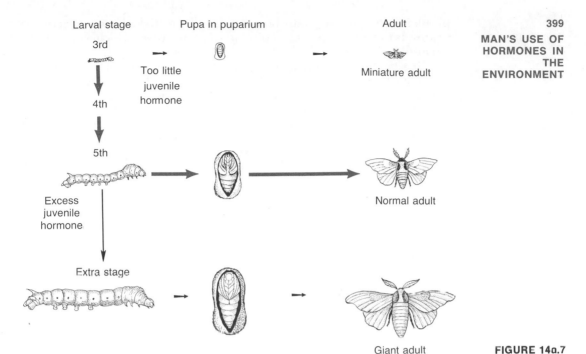

FIGURE 14a.7

Control of growth and molting in the silkworm, *Bombyx mori.*

Normally, larvae molt 5 times, then undergo pupation and emerge as normal-sized adults (pathway shown by colored arrows).

If gland producing juvenile hormone is removed during third larval stage, precocious pupation occurs, and a miniature adult results (upper line). (Too much ecdysone and too little juvenile hormone.)

If extra juvenile hormone is given to fifth-stage larva, it molts to giant larva, which may later pupate and produce giant adult (lower line). (Too much juvenile hormone and too little ecdysone).

many insects. Thus, while juvenile hormone is required at certain stages of development, it is lethal at others.

Juvenile Hormone. The potential of juvenile hormone for control of insect pests is apparent: it is a natural substance produced by insects, which is not toxic even to insects themselves; it simply "derails" their normal development if present in excess. It is not known to be toxic for, or otherwise to affect, any other variety of organism.

What is this potent substance? Although its existence was predicted from experiments, it was not until 1967 that juvenile hormone was finally purified and chemically identified (Fig. 14a.8A). So far, however, only two species of moths have been found which contain enough hormone to permit its extraction. Although the hormone isolated from them is potent on all other insects tested, it may not be chemically identical with juvenile hormone in all species.

It would require impossibly huge quantities of moths to obtain enough juvenile hormone for use in pest control; if it is to be useful, other sources must be found, either natural or synthetic. Although at first glance it appears to be a simple molecule, efforts to synthesize it have so far failed. One synthetic compound, farnesol, exists which has juvenile hormone activity in many insects but is far less active than pure juvenile hormone (Fig. 14a.8B).

Both juvenile hormone and farnesol, were they to become available in large quantities, would nevertheless suffer from a major drawback also shared by today's pesticides. They would act non-specifically on *all* insects, instead of only on those which are pests. Widespread use of such potent chemicals would kill beneficial insects and could easily cause environmental problems even worse than those we face from today's non-selective pesticides. Is there any hope of finding *specific* hormones, which will act only on harmful pest species?

Specific Juvenile Hormone Analogues. During the early 1960's, before juvenile hormone had been isolated and identified, a Czechoslovakian scientist visiting Harvard brought his favorite species of bug with him and attempted to raise it as he did in Prague — on moist paper towels. But at Harvard, instead of developing normally from fifth stage larvae to adults, the bugs always molted to sixth and sometimes seventh stage larvae, which failed to become adults and died (Fig. 14a.9). Excess juvenile hormone in the diet was suspected — and the effect was quickly traced to the paper towels.

For a time, juvenile hormone activity seemed ubiquitous — it was found in The New York Times and even in paper used to print scientific journals! It seemed wildly improbable that so specific a substance as an insect hormone should be so widely distributed in the environment. Two facts about this amazing discovery soon became clear: only paper or other products

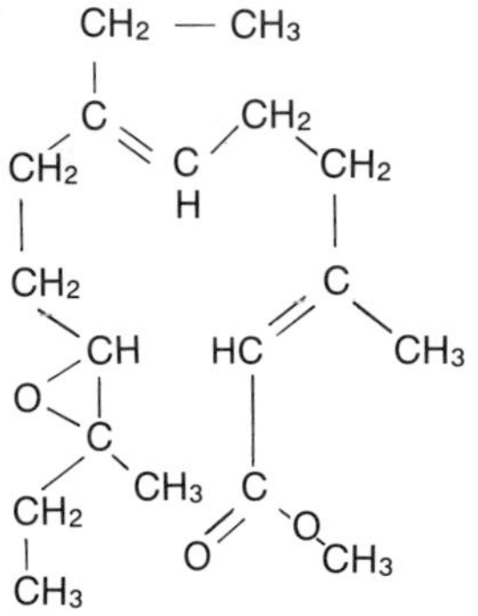

Juvenile hormone

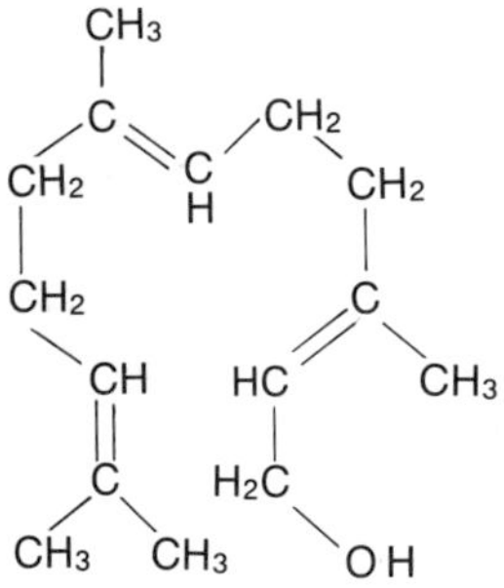

Farnesol

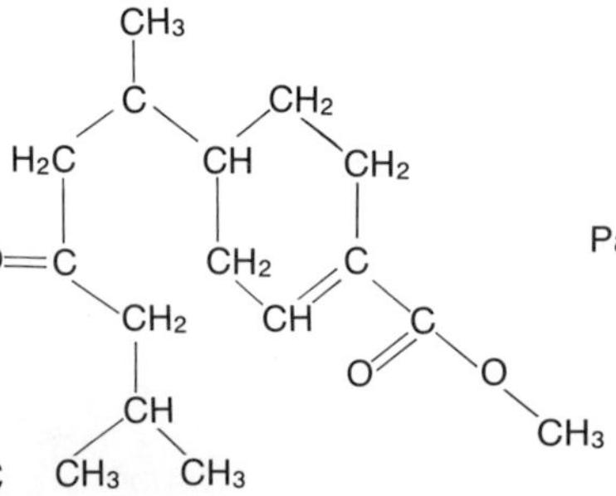

Paper factor

FIGURE 14a.8

Chemicals with juvenile hormone activity.
   A. Juvenile hormone isolated from cecropia moth — active on virtually all insects.
   B. Synthetic hormone, farnesol — active on virtually all insects but far weaker than juvenile hormone.
   C. Paper factor isolated from balsam fir — active only on pyrrhocorid bugs.

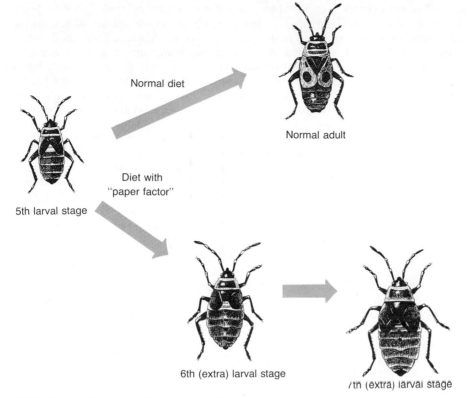

Normal diet

Normal adult

Diet with
"paper factor"

5th larval stage

6th (extra) larval stage

7th (extra) larval stage

FIGURE 14a.9

Effect of paper factor on the insect *Pyrrhocoris apterus.*
Normally, fifth-stage larva molts into adult. If fed on American paper towels made from balsam fir, it molts to a sixth-, and sometimes again, to a seventh-stage larva, then dies.

made from the balsam fir, a tree growing in Canada and the northern United States, were effective; only one group of insects was affected—bugs of the family Pyrrhocoridae, which includes some of the most destructive pests of cotton plants. This juvenile hormone analogue (mimic) came to be known as **"paper factor"**; eventually it was isolated from fir trees and chemically characterized; as shown in Figure 14a.8C, it resembles juvenile hormone.

Apparently the balsam fir, which has been on Earth longer than the insects, evolved the biochemical machinery necessary to synthesize paper factor and may thus have protected itself against some early bug resembling today's pyrrhocorid species. Here, then, we have the first case of a hormone-like agent which acts only on one group of pests, which is readily available and which seems not to be toxic for other organisms—the ideal pest control agent.

More recently other plants have been found which produce juvenile hormone analogues which also act on only a limited group of insects. Certain gymnosperms are also known to produce analogues of ecdysone; insects feeding on these trees undergo premature metamorphosis, which interferes with their normal reproduction. The search for more of these naturally occurring hormone analogues continues. However, certain problems, such as maintaining the stability of the hormone analogues in the environment long enough for them to be effective, remain to be solved.

OTHER ALTERNATIVES

In addition to the possible use of specific juvenile hormone analogues, other types of biological controls for insect pests are possible, and several of these have already proved successful against certain insects.

Sex Attractants.   In Chapter 13 we noted that certain female insects emit small quantities of pheromones, which act as sex attractants, causing the males to search out and mate with females. Bombykol, produced by the female of the moth species *Bombyx,* is one such substance. Recently, a specific sex attractant for the common house fly has been not only isolated and identified but also artificially synthesized. These sex attractants are highly species specific; by placing extracts of a specific sex attractant in traps, capture and removal of males is sometimes possible, decreasing the reproductive potential of the pest species. This method has been used successfully to eliminate the oriental fruit fly from the island of Rota in the Pacific, but it is often far less effective than is desired.

Sterilization of Males.   Adult females of many species of insects mate only once, lay their eggs and die. In such cases, it is possible to release large numbers of sterile males into an area; the females mate with them, laying non-fertile eggs, and the population dies out. This method was used with enormous success in southeastern and parts of southwestern United States to eradicate the screwworm fly *(Cochliomyia hominovorax)* — not, as erroneously believed, a blow fly. The maggots of this fly infest cattle and cause serious loss of livestock. Millions of male screwworms were raised to adults, irradiated to sterilize them and released into the infested areas. The eggs laid by the female flies were infertile and could not develop. The program cost $7 million, whereas annual livestock losses were running at about $40 million. No toxic substances were used and no environmental side effects were produced. Unfortunately, this same method has proved far less successful with other insects, such as the pink bollworm.

Predators and Parasites.   In natural environments, each species is kept in check by a variety of factors, including predators which feed on it and parasites which infect it. Insects are no exception. In large fields with single crops, however, predators find no natural nesting sites and pest insects reproduce unchecked, feeding on the all too abundant food supply — the crop itself. In some instances it has been possible to introduce large numbers of predators, particularly predatory insects, to control a given pest. Control of the Japanese beetle, European corn borer and various scale insects has been achieved in some areas by this means. (In a similar way, it is also possible to control specific weeds by introducing insects which feed preferentially on them.)

Many insects are specific targets for certain disease-producing organisms — viruses, bacteria, protozoans and so forth. (Most parasites attack only one or two groups or species of organism.) It is encouraging to note that a particular bacterium, *Bacillus thuringiensis,* has been found useful for control of two destructive moths — the cabbage looper and the alfalfa caterpillar. The bacteria are marketed for use commercially under the trade name "Thuricide." Search continues for specific parasites of other pestiferous organisms.

OUTLOOK FOR THE FUTURE

It is likely that our present crude methods of controlling both weeds and insects by spraying large quantities of non-specific toxic chemicals into the environment will be superseded by more selective and sophisticated approaches. In particular, **integrated pest control,** where several methods are employed simultaneously, will probably be used more and more. Common-sense agricultural practices, such as rotation and diversification of crops, combined with limited use of short-lived chemical pesticides, and bolstered by the introduction of selected parasites, predators or specific hormones, holds great promise for the future. We must expect, however, that until we know a great deal more about the life cycles and interactions of all species in a given environment we will have many failures and will need to modify our ideas and methods repeatedly.

SUMMARY

Chemicals which affect the growth of both plants and animals have become powerful tools used for various purposes by man. Substances such as 2,4-D and 2,4,5-T, which mimic the natural auxin of plants, have been used both as military defoliants and as weed-killers. They preferentially kill broad-leaved plants, which are replaced by grasses. Not only does the extensive use of weed-killers alter the natural ecosystem, depriving animals of food and nesting sites; these compounds in large amounts may also be toxic to animals and, perhaps, to man. In particular, their potential for inducing teratogenous birth defects has not yet been fully assessed.

In recent years, the future possibility of using specific chemicals for control of deleterious insects—which form a small fraction of the total number of insect species—has become apparent. The finding that growth and reproduction of insects is dependent on internal hormones has led to the discovery of naturally occurring chemicals which have hormonal action only on specific insects. The possibility that similar chemicals specific for many other pests will be found holds hope for the future.

Meantime, alternative pest control methods, such as natural predators, irradiation, specific parasites and specific sex attractants, all offer ecologically more sound and less toxic means for preventing decimation of crops and spread of disease. Further research along these avenues holds promise for the future of both man and his environment.

Aaronson, T. 1971. Gamble. Environment *13*(7):20–29. An article on possible teratogenic effects of herbicides and a cost-benefit analysis of their use.

Boffey, P. M. 1971. Herbicides in Vietnam: AAAS study finds widespread devastation. Science *171*:43–47. Editorial describing how a team of scientists first made public the total extent of deforestation in Vietnam.

Carlson, D. A., Mayer, M. S., Silhacek, D. L., James, J. D., Beroza, M. and Bierl, B. A. 1971. Sex attractant pheromone of the house fly: isolation, identification and synthesis. Science *174*:76–78. Studies like this one may lead to discovery of better, less toxic ways of coping with household pests than pest strips, sprays and so forth.

Edwards, C. A. 1969. Soil pollutants and soil animals. Scientific American Offprint No. 1138. Considers effects of pesticides on soil invertebrates and consequences for soil quality.

Epstein, S. S. 1970. A family likeness. Environment *12*(6):16–25. Points to possible similarities between various herbicide preparations and thalidomide in producing teratogenic effects.

Handler, P. H. (ed.) 1970. *Biology and the Future of Man.* Oxford University Press, New York. (Pp. 601–606 and 819–830 cover topics of this chapter.) An excellent, readable book, highly recommended to the interested student.

Orians, G. H. and Pfeiffer, E. W. 1970. Ecological effects of the war in Vietnam. Science *168*:544–554. Two ecologists attempt, in a war-torn country, the first assessment of ecological damage due to deforestation.

Taussig, H. B. 1962. The thalidomide syndrome. Scientific American Offprint No. 1100. Why the possibility of teratogenicity of new chemicals is regarded so seriously is revealed in the case history of this "safe" drug.

Turner, C. D. and Bagnara, J. T. 1971. *General Endocrinology.* (2nd ed.) W. B. Saunders, Philadelphia. An excellent reference on hormones, with explanations of their role in development of insects.

van den Bosch, R. 1969. The toxicity problem—comments by an applied insect ecologist. Chapter 6 (pp. 97–112), *in* Miller, M. W. and Berg, G. G. (eds.) *Chemical Fallout.* Charles C Thomas, Springfield, Illinois. The story of how one non-specific pesticide actually decreased rather than increased crop yields.

Wigglesworth, V. B. 1970. *Insect Hormones.* W. H. Freeman, San Francisco. An excellent account of the endocrinology of insects; the basis for tomorrow's more specific pest control methods.

Williams, C. M. 1967. Third generation pesticides. Scientific American Offprint No. 1078. An easy-to-read account of how specific hormones may aid in the control of pest insects.

15

"As lyke as one pease is to another."
John Lyly (1554-1606)

GENETICS, THE STUDY OF INHERITANCE

Sexual reproduction, by means of recombination of genes from two parents, results in a great variability among offspring. The detailed study of such inheritance has gradually led to an understanding of what genes are and how they act.

In discussing the reproduction of cells and organisms we noted that sexual reproduction provides a far greater variability among offspring than does asexual reproduction. This fact accounts for the widespread occurrence of sexual reproduction: a species is more likely to survive environmental changes if there is a high degree of genetic variability in the population—some members may, by chance, be pre-adapted to meet the new conditions.

In sexual reproduction, characteristics are inherited from both parents. In Chapter 14 we caught a brief glimpse of the way in which, by production of various types of gametes, two parents could produce many genetically different offspring. In this chapter we shall explore further details of the mechanisms by which parents pass on genetic information to the next generation.

THE BEGINNINGS OF MODERN GENETICS

It has long been observed that many inherited characteristics, such as height and hair color, seem to be distributed in a continuous series of gradations throughout a population of people, while others, such as eye color, sex, and blood type, are sharply defined. The inheritance of these discrete traits is thus more easily studied, and if observed over several successive generations of offspring, the pattern by which they are inherited can be understood. This type of observation was first made by an Austrian monk, Gregor Mendel, over one hundred years ago. His careful observations and interpretations opened up the new field of *genetics*—the science of inheritance.

Mendel chose to observe not people, but pea plants, which produce a new generation each year and whose matings one can control. He studied only those characteristics which were clearly different in the several strains of pea plants he had developed, ignoring other, less distinct traits. Some of the characteristics observed by Mendel were

> Red *versus* white flowers;
> Yellow *versus* green seeds;
> Round *versus* wrinkled seeds.

We call the appearance of such traits in an offspring its **phenotype** (from Greek *phainein,* "to show," plus *typos,* "impression"). In the several pea strains Mendel studied, he noted that each strain always bred true for certain characters—that is, it always had the same phenotype. After establishing this, Mendel spent eight years conducting hybridization experiments between his strains and observing how the various characters were inherited in successive generations.

Observation of Dominance. Upon cross-breeding peas having red flowers with peas having white flowers, Mendel observed that in the first hybrid generation (called the F_1 generation) all plants produced red flowers (Fig. 15.1). The characteristic for whiteness seemed to have disappeared. But when he then bred the F_1 generation with itself (self-cross), he found that one-quarter of the offspring in the next, F_2, generation had white flowers; the rest had red flowers.

Two things required explanation: the return of the white character in the F_2 generation; and the 3 to 1 ratio of red to white flowered plants. The character for white flowers clearly had not been lost, but only "suppressed," during the F_1 generation. Mendel reasoned that each pea plant possessed *two* hereditary factors for each character—one from each parent—and that these factors did not become blended in hybrids but instead retained their individual nature. Two such hereditary factors which both affect the same trait—such as red coloring of flowers and white coloring of flowers—are called **alleles** (from Greek *allēlōn,* "of one another"). We shall become more familiar with this term as we proceed. In Mendel's original, true-breeding strains of pea, red flowered plants possessed two genetic units, or alleles, for red flowers, while white flowered plants possessed two alleles for white flowers. In the F_1 hybrid generation, each plant received one red-flower allele and one white-flower allele (Fig. 15.2).

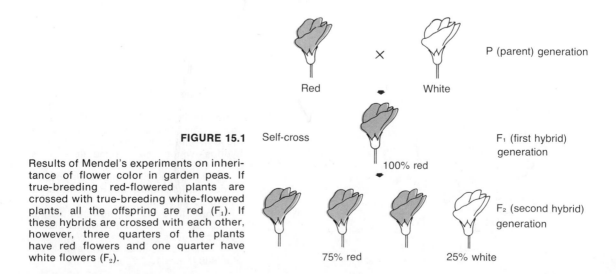

FIGURE 15.1

Results of Mendel's experiments on inheritance of flower color in garden peas. If true-breeding red-flowered plants are crossed with true-breeding white-flowered plants, all the offspring are red (F_1). If these hybrids are crossed with each other, however, three quarters of the plants have red flowers and one quarter have white flowers (F_2).

Red × White   P (parent) generation

Self-cross

100% red

F_1 (first hybrid) generation

F_2 (second hybrid) generation

75% red   25% white

Since all the F_1 plants had red flowers, Mendel further reasoned that the red allele is in some way **dominant** over the white allele, which is therefore **recessive**. By convention, geneticists express dominance and recessiveness for the same trait by capital and small forms of the same letter. Thus, in Figure 15.2, capital C stands for red color, the dominant allele, and small c stands for white color, the recessive allele. In the geneticist's shorthand, the F_1 generation in Mendel's experiments had the genetic constitution Cc. Since C dominates c, all plants have red flowers.

If we assume that each Cc parent passes on, via its gametes, *either one or the other allele* to each offspring, then we can also explain the 3:1 ratio of flower colors in the F_2 generation by examining the probability of different combinations of alleles from two F_1 parents. In our example, Cc combinations happen twice as often as either cc or CC. (Arrows in Figure 15.2 show all ways in which alleles can combine.) Since C dominates c, both CC and Cc combinations will produce red flowered plants. Only cc produces white flowered plants—thus the 3:1 ratio. We call an individual having identical alleles for the same character (CC or cc) **homozygous,** and one having two different alleles (Cc) **heterozygous.** Only a heterozygous individual can produce more than one kind of gamete with respect to the particular trait being observed.

These experiments led to the following generalizations: Traits are inherited in discrete units of inheritance, to which we now give the name **genes;** an individual inherits from each parent one gene for each trait, the two genes being known as alleles; certain alleles are dominant over others when present together. These ideas taken together form Mendel's first law of inheritance, which says that traits remain distinct from one generation to the next and two alleles for the same trait are segregated among offspring in a random fashion. We thus have the concept of an outward appearance, or phenotype, which results from an inner set of genetic traits, the **genotype.** In our example, we cannot outwardly distinguish between CC and Cc red flowered pea plants. They have the same phenotype but a different genotype.

Test Crosses for Genotype. There is a very simple test, often used by geneticists, for determining whether a given phenotype (C__) is due to a homozygous (CC) genotype or a heterozygous (Cc) genotype. The test is to cross the individual having a dominant phenotype (C__) with a homo-

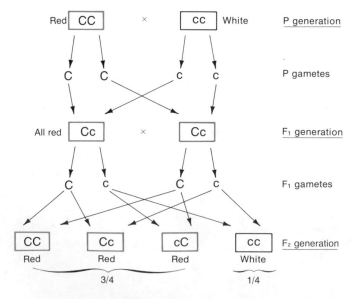

FIGURE 15.2

Interpretation of Mendel's experiments.

True-breeding plants have identical alleles for flower color. Red = CC; white = cc. Gametes of the C parent crossed with those of the c parent produce only Cc F_1 hybrids—which all have red flowers because C is dominant over c.

Self-crossing of F_1 allows all possible recombinations between F_1 gametes as shown, resulting in

$1/4$ homozygous (CC) red-flowered plants.
$1/2$ heterozygous (Cc) red-flowered plants.
$1/4$ homozygous (cc) white-flowered plants.

The ratio of red to white flowered plants is thus 3:1.

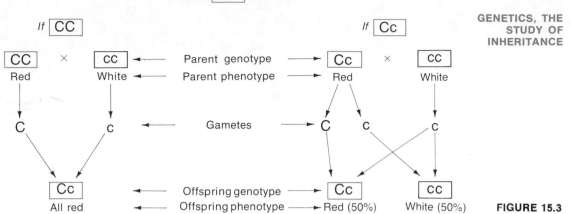

FIGURE 15.3

Homozygous recessive test-cross to determine unknown genotype of individual having dominant phenotype.

A red-flowered pea plant may be either a homozygous (CC) or heterozygous (Cc) genotype. A cross with a white homozygous recessive plant (cc) will distinguish these two possibilities. On the left, each parent produces only one kind of gamete, and all offspring are red (Cc). The unknown red-flowered plant was CC. On the right, the red-flowered plant produced two kinds of gametes, C and c; half the offspring have red flowers (Cc), half have white (cc). The unknown red-flowered plant was thus Cc.

zygous recessive individual (cc) and observe the offspring. Such a test cross of red flowering pea plants is shown in Figure 15.3. If the unknown plant is CC, all offspring will be red (Cc); if the unknown plant is Cc, half the offspring will be red (Cc) and half white (cc).

Independent Assortment of Characters. Mendel did not confine his observations on peas to a single trait, however. By recording the inheritance of several discrete traits at the same time, he discovered another genetic principle—the independent assortment of characters. Let us look at his experiments and analyze them.

Among other characters, Mendel studied the distribution of two seed characters, shape (round versus wrinkled) and color (yellow versus green). In earlier studies he had found that round (R) dominates wrinkled (r) and yellow (Y) dominates green (y). When true-breeding strains of plants having both dominant characters (RRYY) were crossed with true-breeding strains having both recessive characters (rryy), the F_1 generation looked like the dominant parents, as expected (Fig. 15.4), but the F_2 generation gave four phenotypes.

The number of plants Mendel obtained of each phenotype is shown in the figure also, and inspection reveals a ratio very close to 9:3:3:1. If the two characteristics were inherited together, one would expect the dominant characters R and Y to always occur together and the recessive characters r and y to occur together, giving an F_2 ratio of 3 R__Y__ (dominant phenotype) to 1 rryy (recessive phenotype), just as occurs in single factor inheritance. The alternative hypothesis, that the two factors are inherited *independently*, can be shown to fit the observed ratio. By drawing a box showing all possible combinations of the four kinds of gametes produced by each parent, and keeping in mind that all kinds of gametes are produced in equal proportions, we can show that such a ratio is indeed expected (Figure 15.5).

Or, for those more at home with mathematical probabilities, we can make the following statements of the expected ratio of phenotypes in the inheritance of two traits:

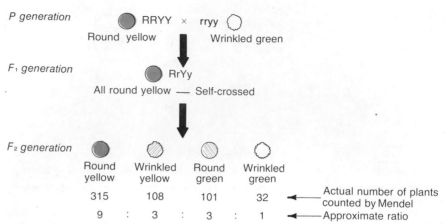

FIGURE 15.4

Mendel's observations on simultaneous inheritance of two characters in pea seeds, shape and color.

Plants with round yellow seeds crossed with plants having wrinkled green seeds produce only plants with round yellow seeds in the F₁ generation. The F₂ generation, however, shows all four possible combinations of the two characters, in the ratios indicated.

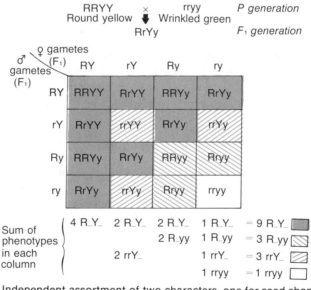

FIGURE 15.5

Independent assortment of two characters, one for seed shape (R = round; r = wrinkled), one for seed color (Y = yellow; y = green). The F₁ hybrids (RrYy) can produce four types of gametes, RY, rY, Ry and ry, which can recombine in 16 ways, as shown in the squares.

The resulting phenotypes are in the following ratio:

9 R_Y_ ▨ (Round yellow)
3 R_yy ▧ (Round green)
3 rrY_ ▨ (Wrinkled yellow)
1 rryy ☐ (Wrinkled green)

If ¾ chance of R__ and ¾ chance of Y__, then ¾ × ¾ =
$\frac{9}{16}$ chance of R__Y__

If ¼ chance of rr and ¾ chance of Y__, then ¼ × ¾ =
$\frac{3}{16}$ chance of rrY__

If ¾ chance of R__ and ¼ chance of yy, then ¾ × ¼ =
$\frac{3}{16}$ chance of R__yy

If ¼ chance of rr and ¼ chance of yy, then ¼ × ¼ =
$\frac{1}{16}$ chance of rryy

In essence we have said that the inheritance of seed shape is independent of the inheritance of seed color and that the probability of two particular characteristics occurring together is the product of the probabilities for the occurrence of each separate character. This independence of the inheritance of separate traits forms the basis of Mendel's second law, the law of independent assortment.

To summarize Mendel's work: long before anything was known of cell nuclei, chromosomes or DNA, Mendel predicted that traits were inherited in discrete units (now called genes); that each trait was controlled by two alleles, one from each parent; that some alleles were dominant over others; and that two different traits could be inherited independently of one another.

THE MODERN EXPLANATION OF MENDEL'S WORK

In the century since Mendel's work, a great number of experiments and observations have provided a much clearer picture of the physical basis of inheritance in terms of DNA molecules, chromosomes and meiosis. Without attempting to place these discoveries in chronological order we can briefly outline our present understanding of how the genotype is inherited and how genes function to produce a particular phenotype.

WHAT IS A GENE?

A gene was first defined as a functional unit of inheritance—that which controls the appearance of a specific trait, such as red or white flower color in pea plants or blue or brown eyes in humans. Later, genes were located at specific sites on chromosomes—the gene **locus** (Latin *locus*, "place"; plural, *loci*). Thus, each pair of homologous chromosomes, one inherited from each parent, bears pairs of homologous loci along the entire length (see Figure 14.16).

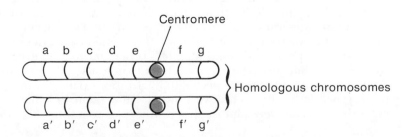

Finally, in the past 20 years or so, it has been shown that a gene is, in fact, a functional region of the DNA of the chromosome, virtually equivalent to the **cistron** described in Chapter 3. We now know that when an activated gene is present in a cell, it directs the synthesis of a specific messenger-RNA which causes synthesis of a particular protein. The final product of a gene is thus a molecule with a unique function to perform.

$$\text{DNA of gene} \longrightarrow \text{m-RNA} \longrightarrow \text{specific protein}$$

Often this protein is an enzyme, perhaps regulating the synthesis of a particular cell product, such as red flower pigment or brown eye pigment. Other genes direct the synthesis of a particular structural protein, such as keratin with regular bends in the molecule, leading to kinky hair. (Actually hair texture is controlled by several genes, as is hair color.)

WHAT IS DOMINANCE?

We now have a partial explanation of the dominance of red flower color over white flower color observed by Mendel. Red flowered plants (C__) have at least one gene capable of producing an enzyme needed for synthesis of red pigment; white flowered plants (cc) are totally deficient in this enzyme. In this case, the recessive condition is due to the absence of a particular enzyme. Either the cistron itself is missing, or the protein it synthesizes fails to function because of an error in its amino acid sequence (see Chapter 3). As we shall see in the next chapter, certain human diseases are the result of recessive deficiencies in particular enzymes.

We cannot yet explain all dominant and recessive traits quite so simply, however, and so must still use these two rather mysterious terms until a more descriptive explanation becomes available.

HOW INDEPENDENT ASSORTMENT OCCURS

Since the same gene is always located at the same spot on a particular chromosome we can now understand how independent assortment will occur *if two traits are located on different chromosomes*. In discussing meiosis we noted that a diploid germ cell with two pairs of homologous chromosomes would produce four kinds of gametes (see page 370). Just so, in Mendel's experiments the R/r alleles for round and wrinkled seeds are located on one pair of chromosomes, and the Y/y alleles for seed color on another (Fig. 15.6). The F_1 hybrids produce equal numbers of four types of gametes, each containing one of each pair of homologous chromosomes. Compare the four types of gametes shown here with those in Figure 15.5.

Only by a lucky chance did Mendel happen to study traits located on different chromosomes in the pea; had he chosen characters located on the same chromosome, he perhaps would never have discovered the law of independent assortment, nor have realized that traits are inherited in discrete units, which we now call genes.

LINKAGE AND CROSSING-OVER

After Mendel published his results in 1866, they lay unnoticed for over 30 years. Not until 1900, more than 15 years after his death, did three separate investigators unknowingly repeat Mendel's experiments, only to find that the Austrian monk had made the same discoveries years before. In trying to repeat and extend Mendel's observations, however, geneticists soon found that independent assortment of traits did not always occur. Sometimes, instead of the expected 9:3:3:1 ratio, a pair of traits gave a ratio of very nearly 3:0:0:1. The two traits did not seem to behave independently, but instead were almost always inherited together (Fig. 15.7).

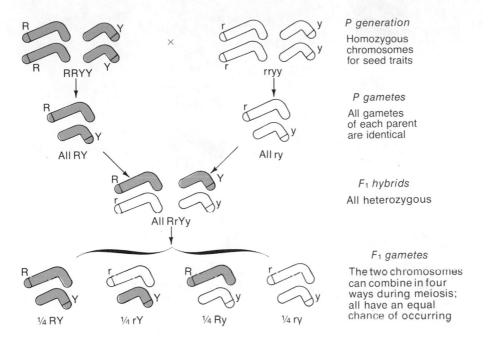

FIGURE 15.6

Independent assortment of characters on two different chromosomes. R allele, round seeds; r allele, wrinkled seeds; Y allele, yellow seeds; y allele, green seeds.
    Gametes of F₁ are of four types: each type of large chromosome can combine with either type of small chromosome—the chromosomes are independently assorted among the gametes.

	AABB	×	aabb		P generation
		AaBb Self-cross			F₁ generation

	A＿B＿		A＿bb		aaB＿		aabb	F₂ generation
Observed per cent	73		2		2		23	
Approximate ratio	3	:	0	:	0	:	1	
Ratio *expected* if independent assortment	9	:	3	:	3	:	1	

FIGURE 15.7

Absence of independent assortment of two traits: instead of the expected 9:3:3:1 ratio, a ratio of *approximately* 3:0:0:1 is found. The traits A and B are *almost* always inherited together as a single unit.

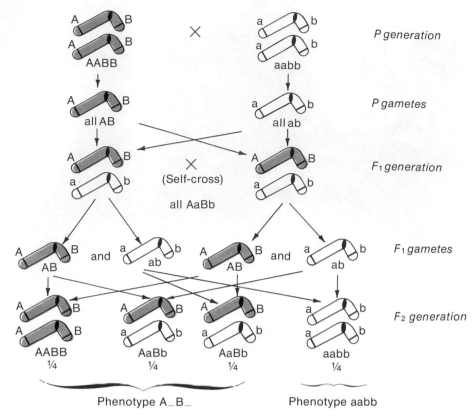

FIGURE 15.8

Phenotype A_B_   Phenotype aabb

Linkage of two traits on a single chromosome.
    The F₁ generation all have the phenotype A_B_. The gametes of each parent of the
F₁ generation are of two types—AB or ab. These randomly recombine to give one quarter
AABB, one half AaBb and one quarter aabb offspring. Unlike independent assortment
of two traits on different chromosomes, the dominant genes A and B are inherited together, as are the recessive genes a and b.

An explanation was not long in appearing: many genes are located in
linear sequence along the length of each chromosome, and when two traits
occur on the same chromosome they are **linked** together and behave as a
single unit (Fig. 15.8).
    We have yet to account, however, for the occasional appearance of
A_bb and aaB_ phenotypes in our example of Figure 15.7; in a few offspring, linkage has apparently not occurred, the traits behaving as if they
were on different chromosomes. It is the mark of a good scientist to think
twice before dismissing unexplained data, since often the unexpected leads
to new discoveries. In this case, the unexpected observations led to the discovery of crossing-over.
    Recall that during the first meiotic division the homologous chromo-
somes join together to form **tetrads.** Investigators had noted that sometimes there are "X-shaped" regions in these tetrads, which they called **chias-
mata** (singular, chiasma; *chi* is the Greek letter X). It was eventually shown
that a chiasma is a point at which two chromatids within the tetrad have
broken apart and rejoined with the opposite strands of DNA. This phenomenon is known as **crossing-over** (Fig. 15.9).
    In our example, the resulting chromatids in the gametes will be of four
types: the majority will be AB and ab, produced from intact chromatids;
there will also be a few Ab and aB gametes from the crossed-over chromatids. When an ab chromosome is combined with either an Ab or aB

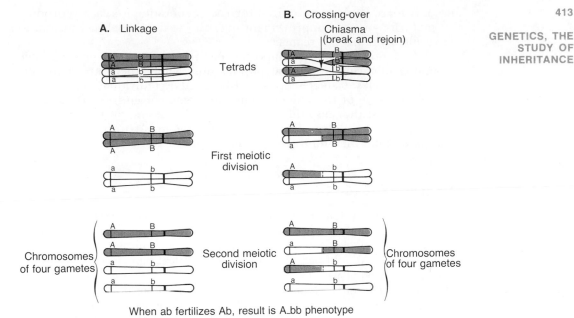

A. Linkage

B. Crossing-over

Chiasma
(break and rejoin)

Tetrads

First meiotic
division

Chromosomes of four gametes

Second meiotic
division

Chromosomes of four gametes

When ab fertilizes Ab, result is A_bb phenotype
When ab fertilizes aB, result is aaB_phenotype

FIGURE 15.9

Linkage and crossing-over in tetrads of an F_1 germ cell heterozygous for two characters on the same chromosome.

A. Linkage. Usually the four chromatids of the tetrad remain intact and linked genes remain together in gametes.

B. Crossing-over Sometimes two chromatids exchange equivalent parts through chiasma formation. Four types of gametes are formed: AB, aB, Ab, and ab.

chromosome during fertilization, the occasional phenotypes A__bb or aaB__ are produced.

It is thought that crossing-over occurs about equally often at any point along the chromatids of a tetrad. Thus, if the loci for A/a and B/b are far apart on a chromosome, the chance that crossing over will occur somewhere between them is high, and the phenotypes A__bb and aaB__ will occur fairly often (Fig. 15.10A). In contrast, if two loci are close together (A/a and C/c), crossing-over between them will occur infrequently, and A__cc and aaC__ phenotypes will be rare (Fig. 15.10B).

By comparing the relative frequency of crossing-over between a whole set of traits linked on a single chromosome, it is possible to make a chromosome map showing the linear relationships of the traits along the chromosome. Such chromosome maps have been constructed for hundreds of inherited traits on the geneticists' favorite animal, the tiny fruit fly, *Drosophila*. This fly, often seen hovering over bananas or other soft fruit, is especially useful because it can be easily raised in the laboratory,

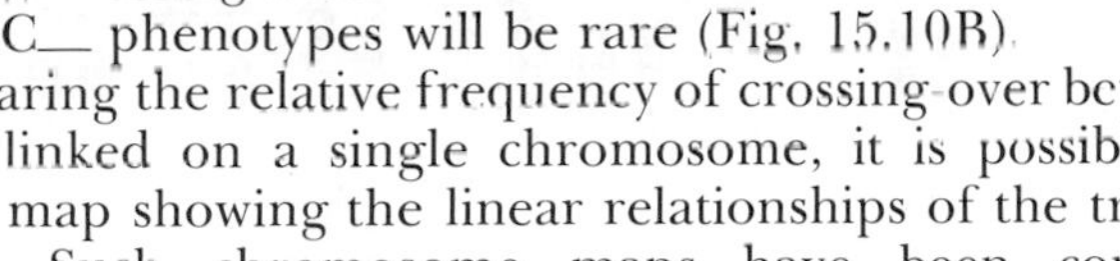

A/a   B/b

A

A/a C/c

B

FIGURE 15.10

Crossing-over frequency depends on distance separating two traits on a chromosome.

A. Loci for traits **A/a** and **B/b** are widely separated, and crossing-over is frequent. (Note: **A/a** = **A** *or* **a** gene at this locus).

B. Loci for traits **A/a** and **C/c** are close together; crossing-over between them is rare.

and only three weeks are required for each generation to reach reproductive maturity—a great improvement over the year Mendel had to wait to get results from his pea plants.

Crossing-over is of great importance in increasing the genetic variability among offspring in natural populations, where there are many heterozygous alleles. Gametes produced by a single individual are different not only because of the independent assortment of entire chromosomes during meiosis; they also differ as a result of crossing-over between two chromatids during tetrad formation of homologous chromosomes. Each chiasma doubles the number of possible types of gametes formed.

THE INHERITANCE OF SEX

We have seen that equal sets of chromosomes are inherited from the female and the male parent, each gamete contributing a complete haploid (1N) set. Thus, if we examine the chromosomes from a dividing diploid (2N) cell, there should be a double set of identical pairs of chromosomes. This is in fact what we find, with one important exception. In one or the other sex of a given species, one pair of chromosomes does not appear to match up very well; the "homologous" chromosomes do not seem to be homologous. In birds it is the female in which there is an odd, unmatched pair of chromosomes; in the fruit fly, *Drosophila,* and in man, it is the male where this occurs. Let us consider the case of man.

In the human diploid cell there are 46 chromosomes, or 23 pairs. Of these, the mates of 22 of the pairs are identical in appearance in both sexes as can be seen by the corresponding bands on homologous chromosomes (Fig. 15.11). But the chromosomes of the 23rd pair, although identical in the female, are different in the male. This 23rd pair are known as the **sex**

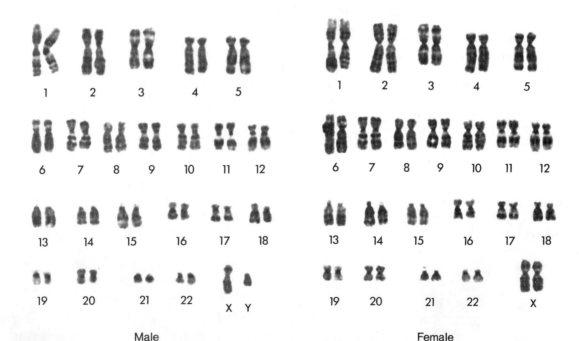

Male   Female

FIGURE 15.11   Human chromosomes, cut out of photographs of stained dividing cells and arranged in homologous pairs. A new staining technique permits pairing of homologous chromosomes according not only to size but also to their distinctive banding. (Photographs courtesy of K. M. Taylor and M. G. Brown, Biology Department, San Diego State University.)

chromosomes. In the human female, the two identical sex chromosomes are called **X-chromosomes.** In the human male, there is only one X-chromosome; its shorter partner is called a **Y-chromosome.** With respect to sex chromosomes, then, human females are XX, human males, XY.

The sex chromosomes in fact determine the sex of an offspring; the sex of a developing human embryo depends on whether the egg was fertilized by a sperm which contained an X or a Y chromosome. During gamete formation in the human male, each sperm receives one member of each of the 22 identical pairs of chromosomes, and *either* an X *or* a Y chromosome. In humans, the male is the **heterogametic** sex (the sex whose gametes do not all contain identically appearing sets of chromosomes); all eggs, on the other hand, contain one X-chromosome. If an egg is fertilized by a sperm with a Y-chromosome it develops into a male (XY); if it is fertilized by an X-chromosome bearing sperm, it becomes a female (XX) (Fig. 15.12).

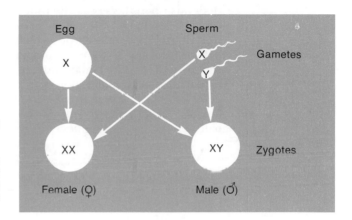

FIGURE 15.12

Inheritance of sex in man.
The male is the heterogametic sex in humans, producing half X-sperm and half Y sperm. Females produce only X-eggs. The sex of an embryo depends on which type of sperm fertilizes the egg.

SEX-LINKED TRAITS

The sex chromosomes bear the genes which determine the sex of an embryo. They also carry other genes which have nothing to do with the determination of maleness or femaleness. Traits controlled by such genes are called **sex-linked** characters. In humans, the Y-chromosome is considerably smaller than the X-chromosome; thus there are numerous genes on the X-chromosome which have no homologous mates on the Y-chromosome. The human male is totally dependent on the genes he inherits from his mother for these traits; he inherits only one allele for most sex-linked characters, those occurring on the X-chromosome.

This has an interesting result for the inheritance of certain recessive characters whose loci are on the X-chromosome, such as color blindness and hemophilia. Let us consider the case of color blindness. The ability to distinguish colors, especially red and green, depends on the synthesis of a particular visual pigment in certain cells of the retina. The gene which directs the synthesis of this pigment is a dominant gene, C, located on the X-chromosome. The recessive gene, c, is unable to produce the appropriate messenger-RNA. A female heterozygous for these genes—having C on one X-chromosome and c on the other—has normal color vision; but she produces two kinds of eggs—C and c. If she marries a man with normal color vision, he *must* possess the dominant allele, C, on his one X-chromosome. If this couple produces a daughter, she *must* receive a C gene from her father, and hence will have normal color vision, whether she receives C or c from her mother. A son, however, inherits no X-chromosome and hence no color vision gene from his father; if he inherits C from his mother, he will have normal vision; but if he inherits c, he will be color blind.

A pedigree of a family in which color blindness occurs is shown in Figure 15.13. Phenotypes are indicated by shapes and colors of symbols representing members of the family; genotypes are given beneath each individual. In some of the females a complete genotype cannot be assigned from the information available. Such sex-linked recessive traits clearly occur far more frequently in males than females. It may prove useful to copy out the pedigree given here and attempt to assign genotypes yourself; or better yet, if you know someone who is color blind, construct a pedigree from what is known of the phenotypes in his family tree.

SEX PENETRANCE OF TRAITS

Certain traits, although not located on the sex chromosomes, have different phenotypes in males and females; the expression of the genotype depends upon the sex of the individual. One particularly common case is a type of baldness, known as pattern baldness. This type of baldness is inherited (unlike other types which result from various diseases) and is controlled by a single gene. In females, the gene for baldness behaves like a recessive, b, and a woman must inherit two such genes (bb) to exhibit pattern baldness. In males, however, baldness occurs even if only one baldness gene is inherited; a man with the genotype Bb will be bald; only if a man has the genotype BB will he retain his hair (Fig. 15.14). It takes two baldness genes in females before the genotype is expressed, or "pene-

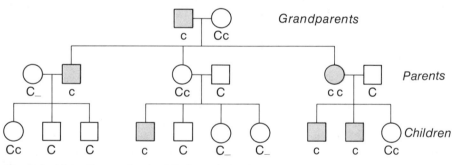

FIGURE 15.13

Sex-linked inheritance of color blindness in humans:

c	color blind male	C	normal male
cc	color blind female	C_	normal female

= marriage

= children

Note that in males, phenotypes and genotypes are identical, because there is only one X-chromosome. In some phenotypically normal females, information in the pedigree allows us to decide if she is a carrier of the c-gene for color blindness. Is it possible for a son to inherit color blindness from his father or his paternal grandfather?

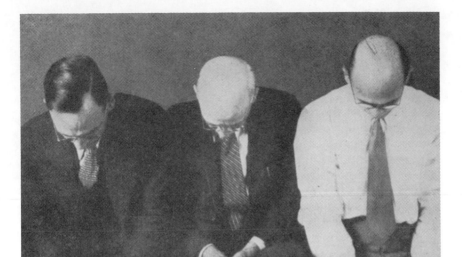

FIGURE 15.14

The inheritance of pattern baldness.
Two sons on either side of bald father (center) show different inheritance. Son on left is BB, son on right is Bb, the b-gene (for baldness) presumably being inherited from his father. (From Sinnott, E. W., Dunn, L. C., and Dobzhansky, Th. 1950. *Principles of Genetics.* McGraw-Hill, New York.)

trates," but only one in males; hence baldness is far more common in men.

One can think of many other traits which depend upon sex — distribution of body and facial hair, depth of voice, breast development, shape of pelvis, height. The *expression* of the genes controlling such sex-related characteristics depends on the presence of sex hormones, produced by the gonads — which in turn depend on sex-determining genes on the X- and Y-chromosomes.

BLENDED INHERITANCE

Not all sets of alleles show the sort of dominant/recessive interactions described above, which result in distinctly different phenotypes and are therefore easily studied. In many cases of heterozygous inheritance, both alleles contribute equally to the phenotypic expression of a particular character; this is known as **blended inheritance.**

Unlike the garden pea, whose heterozygous red flowers (Cc) are indistinguishable from homozygous red flowers (CC), many other hybrid flowers have intermediate colors. In the Japanese four o'clock, in the sweet pea and in the snapdragon, crosses between plants with white and red flowers produce plants with pink flowers (Fig. 15.15). A similar type of inheritance is followed in the production of various types of human hemoglobin, a subject considered further in the next chapter. Segregation of traits in offspring, however, follows the same rules as for complete dominance, and the two genetic alleles retain their individual identity.

POLYGENIC INHERITANCE

The traits studied by Mendel, and several other traits we have mentioned, have been discrete traits dealing with a phenotypic character con-

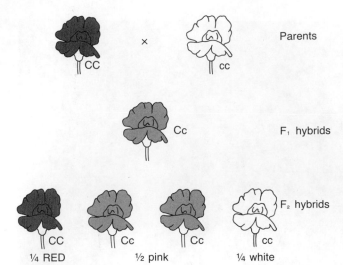

FIGURE 15.15

¼ RED ½ pink ¼ white

Blended inheritance of flower color in the snapdragon. The amount of pigment is proportional to the number of C-genes present: CC = red; Cc = pink; cc = no pigment (white). Genotype ratios are the same whether there is blended inheritance or dominance in expression of traits.

trolled by a single gene. On the other hand, many inherited traits are under the control of several genes, all of which contribute to the final phenotype. One can think of many such traits in humans, including height, skin color, intelligence and so on. Although children tend to be like their parents, they may occasionally differ greatly.

Consider for a moment the genes for height. About 10 gene loci are involved, which have additive effects. Curiously enough, the dominant gene in each case is for shortness. If two medium-sized people, heterozygous for all ten loci, marry, they will mostly produce medium-sized children; but about one time in a million such couples will produce a child that is homozygous dominant for all loci and is extremely short, and just as often, one that is homozygous recessive for all loci and is extremely tall. An example of how this could happen is shown in Figure 15.16, where it is assumed that the 10 genes are not linked and segregate independently during meiosis.

Polygenic inheritance thus produces only a few extreme phenotypes at each end of a continuous scale of variation. The greatest number of individuals are clustered around the average value, and the population as a whole exhibits a normal distribution, as discussed in Chapter 1a. We also saw in that chapter that, in the case of human populations, it is extremely difficult to sort out the contributions of genetic factors to a particular trait such as height from those of non-genetic or environmental origin. In Chapter 21a we shall see that the problem is even more difficult when one considers the inheritance of intelligence.

GENETICS OF POPULATIONS

So far we have considered genes only in relation to inheritance in individuals. Let us now briefly examine the genetics of populations, upon which the processes of evolution act to produce gradual changes in species over many generations.

Let us imagine a population of blue-eyed human beings living on an island. All these people are homozygous for the recessive gene for eye color, bb. Let us further suppose that an equal number of brown-eyed people, all homozygous for the dominant character BB, are shipwrecked on the island, and that the two populations intermarry in a completely random fashion. According to Mendelian laws, $3/4$ of the next generation will have brown eyes—$1/4$ BB and $1/2$ Bb—and $1/4$ will have blue eyes (bb). If random mating continues, this ratio will remain constant generation after generation.

We can perhaps think of genotypes in our hypothetical population by means of the following analogy. All sperm from each generation are placed in one drum. Half of these will carry gene B, half gene b. This comes about in the following way:

Proportion of Individual Genotypes in Population	Contribution to Total Sperm in Population	
	B sperm	*b sperm*
$1/4$ BB $=$	$1 \times 1/4 = 1/4$	$0 \times 1/4 = 0$
$1/2$ Bb $=$	$1/2 \times 1/2 = 1/4$	$1/2 \times 1/2 = 1/4$
$1/4$ bb $=$	$0 \times 1/4 = 0$	$1 \times 1/4 = 1/4$
	$B = 1/2$	$b = 1/2$

A similar situation prevails for eggs placed in a second drum; half are B and half are b. There is a 50 per cent chance that a B sperm will be drawn

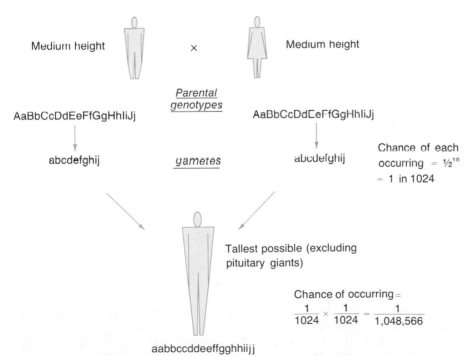

Medium height × Medium height

Parental genotypes

AaBbCcDdEeFfGgHhIiJj → abcdefghij

gametes

AaBbCcDdEeFfGgHhIiJj → abcdefghij

Chance of each occurring $= 1/2^{10}$
$= 1$ in 1024

Tallest possible (excluding pituitary giants)

Chance of occurring $= \dfrac{1}{1024} \times \dfrac{1}{1024} = \dfrac{1}{1,048,566}$

aabbccddeeffgghhiijj

FIGURE 15.16

The chance of 10 selected non-linked alleles for height all occurring in the same gamete from a heterozygous parent is $(1/2)^{10}$, or 1 in 1024. The chance that union will occur between a sperm and an egg both having the *same* genetic constitution for all 10 genes is about 1 in a million!

from the first drum, and a 50 per cent chance that a B egg will be drawn from the second drum; the chance of BB is therefore 25 per cent.

$$\begin{array}{c} \text{♂} \quad \text{♀} \\ B \times B = BB \\ \tfrac{1}{2} \quad \tfrac{1}{2} \quad \tfrac{1}{4} \\[1em] B \times b = Bb \\ \tfrac{1}{2} \quad \tfrac{1}{2} \quad \tfrac{1}{4} \\[1em] b \times B = Bb \\ \tfrac{1}{2} \quad \tfrac{1}{2} \quad \tfrac{1}{4} \\[1em] b \times b = bb \\ \tfrac{1}{2} \quad \tfrac{1}{2} \quad \tfrac{1}{4} \end{array} \qquad Bb = \tfrac{1}{2}$$

This constancy of alleles in a randomly mating population is known as the **Hardy-Weinberg law.** If we know the number of homozygous recessives in such a population, we can immediately calculate the total number of recessive genes which must be present to produce them.

Let us consider a population in which one per cent of the people are albinos. Albinism is a recessive trait, aa, in which an enzyme necessary for production of skin and hair pigment is lacking. We can establish a frequency table, similar to that of Figure 15.5, of possible combinations of eggs and sperm.

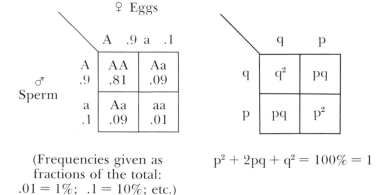

(Frequencies given as fractions of the total: .01 = 1%; .1 = 10%; etc.)

$$p^2 + 2pq + q^2 = 100\% = 1$$

According to the laws of probability, in order for **aa** to occur in one per cent of the offspring, 10 per cent of the gametes must have an **a** gene; the remaining 90 per cent of gametes must therefore have an alternative **A** gene, which is necessary for pigment formation.

To generalize, we can let the fraction of recessive phenotypes in a population be p^2 (.01 in our example); the fraction of recessive genes is $\sqrt{p^2}$, or p (0.1). The remaining genes must be 1−p (1−0.1), or q (0.9). Homozygous dominant phenotypes will occur with a frequency q^2 (.81) and heterozygous dominants with a frequency 2pq (.18). Thus $p^2 + 2pq + q^2 = 1$. The rules for genotype frequencies in randomly mating populations are thus identical with the laws of probability.

It is evident that the number of recessive genes in a population is far greater than the number of recessive phenotypes appearing. This is of especial importance in considering the frequency of recessive genes in human populations. In the next chapter we shall consider some of the genetic diseases occurring in people which result from homozygous recessive genotypes for certain traits.

In randomly mating populations gene frequencies remain constant if no other factors intervene. In real situations, this is seldom the case; there are four important ways in which gene frequency can be altered.

1. Differential migration: Suppose in our case of blue-eyed and brown-eyed populations on an island that, after a few years, many of the shipwrecked brown-eyed people became homesick and left the island. They would remove most of the brown-eyed genes, perhaps leaving behind a few, nearly grown, brown-eyed offspring.

2. Mutation: Once in a while, spontaneous mistakes in copying of DNA occur; these are known as **mutations.** In our island example, let us suppose that mutations from B to b occur more often than the opposite mutation, from b to B. Then b genes will become more numerous, and the proportion of blue-eyed people will increase.

3. Genetic drift: In small populations, by chance one gene is sometimes reproduced more often than another. Suppose there are only 10 families on our island, and in a particular generation brown-eyed parents produce only one child each, and blue-eyed parents produce three children each. Again, the b genes will become more numerous. The laws of probability do not always hold for small populations.

4. Selection: If one gene is more favorable than another for survival and reproduction, then it will become dominant in successive generations because it has adaptive value; it is selected for. Let us suppose our imaginary island is in the tropics and that blue-eyed persons must squint more and are more susceptible to eyestrain and poor vision than are brown-eyed people. If this prevents blue-eyed parents from hunting and feeding themselves and their children as successfully as do the brown-eyed people, fewer blue-eyed children will survive to maturity. The frequency of the b gene will tend to decrease. We shall further consider the forces of selection in Chapter 21.

SUMMARY

During sexual reproduction, each individual receives two genes for each trait, one from each parent, which are known as alleles. The alleles may be the same (homozygous condition) or different (heterozygous condition). In blended inheritance, both alleles contribute to the appearance of the trait (phenotype). In other cases, one allele is dominant over the other; the recessive gene is not expressed. As Mendel showed, recessive characters remain as distinct units of inheritance, even if not expressed in a particular generation.

Genes on different chromosomes are inherited independently, according to the laws of chance, because homologous chromosomes, in turn, are randomly distributed among gametes during meiosis. In contrast, genes on the same chromosome tend to be inherited together—they are linked traits. Crossing-over between chromatids during tetrad formation, however, sometimes allows genes on the same chromosome to be inherited independently. The distance between two loci on a chromosome is inversely proportional to how closely they are linked.

The inheritance of sex in many organisms is dependent on a single, unmatched pair of sex chromosomes consisting of the X- and the Y-chromosome. In man, the male is the heterogametic (XY) sex, producing X- and Y-sperm; all eggs have an X-chromosome, since females have an XX constitution. The large X-chromosome contains many genes not found on the small Y-chromosome. Traits controlled by these genes are called sex-linked — examples of two such recessive traits being color blindness and hemophilia. Other traits, such as pattern baldness and depth of voice, are dependent on sex hormones for their expression.

The frequency of genes in a large, randomly mating population remains constant from one generation to the next, providing no external factors are at work. This is called the Hardy-Weinberg law. In small populations, the laws of probability tend to break down, and genetic drift may occur. Other factors producing changes in gene frequency in a population are differential migration, unequal mutation rates, and selection of more favorable genes.

Certain traits are controlled by several genes, and when these have additive effects, the trait shows a continuous, normal distribution in the population. Examples of such polygenic characters are height, skin color and intelligence.

**READINGS AND
REFERENCES**

Benzer, S. 1962. The fine structure of the gene. Scientific American Offprint No. 120. A summary of the studies on chromosomes of viruses which have revealed much about how genes function.

Carlson, E. J. 1966. *The Gene: A Critical History.* W. B. Saunders Co., Philadelphia. A historical survey of genetics from Mendel to the present.

Moore, J. A. 1972. *Heredity and Development.* 2nd ed. Oxford University Press, New York. The fundamentals of genetics presented in their most up-to-date form, followed by the role of genes in development.

Stern, C. 1960. *Principles of Human Genetics.* 2nd ed. W. H. Freeman Co., San Francisco. An excellent text which includes principles of population genetics.

Sturtevant, A. H. 1965. *A History of Genetics.* Harper & Row, New York. A most readable account of classical genetics.

Taylor, J. H. 1965. *Selected Papers on Molecular Genetics.* Academic Press, New York. A collection of original research papers presenting the classical evidence for the structure and function of DNA, chromosomes and gene products.

Watson, J. D. 1970. *Molecular Biology of the Gene.* 2nd ed. W. A. Benjamin, New York. A most clearly written presentation of what is known about the functions of chromosomes, genes and their products. Highly recommended to the interested student.

MUTATION, RADIATION DAMAGE AND GENETIC RISK

The incidence of both cancer and inherited diseases due to deleterious mutations can be expected to rise as man makes increasing use of radiant energy. The public must decide what total risk is acceptable and which of the alternative benefits are most desirable.

In Chapters 14 and 15 we have seen that the genetic material, DNA, is important for normal development of an embryo. Changes in the structure of DNA produce genetic variations called **mutations.** Although very infrequently a mutation confers positive survival value on an individual, most mutations do not. Many are relatively neutral; for instance, a mutation from a curly-haired to straight-haired gene would have little effect on one's chances of survival in today's world. However, a few mutations are deleterious, producing severe abnormalities or even death. The latter are known as **lethal mutations.** This chapter is concerned mainly with some of the causes of mutations in humans and with the various types of genetic disease which result.

SPONTANEOUS MUTATIONS

During the reproduction of DNA prior to cell division, mistakes or mutations sometimes occur spontaneously. We say "spontaneously" because we do not know whether all such mistakes are caused by some unknown environmental factor—such as chemicals found in the environment or natural background radiation—or whether they sometimes result merely from chance biochemical errors during chromosome duplication. The spontaneous mutation rate in man varies from one gene to another, ranging from once in every thousand to once in every 100,000 times a gene is reproduced. Although the exact number of genes in man is not known (estimates range from 25,000 to 200,000) it is thought that, on the average, each of us carries at least one or two mutations not present in our parents.

THE FREQUENCY OF LETHAL GENES

Lethal mutations which are dominant do not accumulate in the gene pool of a population because the individual dies before reproducing; death usually occurs even before the affected fetus is born. We are thus mainly concerned with recessive mutations, present in unsuspecting parents, which in a homozygous condition are debilitating or lethal to a child. Of the numerous genes in the chromosomes of one individual, about 1600 are known to be capable of deleterious recessive mutations. Thus, even though the spontaneous mutation rate per gene is low, on the average we each are heterozygous carriers of six to eight recessive lethal genes in our germ cells, mostly inherited from our parents.

Recessive lethal mutations usually have little overt effect on a heterozygous carrier, and hence they tend to accumulate in the population. They do not increase indefinitely, however. A certain number of mutations back to the normal gene also occur. Moreover, even in the heterozygous individual, recessive lethal mutations often have subtle effects, such as reducing fertility; they are not transmitted to the next generation at the same rate as the normal gene. Each recessive lethal mutation, then, exists at some equilibrium frequency, q, in the population; q values for lethal genes are mostly quite low, less than one per cent.

What are the chances that two carriers of the *same* lethal gene will marry? If the frequency, q, of the lethal gene in a population is, say .005 (0.5 per cent), then according to the Hardy-Weinberg Law the frequency of carriers, 2 pq, is $(2)(.995) \cdot (.005) = .0095$, or about one per cent. In a large community in which marriages are occurring randomly, the chance that two carriers will marry is thus quite low, about 1 in 10,000 (.01 times .01). But marriages between closely related persons, such as first cousins, greatly increase the chance that both parents will have inherited the same deleterious gene from a common grandparent. Many human societies long ago realized the genetic dangers of such inbreeding and established rules prohibiting intermarriage between closely related individuals.

It is also true that some cultural groups which share a common genetic heritage often have higher q values for certain lethal genes than are found in the general population. Since such people tend to intermarry for cultural reasons, the incidence of those specific genetic diseases tends to be higher among them. For example, Tay-Sachs disease — a particularly severe lethal disease of babies — is common among Ashkenazic Jews, those of northern European origin.

GENETIC DISEASES

In the following discussion we shall consider three genetic diseases of varying degrees of severity, each of which is of interest because of the way it is inherited: hemophilia, sickle-cell anemia and Tay-Sachs disease.

HEMOPHILIA

Hemophilia, like color blindness, is a sex-linked recessive trait located on the X-chromosome. It thus occurs almost exclusively in males and is passed from a carrier mother to her son. (Female hemophiliacs are extremely rare, the first authentic case being described in 1951.) The disease is characterized by excessive bleeding due to a defect in the blood-clotting mechanism; a blood globulin necessary for activation of thrombokinase is absent. Hemophiliacs are often severely handicapped because even small wounds fail to heal. Although hemophilia is not necessarily lethal in humans, it greatly reduces the chance of survival and prevents an active life. If affected males live to reproductive age, they can transmit the defective gene to their daughters (who then are carriers of the disease) but not to their sons.

The most famous family history of hemophilia occurred among the offspring of Queen Victoria of England; the royal children married into most reigning families of Europe (Fig. 15a.1). Since there was no indication of hemophilia among Victoria's ancestors, it is likely that either the queen

FIGURE 15a.1

Queen Victoria of England and her descendents, photographed in 1894 (see pedigree in Figure 15a.3): a family history of hemophilia.

2. Queen Victoria, a mutant carrier of hemophilia.

3. Her daughter Victoria, Empress of Germany, a possible carrier.

7. Her son Edward, later King of England, normal.
   (Her daughter Alice, a carrier, is not shown.)

9. Her son Arthur, normal.

8. Her daughter Beatrice, a carrier.

1. Her grandson Kaiser Wilhelm II of Germany, a normal son of Victoria.

6. Her granddaughter Irene, a carrier daughter of Alice; married Prince Henry of Prussia and bore two hemophiliac sons (not shown).
   (Her grandson Frederick William, a hemophiliac son of Alice, is not shown.)

4. Her granddaughter Alix, a carrier daughter of Alice; married Nicholas II, Tsar of Russia (standing to her right) and bore a hemophiliac heir, Alexis (not shown).
   (Her granddaughter Alice, a carrier daughter of Leopold, is not shown.)

5. Her granddaughter Victoria Eugenie, a carrier daughter of Beatrice; married Alfonso XIV, last King of Spain, and bore a hemophiliac heir.

10. Her granddaughter Marie, a possible carrier.

11. Her granddaughter Elizabeth, a possible carrier.
   (Her grandsons Leopold and Maurice, both hemophiliac sons of Beatrice, are not shown.)

(From The Manseli Collection. *In* Kogan, B. A. 1970. *Health: Man in a Changing Environment.* Harcourt, Brace & World, New York.)

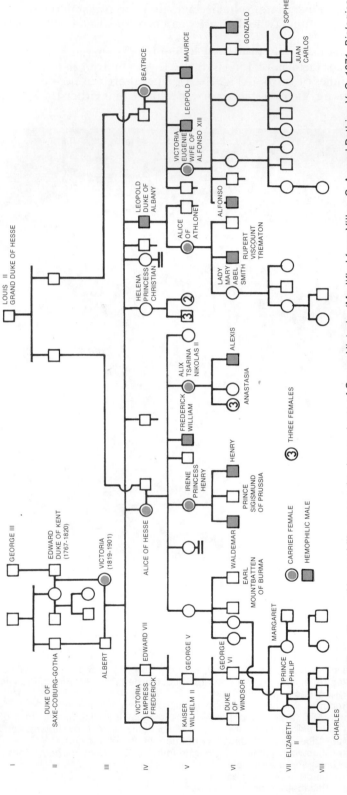

FIGURE 15a.2   The pedigree of hemophilia among the descendants of Queen Victoria. (Modified from Villee, C. A., and Dethier, V. G. 1971. *Biological Principles and Processes.* W. B. Saunders Co., Philadelphia.)

herself or her mother was a mutant. The pedigree of Victoria's family is shown in Figure 15a.2. Note that both the heir apparent of the last Tsar of Russia and the heir apparent of the last King of Spain were hemophiliacs. It is interesting to speculate on the role the disease may have played in shaping the fate of the royal houses of Europe early in this century.

SICKLE-CELL ANEMIA

This trait, unlike hemophilia, is not sex-linked. It is characterized by a tendency of the red blood cells to be fragile and to collapse into a sickle-shape (Fig. 15a.3); they lose their ability to transport oxygen, and severe anemia develops which usually results in death at an early age.

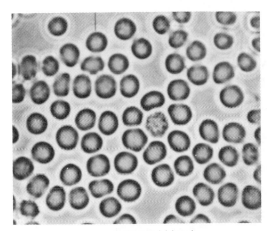

A Normal blood   B Sickle-cell anemia

Comparison of normal red blood cells (A) with those in the genetic disease, sickle-cell anemia (B).
    The shrunken cells are unable to carry oxygen efficiently, and early death is common. (A, courtesy of Dr. A. C. Allison; B, courtesy of Carolina Biological Supply Co.)

FIGURE 15a.3

The cause of sickle-cell anemia has been traced to a mutation in the gene controlling the synthesis of hemoglobin. Normal adult hemoglobin (Hb^A) consists of four globular polypeptide chains. There are two identical alpha (α) chains and two identical beta (β) chains (Fig. 15a.4A). The synthesis of each type of chain is controlled by a separate gene. In sickle-cell anemia, the β gene mutates to form another type of polypeptide chain, the S chain (Fig. 15a.4B), differing from the normal β chain in only a single

FIGURE 15a.4

Normal and sickle-cell hemoglobins.
    A. Normal adults (AA) are homozygous for the gene producing normal β chains, and have only Hb^A in their blood.
    B. Homozygotes (SS) for sickle-cell hemoglobin produce only S chains instead of β chains, and have only Hb^S in their blood. They suffer severe anemia and sickling of red blood cells that are usually lethal.
    Heterozygote carriers (SA) have about 55 per cent Hg^A and 45 per cent Hb^S Their blood has a reduced—but usually not lethal—oxygen-carrying capacity.

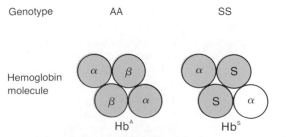

Genotype   AA   SS

Hemoglobin molecule

Hb^A Hb^S

A. Normal hemoglobin   B. Sickle-cell hemoglobin

amino acid—an example of how critical the precise amino acid sequence can be to the function of a protein. This hemoglobin, called Hb^S, is less able to carry oxygen than is Hb^A.

In persons homozygous for the abnormal gene (SS), no normal β subunits are produced; all hemoglobin is of the Hb^S type, and sickling is severe, often causing death, as we have noted. In heterozygous carriers (SA) both β and S genes function (an example of blended inheritance) SA individuals have about 55 per cent Hb^A and 45 per cent Hb^S. Under normal conditions, the oxygen-carrying capacity of the blood of SA persons is adequate, and their red blood cells show little or no tendency to sickle. But under stress, such persons are at a severe disadvantage; during heavy exercise or after moving to high altitudes—whenever oxygen is less available—their red blood cells tend to collapse. In fact, sickling can be induced in a drop of blood from an SA carrier by removing oxygen. This provides a rapid diagnostic test for carriers of the sickling trait.

One would thus expect that not only persons with sickle-cell anemia but also carriers of the gene would be at a reproductive disadvantage relative to persons homozygous for the normal allele, AA; if SA or SS individuals die early and therefore reproduce less often, then the frequency of the gene should be very low indeed. In most parts of the world, sickle-cell disease is extremely rare; but in some areas, notably central Africa, the incidence of SA heterozygotes may be as high as 40 per cent. In these regions, there must be a selective advantage to the S gene.

In comparing the worldwide occurrence of sickle-cell trait with that of malaria, the greatest killer among the infectious diseases, one is struck by the great similarity in their distributions (Fig. 15a.5). In fact, SA heterozygotes are about twice as resistant as AA homozygotes to severe infections of malaria. Thus, even though SS is lethal, SA confers a selective advantage in malarial areas.

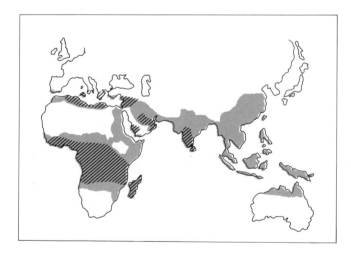

FIGURE 15a.5

Distribution of severe form of malaria (colored area) and of sickle-cell anemia in man (striped).

Persons of African descent in America still show a moderately high incidence of sickle-cell trait (SA genotype). It is estimated that intermarriage with other races over the years would naturally have decreased the frequency of SA carriers from an original level of about 22 per cent down to 15 per cent; the frequency actually found, however, is only 9 per cent. In the absence of malaria in the United States, there is no advantage to being an SA heterozygote—on the contrary, it is a disadvantage. Thus the S gene has been gradually disappearing during the 300 years since inhabitants of Africa were forcibly transported to this country; its decrease over several generations was probably hastened by the selective disadvantage to SA heterozygotes of forced labor, during which bodily demands for oxygen would have been high.

TAY-SACHS DISEASE

There are many genetic diseases which are fatal to infants or young children; among these are cystic fibrosis (a disease in which the sweat glands, salivary glands and certain other glands produce abnormal secretions), one type of muscular dystrophy (progressive degeneration of voluntary muscles), diabetes (high blood sugar due to absence of insulin) and Tay-Sachs disease (progressive brain deterioration). Diabetes can now be treated with insulin injections, and cystic fibrosis patients are sometimes kept alive by careful medical attention; but muscular dystrophy and Tay-Sachs disease are inevitably fatal.

All these diseases are due to the presence of a double dose of a particular defective recessive gene in the baby. Parents do not show any symptoms and are usually unaware that they might be carriers of a defective gene unless a relative has had the disease. Only after producing a normal-appearing, alert baby which gradually becomes more and more passive, retarded and eventually unable to move do many carriers of Tay-Sachs disease discover their genetic constitution (Fig. 15a.6).

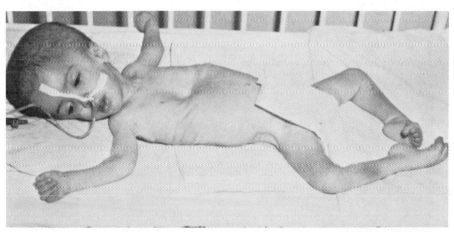

FIGURE 15a.6

A three year old child with Tay-Sachs disease. Enlarged head and atrophied muscles are characteristic. (From Volk, B. W. (ed.) 1964. *Tay-Sachs Disease.* By permission. Grune & Stratton, New York.)

The cause of Tay-Sachs disease is now known to be the deficiency of a certain enzyme which breaks down GM_2 ganglioside, a normal body chemical. This substance forms an important part of cell membranes, especially in the nervous system. Normally, more of this chemical is synthesized than can be used, and the excess is destroyed by the specific enzyme. Tay-Sachs babies do not possess the gene necessary to direct synthesis of this enzyme. In the total absence of the enzyme, GM_2 ganglioside accumulates in nerve cells, especially those in the brain, forming large concretions which gradually increase in size and destroy the neurons (Fig. 15a.7). Death usually occurs before the age of five years. Carrier parents produce less enzyme than normal but enough to prevent the disease.

Until recently, there was no means of counseling prospective parents who had a family history of Tay-Sachs disease as to whether they might be carriers, deficient in one of the two alleles controlling synthesis of the critical enzyme. In the past few years, however, it has been discovered that small amounts of this enzyme occur in the blood; normal homozygous people (EE) have four times more enzyme in their blood than do deficient heterozygotes (Ee). A simple blood test can identify carriers.

Even so, many couples who know they are carriers still wish to have children of their own, and since each child conceived has a 3:1 chance of being phenotypically normal, some parents are willing to take the risk. It is now possible, however, to eliminate producing a Tay-Sachs baby altogether.

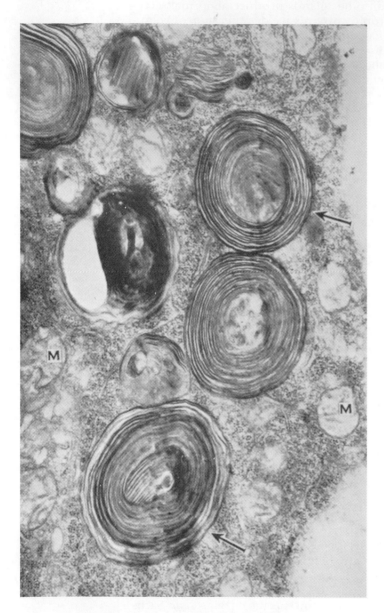

FIGURE 15a.7

Electronmicrograph of part of brain neuron from child dying of Tay-Sachs disease.

Cytoplasm is filled with concretions of GM_2 ganglioside which interfere with cell function (arrows). *M*, mitochondrion. (From Volk, B. W. (ed.) 1964. *Tay-Sachs Disease*. By permission. Grune & Stratton, New York.)

Surrounding the fetus there is a fluid-filled sac, the **amnion,** which contains a few suspended cells (Fig. 15a.8). These cells are genetically the same as those of the fetus, not the mother. By about the fourth month of pregnancy, the fetus and its surrounding membranes have grown sufficiently large that a small sample of the amniotic fluid may be removed through the mother's abdominal wall—a procedure known as **amniocentesis.** The cells it contains are cultured in little vials, where they grow and divide. When enough cells accumulate, they can be tested for presence or absence of the critical enzyme. This test takes several weeks to complete but allows sufficient time for a therapeutic abortion to be carried out within the legal time limit now established in many states, if the parents so desire. The chances are 2:1, however, that a healthy baby born to two heterozygous parents will also be a carrier.

Amniocentesis can also be used to detect Rh factor incompatibility between mother and fetus as well as several other genetic diseases. (The Rh factor, like other red blood cell antigens, is inherited; if the mother is Rh-negative—has no Rh-antigen—she will form antibodies against the Rh-positive red blood cells of a fetus who inhertied the gene from its father. The antibodies cause agglutination of the infant's red blood cells. This subject is discussed further in Chapter 22.) The search is now under way for further biochemical tests which will signal the presence of other lethal mutations early in pregnancy. Much unnecessary suffering of parents and children may thus be alleviated.

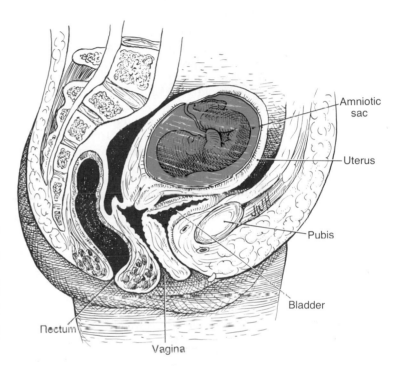

FIGURE 15a.8

Size and position of a four-month human embryo.

A sample of amniotic fluid, the cells of which have the same genetic constitution as the fetus, can be analyzed for certain genetic defects. If these are present, therapeutic abortion can be carried out early in pregnancy.

In the case of those genetic diseases where a biochemical deficiency is uncovered, it is hoped that the missing chemical can be replaced by injection, just as diabetes is now treated with insulin. Much research still remains to be done, however, since the specific causes of only a few of the 1600 genetic diseases which occur in man have yet been discovered.

RADIATION AS A MUTAGEN

For about half a century it has been known that X-rays and the energy emitted by the decay of radioactive elements can be dangerous to man, animals and plants. At extremely high doses of radiation, there is massive destruction of the central nervous system and death is instantaneous, but most of the effects of radiation are more gradual and usually are due, at least in part, to irreversible genetic damage to individual cells. Acute exposure to large quantities of radiation may cause severe body burns and may also result in radiation sickness, characterized by nausea, vomiting, destruction of the blood-forming centers and lesions of the intestine. It can lead to death. A single low exposure or prolonged chronic exposure may not produce immediate symptoms, but permanent damage to body cells can occur as the result of mutations. The spontaneous mutation rate is speeded up. Radiation is thus *mutagenic.*

When mutations occur in the germ cells of the gonads, they can be passed on to future generations and so accumulate in the population. Mutations produced anywhere in the body may result in cancer. If they interfere with the normal genetic controls within a cell which tell it how often to divide, rapid uncontrolled cell division follows. All three types of radiation effect—deaths from radiation sickness, an increase among individuals in mutations leading to cancer and an increase in genetic defects among children born shortly afterwards—were observed in the Japanese cities of Nagasaki and Hiroshima following the bombings of 1945.

TYPES OF RADIATION

Radiant energy falls into two broad categories—electromagnetic radiation, of which light is the most familiar example, and particulate radiation, produced mainly from disintegration of radioactive elements. As we shall see later in this chapter, it is the former which is of most concern to many of us today. First, however, let us briefly consider the nature of these two types of radiant energy.

Electromagnetic Radiation. Radio waves, visible light, ultraviolet light and X-rays are all related forms of energy—all belong to the *electromagnetic spectrum* (Fig. 15a.9). In fact, there is a continuous series, from the very long waves at the radio frequency end of the spectrum to the extremely short gamma rays at the other end. All are emitted in discrete energy "packets" called *photons,* as we have already described for visible light (for a review of wavelength, see Chapter 8). Gamma rays and X-rays are similar forms of radiation. Man-made X-ray machines are capable of producing a slightly broader range of radiant energy than is found among most naturally occurring X-rays and gamma rays. It will be noticed that the shorter the waves, the more energy they contain (photon energy is measured in units called electron volts). High energy rays, such as X-rays and gamma rays, and some low energy rays, such as microwaves, are capable of penetrating deep into living tissue, whereas ultraviolet light and visible light irradiate only the surface of the skin.

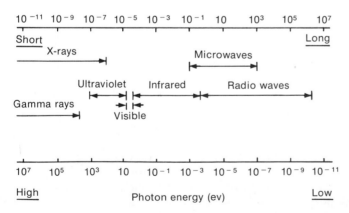

FIGURE 15a.9

The electromagnetic spectrum: wavelengths (in centimeters) and photon energies (in electron volts) of major types of electromagnetic radiation.

Note how narrow the region of visible light is compared to the total spectrum. The shorter the wavelength, the more energy per photon. (After Casarett, A. P. 1968. *Radiation Biology.* Prentice-Hall, Englewood Cliffs, New Jersey.)

Particulate Radiation. In addition to electromagnetic radiation, which may be thought of as infinitely tiny packets of vibrating energy, there is also *particulate radiation,* tiny particles of matter released from the disintegration of radioactive elements. Most elements found in nature are not radioactive. The protons and neutrons in the nuclei of their atoms are bound tightly together by attractive forces. Such nuclei remain stable indefinitely at the temperatures found on Earth. (For a review of the structure of atoms, see Appendix I.) Certain elements exist, however, in which the binding forces in the nucleus are insufficient and the nucleus is unstable or **radioactive.**

All of the heavy elements with more than 83 nuclear protons are radioactive (for example, uranium, thorium and radium), as well as some naturally occurring isotopes of the lighter elements which have unstable ratios of neutrons to protons (a common example is carbon 14). Artificial radioactive isotopes of most of the lighter elements have also been manufactured by man. From time to time these unstable nuclei disintegrate—the more unstable they are, the faster this happens. During **radioactive disintegration,** energy is emitted from the atom. Frequently this energy is in the form of tiny particles released from the nucleus with great force. Sometimes gamma rays are also emitted.

INTERACTION OF RADIANT ENERGY WITH MATTER

Whenever any radiant energy, be it electromagnetic or particulate, passes through matter, it collides with atoms and molecules in its path. In doing so, it gives up part of its energy to those atoms. Relatively low energy radiation, such as visible light or infrared radiation, has only a small effect; it causes the atoms or molecules to vibrate a bit faster, and so their temperature increases slightly. One thus feels warmed by the infrared rays of the sun.

High energy radiations, however, may have more penetrating effects on the matter they hit. Precisely how much effect is produced is determined not only by the energy and type of radiation but also by the nature of the matter through which it travels; a unit volume of air absorbs less energy than water, which in turn absorbs far less than lead. Sometimes chemical bonds are broken; sometimes free radicals are formed; sometimes an atom of one element is converted to another. Details of these interactions are numerous and complex and beyond the scope of our discussion. We shall consider only one example of an interaction of radiant energy with matter, namely the formation of ions, since this is especially important in living cells.

Ions can be produced by the interaction of any type of radiation with matter, *providing enough energy is present.* Let us consider the case of a diagnostic X-ray striking an oxygen atom in a cell (Fig. 15a.10). The initial X-ray photon strikes an orbital electron, ejecting it from the oxygen atom, which now has a deficiency of electrons and becomes an oxygen ion, O^+.

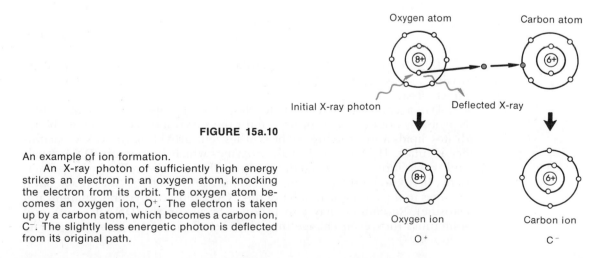

FIGURE 15a.10

An example of ion formation.

An X-ray photon of sufficiently high energy strikes an electron in an oxygen atom, knocking the electron from its orbit. The oxygen atom becomes an oxygen ion, O^+. The electron is taken up by a carbon atom, which becomes a carbon ion, C^-. The slightly less energetic photon is deflected from its original path.

The electron is then taken up by a nearby atom—in this case a carbon atom, which now becomes a carbon ion, C^-. In our example, the photon, having lost but part of its energy to the electron, is deflected and may repeat this process if it strikes an orbital electron in another atom. The ions thus formed, one positive and one negative, are highly energetic and unstable. Unlike ordinary salt ions found in cells, they readily react with one another or with other nearby atoms and molecules, causing further changes in them.

IONIZING RADIATIONS AND LIVING CELLS

The amount of damage done to cells by radiations capable of producing ion pairs (ionizing radiation) depends on how much energy is absorbed by a particular tissue. Bone, for example, is a dense tissue and hence will absorb more of the energy reaching it than will muscle, kidney or brain. Before it can be absorbed, however, the radiation must penetrate to the tissue. Particulate radiation from radioisotopes cannot penetrate deeply into matter: even the most energetic particles are absorbed by a pane of glass. Such radiation reaches only the outer tissues of the body unless the radioisotopes find their way inside through wounds or in food, water or air. Polonium 210, for example, may reach the lungs in tobacco smoke, and strontium 90 and iodine 131 are ingested in milk contaminated by fallout.

Gamma rays and X-rays, on the other hand, are able to penetrate tissues more easily, and the depth to which they penetrate depends upon their energy and the tissues they encounter. Low energy X-rays penetrate less than do high energy X-rays, which may pass clear through the body with little loss of energy. Paradoxically, however, low energy radiation gives up a greater proportion of its energy than does high energy radiation. For taking diagnostic X-rays of the skeleton, for example, an energy should be used which will be maximally absorbed by bone and minimally absorbed by soft tissues, thus giving the clearest X-ray with the least exposure.

Molecules Affected. Once radiant energy reaches a cell, how does it produce its damaging effects? This is a question for which we do not yet have a complete answer. It is believed that the main effects are the result of ion pair formation. Because water is the most common molecule within cells, most ion pairs formed are water ions: H_2O^+ and H_2O^-. These ions are extremely unstable, readily breaking down into other unstable entities called *free radicals*—namely, hydrogen radicals H·) and hydroxyl radicals (OH•). (For a description of free radicals, see Chapter 9.) These free radicals often recombine to give a molecule of ordinary water, but sometimes they react with other molecules in the cell, exchanging electrons and disrupting the normal molecular structure. This is particularly damaging in the case of large, functional macromolecules such as carbohydrates, proteins and nucleic acids. This interaction between free radicals produced from water and critical molecules within a cell is known as the *indirect effect* of radiation.

It is also possible for ionizing radiation to hit a macromolecule directly during its passage through a cell, causing it to ionize—this is known as a *direct effect.*

Damage to carbohydrate molecules, lipids and most enzyme molecules is probably not of great significance for the survival of a cell. New carbohydrates, lipids and proteins can be synthesized, providing the DNA machinery is intact. If DNA itself or the enzymes which function in DNA repair and duplication are damaged, then the effect on the cell may be permanent. It is known that ionizing radiations are capable of causing chemical changes in the all-important bases of DNA (adenine, thymine, guanine and cytosine). Radiations may also produce breaks in the DNA threads, and sometimes they even change the structure of an entire chromosome (Fig.

15a.11). Once such a change has occurred the cell is seldom able to correct it.

Severe genetic damage frequently causes death of a cell, which may then be replaced by division of an unaffected neighboring cell. But sometimes a damaged cell does not die but remains permanently changed—it has undergone a mutation, which will be inherited by all its daughter cells. The mutant cell may become malignant (cancerous). If the affected cell is a germ cell, the mutation may be transmitted to future generations.

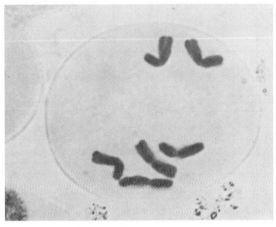 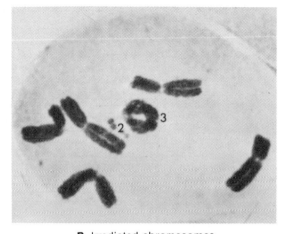

A. Normal chromosomes   **B.** Irradiated chromosomes

Chromosomes of the microspores of the spiderwort, *Tradescantia*.   **FIGURE 15a.11**
    A. Normal chromosomes.
    B. Abnormal chromosomes, after irradiation. Note abnormal ring (3) and fragments (2).
    (Courtesy of Dr. Henry E. Luippold.)

RADIATION DOSAGE

As we have seen, not all types of radiation produce similar effects. The end result depends on many factors, of which the most important are the nature and energy of the radiant particle or ray and the type of tissue with which it interacts. Health physicists—those concerned with the biological effects of radiation—have established a unit that takes into account the biological effectiveness of radiant energy in man; it is known as the **rem,** or *r*oentgen *e*quivalent for *m*an. A roentgen tells us only how much radiation there is and not how it is absorbed or how it affects man. A rem, however, indicates a standard amount of biological damage resulting from various sources. This is important, since it is believed that the effects of all types of radiation accumulate over one's entire lifetime.

"SAFE" LEVELS OF RADIATION

How much radiation may a person safely receive? This is a very difficult question to answer, since many factors are involved. One of these is the rate at which radiation is received. For example, a *single* (acute) exposure to 100 rem of X-rays will cause moderate radiation sickness in about 60 per cent of people, whereas the same dose spread over an entire lifetime does not produce illness, but will probably shorten one's life by about one per cent (or 0.7 year, on the average) (colored arrows in Figure 15a.12). Some effects of radiation, such as those on mutation rates and on life span, are believed to be linear with dose—there is probably *no lower limit* below which damage does not occur (straight lines in Figure 15a.12). It is difficult to obtain proof of this, however, since large amounts of statistical data are required. Other effects, such as radiation sickness, only occur *above a threshold* dose (curved lines in Figure 15a.12).

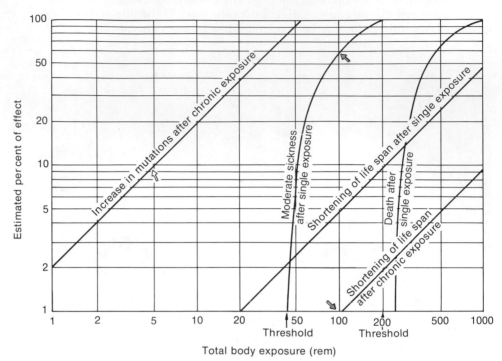

Total body exposure (rem)

FIGURE 15a.12   Relation between radiation dose and per cent of effects of radiation. (The double log scale is used so that the curves can be conveniently placed on a small piece of paper.)
Mutations and life span shortening appear to be linearly related to cumulative dose; radiation sickness and death after a single exposure have threshold dose levels. (After Morgan, K. Z. 1971. Health physics tolerance levels. Health Physics Division, Oak Ridge National Laboratory, Oak Ridge, Tennessee.)

The lines drawn in Figure 15a.12 are only estimates—we do not yet have enough information on man to be more precise. Considerable disagreement exists among scientists as to the accuracy of these values. Some believe the dangers of increased mutations to be far greater than shown in our figure. The estimates shown, however, have served as guidelines for several national and international agencies in setting limits to the amount of man-made radiation which should be permitted in the environment. The permissible limit currently set by these agencies states that the population *as a whole* should not receive more than 170 mrem per person per year (mrem = millirem, or 1/1000 rem). Over 30 years, this amounts to five rem. The limits also state that no individual should receive more than 500 mrem per year. This allows for variations in exposure among the population to achieve the 170 mrem average for the whole population. Note that these limits take no account of medical exposure! We shall return to this point shortly.

Study of Figure 15a.12 shows that a dose of five rem would not be expected to affect the average life span significantly, but it would increase the mutation rate in the population by about 10 per cent (open arrow). This would in turn increase the incidence of radiation-induced cancer and genetic disease. In concrete terms, it is estimated that for each rem accumulated on the average in the whole population, 40 new cases of fatal cancer per million people will occur, and about 20 babies per million born in the next generation will have new genetic mutations. These estimates are from the International Commission on Radiological Protection, Publication 8 (1966), *The Evaluation of Risks from Radiation*, and Publication 14 (1969) *Radiosensitivity and Spatial Distribution of Dose* (Pergamon Press, New York). The mutations mentioned include cleft palate, harelip, sex-linked recessive traits (e.g., hemophilia) and diseases involving too many or too few chromosomes, such as mongolism (Down's syndrome). After irradiation, chromosomes often become "sticky" and fail to separate cleanly during meiosis, giving rise to gametes with abnormal numbers of chromosomes.

EXISTING LEVELS OF RADIATION

There are three major sources of radiation to which people are exposed to varying degrees: natural or background radiation; fallout from bomb tests, power plants and industry; and medical and dental exposure. Only the last two are man-made. Let us consider each of these sources in turn.

NATURAL RADIATION

We are all exposed to natural sources of radiation over which we have almost no control; this has been true since life began. Such radiation is probably the cause of at least some spontaneous mutations. The average background radiation throughout the world is about 100 to 110 mrem per year but is considerably higher in some areas. It comes from two main sources, cosmic and terrestrial.

Cosmic rays are a mixture of high energy radiations emitted by stars and the sun, some of which easily pass through the atmosphere to the Earth's surface. They account for exposures of about 30 mrem per person per year, but this is greater at high latitudes and high elevations. People living in Denver receive about twice as much cosmic radiation as people living at sea level.

The Earth's crust contains various natural radioactive elements which provide another 70 mrem per year exposure on the average. Potassium 40 and carbon 14 exposure are received mainly through ingestion of water and food. Other elements found in the soil—radium, polonium and thorium—emit high energy radiation which penetrates the body surface or enters in drinking water. These elements are not uniformly distributed on the Earth's surface. In the midwestern United States, elevated levels of radium in drinking water result in two to three times the amount normally found in teeth. In the arctic, polonium is found in lichens, which are eaten by reindeer and caribou, the main food of Eskimos; polonium levels in Eskimos are some 10 times greater than in people living elsewhere.

MAN-MADE FALLOUT

Dispersal of man-made radioactive elements in the environment results from two sources: nuclear test explosions and industry. For people living in the United States, these made an *average* per capita contribution in 1970 of about 2 mrem per year and 1 mrem per year, respectively—very small amounts when compared with the 100 mrem per year from natural sources.

Fallout, however, is not uniform over the Earth's surface; air currents and precipitation patterns cause far greater amounts of fallout in the arctic, and once again the Eskimos face a much higher risk than do other people. Another factor is the nature of the radioactive elements released during nuclear reactions. Of particular importance are strontium 90 (^{90}Sr) and iodine 131 (^{131}I), which are both concentrated in milk. Our greatest milk drinkers are children; hence they face the greatest risk. Iodine 131 accumulates in the thyroid gland, where it may produce tumors; ^{90}Sr accumulates in bone, causing leukemia and bone cancer. Although ^{131}I rapidly decays and disappears from the environment, ^{90}Sr decays very slowly—28 years are required for half of it to disappear. It will be 200 years before what has accumulated in the environment has decayed to a barely detectable level.

As long as atmospheric nuclear testing remains at a minimum (France and China are the only two countries which have not yet accepted the moratorium), fallout from this source is unlikely to significantly affect the world's population as a whole, although small areas may still be at considerable risk. Hopefully this source will soon disappear altogether.

Nuclear power plants at present release negligible amounts of radioactivity compared to natural background radiation. It is, however, recommended that future plants be sited, as are today's, far from urban centers, since there is always the remote chance of a nuclear accident occurring. This is especially important for breeder reactors, the next type proposed to be built (see Chapter 7a). These reactors will create more radioactive matter than they consume, and this matter will consist of more dangerous radioactive isotopes than are produced by other types of reactors.

MEDICAL EXPOSURE

Radiation is used in medicine for two purposes: diagnosis and therapy. For example, it is used diagnostically to X-ray teeth, bone, stomach, lungs; or to assay thyroid gland activity with ^{131}I. It is used therapeutically to treat cancer. In the latter instance, treatment with radium needles or massive doses of X-rays is undertaken to kill cancer cells and to save or prolong the life of the individual patient; the attendant risks of causing another cancer or of inducing radiation sickness are known by both doctor and patient, and the decision is made between risk and benefit to a single individual.

In the diagnostic use of X-rays, however, the distinction between benefits and risks to the patient is less clear-cut; yet this is by far the largest source of exposure to man-made radiation for the average person living in advanced countries. In our discussion we shall limit ourselves to doses to the gonads, since such exposures are of greatest importance in assessing

Population	Dose Equivalent(mrem/yr)
Buenos Aires, Argentina	37
Denmark (1956)	22
Hamburg, Germany	17
France	58
Rome, Italy	43
Japan (1960)	39
Leiden, Netherlands	6.8
New Zealand (1963)	12
Norway	10
Sweden (1955)	38
Switzerland	22
Alexandria, U.A.R.	7
Cairo, U.A.R.	7
United Kingdom (1957)	14
United States (1964)	55

TABLE 15a.1 GENETICALLY SIGNIFICANT DOSE TO PEOPLE IN VARIOUS ADVANCED COUNTRIES FROM MEDICAL DIAGNOSTIC X-RAYS*

*It is estimated that the average dose to most organs of the body is two to three times the genetically significant dose. (From Morgan, K. Z. 1971. Health physics tolerance levels.)

permanent damage, resulting from mutations, to the population as a whole. A consideration of the chances of producing cancer as a result of various types of X-ray diagnosis, while of great importance to the individual, would require far more space than is here available.

The average exposure to the gonads of people of reproductive age living in the United States in the year 1964 was 55 mrem. This is higher than for any other advanced country in the world except France (Table 15a.1) and is already half that received from natural background! Moreover, the dose varies enormously from one clinical facility to another (Table 15a.2). Many people receive far more than "average" doses.

TABLE 15a.2
RANGE
OF GONAD DOSE
FOR VARIOUS TYPES OF
DIAGNOSTIC
X-RAYS*

Region Examined	Gonad Dose (mrem/X-ray)	
	Male	Female
Chest	0.1–6	0.1–15
Stomach	3–123	6–1108
Kidney	423–3700	403–1940
Large intestine	25–1310	20–2530
Spine (lower)	16–767	47–700

*From Casarett, A. P. 1968. *Radiation Biology.* Prentice-Hall, Inc., Englewood Cliffs, N.J., p. 330.

Is this amount of exposure really necessary? The answer is "no"! Dr. Karl Morgan, Director of the Health Physics Division of the Oak Ridge National Laboratory, has pointed out many ways in which exposure could be decreased tenfold without in any way curtailing the number and quality of X-rays taken. Five of the most important follow:

1) Require licensing of all X-ray technicians. At present, only New York, New Jersey and California require that a technician be certified. In other states, anyone can own and operate an X-ray machine, regardless of his level of training!

2) Update all X-ray machines so that a narrow, well directed beam is used (Fig. 15a.13). Many machines irradiate a large fraction of the patient when only a small area need be exposed. Mobile chest X-ray

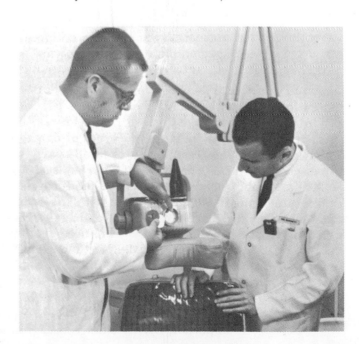

FIGURE 15a.13

Insertion of a lead collimator into a dental X-ray machine.

The device directs X-rays only at the film, greatly cutting down exposure to other tissues, such as the thyroid gland. (Bureau of Radiological Health – Public Health Service.)

units are especially remiss. Persons waiting in line may receive almost as much exposure to X-rays as they do in front of the machine!

3) Shield those areas of the patient which do not need to be exposed with lead aprons or frames.

4) Use faster X-ray film so lower doses are required to obtain the X-ray.

5) Record exposure of individual patients in a permanent record. At present, a doctor has no idea of how many X-rays his patient has already had.

APPORTIONING THE PERMISSIBLE MAN-MADE DOSE

If we agree—and not all scientists do—that 170 mrem per person per year is an acceptable increase over background radiation for the population at large, how shall we apportion that dose? *At present, it does not even include medical X-rays,* which Dr. Morgan and others say emphatically should be included. Nor does it include exposure from non-ionizing radiations such as those from lasers, microwave ovens, radar and so forth. At the moment, the risks from these to the average person are probably small; few people as yet own microwave ovens or are exposed to lasers or radar transmitters. But as more beneficial uses for various types of radiation are found and these become more widespread, some limitations will have to be imposed. Figure 15a.14 provides a summary of the various benefits and risks currently to be derived from the use of radiant energy.

Man-made radiation has provided many benefits for man. It also has been and is now being frequently misused. One college in the United States X-rays the throats of its voice students weekly to determine if they are using their vocal cords properly! In the near future, as the 170 mrem limit is approached, decisions must be made, by the public or their elected representatives, as to which benefits are to take precedence—medical use, nuclear power, entertainment—in utilizing the limited amount of radiant energy the risks of which we are willing to accept.

SUMMARY

Mutations, or permanent changes in the genetic information, are only seldom advantageous and are sometimes lethal. Spontaneous mutations occur with low but measurable frequencies and eventually reach an equilibrium level in the population. A small proportion of human genes are capable of recessive mutations which are highly deleterious or lethal in the homozygous condition, and we each carry six to eight of these.

For only a small fraction of the 1600 identified genetic diseases do we have a biochemical explanation; some, such as diabetes, can be treated, while others, like Tay-Sachs disease, cannot. Sex-linked diseases, such as hemophilia, are almost entirely restricted to males. Other diseases, like sickle-cell anemia, are prevalent in certain areas because the mutation has selective advantage in a particular environment—in this case it confers resistance to malaria. Tay-Sachs disease is one in which a critical enzyme is missing. Although incurable, it is now possible to detect potential carriers by simple blood tests and to diagnose the presence of the disease in an unborn fetus by amniocentesis; this may be followed by therapeutic abortion if desired.

The incidence of genetic diseases can be increased by mutagenic agents, of which radiation is an important example. Formation of ion pairs by radiant energy causes disruption of macromolecules, including the all-important DNA. Average background radiation in the United States is

BENEFITS

Good Health—Entertainment—Employment—
Modern Conveniences—Affluent Society

Medical X-Ray
Treatment of diseases
Medical diagnosis

Color Television
Educational programs
News and political information
Entertainment

Nuclear Power
Cheap electricity
Reduction in air pollution

Industrial X-Ray
Locate metallurgical flaws (prevent accidents)

Microwave Ovens
Safer and quicker preparation of food

Ultraviolet Radiation
Destroy bacteria in operating rooms

Radar
Improved and safer air transportation

Laser
Improved communications

Radioisotopes
Medical diagnosis and treatment,
    power source for heart pump, etc.

RISKS

Sickness and Suffering—Deformities and Physical
Handicaps—Life Shortening and Early Death

Cancer
Leukemia
Central nervous system cancer
Bone Tumor
Thyroid cancer
Lung carcinoma

Eye Cataract

Life Shortening

Damage to Unborn Children
Mongoloidism
Microcephaly (abnormal
    smallness of head)
Various forms of cancer

Genetic Mutations
Fetal and infant deaths
Deformities (physical
    and mental)

The benefits and risks of radiation.   **FIGURE 15a.14**
    The public (or its elected representatives) must decide how much total risk it wishes
to take and how this risk should be divided among the various benefits. The decision will
become ever more difficult as more potential benefits are discovered. (From Morgan, K. Z.
1971. Environment 13(1):34.)

about 100 mrem per year and probably accounts for a fraction of spontaneous mutations. On the assumption that any increase in radiation will increase mutation rates, a permissible limit to man-made radiation exposure to the average population has been arbitrarily set at 170 mrem per year. Paradoxically this limit does not now include medical exposure, which presently is 55 mrem per average person per year—far greater than fallout from bomb testing, power plants and industry combined. This amount could be drastically reduced by increased training of X-ray technicians and technologists and by updating of X-ray equipment throughout the country without any decrease in number and quality of X-rays taken. There is also great scope for legal prevention of current misuse of X-rays.

Radiant energy has the potential for greatly benefiting man while also giving rise to calculated health risks. The public must decide what total risk it will accept and which of the alternative benefits are to be derived from the use of radiation.

READINGS AND REFERENCES

Aaronson, T. 1970. Mystery. Environment *12*(4):2–10.

Aaronson, T. 1970. Out of the frying pan. Environment *12*(5):26–31. Two articles pointing out unknowns in health safety factors in the use of microwaves for radar, cooking and so forth.

Allison, A. C. 1956. Sickle cells and evolution. Scientific American Offprint No. 1065.

Allison, A. C. 1964. Polymorphism and natural selection in human populations. Cold Spring Harbor Symposia in Quantitative Biology *29*:137–147. A lay article and a more technical discussion of the relation between sickling trait and resistance to malaria.

Boffey, P. M. 1971. Radiation standards: Are the right people making decisions? Science *171*:780–783. An excellent analysis and summary of current debate.

Casarett, A. P. 1968. *Radiation Biology*. Prentice-Hall, Englewood Cliffs, N.J. A clearly written textbook for those interested in more detail on the subject.

Friedmann, T. 1971. Prenatal diagnosis of genetic disease. Scientific American Offprint No. 1234. Describes techniques of amniocentesis and biochemical screening.

Gofman, J. W. and Tamplin, A. R. 1970. *Population Control Through Nuclear Pollution*. Nelson-Hall, Chicago. Two radiation biologists who are among the most ardent critics of our present safeguards over radiation present their opinions.

Holcomb, R. W. 1970. Radiation risk: a scientific problem? Science *167*:853–855. Discusses difficulties in scientifically predicting population effects of low doses of radiation.

Morgan, K. Z. 1971. Health physics tolerance levels. A paper presented by the Director of the Health Physics Division, Oak Ridge National Laboratory, Oak Ridge, Tennessee in January, 1971 to a symposium, Environmental Aspects of Radiation Sciences, at Oak Ridge Associated University.

Morgan, K. Z. 1971. Never do harm. Environment *13*(1):28–38. Two articles by a foremost U.S. authority on health physics, clearly and honestly presenting the present hazards from radiation, especially medical diagnostic misuse.

Nelson, B. 1971. Mobile TB X-ray units: An obsolete technology lingers. Science *174*:1114–1115. Interesting social reasons exist for retention of these outmoded and dangerous units.

Volk, B. W. (ed.) 1964. *Tay-Sachs Disease*. Grune & Stratton, New York. Details the cause and detection of this inherited disease.

Wiesenfeld, S. L. 1967. Sickle-cell trait in human biological and cultural evolution. Science *157*:1134–1140. Considers sociological aspects of inheritance of a single trait.

16

AN INTRODUCTION TO ECOSYSTEMS

The science of ecology deals with the total pattern of complex interactions among populations of organisms and between them and their environment—that is, with the study of ecosystems.

So far in this book we have focused our attention mainly on individual organisms: how they are constructed, how they obtain and utilize nutrients, how they maintain themselves and reproduce. We have also examined the principles of inheritance, noted the factors leading to variability among offspring and considered the genetics of populations. Logically we should next look at the importance of this genetic variability in determining which offspring in a population will survive and what types of organisms one might thus expect to find on Earth. In other words, we should concern ourselves with the process of evolution. But although our study of the internal workings of an organism allows us to comprehend why some inherited traits are of value, we still lack information about environmental factors to which an organism must be genetically adapted if it is to survive. To understand the importance of such factors we must have some knowledge of the totality of the environment in which an organism exists, both in its physical and biological aspects.

In Chapter 2, we touched upon the more important physical factors in the major environments of the world. In this chapter, we shall present a broad view of environmental interactions and, in later chapters, consider specific aspects of these interactions. Thus armed, we shall eventually tackle the problem of evolution itself and consider what role man may play in this process.

ECOLOGY AND ECOSYSTEMS

"Ecology" has become a household word frequently used in connection with antilitter and recycling campaigns. It is often misused by politicians and others when they really mean to say "environment." **Ecology** is a science; the word is derived from the Greek words *oikos*, "house," and *logos*, "word" or "discourse." Each derivative, however, is used in a broad sense — *oikos* connoting habitat or environment, and *logos*, doctrine or science; hence the science of the environment. For a given organism, the environment consists of all the surrounding physical and biological factors with which it interacts. In theory the total environment of each of us is the world, itself — the **biosphere.** In practice, we recognize subunits of the biosphere, which we call **ecosystems,** each of which has certain physical and biological characteristics.

PHYSICAL FACTORS

Among the main physical components of an ecosystem are such climatic factors as solar radiation, temperature and rainfall and such chemical factors as acidity, salinity and the availability of inorganic nutrients needed by plants. In some environments, other factors are of importance. In the ocean, for example, tides, currents, upwelling, and silt and freshwater run-off from land all play a role. On land geological and geographical features strongly influence the abundance and distribution of organisms. Geologically determined soil characteristics, such as acidity, porosity, nutrient supply and drainage, in turn affect the types of plants that will thrive in an area. Geographical features, such as mountains, rivers, valleys and plains influence the patterns of climate and weather, the temperature and the amount of light and rainfall in a region.

We have not space to enumerate all the physical factors found in the many types of ecosystems. It is important to realize, however, that many physical factors are to some degree affected by the living organisms with which they interact. Soil quality is a good example of such an interaction between physical and biological factors. The chemistry and structure of the parent soil determines which plants and soil organisms can live in it in the first place. Plant roots, in turn, hold the soil in place, slowing down erosion processes. Humus in the soil, formed from dead organisms by soil bacteria, fungi and invertebrates, retains nutrients that would otherwise wash away and thus provides a steady supply of these nutrients to the roots of plants. Soils which do not support vegetation, including those despoiled by man, become subject to rapid erosion.

BIOLOGICAL FACTORS

The biological factors in an ecosystem are the totality of all living organisms and their organic by-products. This includes not only the larger plants and animals that are readily observed, but all the microscopic organisms as well — bacteria in the soil or sediments, in the digestive tracts of animals or infecting plants and animals; plankton that live in water; viruses that cause disease; fungi that live in the soil or on decaying organisms; mites that feed on bacteria and fungi, and so on.

Each kind of living organism found in an ecosystem is called a **species.** We can anticipate our discussion of this term in Chapter 21 by saying that a species comprises all those organisms which are sufficiently alike genetically so as to make them distinguishable from all other types of organisms. Within an ecosystem, there usually exists a wide variety of species, all of which interact directly or indirectly with each other.

All individuals belonging to a particular species in a given ecosystem form a **population.** We thus speak of the population of red-legged frogs in a pond, of the population of the blue-green alga, *Anabena,* in a eutrophied lake or of the population of herring in the North Sea. Within an ecosystem there is a great number of populations of different species, all interacting with one another as a **community** and with the physical environment as well (Fig. 16.1). A very complex picture indeed!

FIGURE 16.1

A community of organisms in the African savanna. The tropical environment, characterized by a prolonged dry season, supports various species of grasses, scattered deciduous trees, and palms near water holes. Shown in addition are zebras and wildebeests,
two of the many herbivores of the savanna. Add also, in your imagination, the lions and
jackals, the burrowing rodents, the birds and insects, the soil and water micro-organisms
and the plant and animal parasites to form a more complete picture of the total ecosystem.
(Photo courtesy of Herbert Lang, from Villee, C. A., and Dethier, V. G. 1971. *Biological
Principles and Processes.* W. B. Saunders Co., Philadelphia.)

At this point let us briefly consider a concept about populations which
we shall explore in greater detail in Chapter 19—namely, the regulation of
population size. Organisms generally tend to produce more offspring than
actually grow up and become reproductive themselves. A female salmon or
frog lays many eggs, but in a stable situation on the average only two of
these grow up to replace the parents. Each species has a certain **reproductive
potential**—a maximum number of offspring to which it can give birth under
optimum conditions. There are, however, many reasons why an offspring
may die before reaching adulthood. It may be killed by a frost, it may starve
or be eaten by a predator or it may die from disease. Taken together, these
external factors causing death are called the **environmental resistance.** The
actual size of a population at a given moment will be determined by the balance between its reproductive potential and the environmental resistance.
To fully describe the working of even the simplest ecosystem, the ecologist
must assess the relative importance of all the factors acting on each of the
populations within it—a formidable task, and one which has yet to be accomplished, even for the simplest ecosystems occurring in nature.

THE STRUCTURE OF ECOSYSTEMS

Since all organisms contain organic molecules and, with the exception
of a few bacteria, only the photosynthetic plants are capable of synthesizing
these from inorganic matter, every ecosystem ultimately depends on green
plants for its existence. These photosynthetic autotrophs, described in
Chapters 8 and 9, are thus the primary producers in an ecosystem. They
utilize energy from solar radiation for synthesis of organic molecules from
carbon dioxide, water and nutrient minerals. Part of the energy absorbed
by plants is used for their own maintenance, the rest going into the synthesis of plant tissue or **biomass.** The resulting increase in plant biomass is

often referred to as "net photosynthesis"—that which is available for support of the rest of the ecosystem.

Plant biomass serves as food for **herbivores**—animals feeding on plants. Herbivores are thus the primary (first-level) consumers. They, in turn, become food for **carnivores,** who are secondary (second-level) consumers. Each of these steps in the transfer of the original plant biomass through an ecosystem is known as a **trophic,** or feeding, level. This type of feeding sequence forms a food chain, or more accurately, a **food web,** since omnivorous species, such as man, may feed upon anything from plants to top-level carnivores.

One last category in the food web requires our attention—namely, the **decomposers.** Plants lose leaves or die; animals excrete wastes and, on dying, become carcasses. Eventually all these dead materials, if not eaten by scavengers, are decomposed, their organic matter being converted back into organic nutrients. The major decomposers are bacteria and fungi living in sediments or soils where organic debris accumulates. A summary of these trophic levels is given in Figure 16.2.

A study of Figure 16.2 reveals that the basic elements of living systems, carbon, oxygen, nitrogen, phosphorus and so forth, are originally obtained in the form of minerals (CO_2, H_2O, NO_3^-, $PO_4^\equiv$) by plants, They then pass from one trophic level to the next in the form of organic molecules. Eventually the decomposers return them, as mineral nutrients, to the air, water and soil, to be used again by plants. Mineral nutrients are thus **recycled** over and over again within an ecosystem. This important process is considered in detail in Chapter 18.

As we saw in our study of nutrition (Chapter 9) all organisms utilize food for two purposes: (1) to maintain themselves and (2) for growth of new tissue. Both processes utilize energy made available from the splitting of ATP. As noted in Chapter 8, the overall process of respiration during which ATP is synthesized is only slightly less than 50 per cent efficient. Energy not captured in ATP molecules is lost as metabolic heat.

The original energy source for the entire ecosystem is the sun. As shown in Figure 16.2, a fraction of this solar energy, initially captured during photosynthesis, is dissipated at each trophic level as waste metabolic heat. Note that during the final conversion of organic matter to mineral nutrients by decomposers, the last of the chemical energy stored in the organic molecules is converted to heat. Energy thus flows *only once* through an ecosystem. The important consequences of this are considered in Chapter 17. Here we need only stress the importance of green plants, the primary producers, in the functioning of any ecosystem. Let us next consider a few concrete examples of ecosystems.

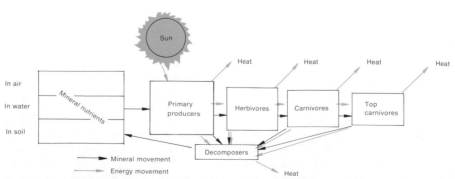

FIGURE 16.2

A summary of energy and nutrient flow through an ecosystem. Energy is gradually lost as heat, owing to respiration by organisms at each trophic level, whereas nutrients are recycled.

TYPES OF ECOSYSTEMS

In Chapter 2 we saw that there are four major environmental categories—marine, brackish, freshwater and terrestrial. Each of these may be further subdivided into smaller categories. For example, a freshwater environment may be a large lake, a pond, a river or a stream. Each subtype has a typical set of physical and biological features to distinguish it and thus is a discrete ecosystem.

To delimit the boundaries of an ecosystem is somewhat arbitrary, however, for each region or area interacts with neighboring regions. The edge of a lake, for example, often is taken as a convenient boundary; yet the lake receives water containing dissolved nutrients, bacteria and other organisms from the surrounding land, and it exports water downstream which again contains inorganic and organic matter.

TERRESTRIAL BIOMES

We have discussed in Chapter 2a the nature of one freshwater environment—a lake—and shall consider later (Chapter 17a) the nature of the marine environment. In this chapter, therefore, let us examine some of the types of terrestrial environments, especially those characteristic of North America.

The main subdivisions of the terrestrial environment are known as **biomes.** Each biome is characterized by the major type of vegetation it will support *under stable conditions*, known as the **climax vegetation.** Obviously, after a flood, a fire, or interference by man, conditions are temporarily changed in an area. The vegetation it then supports will be different from the climax type, but will gradually change toward it in a sequence of **successional stages**. An example of ecological succession is given later in the chapter.

The life form of the vegetation in a particular biome reflects the major features of soil and climate. The type of vegetation, in turn, determines the kinds of habitats available to animals within the biome and thus what types of animals will be found there. A biome, then, is a major type of ecosystem characterized by a typical community of plants and animals. The most important biomes of the world are shown in Figure 16.3. Their boundaries are determined by several factors, some of which, such as latitude and topography, are apparent in the figure. A map of world soil types, superimposed on the biome map, would show a quite similar pattern, underlining the role of soil quality in the distribution of vegetation. Climatic factors are also of great importance.

Tundra. In the most extreme latitudes, or above timberline in high mountains, one finds the characteristic arctic-alpine biome known as **tundra** (Fig. 16.4). Cold air and acid soils are primary physical characteristics. The climax vegetation consists mainly of perennial herbaceous plants, dwarf shrubs and trees, and lichens (reindeer moss). In the arctic, the ground is permanently frozen all year round. Only the top few inches thaw in summer, producing numerous ponds and bogs in low-lying areas and a rather water-logged soil. This year-round "permafrost" precludes the growth of the large root systems required to support trees of substantial size. The growing season in the arctic is limited to about two months. Relatively few species of animals thrive in the tundra—mosquitoes and midges abound in the ponds, small herbivores such as lemmings (which are rodents) and ptarmigans (grouse-like birds) are common, and a number of predatory birds and a few large mammals, such as reindeer or caribou, musk ox, arctic hare and arctic fox, are found. Migratory summer birds are also common. Because the tundra is relatively simple compared to other ecosystems it has been intensively studied by ecologists.

Boreal Coniferous Forests. The "Great North Woods" or **taiga** covers much of central Canada and extends into the mountainous regions of the

United States (Fig. 16.5). Cool climates, acid soils and plentiful rainfall are typical. The climax vegetation consists mainly of evergreen conifers, such as fir, spruce and pine, with some deciduous aspen and birch intermingled. Because the leaves of these coniferous trees are retained all year, most of the forest floor is constantly shaded and the growth of "understory" shrubs is consequently limited. Ground litter tends to accumulate. The number of animals here is far greater than in the tundra, there being many species of birds and mammals and numerous small, soil-dwelling animals. Outbreaks of insect pests are frequent, but the forests regularly recover under natural conditions.

In the taiga, more water falls as rain than is lost by evapotranspiration from the leaves and soil. The resulting heavy run-off of surface and ground waters carries mineral nutrients out of the ecosystem (Fig. 16.9A). The nutrient-poor acid soil restricts the climax vegetation to those plant species that can withstand such conditions—namely, the conifers.

Temperate Rain Forests. Along the northwestern coasts of America, from Alaska to northern California, are found the temperate equivalent of tropical rain forests. Rainfall and fog in these areas combine to produce a high degree of humidity all year around. The Olympic Peninsula near Puget Sound is famed for its thick stands of hemlock and fir, draped with mosses, and for a rich understory vegetation (Fig. 16.6). Farther south, the Douglas fir and magnificent redwoods characterize this biome.

Temperate Deciduous Forests. In still lower latitudes of North America, especially in the eastern United States, one finds forests composed mainly of such deciduous trees as maple, beech and oak (Fig. 16.7). The climate is moderate, with cold winters and more or less uniform rainfall throughout the year. The understory is generally well developed, with

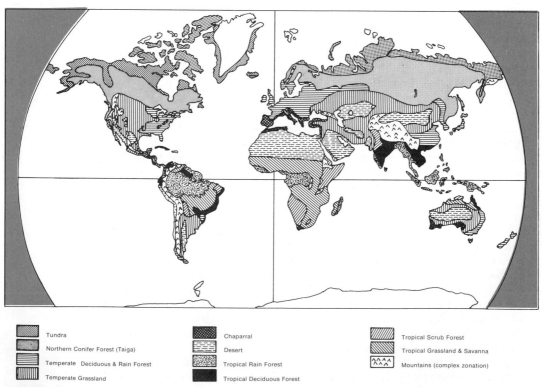

Tundra	Chaparral	Tropical Scrub Forest
Northern Conifer Forest (Taiga)	Desert	Tropical Grassland & Savanna
Temperate Deciduous & Rain Forest	Tropical Rain Forest	Mountains (complex zonation)
Temperate Grassland	Tropical Deciduous Forest	

FIGURE 16.3   Map of the major biomes of the world. Note that the boundaries are to some extent determined by latitude and topography. Other important factors are climate and soil. (From Odum, E. P. 1971. *Fundamentals of Ecology.* W. B. Saunders Co., Philadelphia.)

FIGURE 16.4

Tundra of the arctic coastal plain near Point Barrow, Alaska. Note the extremely low vegetation. The troughs are created by ice wedges which force the surrounding ground upward into raised polygons. (From Odum, E. P. 1971. *Fundamentals of Ecology.* W. B.
Saunders Co., Philadelphia.)

Boreal coniferous forest or taiga. Note the dense canopy of leaves which shade the forest
floor all year round. (U.S. Forest Service photo.)

FIGURE 16.5

FIGURE 16.6

Temperate rain forest in the Olympic National Forest, Washington. Mosses drape the large trees, and there is abundant understory growth. (From Odum, E. P. 1971. *Fundamentals of Ecology.* W. B. Saunders Co., Philadelphia.)

many early-flowering shrubs and perennials which reach their peak of flowering before the trees are in full leaf. Much of this biome has been greatly modified by man for agricultural and urban development.

Subtropical Evergreen Forests. In the warmer southeastern United States, small areas of stable, subtropical, broadleaved evergreen forest exist. In much of this region, changing soil conditions and frequent floods

FIGURE 16.7

A virgin stand of temperate deciduous forest in North Carolina. Note abundant understory vegetation. (From Odum, E. P. 1971. *Fundamentals of Ecology.* W. B. Saunders Co., Philadelphia.)

FIGURE 16.8

Grassland biome in a wildlife refuge in Montana with a small herd of pronghorn antelope, a native herbivore. (From Odum, E.P. 1971. *Fundamentals of Ecology*. W. B. Saunders Co., Philadelphia.)

and fires prevent the establishment of a stable climax vegetation over wide areas. Rather, a mosaic of various ecosystems is usually found. In the more northern regions, oak, bay and magnolia predominate, while in southern Florida, clumps of strangler-fig, mahogany and palms occur, with mangrove swamps along the coastal areas.

Grasslands.   The great prairies of the Midwest (Fig. 16.8) have a moderate climate but one which fluctuates annually between temperature extremes. Precipitation is moderate, from 10 to 30 inches per year. In the eastern plains, where rainfall is relatively more plentiful, the climax vegetation consists mainly of grasses which grow four to six feet tall, such as the two species of bluestem. In the drier western plains, the grasses tend to grow only six to 12 inches in height and include buffalo grass and various bunchgrasses. Few original grassland areas remain, and the herds of grazers once common on them have also largely disappeared. Small rodents and birds, however, are still prevalent.

Over most of the prairie, potential evapotranspiration rate exceeds precipitation; ground water is thus constantly being brought to the surface, especially by the roots of grasses, which may extend as deep as six feet. In contrast to the situation in the boreal coniferous forest, this results in an accumulation of nutrient ions at the surface of the soil (Fig. 16.9B). As a

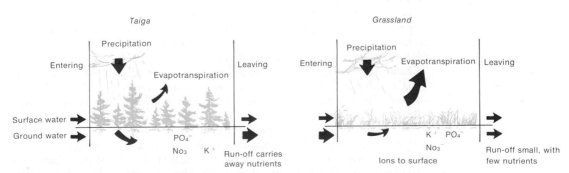

Comparison of the hydrologic cycle and its effect on soil quality in two biomes, taiga and grassland.

FIGURE 16.9

FIGURE 16.10

Desert biome in Arizona. Note small sclerophyll leaves on shrubs and wide spacing of plants. (From Odum, E. P. 1971. *Fundamentals of Ecology.* W. B. Saunders Co., Philadelphia.)

result, grasslands of the Midwest are among the most fertile soils in the world.

Desert.   Large areas of the western and southwestern United States and parts of Mexico receive less than 10 inches of rainfall per year and are typical deserts (Fig. 16.10). Temperatures are generally high during the day and low at night, but the limiting factors for plant growth are low humidity and precipitation. Climax plants in the desert are mainly of two types—shrubs with small, hard leaves, such as creosote and sagebrush, and the fleshy succulents, or cacti; both are adapted for water conservation. Cacti have virtually no leaves, using their swollen stems for photosynthesis and also as sites for water storage. The reduced leaves of the shrubs are an adaptation for decreasing the rate of water loss by transpiration, an adaptation which also results in a slower growth rate. In addition, these leaves, instead of requiring large quantities of water to maintain internal turgor pressure for stiffness, are supported by hardening of the external layers, which are known as **sclerophyll tissue.** Besides shrubs and succulents, deserts are characterized by a multitude of flowering annuals. The seeds of these species germinate only after the winter rains, and their entire life cycle is generally completed in three to four weeks.

Despite its barren appearance, the desert is not devoid of animal life: insects, reptiles, rodents and birds are common, and each species shows special adaptations for conserving water.

Chaparral.   There is a narrow zone along the Pacific coast of California and northern Baja California known as **chaparral** (Fig. 16.11). Rainfall is low and is confined to the winter months, but it is slightly greater than in the deserts. Humidity is maintained by moist air from the ocean. Temperatures are moderate to subtropical, but without extremes.

The climax vegetation consists primarily of large shrubs—chamise, manzanita, *Ceanothus* and others. Like the desert shrubs, these plants have water-conserving mechanisms; but since the climate is less demanding, these adaptations are less exaggerated. Typically, the shrubs have larger leaves than their desert counterparts, but they are still strengthened by hardening substances. These plants are therefore known as broad-leaved sclerophylls.

COMMUNITIES AND HABITATS

No large biome is uniform throughout. Local variations, due to topography, soil type and other factors, occur from one acre to the next. Lakes are found in the Canadian taiga; rivers flow through the midwestern plains; springs support oases in the deserts. Within each biome, one can thus recognize many subdivisions, or ecosystems, each with its own characteristic vegetation and animals. As noted above, whatever limits we choose to set upon a particular ecosystem, we can recognize the existence within it of a characteristic group of plants and animals, the **community.** The individuals of the various species of such a community interact with one another in a tightly-woven fashion.

Each species within a community physically occupies a particular part of the ecosystem, known as its **habitat.** Let us take a woodpecker, by way of example. It is never found underground and seldom on the ground; it spends most of its life in trees, flying from one to another. It nests in a hole usually excavated in the trunk of a dead tree; it searches for food over the branches and bark of other trees; it perches on prominent branches to display its conspicuous plumage to other woodpeckers. The woodpecker thus inhabits a particular part of its total environment.

In addition to the *places* where it is found, we can also assign to the woodpecker certain *functions* in the ecosystem of which it is a part. For instance, it eats bark insects, helping to control plant pests; it also feeds on berries and acorns, assisting in their dispersal; it helps to recycle nutrients by excreting its wastes; it hastens the destruction of dead tree trunks and so

Chapárral in southern California. In this broad sclerophyll biome, the shrubs have larger leaves than do those in the desert. Note also that the shrubs are taller and more closely spaced. (U.S. Forest Service photo.)

FIGURE 16.11

forth. The totality of the functions performed by an organism, such as the woodpecker, is often referred to as its "ecological niche"—an unfortunate term, since niche implies to most of us a sort of cubbyhole or nook and hence is easily confused with habitat. Although a squirrel and a woodpecker may share essentially the same habitat, they certainly do not fulfill the same functions in an ecosystem! Nor does any other organism in the same ecosystem perform exactly the same functions as the woodpecker; its role is unique.

Insofar as it has been possible for ecologists to study the roles of different species, this important generalization, known as **Gause's competitive exclusion principle** seems to be followed. Put in other words, this principle says that no two species can perform exactly the same function in a given ecosystem. When the functions of two different species begin to overlap significantly, competition between them develops, and the least well adapted species is eliminated.

We can thus envision a range of complexity in ecosystems. There are relatively simple ecosystems, with only a few species of plants, which in turn provide only a limited number of habitats for animals; there are relatively few functions to be performed within such communities. In general, such simple ecosystems are characteristic of extremely cold or dry climates and also of some of the artificial ecosystems produced by man—large fields devoted to a single crop or forests planted with but one kind of tree. In the arctic tundra, the deserts or a midwestern cornfield, **species diversity** is relatively low. In such simple ecosystems, the number of interactions between species is often insufficient to maintain stability. Thus, an unusually cold or dry year, or the introduction of a new pest, may have a catastrophic effect. The more complex an ecosystem is—that is, the greater its species diversity—the greater its stability and its ability to withstand periods of stress.

ECOLOGICAL SUCCESSION AND SPECIES DIVERSITY

So far we have concerned ourselves with relatively stable communities, in which climax vegetation is present and each species plays a well-defined role in relation to the rest of the community. But occasionally there are cataclysms—either natural or man-made. Fires, floods and avalanches can occur; or man may bulldoze the landscape, abandon unproductive fields or create a new lake behind a dam. The original ecosystem is destroyed, and a new one gradually takes its place. In such cases one can observe a sequence of biotic stages as a new ecosystem develops on an initially barren site. This is known as **ecological succession.** Of special interest are islands newly formed by volcanic eruptions; if the observer is careful not to introduce new species inadvertently, he can record colonization by new species through natural means and describe the interactions of the various populations as more and more species arrive. In such a way it is possible to gain insight into the factors which lead to the eventual stability of a complex ecosystem. Other examples also offer themselves to study, however.

SUCCESSION IN BEACH PONDS

Along the south shore of Lake Erie lies a four mile peninsula known as Presque Isle. Not infrequently during winter storms, a sandbar is formed, cutting off a small pond at the edge of the lake (Fig. 16.12A). Some of these ponds, which are usually 100 to 200 meters long, 10 to 20 meters wide and about a meter deep, persist long enough for vegetation to become established. If this happens, they then gradually fill in and become part of the terrestrial environment. An intermediate stage in this process is shown in Figure 16.12B.

A

B

Formation of beach ponds on the shores of Lake Erie at Presque Isle, Pennsylvania.   **FIGURE 16.12**
    A. A recently formed pond, less than a year old.
    B. A pond which was formed about 50 years ago. (From Edward J. Kormondy, *Concepts of Ecology,* © 1969. By permission of Prentice-Hall, Inc., Englewood Cliffs, N.J.)

Observations of the changes in the major plant species during the developmental stages of such a pond reveal some interesting points. The changes themselves are shown in the drawings of Figure 16.13.

The first, most obvious change is in the total amount of biomass associated with the pond. This is perhaps most striking in the photographs of Figure 16.12. Second, the pond gradually fills in with dead organic matter, altering both its physical shape and its nutrient content. Third, there is an increase in species diversity—the total number of species. As the pond ages,

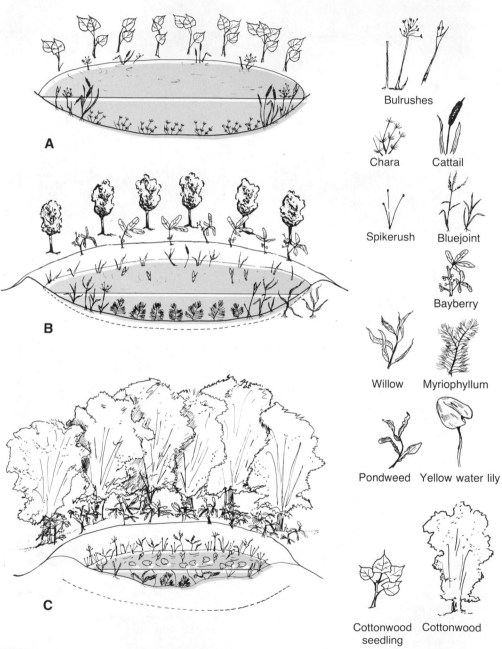

FIGURE 16.13   Successional stages of the aging of beach ponds at Presque Isle, Pennsylvania.
    A, four years. *B*, 50 years. *C*, 100 years of age. Note the changes in kind, distribution and abundance of different species and the accumulation of organic matter filling in the bottom. (After Kormondy, E. J. 1969. *Concepts of Ecology.* Prentice-Hall, Inc. Englewood Cliffs, New Jersey.)

not only does the number of plant species increase (Fig. 16.13), but there is also an increase in both the microscopic plankton species and the animal species associated with the pond. Each new type of plant provides habitats for a new group of animals, and so the changes in plant species are followed by a succession of animal species. Fourth, there is a change in species composition. New species are added, and some early species disappear as the environment gradually changes. Eventually the pond is filled in completely; it then goes through a succession of terrestrial stages, culminating in the establishment of a highly diverse climax community, the beech-maple forest characteristic of that part of Pennsylvania.

Two important properties of ecosystems are seen in our example. Perhaps the more obvious is that organisms, as a result of their own presence, alter the non-living environment around them. They remove some chemicals and add others; they change the soil or sediment quality; they affect water and air currents, and so on. Each change not only acts back on the causative species itself, increasing or decreasing its chance for survival; it also creates new habitats and functions for other species, at the same time perhaps destroying old ones. Not infrequently, a species highly successful during the early stages of succession is unable to live in new conditions which were partly brought about by its own presence.

The second important property is the increase in species diversity as a community develops towards a stable, climax condition—one which then remains relatively constant over long periods of time. A high degree of correlation between diversity and stability has been observed so many times in a wide variety of ecosystems that ecologists are convinced that the relationship is real. But it is difficult to explain which is cause and which effect: that is, does diversity create stability by buffering the ecosystem against sudden changes, or does stability create an increase in the number of permanent habitats and functions which may then be filled by new species?

It is known that species diversity generally decreases as a result of many human activities—monocrop agriculture, some types of pollution of streams and bays and so on—and that in these instances large fluctuations of populations are common. Insect pests may multiply rapidly and destroy an entire crop, for example. Although similar attacks occur in natural ecosystems, the whole system is not destroyed; eventually the damaged plants are replaced and the community returns to normal. In diverse communities, there appears to be a natural buffering action.

SUMMARY

Ecology, the science of the environment, is a study of ecosystems, or the totality of the reciprocal interactions between living organisms and their physical surroundings. Among the more important physical factors are temperature, light, water and inorganic nutrients. Biological factors include all interacting species of organisms, from macroscopic plants and animals to microorganisms and non-living organic molecules. The size of a population of any given species is determined by a balance between its reproductive potential and the environmental resistance.

The structure of an ecosystem ultimately depends upon the photosynthetic plants which support it. These primary producers provide both energy and biomass for the sustenance and growth of all other consumers—herbivores, carnivores and decomposers. In any ecosystem, energy is utilized only once, gradually being dissipated as heat during respiration at each trophic level; nutrients, in contrast, are constantly recycled.

A great variety of ecosystems exist in the world. Those on land are referred to as biomes, of which the characteristic ones in North America are tundra, taiga, temperate rain forests, deciduous forests, subtropical evergreen forests, grasslands, desert and chaparral. In general, the patterns of biomes are delimited by latitude, topography, climate and soil. Each biome is characterized by a climax vegetation.

Within an ecosystem there is a specific community of interacting organisms. Each species of organism within such a community possesses a particular habitat and carries out a specific environmental function. Each species is highly adapted to its unique function, and, because of competition between species whose activities overlap, no two species perform exactly the same function.

Occasionally natural or man-made cataclysms occur which disrupt a stable ecosystem, and a new one gradually takes its place. It is then possible to observe a chronological succession of species, each altering the environment. With time, early colonists are often replaced, and species diversity increases until eventually a stable community is achieved. It is hypothesized, but not yet proved, that species diversity leads to community stability in a self-reinforcing manner.

**READINGS AND
REFERENCES**

Hazen, W. E. 1970. *Readings in Population and Community Ecology.* 2nd edition. W. B. Saunders Co., Philadelphia. An excellent collection of original papers thoughtfully selected to introduce the student to major concepts of interactions within ecological communities.

Hutchinson, G. E. 1959. Homage to Santa Rosalia, or Why are there so many kinds of animals? American Naturalist 93:145–159. A well-reasoned and entertaining argument about the causes of species diversity.

Kormondy, E. J. 1969. *Concepts of Ecology.* Prentice-Hall, Inc., Englewood Cliffs, N. J. A highly readable introduction to the general subject of ecology.

Kormondy, E. J. 1969. Comparative ecology of sandspit ponds. American Midland Naturalist 82(1). An original work dealing with successional stages in beach ponds.

Odum, E. P. 1971. *Fundamentals of Ecology.* 3rd edition. W. B. Saunders Co., Philadelphia. A comprehensive treatise and basic reference text covering all aspects of ecology.

ENERGY FLUX AND FOOD CHAINS

 17

The amount of energy captured by plants determines the productivity of an ecosystem, while the pattern of energy flow through the food chain determines the biomass found in various trophic levels. An understanding of these parameters is important for solving the world food problem.

The initial source of all energy in any ecosystem is the photosynthetic activity of green plants. The chemical energy trapped by them is gradually distributed through the food web, until it is all eventually dissipated as heat, through the respiration of the various organisms. This chapter is concerned with how that energy is captured, transferred and finally dissipated.

As a first approximation, it is useful to consider an ecosystem as a single unit, a sort of giant organism, which has energy and nutrient requirements similar to those of an individual. The parallel is illustrated here:

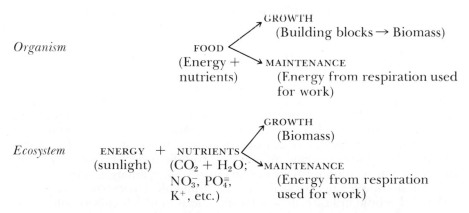

Available energy is on the left; the use to which it is put is on the right. Just as in animals, if more food is eaten than is needed for maintenance, growth occurs; so with an ecosystem, if photosynthesis outstrips respiration, there will be an increase in biomass. If photosynthesis exactly equals

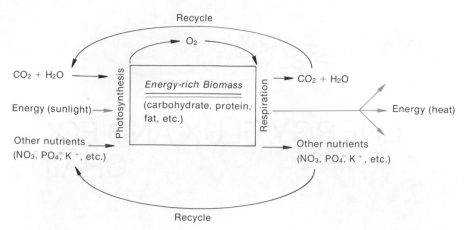

FIGURE 17.1

The metabolism of a whole ecosystem.

Photosynthesis, on the left, produces the entire energy-rich biomass of the ecosystem. Respiration, on the right, releases energy for maintenance. Minerals and gases are recycled, but energy is lost as heat.

respiration, however, then the biomass will remain constant, and the ecosystem is in a **stable state.** If photosynthesis is less than the total respiration of the ecosystem, the total biomass will decrease. We can express this in simple shorthand fashion, as follows, letting P stand for total photosynthesis and R for respiration:

If $P > R$, biomass increases ($>$ = greater than)
If $P = R$, biomass remains constant (stable state)
If $P < R$, biomass decreases ($<$ = less than)

The net changes going on in an ecosystem are summarized in Figure 17.1. Photosynthesis on the left produces energy-rich biomass; this in turn is converted, through respiration, back into the original nutrients. Energy is lost as heat.

THE STRUCTURE OF FOOD CHAINS

The diagram given in Figure 16.1 of the stages by which nutrients and energy flow through the various trophic levels of a food chain was oversimplified for the sake of clarity; it gives the impression that all organisms can be placed in discrete trophic levels. Although all four main trophic levels—herbivores, carnivores, top carnivores and decomposers—are found in almost every ecosystem, the assignment of a particular species to one or other trophic level is often difficult. An owl, for example, when feeding on a seed-eating rodent is a first-level carnivore, but when eating an insectivorous lizard, is a top carnivore. Man feeds at all trophic levels—eating a hamburger places one in at least two trophic levels simultaneously! Many insects change from one trophic level to another during development, feeding on plants as larvae and on animals as adults.

These complexities make it extremely difficult to construct a real food web which would indicate all possible directions in which energy could flow through an ecosystem. A hypothetical food web is shown in Figure 17.2, and an attempt to construct a real one for a deciduous forest in Illinois is shown in Figure 17.3. Note how much more complex the real system is.

Yet even the real food web does not present the full picture, for only the larger animals are shown—the microorganisms have been omitted. Within the gut of each species shown in Figure 17.3 are protozoans, bacte-

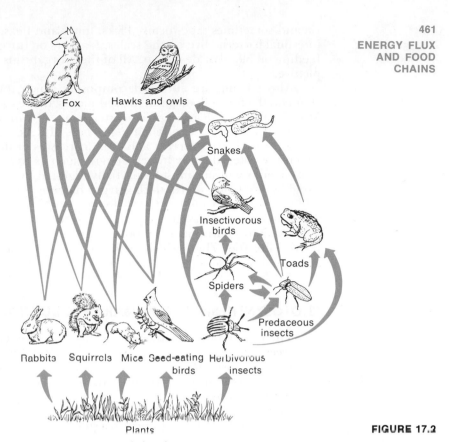

FIGURE 17.2

A hypothetical food web, far simpler than any real situation.

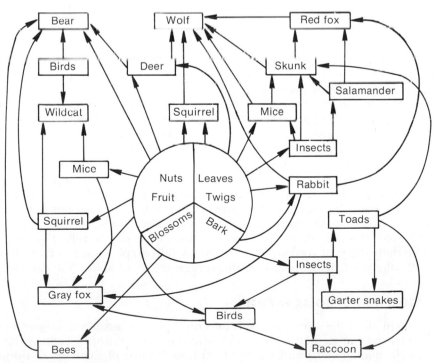

FIGURE 17.3

Diagram of the interrelationships in a food web in a deciduous forest in Illinois. Only the larger species are shown; small animals, microorganisms and parasites, as well as detritus feeders, are omitted.

ria and sometimes tapeworms. Ticks, mites and fleas, each with their own intestinal bacteria, live on the scales, feathers or fur of the larger animals, feeding on blood or dead skin. All of these organisms are missing from the picture.

Also missing are all the decomposers which form the **detritus** food chain on the forest floor (or in the sediments, in the case of a lake or ocean). Most detritus, or dead organic matter, comes from unconsumed parts of plants—leaves, twigs and fallen logs, roots of dead plants and so on. The rest is derived from excreta and from carcasses of dead animals, some of which are eaten directly by scavengers, such as hyenas and vultures. But most dead organisms must be partially digested by microscopic bacteria or fungi before being useful to the larger detritus feeders—earthworms and millipedes on land and worms, snails, clams and so forth in a lake or ocean. These detritus feeders eat the partly decayed matter, bacteria and all, probably obtaining nutrients from both. Ultimately it is the microorganisms in the detritus food chain which, with their complex biochemical machinery, are able to recycle minerals back into the ecosystem.

PHOTOSYNTHESIS AND PRIMARY PRODUCTION

The nature of an ecosystem, insofar as it is determined by its energy resources, will depend upon two things: the total amount of energy captured by photosynthesis and the distribution of this energy among the trophic levels. Our first concern, then, is, How much solar energy is trapped by an ecosystem? This clearly depends not only on the nature of the plants and their ability to capture energy in photosynthesis but also on how much sunlight falls on the ecosystem in question.

SOLAR FLUX

Radiant energy from the sun includes energy of many wavelengths, from short ultraviolet to long infrared. This energy is continuously emitted at a constant rate and, at the distance the Earth is from the sun, amounts to a total of two calories per square centimeter per minute (2 cal/cm²/min). This is known as the **solar flux.** Remember that a calorie is a unit of energy; it is convenient to use it here, since it can be directly compared with calories stored in chemical bonds as biomass. A square centimeter is about the same area as one-sixth of a square inch.

Due to the tilt of the Earth's axis and also to its slightly elliptical path, not all parts of the Earth receive equal amounts of radiation. The solar flux falling on a particular spot depends on its latitude and on the time of year, as seen in Figure 17.4

The values in Figure 17.4, however, take no account of energy reflected upward on cloudy days, which ranges from 50 to 80 per cent. On the average, only about half of the sunlight reaching the Earth's atmosphere passes through it to the surface, and much of this does not fall on plants. Some strikes areas of sparse vegetation, such as deserts, and is reflected upward; about 75 per cent falls on the highly reflective oceans. The daily input of sunlight which actually falls on plants varies from 100 to 800 cal/cm² of plant surface, with average values of about 350 cal/cm².

ABSORPTION OF SOLAR ENERGY

Not all of the sunlight which falls on a plant is absorbed, however. As we saw in Chapter 8, the absorption spectrum of chlorophyll shows that it absorbs mainly violet and red light. Although most plants have additional pigments (the orange carotenoids and yellow xanthophylls, for instance,

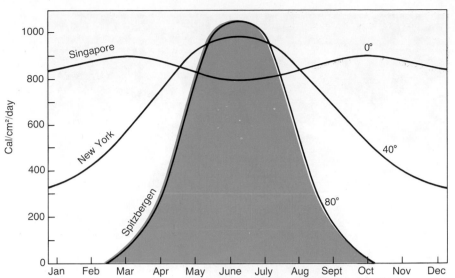

FIGURE 17.4

Total daily solar radiation received at various latitudes during different months of the year. The annual radiation at a particular latitude is the total area under the appropriate curve (that for 80°N is shown in color).

which are able to absorb other wavelengths of light and transfer the energy to chlorophyll), most plants are able to absorb only about 50 per cent of the total radiant energy falling on them. The remaining light either is reflected from the surface of the plant or passes on through it.

PRIMARY PRODUCTIVITY

Not all of the radiant energy absorbed by a plant is converted into useful energy—energy which goes into synthesis of ATP or of reduced organic molecules. In fact, only a small fraction of absorbed sunlight is captured for useful work; the rest is lost as heat without ever being used by the ecosystem. Our next question, then, is: How efficient are primary producers at converting light energy into useful energy?

Gross Primary Productivity. In order to discover how much total energy plants capture, we must consider both the growth (synthesis of new plant biomass) and the energy required for the maintenance of the plant. Taken together, these constitute the total photosynthesis or **gross primary productivity.**

Accurately measuring the gross productivity of a given ecosystem is a difficult matter, however. If we assume that the total energy requirements for maintenance of an ecosystem are constant day and night, then by measuring its total respiration (R) we can estimate how much energy is being used by it. Respiration is most easily estimated by measuring the utilization of oxygen in the absence of light (where no photosynthesis occurs). If we then simultaneously measure oxygen produced by photosynthesis during the day in the same ecosystem, we will have an estimate of its gross primary productivity (P). Measurements of this type are relatively simple to make on an aquatic ecosystem such as a lake or the open ocean, where most of the primary producers are phytoplankton (Fig. 17.5).

To accomplish this, water collected at different depths is placed into pairs of bottles; one bottle of each pair is dark (for measuring respiration only), one is transparent (for measuring photosynthesis and respiration simultaneously). The pairs of bottles are then returned to the depths from

FIGURE 17.5

Measuring gross primary productivity of a lake, using pairs of light and dark bottles. Each pair of bottles, filled with water from a known depth, is lowered on a string and left in place 24 hours. The white jug acts as a float. Afterward, changes in dissolved oxygen content of the water in each pair of bottles are used to calculate gross production, as described in the text. (From Odum, E. P. 1971. *Fundamentals of Ecology.* W. B. Saunders Co., Philadelphia.)

which the water samples were taken and are left for 24 hours. During this time, the microscopic plants and animals in the bottles use up oxygen in both bottles, while plants in the transparent bottle photosynthesize during the day, when light is present. The change in oxygen content of the water in each bottle is then measured and converted to grams of oxygen per cubic meter of water ($g\ O_2/m^3$) (Table 17.1). The data shown in the table can then be used to calculate the gross productivity, P. The change in the dark bottle represents total community respiration, R, while that in the light bottle represents gross productivity, P, minus simultaneous respiration, R, or P−R. By adding back the energy lost through respiration, R, we obtain P, gross productivity.

**TABLE 17.1
MEASUREMENT
OF GROSS
PRODUCTIVITY
IN A LAKE**

Depth	Oxygen Change $g\ O_2/m^3/day$		Community Respiration $g\ O_2/m^3/day$	Gross Productivity $g\ O_2/m^3/day$
	Light Bottle	Dark Bottle		
	$P-R$	$-R$	R	$[(P-R)+R]=P$
Top m³	+3	−1	1	+4
2nd m³	+2	−1	1	+3
3rd m³	0	−1	1	+1
bottom m³	−3	−3	3	0
Total for entire water column ($g\ O_2/m^2/day$)*			6	+8

*Since the oxygen produced is directly proportional to the energy captured during photosynthesis, the value for grams of oxygen/m²/day can easily be converted into calories/m²/day. (After Odum, E. P., 1971. *Fundamentals of Ecology.* W. B. Saunders Co., Philadelphia.)

Several things may be noted in our example. First, P is greater than R for the system measured. Either excess biomass is being produced or not all of the respiration has been measured. During the long days of summer, for P to be greater than R is common, whereas in winter, P may become less than R. It is also likely that respiration is occurring among the decomposers in the lake sediments which was not measured. Nor has the respiration of large animals such as fish been included. Both would tend to bring R closer to P, which is characteristic of a stable ecosystem.

We can also note that photosynthesis is highest near the surface, where light concentration is highest, whereas respiration is greatest at the bottom, where the highest density of heterotrophs (animals and decomposers) must therefore occur.

Measurement of gross primary productivity on land is much more difficult, although new techniques are being worked out. One method, similar in principle to that using the light and dark bottles in a lake, is shown in Figure 17.6. Air is run through a plastic bag surrounding the plants in question, and the change in its carbon dioxide content is measured with an infrared gas analyzer. During the day, photosynthesis removes CO_2, while respiration simultaneously releases some of it (P−R); at night (or when the bag is covered with a black cloth) only carbon dioxide released by respiration (R) is measured. Again, adding back R to P−R gives P, gross primary productivity. A difficulty of this method is that respiration by plant roots is ignored. Also, it is limited to small plants; it is obviously awkward to put a plastic bag around a redwood tree!

A great many other methods exist for measuring gross primary productivity in an ecosystem, but all of them have some drawbacks. Nevertheless, it is possible to estimate the efficiency of various ecosystems in converting absorbed sunlight into useful energy. On a year-round basis, the mean value for gross primary productivity throughout the biosphere is *0.4 per cent*. That is, for every 100 calories of energy absorbed by all plants, 0.4 calorie is converted to gross production. A very low value indeed! During the peak of the summer growing season, however, many plants are capable of far greater efficiencies. In certain crop plants—sugar cane, irrigated maize, sugar beets—up to *15 per cent* of the absorbed sunlight is converted to gross productivity. It is thus apparent that *productivity is not always limited solely by light but is also limited by available nutrients, temperature and water.*

FIGURE 17.6

Measuring gross primary production in a field. Air is passed through the plastic bag, and the change in carbon dioxide is measured with an infrared gas analyzer. The experiment is repeated while the bag is covered with a black cloth. Calculation of gross productivity is similar to that employing light and dark bottles in an aquatic ecosystem. (From Odum, E. P. 1971. *Fundamentals of Ecology*. W. B. Saunders Co., Philadelphia.)

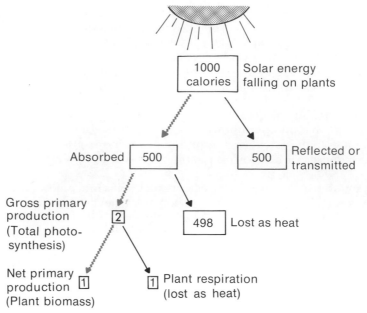

FIGURE 17.7

Summary of the efficiency of net primary production (plant biomass) relative to total solar flux. Numbers are based on mean annual efficiencies for the whole biosphere. Only 50 per cent of the light falling on leaves is absorbed; only 0.4 per cent of this is utilized for photosynthesis (gross primary production). Only 50 per cent of this is converted to plant biomass (net primary production).

It is important to remember, however, that other factors may affect efficiency, especially at the level of photosynthesis. Low temperature and too little water or nutrients restrict photosynthesis.

Net Primary Productivity. It is not the gross productivity, however, which determines the size of an ecosystem, but the **net primary productivity**—that is, how much plant biomass is available for food for all the heterotrophs:

Gross Primary   Net Primary   Plant
Productivity = Productivity + Respiration
(total photosynthesis)   (growth of plant biomass)   (plant maintenance)

The respiration of plants on the average consumes about 50 per cent of the gross production resulting from photosynthesis. The remainder shows up as an increase in plant biomass. The average efficiency of *net production* for the whole biosphere is thus reduced to 0.2 per cent of the energy absorbed by plants, or 0.1 per cent of the solar radiation falling on them. All these losses are summarized in Figure 17.7.

NET PRODUCTIVITY OF MAJOR ECOSYSTEMS

We are now in a position to compare the net primary productivity of some of the major types of ecosystems in the world. A list of the estimated annual productivity of various ecosystems is given in Table 17.2. The most fertile sites of each type of ecosystem have been chosen, so the figures are maximum rather than average values. The reader should also realize that these numbers are constantly being revised as new data are acquired.

From Table 17.2 we can draw some inferences about the factors which limit net productivity in various ecosystems. First of all, latitude plays an important role. Near the equator, there is more light per year (Fig. 17.4), and the average temperature is higher. Since enzyme-catalyzed reactions, including those of photosynthesis, are temperature dependent, one would expect higher photosynthetic rates at higher temperatures, all other factors being equal. Comparison of productivity in similar ecosystems in temperate and tropical climates illustrates the importance of light and temperature as limiting factors.

Nutrients are also an important factor in productivity, as shown by two examples in the Table. A polluted lake—one containing excessive amounts of nitrate (NO_3^-) and phosphate ($PO_4^{\equiv}$) (see Chapter 2a)—has a higher productivity than a non-polluted lake. Likewise, terrestrial environments have a higher productivity than do most aquatic ecosystems at the same latitude. This is mainly due to the higher concentrations of mineral nutrients in soils than in large bodies of water. Oceans and lakes are generally nutrient-poor ecosystems. Important exceptions are estuaries and coral reefs, which are as productive as the most fertile terrestrial ecosystems. Agricultural plants tend to have a high productivity for two reasons. First, they have been selected by man for a high photosynthetic efficiency; and second, they are generally grown on nutrient-rich alluvial plains, with high evapotranspiration rates which bring nutrients to the surface.

A third factor affecting net primary productivity is water. Deserts have the lowest productivity of all ecosystems, although neither light, temperature nor nutrients are limiting.

It is interesting to note that although the oceans constitute almost 75 per cent of the Earth's total surface, only about half of the world's total primary productivity occurs there. We shall further consider this point, as well as the potential for increasing the productivity on land as a means for feeding the growing human population, in the next chapter.

TABLE 17.2
ESTIMATES OF
ANNUAL NET
PRIMARY
PRODUCTIVITY
OF FERTILE
SITES*

Type of Ecosystem or Producer	Climate	Productivity ($kcal/m^2/yr$)
Desert	arid	400
Phytoplankton		
Ocean	–	800
Lake	temperate	800
Polluted lake	temperate	2400
Submerged macrophytes		
Freshwater	temperate	2400
Freshwater	tropical	6800
Marine	temperate	11,600
Marine	tropical	14,000
Forest		
Deciduous	temperate	4800
Coniferous	temperate	11,200
Rain forest	tropical	20,000
Agricultural plants		
Annuals	temperate	8800
Perennials	temperate	12,000
Annuals	tropical	12,000
Perennials	tropical	30,000
Swamps and marshes		
Salt marsh	–	12,000
Reedswamps	temperate	17,100
Reedswamps	tropical	30,000

*After Kormondy, E. J. 1969. *Concepts of Ecology.* Prentice-Hall, Inc., Englewood Cliffs, N. J., p. 19.

ENERGY FLOW IN AN ECOSYSTEM

Having considered in some detail the factors affecting the primary productivity of an ecosystem, let us now consider how that energy is distributed among the several trophic levels and eventually dissipated by them.

Just as only a fraction of the energy absorbed by green plants is utilized for maintenance and growth, so also is only a part of the chemical energy consumed by each trophic level actually assimilated. Of this part, a large fraction is utilized for maintenance, and the remainder goes towards growth. The unassimilated food (feces), still containing its energy-rich bonds, is later utilized by other members of the community. If we equate maintenance with respiration, then we can estimate both the calories expended by each trophic level to keep itself functioning and those which are left over for formation of biomass.

On paper this all sounds easy enough; one simply collects all the herbivores in a community, measures their biomass, respiration rate and fecal production, and calculates how much food they have therefore consumed. One then repeats this procedure for each trophic level, and comes out with an energy flow diagram. In practice, however, matters are not that simple. It is difficult to estimate, even for an animal as large as a cow, its total growth rate, its average respiration and how much organic matter is excreted in its feces or left behind when it dies. To make the same estimates for a population of birds or squirrels or mice is even more difficult, since one must often guess at how many are hibernating, how many are born and die, and so on, during a year. To make accurate estimates for the populations of decomposers—the bacteria and fungi—is nearly impossible; it is even more difficult to determine their metabolic activities during a full year. Yet these are just the sort of measurements that are required if we are to accurately determine the rate at which energy flows through an ecosystem and, hence, the rate at which biomass is synthesized at each trophic level. These are not merely academic questions, however; the production of food (biomass) for human consumption depends precisely on such calculations.

ENERGY FLOW IN A SIMPLE ECOSYSTEM

In a few simple ecosystems it has been possible to measure the energy flow between trophic levels. One such case is the famous tourist attraction of Silver Springs, Florida, studied by H. T. Odum. Odum made estimates of the energy flowing from one trophic level to another and of the way the energy was utilized at each level—either for maintenance (respiration) or for growth (biomass). A summary of his calculations is shown in Figure 17.8, in energy units of kilocalories per square meter per year (kcal/m²/yr; a kilocalorie [kcal] = 1000 calories).

First of all, Odum noted that only about a quarter of the light falling on the green plants in this ecosystem is absorbed (410,000 kcal); of this, less than five per cent is captured during gross photosynthesis (20,810 kcal). He also noted that more than half of the latter is lost as plant respiration (11,977 kcal). The remainder (8833 kcal) becomes part of the plant biomass. Since the ecosystem is in a stable state ($P = R$), the newly synthesized plant biomass is utilized by the rest of the ecosystem. Odum also assumed that each trophic level was in an equilibrium state—that is, its biomass remained constant. Therefore the energy entering a trophic level had also to leave it, either in the form of respiration (for maintenance) or as food for another trophic level.

Odum also took into account external sources of energy besides photosynthesis—namely, organic molecules imported into the springs from elsewhere (406 kcal). (In fact, this represents food thrown to turtles by tourists, since the spring water itself contains no organics.) He noted, too, the export of organic matter in the stream leaving the springs (2500 kcal). In this way he was able to make an "energy balance sheet" for the entire ecosystem, accounting at each step for the loss of energy (as heat) or its transfer (as biomass) to the next trophic level.

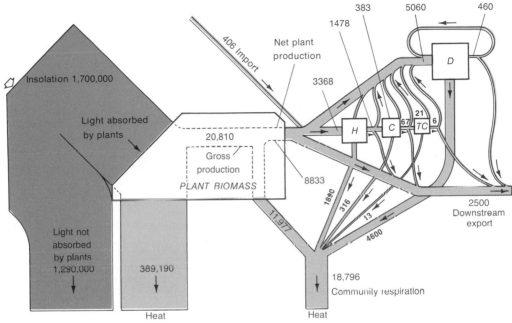

Diagram of energy flow through a simple ecosystem at Silver Springs, Florida. This ecosystem is in a state of equilibrium—the biomass of each trophic level remains constant. Note that, at each trophic level, incoming energy is used either for maintenance (respiration) or for food by a higher trophic level. Excess energy leaves the ecosystem as organic matter exported downstream.

Numbers refer to kilocalories per square meter per year. H, herbivores; C, carnivores; TC, top level carnivores; D, decomposers.

FIGURE 17.8

ENERGY FLOW AND THERMODYNAMICS

If we compare gross energy production in Silver Springs (20,810 kcal) with the sum of the energy used in community respiration (18,796) and the *net* exported (2094), we see that they are virtually equal. (Presumably the slight excess utilization (80 kcal) caused a small decrease in total biomass during the year of Odum's observations.) Thus the First Law of Thermodynamics, which says that energy is neither created nor destroyed, has been obeyed. All energy is accounted for.

Likewise, the Second Law of Thermodynamics is also obeyed. During each transfer of energy, a fraction is dissipated as heat—no energy transfer is 100 per cent efficient. This heat arises from respiration (see Chapter 7).

The food chain Odum studied was remarkably efficient. In most food chains, about 90 per cent of the energy transferred at each step is either lost through respiration or is never assimilated in the first place, and is passed as feces to the decomposers. Only about 10 to 20 per cent is available for growth. On the whole, herbivores tend to be somewhat less efficient than carnivores at converting food into biomass.

ENERGY FLUXES IN DIFFERENT TYPES OF COMMUNITIES

We now have two parameters by which to compare various ecosystems: (1) the net productivity and (2) the ratio between total productivity and respiration. The latter requires further explanation. In Odum's study of Silver Springs, he found a slight excess of productivity over respiration (P > R); this was compensated by an export of energy-rich organic matter. Had the organic matter not been exported it might have accumulated as detritus and perhaps eventually have become fossil fuel (P >R still); or it might have been oxidized by decomposers (P = R). It is also possible to imagine an ecosystem where gross primary production is less than total community respiration (P<R). In such a case, chemical energy must be imported from outside the ecosystem to maintain it in a steady state. Some examples of these various types of ecosystems are shown in Figure 17.9.

Ecosystems lying on the line P = R are generally stable, whether they have a high productivity (coral reefs) or a low productivity (deserts). Ecosystems where P>R (above the line) are increasing in biomass or stored organics. This situation is characteristic of young, growing communities, such as recently planted forests or fields. It also must have characterized the giant fern forests of the Carboniferous period some 300 million years ago; during that time, productivity greatly exceeded respiration, leading to the accumulation of organic matter which was transformed into the great coal deposits.

Ecosystems where P<R (below the line) depend upon imported organics for maintenance of their biomass. It is no coincidence that two of the cases shown in Figure 17.9 are the result of man-made pollution—eu-

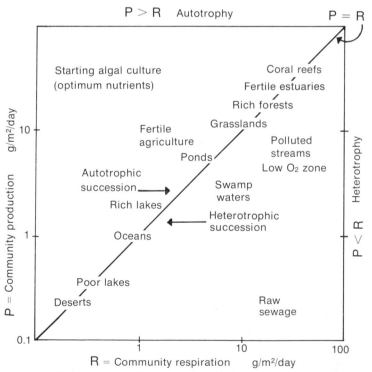

FIGURE 17.9

Classification of communities according to balance between gross primary production (P) and community respiration (R). Communities lying on the line *P = R* are generally stable, although productivity may vary enormously. Communities where P is greater than R are increasing in biomass or organic detritus; those where P is less than R are either decreasing in total organic matter or living off imported organics.

trophied streams and sewage. Man tends to collect his wastes in one spot and dispose of them in an ecosystem which does not have the photosynthetic capacity to recycle them.

DISTRIBUTION OF BIOMASS IN ECOSYSTEMS

From our earlier discussion, it is apparent that the energy available to produce each trophic level in a food chain is about one-tenth that in the preceding level. Thus, if 10,000 calories of plant are available, they can make about 1000 calories of herbivores, which in turn can make about 100 calories of first level carnivores, and so on. This point is clearly made in the biomass pyramids of Figure 17.10. At each successive trophic level, the biomass is about one-tenth that of the level on which it feeds.

A. Hypothetical biomass pyramid

Boy — 100   C = Carnivore

Calf — 1,000   H = Herbivore

Alfalfa — 10,000   P = Primary producer

B. Actual biomass pyramids

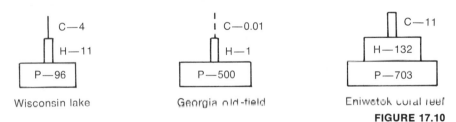

C—4   C—0.01   C—11

H—11   H—1   H—132

P—96   P—500   P—703

Wisconsin lake   Georgia old-field   Eniwetok coral reef

FIGURE 17.10

Biomass pyramids:
    A. Hypothetical biomass pyramid, in pounds; ten per cent of the energy at each trophic level is available to the next one.
    B. Actual biomass pyramids; numbers are in grams of dry weight per square meter. Note that there is more plant biomass in each case than there is herbivore biomass, and that there is more herbivore biomass than carnivore biomass. (An old-field is former farmland returned to a natural state.)
    C, carnivores; *H,* herbivores; *P,* primary producers.
    (After Odum, E. P. 1971. *Fundamentals of Ecology.* W. B. Saunders Co., Philadelphia.)

In some ecosystems, however, an apparently incongruous situation exists. The biomass in the English Channel, for instance, is four times that of the primary producers (Fig. 17.11). At first this seems to be a refutation of all the laws of thermodynamics—energy (or biomass) seems to have been created out of nothing.

The explanation lies in the fact that biomass pyramids only show where energy is deposited in an ecosystem *at a given moment.* They tell us nothing about the *rates* of energy flow between trophic levels. Let us use the analogy of water running down a stream. The same amount of water entering the stream at one point leaves it lower down (ignoring additions or losses along the way). In narrow regions, the total amount of water at a given moment is small, but the flow rate is rapid; in wide regions, there is much water, but it flows slowly. In ecosystems like that of the English Channel, the biomass of the primary producers is small in volume, but the energy flow rate through this trophic level is very fast indeed. This is due to the rapid growth and reproductive rates of the small phytoplankton which are the basis of the food chain. During a year, far more phytoplankton than herbivore biomass is produced, but since energy flows only slowly through the herbivore trophic level, it tends to accumulate there as herbivore biomass.

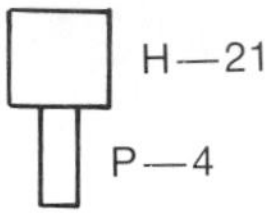

FIGURE 17.11

H—21

P—4

English Channel

An inverted biomass pyramid occurring in the waters of the English Channel. Numbers are in grams of dry weight per square meter. The instantaneous picture of biomass does not give us information about the *rates* of energy flow between trophic levels.

H, herbivores; *P,* primary producers.

(After Odum, E. P. 1971. *Fundamentals of Ecology.* W. B. Saunders Co., Philadelphia.)

Or let us take an example on land. It is entirely possible that a fertile pasture which is constantly grazed by livestock will always have less plant biomass at a given moment than a neighboring ungrazed pasture of low productivity. Yet during the year the total plant biomass produced by the fertile pasture may far exceed that of the unfertile one.

Thus, although such pyramids are useful for gaining understanding of the sequence of energy flow in an ecosystem, they do not represent a true picture of the process. Far more useful are energy flow diagrams, such as that in Figure 17.8, where rates of energy transfer are what is actually being measured.

SUMMARY

The ultimate source of useful energy in an ecosystem is photosynthesis. The energy captured may be used either for maintenance of organisms (respiration) or for synthesis of new biomass (growth). If gross photosynthesis (P) exceeds community respiration (R) biomass increases; if P = R, biomass remains the same; if P<R, biomass decreases. Energy is transferred through various trophic levels, known collectively as food chains or, more correctly, food webs, since some organisms belong to several trophic levels.

Gross photosynthesis depends on the amount of solar energy captured by plants. Only about 0.2 per cent of the light falling on plants is utilized for gross photosynthesis, and about half of this is required for plant respiration. The remaining energy goes into synthesis of plant biomass—net photosynthesis—which is then available to other organisms. Net photosynthesis varies among ecosystems and is limited not only by the amount of light but also by temperature and by nutrient and water availability.

Measuring energy flow through an ecosystem requires a knowledge of the rates of transfer of biomass from one trophic level to the next and of the efficiency of its utilization by each group of organisms. This implies an ability to estimate the biomass (population sizes), respiration and excretion of each trophic level. In relatively simple, stable ecosystems where measurements are possible, the distribution and dissipation of energy can be estimated with reasonable accuracy. Such measurements indicate that the efficiency of energy transfer between trophic levels is between 10 and 20 per cent.

Construction of biomass pyramids is an alternative way of representing energy flow through an ecosystem, but because these are instantaneous representations, which do not take into account the flow rates of energy between trophic levels, they can be misleading.

(In addition to specific chapters in the general texts listed in Chapter 16.)

Lindemann, R. L. 1942. The trophic-dynamic aspect of ecology. Ecology *23*:399–418. The first attempt to analyze in detail the energy flow through a small, circumscribed ecosystem.

Turner, F. B. (ed.) 1968. Energy flow and ecological systems. (Symposium.) American Zoologist *8*:10–69. A collection of six readable papers intended as a refresher course for biologists needing up-dating in the energetics of ecosystems.

Westlake, D. F. 1963. Comparisons of plant productivity. Biological Reviews *38*:385–425. Describes ways of determining primary productivity in various types of plant communities and compares data from different communities.

Woodwell, G. M. 1970. The energy cycle of the biosphere. Scientific American Offprint No. 1190. Considers further some implications of the principles outlined in this chapter.

MAN'S POSITION IN THE FOOD CHAIN

The prime concern of a great part of the world's population is obtaining enough of the right sort of food to eat. Nevertheless, increasing food production without due consideration to ecological principles may create more problems than it solves.

This chapter is concerned with applying some of the principles gained in the last chapter to the very real problem which exists today—namely, how to feed the world's burgeoning and already hungry human population. We have not space—nor is it our intent—to consider in detail all possible food resources and the social and economic factors affecting them, although these are obviously of great importance. Our discussion will be mainly concerned with the ecological position of man as a consumer in the biosphere and with the ecological impact of expanding his food resources. Many of the suggested solutions which we shall evaluate in this chapter are summarized in the 1967 report of the President's Science Advisory Committee Panel on the World Food Supply, entitled *The World Food Problem*. It is a comprehensive and basic source of information.

THE PROBLEM

In Chapter 9 we outlined the basic food requirements of animals generally—calories, organic precursor molecules such as amino acids, and micronutrients such as iron and vitamins. All three are essential for a healthy organism, including man, but we shall be concerned in this chapter primarily with the first two. Man obtains his calories mainly from staple foods—starches such as wheat, rice, corn and potatoes. An active adult needs about 3000 kilocalories per day, a sedentary adult about 2300. The major organic precursor molecules required are amino acids, especially the 10 essential ones which we cannot synthesize for ourselves. These are obtained from foods with a high protein content—meat, fish, poultry, eggs, milk, cheese and protein-rich seeds, like peas and beans. Only those proteins, or combinations of proteins, which contain adequate amounts of all 10 essential amino acids are of value in meeting dietary protein requirements. These are known as **complete proteins.** Although whole grain cereals contain considerable amounts of proteins, these are generally incomplete; animal proteins and certain seeds and nuts are our major sources of complete protein. The minimum adult daily requirement for complete proteins to maintain good health is between 60 and 70 grams, more being required by adolescents and by pregnant and nursing women.

It is estimated that, at least as of 1966, there was enough food in the world to supply everyone alive with enough calories and protein! Yet because of unequal distribution, spoilage and waste, approximately two-thirds of the world's population is either undernourished—too few calories—or malnourished—not enough vitamins and protein (Fig. 17a.1). Meanwhile, obesity and heart disease are becoming increasing medical problems in the well-fed (overfed?) segment of the world population.

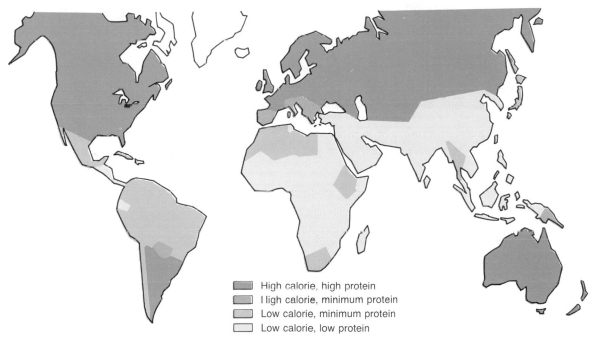

　　 ▊ High calorie, high protein
　　 ▊ High calorie, minimum protein
　　 ▊ Low calorie, minimum protein
　　 ☐ Low calorie, low protein

The geography of hunger: the unequal distribution of food resources. (After Ehrlich, P. R., and Ehrlich, A. H. 1970. *Population, Resources, Environment.* W. H. Freeman, San Francisco.)

FIGURE 17a.1

Growing children and pregnant and nursing mothers are especially affected by protein deficiencies. As a result, infant mortality is higher among poorly nourished populations. In the United States, about 25 in every 1000 children born alive die in their first year, usually because they are premature or have congenital defects. In underdeveloped countries, the rate is 50 to 100 deaths per 1000 live births, the increase being due mainly to malnutrition. Death rates from childhood diseases, such as measles, are 10 times higher among poorly fed populations.

A major childhood disease directly resulting from protein deficiency is **kwashiorkor**—a West African word meaning "sickness the child develops when another baby is born," i.e., when it is weaned and fed a starchy, low protein diet. Words for the same syndrome occur in many languages. A swollen stomach, retarded growth and listlessness characterize the disease, which is fatal in severe cases (Fig. 17a.2). Children who survive often are permanently retarded mentally as well as physically. It is estimated that moderate to severe protein malnutrition affects 50 per cent of children in some underdeveloped countries. Numerous vitamin deficiencies also are found among the undernourished and malnourished populations of the world; all these nutritional diseases not only cause suffering of themselves but also decrease resistance to infectious diseases as well.

FIGURE 17a.2   Symptoms of kwashiorkor, or protein-deficiency disease, in African children.
     In addition to listlessness and mental retardation, such children are readily susceptible to other diseases.

NEEDS FOR THE FUTURE

Were the population of the world to remain static, it is conceivable that better distribution of today's food resources would solve most of the world food problem. A leveling off of population growth, however, does not yet seem to be on the horizon. If today's present birth rate prevails, it is estimated that by 1985 world food requirements will have increased by 52 per cent. Even if there is a 30 per cent *decline* in fertility resulting from widespread acceptance of family planning, the need for food will still have increased 43 per cent, since babies already born will be reaching their maximum food requirements. In less than 15 years, 50 per cent more food must be made available to the world population if famine, malnutrition and their attendant diseases are not to become even more widespread than they are today.

Let us consider some suggested ways of increasing world food supply and examine their likely effects on the ecological health of the world.

INCREASED FOOD GATHERING

Prior to becoming herdsmen and farmers, men obtained food as do other animals, by hunting, fishing and gathering plants. As long as man did not remove more of the animal or plant crop than could be replaced by natural means, his supplies were continuous. All he had to do was apply his energies and skills to collecting food as needed. Today, only a small part of our diet is obtained in this way, primarily fish and shellfish. It is possible, however, that substantial additions to world food resources could be made by increasing our efforts in gathering "natural" foods.

WILDLIFE RESOURCES

One untapped resource suggested by the President's Science Advisory Committee is wildlife, especially African game animals. Much of eastern Africa has been colonized by European farmers who imported domesticated cattle. The lands they fenced off became unavailable to game, and many wild species have suffered severe population decreases. Yet areas remain which are unsuited for cattle and where game still abound. One study made in Rhodesia suggests that it may, in fact, be both cheaper and more productive to harvest wild game than to grow domesticated animals. An area of 50 square miles was observed for several years to obtain an estimate of the annual production of game. This number of animals was then harvested, yielding 118,000 pounds of meat which was sold at a profit of $8900. It was estimated that the same land fenced for raising domesticated cattle would have yielded 95,000 pounds of meat, but at a profit of only $1416! The difference in profit is due to the extra labor and other costs of producing cattle. Obviously in some ecosystems we cannot, at the moment, improve on nature, even with technological inputs. It seems likely that by employing sound wildlife management, the nutrition of parts of Africa could be improved while yet preserving a healthy ecosystem!

MARINE FISHERIES RESOURCES

The main natural food resource which holds some promise for the future, however, is the sea. It has been highly touted by some as the panacea for all the world's food problems. As we shall see, this simply is not true. In fact, if we are to improve our food intake from the sea significantly, we shall have to begin fishing in some rather distant places.

A general impression of the economy of the sea may be gathered from Figure 17a.3. The shoreline, especially the western coasts of the major continents, is an area where upwelling occurs (see Chapter 2). When surface waters are blown out to sea by offshore winds, they are replaced by nutrient-rich waters from below that contain dissolved nitrates and phosphates. Coastal waters are further enriched by rivers carrying dissolved nutrients leached from the land. The productivity of the phytoplankton—single-celled algae—in these regions is relatively high. Seaweeds growing along the shore also contribute to the primary productivity of coastal waters.

In the open ocean the only primary producers present are phytoplankton. Because the open seas are generally deficient in nutrients, they have a low productivity and are often referred to as "biological deserts." The phytoplankton are generally eaten by tiny zooplankton (Fig. 17a.3), the main herbivores of the open ocean. Zooplankton are small animals which live both in the upper, euphotic zone and in the deeper waters. Included are small crustacea and other small animals that spend their entire lives suspended in the water; also included are the larvae of larger animals which later metamorphose and settle on the bottom as adults. The smaller zooplankton are eaten by larger zooplankton or by small fish, the first-level carnivores; these are in turn eaten by larger fish, the second-level carnivores. If we recall that only about 10 to 20 per cent of the energy is transferred at each trophic level, then large fish in the open ocean must be very scarce indeed. Any unused food—dead algae, zooplankton and larger animals—slowly falls to the bottom of the ocean, where it becomes available to benthic animals and decomposers in the sediments.

478

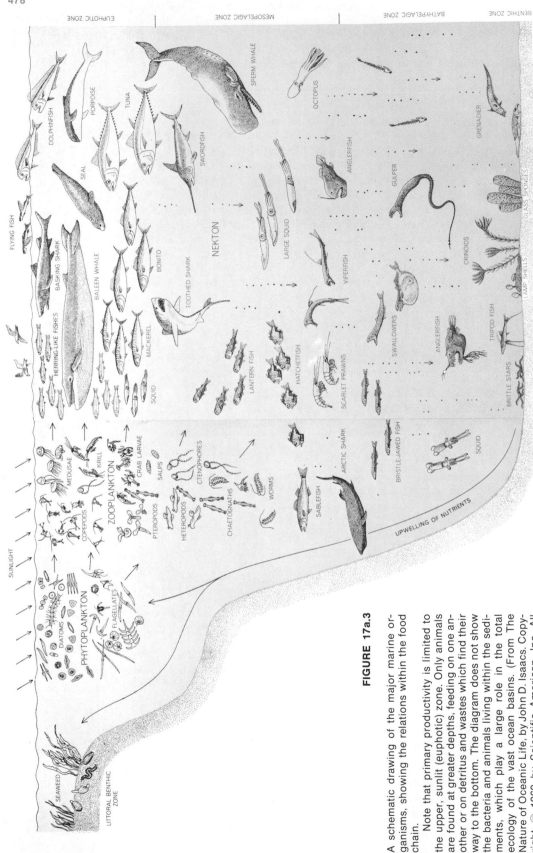

FIGURE 17a.3

A schematic drawing of the major marine organisms, showing the relations within the food chain.

Note that primary productivity is limited to the upper, sunlit (euphotic) zone. Only animals are found at greater depths, feeding on one another or on detritus and wastes which find their way to the bottom. The diagram does not show the bacteria and animals living within the sediments, which play a large role in the total ecology of the vast ocean basins. (From The Nature of Oceanic Life, by John D. Isaacs. Copyright © 1969 by Scientific American, Inc. All rights reserved.)

Estimating Total Sustainable Yield. We are interested here not in the number of fish in the sea, but rather in how many can be harvested without depleting the supply—the total **sustainable yield.** Estimates of the total sustainable yield of edible fish in the sea vary enormously; they range from 55 to 2000 million tons per year. The very vastness of the seas and the few data that exist make such extrapolations extremely difficult.

An attempt to resolve these discrepancies was undertaken by John Ryther, a marine ecologist at the Woods Hole Oceanographic Institution in Massachusetts. He made pertinent calculations about the total available fish in the sea based on closely argued calculations of primary productivity, number of trophic levels in food chains and efficiency of energy transfer at each trophic level. His estimates seem to offer the best we now have of the potential food resources of the sea.

Ryther divided the oceans into three provinces—the open ocean, ordinary coastal regions and regions of maximum upwelling (western coasts of Africa and of North and South America). He then estimated for each province the area, the primary productivity (expressed as carbon fixed by photosynthesis per year), the number of trophic levels, the efficiency of energy transfer between trophic levels and, finally, the fish production. The results of his calculations are shown in Table 17a.1.

TABLE 17a.1
ESTIMATED FISH
PRODUCTION IN
THREE OCEAN
PROVINCES*

Province	Percentage of Ocean	Primary Production (Tons Organic Carbon Per Year)	Number of Trophic Levels	Transfer Efficiency (Per Cent)	Fish Production (Millions of Tons Fresh Weight Per Year)
Oceanic	90	16.3 billion	5	10	1.6
Ordinary coastal	9.9	3.6 billion	3	15	120
Upwelling coastal	0.1	0.1 billion	1½	20	120
Total	100				241.6

*After Ryther, J. H. 1969. Science *166*:72–76.

Let us analyze Ryther's calculations, for they are instructive in understanding the role of energy transfer through trophic levels in the production of food for man. Consider first the total production of fish in the open ocean (oceanic province). This province comprises 90 per cent of the total surface of the ocean, and its phytoplankton synthesize some 16 billion tons of organic matter per year. However, phytoplankton in the open ocean tend to be very small. Few fish have filtering mechanisms capable of trapping them. In fact, only the smaller zooplankton can efficiently feed on phytoplankton in the open ocean, and they, also, are too small to be used as food by most fish. In fact, Ryther calculates that at least *four transfers* (five trophic levels) are required before energy trapped by phytoplankton in the open ocean can be utilized by fish big enough to be caught in the nets of fishermen. He also assumes that because organisms are not very concentrated in the open ocean, the efficiency of energy transfer is only 10 per cent. Many algae and zooplankton are not consumed but instead die and fall to the bottom as detritus. The annual fish production in the entire open ocean is thus estimated to be 1.6 million tons.

The ordinary coastal waters occupy about 9.9 per cent of the ocean, but they are richer in nutrients, and therefore the plankton have a higher annual productivity rate per unit area—about twice that of the open ocean. These plankton are also larger, and hence fewer trophic levels are required before the photosynthetic energy is transferred to fish. Ryther assumes that the higher density of biomass results in greater efficiency of energy transfer. Relatively more food is eaten than dies and sinks to the bottom. The ordinary coastal waters thus produce far more fish per year than the open ocean—about 120 million tons.

Upwelling areas, although they constitute only 0.1 per cent of the total ocean, have an extremely high primary productivity due to the availability of nutrients. Moreover, phytoplankton in these areas are relatively large and tend to clump together in gelatinous masses which are easily eaten by large animals, including some fish. Thus the number of trophic levels is greatly reduced. Moreover the high concentration of food organisms results in a greater efficiency of energy transfer between trophic levels; animals expend less energy obtaining food. Consequently, upwelling areas produce about 50 per cent of the world's marine fish!

By means of these calculations, Ryther estimates that some 240 million tons of fish are produced annually, almost entirely near the continental margins. This explains why countries with a large fishing industry have been recently expanding their territorial limits from 3 to 12—or more—miles. Mankind now takes about 55 to 60 million tons of fish per year from the ocean. We cannot, however, harvest the whole of the annual 240 million tons produced annually. Some fish will inevitably be eaten by other predators—guano-producing birds, tuna, squid, sea lions and so forth. Allowing for losses to them, and leaving behind enough fish to produce the next year's crop, Ryther estimates we could safely increase our total catch to a maximum sustainable yield of about 100 million tons per year, or almost twice its present level.

There is one snag, however. Most of today's marine fisheries are located in small areas of the ocean—the North Atlantic, the western Pacific and the coast of Peru. There is evidence that these are already nearly maximally exploited; in the North Atlantic the maximum sustainable yields have already been exceeded and the annual catches are now falling. To increase our present catch we must utilize new fishing grounds, especially the distant waters of the Antarctic, which are known to be highly productive but as yet almost untapped fisheries.

Recent discoveries that phytoplankton may be more numerous in the open ocean, and occur at greater depths than was generally supposed at the time Ryther made his calculations, may mean that twice as many fish are produced in the open oceans as he estimates. Even if this is true, it still leaves the open ocean as a relatively infertile region compared to the coasts (see Table 17a.1). The energy expended in catching such sparsely located animals may well exceed their value as food.

FRESHWATER FISHERIES RESOURCES

It has been estimated that natural populations of freshwater fish could supply about 20 per cent more fish than at present before reaching a maximum sustainable yield. However, if industrialized farming practices similar to those used in western countries become prevalent throughout the world, much of this resource may be made less productive through pollution, as has already happened in the United States.

The potential for increasing food supplies from natural resources is thus rather limited, and it is probable that the major increases will come from managed resources—from agriculture and from aquaculture, the cultivation of fish and shellfish in managed ponds, streams or other aquatic impoundments. Let us now turn our attention to some of the proposals for extending and increasing the yield of food by these means.

FOOD CULTURE—NEED FOR ENERGY INPUT

Before considering some specific proposals for increasing world food supplies by the application of culturing techniques, we should examine how it has been possible for man to "improve" on nature. As we saw in the preceding chapter and also in our study of the productivity of the oceans, a natural ecosystem depends ultimately on the sun for energy to support all the functions it carries out — synthesis of biomass at each trophic level, maintenance of individual organisms and, not least, *the eventual recycling of nutrient minerals.* The total productivity (calories per acre, for example) may be limited by light itself or by nutrients, water supply, temperature or some other limiting factor. Every stable climax ecosystem therefore achieves a maximum productivity — or maximum turnover of solar energy — in accordance with the limiting conditions within which it functions. *In order for man to increase productivity in any system, he must supply the limiting factors, and this requires input of some form of extra energy.* This is precisely what man does when he cultivates the fields or the ponds or the forests — he adds energy over and above that coming from the sun.

In early agricultural practices — what is now called "subsistence farming" — the extra energy was provided by man himself and his domesticated animals. Man learned to add his own energy to the ecosystem in ways which would benefit himself rather than some other species. He cleared land of its natural climax vegetation; he mulched and fertilized, speeding up the natural recycling of nutrients. He selected species of benefit to himself, eventually creating domesticated strains of grains and livestock. He irrigated, often changing the local hydrologic cycle. Sometimes in the process, he destroyed ecosystems which once supported him (the deserts of the southern Mediterranean coasts are a case in point). Other times, he learned to manipulate ecosystems so that they became relatively stable as long as he continued to add his own energy to them (the medieval fields of most of

Cattle in a midwestern feedlot.
    Food and water are supplied through addition of external energy; the animals expend no energy themselves on foraging. At present, wastes are not extensively recycled, and often they pollute local water resources. (One cow produces waste equivalent to that of 10 people.) When waste recycling becomes mandatory during the next decade, even further external energy will be required to maintain cattle in such high-yield feedlots. (United Nations photograph, from Wagner, R. H. 1971. *Environment and Man.* W. W. Norton & Co., New York.)

FIGURE 17a.4

Europe are an example). So long as man depended upon his own energies or those of his beasts, however, he was not essentially different, ecologically speaking, from the ants which maintain aphid colonies to increase their food resources. The ultimate energy source was still the sun.

But in the past 200 years or so, man has begun to add external energy—obtained from the burning of fossil fuels or, more recently, from nuclear power—to the ecosystem and has thereby greatly increased the productivity of his crops. One example may be used to clarify the point. Semi-domesticated cattle, such as the longhorns of the wild West, ranged widely for their food. The annual roundup and drive to the railroads collected the herbivore biomass together for redistribution to the cities via the Chicago stockyards. A small amount of external energy was involved in the collection and redistribution processes—mostly the burning of coal to construct the railways and run the locomotives. Today, however, cattle in the United States range less widely for forage, and are finally fattened in feedlots where they expend virtually no energy to obtain food and water (Fig. 17a.4). The energy they would have used is supplied by man from external sources.

Cattle feed, mainly corn, is grown on farms where gasoline-powered tractors are used to till, weed and harvest. Insecticides, herbicides and fertilizers, produced in chemical plants powered by fossil fuels and distributed to the farmer by fossil fuel powered trains and trucks, are used to maximize the production of corn. Water may be supplied to the cattle by pumps, driven by gasoline or electric engines. Synthetic hormones, whose manufacture requires energy, are often injected into cattle to increase meat production. No doubt we have omitted many other non-solar energy inputs which go into the production of animal protein in the industrialized farming of developed countries—but nevertheless, the point should be clear. If man is to significantly increase his food resources, he must add external energy to the ecosystem upon which he depends. Modern farmers depend as much on urban industry for agricultural energy subsidies as the inhabitants of cities depend on agriculture for food; the relationship is a reciprocal one.

SOLUTIONS AND PROBLEMS

It is evident that, to adequately feed the increasing human population in the next few decades, both more and better food will need to be produced. Wherever possible, it should be produced near the consumer population, to minimize spoilage, storage and distribution costs. In general, this means increasing the productivity of existing croplands in the underdeveloped countries and, to a lesser extent, increasing cropland acreage. In most cases, this does not mean merely exporting seeds, chemicals and tractors to these countries in an attempt to introduce Western technology throughout the world. For one thing, many farming practices in the developed countries are ecologically unsound, as we shall see. In addition, most underdeveloped countries lie within the tropics, and innovations must be tailored to suit local climates, soils and so forth. Attempts to apply temperate farming practices in such areas have frequently met with disaster, as noted in Chapter 8a in our discussion of laterite formation. We shall return to this point in Chapter 18.

A great variety of possibilities exist, however, for increasing world food production. Let us examine a few of these and consider possible problems they may create.

AQUACULTURE

One of the most important means for increasing animal protein foods is aquaculture, the growing of fish in freshwater or brackish water ponds or other impoundments. This technique is already highly developed in parts of China, Japan and Southeast Asia. It could readily be extended elsewhere, especially to the tropics, where fish grow rapidly all year round; transportation facilities are not needed, since ponds can be located near centers of population.

In essence, the pond is fertilized with crop and animal wastes, producing a rich growth of algae and water plants. These are fed upon directly by herbivorous and detritus-eating fish such as carp, milkfish, mullet and shrimp. The most efficient harvest is obtained by draining the ponds; the fish are collected and either eaten directly or made into concentrated sauces preserved with salt. Accumulated sediments are removed regularly and used to fertilize the fields. Annual yields of 2000 to 5000 pounds or more of fish per acre can be obtained from such ponds (Table 17a.2).

TABLE 17a.2
PRODUCTIVITY
IN FISH
PRODUCING
ECOSYSTEMS*

Ecosystem and Trophic Level	Harvest — pounds/acre/year
I. Unfertilized Natural Waters	
Mixed carnivores, natural population	
World marine fishery (average)	1.5
North Sea (a coastal fishery)	27.0
Great Lakes (large lakes)	1–7
U.S. small lakes (small lakes)	2–160
Stocked carnivores	
U.S. sport fish ponds	40–150
Stocked herbivores	
German fish ponds	100–350
II. Naturally Fertilized Waters	
Plankton-feeding carnivores (anchovies)	
Peru coastal upwelling	1500
III. Artificially Fertilized Waters	
Stocked carnivores	
U.S. sport fish ponds	200–500
Stocked herbivores	
Philippine marine ponds	500–1000
IV. Artificially Fertilized Waters plus Added Food	
Stocked carnivores	
U.S. small sport fish ponds	2000
Stocked herbivores	
Hong Kong	2000–4000
South China	1000–13,500
Malaya	3500

*Adapted from Odum, E. P. 1971. *Fundamentals of Ecology.* 3rd ed. W. B. Saunders Co., Philadelphia.

TABLE 17a.3
EFFICIENCY OF
FEED
UTILIZATION BY
VARIOUS
ANIMALS*
(Based on 100
Pounds of Food
Intake)**

Species	Live Weight Gain (Pounds)	Protein Gain (Pounds)
Cattle	16	2.6
Sheep	19	2.2
Hogs	29	3.0
Poultry	36	10.1
Catfish	72	11.8
Trout	58	7.5

*Adapted from *The World Food Problem.* 1967. Volume II, p. 352.
**The warm-blooded animals were all fed similar diets.

In the United States, where dietary protein is not limiting, ponds are cultivated for sport fish; productivity in them is lower, since bass and other sport fish are carnivores, existing at one higher trophic level than the herbivores of Asiatic ponds. Such ponds are also less heavily fertilized than those in Asia.

A particular advantage of fish culture (which is shared by poultry) is the high efficiency of production of edible protein (Table 17a.3). Whereas only about three per cent of the food eaten by cattle, pigs or sheep is converted to protein, about 10 per cent of that eaten by fish and poultry becomes protein. Moreover, fish can utilize spoiled grain, cotton seed and other organic resources which are not directly useful to man. It is estimated that by the year 2000, fish culture could provide some 19 million tons of fish annually for human consumption, supplying about one-fifth of the projected world fish requirements.

PRODUCTION OF PROTEIN ON LAND

The annual marine fish catch for 1967 was about 60 million tons net weight, and the freshwater catch was about 15 million tons. Together these amount to around 7 to 8 million tons of edible protein. By comparison, the total production of animal protein on land in 1964 was 22 million tons, about half of which was milk (Table 17a.4).

Commodity	Protein Production (10^6 metric tons)
Eggs	1.6
Beef and veal	5.3
Pork	3.0
Poultry	1.4
Cow's milk	11.0
Total	22.3

TABLE 17a.4 WORLD PRODUCTION OF ANIMAL PROTEIN IN 1964*

*Adapted from *The World Food Problem*. 1967. Volume II, p. 340.

Although animal protein is superior in quality to plant protein, its production is far less efficient in yield of edible protein per acre of land because at least one extra step in the food chain is required (Fig. 17a.5). Beef and veal, which account for half the world's meat production, are the least efficiently produced of all. They also compete directly with man for edible foods, such as corn and fish—about half the Peruvian anchovy catch, the richest fishery in the world, is not used to feed man directly but to fatten livestock and poultry! While two-thirds of the world's population is deficient in protein, the other third is consuming large quantities of inefficiently produced meat. This imbalance is due to both the subsistence level of agriculture in much of the world and the economic wealth which allows developed countries to import animal protein for feed.

It is evident in Figure 17a.5 that grains and seeds give higher per acre yields of dietary protein than do animals. This is especially true of the legumes—peas, beans and soybeans. In fact, grains and seeds supply 70 per cent of man's protein intake on a worldwide basis. Much of the protein in soybeans, and in other oil-producing seeds such as peanuts and cotton seeds, is not used at present for human consumption, however—like the Peruvian anchovies, it is fed to cattle instead! As the report on the world food problem states, "one of the greatest opportunities for increasing the world protein supply, without increasing agricultural production, is to utilize the presently available oilseed resources directly for humans." The major difficulties with such an approach are the unpalatability of these proteins and also the presence of toxic substances. The latter can be easily removed by processing, but the former objection will require new food technologies—adding flavor and texture in manufactured products.

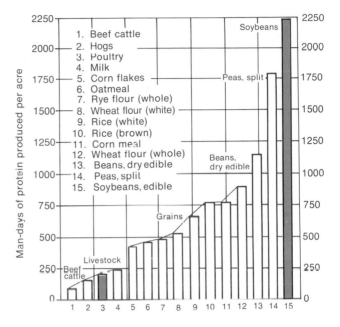

FIGURE 17a.5

The efficiency of protein production of various foods, expressed as man-days of protein produced per acre. A man-day of protein is the amount required by one adult man per day.

Note that one acre that is used to grow *soybeans* will feed about 10 times more people than one acre used to grow food for *poultry*. (After *The World Food Problem*, Vol. II.)

Another means for improving the dietary quality of plant proteins is the addition of specific amino acids to flour made from various grains. Several essential amino acids, including threonine, tryptophane and especially lysine, are often insufficient in the incomplete proteins of such foods. If all the essential amino acids are not present simultaneously, body proteins cannot be made, and much of the incomplete protein nitrogen is excreted unused. When the lacking amino acids are added to corn, wheat or rice

	Amount Fed (Milligrams Nitrogen/ Kilogram/Day)	Per Cent Nitrogen Retained
Protein Source		
Skim milk (complete protein control)	454	18.6
Corn masa (incomplete protein)	461	2.3
Corn masa + lysine + tryptophane (supplement)	464	17.8
Skim milk (complete protein control)	310	24.8
Wheat flour (incomplete protein)	328	8.2
Wheat flour + lysine (supplement)	335	17.9
Skim milk (complete protein control)	317	28.1
Rice (incomplete protein)	320	18.7
Rice + lysine (supplement)	320	24.7

TABLE 17a.5
EFFECT OF ADDING ESSENTIAL AMINO ACIDS TO DEFICIENT CEREAL PROTEINS FED TO CHILDREN ON THEIR ABILITY TO UTILIZE PROTEIN* (Measured as Per Cent Nitrogen Retention)

*Adapted from *The World Food Problem*. 1967. Volume II, p. 318.

diets, the amount of protein nitrogen retained by the body and converted into body protein is greatly increased—thus, *all* the amino acids become available for nutrition and growth if the limiting ones are supplied in sufficient amounts. This is clearly shown by the experiments summarized in Table 17a.5. The addition of essential amino acids to grains can be accomplished much more cheaply than can the provision of animal protein.

It is also possible that the quality of grains can be improved by genetic selection of strains with more complete amino acid complements. So far, the greatest emphasis has been on developing pest-resistant, high calorie-yielding grains. But experiments with corn and other grains have shown that improvements in quality as well as quantity are possible. A strain of maize known as "opaque-2" bears a mutant gene which increases its lysine content. This maize is far superior to ordinary corn in supporting the growth of experimental rats. Unfortunately, between 10 and 20 years are required for commercial development of a new strain and its widespread cultivation.

INCREASING CALORIE PRODUCTION

So far, except for aquaculture, we have said little about the means by which more food can be produced in the underdeveloped countries. There are at least four possibilities, in addition to increasing acreage under cultivation: irrigation, introduction of high-yield crop varieties, increased use of chemicals and increased use of machinery. We have considered some of the problems of irrigation projects in Chapter 2a and will examine a specific case, the Aswan High Dam, in Chapter 19a. For the moment, we can say that irrigation can substantially increase crop yields but may bring with it problems of its own unless carefully planned.

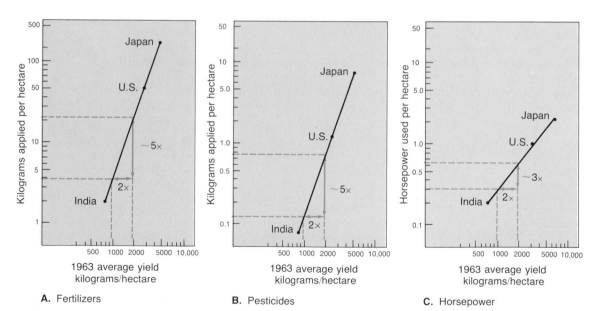

A. Fertilizers **B.** Pesticides **C.** Horsepower

FIGURE 17a.6   Diminishing returns in the application of external energy to food crops. Annual yield is expressed in kilograms dry weight of food per hectare (a hectare = 2.47 acres).
A. Fertilizers. Doubling the yield requires a fivefold increase in application. Quadrupling the yield thus requires 25 times as much fertilizer.
B. Pesticides. Doubling the yield requires a fivefold increase in application.
C. Horsepower. Doubling the yield requires a threefold increase in use.
(Modified from Odum, E. P. 1971. *Fundamentals of Ecology.* W. B. Saunders Co., Philadelphia.)

In the past two decades great advances have been made by plant genet-icists in the development of new high-yield varieties of grain, especially wheat and rice. We shall consider the history of this development further in Chapter 21a. These new grains, if given sufficient artificial fertilizer and irrigation water, are capable of greatly increasing the yields per acre—lead-ing to the recent Green Revolution. In fact, worldwide production of rice by early 1972 had reached such proportions that India and Japan, once rice importers, are, for the moment, self-sufficient—Japan is even an ex-porter of rice. This has led to a dramatic fall in the price of rice and por-tends economic disaster in the traditional rice-exporting countries of southeast Asia. The economics of solving the world food problem is ob-viously complexly intertwined with other areas of economics and politics.

In order to achieve the high crop yields characteristic of the Green Revolution, however, a great deal of extra energy has had to be added in the form of fertilizer, pesticides and horsepower, as shown in Figure 17a.6.

TABLE 17a.6
RELATIONSHIP BETWEEN
QUANTITY OF ARTIFICIAL
FERTILIZER APPLIED AND
CORN CROP YIELD IN
ILLINOIS*

Year	Tons of Fertilizer Applied in State of Illinois	Average Yield, Bushels Per Acre
1945	10,000 ⎱ 10×	50 ⎱ 1.4×
1958	100,000 ⎰ 5×	70 ⎰ 1.3×
1965	500,000	90

*Based on data from Commoner, B. 1971. *The Closing Circle: Nature, Man and Technology.* Alfred A. Knopf, New York, pp. 84–85.

Each doubling of crop yield requires an addition of five times as much fer-tilizer, or pesticides, and three times as much horsepower. Thus, agro-in-dustry in Japan produces about seven times the yield of food per acre, while using about 100 times the resources and energy that are used in India. It has been estimated that, without the addition of fertilizers, especially those containing phosphorus, world agriculture today could ad-equately feed only about 2 billion people. We are already heavily commit-ted to subsidizing our food production.

The environmental backlash of industrialized farming as practiced in developed countries is now being felt. We have already noted the wide-spread effects of DDT and other long-lived pesticides throughout the world. Artificial fertilizers also cause problems. Unlike natural fertilizers such as mulch and manure, which contain organically bound nitrogen that is slowly made available to plants by soil microbes, artificial fertilizers are readily washed out of the soil, and far more of them must be applied than is actually utilized by the crop plants. This results in diminishing returns, as shown for corn production in the state of Illinois over a 20 year interval. In this case, a 50-fold increase in fertilizer barely doubled the yield (Table 17a.6).

In addition to causing eutrophication of streams and lakes, the excess nitrate from fertilizers seeps into ground water and reservoirs. Decatur, Il-linois, a city of 100,000 people, annually faces excess nitrate in its drinking water after the winter snow leaches nitrate from the surrounding farms. Isotope fingerprinting, similar to that used for identifying the source of en-vironmental lead (Chapter 9a) has shown that the offending nitrogen comes from artificial fertilizers. Nitrate itself is not toxic, but a type of intes-tinal bacteria which is common in young children converts nitrate to nitrite. Nitrite in turn reacts with hemoglobin, forming methemoglobin, which is incapable of carrying oxygen (see Chapter 9a). Labored breathing results, and an affected infant may turn blue and die of asphyxiation. Nor is Decatur the only city so affected; deaths from infant methemoglobinemia have occurred in California's central valley and in other farming centers in the United States, Europe and Israel.

This is but one example of the multitude of ecological problems created by our present agricultural technology. In striving for short-term gains, long-term debts have been incurred. It is evident that, although much extra-solar energy has already gone into industrialized agriculture, there have been too many economic shortcuts. Soil, air and water pollution are the price. To rectify this situation, even more energy will be required in the future to maintain present yields—recycling of organic wastes from feedlots and urban sewage plants to the land, purifying and recycling water supplies, using more specific pest controls, introducing strip cropping and crop rotation (Fig. 17a.7) and so on.

FIGURE 17a.7   Strip cropping and crop rotation being practiced on rich agricultural land. Such practices insure constant fertility and reduce the need for pesticides, fertilizers and other artificial inputs. (From USDA—Soil Conservation Service.)

To export wholesale our present farming methods to the underdeveloped countries, even taking into account climatic, economic and social differences, will only spread the ecological problems and incur a further debt. Nevertheless, the problem is urgent, and many will argue that such measures are essential to prevent worldwide famine. It is to be hoped that greater emphasis will be placed on more efficient distribution and equitable utilization of food resources already available. Meantime, there is great scope for research into and application of ecologically sound principles for increasing the world's food supply. No solution can be successful, however, until the rate of population increase is drastically slowed. We shall return to this issue in Chapter 22a.

SUMMARY

It is estimated that today's global production of both calories and proteins is sufficient to feed the world's population were food more equitably distributed. This will soon not be true if population growth continues at its present rate. About two-thirds of the world's population is malnourished, leading to kwashiorkor, disability and disease.

Partial solution may come from increased food gathering. Utilization of wild game may be nutritionally and ecologically beneficial in some areas. The marine fisheries are not likely to be the panacea many believe; the open ocean produces few fish because of lack of nutrients, a long food chain and inefficient energy transfer. Many coastal areas and regions of upwelling, where most of the world's fish are produced, are already over-fished; to double our present catch and thus reach the maximum sustainable yield will necessitate exploitation of remote regions.

The cultivation of food requires extra energy input into an ecosystem—the addition of nutrients, limitation of parasites, pests and competitors, irrigation and so forth. Aquaculture could be extensively increased, especially in underdeveloped countries, to provide significant amounts of animal protein. Although animal protein is generally more nutritious than plant protein, it is ecologically expensive to produce, feeding only about one-fifth or one-tenth as many people per acre. Much plant protein could be made more nutritious by addition of essential amino acids.

To increase calorie production also requires the addition of horse-power and chemicals, five times as much of these being needed to double productivity. Modern agricultural practices often create environmental problems which threaten the health of man and the biosphere and therefore should not be expanded unmodified. Even greater energy inputs will be needed to maintain productivity without destroying the ecosystem. Food production in the future must adhere to sound ecological principles.

Bardach, J. 1968. *Harvest of the Sea.* Harper & Row, New York. A somewhat more optimistic appraisal of marine food resources than given here.

Borgstrom, G. 1968. *Too Many.* Collier-MacMillan, Toronto. A considered discussion of the limits to world food production.

Brown, L. R. 1970. *Seeds of Change: The Green Revolution and Development in the 1970's.* Frederick A. Praeger, New York. The former administrator of the International Agricultural Development Service analyzes the impact of the Green Revolution.

Commoner, B. 1971. *The Closing Circle: Nature, Man and Technology.* Alfred A. Knopf, New York. A book for the lay reader describing man's technological impacts on the ecosystem, including those from agro-industry.

Dasmann, R. F. 1964. *African Game Ranching.* Pergamon-Macmillan, New York. Experiments show that African game produce as much meat per acre as cattle, but at less cost and without destroying the environment.

Ehrlich, P. R. and Ehrlich, A. H. 1970. *Population, Resources, Environment: Issues in Human Ecology.* W. H. Freeman & Co., San Francisco. Considers, somewhat pessimistically, the population problem and the chances of feeding a hungry world.

Kohl, D. H., Shearer, G. B. and Commoner, B. 1971. Fertilizer nitrogen: contribution to nitrate in surface water in a corn belt watershed. Science *174*:1331–1334. A research article pinpointing fertilizer nitrogen as the source of nitrate in drinking water of farm communities.

Ladejinsky, W. 1970. Ironies of India's green revolution. Foreign Affairs *48*:758–768. Exposes some of the economic and social dislocations of surplus food production.

Mertz, E. T., *et al.* 1965. Growth of rats fed on opaque-2 maize. Science *148*:1741–1742. Original data demonstrating nutritional value of a new genetic strain of corn.

Odum, E. P. 1971. *Fundamentals of Ecology.* W. B. Saunders Co., Philadelphia. In chapters 3 and 15 an eminent ecologist explains the ecological necessity for extra-solar energy input to increase crop yields and some of the attendant dangers in taking economic short cuts.

Ryther, J. H. 1969. Photosynthesis and fish production in the sea. Science *166*:72–76. A foremost marine ecologist shows why the ocean does not offer unlimited supplies of food for man.

The World Food Problem. A Report of the President's Science Advisory Committee Panel on the World Food Supply. 1967. 3 volumes. Superintendent of Documents, Washington, D.C. A comprehensive analysis of the practicability of various means of feeding an ever-increasing population.

NATURAL RECYCLING

Within the biosphere there is a more or less constant recycling of nutrients, which is accomplished by expenditure of a certain fraction of the energy flowing through various ecosystems. Man's technology has altered these cycles, sometimes locally, sometimes on a global basis.

Aside from minor losses to outer space and a small number of transmutations due to radioactive disintegrations, all of the atoms originally present on our giant spaceship Earth are still here. The atoms of the 30 to 40 elements which go to make up living organisms have been used over and over again, moving between the living and non-living worlds in an endless process of natural recycling—what have come to be known as *biogeochemical cycles.* All living organisms are in a state of dynamic equilibrium with their non-living surroundings, and there is continuous interaction between them. Not only does the non-living environment determine the nature and abundance of the organisms it supports, but the organisms in turn alter their non-living surroundings.

In this chapter we shall concern ourselves with only a few of the major elements—hydrogen and oxygen (combined as water), carbon, nitrogen and phosphorus. But the reader should keep in mind that less abundant elements—sulfur, potassium, calcium and so forth—as well as certain small organic molecules, such as vitamins, also undergo continuous cycling which is of equal importance to the health of the biosphere. At certain points in the cycle of each element, technological man is having an impact, as we shall see, and there is much to be learned from a study of natural recycling.

SOME GENERALIZATIONS

Prior to discussing the details of individual cycles let us examine some general concepts about biogeochemical cycles. In the non-living environment there is usually a large reservoir or storage depot of a given element. In the case of nitrogen, for example, this is the air; in the case of phosphorus, it is the Earth's crust. There is also a more "fluid" compartment—one which exchanges the element readily with living organisms. In the case of nitrogen, this is mainly inorganic nitrate in soil or water;

491

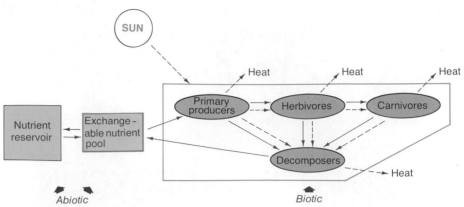

FIGURE 18.1

Schematic diagram of the three major compartments of a recycling nutrient (solid arrows) in an ecosystem.

There are usually two abiotic (non-living) compartments: the large, relatively unreactive nutrient reservoir, and the smaller, exchangeable pool. The latter readily interacts with the biotic (living) compartment, through which nutrients are recycled.

This diagram, which also shows one-way flow of energy through the system, may be compared with that of Figure 16.1.

phosphorus exchanges with organisms mainly as dissolved phosphates. We can thus recognize at least three compartments for any element: the reservoir, the non-living exchange fraction or "pool" and the living organisms themselves (Fig. 18.1). Each of these may be further subdivided, as we shall see below, and organisms may convert an element from one mineral form to another.

Although it is of importance to know the total amount of an element in each of its compartments, of greater interest is the rate at which it moves from one compartment to another. For example, since nitrate is often the limiting factor for plant growth (Chapter 17), it is of less importance to know how much nitrogen gas is in the atmosphere than it is to have an estimate of the rate at which certain soil microorganisms convert atmospheric nitrogen to forms usable by plants.

It is also useful to keep in mind that certain steps in each biogeochemical cycle are energy-yielding reactions, whereas others require energy. In general, conversion from organic to inorganic forms is accomplished with the release of energy, while the reverse steps require input of energy, either from photosynthesis or from oxidation of organic molecules, that is, from cellular respiration.

In considering biogeochemical cycles of individual elements it is necessary to realize that the cycles overlap at certain points. This is especially true of oxygen, which forms complex ions or molecules with many other elements: for example, hydrogen (H_2O), nitrogen (NO_3^- and NO_2^-), carbon (CO_2), phosphorus ($PO_4^\equiv$ and P_2O_3 — phosphorus trioxide, the common soil form of P) and sulfur (gaseous SO_2, and $SO_4^\equiv$). Although we shall not consider the oxygen cycle separately from that of water, the reader should keep in mind the importance of the interactions of oxygen in these other cycles, as well as the significant roles of molecular oxygen (O_2) in respiration of plants and animals discussed in Chapter 7.

THE HYDROLOGIC CYCLE

Although water is a compound of two elements, hydrogen and oxygen, it is convenient to consider the cycling of water rather than of its constituent atoms since only a very small proportion of the water molecules are chemically degraded during the cycling process. In addition to circulating water, the hydrologic cycle also serves an important function in distributing dissolved gases, ions and organic molecules and thus participates in the cycling of most other elements.

In Chapter 2 we considered the daily global turnover of water and in Chapter 2a, some of the movements of water on land, both natural and man-made. We noted that the main driving force of the water cycle is the solar-powered evaporation of water from the oceans and its transfer to land as precipitation. The major role played by living organisms in cycling water is through *transpiration* of terrestrial plants (evaporation of water from leaves). Transpiration often significantly affects the rates of water run-off on land as well as insuring cycling of nutrients. Because transpiration is so important in terrestrial ecosystems, let us examine its functions in more detail. These are summarized in Figure 18.2. It is important to keep in mind that whenever water evaporates—whether from oceans, soil or leaves—energy is required to convert it from a liquid to a gas; this energy is almost always solar energy—that is, heat from the sun.

THE ROLE OF TRANSPIRATION

Photosynthesis in most terrestrial plants requires the transpirational movement of large volumes of water. In fact, 97 to 99 per cent of water entering the roots of plants is lost by evaporation from the leaves; the rest is

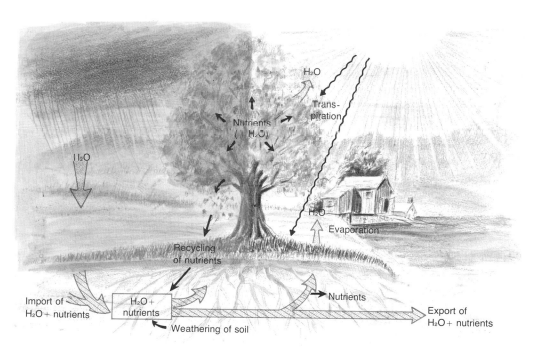

FIGURE 18.2

The role of evapotranspiration in nutrient retention and utilization.
Rainwater dissolves nutrients from the soil; together they are absorbed by the roots of vegetation and transported upward owing to transpiration. Nutrients (and a small amount of water) are retained in the plant, permitting growth to result from photosynthesis. Eventually, nutrients are recycled in decaying vegetation. Evaporation directly from the soil leaves nutrients behind. Excess water run-off from the ecosystem removes nutrients. If this run-off exceeds nutrient import, the ecosystem becomes impoverished. Removal of vegetation prevents transpiration, and the excess water run-off leads to increased nutrient loss.

utilized in photosynthesis and water storage within the plant. In heavily vegetated ecosystems, such as grasslands and forests, from 50 to 90 per cent of the total solar energy involved in evaporation of water is accounted for by transpiration from plants, the remainder being water evaporated from soil.

Although often regarded as an unwanted stress since it greatly increases water requirements of crops, transpiration is essential for plant growth. About 1000 grams of water must be transpired for every two grams of dry plant tissue synthesized. As we saw in Chapter 9, transpiration is a major force in moving water and nutrients up the xylem and distributing them throughout the plant. Under conditions where transpiration is inhibited (in deserts where groundwater is limited or in cloudy rain forests where high humidity suppresses evaporation from leaves), plant growth is slow. Rates of plant growth are often closely correlated with transpiration rates.

As we also saw earlier, in our discussion of grasslands (Chapter 16), transpiration increases the retention and availability of dissolved nutrients in the soil. Nutrients are brought to the surface by capillary water movements and, while incorporated into plants, cannot be leached from the ecosystem through groundwater run-off. The trapping of nutrients in plants is a major means of retaining them so that they can be recycled over and over again.

Plants also, through transpiration, reduce the amount of water run-off in an ecosystem and so decrease nutrient loss due to erosion. This was clearly demonstrated experimentally in forested areas in New Hampshire. In an untouched control forest, nutrient losses from the ecosystem were monitored by measuring the quantities of nutrient ions leaving the watershed in effluent streams; losses under these natural conditions were exceedingly small. The vegetation in a second, experimental forest was felled but left lying on the ground, and further growth was suppressed by spraying with herbicides. In this experimentally denuded forest, water run-off and nutrient loss were far greater than in the control forest. The only difference between the control and experimental forests was the presence of living plants. The results clearly show that transpiration, by decreasing water run-off, reduces leaching of nutrients. Prolonged denuding of soils thus leads to loss of nutrients, erosion and formation of wasteland (Fig. 18.3).

THE CYCLE OF METABOLIC WATER

So far we have been considering the water molecules which are not incorporated permanently into living organisms. Most of the water involved in plant transpiration, for instance, does not become a part of the living system. Only a tiny fraction of the cycling water molecules actually form a part of living organisms, but these are of great significance to life processes. From 70 to 95 per cent or more of the biomass of living organisms is composed of water molecules; moreover, water is incorporated into organic molecules during photosynthesis and released again during respiration, with the simultaneous production and utilization of molecular oxygen.

$$H_2O \quad + \quad CO_2 \quad \underset{\text{respiration}}{\overset{\text{photosynthesis}}{\rightleftharpoons}} \quad [CH_2O] \quad + \quad O_2$$

(water)   (carbon dioxide)   (carbohydrate)   (oxygen)

A brief summary of the cycling of water in living organisms, as shown by Figure 18.4, will refresh the reader's memory regarding these important

FIGURE 18.3

An example of needless destruction in a logging operation leading to nutrient loss and erosion from exposed soil. (From Odum, E. P. 1971. *Fundamentals of Ecology.* W. B. Saunders Co., Philadelphia.)

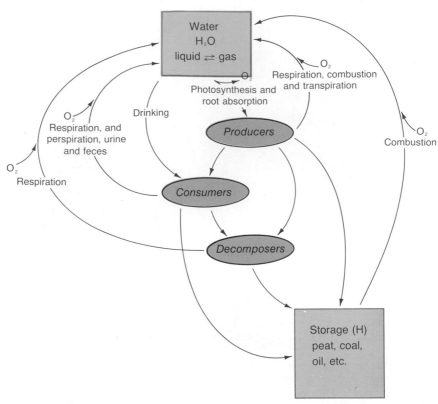

FIGURE 18.4

The cycle of water in living organisms.
    Between 70 and 95 per cent of an organism is composed of water; organic molecules are also synthesized from water during photosynthesis. Aside from transpiration, the main recycling of water occurs as a result of respiration and combustion. Note that oxygen is released from water during photosynthesis and recombined to form water during respiration and combustion. Hydrogen in unoxidized organic matter may be stored as fossil fuels (peat, coal, oil and so forth), to be burned subsequently by man, again forming water.

processes and also aid him in fitting photosynthesis and respiration into the overall cycling of elements in the global ecosystem.

    A small fraction of the water entering plants is converted into organic molecules by photosynthesis, with the release of oxygen. Both molecular water and organic molecules flow through the food chain, some of the latter being reconverted to water through respiration at each step. Animals also cycle water by drinking and excretion. Unburned organic molecules, which contain hydrogen originally obtained from water, may become stored as fossil fuels. If these are burned later by man, their hydrogen is combined with atmospheric oxygen to form water, which is thus restored to the global ecosystem.

    We have already discussed in Chapters 2, 2a and 8a some of the ways in which man has affected or may affect the hydrologic cycle, and we shall consider some effects of a specific case, the Aswan Dam, in Chapter 19a.

THE CARBON CYCLE

    The carbon of all organic molecules comes ultimately either from carbon dioxide in the air or from bicarbonate ion dissolved in water. A review of the way carbon dioxide and bicarbonate ion are interconverted is shown in Figure 18.5, together with the formation of the main long-term storage form of carbon, calcium carbonate. Although all of the reactions shown in the diagram are reversible, for all intents and purposes calcium carbonate buried in the deep layers of sediments is unavailable. It remains buried for millions of years, and thus represents an inaccessible reservoir. The exchangeable carbon dioxide pool is divided between the two per cent which is free in the atmosphere and the 98 per cent dissolved in oceanic waters. The oceans thus act as a great buffer or stabilizer for atmospheric carbon dioxide, helping to maintain it at a constant level. We shall consider man's effects on this balance shortly.

The carbon cycle is shown schematically in Figure 18.6. The major cycle (heavy arrows) involves the incorporation or fixation of carbon by photosynthetic organisms, and its stepwise release through respiration at each step in the food chain. A much smaller amount of carbon dioxide is also fixed by non-photosynthetic, energy-requiring biochemical pathways, and this also is recycled via respiration. Carbon is recycled fairly rapidly by these processes, it being estimated that all atmospheric carbon dioxide cycles through the biosphere once every 20 to 30 years.

A much slower cycling of carbon involves its long-term storage in the Earth's crust, which contains more than 99 per cent of the world's total carbon. About 98 per cent of this stored carbon exists in the form of oxidized carbonates, mainly limestone which forms from precipitation of calcium carbonate in the oceans and from sedimentation of carbonate-containing shells. The remaining two per cent exists as reduced organic carbon; of this, one-third is concentrated in fossil fuels (coal, peat, oil) and two-thirds is widely dispersed in shales and other rocks.

The stores of reduced organic carbon were gradually accumulated at periods in the past when global photosynthesis slightly exceeded respiration. This also resulted in a simultaneous accumulation of atmospheric oxygen, which accounts, at least in part, for the high level of oxygen in the air today (a point we shall return to in Chapter 20).

On the whole, however, carbon losses through chemical precipitation, shell sedimentation and fossil fuel formation have been approximately equaled by natural weathering and volcanic emission. It is thought that, prior to the industrial revolution, the concentration of atmospheric CO_2 remained fairly constant for thousands of years due to two great buffering effects. One of these is the capacity of the ocean to dissolve and release CO_2,

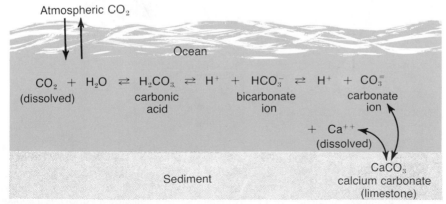

FIGURE 18.5

The major inorganic reactions of carbon dioxide.

Atmospheric CO_2 dissolves readily in water, where it combines with water molecules to form carbonic acid. This dissociates to hydrogen and bicarbonate (HCO_3^-) ions. The latter can dissociate further to another hydrogen ion and carbonate ($CO_3^=$) ion. This combines readily with dissolved calcium ions (Ca^{++}) to form a precipitate, calcium carbonate or limestone ($CaCO_3$).

All of these reactions are reversible.

already mentioned. The other is the fact that photosynthesis, especially on land, is limited in many areas by the availability of carbon dioxide. Florists sometimes release carbon dioxide inside greenhouses to promote plant growth! Thus, as atmospheric CO_2 increases, photosynthesis is accelerated, again decreasing the atmospheric level; conversely, when CO_2 levels in the air fall, photosynthesis is slowed. Plants thus exert a negative feedback control on atmospheric CO_2.

In recent times, as we noted earlier, burning of fossil fuels by man has exceeded the rate at which CO_2 buffering can be accomplished by the plants and the oceans, and CO_2 is accumulating in the atmosphere (Fig. 18.7). About six billion tons of CO_2 are released into the atmosphere each year from homes, factories and power plants. In addition to this, clearing and plowing of land for agriculture hastens oxidation of humus in the soil and adds another two billion tons of CO_2 to the atmosphere annually. A

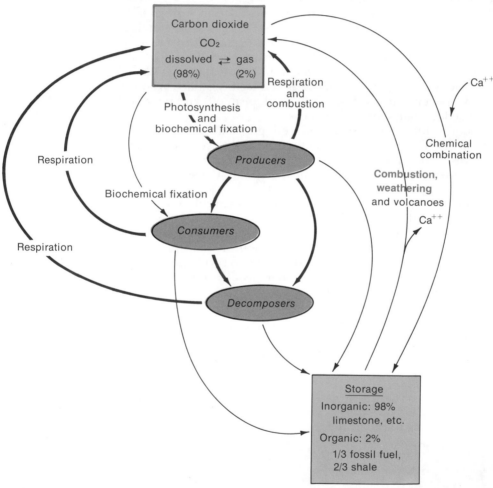

FIGURE 18.6   The carbon cycle.

Dissolved bicarbonate and atmospheric carbon dioxide are incorporated by producers during photosynthesis and are transferred through the food chain. At each step, some is returned to the exchangeable fraction by respiration. A small amount of CO_2 is also fixed by energy-requiring biochemical pathways. Weathering of limestone and shale rocks, volcanic emissions and man's combustion of fossil fuel return stored carbon to the atmosphere and oceans. Dissolved carbonate continuously forms new limestone in oceanic sediments by precipitation with calcium ions and by accumulation of carbonate-containing skeletal matter from dead organisms.

Above, the heavier arrows indicate routes of greatest turnover. Blue lettering indicates points where the activities of man affect the cycle.

century ago, atmospheric CO_2 concentration was only 290 parts per million; today it is 320. It is estimated that by the year 2000, atmospheric CO_2 will have reached about 380 parts per million, an increase of nearly 50 per cent over what it was a century ago. Some of the possible effects of such an increase have already been discussed in Chapter 7a.

FIGURE 18.7

Recent changes in atmospheric carbon dioxide measured at Mauna Loa Observatory in Hawaii, which is free from effects of local industrial pollution. Note the normal seasonal fluctuations: CO_2 rises with the general increase in respiration in the spring and falls at the end of the major photosynthetic period in the autumn. The overall upward trend is shown by the solid line. (After Bolin, B. 1970. The carbon cycle. In *The Biosphere*. Freeman, San Francisco.)

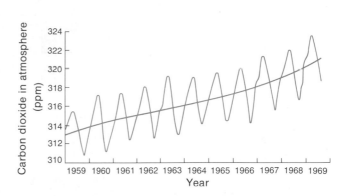

THE NITROGEN CYCLE

The major reservoir for the element nitrogen is the atmosphere. Unlike carbon dioxide and water, however, gaseous nitrogen, N_2, is not directly usable by most green plants. It enters the biosphere only after chemical conversion, a process known as **nitrogen fixation** (Fig. 18.8).

Less than 10 per cent of nitrogen fixation results from electrical discharges and photochemical reactions in the atmosphere, during which gaseous nitrogen and oxygen are combined chemically to form nitrate. The remainder is due to the action of nitrogen-fixing microorgansims. The most familiar of these are the root nodule bacteria associated with leguminous plants (peas, beans, alfalfa, clover). The bacteria invade the root hairs, which are then stimulated to grow into nodules wherein bacteria incorporate nitrogen into organic molecules (Fig. 18.9). Legumes are thus extremely useful crops for returning nitrogen to the soil, since the nodule bacteria fix far more than is used by their host plants.

Free-living soil bacteria are also capable of fixing nitrogen and so are certain blue-green algae. Most nitrogen fixation in aquatic environments is probably due to these primitive photosynthetic organisms. Recent investigations have revealed that far more types of microorganisms than previously believed are capable of nitrogen fixation.

Once nitrogen has been converted into a form usable by plants, it can be taken up and synthesized into amino acids, proteins and other nitrogen-containing compounds. These are transferred through the food chain, finally to be released by decomposers as small organic molecules. These in turn are converted by other bacteria to ammonia.

Although some plants are capable of utilizing ammonia directly as a nitrogen source, most grow best on nitrate. Because nitrogen can occur in so many oxidation-reduction states (NH_3, NO_2^-, NO_3^-) several transformations are required to complete the cycle in the soil. For this, two more groups of bacteria are required: the nitrite bacteria, which oxidize ammonia to nitrite, and the nitrate bacteria, which oxidize nitrite to nitrate. Both groups of bacteria, incidentally, obtain energy during these oxidation steps. This two-step conversion of ammonia to nitrate is called **nitrification.** To further complicate the picture there is yet another group of soil bacteria, the nitrogen-reducing bacteria, which reverse this process; they reduce nitrate to nitrite and ammonia.

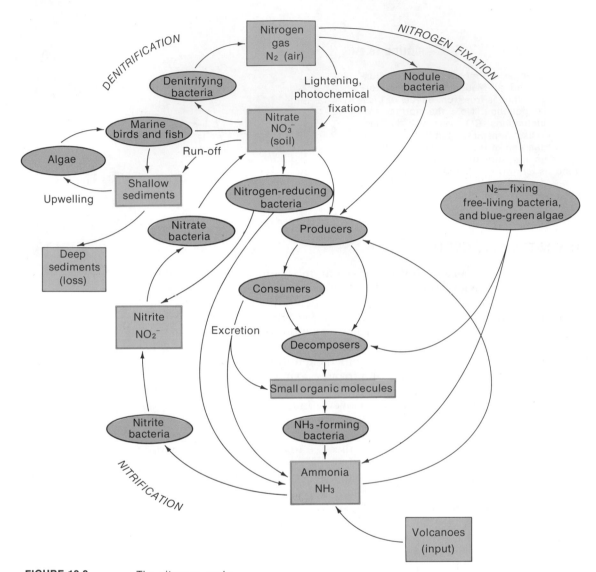

FIGURE 18.8   The nitrogen cycle.
        The storage reservoir is atmospheric N_2. This is converted by bacteria and blue-green algae into forms which can be used by green plants (nitrogen fixation). The main exchangeable pool is NO_3^- in soil, fresh water and oceans. Recycling of nitrate occurs both on land and in the sea. Decomposers release small organic molecules from the food chain, whose nitrogen is sequentially converted to NH_3, NO_2^- and NO_3^- by various soil bacteria. Nitrogen-reducing bacteria partially reverse this process. Nitrogen is returned to the atmosphere by denitrifying bacteria. Nitrogen is returned from sea to land by excreta from birds and man's catch of fish. Nitrogen lost to deep sediments is approximately balanced by ammonia released from volcanoes.

Unless nitrate in the soil is rapidly taken up by organisms it is readily leached out by rainwater; as a result, a considerable fraction of nitrogen fixed on land finds its way into the sea. This partly accounts for the high productivity of coastal waters, which receive nutrient rich run-off from the land. After cycling in the coastal waters, some of this nitrogen is eventually returned to the land, either in the excreted wastes of marine birds or as fish caught and consumed by man. That which accumulates in the deep sediments, however, is lost for literally millions of years, until uplifting of the Earth's crust returns it to the cycle. To a large extent, this loss is balanced by the addition to the cycle of nitrogen, in the form of ammonia, discharged by volcanoes.

Finally, nitrogen is returned to the atmosphere by the action of denitrifying bacteria, yet one more group of microorganisms found in soil. This final process is called **denitrification.**

The nitrogen cycle is far more complex than those of most other elements because nitrogen exists in so many inorganic forms. Nitrogen recycling is wholly dependent on the activities of soil microorganisms, particularly bacteria, to convert it from one form to another. Although there is no clear-cut evidence to date that the heavy use of pesticides adversely affects soil bacteria directly, it is known that they destroy many other soil organisms, such as earthworms, roundworms, mites and insects. We are only beginning to comprehend the complex relationships among all the organisms in the soil microenvironment, but experience from larger ecosystems suggests that effects on one species are usually translated indirectly to others. Although soil invertebrates contribute an insignificant fraction to the total respiration in a soil sample, they appear to have an important, although as yet unknown, relationship with their associated microorganisms.

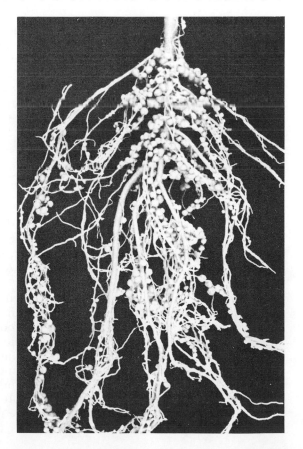

FIGURE 18.9

Photograph of roots of a legume (bird's-foot trefoil), showing nodules. (Courtesy Nitragin Co., Milwaukee, Wis.)

For example, killing the invertebrates (but not the soil bacteria) was found to decrease the rate of nutrient recycling in an experimental soil sample (Fig. 18.10). Thus, in the absence of associated animals, soil microbes cannot perform as effectively.

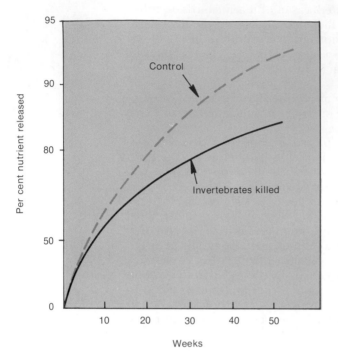

FIGURE 18.10

Release of nutrients from soil samples in which invertebrates were present (dashed line) and in which they had been killed (solid line). Soil microbes appear unable to function as effectively in the absence of the invertebrates. (After Odum, E. P. 1971. *Fundamentals of Ecology.* W. B. Saunders Co., Philadelphia.)

Depletion of soil nitrogen following intensive agriculture is a common occurrence. As we saw in the preceding chapter, fertilizing with excessive amounts of artificial nitrates leads to contamination of water supplies as well as unwanted eutrophication of rivers and lakes. A better solution is crop rotation, in which legumes are regularly alternated with other crops. Extensive recycling of animal and human wastes, from feedlots and sewage plants, would also help restore nitrogen to the soil while reducing current water pollution problems.

THE PHOSPHORUS CYCLE

The main reservoir of phosphorus is the soil and rocks of the Earth's crust, from which it is gradually released by natural weathering processes and erosion resulting from man's activities. Phosphorus is released and utilized as inorganic phosphate ion ($PO_4^{\equiv}$), the main exchangeable form. The phosphorus cycle, shown in Figure 18.11, is relatively simple compared to that of nitrogen.

As with other nutrients, the exchangeable inorganic phosphate is taken up by primary consumers, incorporated into organic molecules, and transferred through the food chain. The decomposers, however, do not release inorganic phosphate; instead, they release large quantities of soluble organic phosphate compounds, which cannot be utilized directly by plants. The dissolved organic phosphates are recycled to phosphate ions by the action of phosphatizing bacteria. The organic phosphate pool in aquatic ecosystems may often be far larger than the inorganic pool of phosphate ions because the rate of conversion by phosphatizing bacteria is usually slower than the rate of phosphate utilization by plants. Indeed, the

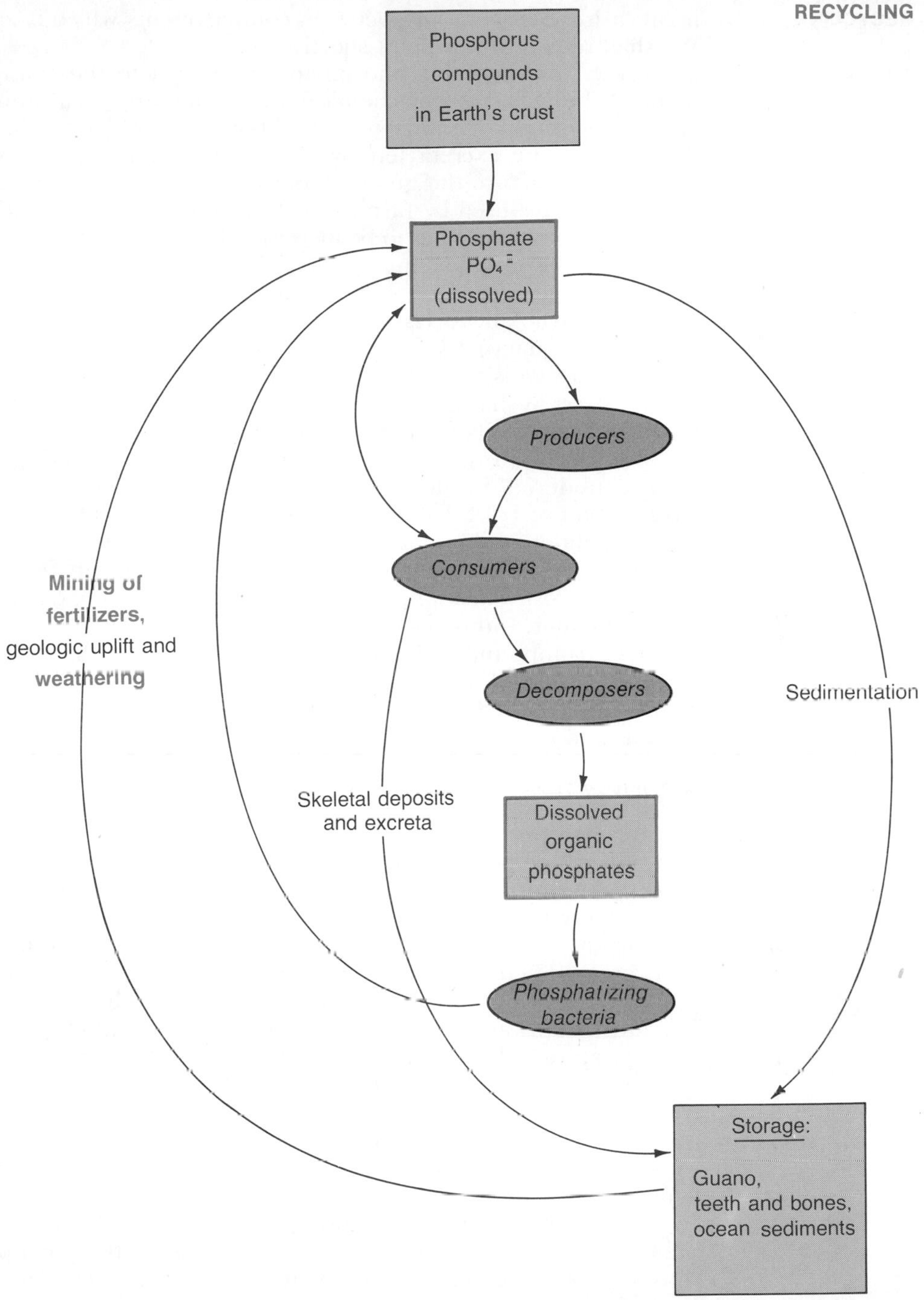

FIGURE 18.11

The phosphorus cycle.

   The major reservoir of phosphorus is the Earth's crust, which is converted by weathering to $PO_4^=$. This is the form in which phosphorus is incorporated into organic molecules by producers and is transferred through the food chain. Decomposers release mainly organic phosphorus compounds which are converted to $PO_4^=$ by phosphatizing bacteria. Storage occurs from chemical precipitation as well as accumulation of excreta and skeletal matter.

   Natural geological activity returns stored phosphorus to the cycle. Man hastens this process by bulldozing and plowing, and by mining fertilizers, as shown in color.

phosphate cycle is a prime example of a situation in which little can be determined from measuring the quantities of the element in each compartment: it is the *turnover rate* between compartments which is all-important. We shall return to this point shortly.

Freely exchangeable phosphate is removed to the storage compartment of the cycle both by chemical sedimentation, as calcium phosphate, and by deposition of animal teeth and bones and excreta. The huge **guano** deposits are the excreta left by shore birds along sea cliffs. The most famous of these are the guano deposits along the coasts of Ecuador and Peru, which are used by farmers as fertilizer. If the rich anchovy fishery along that coast continues to be increasingly exploited by man, the food of these shore birds will be in short supply, and the all-important guano deposits will decline. Thus man, in seeking to augment his food supplies from one source (the anchovies), may be depleting them elsewhere (the fields fertilized by guano).

Weathering is the most effective natural agent releasing phosphorus. This process is hastened not only by man's use of guano and other phosphorus deposits for fertilizer, but also through farming and clearing of land, which powders the soil and exposes it to weathering. We have already considered the effects of bulldozers on the phosphorus-induced eutrophication of Lake Tahoe in California (Chapter 2a).

In addition to these practices man has also interfered with the normal phosphorus cycle by adding large quantities of phosphate-containing detergents to freshwater supplies. TSP (trisodium phosphate, or Na_3PO_4) is a prime offender. Industrial wastes and phosphate-containing artificial fertilizers have compounded the problem. Wherever phosphate is limiting in an ecosystem, its addition leads to increased plant growth. All too often this results in unwanted algal blooms in previously clear and attractive lakes, streams and lagoons.

TURNOVER TIMES OF PHOSPHORUS

Because phosphorus in the exchangeable fraction is a nutrient which is rapidly cycled between organisms and their environment, let us use it to explore the relation between exchange rate or **turnover time** and productivity. In many ecosystems, especially aquatic ones, phosphate tends to be a limiting nutrient. If more phosphate is added, productivity increases. However, as we noted above, simply measuring the amount of dissolved $PO_4^{\equiv}$ present at a given moment tells us little about productivity.

Most plants are capable of rapidly incorporating $PO_4^{\equiv}$ even when it is present at very low concentrations. In proportion to its concentration as phosphate in the environment, phosphorus has a higher concentration in living organisms than does any other element. Thus it is not the momentary concentration of available phosphate which is important but the rate at which it *becomes* available — or, in other words, the rapidity with which it is recycled.

Upon adding radioactive phosphate ($^{32}PO_4^{\equiv}$) to a sample of lake or ocean water, one can follow its rate of removal by the phytoplankton and thus estimate the turnover time of $PO_4^{\equiv}$ in the exchangeable fraction. Data obtained from such experiments on lake and ocean waters are given in Table 18.1.

	Summer	Winter
Lake	1–7 minutes	30–196 hours
Ocean	1–56 hours	13–166 hours

TABLE 18.1
TURNOVER TIME
OF PHOSPHATE
IN LAKES AND
OCEANS*

*Based on data from Kormondy, E. J. 1969. *Concepts of Ecology*. Prentice-Hall, Inc., Englewood Cliffs, N. J.

Note that in the summer, when $PO_4^{\equiv}$ concentrations in a lake are lowest owing to the rapid growth of algae, turnover time is extremely short —from one to seven minutes. A $PO_4^{\equiv}$ ion released by phosphatizing bacteria remains free for a very short time indeed! In the winter, turnover times are longer; utilization by plants is slower owing to less light and lower temperatures. Phosphate concentrations thus tend to increase in winter.

In the oceans, turnover time is longer than in summer lakes and shows little change with season. Ocean temperatures vary less during the year than do those in lakes, which may account for the smaller seasonal differences in ocean waters. Recent evidence also suggests that nitrate, rather than phosphate, may be the limiting nutrient in the ocean (at least in coastal waters), hence explaining the slower summer turnover of phosphate in ocean waters.

RECYCLING IN THE TROPICS

As we noted in the last chapter, it is often impractical to transport temperate agricultural practices to the tropics—they simply don't work. One reason for this is that nutrient recycling patterns in the tropics are rather different from those in temperate latitudes. Figure 18.12 compares the distribution of organic carbon in temperate and tropical forest ecosystems.

In a temperate forest more than half the organic carbon exists either as litter on the forest floor or as humus in the soil. Not so in a tropical forest, where almost all the organic carbon is in the living biomass. The reason is simple. At high temperature and humidity, decay processes are rapid; the heavy tropical rains soon leach soluble molecules from the soil unless they are taken up by organisms. If we were to examine the soil in a tropical forest, we should find that much of the organic carbon does not occur as humus but is incorporated in an extensive network of soil fungi closely associated with the roots of trees. What is true for carbon is also true for the other nutrients in a tropical forest. Very little nutrient matter exists outside the living organisms. Litter and humus do not accumulate. It is almost as though each organism passed its nutrients directly on to the next without ever releasing them into a non-living compartment. The stability and continuing productivity of tropical ecosystems is in large part due to the great diversity of plants and animals, interacting with one another in a constant process of recycling. The soils themselves have virtually no nutrients.

This difference between nutrient cycling in temperate and tropical regions has important practical consequences for tropical agriculture. In general, the temperate zone practice of clearing large areas and planting

FIGURE 18.12

The relative distribution of organic carbon in two types of forest. In northern coniferous forests, the soil and litter contain most of the organic carbon, but in a tropical forest, it is mostly in the trees themselves. (After Odum, E. P. 1971. *Fundamentals of Ecology*. W. B. Saunders Co., Philadelphia.)

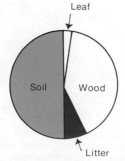

Northern Coniferous Forest

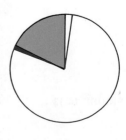

Tropical Rain Forest

them in single crops year after year soon leads to deterioration of tropical soils, since nutrients are rapidly leached away by heavy rains. Even plantations of tropical bananas, which initially produce highly marketable fruit, eventually deteriorate because of insufficient species diversity to maintain nutrient recycling. The native practice of mixed-crop agriculture, or **polyculture,** although less efficient, can be maintained on the same plot of ground for long periods of time (Fig. 18.13). The cultivation of many plant species side by side mimics the natural species diversity and thus helps to maintain natural recycling.

Native agriculture, however, is generally of a shifting type: a plot is cleared and used only for a temporary period and then allowed to return to its natural state. When the more tropical plants, such as bananas, yams and sugar cane, are replaced by polycultures of seed crops (corn, beans and squash) a particular area can be farmed for only a short period, perhaps two to five years. In certain areas of the Philippines, where rice has been grown on the same small jungle clearings for centuries, it is thought that the neighboring forests make an essential contribution to the stability of the overall ecosystem.

SUMMARY

The elements found in living organisms are constantly being cycled between the living and non-living worlds. The abiotic part of a biogeochemi-

FIGURE 18.13

Polyculture of tropical plants in a forest clearing in Costa Rica. Resembling a jungle itself, this garden contains bananas, sugar cane, young papayas and other plants used for food and medicine. (From Ehrenfeld, D. W. 1970. *Biological Conservation.* Holt, Rinehart and Winston, New York.)

cal cycle is usually subdivided into a large storage reservoir and a small exchangeable pool which interacts with the biotic compartment.

The hydrologic cycle includes not only the geochemical recycling of fresh water but the transpiration of water by terrestrial plants, important for distribution of nutrients within them, and the photosynthesis-respiration cycle within the biosphere. Most environmental water is directly available to organisms.

Most carbon is stored in limestone. Gaseous or dissolved carbon dioxide readily exchanges with organisms via photosynthesis and respiration. Combustion of fossil fuel by man is increasing atmospheric CO_2, despite the ocean's buffering capacity.

The major nitrogen reservoir is the atmosphere; gaseous nitrogen is made available to the biotic compartment mainly by nitrogen-fixing microorganisms, including nodule bacteria of legumes. Recycling of nitrogen within the exchangeable compartment depends on several groups of soil and sediment bacteria. Agriculture tends to deplete soil nitrogen, and pesticides may slow recycling of soil nitrogen by killing invertebrates.

The Earth's crust is the main reservoir of phosphorus, which is often released at high rates through man's activities. Inorganic phosphate, which forms the primary exchangeable fraction, is rapidly taken up by plants, especially during the peak of the growing season. It is the turnover rate of phosphorus rather than its concentration which determines productivity.

The mechanisms of recycling described do not necessarily apply in tropical terrestrial ecosystems, a fact which often makes temperate agriculture impractical there.

READINGS AND REFERENCES

(In addition to the general texts listed in Chapter 16.)

Harris, D. R. 1972. The origins of agriculture in the tropics. American Scientist 60:180–193. An excellent discussion of why native shifting, polycultural practices are ecologically more effective in the tropics than mechanized monoculture.

Johnson, F. S. 1970. The balance of atmospheric oxygen and carbon dioxide. Biological Conservation 2:83–89. Describes the distribution of carbon compounds in the Earth's crust and the history of atmospheric gases.

McNeil, M. 1972. Laterite soils in distinct tropical environments: Southern Sudan and Brazil. Pp. 591–608, in Farvar, M. T. and Milton, J. (eds.) The Careless Technology. Natural History Press, Garden City, N. Y. Explains how the natural ecosystems of the lush tropics, dependent on a delicately balanced natural cycling of water and nutrients, are easily disrupted by man.

Phillips, J. 1972. Problems in the use of chemical fertilizers. Pp. 549–566, in Farvar, M. T. and Milton, J. (eds.) The Careless Technology. Natural History Press, Garden City, N. Y. Analyzes how additions of mineral nutrients affect natural cycles in various soils.

Plass, G. N. 1959. Carbon dioxide and climate. Scientific American Reprint No. 823. W. H. Freeman & Co., San Francisco. Explains how man-made increase of atmospheric CO_2 can affect climate.

The Biosphere: A Scientific American Book. 1970. W. H. Freeman & Co., San Francisco. A collection of excellent articles on global energy flow and nutrient recycling and on man's impact on these.

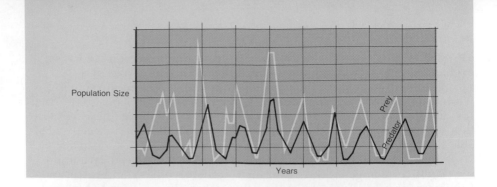

Population Size

Prey

Predator

Years

19

POPULATION SIZE AND COMMUNITY INTERACTION

The number of individuals of a given species found in an ecosystem is a compromise between the number of offspring produced and the limitations placed on their survival by the environment.

We have described an ecosystem as a community of many types of organisms—plants, animals, decomposers and parasites—interacting with one another and with the physical environment within a more or less circumscribed area, such as a field, forest, lake or estuary. So far we have mainly been concerned with the flow of energy through such an ecosystem and with the recycling of nutrients within it. Let us now turn to the interactions among the various organisms within an ecosystem.

We cannot possibly describe all of the interactions taking place between all of the individual organisms in an ecosystem—there are simply too many individuals to study each in detail. Instead, we attempt to study the interactions of **populations** of organisms, where a population includes all those individuals belonging to the same type or **species.** In anticipation of Chapter 21 we can define a species as all those individuals which are so similar that they are capable of interbreeding. In order to begin our study, it is first necessary to examine some of the general properties of populations.

POPULATION GROWTH CURVES

Every species of organism we might choose to look at has a far greater capacity for reproducing itself than is necessary simply to replace the parents. A stalk of wheat or an apple tree produces far more seeds than will ever develop into mature plants; a female frog produces hundreds of eggs each year; a human female is capable of bearing 20 or more children during her reproductive years. Under optimum conditions of food, habitat and climate, adult organisms can obviously produce great numbers of off-spring. This maximum reproductive capacity of a species is known as its **reproductive potential.**

Let us take a hypothetical example. A farmer builds a brand new pond, which initially is colonized by a single pair of frogs, a population size of two. Let us assume that at first conditions for reproduction and survival of offspring are optimum—there is plenty of food, water and space, and no predators or diseases exist. The frogs will reproduce at their maximum rate, which is far greater than that necessary to merely replace the parents when they die, and the population in the pond will tend to grow logarith-mically (dotted line in Figure 19.1). The population grows in the same way as a bank account earning compound interest.

FIGURE 19.1

Hypothetical case of growth of a frog popu-lation in a pond, starting with one pair.

Dotted line is theoretical maximum in-crease in population resulting from repro-ductive potential. Note its logarithmic na-ture. Solid line is S-shaped curve produced by environmental resistance. The population eventually reaches an equilibrium level, called the carrying capacity of the environ-ment.

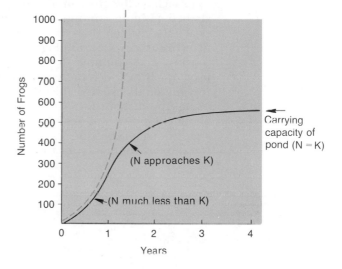

We can generalize this to any population we choose. If **N** is the popula-tion size at a given moment (analogous to the balance in a savings account), and **r** is its excess reproductive potential (analogous to the compound inter-est rate), then the rate of growth of the population under optimum condi-tions will be equal to $r \cdot N$ (capital times interest rate). The main difference between the growth rate of a bank account and that of a population living under optimum conditions is in the interest rate—annual rates for bank accounts are about six per cent, while they may be several thousand per cent for populations of organisms! Without limiting factors, populations would soon reach astronomical sizes.

It is apparent that our assumptions about unlimited food, water and places to live and about the absence of predators and diseases are unreal. Not all offspring survive. Some find no food, others are eaten by predators, while still others die of infection or from inability to find a suitable place to live. Few organisms in nature die of "old age." There are many causes of premature mortality which can be summed under the single heading, **envi-ronmental resistance**—the totality of factors other than aging which pro-duce death among members of a population. In the case of our frog popu-lation, instead of increasing indefinitely, the population size will gradually approach some equilibrium level, as shown by the solid line in Figure 19.1. The final size of a population is thus determined by a *balance* between its reproductive potential and the environmental resistance. For a particular species in a given situation, this equilibrium population size is known as the **carrying capacity** of the environment, denoted by **K.**

In our example, the initial population, **N,** of two frogs in the pond is far below the carrying capacity, **K,** and so the population grows logarithmically at first, at a rate **r · N.** But as **N** increases and approaches **K,** the growth rate slows down, resulting in the S-shaped growth curve indicated by the solid line. Finally, when **N = K,** the population no longer grows, and is now in an equilibrium state with respect to its environment. Despite the great number of eggs and tadpoles produced each season, only enough survive to replace the parents. Environmental resistance is limiting the population to a more or less constant number.

Not all populations found in nature maintain a constant size, however. Seasonal fluctuations commonly occur. For example, many species of small rodents, for which the time elapsing between conception and adulthood is often less than two months, have large populations in the spring and summer when food is plentiful, but decline in late fall and winter when food is scarce and environmental resistance is high. Other organisms, instead of approaching the environmental carrying capacity gradually in the usual S-shaped fashion, continue to increase logarithmically until they reach a peak, after which there is an abrupt decline or "crash." Such a J-shaped growth curve is shown in Figure 19.2. In the case shown here, phytoplankton in a lake continue to grow logarithmically until all the nutrients are used up, after which there is a precipitous fall in population size. The phytoplankton themselves alter the carrying capacity of their environment by using up nutrients faster than they can be recycled; the plants thus bring about their own population crash. J-shaped growth curves are typical of species which reproduce rapidly and are greatly affected by seasonally fluctuating environmental factors, such as light, temperature and rainfall.

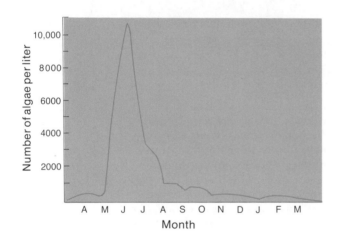

FIGURE 19.2

J-shaped growth curve of phytoplankton population. After the summer bloom, when nutrients suddenly are used up, there is a sudden crash. (After Kormondy, E. J. 1969. *Concepts of Ecology.* Prentice-Hall, Englewood Cliffs, New Jersey.)

POPULATION AGE STRUCTURE

So far we have considered only the total number of individuals in a population without regard to their age distribution. In fact, the actual number of offspring produced will depend upon the number of reproductive individuals present in the population, and this, in turn, is related to life expectancy.

LIFE SPAN AND LIFE EXPECTANCY

Life span or the maximum length of life of an individual organism varies greatly from one species to the next. In general, the larger the organism, the longer its maximum life span. Redwood trees may live 2000 years; man about 110 years; the American robin about seven years; and the water flea, *Daphnia*, about seven weeks. Aging thus occurs at different rates in different species. The causes of aging are poorly understood despite much study. It is partly due to accumulated mutations in the DNA of non-germ cells in the body; these interfere with normal gene function, leading to cell death.

Life expectancy depends on the degree of environmental resistance. Most populations suffer large losses during the early stages of life, and therefore an individual has a short life expectancy at birth. *Survivorship curves,* which relate life expectancy to life span, allow a graphic representation of the stages in the life span when mortality is greatest. In Figure 19.3, survivorship curves for three species are shown: man (living in the United States), the American robin, and the blacktail deer. The number of survivors, plotted on a logarithmic curve, is shown as a function of the total maximum life span. The average or mean life expectancy at birth is that age to which half the individuals born survive—thus for each 1000 births it is the age at which there will be 500 survivors. This is indicated by the colored horizontal line in the figure.

FIGURE 19.3

Survivorship curves for three species: man (in U.S.), robin, blacktail deer.

Arrows point to age relative to total life span to which 50 per cent of individuals born survive (mean life expectancy at birth). (After Kormondy, E. J. 1969. *Concepts of Ecology.* Prentice-Hall, Englewood Cliffs, New Jersey.)

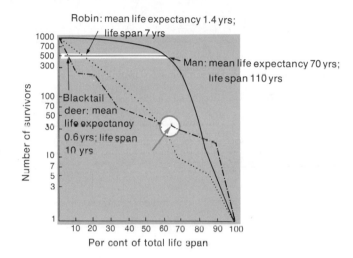

Whereas half the people born in the United States survive 70 years, or 65 per cent of the maximum human life span, only about 3 per cent (30 out of 1000) of robins and blacktail deer survive the same fraction of their respective life spans (colored arrow). Note that deer have an especially high mortality rate during the first year of life, so that the average fawn at birth has a life expectancy of only 0.6 year. It is probable that survivorship curves of early prehistoric man were more like those of the deer and robin. However, despite an enormous increase in human life expectancy through improved agriculture, medicine and sanitation, there has been virtually no increase in the maximum life span of man.

AGE CLASSES AND AGE PYRAMIDS

Ecologically speaking, an individual falls into one of three age classes: prereproductive, reproductive or postreproductive. By counting the number of individuals in each and constructing a pyramid of age classes, one can quickly visualize the age structure of a population (Fig. 19.4). A population which is expanding often has a broad-based pyramid, one which is stationary may be bell-shaped and one which is diminishing may have an urn shape.

Human populations provide interesting examples of all three types of pyramids (Fig. 19.5). Korea, in 1940, had a rapidly expanding population, with a great preponderance of prereproductives. Similar pyramids characterize most populations in Africa, southern Asia and Latin America today. Such distributions are the result of a continuing high birth rate coupled with a sharp decline in infant mortality due to modern medical care. As birth control spreads and life expectancy increases, the curves will become more bell-shaped, approaching the stationary pyramid characterized in our diagram by the Swiss population in 1947. An example of a diminishing population occurred in Sweden in the middle 1930's.

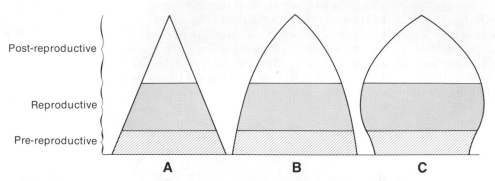

FIGURE 19.4

Age pyramids for three types of populations.
A. Broad-based pyramid indicates an expanding population.
B. Bell-shaped pyramid indicates a stable population.
C. Urn-shaped pyramid indicates a diminishing population.
(After Kormondy, E. J. 1969. *Concepts of Ecology.* Prentice-Hall, Englewood Cliffs, New Jersey.)

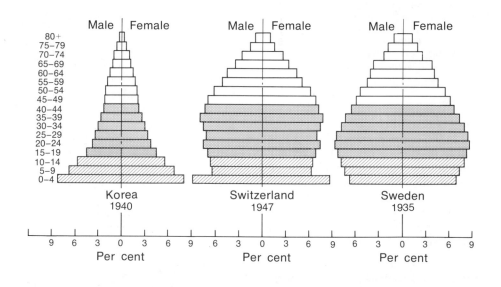

FIGURE 19.5

Pyramids of age structure among different human populations: left, an expanding population; center, a stationary population; right, a diminishing population. (After Petersen, W. 1969. *Population.* Macmillan, New York.)

FACTORS AFFECTING REPRODUCTIVE POTENTIAL

We can next ask, What factors affect the reproductive potential of a population—that is, what determines the number of offspring produced? We must here arbitrarily define what we mean by offspring. For convenience, most biologists measure **natality,** or the number of live births, hatched eggs or sprouted seeds produced by the parents in a given period of time. Eggs laid or seeds shed are also a measure of reproductive potential, but they are often much harder to count.

Factors affecting natality may be divided into non-living or physical factors, on the one hand, and those which are due to living agents—that is, biological factors—on the other.

PHYSICAL FACTORS

One important environmental factor affecting the reproductive capacity of many organisms is temperature. Moderate rises in temperature will increase the rates of enzyme catalyzed reactions—until the point is reached at which the specific structure of the protein enzymes is destroyed as a result of heat denaturation (akin to cooking an egg white or boiling surgical instruments to sterilize them). We have also noted that some organisms can adapt to gradual changes in temperature but are killed by rapid fluctuations such as may occur in cooling waters flowing from power plants (see Chapter 7a). Not all species are capable of wide temperature tolerance, however, and reproductive capacity is often quite sensitive to temperature fluctuations. In Figure 19.6 we see the effect of temperature on the reproductive rate of an insect, the rice weevil. A temperature near 30° C results in more offspring than either a lower or a higher temperature.

FIGURE 19.6

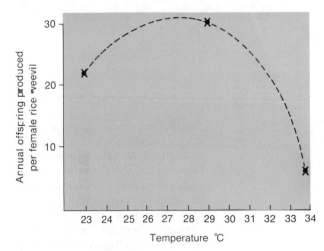

Effect of temperature on reproductive potential of rice weevil. **x** = actual data points; dotted line is a possible curve connecting the points. (Based on data of Birch, L. C., *in* Kormondy, E. J. 1969. *Concepts of Ecology.* Prentice-Hall, Englewood Cliffs, New Jersey.)

In addition to temperature, one can think of many other physical environmental factors affecting natality. Lack of mineral nutrients, light or water may all diminish the reproductive capacity of plant populations, for example, and the eggs of many marine invertebrates fail to develop into larvae in sea water diluted by freshwater run-off.

BIOLOGICAL FACTORS

There is also a variety of biological factors which can affect reproductive potential. Some of these arise from the nature of the reproductive individual itself. The age of a reproductive female animal has often been found to affect her reproductive capacity. The water flea, *Daphnia*, has been well studied in this respect. The adult lifetime lasts about five weeks, during which females produce offspring daily. Maximum reproductive rate occurs between the ages of 14 and 17 days (Fig. 19.7).

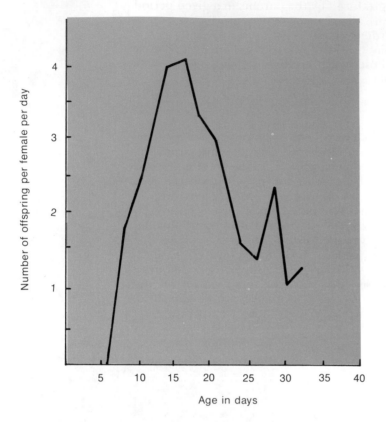

FIGURE 19.7

Effect of age on reproductive rate of the female water flea, *Daphnia*.
Under optimal conditions, the females produce a maximum number of offspring between the ages of 14 and 27 days. (After Kormondy, E. J. 1969. *Concepts of Ecology*. Prentice-Hall, Englewood Cliffs, New Jersey.)

Self-regulation Within Populations.   Among the biological factors affecting reproductive potential, a great many are population-dependent and help to maintain the population at an equilibrium size. Several examples of such self-regulation are given below, and in each instance there is an element of negative feedback: as the population increases, reproductive potential of the population as a whole decreases. Sometimes this occurs by reduction of the reproductive capacity of all individuals, sometimes by restriction of reproduction to only selected individuals.

Individual crowding is one such factor. As populations of *Daphnia*, for example, become more and more crowded, the number of offspring produced per female declines (Fig. 19.8). *This occurs even when food is abundant.* Mice show a similar decrease in reproductive rate under crowded conditions. In this case, it is believed that social stress stimulates the adrenal cortex to secrete hormones, one of the effects of which is to reduce the size of the gonads, resulting in fewer offspring.

Another factor influencing reproductive potential in populations is social interaction. Only one of many potentially reproductive females becomes capable of laying eggs in populations of highly social insects such as bees and termites. Once a "queen" has been selected, she produces chemicals which suppress reproductive maturation in her sisters. They function to nourish her, and the entire colony becomes a single reproductive unit centered around one individual.

TERRITORIALITY. In many species of animals, especially fish, birds and mammals, territorial behavior is a common form of self-regulation of population size. In some, territories are maintained throughout the year; in others, only during the breeding season. Territoriality is exhibited most often by the males of a species, who establish territorial boundaries which other males are not allowed to cross. Male birds usually mate with only one female, whereas in many mammals several females are accepted within the territory of a single male. Only those individuals possessing territories reproduce; the unlucky ones—usually weaker or less well adapted to the environment—may wander off in search of their own territories and mates.

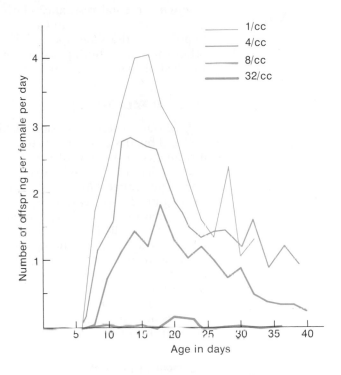

FIGURE 19.8

Effect of population density on reproductive rate in the water flea, Daphnia.

At low densities, females have a maximum reproductive capacity between the ages of 14 and 17 days. Crowding significantly decreases this capacity, and it also tends to increase the age at which the maximum occurs. (After Kormondy, E. J. 1969. *Concepts of Ecology.* Prentice-Hall, Englewood Cliffs, New Jersey.)

Among animals exhibiting territorial behavior, the number of offspring produced in a given ecosystem is relatively constant and depends on the number of territories available. Territorial size may be determined by food supply, nest sites and other environmental factors. An individual unsuccessful in obtaining a territory one year may gain one the year following and so enter the breeding population (Fig. 19.9). Thus, although the total population may fluctuate yearly, the number of breeding pairs or groups remains fairly constant from year to year. It is thought that territorial be-

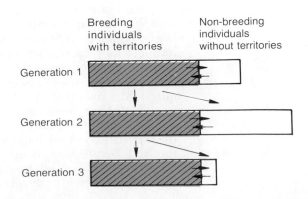

FIGURE 19.9

Territorial behavior results in a constant size of breeding population in many species of animals, since only individuals holding territories mate. A few individuals may lose or gain territories during the breeding season.

havior insures that those offspring which are produced have a maximum chance of survival and that their numbers are therefore greater than would be the case if the entire adult population were to reproduce and their offspring were to compete simultaneously for limited resources in the environment.

FACTORS RELATED TO ENVIRONMENTAL RESISTANCE

Almost all factors affecting the mortality rate in a population are external to the individual organism itself and therefore fall within the category of environmental resistance. Three exceptions are the rate of aging, genetic defects (discussed in Chapter 15a) and destruction of self or of other members of the same species. The latter is relatively rare among animals other than man. Both physical factors and biological factors can affect the survival of living organisms.

PHYSICAL FACTORS

In our earlier discussion of biomes (Chapter 16) we noted that such physical factors as temperature, rainfall, sunlight and so on are major determinants of the type of vegetation found in different parts of the world. Each variety of plant survives best in a particular environment, a fact, unfortunately, that Americans tend to overlook: in our attempts to emulate the beautiful lawns and gardens of the English countryside, we sometimes forget that we are working with quite different climatic and soil conditions. Likewise, the distribution of animals is often limited directly by physical factors; whereas optimum temperatures may stimulate reproduction, as in the rice weevil, extremes are often lethal. Each species tolerates a specific temperature range which limits its distribution. The same is true for other physical factors, such as salinity of water, oxygen availability and so forth. In the next chapter we shall consider a specific instance where the activities of man have affected the physical factors of an ecosystem and hence the populations of organisms which exist in it.

BIOLOGICAL FACTORS

We now turn to a consideration of the interactions between members of different species within an ecosystem, or how populations interact within a community. The many possible types of interactions can be broadly grouped into predator-prey relationships, parasitism and direct competition.

Predator-Prey Relationships. Virtually every organism has a predator which utilizes it for food. Obviously, the size of a prey population is, to a greater or lesser degree, regulated by its predators. Conversely, the size of a predator population is determined by how much prey is available.

Observations on predator-prey relationships are made most easily in relatively simple ecosystems such as the arctic tundra. In the arctic, herbivore populations often exhibit long-term cyclic fluctuations of great magnitude, with peaks occurring once every three to four years in some species and every nine or ten years in others. One case is the snowshoe hare, where the population rises rapidly for several years to a maximum and then quickly declines (Fig. 19.10). The lynx, whose major food source is the snowshoe hare, undergoes similar population cycles which lag slightly behind those of the hare. It is thought that availability of food determines the fluctuations in the lynx population, but the reciprocal effect, control of the hare by the lynx, is not clear-cut—hare populations decline precipitously even in areas where lynx are absent.

Another arctic herbivore famed for its striking population fluctuations is the lemming. Every three or four years the population builds up dramatically, only to crash and virtually disappear again. Among the explanations offered to explain these crashes are depletion of food resources, control by predators, self-regulation leading to diminished reproductive potential, and migration. None of the first three seems to apply, at least among lemming populations in Norway. Observations there indicate that these rodents are highly anti-social, exhibiting negative behavior when crowded together (although wounds are never inflicted in the wild). The losers of

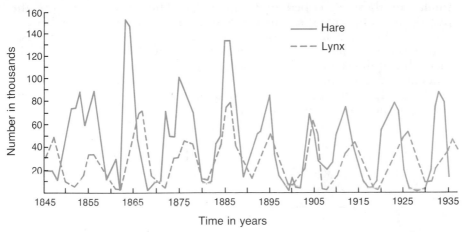

Cyclic fluctuations in population size of the snowshoe hare and its predator, the lynx. Note that the cycles of the predator population lag slightly behind those of the prey. Cycles occur at approximately 10 year intervals. (After Odum, E. P. 1971. *Fundamentals of Ecology*. W. B. Saunders Co., Philadelphia.)

FIGURE 19.10

these continuous battles eventually migrate in great hordes and, unless they find suitable habitats, die off. Such peculiar behavior seems designed to insure that not all food is eaten and that a few individuals survive to reproduce the next year. Meantime, predators of the lemming also increase in numbers until the crash comes. One of these, the snowy owl, emigrates southward in great droves when the lemming population crashes and becomes a common sight in some cities about once every four years.

Another species well known for its cyclical population increases and ensuing migrations is the Eurasian migratory locust. Normally restricted to desert regions, this grasshopper periodically undergoes spectacular population increases. As the population density builds up, the adult animals change form, developing long wings. Hordes of them then spread into cultivated areas, eating everything in their path. Such migrations occur about once in every 40 years.

Population explosions of the types described above are not well understood, but they seem to be characteristic of animals living in simple, unstable ecosystems. Perhaps they are triggered by temporary increases in food available to the herbivore population during favorable years, which lead to a secondary increase in predator populations. Much study remains to be done on these highly fluctuating populations.

CHEMICAL INTERACTIONS. Both plants and animals have developed specific defenses against excessive predation. Some of these are considered in Chapter 21, and we shall limit ourselves here to a few examples of specific interactions between prey and predator populations, many of which are chemical in nature.

Many plants have evolved mechanisms for defending themselves against herbivorous predators. Thorns, spines, toxins and specific insect hormones (see Chapter 14a) are all examples. The image of a cow content-edly feeding on buttercups is a misconception propagated by the advertis-ing world. Buttercups and other members of the genus *Ranunculus* produce unpalatable substances which cause them to be avoided by grazers unless no other food is available.

Of particular interest is the production of distinctly toxic substances by certain plants. Members of the chrysanthemum family, for example, produce insecticidal chemicals called pyrethrins which deter predation by most insects (Fig. 19.11). The larvae of a few species of butterflies and moths, however, have evolved detoxifying enzymes in their digestive tracts, similar to those described in Chapter 3a. The enzymes destroy the pyrethrins and permit these insects to feed on the plants. Certain chrysan-themums, however, have subsequently developed another chemical, sesa-min, which is an inhibitor of these detoxifying enzymes. Sesamin thus limits the ability of the larvae to feed on the plants. We are witnessing here three sequential stages in the evolution of "chemical warfare" between predators and prey. Each strategy is designed to increase the population of one species at the expense of the other.

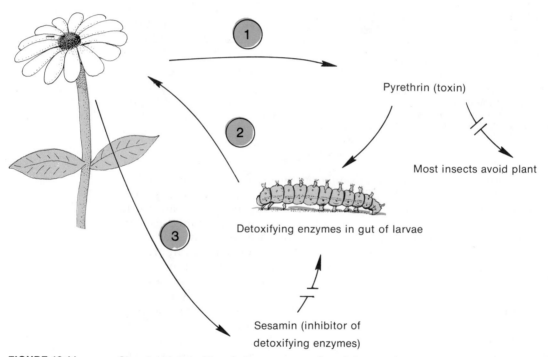

FIGURE 19.11   Chemical interactions between prey and predator species.
   Many species of chrysanthemum produce pyrethrins (*1*), natural insecticides which deter most insect predators (interrupted arrow). Larvae of some butterflies and moths contain detoxifying enzymes (*2*) in their digestive tract which destroy the pyrethrins, allowing them to feed on the plant. Some chrysanthemums have evolved an inhibitor of these enzymes (interrupted arrow), sesamin (*3*), which helps protect them from moth larvae.

Another, somewhat different chemical relationship exists among cer-tain pine trees, the bark beetles which feed on them and the predators of the bark beetles. The sap of many trees contains odoriferous chemicals. When the *Ips* beetle locates its host tree, the ponderosa pine, it begins to feed on the sap and converts the scented chemicals into specific, volatile substances known as terpenes. The excreted terpenes attract other members of the same species to the food source, thus insuring mating and reproduction of the *Ips* beetle. However, a second, carnivorous beetle which feeds on *Ips* is also attracted by these same terpenes. Evidently this scent is also used as a cue by the predatory beetle for locating high densities of its prey!

There are many other known examples, and probably many yet to be discovered, of chemical interactions between prey and predator species which play a role in regulating the populations of each.

Symbiotic Relationships. Another type of population interaction occurs when two species of organisms live in very close association with one another, a condition known as **symbiosis.** Symbiotic interactions are usually obligatory for one or the other, and sometimes for both, of the species involved. Three somewhat arbitrary categories have been established to describe different types of symbiotic relationships based on the degree of benefit or harm to each of the two species. These are **mutualism** (both species benefit); **commensalism** (one benefits, the other is unaffected); and **parasitism** (one benefits, the other is harmed). Examples are shown in Figure 19.12).

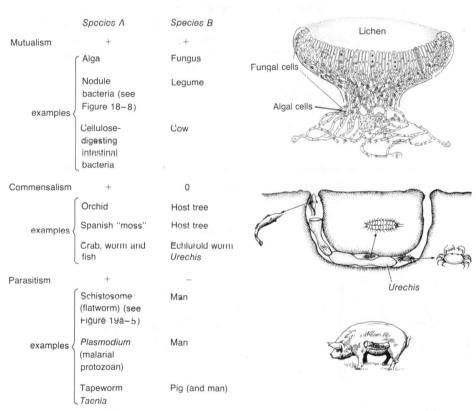

FIGURE 19.12

Three types of symbiotic relationships.

Mutualism. Both species benefit. In some cases the relationship is obligatory, in others it is not. Photosynthetic algae live between the hyphae of the fungus in lichens.

Commensalism. One species provides a habitat for a second species, but the former is neither benefited nor harmed by the association.

Parasitism. The parasite obtains shelter and, most importantly, food from the host, which suffers harm from the relationship.

MUTUALISM. In lichens, algal cells interspersed between the fungal cells obtain protection and moisture while providing photosynthetic nutrients to the heterotrophic fungus—both species benefit from this mutualistic relationship. Neither population could exist without the other, and hence the size of each is determined by that of the other.

COMMENSALISM. **Epiphytes** are plants which live in a commensal relationship on another plant, deriving support from it but obtaining their water and food from moisture and dust in the air. Examples are the epiphytic orchids and ferns and Spanish "moss," which really belongs to the pineapple family. The population of host trees is unharmed unless the leaves become shaded or the branches break from the weight of the epiphyte, but the population of commensals is entirely dependent on that of the host species. Another host to commensal species is the "fat innkeeper," *Urechis,* a marine echiuroid worm; it regularly shares its burrow in the sand with a crab, a worm and a fish.

PARASITISM. Many organisms are parasitic: some fungi, all viruses, many bacteria, protozoa, flatworms, roundworms and so forth. In general, the parasite population exists at the expense of the host population. The adult tapeworm of the pig, living in its intestine, merely feeds on food digested by the host's enzymes; but the larvae burrow into the host's muscle, which, if it is eaten by man, may infect him also. The larvae of a roundworm which may also infect pork muscle causes another disease in man, **trichinosis.**

Although all types of symbiosis are ecologically important, man is generally most concerned with parasitic infestations of himself or species he values. In nature, a stable equilibrium is generally reached between the populations of host and parasite, since if the parasite killed all its hosts it would cause its own extinction. Fluctuating cycles, similar to those shown for the prey-predator relationship in Figure 19.10, may occur. Quite often, the control of the host population by its parasite prevents the former from exceeding the carrying capacity of the environment, thus proving ultimately beneficial to the general host population, if not to the infected individual.

When man intervenes, however, natural population balances are frequently upset; often excess numbers of hosts are available, as in crowded cities, or man inadvertently introduces a new parasite to which the host species has no resistance. Two examples will serve to illustrate these points.

Bubonic plague or Black Death has historically been one of the most devastating diseases of mankind and has probably been common since man

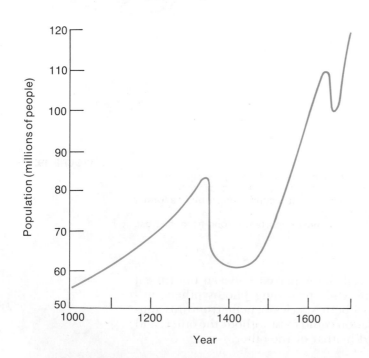

FIGURE 19.13

Estimated European population during the Middle Ages. Bubonic plague caused the two large dips in the fourteenth and sixteenth centuries.

first began to inhabit densely populated cities. Continuous records of its occurrence go back to the third century B.C. Europe suffered several prolonged periods of plague; one in the sixth century decimated the population of Constantinople, to be followed by others in the fourteenth and seventeenth centuries (Fig. 19.13). The great plague of London in 1665 was only halted by the fire of 1666 which destroyed much of the city. The disease is still prevalent in India, Africa and parts of South America, and there have been outbreaks in cities of the United States within this century.

Plague is a native or endemic disease among rodent populations. It occurs not only among rats living in association with human habitations, but also in populations of wild squirrels, chipmunks, prairie dogs and other species. The wild populations serve as reservoirs for spread of the disease to cities. The causative agent of the disease is a bacterium, *Pasteurella pestis*, which is transmitted from one rat to another by certain fleas, mainly those of the genus *Xenopsylla;* a flea thus acts as a carrying agent or **vector** (Fig. 19.14). When rat populations increase as a result of lack of sanitation, infected rat fleas frequently bite humans. The bacteria multiply in the blood stream and spread to all organs; enormously swollen lymph nodes (buboes) are characteristic; lungs are frequently infected, allowing direct

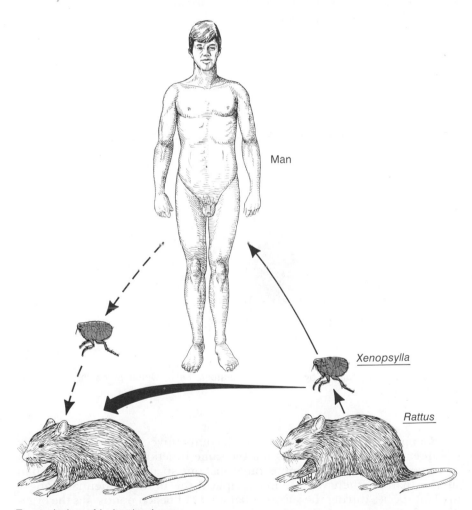

Man

Xenopsylla

Rattus

FIGURE 19.14

Transmission of bubonic plague.
    The major host of the bacterium, *Pasteurella pestis,* is the rat; the vector is the rat flea, which occasionally bites man. (Heavy arrow indicates major route; dotted arrows, possible route.)

man-to-man infection via sputum. Death occurs in more than half the cases unless antibiotics are available. It is probable that more than 150 million people have died of plague throughout history—not many in comparison to today's world population of 3.7 billion, perhaps, but remember that, until recently, human populations were sparse.

A disease of trees, also transmitted by a vector, is Dutch elm disease. It is produced by a fungus, *Ceratostomella,* which plugs the xylem and phloem of the elm, causing death. The fungus is transferred from one tree to the next by the bark beetle, *Scolytus* (Fig. 19.15). Imported from Europe to the East Coast some decades ago, the disease has destroyed most elms as far west as Denver. Extensive, heavy spraying with DDT and other insecticides to kill the beetle has slightly slowed but not halted the spread of the disease while simultaneously killing thousands of robins and other birds. Another fungus, accidentally introduced from China in 1904, has virtually wiped out the great American chestnut, once an important part of the eastern deciduous forests. In both cases, the fungus in its native habitat had established a natural equilibrium with its host population, but American trees, without any defense mechanisms, quickly succumbed. Just as animals evolve specific defense mechanisms against parasites (see Chapter 10), so do plants; but the development of genetically resistant strains requires many generations—time not available to organisms when man suddenly exposes them to a foreign parasite.

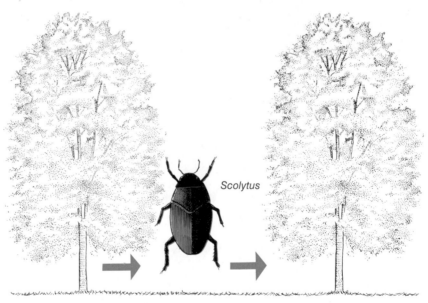

Scolytus

FIGURE 19.15

Transmission of Dutch elm disease by the bark beetle, *Scolytus.*
    The disease is caused by a fungus, *Ceratostomella,* which plugs the xylem and phloem.

Competition. According to the competitive exclusion principle (Chapter 16), no two species fulfill the same function in an ecosystem. In many instances it can be demonstrated that one species is better adapted to survive than another. We saw an example of this in the sequential changes in plant species during the successional stages of beach ponds—as the organisms altered their environment, new species became favored and others were excluded.

Some species of plants are able to actively exclude the growth of other species by exuding noxious substances from their leaves, stems or roots—a phenomenon known as **allelopathy** (Greek *allēlon*, "other," plus *pathos*, "suffering"). These volatile substances fall on the ground near such plants, suppressing germination and growth of other species and insuring these plants' own roots of a plentiful supply of water and nutrients (Fig. 19.16). In the same way, many soil fungi release antibiotics (now made use of by man) which suppress bacterial growth and thus lessen the competition for a limited food supply of detritus. Some soil bacteria have in turn evolved enzymes to destroy the antibiotics. Many other examples of competition between species have been discovered, which favor the survival of one population at the expense of another.

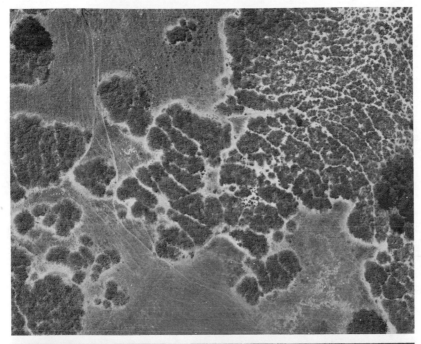

FIGURE 19.16

Allelopathy between plant species.
Upper picture shows aerial view of chaparral invading grassland. Lower picture shows ground-level view of the same. The chaparral plants exude volatile toxins which keep area between *A* and *B* free of all plants. The region between *B* and *C* has fewer species and smaller plants than the surrounding grassland. (From Odum, E. P. 1971. *Fundamentals of Ecology.* W. B. Saunders Co., Philadelphia.)

SUMMARY OF INTERACTIONS BETWEEN SPECIES

By way of a summary and as a convenience to the reader, we have listed here the positive and negative ways in which a given population may interact with another species. This list comprises the various types of interactions observed in nature.

INTERACTIONS INCREASING THE SIZE OF A POPULATION
Preying on other species
Parasitizing another species
Commensalism with another species
Mutualism with another species

INTERACTIONS DECREASING THE SIZE OF A POPULATION
Predation by another species
(Compensated by escape, defense, camouflage, noxious or toxic chemicals, etc.)
Parasitism by another species
(Compensated by genetic resistance, phagocytosis, antibiotics, antibodies)
Competition with another species
(Compensated by genetic adaptation to another habitat or by detoxification of inhibiting substances)

One important result of these various types of interactions between organisms is that diverse communities, with many species, tend to be stable. Each species interacts, directly or indirectly, with so many others that there is an enormous number of feedbacks—checks and balances—which maintain a nearly constant population size. Man tends to oversimplify ecosystems, making them vulnerable to competition, predation or parasitism. For instance, removing predators (shooting hawks) may increase competition among different prey species (rodents). This can result in a decrease in the variety of prey species—one species may then take over and become a pest. We shall consider some specific effects of man's activities on population interactions in the next chapter.

SUMMARY

The size of a population is determined by a balance between births (reproductive potential) and deaths (environmental resistance). The maximum population attainable is set by the carrying capacity of the environment. Some species exhibit an almost constant population size in nature, while others undergo rapid growth followed by a crash (J-shaped growth curves).

The age structure of a population is reflected in survivorship curves which relate life expectancy to life span. In human populations, age pyramids provide pictorial information about the growth or decline of a population.

Reproductive potential may be affected by factors such as temperature, age, and population density. Density-dependent factors include stress, social structure and territorial behavior. Environmental resistance is determined both by physical factors, such as temperature, rainfall, sunlight and so forth, and by biological factors like predation, parasitism and competition between species. Many examples exist of evolutionary adaptations by a species to a particular aspect of environmental resistance. Such adaptations are particularly observable in interactions between species.

Barduducci, T. B. 1972. Ecological consequences of pesticides used for the control of cotton insects in Cañete Valley, Peru. Pp. 423–438, *in* Farvar, M. T. and Milton, J. (eds.) *The Careless Technology*. Natural History Press, Garden City, N. Y. The pest became resistant, its predators were wiped out and the crop failed. Re-establishment of predator-prey control and other ecologically sound techniques reversed the disaster.

Boughey, A. S. 1968. *Ecology of Populations*. Macmillan, New York. A standard text on this subject.

Clough, G. C. 1965. Lemmings and population problems. American Scientist *53*:199–212. Field observations in Norway indicate crashes may be mainly due to massive migrations.

Hairston, N. G., Smith, F. E. and Slobodkin, L. B. 1960. Community structure, population control, and competition. American Naturalist *94*:421–425. A brief, controversial article proposing that, unlike all other organisms, herbivore populations are not limited by resources but by predators

Huffaker, C. B. 1958. Experimental studies on predation: dispersion factors and predator-prey oscillations. Hilgardia *27*:343–383. Simple laboratory experiments demonstrating important aspects of predator-prey interactions between two mite populations living on oranges.

Langer, W. L. 1964. The black death. Scientific American Offprint No. 619. A fascinating account of the role of bubonic plague in human history.

MacArthur, R. H. and Connell, J. H. 1969. *The Biology of Populations*. John Wiley and Sons, New York. A concise treatment of the subject.

Whittaker, R. H. and Feeny, P. P. 1971. Allelochemics: Chemical interactions between species. Science *171*:757–770. Detailed account of a whole host of chemically mediated relationships between various species of plants and animals.

**READINGS AND
REFERENCES**

19a

THE ASWAN DAM AND COMMUNITIES OF THE NILE

In trying to feed his own growing population man often disturbs the interdependent nutrient cycles and population balances between other species. As in the case of the Aswan Dam, this often results in an ecological boomerang striking back at man himself.

The human population since its first appearance some million years ago has not grown at a constant rate, but has shown three large surges, each the result of important changes in human life styles (Fig. 19a.1). The first surge resulted from the evolution of cultural patterns; the development of language and resulting social interactions improved food gathering techniques and community defenses against enemies. The second surge resulted from the advent of agriculture and husbandry, and the third from the industrial and medical revolutions of the last centuries. This last surge is still in progress in many underdeveloped countries, where attempts are

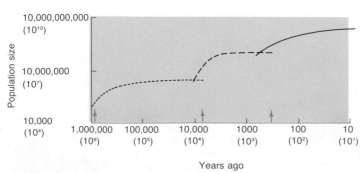

FIGURE 19a.1

Stages in the growth of the human population since its earliest appearance, plotted on a log-log scale.

Note three surges, coinciding with the cultural, agricultural and industrial-medical "revolutions," as indicated by arrows. (After Ehrlich, P. R., and Ehrlich, A. H. 1970. *Population, Resources, Environment.* W. H. Freeman, San Francisco.)

being made to bring the benefits of modern technology to largely agrarian peoples.

Modern communications have spread awareness of the great economic disparities which exist among countries; the result is increasing social and political pressure for rapid improvements which, if not properly managed, may cause as many problems as they solve. In this chapter we shall consider in detail the effects of one such attempt: the construction during the 1960's of the Aswan High Dam in Egypt. This is, to date, the largest dam of its kind ever built, costing over a billion dollars. Although many other examples could be chosen, this single case embodies several far-reaching ecological effects which would have been foreseen had the principles of ecology rather than of short-term economics and politics been followed. As we shall see, it is possible that the damage done to the ecosystems associated with the Nile and consequent human hardships will outweigh the benefits of the dam to the Egyptian population.

In this chapter we shall pay particular attention to the way in which man's manipulation of one physical factor in the environment—water supply—has affected the balance of populations of various organisms. Many of the principles covered in the preceding chapter are applied here to a real situation.

EGYPT AND THE HIGH DAM

The United Arab Republic is a desert country. Annual rainfall is less than 10 inches; temperatures are between 80 and 90° F in summer, and 60 and 70° F in winter. The 35 million inhabitants are almost entirely dependent on the fertility of their only river, the Nile, and more than 32 million of them live in the wide valley formed by the annual floods (Fig. 19a.2).

The Nile is the longest river in the world, arising in the mountains surrounding the southern Sudan and the Ethiopian highlands, where rainfall is 40 to 80 inches per year. Unlike temperate latitudes, however, these mountains have relatively high winter temperatures (35 to 50° F), and so the water is not stored as snow. During the rainy season, from August to November, great floods rush from the mountains, carrying mud and nutrients with them. Prior to construction of the High Dam, these annual floods deposited a rich layer of silt in the Nile Valley, which has maintained its high fertility despite millennia of agriculture. The floods also replenished the groundwater supply. The silt carried nutrients into the eastern Mediterranean, which supported an abundant fishery.

Crops grown along the Nile include sugar cane, cotton, tobacco and wheat, with corn and rice being raised in the highly fertile delta. Camels and dates are harvested around the oases and along the coasts. Until the High Dam was built, agriculture was limited both seasonally and in extent by the flooding of the river. Only those areas flooded or near the flood plain could be cultivated, and these were sown only once a year, just after the floods, while moisture remained in the soil. Earlier in the century, several smaller dams were built along the Nile, three in the Aswan region, to help control the destructiveness of the floods and to increase the area under cultivation by supplying irrigation water. In addition, temporary earthen dams blocking the two main mouths of the Nile have long been used to prevent loss of water into the Mediterranean Sea, thus allowing year-round irrigation of the delta. When the floods reached the delta, first one, then the other dam was opened.

INTENT OF THE HIGH DAM

It was the intent of the late Gamal Abdel Nasser that the Aswan High Dam should be the turning point in the Egyptian economy—increasing ag-

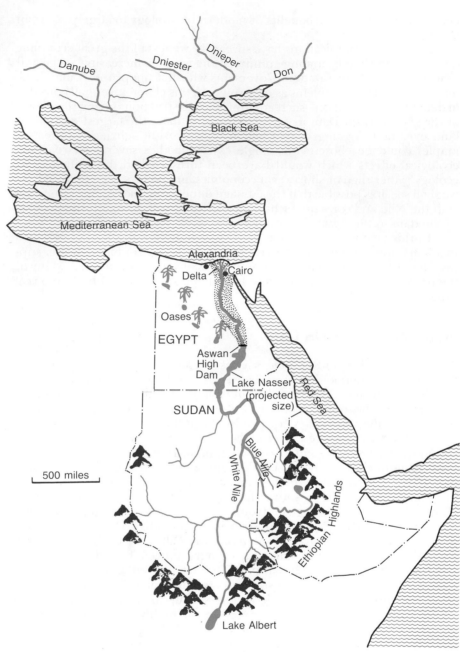

FIGURE 19a.2   The geography of the Nile and the eastern Mediterranean.
    The shaded area indicates the flood plain of the Nile, where most of Egypt's popula-
tion lives. Note the projected size of Lake Nasser behind the Aswan High Dam; the lake
is only about half full at present.

ricultural productivity and providing power for industrialization (Fig.
19a.3). Another 1.3 million acres were to be added to the six million already under cultivation, and year-round agriculture was to be introduced
over much of the farmland. The 12 generators of the dam were to produce
10 billion kilowatt hours annually, providing the average Egyptian with
about one-twentieth the electrical power now available per person in the
United States. It was expected that, following the closure of the dam in
1964, Lake Nasser would have filled by 1970 and begun providing its max-

imum benefits soon after. These were to have repaid the costs of the dam in two years. None of these expectations has been fully realized and some may never be.

Prior to its construction, some objections to the dam were raised. Particularly opposed was the distinguished Egyptian hydrologist, Dr. Abdel Aziz Ahmed, who foresaw a number of the disastrous consequences, particularly those involving loss of silt deposition and changes in underground water flow. His warnings were met by his summary dismissal. Unfortunately, not only have his predictions come true, but there have also been many other dire results—all of which could have been foreseen had the ap-

The release of Nile river water from the Aswan High Dam. (UPI photo.)   **FIGURE 19a.3**

propriate ecological studies been carried out. In essence, the quick gains expected from the High Dam have not yet been realized while the ecological backlash is already being felt. Most of the problems involve drastic fluctuations in populations of plants and animals resulting from a profound disturbance of the normal ecological balances. In general, these fall into three areas: effects on agriculture; effects on fisheries; and effects on public health. We shall consider each of these in turn.

AGRICULTURAL PRODUCTIVITY

Agriculture, the mainstay of the Egyptian economy, was to have been greatly expanded by the introduction of year-round irrigation and utilization of previously arid land. The effective increase in arable acreage from these two sources has been about 30 per cent, but the projected increase in productivity has not been realized for several reasons.

LOSS OF NUTRIENTS

As we have seen over and over again, plant growth is limited not only by availability of water but also by nutrients. These used to be supplied by the annual deposits of Nile silt. The Nile today, sparkling clear, provides virtually no nutrients—the silt is trapped in Lake Nasser, which, in a few centuries, will become filled with it.

To replace these nutrients, costly artificial fertilizers are being applied which, as we have noted, cause ecological problems of their own. The use of highly productive strains of grain ("miracle grains") has been introduced to increase yield; these also require addition of extra fertilizer. Over two billion tons of fertilizer are being used annually, two-thirds of which is needed just to compensate for lost natural fertility. The annual cost is 100 million dollars—one-tenth the cost of the dam and an expense that will recur annually. Even so, soil fertility is decreasing, and with it, productivity per acre.

Of the 1.3 million acres which were to have been reclaimed from the desert, only about 300,000 are now being converted to farmland. Once again, water alone is not enough. Mechanized transport of silt to the deserts has been costly. It is estimated that a third of a billion dollars has already been expended in reclaiming desert lands.

HYDROLOGIC PROBLEMS

The hydrologic consequences of building the Aswan Dam, many of which were predicted by Dr. Ahmed, have been enormous. Far less water is being stored than was predicted—in fact, Egypt may end up with a net loss of water! Lake Nasser should have been full in 1970; by 1971 it was less than half full. Evaporation from the lake was underestimated by about 50 per cent. Large bodies of water such as Lake Nasser, already over 100 miles long, create temperature gradients in the overlying air resulting in high wind velocities. Engineers took into account temperature and humidity but forgot to allow for increased winds which greatly accelerate evaporation. Another factor incorrectly foreseen was the continuing seepage of water from the bottom of the lake. The porous Nubian sandstone underlying Lake Nasser was supposed to have become quickly plugged with silt. The silt, however, has settled mainly in the old river channel in the center of the lake, and leakage continues unchecked around its enormous perimeter. No one knows how long it will take for the lake finally to fill to capacity; estimates range from 12 to 200 years, by which latter time the lake will be substantially full of silt rather than water.

One result of seepage from the lake is a reversal of groundwater flow throughout much of Egypt. Groundwater from the western deserts used to seep toward the Nile Valley, augmenting the flow of the river. High hydrostatic pressures generated by the lake have reversed this, so that the valley is now losing groundwater to the deserts rather than gaining it. Water meant to be used for irrigation and power generation is being lost to the uninhabited wilderness. Such a paradox should have been foreseen, since it was already known that the largest of the earlier Aswan reservoirs, while storing five billion cubic meters of water per year, was causing an annual loss, through seepage and reversed underground water flow, of 12 billion cubic meters. The combined losses from Lake Nasser due to evaporation, seepage and reversed groundwater flow almost equal the total annual discharge of the Nile into the Mediterranean prior to construction of the dam. Thus, the dam is squandering most of the water it was supposed to preserve!

The changed water economy of the Nile Valley has resulted in yet another problem. The annual floods once washed away salts accumulated at the soil surface owing to the high rates of evaporation in the tropical climate. Irrigation hastens the rate of salt accumulation unless extra amounts of water are used to flush the salts away. As we saw in Chapter 2a, this requires construction of tile drainage to remove the excess water. Tiles are now being installed in a million acres of the fertile delta, which has always been subject to this problem. But the rest of the valley is also being affected, and installation of adequate drainage throughout would cost a billion dollars—the same as the High Dam itself—even if sufficient extra irrigation water were available. It is likely that with time more and more land will be lost to agriculture through salt accumulation, offsetting gains from the introduction of irrigation.

One further result of impounding the Nile is the problem of erosion. The silt-free waters flow faster and are undermining the foundations of the downstream dams, 550 bridges and the river banks. There is also an annual loss from erosion of the rich agricultural lands of the delta (Fig. 19a.4). Without its yearly supply of silt the coastline is now receding noticeably—as much as several yards per year in some places. Rice raised in the delta is a major dollar-earning export for Egypt, and hence loss of this prime land is economically significant.

FIGURE 19a.4

View of the Nile Delta; photo taken during the Gemini missions.
The Red Sea is on the upper right, the Sinai peninsula on the upper left. Erosion of the Delta is now occurring in the absence of silt deposition, as is depletion of the adjacent fisheries. (Courtesy of NASA.)

FISHERIES

One of the great benefits of Lake Nasser was to have been the generation of a new fishery, yielding ultimately some 12,000 to 16,000 tons of fish yearly—a boon to a country where annual consumption of animal protein is less than 25 pounds per person, or less than one-fourth of what it is in the United States. The number of fish taken so far has been disappointingly small, amounting to only 4500 tons per year. This is partly a result of the paucity of fishermen; since the shoreline is still rising, no permanent settlements are possible, and only a few people have been attracted to the area. Little is known about the dynamics of fish populations in large artificial lakes, so it is impossible to predict what the eventual sustainable yield in

Lake Nasser will be. Most experience with other large impoundments suggests that there is often an early peak in production due to sudden utilization of nutrients, followed by a rapid decline to a modest level as the ecosystem settles into a steady state. Such was the case in the lake behind Zambia's great Kariba Dam, where catches of over 10,000 tons per year have now dwindled to less than 2000 tons.

MARINE FISHERIES

Disappointing as the lack of fish in Lake Nasser may be, of far more consequence has been the loss of much of the fishery of the eastern Mediterranean. The Mediterranean is an impoverished sea; there is virtually no upwelling, and much of its nutrient rich sediment is lost to the Atlantic as the deeper waters flow out over the sill at the Strait of Gibraltar. Fishermen remove a large fraction of nutrients which might otherwise be recycled, and the only significant inputs are the rivers. Of the great rivers supplying water to the eastern Mediterranean, the Don, Dnieper, Dniester and Danube flow first into the Black Sea, where much of their nutrients are deposited as sediments or removed by man. The Nile, prior to the High Dam, was therefore the main source of nutrients for that end of the Mediterranean (Fig. 19a.2). During the annual floods, its silt supplied nutrients for a great phytoplankton bloom which supported the entire aquatic food chain. The effects extended for hundreds of miles. Since the dam, all this has drastically changed.

During 1962, three years before the closing of the dam, the total commercial fish production from the Mediterranean Sea was 38,000 tons. By 1967, this had fallen precipitously. Catches of sardine dropped from 18,000 tons to 500 tons in the same period and disappeared completely in 1971. Not only Egypt but other countries — Greece, Israel, Lebanon — were drastically affected. At least 50 per cent of the total fishery has disappeared; nor will the smaller and uncertain catches in Lake Nasser be significant compensation for the losses in dietary protein suffered by the peoples of the Near East.

PUBLIC HEALTH EFFECTS

The High Dam is causing yet one other unexpected detrimental effect on Egypt's population — namely, a decline in the general health of the people through spread of debilitating diseases. The most important of these is **schistosomiasis** — also known as bilharzia after Theodor Bilharz, who first described the infective agent. The disease has long been known in the Nile Valley and other tropical regions, but was not prevalent until extensive irrigation was introduced at the beginning of this century. Literally millions became affected by it, and it has spread even more since construction of the High Dam. Schistosomiasis results from infection with a flatworm belonging to the group known as trematodes or flukes. The most common species in Egypt are *Schistosoma hematobium* and *Schistosoma mansoni.* Like many parasites, the schistosomes have a complex life cycle which enables them to reproduce in great numbers and disperse themselves widely, thus insuring survival of the species (see special section on page 534).

EPIDEMIOLOGY SINCE THE HIGH DAM

Prior to construction of the Aswan High Dam, approximately 47 per cent of Egyptians were affected by schistosomiasis. Incidence was highest among peasants, especially in perennially irrigated regions around the delta. This high rate of infection is due to the use of open water for drinking, bathing and laundering clothes and as latrines. Children daily wade in the local ditches to tend the water buffalo. The Moslem practice of *wadu,* or washing the body before daily prayers, exacerbates the situation. The annual floods, however, once tended to suppress the population of snails, thus reducing the incidence of infection. Silt-laden flood waters swept the snails away, and annual desiccation prevented their multiplication throughout the year.

The recent construction of extensive year-round irrigation ditches has effected an enormous change in the snail populations. The clear, slow-flowing waters in the ditches provide optimum habitats for the host snails, whose numbers have consequently increased greatly. The shallow waters of the extensive shores of Lake Nasser also provide an excellent breeding ground for snails. As a result, infection today occurs in areas where it was previously unknown. The overall incidence of schistosomiasis in the rural population is now estimated to be 80 per cent, with 90 per cent of agricultural workers infected. There are probably three million new cases since the dam was closed in 1964, at a cost to the Egyptian economy of some 80 million dollars a year.

CONTROLLING SCHISTOSOMIASIS

At present, there is no certain cure for the disease. The schistosome is a very successful parasite, since it resists antibodies produced by the host and does not often kill its host outright. Various drugs have been developed but none is as yet truly effective. One promising drug, hycanthone, produces only temporary remissions. It has been recently discovered, moreover, that schistosomes readily develop resistance to hycanthone, the reason being that the drug is an excellent mutagen. Apparently, a few mutant adult worms are capable of detoxifying the drug and subsequently lay eggs which are genetically resistant to it. It is evident that medical technology has much to learn concerning therapeutic measures for this disease. Even if a person is cured through treatment, reinfection is likely the next time he contacts snail-infested waters.

Molluscicides (snail killing chemicals such as copper sulfate and sodium pentachlorophenate) are used extensively but are only marginally successful. When used in high concentrations, they are toxic to other organisms, including fish and man. It has also been proposed to introduce species of snail-eating fish into the irrigation ditches for biological control of the snails. Effectiveness of this method remains to be demonstrated.

Obviously the most effective measure would be to change the habits of the rural populace. Introduction of piped water for domestic use and of septic tanks, and the construction of village swimming facilities to supplant the recreational aspects of the irrigation ditches, would all do much to decrease infection. Such innovations require changes in cultural habits. A decade ago, the World Health Organization built latrines in a highly infested part of Egypt, but these were largely unused. For one thing, they rapidly filled with water and, in the high temperatures, quickly became putrid. Also, the peasants were unwilling to give up their ablutions before prayer. Whether the expense of providing clean water can be met and the necessary cultural changes brought about remains to be seen.

OTHER DISEASES

The extensive irrigation ditches and the waters of Lake Nasser provide breeding grounds for at least two insects which carry diseases: the black fly, which produces trachoma, a disease of the eye which can lead to blindness; and the *Anopheles* mosquito, which transmits malaria. The latter is of great potential danger to Egypt, since malaria occurs in areas of the Sudan only

50 miles south of Lake Nasser. In 1942, the mosquito invaded Egypt far enough to infect a million persons, killing a tenth of them. The swampy shores of Lake Nasser offer an ideal breeding ground and hence a good focus for spreading the mosquitoes and the disease further northward. Present attempts at several check points to prevent ingress of the vector on travelers coming from the Sudan may or may not prove sufficient to keep out the disease.

The tragedies surrounding the Aswan High Dam are but an example of the effects of man's attempts to "improve" on nature *before* he understands the ecological implications of his acts. In Chapter 2a we noted some

THE LIFE CYCLE OF SCHISTOSOMES

Most parasites are able to infect only a limited number of host species, where they find the optimum conditions necessary for their survival and reproduction. Under natural conditions, suitable hosts are met with only occasionally. In order to insure their survival as a species, many parasites must therefore reproduce in large numbers and become widely dispersed. One means of attaining dispersal is to infect, during part of the life cycle, a second, intermediary host, which also must be one of a limited number of species. In the course of evolution, a great many parasites have thus developed complex life cycles, often involving two or more distinct host species, in each of which they spend part of their lives. The life cycle of the blood flukes or schistosomes offers a good example of the complexity typical of parasitic existence. The schistosomes spend the adult part of their lifetime in man and the larval stages in certain species of aquatic snails, which thus act as intermediate hosts and vectors of the disease. The complete life cycle is shown in Figure 19a.5.

The adults live in the blood vessels of humans, either in the blood vessels of the bladder or in the portal veins carrying blood from the intestine to the liver, which are especially rich in digested foodstuffs. The genus *Schistosoma* is unusual among trematodes in having separate sexes which are strikingly different in appearance. The male is larger, flatter and slightly shorter than the female and forms a groove in which the female resides. Each adult bears suckers at the front end by which it attaches to the walls of the blood vessel. After the eggs of the female are fertilized she leaves the male and migrates to smaller blood vessels, where the eggs are laid. The eggs are heavily encapsulated and eventually reach the exterior by way of either the bladder or the intestine. During this time, the egg, still within its capsule, develops into a microscopic multicellular larva, the **miracidium**.

Once excreted, the encapsulated miracidium remains viable up to two years provided it neither freezes nor dries out. Under appropriate conditions of moisture, light and temperature, the larva hatches and swims actively by means of numerous cilia until it finds a host snail. It then bores into the snail's tissues, where it undergoes two **sporocyst** generations. This is a form of asexual (mitotic) reproduction in which the miracidium becomes converted into a sac-like organism containing many germ cells. Each of these develops into a second sporocyst and the whole process is repeated. These two stages serve to greatly increase the number of individuals. The offspring of the second sporocyst develop into yet another larval stage, the **cercaria.** The cercariae emerge from the host snail and swim about by means of a forked tail. Upon contacting human skin, they attach by suckers, lose their tail and digest the skin, thus penetrating into the circulatory system. During the few days they spend in the general circulation the cercariae grow into adults and finally settle down in the bladder veins or portal veins. The complexity of this propagative mechanism is a strategy of nature designed not to confuse students but rather to insure maximum survival of the parasite population.

Infection of man usually occurs via the skin while wading in snail-infested waters or by drinking such water, in which case cercariae penetrate the mucus membranes of mouth and throat. Symptoms include weakness, abdominal pain, headache and fever, and in some cases effects on the central nervous system are evident. The age groups most susceptible to infection are children and adolescents. The disease is seldom a direct cause of death but usually leads to malaise and general disability, together with a decreased resistance to other diseases. Infected adults frequently are able to work only a few hours a day. Life expectancy among highly infected populations is greatly reduced, being but about 25 years in some areas. Side effects include anemia and mental retardation. Polyps sometimes develop in the bladder, owing to the presence of encapsulated schistosome zygotes, and these may become cancerous. Adult schistosomes live up to 20 or 30 years in the human host despite the presence of demonstrable antibodies.

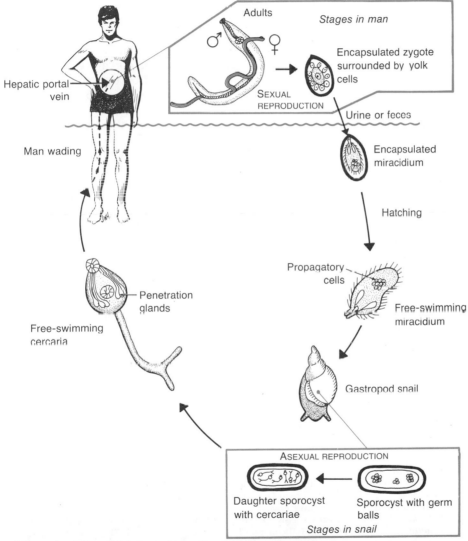

FIGURE 19a.5

Life cycle of the schistosome (parasites not shown to scale).
     The adults live in the hepatic portal vein of man. The encapsulated egg develops into a miracidium which is excreted. Under optimal conditions it hatches and infects a snail. In this intermediate host, two sporocyst generations occur; eventually numerous cercaria larvae are released which penetrate the skin of man. In the human circulation, cercariae develop into adults which may persist for 20 to 30 years.

likely consequences of interbasin water transfer in North America and also the effects of attempts at climate modification. Other projects presently in the planning stages which could have similar far-reaching ecological consequences are (1) the Russian plans to divert rivers from the arctic to central Siberia; (2) plans for agricultural development of the Amazon basin (see Chapter 17a); (3) plans for an oil pipeline from Alaska to the conterminous United States, and (4) strategies for blasting a sea-level canal through the Isthmus of Panama. Let us hope the Aswan High Dam is a lesson: technology without ecological understanding is, at best, a gamble, and is likely to create more problems than it solves.

SUMMARY

Egypt, one of the world's underdeveloped countries, is seeking to improve its economy by modernizing both agriculture and industry. The Aswan High Dam, constructed in the 1960's, was designed to play a decisive role in both areas, providing water for the one and power for the other. The subsequent ecological impacts of the dam were either unforeseen or ignored during planning, although all could have been predicted had the effort been made. Interference with natural nutrient and water cycles has had far-reaching effects on the populations of plants and animals in several ecosystems.

Agricultural productivity, expected to rise sharply through increased irrigation, has increased only moderately and then only with the addition of massive amounts of fertilizer to replace lost nutrients. Less water than expected is available, since losses from Lake Nasser through evaporation, seepage and reversed groundwater flow are far greater than predicted.

The great fisheries of the eastern Mediterranean have declined precipitously, since the nutrient-laden silt no longer flows into the sea to support the food chain. Even the most optimistic estimates of the productivity of Lake Nasser's fishery, far from being met at present, could not equal the losses already experienced in the Mediterranean.

A third major effect of the dam has been the widespread increase in schistosomiasis. The rapid rise in the population of the intermediate host snails in the clear, still waters of the new irrigation ditches has increased the population of the infective flatworm causing the disease, for which there is no effective cure.

All of these detrimental side effects of the dam are costly and are only partially reversible. It is likely that the High Dam will have created more problems than it solves.

READINGS AND REFERENCES

Deevey, E. S. 1960. The human population. Scientific American Offprint No. 608. The historical growth of the human population is set against the limits of Earth's capacity to support it.

George, C. J. 1972. The role of the Aswan High Dam in changing the fisheries of the southeastern Mediterranean. Pp. 159–178, *in* Farvar, M. T. and Milton, J. (eds.) *The Careless Technology.* Natural History Press, Garden City, N. Y. Details the sudden, drastic decline of fish due to loss of Nile nutrients.

Greany, W. H. 1952. Schistosomiasis in the Gezira irrigated area of the Anglo-Egyptian Sudan. Annals of Tropical Medicine and Parasitology *46*:250–267, 298–310. This earlier irrigation scheme on the upper Nile gave clear warning of the increased infection that would occur in Egypt.

Hughes, C. C. and Hunter, J. M. 1972. The role of technological development in promoting disease in Africa. Pp. 69–101, *in* Farvar, M. T. and Milton, J. (eds.) *The Careless Technology.* Natural History Press, Garden City, N. Y. The nutrition and health of much of Africa is declining rapidly as a result of technological "improvements" in agriculture.

Kassas, M. 1972. Impact of river control schemes on the shoreline of the Nile Delta. Pp. 179–188, *in* Farvar, M. T. and Milton, J. (eds.) *The Careless Technology.* Natural History Press,

Garden City, N. Y. As the coast erodes, the large brackish lakes of the delta will become saline, destroying thousands of acres of fertile farmland.

Rogers, S. H. and Bueding, E. 1971. Hycanthone resistance: Development in *Schistosoma mansoni*. Science *172*:1057–1058. Experiments demonstrate that the drug used to treat schistosomiasis may itself induce formation of resistant forms.

Shiff, C. J. 1972. The effects of molluscicides on the microflora and microfauna of aquatic systems. Pp. 109–115, *in* Farvar, M. T. and Milton, J. (eds.) *The Careless Technology*. Natural History Press, Garden City, N. Y. Although new molluscicides kill mainly snails, fish are also highly affected, including species grown for food in aquaculture ponds.

Sterling, C. 1971. The Aswan disaster. Environmental Journal, Nat'l Parks and Conservation Magazine *45*(8):10–13. A journalist discusses many facts related to the Aswan Dam.

van der Schalie, H. 1972. World health organization project Egypt 10: A case history of schistosomiasis. Pp. 116–136, *in* Farvar, M. T. and Milton, J. (eds.) *The Careless Technology*. Natural History Press, Garden City, N. Y. Vividly describes the development and extent of schistosomiasis in Egypt and the Sudan and reasons for virtually complete failure of control attempts.

Zilberg, B., Saunders, E. and Lewis, B. 1967. Cerebral and cardiac abnormalities in Katayama fever. South African Medical Journal *41*:598–602. Case histories indicate schistosomes may also affect the brain and spinal cord, especially in children.

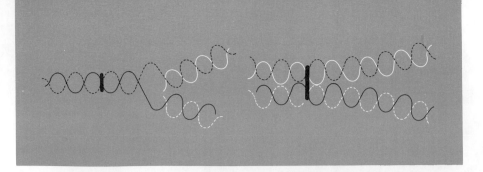

20

THE ORIGINS OF LIFE

In the past decade speculations about how life originated on Earth have become more and more scientifically plausible. It is generally agreed that a major result of the evolution of self-reproducing organisms was the irreversible change they wrought on their environment.

One of the most apparently impractical preoccupations of man is his age-old curiosity about his own origins. Every culture has invented a story of the creation of the world, living organisms and man. The possibility of supernatural creation, since it is not subject to experimental testing, falls outside the domain of scientific argument. As we continue to gather information about the probable conditions existing on our primitive Earth, however, scientific explanations for the origin of life become ever more credible.

As we shall see in the next chapter, once living organisms had appeared on Earth, their further evolution from simple to complex forms, leading eventually to man, can be explained by the theory of natural selection first propounded by Darwin and Wallace—a theory now widely accepted by biologists. It is, rather, the initial appearance of the first organisms that is the subject of this chapter. Some have suggested that life arose elsewhere in the universe and was transported to Earth as bacterial spores; but that idea only transfers the problem and does not answer it. Since it is believed that other planets which would be capable of supporting life would have had a history similar to that of Earth, it is simpler to assume that life evolved from inanimate matter found on Earth, where all of the necessary conditions probably existed.

Most biologists believe that after life first evolved, the possibility for further abiogenic (non-living) origins disappeared. Since Pasteur's demonstration that sterilized media, no matter how rich in organic molecules, does not generate life, the idea of spontaneous generation of organisms *under conditions of the modern Earth* has all but disappeared. At the end of this chapter we shall briefly consider the arguments for neobiogenesis, or the spontaneous generation of life in the presence of living organisms. First, however, let us trace the stages by which it is hypothesized that life began *under conditions existing on the primordial Earth.*

THE NATURE OF THE PROBLEM

No one knows precisely what the Earth was like when it was formed some 4½ billion years ago, but some reasonable possibilities can be envisioned from what we do know of the Earth's formation. When these hypothetical conditions are duplicated in the laboratory we can observe what might have happened. Many experiments of this nature have been conducted, and many interesting results have emerged which allow us to construct a tentative hypothesis about how life originated. As more information is gathered and new ideas are formed and tested, the hypothesis will no doubt be modified, or it may be discarded altogether and replaced with a new one.

In attempting to bridge the gap between the non-living and living worlds, it is desirable to define the fundamental characteristics of life. A working definition of life might be "the organization of organic molecules into self-sustaining entities which are capable of growth and self-duplication." In fact, it is probable that when the first forms of life were emerging no sharp distinctions between living and non-living existed. (Nor do they today, since scientists find it difficult to classify viruses as living or non-living—see Chapter 4.) One thing which is certain is that the evolution of life required a source of energy. Any theory must therefore account for the formation of organic molecules, for their organization into entities separate from their environment and for a source of energy for their growth and self-duplication.

CONDITIONS ON THE PRIMITIVE EARTH

The first questions concerning the origin of life which scientists must try to answer are, How was the Earth formed? What conditions existed? What consequences would these conditions have for the origin of life?

THE FORMATION OF SUN AND EARTH

Stars like our sun are now thought to have discrete lifetimes. They are "born" from the condensation of clouds of atoms, mostly hydrogen, which exist in space (Fig. 20.1). When such a cloud is dense enough, gravitational forces between the atoms cause them to fly together. During this acceleration they gain energy, and eventually an intensely hot star is born. The energized hydrogen atoms interact in a type of fusion process which generates helium; enormous amounts of energy are released, which are radiated as electromagnetic energy—heat, ultraviolet and visible light and gamma rays. Slowly, as the star ages, it expands, becoming a so-called red giant. Just before it expires, a star gains a new lease on life, burning brightly as a white dwarf, and producing a multitude of other elements. Then it either explodes or fades away.

Our sun is about 5 billion years old and is middle-aged. In another 5 billion years it will become a red giant, heating the Earth's surface to 2200° centigrade; a few millennia later it will have completed its life. Since large stars burn up faster than small ones, we are fortunate that our sun is only a middle-sized star.

The debris left when a star expires consists not only of hydrogen and helium but also of the other elements formed during its declining years. These various atoms float in space, contributing to the cosmic dust from which Earth and the other planets are believed to have been formed. The great variety of elements found on Earth are thus possibly the remnants of stars long since gone. At the time when the sun was forming, other cosmic debris lying within its influence condensed to form the planets. The small mass and warmer temperatures of some of the planets such as Earth proba-

FIGURE 20.1

A cloud of cosmic gas and dust in the constellation Sagittarius, in which new stars are forming.

The hydrogen gas emits light due to its excitation by radiation from neighboring stars. The small dark areas are believed to be new stars in the process of condensation. (Photograph courtesy of the Hale Observatories.)

bly resulted in the loss of much of the lighter elements, hydrogen and helium—the commonest gases in space. Earth thus acquired disproportionate amounts of the heavier elements. As the gaseous elements and dust particles condensed, they inevitably collided with one another, producing small molecules.

At some time in its early life, the interior of the Earth became very hot—either owing to heat generated during its condensation or, later, through accumulation of energy from radioactive disintegration within (radiogenic heat). Most of its rocks melted, and the heavier elements such as iron and nickel gravitated to the center, forming a dense core. The lighter atoms and molecules—those common to living cells—remained near the surface, forming the crust and later the primordial atmosphere.

THE PRIMORDIAL ATMOSPHERE

The first atmosphere surrounding our planet is generally believed to have been emitted from volcanoes and cracks in the Earth's crust, but there is disagreement as to its composition. It certainly contained enormous quantities of water vapor, which, as the surface gradually cooled, condensed to form the primitive oceans. Some think the remaining gases were mainly hydrogenated compounds, such as methane (CH_4), ammonia (NH_3) and hydrogen itself (H_2); others claim that the gases were less hydrogenated, consisting mainly of carbon dioxide (CO_2), nitrogen (N_2) and a little carbon monoxide (CO), with only traces of hydrogen and perhaps methane. Cyanide (CN) was also probably present. In either case, there was virtually no free oxygen, which forms a fifth of today's atmosphere.

In the prebiotic world, the only source of free oxygen would have been that released by *photolysis* or ultraviolet (UV) splitting of water vapor in the upper atmosphere:

$$UV\ light$$
$$2\ H_2O \longrightarrow 2\ H_2 + \mathbf{O_2}$$

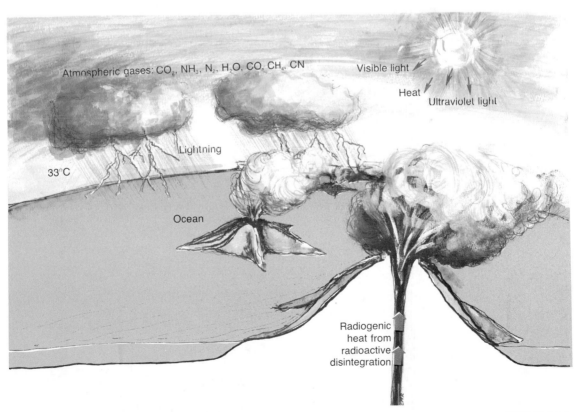

A diagram of the Earth about 3.5 billion years ago.
   The atmosphere contained mainly CO_2, N_2 and H_2O, with several other gases also present, but no O_2. The surface temperature was hot; shallow oceans formed. Radiant energy from the sun, radiogenic heat from the interior and lightning all provided potential energy sources for synthesis of organic molecules.

FIGURE 20.2

We know from the composition of rocks erupted from the Earth's interior by modern volcanoes that the Earth's rocks were initially unoxidized, yet all exposed rocks today are oxidized. Thus any free oxygen generated by photolysis must have quickly combined with the rocks of the crust and with the poorly oxidized volcanic gases. The early Earth therefore had a more or less oxygen-free or *reducing atmosphere.*

A summary of our picture of the primitive Earth is shown in Figure 20.2: a thin layer of oceans; a small land mass, mostly volcanic; and an atmosphere free of oxygen, containing much water vapor. Various sources of energy were present — radiogenic heat from the center of the Earth, ultraviolet, visible and infrared radiation from the sun, and electricity from the continuing thunderstorms. The average temperature was perhaps 33°C — considerably warmer than today's 15°C average. This stage was reached about 3½ billion years ago.

FORMATION OF ORGANIC MOLECULES

If our picture of the primitive Earth is correct, what consequences would there have been for the generation of life? Despite their disagreements about the early atmosphere of Earth, all scientists agree that organic molecules must have been formed before life appeared. A great many experiments have been performed using conditions simulating those believed to have existed some 3½ billion years ago. One of the first was conducted in 1952 by Stanley Miller while still a graduate student at the University of Chicago, using the apparatus shown in Figure 20.3. Starting with boiling water and the gases methane, ammonia and hydrogen, Miller subjected the continuously cycling mixture to an electrical discharge. After a week he opened the apparatus and analyzed its contents. The water contained various amino acids, the building blocks of proteins. Since then, many similar experiments have been conducted, using slightly different combinations of gases and various sources of energy. In all cases, organic molecules have been obtained — organic acids, amino acids, nitrogenous bases (which form nucleic acids in cells). It is therefore possible that these same molecules were formed in the primitive atmosphere and that, washed to the surface by rain, they accumulated in the primordial oceans.

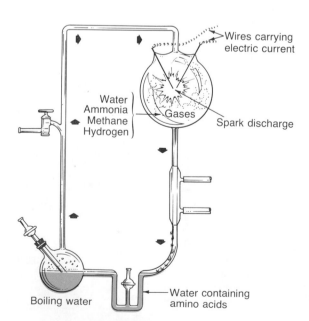

FIGURE 20.3

Apparatus used by Miller in the first experiments generating organic molecules from a mixture of gases probably present on the primitive Earth. Energy was generated by electric discharge. (After Jastrow, R. 1967. *Red Giants and White Dwarfs.* Harper & Row, New York.)

THE STEPS TO LIFE

Having demonstrated the possibility of abiogenic formation of small organic molecules, we must next attempt to explain how they became macromolecules, how these aggregated together and, finally, how the aggregate became self-supporting.

SYNTHESIS OF MACROMOLECULES

In many of the experiments just mentioned, not only are free amino acids present, but there are also protein-like polymers. In other experiments, heating amino acids to temperatures greater than boiling, as might occur under high pressures in underground hot springs, has yielded proteins. Clay particles and certain simple chemicals have also been shown to function catalytically in the polymerization of macromolecules. Like small organic molecules, macromolecules can be formed under abiogenic conditions such as might have existed on the primitive earth.

THE PROTOBIONTS

What was the fate of these primitive macromolecules? On today's Earth they would have been rapidly consumed by bacteria or gradually oxidized by molecular oxygen; but neither of these was present at that early time. Some macromolecules undoubtedly were destroyed by the very ultraviolet irradiation which brought them into existence. Remember that there was no atmospheric oxygen yet, and therefore no ozone, which is a photoproduct of oxygen; ozone is the main absorbent in today's atmosphere of the sun's ultraviolet light. The macromolecules of today's living organisms are highly sensitive to UV, and it is likely that many primitive macromolecules were as well. But what of those which found a spot protected from UV, such as in the subsurface waters or sediments of the shallow seas? How did they come together to form aggregates—the **protobionts** or pre-cells?

It has long been known that aqueous solutions of macromolecules, under the right physical and chemical conditions, tend to separate spontaneously into two phases. One phase consists of droplets containing most of the macromolecules; the other, continuous phase contains mostly ions and small molecules dissolved in water (Fig. 20.4). The macromolecules, under these conditions, have a greater affinity for one another than for the water molecules; they aggregate together. Such two-phase systems are called **coacervates.**

FIGURE 20.4

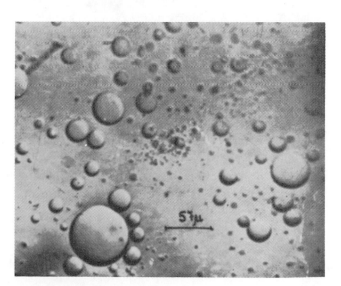

Coacervate droplet formation.
    Two types of macromolecules, gelatin and gum arabic, were mixed together with ions and water. Small droplets spontaneously separated out which contained most of the macromolecules; ions and water form the continuous phase in between. Note the sharp boundaries in this two-phase system (magnification about 200 ×). (From Oparin, A. I. 1964. *The Chemical Origin of Life.* Charles C Thomas, Springfield, Illinois.)

Russian biochemist A. I. Oparin conducted many experiments on coacervates formed from various mixtures of different kinds of macromolecules, including proteins, nucleic acids and chlorophyll. In one experiment he included in the system an enzyme which was capable of catalyzing the synthesis of another macromolecule. When the monomer of this second macromolecule was then added to the continuous phase of the coacervate suspension, the droplets were observed to grow (Fig. 20.5). Monomers diffused into the droplet, where they were catalytically added onto the macromolecules, causing the droplet to increase in size. Sometimes the droplets fragmented into two or more "daughter" droplets which also grew as long as they each contained enzyme molecules (Fig. 20.5C).

Similar phenomena have been observed by other workers, and the resemblance to living cells is sometimes remarkable (Fig. 20.6). Obviously chemical reactions go on within these droplets which are different from those in the surrounding medium. Coacervates clearly may represent a stage which actually occurred during the origin of living cells—a stage during which innumerable combinations of macromolecules were constantly being formed and reformed. One can imagine that several catalysts eventually came together in one droplet, allowing a complete biochemical pathway such as glycolysis to occur. Such a droplet would have a great ad-

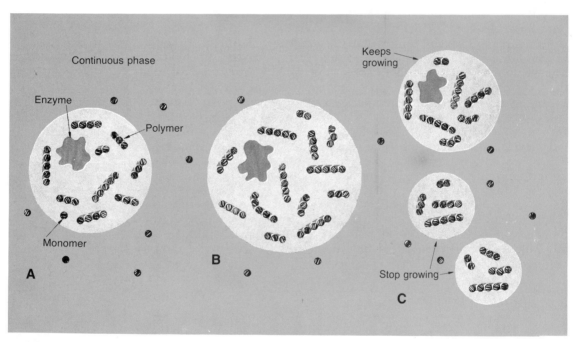

FIGURE 20.5   Growth of coacervate droplet containing polymerizing enzyme.
    A. As monomer diffuses in from continuous phase it is added to polymer chain by enzyme.
    B. The droplet grows and may fragment.
    C. Only daughter droplets containing an enzyme molecule continue to grow.

vantage, and could grow faster than its neighbors—until, of course, it lost one of its enzymes. So far our hypothesis has made no provision for replacing or duplicating molecules.

SELF-REPLICATION

The only molecules known to be capable of reproducing themselves are the nucleic acids—in particular, DNA. Experiments with isolated fragments of DNA from viruses which infect bacteria have shown that at least 500 subunits (monomers) must be present in a fragment for it to duplicate itself. Once nucleic acid molecules of that size appeared on the primitive Earth, they could have acted as templates or models for synthesis of their own complementary, mirror images. This "naked gene" would have become a focus for growth and reproduction of itself, aided by the presence of high concentrations of nucleic acid precursors and ATP in its environment. Perhaps it would gradually have attracted other macromolecules or come to reside in a coacervate droplet, thus resembling a modern virus (Fig. 20.7). Modern viruses are incapable of self-replication, however, even in an enriched broth—at least within the time periods allowed them by the experimenters. They must borrow the protein synthesizing machinery of a living host cell. Therefore, today's viruses could not have been the primordial organisms in the long road of life.

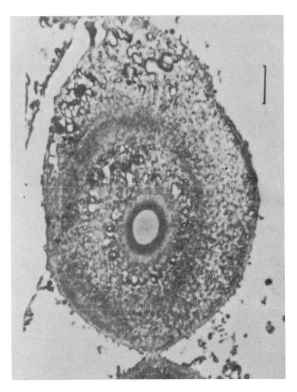

FIGURE 20.6

Resemblance of non living droplet of protein like material to a living cell. The central space is not a nucleus but a cavity. These droplets form when heated solutions of protein-like molecules are cooled. (From *Origin of Life* by John Keosian (2nd ed.) © 1964 by Litton Educational Publishing, Inc. Reprinted by permission of Van Nostrand Reinhold Company.)

We can imagine that in the primordial world, as more and more self-duplicating genes appeared, they began to compete with each other for an ever diminishing environmental supply of precursors and ATP. Those which could synthesize faster became predominant. A gene associated by chance with an enzyme which catalyzed its self-replication would have a great advantage—until the enzyme was lost or destroyed. A gene which also contained the correct code for synthesizing the appropriate enzyme would have a decisive advantage. Thus we see that catalysts are essential in the competition for limited resources, but only genes can reproduce both themselves and new catalysts. Once this interrelationship was established, the genetic code had come into being—and life as we know it may be said to have evolved on Earth.

Experiments designed to show how this last, crucial step may have come about have not so far been devised, although there is much evidence, especially from studies on viruses, which suggest it is not improbable.

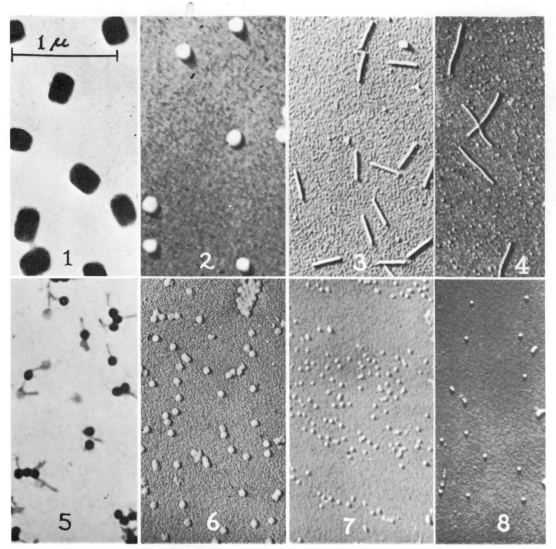

FIGURE 20.7 A variety of modern viruses: these contain DNA enclosed in a protein coat. The earliest self-replicating organisms may have been similar in size and composition, but must have including machinery for self-replication. (Courtesy of C. A. Knight, from Villee, C. A., and Dethier, V. G. 1971. *Biological Principles and Processes.* W. B. Saunders Co., Philadelphia.)

FURTHER DEVELOPMENTS

The conditions under which life first evolved were unfavorable to modern organisms. High intensities of ultraviolet light and the presence of carbon monoxide and cyanide would destroy most life today. Our earlier studies have also shown us that living organisms interact with and change their environment. This has been true since the earliest appearance of catalytic and self-reproducing molecules. Let us next consider some of the ways in which protobionts and the earliest cells would have modified their surroundings irreversibly.

THE NEED FOR ALTERNATIVE ENERGY SOURCES

As we have seen, organic molecules including ATP were originally formed abiogenically and gradually accumulated in the environment. The first organisms were therefore **heterotrophs** (see Chapter 9). Once these primitive organisms began to use the available molecules for their own growth, however, concentrations would have rapidly decreased. This would have been especially true of ATP, the major chemical energy source available for molecular synthesis. The temperature of the Earth was also falling, resulting in a decreased rate of all reactions. A premium was therefore placed on being able to obtain extra energy—and those primitive organisms capable of releasing energy from the organic molecules in their environment would have had a great advantage.

The reactions of glycolysis, during which sugar is degraded into lactic acid or alcohol (see Chapter 7) are found in almost all living organisms, from the most primitive bacteria to man. For this reason, glycolysis is believed to be one of the most ancient of all metabolic pathways. It provides a steady supply of ATP, as long as organic molecules are available.

A few other organisms, the **chemosynthetic autotrophs,** have evolved the ability to obtain energy from available inorganic sources. The bacteria belonging to this group obtain energy by the oxidation of iron from Fe^{++} to Fe^{+++}, by the oxidation of sulfur to sulfate or by the oxidation of ammonia to nitrate. It is probable that this type of energy utilization was developed early in the evolution of life. However, since reduced iron, ammonia and sulfur have a limited distribution, they now support these unusual bacteria in only a few places.

THE EVOLUTION OF PHOTOSYNTHESIS

Once glycolysis evolved, the utilization of abiogenically produced organic molecules would have increased rapidly, eventually exceeding their rate of formation. As these molecules disappeared from the environ-

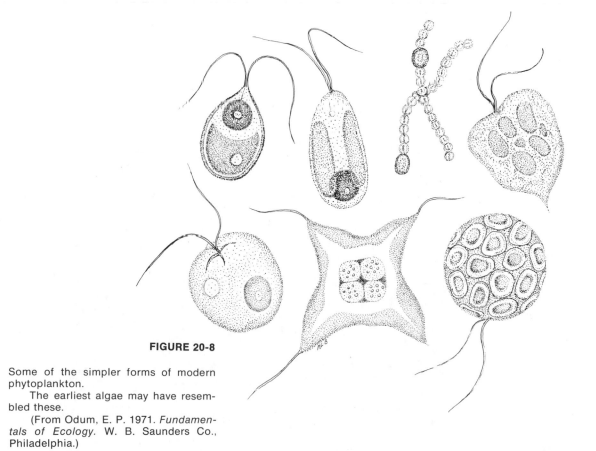

FIGURE 20-8

Some of the simpler forms of modern phytoplankton.

The earliest algae may have resembled these.

(From Odum, E. P. 1971. *Fundamentals of Ecology.* W. B. Saunders Co., Philadelphia.)

ment, there was once again a premium on being able to utilize an alternative source of energy.

Visible light, formerly of little use as an energy source, suddenly became important. Single-celled organisms in the oceans which contained pigments capable of absorbing light energy may gradually have evolved the ability to utilize this energy for synthesis of ATP; thus cyclic photophosphorylation came about (see Chapter 8). It is likely that ATP synthesis was the first use to which chlorophyll was put, for there would still have been abiogenically formed organic molecules in the environment to serve as building blocks for macromolecular synthesis and growth.

But as these also began to disappear, the ability to carry out non-cyclic photophosphorylation would have been of great advantage and become widespread—thus the first phytoplankton appeared (Fig. 20.8). Pho-

FIGURE 20.9   Rock formation in Wyoming 2 to 3 billion years old which contains traces of primitive algae. (From Jastrow, R. 1967. *Red Giants and White Dwarfs.* Harper & Row, New York.)

tosynthesis, utilizing as it does a continuing supply of energy from the sun to produce organic molecules, allowed living organisms finally to become independent of abiogenically formed food supplies. Judging from the age of ancient rocks in which remains of primitive algae have been found, this change took place approximately 3 billion years ago (Fig. 20.9).

CONSEQUENCES OF PHOTOSYNTHESIS

As we have repeatedly observed, one product of photosynthesis is oxygen. The evolution of photosynthetic organisms is believed by some scientists to have caused the Earth's atmosphere to change from a reducing to an oxidizing condition through the release and gradual accumulation of oxygen. Others disagree, claiming that not enough reduced carbon exists on Earth to account for the large amount of oxygen in today's atmosphere.

The Source of Atmospheric Oxygen. It may be of interest to contemplate some of the reasons why there is such great disagreement among scientists about the source of the oxygen in today's atmosphere. Some knowledge of this is essential in predicting what effects industrial man may have on global oxygen supplies, a matter about which scientists are also in disagreement. We shall deal only briefly with the subject here; further readings are given at the end of the chapter. Figure 20.10 summarizes the major known pathways of oxygen in the world today, excluding contribu-

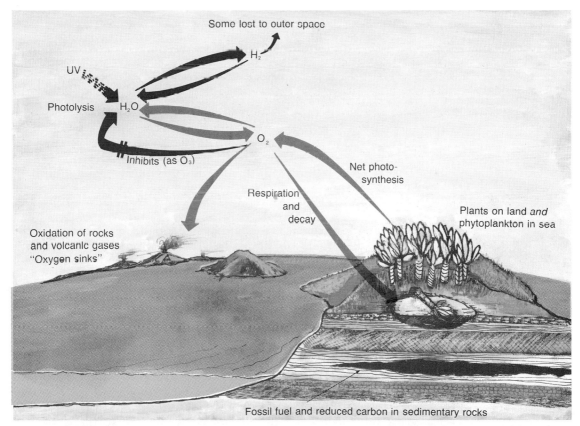

The balance of atmospheric oxygen.

FIGURE 20.10

Formation of O_2 by photolysis of water in the upper atmosphere slows down as oxygen, and hence ozone, accumulate. Photosynthetic production yields a net increase only if the plants are not respiring. Oxygen is constantly removed by oxidation of rocks and volcanic gases.

Initial oxygen accumulation must have resulted from an increased rate of production —either photolysis or photosynthesis in excess of respiration—or from a decreased rate of loss to oxygen sinks.

tions by man. It should be recognized that the trees pictured are meant to represent not only terrestrial plants but all phytoplankton and macroalgae in the sea as well.

As we noted earlier, before photosynthetic organisms appeared, the only source of atmospheric oxygen was photolysis of water vapor in the atmosphere by ultraviolet light. Most of the hydrogen and oxygen recombined; but some hydrogen, being light, escaped to space, leaving the heavier oxygen behind. It is thought that most of the oxygen so formed was utilized in the oxidation of volcanic rocks and gases, at least during the first 3 billion years of the Earth's history, when volcanoes were more active than they are now.

The development of photosynthesis some 3 billion years ago provided a second source of oxygen. At that time there were no organisms capable of using the oxygen; the remains of the dead phytoplankton, with only anaerobic decomposers to partially destroy them, would have been incorporated into the ocean sediments as reduced carbon. It is assumed that the first oxygen formed by photosynthesis was used up in the oxidation of surface rocks but that eventually enough formed to accumulate in the atmosphere. When aerobic respiration evolved later on, however, much of the oxygen subsequently produced by plants was re-used by them and by consumers. Eventually a new balance of atmospheric oxygen would have been reached—namely, that now found in today's air.

If photosynthesis has indeed significantly added to atmospheric oxygen over the millennia, there must exist at least an equal amount of reduced carbon—the debris of unused plants. Most estimates agree that at least six times more reduced carbon is present on Earth than would be necessary to account for today's atmosphere by the photosynthetic equation:

$$CO_2 + H_2O \longrightarrow \overset{\textstyle \text{atmospheric oxygen}}{\underset{\textstyle \text{reduced carbon}}{[CH_2O] + O_2}}$$

The main point of disagreement is that this may not be enough to have compensated over the millennia for the continuing loss of oxygen to "oxygen sinks"—the oxidation of eroded rocks and, especially, of volcanic gases.

The greatest problem is that no accurate estimates exist for any of the parameters involved. No one knows how much oxygen has been and is being formed by photolysis or lost to oxygen sinks. Some think these may have changed greatly with time. Even estimates of stored reduced carbon and of today's rates of global photosynthesis vary greatly. Whatever the initial origins of atmospheric oxygen, which began to accumulate about 2 billion years ago, it has had some significant results for life on Earth.

Advent of Respiration. Whether or not photosynthesis alone was responsible for the increase in atmospheric oxygen, the appearance of oxygen in the environment permitted the evolution of aerobic respiration. Initially, respiratory enzymes such as the cytochromes probably functioned as a protection against the presence of molecular oxygen, which in its free form is detrimental to life. Once the ability evolved to utilize the energy released from oxidation of reduced molecules for ATP synthesis and cell work, a rapid increase in the development of consumer organisms, and especially multicellular animals, became possible. This is believed to have occurred 600 million years ago, when atmospheric oxygen was about one-hundredth its present level (see Figure 20.11).

Oxygen, Ozone and Ultraviolet Radiation. A further important consequence of the increase in atmospheric oxygen is the formation of ozone in the upper atmosphere. As noted in Chapter 9a, ultraviolet light reacts with molecular oxygen (O_2) to form ozone (O_3). Ozone strongly absorbs UV light and thus slows down the rate of photolytic production of oxygen (Fig. 20.10). It also prevents much UV from reaching the Earth's surface.

Since ultraviolet light was probably an important source of energy for abiogenic synthesis of organic molecules, the increase in atmospheric oxygen greatly suppressed the amount of organic molecules formed in this way. Ultraviolet light is also destructive of living organisms; until an ozone screen formed in the upper atmosphere, life must have been restricted to deeper waters, sediments and perhaps protected crevices on land. Only after an ozone layer was present could life have invaded the exposed continents. There is concern today that supersonic transports will cause a decrease in the ozone shield along their routes, creating gaps several miles wide through which harmful ultraviolet radiation could reach the Earth's surface.

RESULTS OF THE ORIGIN OF LIFE

Once life began to evolve on Earth, it changed the environment in ways which preclude a repetition of the same events. It is probable that non-living processes alone would have brought about a different set of changes.

Many scientists think that during the period of some billion years or so when organic molecules and primitive organisms existed side by side, a great many forms of "life" must have evolved, some more successful than others. Whether today's living organisms have derived from only one or from several of the successful protobionts cannot be answered. Certain similarities shared by all life suggest a single origin to some, while others argue that the prevailing conditions would have favored parallel evolution in many places simultaneously.

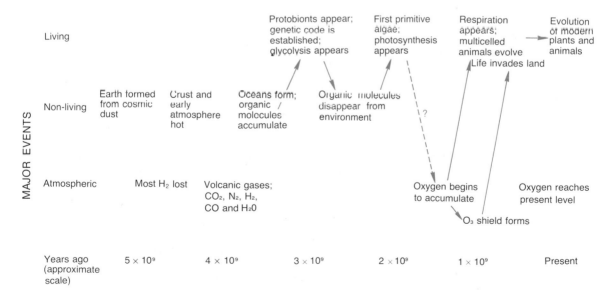

The major events hypothesized to explain the origin of life on Earth.   **FIGURE 20.11**
    The time scale is only approximate with respect to some events. Arrows indicate what are believed to be cause and effect interactions. Many other important events, such as the formation of sedimentary rocks, changes in the shape of the ocean beds and formation of the continents are omitted.

There is little disagreement, however, that once life began to alter the Earth's environment, no new life could arise abiogenically or spontaneously. The genetic advantage gained by living organisms is considered too great for the evolution of new forms except from those already living — the subject of our next chapter. A few scientists have argued otherwise — that neobiogenesis is continuingly possible. One suggestion is that, from the organic remnants of dead organisms, a new primitive species might arise capable of existence as a parasite within an existing species. Such a suggestion ignores the defense mechanisms possessed by organisms against unfamiliar invaders; it also ignores the fact that parasitic organisms can almost always be shown to have evolved from ancestors more complex than themselves. While it is true that DNA obtained from dead bacteria has been experimentally transferred to living bacterial cells, causing genetic changes in them, it would seem a matter of semantics to claim that neobiogenesis has taken place.

SUMMARY

Figure 20.11 provides a summary of the chronological events which are hypothesized to have taken place during the origin of life on Earth. Note that many of these events are believed to have set the stage for subsequent developments; thus the evolution of both the non-living and living components of the biosphere wrought irreversible changes which precluded a second origin of life by the same stages.

READINGS AND REFERENCES

Broecker, W. S. 1970. Man's oxygen reserves. Science *168*:1537–1538. A professor of Earth science claims that there is too much oxygen for man to deplete it seriously.

Gas exchange. (Letters to the Editor). Environment *12*(10:39–45 (1970). Comments for and against Broecker's ideas are aired.

Jastrow, R. 1967. *Red Giants and White Dwarfs — The Evolution of Stars, Planets and Life*. Harper & Row, New York. A readable lay book emphasizing the formation of Earth, sun and stars.

Johnson, F. S. 1970. The balance of atmospheric oxygen and carbon dioxide. Biological Conservation *2*(2):83–89. Argues that photosynthesis was the main source of atmospheric oxygen; difficult reading.

Kenyon, D. H. and Steinman, G. 1969. *Biochemical Predestination*. McGraw-Hill, New York. A readable paperback going into more depth on the various topics of this chapter.

Keosian, J. 1968. *The Origin of Life*. 2nd ed. Reinhold, New York.

Keosian, J. 1970. Neobiogenesis. Pp. 604–606, *in* Gray, P. (ed.) *The Encyclopaedia of the Biological Sciences*, 2nd ed. Van Nostrand-Reinhold, New York. The preceding two references expound what might be called "the minority opinion" about the consequences of the first origins of life.

Oparin, A. I. 1964. *The Chemical Origin of Life*. Charles C Thomas, Springfield, Illinois. The originator of many contemporary ideas about the origin of life provides a readable account of his theories.

VanValen, L. 1971. The history and stability of atmospheric oxygen. Science *171*:439–443. A highly technical argument about oxygen balance; points out uncertainties in our estimates; many useful references.

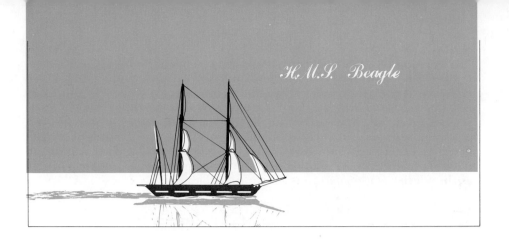

EVOLUTION—THE ORIGIN OF NEW SPECIES

When Darwin set sail upon the Beagle some one-hundred and fifty years ago, almost everyone believed either in spontaneous generation or divine creation of all species. Himself a strongly religious man, Darwin could not, however, dismiss the evidence before his eyes that evolution of species one from another has, in fact, occurred.

When we look at the plants and animals about us we take for granted that there is a great diversity of types: there are many kinds of trees, hundreds of flowering plants, a variety of birds, of insects—indeed, of all groups of organisms. We also have no difficulty distinguishing a pine tree from a cherry tree, nor a parrot from a robin. Each type of organism has a specific set of traits by which we distinguish it from all other types. It is also often possible, however, to recognize basic similarities among groups of organisms—all birds, for instance, characteristically have wings, beaks and feathers. The systematic grouping of organisms into related categories forms the basis of a branch of biology known as taxonomy.

The basis of our modern system of classification began with the Swedish botanist, Carolus Linnaeus, in the mid-eighteenth century. He grouped organisms into convenient categories, primarily based on their degree of structural similarity. As the concept of evolution and the sequential development of organisms one from another became accepted, Linnaeus' original groups were modified to take the fact of evolution into account. But even our present system, briefly outlined in Appendix II, is constantly being revised. Today taxonomists recognize several levels of relatedness among the enormously varied types of living and fossil organisms. Within the broadest, most general categories are grouped many diverse organisms which share only one or two major traits. The most inclusive categories are the **kingdoms,** of which five are now generally recognized—the Monera (bacteria and blue-green algae), the Protista (single-celled protozoans), the Fungi (yeasts and molds), the Plants and the Animals. Within each kingdom are large subdivisions, the **phyla** (singular, phylum); within each phylum are several **classes,** and so on. Each new subdivision includes a smaller group of more and more similar organisms, until in the last category—the **species**—the organisms all appear so similar that it is often difficult to distinguish one individual from another. The classification of man—from his membership in the Animal kingdom to his unique position as the species *Homo sapiens*—is shown in Table 21.1.

Kingdom	Animalia	(animals)	**TABLE 21.1 THE CLASSIFICATION OF MAN**
Phylum	Chordata	(with notochord)	
Class	Mammalia	(with hair and mammary glands)	
Order	Primates	(with opposable thumb and binocular vision)	
Family	Hominidae	(with erect posture)	
Genus	*Homo*	(man)	
Species	*sapiens*	(wise) (?)	

THE CONCEPT OF SPECIES

Within many species, however, we frequently recognize consistent varieties or races. This is especially true among those which have been bred by man; there are many varieties of corn, cattle, sheep, dogs and so on. Yet each variety, although often remarkably distinct, is still only a variety and not a true species. What, then, is the criterion by which we distinguish a variety from a species? The criterion which has been chosen has considerable biological significance, as we shall see later in this chapter. Two populations of closely related organisms are considered as two separate species if they are unable to interbreed successfully—that is, produce fertile offspring. Various types of cattle (varieties of the same species) can be interbred, but cattle and sheep (different species) cannot. The horse and donkey are different species; and even though they produce a vigorous offspring, the mule is sterile.

THE DARWINIAN THEORY OF EVOLUTION

Until 150 years ago, it was generally held that each species of organism had been separately created—and, once formed, did not change from one generation to the next. But early in the nineteenth century, a few people began to question these ideas. Evidence from the fossil record was beginning to accumulate, showing two things: first, that many species had once lived but were now extinct; second, that among certain groups of extinct species one could observe sequential stages of body characters which led directly to those found in living forms. Many of the extinct types, however, were so different from living forms that no one claimed them to be mere variants of contemporary species; all agreed they were distinct species. Excellent examples of such sequences are to be found in the fossil records of the elephant and the horse (Fig. 21.1).

The evidence of the fossil record could be interpreted in two ways. Either each fossil species arose independently at some time in the past (either through divine creation or spontaneously from non-living matter), or there has been a gradual evolution from one species to another. Spontaneous generation was long believed to occur; the sudden appearance of tadpoles in an apparently lifeless pond, of maggots in meat exposed to the air or of "worms" in a barrel of rainwater seemed sufficient evidence. Despite the experiments of Redi, an Italian scientist who demonstrated about 1680 that maggots did not develop in meat if it was protected from flies, the concept of spontaneous generation was widely accepted even 150 years ago. Even more widely accepted was the Biblical account of the original creation of species, which places the origin of life at only a few thousand years ago. Such was the general climate of opinion when a young English naturalist, Charles Darwin, undertook a voyage of nearly five years, during which he made observations which led him to formulate the theory of evolution of one species from another.

THE VOYAGE OF THE BEAGLE

In 1831, Darwin, a young man of 22 who was a dropout from medical school and felt formal education a waste of time, was taken on board the survey ship, *H.M.S. Beagle,* as ship's naturalist (Fig. 21.2). At that time, the concept of evolution had already begun to gain favor in certain circles in Europe, and one reason for Darwin's being invited on the voyage was to obtain geological evidence to refute these views. Young Darwin was then also a believer in the Biblical account of creation. But the seeds were planted. His grandfather, Erasmus Darwin, had held evolutionary theories. Also, Darwin took with him on the voyage a copy of the first volume of Sir Charles Lyell's *Principles of Geology,* which proposed the then revolutionary view that the Earth's physical features had developed not in a few thousand years—as the Bible suggested—but over immense periods of time. These, together with his own extraordinary powers of observation, led him eventually to discard his creationist beliefs and, more than 20 years after the voyage, to propose his theory of evolution in a book called *The Origin of Species by Means of Natural Selection or the Preservation of Favored Races in the Struggle for Life.*

The voyage of the *Beagle* took Darwin to a great many places—both coasts of South America, New Zealand and Australia, as well as many isolated groups of islands in the Atlantic and Pacific oceans. Among the pages of his diary of the trip are to be found not only his general observations on the geology, plants, animals and people of each area, but also special insights which later formed part of the argument for the *Origin of Species.* Let us examine a few examples of these insights.

Excess Reproductive Capacity and Competition for Survival. In the intertidal waters of the Falkland Islands and Tierra del Fuego at the southern tip of South America, there exists a shell-less mollusc, or sea-slug (Fig. 21.3), about which Darwin made the following observations:

> I was surprised to find, on counting the eggs of a large white Doris (this sea-slug was three and a half inches long), how extraordinarily numerous they were. From two to five eggs (each three-thousandths of an inch in diameter) were contained in a spherical little case. These were arranged two deep in transverse rows forming a ribbon. The ribbon adhered by its edge to the rock in an oval spire. One which I found, measured nearly twenty inches in length and half in breadth. By counting how many balls were contained in a tenth of an inch in the row, and how many rows in an equal length of the ribbon, on the most moderate computation there were six hundred thousand eggs. Yet this Doris was certainly not very common: although I was often searching under the stones, I saw only seven individuals. *No fallacy is more common with naturalists, than that the numbers of an individual species depend on its powers of propagation.*

The tendency of all animals to produce far more offspring than can possibly survive was also noted by Darwin in the following passage, in which he is discussing replacement of extinct species of animals in South America by closely related living species.

Nevertheless, if we consider the subject under another point of view, it will appear less perplexing. We do not steadily bear in mind, how profoundly ignorant we are of the conditions of existence of every animal; nor do we always remember, that some check is constantly preventing the too rapid increase of every organized being left in a state of nature. The supply of food, on an average, remains constant; yet the tendency in every animal to increase by propagation is geometrical; and its surprising effects have nowhere been more astonishingly shown, than in the case of the European animals run wild during the last few centuries in America. Every animal in a state of nature regularly breeds; yet in a species long established, any *great* increase in numbers is obviously impossible, and must be checked by some means. We are, nevertheless, seldom able with certainty to tell in any given species, at what period of life, or at what period of the year, or whether only at long intervals, the check falls; or, again, what is the precise nature of the check. Hence probably it is, that we feel so little surprise at one, of two species closely allied in habits, being rare and the other abundant in the same district; or, again, that one should be abundant in one district, and another, filling the same place in the economy of na-

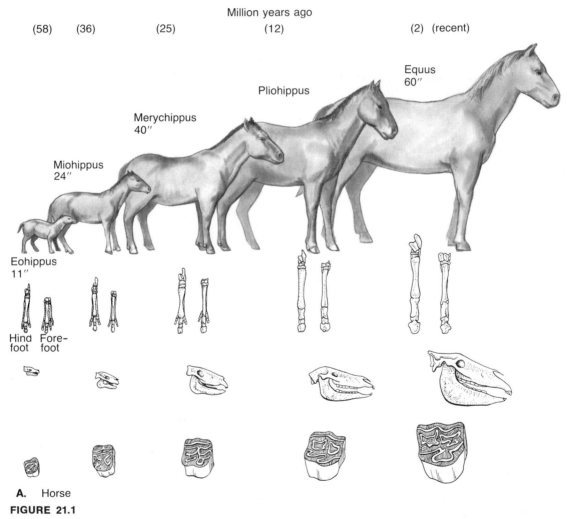

Million years ago

(58)　(36)　　(25)　　　(12)　　　　(2)　(recent)

Equus
60″

Pliohippus

Merychippus
40″

Miohippus
24″

Eohippus
11″

Hind　Fore-
foot　foot

A.　Horse

FIGURE 21.1

Illustration and legend continued on opposite page.

ture, should be abundant in a neighbouring district, differing very little in its conditions. If asked how this is, one immediately replies that it is determined by some slight difference in climate, food, or the number of enemies; yet how rarely, if ever, we can point out the precise cause and manner of action of the check! We are, therefore, driven to the conclusion, that causes generally quite inappreciable by us, determine whether a given species shall be abundant or scanty in numbers.

In this statement Darwin implies that overproduction of offspring is typical and that some environmental check eventually comes into play to control population growth; he also notes that such checks will have different effects on two species "filling the same place in the economy of nature" such that one species thrives while the other becomes rare. The idea of competition in the struggle for existence is clearly beginning to form in Darwin's mind. The next step is to discover why one species should succeed where another fails.

Adaptive Fitness.   The answer to this question came to Darwin slowly. On his visit to the Galápagos Islands—a small group of volcanic islands situated about 500 miles west of Ecuador, which Darwin knew to be of rela-

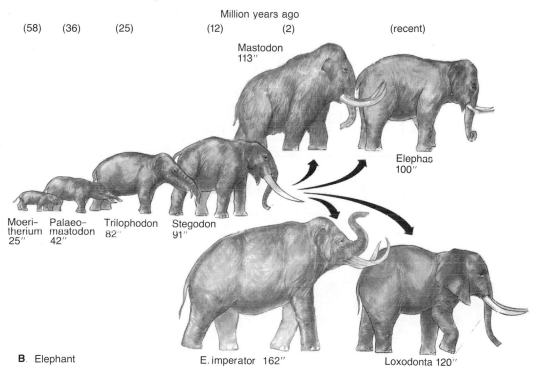

B. Elephant

Continued

FIGURE 21.1

Fossil evidence for evolution of two lines of mammals. The geologic period in millions of years, prior to the present, when each species lived is shown in parentheses; also, the shoulder-height of each species is given in inches.

A. The horse. Note increase in size from forest-dwelling *Eohippus* to modern, plains-dwelling *Equus*. Sequential changes in foot bones (note reduction in number of toes), skull (elongation of snout) and grinding surface of upper molar (enamel pattern indicated in black) are shown.

B. The elephant. Note gradual increase in body size and development of trunk and tusks. The primitive *Moeritherium* evolved in Egypt, but later forms migrated to other continents. Two million years ago, four species of elephants inhabited North America. Of the two living genera of elephants, *Elephas* is found in India and *Loxodonta* in Africa.

(After Storer, T. I. 1943. *General Zoology*. McGraw-Hill, New York.)

FIGURE 21.2

Charles Darwin as a young man,
four years after the return of the
Beagle. (From Engel, L. (ed.) 1962.
The Voyage of the Beagle, by
Charles Darwin. Doubleday & Co.,
New York.)

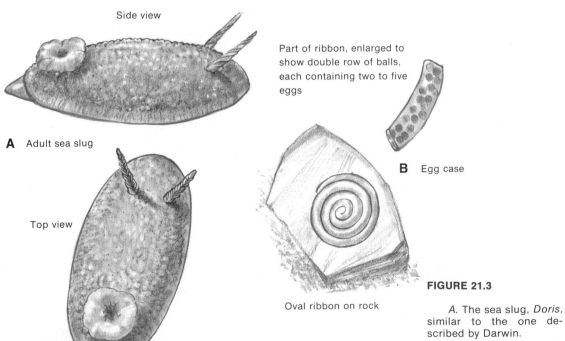

Side view

Part of ribbon, enlarged to
show double row of balls,
each containing two to five
eggs

A Adult sea slug

B Egg case

Top view

Oval ribbon on rock

FIGURE 21.3

A. The sea slug, *Doris,*
similar to the one de-
scribed by Darwin.
B. Its egg case, con-
taining hundreds of thou-
sands of eggs.

The 13 finches found on the Galápagos Islands.
 1–6, ground finches (Geospiza); *7–12*, tree finches (Camyrhynchus); *13*, warbler finch (Certhidea).
 Compare the following finches with those illustrated in Darwin's woodcut (Fig. 21.5): *2, Geospiza magnirostris* (Darwin's no. 1); *3, Geospiza fortis* (Darwin's no. 2); *6, Geospiza fuliginosa* (Darwin's no. 3); *13, Certhidea olivacea* (Darwin's no. 4).
 (From Keeton, W. T. 1967. *Elements of Biological Science.* W. W. Norton & Co., New York.)

FIGURE 21.4

tively recent geological origin—Darwin was struck by the relatively few species of plants and animals as compared with the American mainland. In particular, he noted that fully half of the 26 species of land birds on the islands were a closely related group of finches. Altogether, Darwin recognized four genera:

Cactornis	Two species, inhabiting cactus trees (#7 and 12)
Geospiza	Six species, inhabiting the ground (#1–6)
Camarynchus	Four species, inhabiting mainly trees, rather than ground, as Darwin thought (#8–11)
Certhidea	One species with a warbler-like bill, inhabiting the trees (#13)

The 13 species are shown in Figure 21.4. Today, the genus *Cactornis* is included with *Camarynchus*. Darwin wrote about these finches as follows (the figures referred to being those of Figure 21.5).

The remaining land-birds form a most singular group of finches, related to each other in the structure of their beaks, short tails, form of body, and plumage: there are thirteen species, which Mr. Gould has divided into four sub-groups: All these species are peculiar to this archipelago. . . . The males of all, or certainly of the greater number, are jet black; and the females (with perhaps one or two exceptions) are brown. The most curious fact is the perfect gradation in the size of the beaks in the different species of Geospiza, from one as large as that of a hawfinch to that of a chaffinch, and (if Mr. Gould is right in including his sub-group, Certhidea, in the main group), even to that of a warbler. The largest beak in the genus Geospiza is shown in Fig. 1, and the smallest in Fig. 3; but instead of there being only one intermediate species, with a beak of the size shown in Fig. 2, there are no less than six species with insensibly graduated beaks. The beak of the sub-group Certhidia, is shown in Fig. 4. The beak of Cactornis is somewhat like that of a starling; and that of the fourth sub-group, Camarhynchus, is slightly parrot-shaped. Seeing this gradation and diversity of structure in one small, intimately related group of birds, one might really fancy that from an original paucity of birds in this archipelago, one species had been taken and modified for different ends.

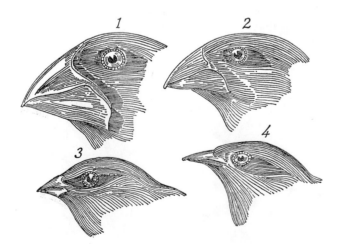

FIGURE 21.5

Beaks of four of the 13 species of finches found by Darwin on the Galápagos. Numbers refer to Darwin's text. (From Darwin's *The Voyage of the Beagle*.)

The fact that there were so many different types of finches on the Galápagos, but that none of them was exactly identical to any finch found on the American continent puzzled Darwin; they were obviously closely related to them. He found a similar state of affairs among the plants on the Galápagos Islands, as shown in Table 21.2.

Only about half the species of plants on the Galápagos were found elsewhere in the world, and some of these had been introduced by man. The rest were unique to the islands. Moreover, most of the indigenous species were confined to a particular island. (Each island in the archipelago originated as a separate volcano and to this day has remained separated by several miles of deep water from its neighbors.) The following paragraph from Darwin's diary sums up the wonder he experienced at the curious distribution of plants and animals on these islands:

The distribution of the tenants of this archipelago would not be nearly so wonderful, if, for instance, one island had a mocking-thrush, and a second island some other quite distinct genus;—if one island had its genus of lizard, and a second island another distinct genus, or none whatever;—or if the different islands were inhabited, not by representative species of the same genera of plants, but by totally different genera, as does to a certain extent hold good; for, to give one instance, a large berry-bearing tree at James Island has no representative species in Charles Island. But it is the circumstance, that several of the islands possess their own species of the tortoise, mocking-thrush, finches, and numerous plants, these species having the general habits, occupying analogous situations, and obviously filling the same place in the natural economy of this archipelago, that strikes me with wonder.

For a long time after returning to England, Darwin pondered the problem of the distribution of species on the Galápagos. He finally realized that only a few types of plants and animals had found their way from other parts of the world to the newly formed islands; each of these few colonizers had subsequently evolved into a variety of new species filling a particular habitat on the separate islands.

In 1838, two years after the voyage, Darwin read *An Essay on the Principles of Population* written in 1798 by Thomas R. Malthus, an English economist. In his book, Malthus emphasized the tendency of all species to increase their numbers geometrically—that is, for each set of parents to produce far more offspring than were necessary to replace themselves. Darwin, as we have seen, had already observed this phenomenon in the case of the sea-slug, *Doris*, and other species. In nature, then, each species produces an abundance of offspring, and selective forces determine which of these are best adapted to survive.

Variation Among Offspring. By this time Darwin had evidently formu-

TABLE 21.2 DISTRIBUTION OF PLANT SPECIES FOUND BY DARWIN ON GALÁPAGOS ISLANDS*

Name of Island.	Total No. of Species.	No. of Species found in other parts of the world.	No. of Species confined to the Galápagos Archipelago.	No. confined to the one Island.	No. of species confined to the Galápagos Archipelago, but found on more than the one Island.
James Island	71	33	38	30	8
Albemarle Island	46	18	26	22	4
Chatham Island	32	16	16	12	4
Charles Island	68	39 (or 29, if the probably imported plants be subtracted)	29	21	8

*From Darwin's *The Voyage of the Beagle*.

FIGURE 21.6 Breeds of pigeons, all but *A* developed by man-made selection.
    A. The wild rock pigeon of Europe, believed to be the ancestor of all domesticated breeds pictured.
    B. Fantail.
    C. Frill back.
    D.Satinette oriental frill.
    E. English pouter.
    F. Pomeranian pouter.
    G. Carrier.
    (From Keeton, W. T. 1967. *Elements of Biological Science.* W. W. Norton & Co., New York.)

lated his hypothesis about natural selection, but he required one further piece of evidence to complete his argument. It was necessary to demonstrate that individual differences do in fact exist among the offspring of a single set of parents. Under natural conditions, the observer is hard-pressed to discover any visible differences among the offspring in a particular generation which would give one individual an advantage over another. To demonstrate the existence of variation, Darwin turned to domesticated species.

Darwin began a careful study of such domesticated animals as cattle and pigeons, in which he observed that man had created innumerable varieties by selectively breeding individuals possessing particular traits. (Fig. 21.6). He reasoned that variations among traits which arose spontaneously in offspring were being perpetuated by man-made selection and transmitted to further generations. Darwin had the vision to realize that the selection practiced by man on domesticated organisms was essentially no different from that occurring in nature. That is, nature "selects" from the overabundance of offspring those individuals which are best adapted to a particular habitat; it is these which survive and reproduce the next generation.

It took Darwin some 23 years after returning from his voyage on the *Beagle* to collect and sort the information needed to support his theory on how new species originated. Perhaps his reticence in publishing his ideas sooner can be explained by the following paragraph taken from the last page of his diary of the voyage:

> In conclusion, it appears to me that nothing can be more improving to a young naturalist, than a journey in distant countries. It both sharpens, and partly allays that want and craving, which, as Sir J. Herschel remarks, a man experiences although every corporeal sense be fully satisfied. The excitement from the novelty of objects, and the chance of success, stimulate him to increased activity. Moreover, as a number of isolated facts soon become uninteresting, the habit of comparison leads to generalization. On the other hand, as the traveller stays but a short time in each place, his descriptions must generally consist of mere sketches, instead of detailed observations. Hence, arises, as I have found to my cost, a constant tendency to fill up the wide gaps of knowledge, by inaccurate and superficial hypotheses.

Darwin did not wish to present his theory to public scrutiny until he had amassed a great abundance of data to support it.

But Darwin was not the only one to have such ideas. Another English naturalist, Alfred Russell Wallace, wrote an essay in 1858 entitled "On the Tendency of Varieties to Depart Indefinitely from the Original Type." Later that year, Darwin and Wallace presented their ideas together before the highly respected Linnaean Society of London. It is Darwin, however, who is mainly remembered, since his book, *The Origin of Species*, published the following year, sets forth the arguments in favor of evolution in such great detail. Wallace himself generously proposed the term **Darwinism** for the theory of evolution.

In the years since Darwin, experiments and observations have further supported his theory and have also provided insights into the mechanisms of evolution. We shall conclude our chapter with a consideration of some of the factors believed to act in the origin of new species.

DEGREE OF FITNESS

We have seen that more offspring are produced in a given generation than can survive to maturity due to limitations of food supply and nesting sites, predation and so on. It is thus apparent that any inherited trait which improves an individual's chance for survival to reproductive age, or its chance of mating or the number of offspring produced as a result of mating, will be transmitted to the next generation; and on the other hand, any trait which decreases the probability of any of these is less likely to be passed on. Each trait can thus be said to have a certain degree of "fitness." If the trait favors survival, the proportion of individuals having that trait will gradually increase in each succeeding generation; if the trait is unfavorable, it will gradually disappear in succeeding generations.

A trait may be unfavorable in one environment, however, but not in another. As we noted in Chapter 15a, the gene for sickle-cell hemoglobin is of great advantage in Africa since it confers resistance to malaria, but it is a disadvantage in malaria-free areas such as the United States, since even carriers tend to be anemic under conditions of oxygen stress. Hence, fitness refers to the adaptive value of a trait in relation to a particular environment. Since environments are constantly changing, so is the fitness of the many inherited traits of a given species—evolution is a continuing process. As Darwin noted towards the end of his voyage:

> Where on the face of the earth can we find a spot, on which close investigation will not discover signs of that endless cycle of change, to which this earth has been, is, and will be subjected?

Virtually all inherited traits of a given species probably have some adaptive value, although occasionally there are useless traits which are ves-

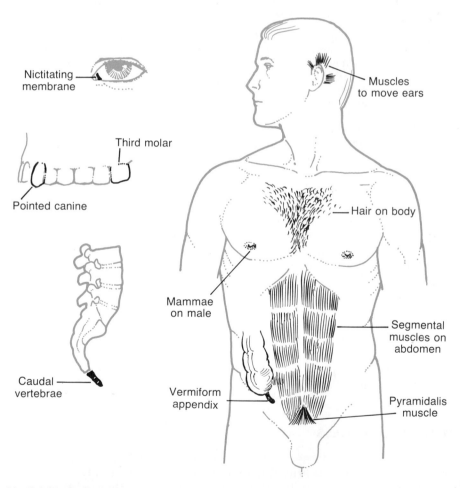

FIGURE 21.7   Vestigial traits in man.
    The inherited characteristics shown had functional significance in some ancestor of ours. Homologous organs with a useful function are still found in other living mammals. (After Storer, T. I. 1943. *General Zoology.* McGraw-Hill, New York.)

tiges from the past. There are several such vestigial characters in man, some of which are shown in Figure 21.7. These are remnants of organs once functional in some ancestor of man; they remain as mute evidence of our evolution from a species different from ourselves.

EXAMPLES OF SPECIFIC ADAPTATIONS

A comprehensive catalogue of the many types of adaptations found in organisms is not possible here, and we shall note only a few examples; others have been discussed in Chapter 19.

ADAPTATIONS FOR FEEDING

Obtaining food is of major importance for survival, especially where competition exists within and between species. The various sizes of beaks found in different species of ground finch on the Galápagos, for example, allows each species to feed on seeds of a given size, thus insuring a sharing of all available food without excessive competition between species.

An animal exhibiting many traits which enable it to feed on food not accessible to most other species is the hummingbird (Fig. 21.8). These tiny, swiftly flying creatures can obtain nectar from flowers that are otherwise utilized only by a few insects. The light weight of their body results from the extreme hollowness of their bones, together with the presence of air sacs which are modified extensions of the lungs filling much of the body cavity. Their ability to hover before a flower is due to the flexible joints of the shortened bones of the forearm, permitting the wings to be moved in a rapid sculling, or fanning motion. The elongate beak can penetrate the deepest flower, and the long, extensible tongue is adapted for seeking out nectar.

FIGURE 21.8

A hummingbird shows several specializations for feeding on nectar of flowers: short wings and lightweight body for hovering; elongate beak and tongue (not seen here) for gathering nectar. (From Jessop, N. 1970. *Biosphere: A Study of Life.* Prentice-Hall, Englewood Cliffs, New Jersey.)

ADAPTATIONS FOR DEFENSE

Protecting oneself from being eaten by a predator has obvious survival value, and various organisms have evolved a variety of defense mechanisms. One solution is to become so large and have such a tough skin that predators are ineffective—such solutions are found in the rhinoceros, elephant and armadillo. Trees have hard bark which serves the same purpose (as well as other functions); many plants are covered with thorns or spines.

The porcupine and hedgehog have hairs fused into sharp bristles. The ability to run very fast is another effective defense made use of by such long-legged animals as horses and gazelles. Production of unpalatable or poisonous secretions is another form of protection. Many plants are highly poisonous—the colorful red and white European toadstool, *Amanita,* is a deadly example. Skin irritants, as found in nettles, poison oak and poison ivy, are effective deterents to predators. Many animals have poisonous venoms used either to capture prey or for defense (or both). Certain marine snails produce some of the most lethal toxins known; some snakes, the scorpions, bees and wasps are all known for their toxic venoms. Other animals are merely unpalatable: many frogs and toads, for example, produce unpleasant, mildly toxic skin secretions; skunks exude noxious odors if alarmed; the monarch butterfly has an unpleasant taste.

Hiding is another, extremely effective defense mechanism. This may be accomplished either by literally withdrawing into a hole or by camouflage. Innumerable animals have evolved body shapes and colors which so match their surroundings that they become invisible to a predator. Mottled salamanders and frogs blend with their background; so do chameleons. The latter may also change color as they move to backgrounds of different shades. The stick insects match in both shape and color the twigs they live on.

Some 150 years ago, dark varieties of a particular light-colored moth,

FIGURE 21.9   Cryptic coloration in an English moth.
    The original light form is all but invisible on a lichen-covered tree in soot-free region of country (left), but contrasts sharply against soot-covered tree (right). Pigmentation in wings of a variety common in the sooty area make it invisible against the blackened tree. (From Keeton, W. T. 1967. *Elements of Biological Science.* W. W. Norton & Co., New York.)

called *Biston betularia*, were rare in England—so much so, that they were prized by collectors of butterflies. As industrialization spread, bringing coal soot with it which blackened the tree bark on which the moths settle, more and more dark forms were found (Fig. 21.9). Today it is rare to find a light-colored individual in the heavily industrialized English Midlands. The ability to form dark pigments in the wings and body is a genetically inherited trait known as **melanism**. It appears that natural selection has been favoring survival of the moths best able to camouflage themselves in their new surroundings. In fact this is the first direct evidence actually obtained to support Darwin's theory that natural selection occurs.

The above are all examples of **adaptive** or **cryptic coloration.**

Sometimes an animal does not mimic its background for concealment but instead obtains the desired result by mimicking a particularly noxious species. Hover flies, for example, have striped bodies resembling wasps, but have no sting of their own. The viceroy butterfly, itself delectable to birds, mimics the appearance of the unpalatable monarch (Fig. 21.10).

We could mention many other types of adaptation found in organisms—such as the adaptations evolved by animals for reproduction in aquatic environments and on land (Chapter 14)—but the preceding should suffice to give the reader an idea of the significance of adaptive traits.

FIGURE 21.10

Mimicry in butterflies
The Viceroy, below, mimics the unpalatable Monarch, above. The two species occur in similar areas. (From Keeton, W. T. 1967. *Elements of Biological Science.* W. W. Norton & Co., New York.)

RESULTS OF SPECIFIC ADAPTATIONS

If we consider the members of any large group of organisms, such as one class within a phylum, we are struck by both the similarities and diversities within the group. Just as Darwin noted that the several species of finches on the Galápagos seemed to have developed from one species "modified for different ends," so we may note that the structures of a common primitive ancestor of any diverse group seem to have been modified for life in a variety of specialized habitats. Let us consider the case of the mammals.

Among the living mammals we find a great many similarities in form and structure: hair, two pairs of appendages, mammary glands and so on. But there is also great diversity, as shown in Figure 21.11. The least specialized of all living mammals are the insectivorous shrews. It is thought that

FIGURE 21.11   Adaptive radiation in mammals.
    The original unspecialized ancestor is thought to have looked like the animal in the center—not too different from today's shrews. Each of the various descendants has become adapted to a specific habitat. (After Villee, C. A., and Dethier, V. G. 1971. *Biological Principles and Processes.* W. B. Saunders Co., Philadelphia.)

the common ancestor of the mammals, an evolutionary modification of a mammal-like reptile, may have looked very much like today's shrews.

The great variety of mammals which has evolved from a single common ancestor exhibit many types of modifications: changes in limb structure, dentition and overall body shape are easily observed. Such diversification from a single ancestor is called **adaptive radiation.** It has occurred in many other groups as well. The diversity among the extinct reptiles is one such case; the diversity of the marsupial mammals in Australia is another. Because Australia separated early from the main land mass of the world, the mammals found there are not closely related to those elsewhere (except for our oppossum); yet the Australian species are adapted to fill almost as wide a variety of habitats as may be found elsewhere.

As one might expect, the same environmental demands can also result in the evolution of similar structures in quite unrelated animals. Many animals have developed wings at various times — insects, some of the extinct reptiles, birds and bats. Many aquatic animals have streamlined bodies, and people often confuse whales and porpoises, which are warm-blooded, air-breathing mammals, with fish. The camera-type eyes of octopus and man are very similar in structure and function, yet their embryonic origin in each species is quite different. This tendency to similarity of structures in quite unrelated species is called **convergent evolution.**

FORMATION OF NEW SPECIES

It is perhaps not too difficult to comprehend how a modern species such as today's horse evolved in a stepwise fashion from some small, primitive, horse-like ancestor. In each generation, the more favored offspring were selected and gradually replaced their less well adapted brothers. But it is more difficult to visualize how several different species of finches could have evolved simultaneously from a single common ancestor on the Galápagos, or how the varied types of mammals arose from a single, unspecialized species of mammal-like reptile.

If we consider a species to include all individuals capable of interbreeding, then there exists the possibility for exchange of genetic information among the entire population. **Gene flow** within the population is complete, producing a single **gene pool.** More favored offspring, wherever they occur within the geographic distribution of the species, will become predominant throughout the whole population (Fig. 21.12A). But if subunits within a population become genetically isolated — if gene flow is interrupted — then favorable adaptations in one subunit will *not* spread throughout the entire population. The subunits begin to diverge from one another and form new species.

GEOGRAPHIC ISOLATION

A major factor leading to formation of subunits within a population is **geographic isolation.** A barrier — a new mountain range, a river gorge or the separation of continents — prevents interbreeding (Fig. 21.12B). Each subunit of the original population continues to evolve independently, being acted upon by its own environmental selection factors. Since it is unlikely that any two environments will be identical, different traits will be selected for in each separate environment.

In the case of the species *Homo sapiens*, environmental factors have produced relatively minor changes in the total genotype. Man's adaptations to environmental differences have been mainly cultural rather than genetic, and reproductive isolation has never evolved — all populations of men can still interbreed.

But for all other organisms, which lack the cultural adaptability of man, geographic isolation of subpopulations results in well-integrated genotypes within each area; each genotype is highly suited to the local environment. A consequence of this genetic specialization is that should the two subpopulations once again come together (Fig. 21.12C), the recombination of the two genotypes in hybrids may produce offspring with a less well integrated set of genes. Such hybrids are less likely to survive and will thus be selected against. In some cases the two parent genotypes are so far out of step with one another that the hybrids die during development. The two subpopulations have become *reproductively isolated*, forming two distinct species.

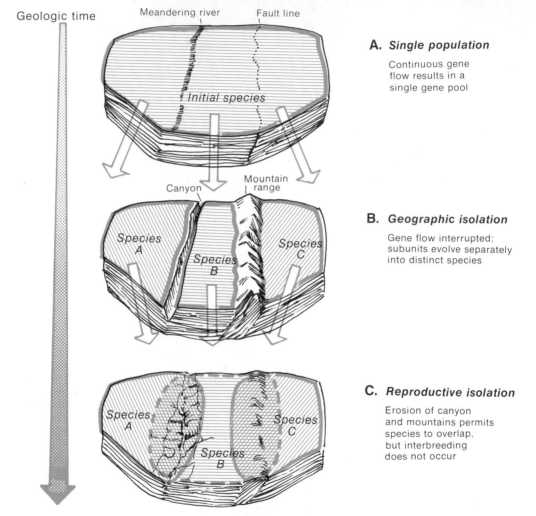

Geologic time

Meandering river   Fault line

A. *Single population*

Continuous gene
flow results in a
single gene pool

Initial species

Canyon   Mountain range

B. *Geographic isolation*

Gene flow interrupted;
subunits evolve separately
into distinct species

Species A   Species B   Species C

C. *Reproductive isolation*

Erosion of canyon
and mountains permits
species to overlap,
but interbreeding
does not occur

Species A   Species B   Species C

FIGURE 21.12   Theoretical stages in the evolution of new species from a single common ancestor.
Prolonged geographic isolation leads to permanent reproductive isolation. Interbreeding cannot occur, even if the new species again overlap.

For many closely related groups of animals it is believed that this is just what has occurred. Races or subunits of a once homogeneous population remained isolated long enough so that they eventually could no longer produce viable offspring. This is happening at present with two races of squirrel inhabiting opposite sides of the Grand Canyon.

Thus, even when the original geographic barriers disappear and the divergent populations once again come into contact, they can no longer interbreed. Gene flow is permanently interrupted (Fig. 21.12C).

STAGES IN THE DEVELOPMENT OF REPRODUCTIVE ISOLATION

We can next list some of the sequential steps that occur during the reproductive isolation of two overlapping populations.

1. Lack of hybrid perpetuation
   a. Hybrid offspring are less well adapted to the environment and succumb more often to natural selection than the parent genotypes; the latter predominate.
   b. Hybrid offspring are sterile, and no longer contribute genes.
2. Lack of viable offspring
   a. Zygote may fail to develop; incompatibility of male and female DNA.
   b. Gametes may fail to fuse; each type of sperm produces digestive enzymes specific for the membranes of the eggs of its own species.
3. Lack of mating
   a. The two populations may occupy different habitats within a geographic area, and hence seldom meet.
   b. Reproduction may occur at different seasons of the year.
   c. Copulation may become physically impossible as a result of changes in anatomy of male or female reproductive organs.
   d. Courtship behavior patterns, which are often inherited, may become different.

The initial cause of speciation is thus the disadvantage of the hybrid genotype. However, if the two parent species are to survive, they must maintain a high reproductive potential (Chapter 19). Any adult which mates with a member of the other population will produce fewer successful offspring and is at a genetic disadvantage. There is therefore strong selection for those parents which do not make the mistake of incorrect mating. This results in the selection of further genetic isolating mechanisms, as shown in our list above, and leads eventually to such very precise isolating mechanisms as anatomical changes in the reproductive tract and, especially, behavioral isolation. Copulation with a mate of the "wrong" species at this stage of isolation is never attempted.

In a few organisms, however, viable hybrids between species are sometimes produced which are larger and more vigorous than either parent. A particularly interesting example of this occurs fairly often among closely related species of plants. If, by chance, *diploid* gametes are formed through failure of the chromatids to separate during meiosis (see Chapter 14), then offspring with a double number of chromosomes are produced. In this condition, known as **polyploidy,** no vital information of either parent species is lost since a complete (2N) chromosome complement of each parent exists in the offspring.

Polyploid hybridization has been utilized in producing more robust species of wheat and other grains. The normal diploid chromosome number in wheat is 14 (2N), but vigorous species with 28 (4N) and 42 (6N) chromosomes have been developed. Another promising new hybrid is a cross between wheat and rye; the cells of these plants contain 6N wheat chromosomes and 2N rye chromosomes.

SUMMARY

Living organisms fall into related groups which may be arranged in a hierarchy of classification from the broadest, all-inclusive category, the kingdom, to the narrowest category, the species. Although varieties may exist within a species, since these can interbreed they are part of the same genetic pool.

A theory explaining the origin of new species was proposed by the English naturalist, Charles Darwin, in 1859, in a book entitled *The Origin of Species by Means of Natural Selection or the Preservation of Favored Races in the Struggle for Life*. The ideas presented in the theory began to form in Darwin's mind 20 years earlier during a five year voyage around the world. He noted that most organisms produce far more offspring than can possibly survive, and concluded that those best adapted to a particular environment would live to reproduce. His observations of the diversity among finches and other species in the Galápagos Islands suggested the evolution from one common ancestor of many types, each filling a distinct role in the environment. Examples of fitness and of special adaptations for survival were abundant. The fossil record supplied him with considerable evidence for the gradual evolution of species, one from another. Darwin obtained evidence for the existence of variation among offspring from man-made selection of varieties of domesticated animals. The final piece of evidence, the actual observation of selective forces acting in nature was not available to Darwin—he assumed this to be the case. Only recently has such an observation been made, in the evolution of industrial melanism among moths in England.

The origin of a new species, as distinct from a variety or race, depends on interruption of gene flow between two subpopulations, thus permitting their independent evolution. This is usually caused by geographic isolation. Once reproductive isolation occurs between two subpopulations, they can be recognized as distinct species.

READINGS AND REFERENCES

Darwin, C. *The Voyage of the Beagle*. Ed. by Leonard Engel. 1962. Doubleday & Company, Garden City, N.Y. (Anchor Books Natural History Library.) Darwin's own account of his experiences, observations and thoughts during his five-year voyage as ship's naturalist aboard *H.M.S. Beagle*.

Darwin, C. 1859. *The Origin of Species by Means of Natural Selection or the Preservation of Favored Races in the Struggle for Life*. John Murray, London. The book in which Darwin expounds in detail his theory of evolution.

Lack, D. 1947. *Darwin's Finches*. Cambridge University Press, New York. (Paperback edition, Harper Torchbooks, 1961).

Lack, D. 1953. Darwin's finches. Scientific American Offprint No. 22. The preceding two references explain in a readable manner the adaptive significance of the variations in finches observed by Darwin in the Galápagos.

Kettlewell, H. B. D. 1956. A résumé of investigations on the evolution of melanism in the Lepidoptera. Proceedings of the Royal Society of London, Series B. *145*:297–303.

Kettlewell, H. B. D. 1958. Frequencies of melanic forms in *Biston*. Heredity *12*:51–72.

Kettlewell, H. B. D. 1959. Darwin's missing evidence. Scientific American Offprint No. 842. The preceding three references describe the actual observation of natural selection occurring in a species of moth.

Mayr, E. 1963. *Animal Species and Evolution*. Harvard University Press, Cambridge, Massachusetts. A detailed presentation of modern evolutionary theory, emphasizing the origin of new species.

Stebbins, G. L. 1966. *Processes of Organic Evolution*. Prentice-Hall, Inc., Englewood Cliffs, New Jersey. A short review of the subject by one of the foremost students of evolutionary theory.

NATURAL SELECTION AND MAN

Man, himself a product of the forces of natural selection, has now become the most potent force on Earth in determining his own further evolution, as well as that of his fellow creatures.

In this chapter we shall consider two very disparate yet closely related subjects, both arising naturally from our discussion of evolution. The first of these deals with the biological nature of man: how he evolved and what significance there is in so-called racial differences. The second subject, and one of far more importance for the future of man, is his role, both past and present, as an evolutionary agent affecting the species around him. We shall consider only highlights of this vast topic, however, noting examples of both unintentional and intentional man-made selection, either for or against survival of various species of plants and animals. Some of the consequences to man of his own actions will become apparent as we go.

MAN'S EARLY EVOLUTION

After publication in 1859 of Darwin's *The Origin of Species*, there was not only a great deal of angry dissent by Creationists, on the one hand, but also much speculation by Darwin's supporters, on the other, as to the nature of the postulated "missing link" between man and the apes. Three years earlier, in 1856, a most unusual skull, obviously human but far heavier than any modern skull, was found in the Neander Valley in Germany. Thomas Huxley, an outstanding zoologist and strong supporter of Darwin, undertook to examine the skull. He found its brain case to have a capacity about the same as that of the average modern man, but it also had certain ape-like qualities—heavy brow ridges, a large jaw and thick bones.

Neanderthal man, as he came to be known, was definitely man, however, not ape (Fig. 21a.1).

Some twenty years later a young Dutchman by the name of Eugène Dubois discovered in Java a rather more ape-like head belonging to a creature whose thigh bone showed that he also stood upright. Although smaller than that of Neanderthal, the brain case was still far greater than that of any living ape. Java man was estimated to be 500,000 years old.

FIGURE 21a.1

Two reconstructions of Neanderthal man.

The earlier one on the left emphasizes ape-like characteristics, the later one a more human aspect. The righthand restoration more nearly agrees with fossil evidence of underlying bone structure. (Courtesy of the American Museum of Natural History.)

This find caused renewed furor among both evolutionists and Creationists—an ape's head on a human body? Surely the human brain must have evolved before the body if indeed evolution occurred at all. The esthetically unacceptable situation posed by Java man may have led to that great hoax, the Piltdown man. In 1908, at Piltdown in Sussex county, England, an amateur paleontologist "discovered" pieces of bone which he deftly fitted together to give an ape's jaw attached to the cranium of a modern man. Although plainly puzzling, since it did not fit the story gradually emerging from other finds throughout the world, this peculiar being was duly listed for years in textbooks as one of our ancestors. Not until 1953 was the hoax uncovered; its perpetrator had artificially "aged" fragments of bone from a modern ape's jaw and a modern man's cranium.

Around 1930, investigations in caves outside Peking turned up another, this time genuine, fossil ancestor of modern man. Peking man was just a little more "human" than Java man and knew the use of fire. But Peking man still was not the missing link; he was clearly an ancient form of man, not an advanced ape (see Figure 21a.2).

Meantime, discoveries in South Africa began to reveal that true apemen had indeed existed. At first, only a few scientists believed in Raymond Dart's interpretation of the bones he found as those of a very ancient, pygmy-sized ape-man. One of these believers was a Scottish physician, Robert Broom, who himself spent more than twenty years searching for bones in South Africa. By 1950, he had found remains of upright-walking apemen at several stages in their evolution, and these were finally accepted by scientists throughout the world as being probable ancestors of man. These various African ape-men lived from about 2 million to 800,000 years ago.

About this time, Dart showed that the earliest of these ape-men were able to use bones and tusks as tools. Among the remains of other animals associated with the pre-human fossils were a preponderance of "useful" leg bones which could have served as weapons. Moreover, the skeletons of the ape-man's prey found beside him showed signs of being broken violently. This fact led some scientists to conclude that even these early ape-men used weapons. In some of the later ape-man caves Broom also found remnants of early tool building, in the form of chipped pebbles the size of a fist. It seems that even our earliest ancestors were tool-makers as well as tool-users.

Meantime, in another part of Africa, Dr. L. S. B. Leakey and his wife, Dr. Mary Leakey, after spending some 24 years studying shaped pebbles and other artifacts left by early man in Tanzania's Olduvai Gorge, finally discovered a skull of an early toolmaker (Fig. 21a.2). This skull resembled the later ape-men of South Africa and was their approximate contemporary, living more than a million years ago. Higher up in the walls of this same gorge the Leakeys found remains of his more human-like descendants who lived some 500,000 years later and were approximate contemporaries of Java man and Peking man. These later men made more sophisticated tools, capable of killing large game. The most likely hypothesis for the origin of man is that he evolved first in Africa as a tool-using, erect-walking ape-man, spreading from there across Eurasia and learning to make ever more sophisticated tools and eventually, to use fire. In the following millennia, subpopulations evolved special traits selected for by local environmental conditions, but gene flow, through migration, continued to maintain a worldwide breeding population throughout the evolutionary history of man (see Chapter 21).

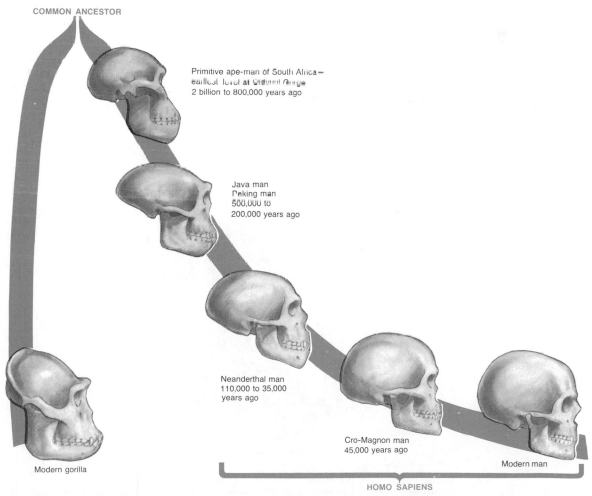

COMMON ANCESTOR

Primitive ape-man of South Africa — earliest level at Olduvai Gorge
2 billion to 800,000 years ago

Java man
Peking man
500,000 to
200,000 years ago

Neanderthal man
110,000 to 35,000
years ago

Modern gorilla

Cro-Magnon man
45,000 years ago

Modern man

HOMO SAPIENS

The evolution of the human skull. Note gradual increase in cranial capacity and reduction of jaw, brow-ridges and other ape-like features. (After Young, L. B. (ed.) 1970 *Evolution of Man*. Oxford University Press, New York.)

FIGURE 21a.2

During the million or so years of man's evolution there were four Pleistocene glaciations interspersed with warm periods. These climatic upheavals have left us with an incomplete fossil record of man's prehistoric existence, a summary of which is shown in Figure 21a.2. It seems likely that Java and Peking man disappeared about 200,000 years ago, being replaced by various populations of modern man, or *Homo sapiens.* One geographic variant of *Homo sapiens* in Europe was Neanderthal man, who populated the area for a time. Gradually Neanderthal man himself was replaced by a more modern variant, Cro-Magnon man, an excellent toolmaker and, perhaps, a direct ancestor of modern man.

CONSEQUENCES OF THE HUMAN CONDITION

There are several externally recognizable features by which man is identified — his upright gait, his lack of body hair and his specialized mouth and tongue, necessary for the development of speech. The most important factor of all, however, is man's large brain, which makes possible the full realization of his other potentials.

Upright Gait.   The advantages of having a pair of grasping limbs available for carrying food or throwing weapons while still retaining mobility

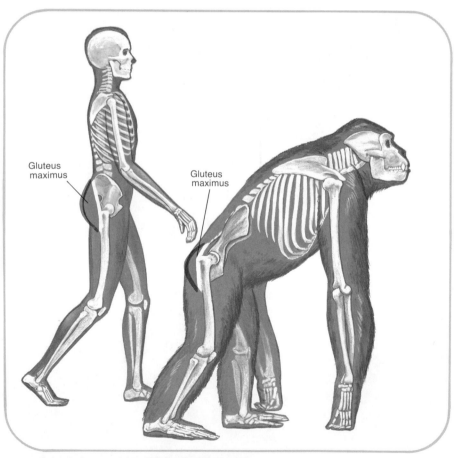

FIGURE 21a.3

Skeletal systems of man (left) and gorilla (right).
    Note differences in joining of head to vertebral column, shape of spine, length of arms and legs, center of gravity and shape of pelvic girdle. The gluteus maximus muscle of man (dark color) is highly developed. It functions in preventing forward collapse of the body, especially during running and climbing.

are obvious. The evolution of an upright posture required a change in skeletal and muscular arrangements, as shown in Figure 21a.3. The attachment of the head to the spine shifted from the back of the skull to its base. The vertebral column became curved, distributing body weight so that the center of gravity was directly over the pelvis rather than in front of it. The shape of the hips was altered, a feature which makes childbirth harder for women than for other mammalian females. In walking, instead of shuffling forward as does an ape—that is, letting himself start to fall and then catching his weight by sliding a foot forward—man moves hips, legs and feet in a striding motion. In man, the buttocks muscle or **gluteus maximus,** is developed as in no other animal. Its function is to maintain an upright stature. Although used minimally in walking, this muscle is essential for climbing over uneven terrain and for running on two legs (a fact one can easily test on oneself).

These skeletal and muscular developments allowed primitive ape-man to utilize a variety of habitats not available to his great ape contemporaries. It also permitted him to travel great distances in search of prey while carrying weapons and perhaps spare food and water. Although he is not the swiftest of animals, man's physical endurance is greater than most, enabling him to track game for days on end (Fig. 21a.4). An upright gait was thus the first major step in the evolution of man.

FIGURE 21a.4

Masai elephant trackers, whose physique is remarkably adapted to following game over long distances. Men who hunt on foot have undergone selection for long limbs and lithe bodies. (Courtesy of the American Museum of Natural History.)

Evolution of the Brain. The original bipedal ape-man of South Africa had a brain scarcely larger than that of other apes and presumably possessed behavioral capacities to match. During the later evolution of man, brain size approximately tripled, most of the increase resulting from additions to the cerebral cortex (see Chapter 13a). The cortex appears to be the region of the brain where finer distinctions of sensory input are made and acted upon, such as those between similarly pitched sounds, shades of color and so forth. The cortex is also the region where conscious association of information occurs, facilitating an awareness of relationships between sensory stimuli. Lastly, it is the center where abstract ideas are formulated into language. All of these functions are essential to the complex behavior of *Homo sapiens.* It is not surprising that large areas of the evolutionarily recent human cortex are associated with those two organs having specifically human functions, the tool-making hand and the language-forming mouth and tongue (refer to Figure 13a.7).

Interestingly enough, the maximum size of the human brain was achieved with the evolution of Neanderthal man some 100,000 years ago. Since then, the brain-size of man has remained constant, and if, as is thought, brain size is related to mental capacity, it seems unlikely that there has been any significant increase in intellectual potential of mankind since

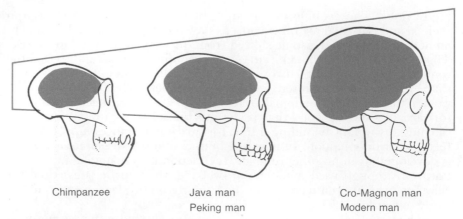

Chimpanzee	Java man	Cro-Magnon man
	Peking man	Modern man

FIGURE 21a.5

Evolution of the human brain (brain size in color).

Brains of earliest South African ape-men were about the size of that of today's chimpanzee. Note that the brain size of Cro-Magnon man, about 50,000 years ago, equaled that of modern man.

(Fig. 21a.5). Further advances are due mainly to accumulated cultural factors.

THE EVOLUTION OF "RACES"

Modern genetics has shown that only a very small fraction of the total number of human genes are responsible for the differences between the so-called races of man and that these differences are mainly adaptations to a particular environment or to a particular cultural pattern suited to that environment. Eskimos, for instance, have an advantage in the cold Arctic because of their short stature and fatty padding (both heat-conserving mechanisms), their truncated noses (less susceptible to frostbite) and their lack of an eyefold (which helps keep the eyeball warm). It is thought that these Mongoloid features evolved about 25,000 years ago among a population of man trapped in northern Asia by the glaciers.

Another example of an adaptation to the environment is the amount of skin pigment. A certain amount of ultraviolet light must penetrate the skin to convert dietary precursors of vitamin D into the active form of the vitamin, thus preventing rickets. On the other hand, too much ultraviolet light is detrimental, causing severe burns and perhaps skin cancer. People living in northern latitudes where light intensity fluctuates throughout the year have pale skins in winter and become pigmented (tanned) in summer after repeated exposure to sunlight. The pigment absorbs excess ultraviolet light and prevents its bad side effects. In contrast, tropical races exposed to continuous sunlight produce pigment all year round, whether sunlight is present or not. Both adaptations are genetically controlled.

Many more examples could be given, all of which would simply add up to the fact that man, like any other creature, has been molded by his environment over thousands of years. As the environment varied from place to place, so did the men—they became distinguishable as so-called "races."

What Is a Race? In the preceding chapter we defined the term *race* biologically as a circumscribed subpopulation of a species; a race is genetically distinct but not yet reproductively isolated. Various people have tried to apply this same concept to mankind. In 1795, a German physician studying skull types decided there were five "races" of men. These soon became identified by their skin color—black, brown, yellow, red and white—a classification still learned by school children today. More recently, anthropologists using more refined criteria have arrived at something between 30 and 34 human "races." Presumably, one could go on modifying criteria indefinitely (Fig. 21a.6).

The so-called races of man, however, are not nearly as genetically distinct from one another as are the well-studied geographical races of other organisms. External "racial" characters—such as skin pigment, size and shape of nose and lips and presence or absence of an eyefold—are not limited to particular subpopulations of man; all of the usual racial characters occur, to greater or lesser extent, among men everywhere, and no one character is limited to a single "race." To take but one example, Hindus of India, with "Aryan" facial bone structure, may have extremely dark skins. There has always been such a great mobility of people that genetically "pure" races of man have never developed; instead there is a continuous gradation from one general type to the next.

The people of the world: How many "races" are there? (U.S. Forest Service photo.) **FIGURE 21a.6**

Trying to make rigid legal classifications about racial differences often leads to biologically meaningless situations. The South African government, for example, found it necessary to classify a dark-skinned child as "colored," even though both her natural parents were classed as "white." She was therefore legally barred from places her parents could enter and was sent to different schools from her "white" brothers and sisters. In that same country, Japanese are legally regarded as "honorary whites"—a status recently awarded them to facilitate trade relations between Japan and South Africa.

The differences between human populations are overwhelmingly cultural and linguistic in origin—that is, they are learned characters, not inherited ones.

What Is Intelligence? Claims of intellectual differences between various groups of people are frequently used as justification for racial discrimination. This immediately requires a long, hard look at the word "intelligence." Webster says it means "the power or act of understanding" — but understanding *what*? What are the criteria of intelligence?

In America we use so-called "intelligence tests" to measure something called an **intelligence quotient** — mental age relative to chronological age, based on a population mean of 100. Such tests, as pointed out by the population geneticist, James C. King, in his book *The Biology of Race*, have been developed and used most extensively in Europe and the United States as a means for predicting the achievement of children in school. Since success in school is highly correlated with adult performance in Western society — in terms of status, earnings and responsibility — it is not surprising that I.Q. tests are regarded as measures of intelligence. However, they make no predictions whatever about the intellectual abilities necessary to survive in other cultural situations, such as in the Arctic or the Australian bush. As Dr. King states, "the elimination of cultural bias from intelligence tests has not been achieved and appears to be a next-to-impossible task."

In 1969, an educational psychologist, A. R. Jensen, published a paper purporting to show that blacks in the United States are genetically inferior to whites in their ability to perform on I.Q. tests. This report was followed by a wave of controversy. There is no doubt that, among humans, intellectual ability is partly inherited; it is also a fact that black Americans, on the average, score lower on I.Q. tests than white Americans. Interestingly enough, neither Jensen nor anyone else has tried to correlate *degrees* of "blackness" (if that is to be our criterion of "race" in the issue of intelligence) with degrees of I.Q. test performance. In other words, all "blacks" from a black ghetto, whatever their genetic makeup, are classified as "black" in terms of their performance on an I.Q. test.

Dr. Jensen has combined these two facts — that intelligence is partly inherited and that persons identified as "black" score poorly on I.Q. tests — in arguing that blacks are genetically limited. Several psychologists and geneticists have strongly disagreed with him, pointing out that aside from the cultural bias inherent in I.Q. tests, general intelligence is more strongly affected by environmental factors than Dr. Jensen admits.

There are at least four factors involved in scoring well on an I.Q. test: (1) the intellectual capacity inherited from one's parents; (2) the amount of stimulation during development. (As we saw in Chapter 13, the development of the nervous system in a young organism is dependent upon environmental stimulation. This is especially true in man, where environmental input during the prenatal (see Chapter 22) and early postnatal years is extremely important); (3) the familiarity of the subject with the cultural biases present in the I.Q. test itself; (4) the subject's culturally determined expectations of performance in a test situation. (A person who expects not to do well often fulfills his own expectations; for example, some students consistently miss significantly more than half the questions on true-false examinations.) The latter three environmental factors were not adequately considered by Dr. Jensen in his estimates of the contribution of environment to I.Q. test performance.

As pointed out in Chapter 1a, any studies on human populations are made difficult by the large number of uncontrolled variables involved. In our consideration in that chapter of factors affecting height, we noted that nutrition, sex, general health and much else could modify the genetic factors inherited from one's parents. In fact, in the past few decades, the average height of many human populations has increased remarkably due to improved nutrition, even though the gene pool has remained constant. This point has been used by many of Jensen's critics as evidence that environmental factors may play a far greater role in development of intelligence than was previously thought. Their argument runs as follows: height, which is inherited through many genes in much the same way as intelligence, can be remarkably affected by environmental factors; thus it also may be possible to increase the intellectual potential of a person significantly by improving his surroundings during childhood, especially during the all-important formative years for the brain, between ages 0 and 5. It is quite possible that the intellectual performance of all people, whatever their inherited capacities, can be greatly improved by a better environment and better schooling.

MAN AS AN EVOLUTIONARY AGENT

We have seen in our examination of ecological principles that no organism in an ecosystem stands alone. To some extent, each individual affects all other members of the community and so exerts a selective force on them. In this man is not different from other organisms, except in *degree*.

There are several ways in which man acts as an evolutionary force in the biosphere, which can be listed in a general way as follows:

1. Direct physical manipulation of species, as occurs in hunting, food gathering, breeding of selected strains, agriculture and so forth;
2. Changing the local environment, as by clearing land, building dams, releasing pollutants and the like.
3. Introduction of foreign species into an environment.

These categories overlap to some extent, but they are still useful guidelines to keep in mind as we examine the history of man's effects on the evolution of his fellow organisms.

THE ROLE OF FIRE

When early man merely hunted and gathered food, he was not the controlling agent in his environment, but rather interacted with it as do other organisms. When he learned to make fire, though, he became capable of significantly altering his surroundings. To provide himself with fuel, he stripped the bark from trees, causing them to die. Clearings were burned in the forests to increase the growth of grass and so provide forage for the wild animals man fed upon. The American Plains Indian burned off large areas of forest and shrub to provide more fodder for wild bison. Fire also provided the means for cooking plants which were otherwise unpalatable.

A particularly interesting speculation about man's use of fire in relation to the disappearance of many of the great Pleistocene mammals was made by the American ecologist Paul Martin. The extinction of species has, of course, gone on continuously since life began, and many animals, large and small, had become extinct before man appeared. As in the case of the dinosaurs, there generally seems to have been some great climatic change which could explain it. As Martin points out, however, the North American camels and horses, the long-horned bison, the mammoths and mastodons of the New and Old Worlds, the wooly rhinocerous of Russia, had all managed to survive several glacial periods, only to disappear toward the end of the Ice Ages. Even large animals in parts of the tropics suddenly disappeared. Moreover, only large mammals were affected; smaller animal species, as well as most plants, survived the climatic changes. What happened?

Martin's theory runs like this: Although the climatic changes occurred more or less simultaneously over the world, the extinctions did not; they spanned a period extending from 50,000 years ago in Africa to 14,000 years ago in Australia and northern Eurasia, 12,000 years ago in America and only 1000 years ago in Madagascar and New Zealand. It is Martin's thesis that the waves of extinction took place following the migrations of Stone Age peoples into new territories (Fig. 21a.7). He postulates that these prehistoric hunters, having developed better weapons and perhaps also by conducting fire drives, slaughtered whole herds of bison or mammoths, sometimes forcing them over cliffs; the theory assumes that to catch the game they needed the hunters had to kill the whole herd—what Martin refers to as "Pleistocene overkill." Eventually the larger animals became extinct, and the human population had to resort to killing smaller, less gregarious species on an individual basis. Although fossil evidence from this period is still scant, what has so far been uncovered does not refute this interesting hypothesis. As we shall see later in the chapter, man's propensity for causing extinction of other species has increased rather than diminished with time.

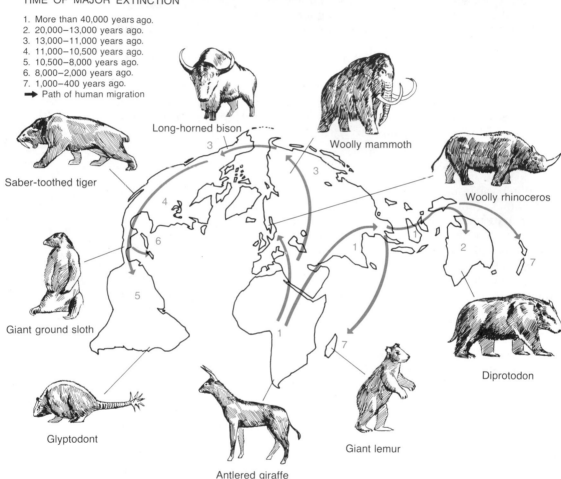

TIME OF MAJOR EXTINCTION

1. More than 40,000 years ago.
2. 20,000–13,000 years ago.
3. 13,000–11,000 years ago.
4. 11,000–10,500 years ago.
5. 10,500–8,000 years ago.
6. 8,000–2,000 years ago.
7. 1,000–400 years ago.
➡ Path of human migration

Long-horned bison

Woolly mammoth

Saber-toothed tiger

Woolly rhinoceros

Giant ground sloth

Diprotodon

Glyptodont

Giant lemur

Antlered giraffe

FIGURE 21a.7   Hypothesized extinction of large Pleistocene mammals following man's footsteps. (After Martin, P. S. 1967. Pleistocene overkill. Natural History *76*(10):32–38.)

THE ROLE OF AGRICULTURE

The earliest agriculture practiced by man probably consisted of the shifting type of polyculture carried out in temporary clearings of tropical forests, as described in Chapter 18. However, even before that, certain wild plants with some usefulness as food probably sprang up as weeds in the disturbed soils associated with human habitations. Most crop plants have evolved from weeds, and weeds are typically derived from plants adapted to disturbed habitats, such as river valleys, rock-slides, deserts, anthills and so forth. It is thought that domesticated wheat is derived from a minor weed-type grain which, through hybridization with related weeds, gradually evolved through selection by man into one of the world's most productive cereals. This probably took place about 10,000 years ago, in the alluvial plains of more temperate regions and especially in the Fertile Crescent of the Middle East, where grains such as wheat and barley were first grown. Oats and rye were originally weeds in these Near Eastern wheat fields but became the predominant grains when agriculture spread to the colder climates of northern Europe, where they thrived better. So great has been man's selective action on grains that it is now impossible to identify with certainty the original wild species of a single domesticated seed plant.

About 10,000 years ago, man also began to selectively breed herd animals—cattle, goats, asses and so forth. It is thought that these were originally raised for milk production, any surplus animals being utilized for meat and hides. Eventually, at least in Europe, man cleared land by burning to produce the grasses needed for his livestock; the excreta of the latter returned nutrients to the soil. The man-made ecosystems of that time, although based on selected strains of plants and animals, were ecologically self-sustaining as long as man tended them.

THE ROLE OF URBANIZATION

Only when agriculture had achieved a certain level of productivity was it possible for a fraction of the population to engage in pursuits other than obtaining food, and thus to found cities. As urban populations grew, they made ever greater demands upon agricultural productivity. Plowing and irrigation provided the extra food needed to support a non-rural population, and new engineering feats were forthcoming to extend the food-producing capability of the farmer. By early historic times, agriculture ceased to be an individual pursuit and evolved a commercial relationship with urban centers. Farmers traded food for manufactured goods. In fact, it is thought that writing originated as an accounting system to keep track of property transactions and taxes. As the demand for food increased, farmers increased their genetic selection of plants and animals and further modified the ecosystem to accommodate them. The ecological results of the Aswan High Dam (discussed in Chapter 19a), although far greater in magnitude, serve as an example of the sort of changes which must have resulted from man's more primitive feats.

THE ROLE OF COLONIZATION

With civilization came empires. Populations grew faster than ever and expanded their territories. And as people moved, they took with them, either intentionally or unintentionally, plants, animals and microorganisms.

Microorganisms. Wherever man moved, he took his favorite microorganisms: yeasts for making alcoholic beverages and raising bread; bacilli for making cheese and yogurt and so on. He also transported, unwittingly, both organisms causing his own diseases, especially contagious intestinal forms such as typhoid and cholera, and non-pathogenic soil organisms.

Plants. Wherever he went, man intentionally transported useful plants—crops, ornamental flowers and shrubs, medicinal herbs. Olive, grapevine, fig and other ancient natives of the Middle East spread thousands of years ago to the Mediterranean countries and but a century or so ago across the world to North America. Many of these species would not have survived in their new habitats if man had not provided for them. Extreme examples are bananas and breadfruit (a tree producing a mealy, carbohydrate-rich pod), both of which have lost the ability to form seeds since becoming domesticated and now survive only because man continues to propagate them from cuttings.

A small proportion of these transported plants have "escaped" from domestication and become wild in their new environments. For example, Kentucky bluegrass and Canadian bluegrass, now "natives," arrived in the New World centuries ago from Europe and originally came from Asia. Likewise, the wild oats, mustard, radishes and fennel covering California's hillsides today are natives of the Mediterranean area. The proportion of wild species actually native to California is quite small.

Animals. As with his domesticated plants, man brings along his domesticated animals, creating environments for their survival wherever he goes. Today's silkworm is an extreme case, for it is so altered by domestication that it can no longer survive in the wild. But in addition to his more obvious imports, man also brings along his camp followers and pests: lice, bedbugs, fleas, cockroaches, silverfish and so forth, which live either on man or in his habitat. Man also transports crop pests; about half of the insect pests at any spot in the world have been brought there by man. So too are the vectors of human and plant diseases spread by man. Recall the decline of chestnut trees in America after introduction of the foreign vector carrying the infectious fungal disease (see Chapter 19). Rapid air transport has speeded up these processes. The African malaria-carrying mosquito was introduced by plane from Africa into Brazil about 1930, and only after 10 years of effort was it finally eradicated.

Opening up waterways also affects distribution of animals. When the Welland Canal was opened over a hundred years ago, allowing ships to by-pass Niagara Falls and so enter Lake Erie, not only did ships come—so did the sea lamprey. This primitive fish lives as a parasite on other fish species, attaching by its oral sucker to the fish's body and feeding on its

FIGURE 21a.8   The Great Lakes lamprey and the damage it causes.
Note the jawless, sucker-like mouth by which the lamprey attaches to another fish, rasping a hole through the skin in order to feed on the host's blood. (From Case, J., and Stiers, V. 1971. *Biology: Observation and Concept.* Copyright by the Macmillan Company, New York.)

blood (Fig. 21a.8). Before being destroyed by a specific chemical in the 1950's, the lamprey all but eliminated the commercially important white-fish and trout from the western Great Lakes. Opening a sea-level canal across Panama, which would permit mixing of species now kept separate, could have consequences many times greater than that of the Welland Canal, and they would be almost impossible to control in the open ocean.

Intentional introductions of exotic wild animals often result in populations of pest magnitude. The muskrat was introduced into Europe from America in 1905 to begin a new fur industry but soon spread out of control and became destructive of crops. The story was repeated with the fur-bearing South American rodent known as coypu or nutria, which is now a pest in rice-growing regions of the southern United States. Four pairs of mongooses, brought to Jamaica from India in 1872 to kill rats in the sugar cane felds, multiplied rapidly and accomplished the job in 10 years; then they turned to toads, lizards, birds and small mammals, virtually annihilating many species. Rabbits, introduced into Australia by a farmer in 1859, increased from 24 individuals to 30,000 in just six years (Fig. 21a.9). At the peak, there were an estimated billion rabbits, competing with the sheep for sparse food in the dry countryside. Eventually, a specific rabbit virus, **myxomatosis,** was introduced among the Australian rabbit population where it proved especially virulent, killing millions of them.

Rabbits at an Australian water hole during a drought. Photograph taken at peak of population, prior to introduction of myxomatosis. (From Wagner, R. H. 1971. *Environment and Man.* W. W. Norton & Co., New York.)

FIGURE 21a.9

THE SPECIES BALANCE-SHEET

Human activities thus result in population increases in some species and decreases in others almost everywhere; man is a highly potent force in the process of natural selection. While causing extinctions of some species, man is also hastening the evolution of new species, as we shall see shortly.

It is enormously difficult to estimate accurately the net effect of man's activities on worldwide species diversity. It is probably safe to say, however, that the surviving species are more dependent on the unstable ecosystems created by man than were those which became extinct.

Endangered Species. The extinction of about 100 species of large Pleistocene mammals was almost insignificant compared to the number of species in danger of dying out today. Direct predation, whether for food, fur or trophies, continues in many areas unchecked. Perhaps the most infamous case is that of the whales. The whaling industry has attempted unsuccessfully to regulate itself through the International Whaling Commission. Not all countries involved in whaling, however, belong to the IWC. Moreover, the self-imposed limits put on the catches of those countries which do belong have been unrealistically high—far above the maximum sustainable yield (defined in Chapter 17a). The first species to decline was the blue whale, the largest animal ever to live on Earth (Fig. 21a.10). In

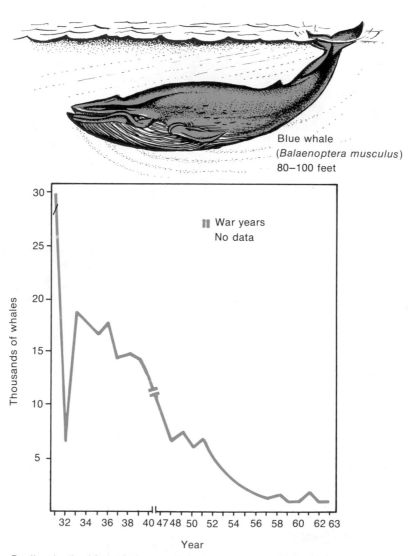

Blue whale
(*Balaenoptera musculus*)
80–100 feet

FIGURE 21a.10

Decline in the blue whale populations, as estimated from annual catches. The number of whales now remaining is believed to be too few to maintain the species. (After Ehrenfeld, D. W. 1970. *Biological Conservation*. Holt, Rinehart and Winston, New York.)

1970, less than 300 individuals were alive—too few, it is thought, to maintain the species. As blue whales declined, the next smaller species, fin whales, were taken, then the Seis and lastly the sperm whales. At present, all whale populations are rapidly declining.

Predator control is another, usually unwise, undertaking by which man threatens species with extinction in the mistaken belief he is protecting himself, his livestock, or a favored game animal (Fig. 21a.11). Bonafide cases of human deaths due to unprovoked wild predators are few; in contrast, that man-made "predator," the automobile, kills many times more people every year than all the world's predators combined, yet there is no outcry to destroy this "species"! In the case of alleged game destruction by predators, after predators have been removed from an area the game population they were supposedly decimating usually does not increase in size—it was declining for other reasons. Pheasants in South Dakota, for example, began to disappear after prime nesting land was mowed upon being withdrawn from the soil bank program. Sportsmen, blaming foxes and coyotes, began a vigorous predator control program, but the pheasant population has not recovered. Instead, those prey species which were being checked from gross overpopulation by the predator—rabbits, rodents and so forth—suddenly increased to become pests. Ranchers in at least one area, Colorado's Yampa Valley, realizing that predators may be more beneficial in rodent control than detrimental in killing livestock, halted a coyote extermination program in order to avoid similar consequences. It is now becoming clear that the diet of most large predators is made up mostly of rabbits, field mice and other small animals, rather than the larger game or livestock species.

FIGURE 21a.11

Mountain lion cornered by bounty hunters. (Carl Iwasaki/Life Magazine. © 1970, Time Inc.)

Perhaps the greatest losses of species result, however, from man-made alterations of the environment. We have already noted that non-specific pesticides seldom eradicate the target pest species, but nevertheless affect the course of evolution. First, they select for resistant mutants among the pest species. Second, if the pesticide is long-lived in the environment, it may kill far distant species by accident. As we saw in Chapter 1, several species of predatory birds are becoming extinct because DDT interferes with their reproductive capacity. But far greater numbers of plant and animal species are declining in a less dramatic fashion as man nibbles away at their natural habitats, damming rivers, draining swamps, replacing scrub and forest with fields and, more recently, with subdivisions. In such cases, the original species decline in numbers, perhaps to become extinct, while other, introduced species increase and fluorish.

Creation of New Species. It is only an instant in the history of the Earth since Darwin provided an explanation of the origin of species in terms of evolution. Although we can surmise that man, in the course of his own evolution, has brought about new species in the past, the time since Darwin is yet too short for any verifiable scientific documentation of the development of distinct new species due to man's activities. The best evidence we have to date is the selection of a strain of darkly pigmented moths in soot-covered areas of industrial England in preference to the more lightly colored strain found elsewhere (see Chapter 21). Man has also inadvertently selected for strains of insects, including malarial mosquitoes which are genetically resistant to DDT and other pesticides, and also for virulent strains of bacteria resistant to antibiotics.

There is also abundant evidence that man accelerates those genetic processes by which it is thought evolution occurs. We already noted in Chapter 15a that man-made radiation accelerates the mutation rate in man. The same radiation, of course, is also mutagenic for other species, providing a greater variety of offspring upon which natural selection can act.

In Chapter 15 we saw that when two parent strains are crossed, the hybrid offspring are much more varied in phenotype than are the parent populations. It is thought that spontaneous hybridization between related strains in nature has been a major means for providing the variable offspring needed for selection to occur. Ordinarily each parent strain is well adapted to its particular habitat, and the hybrids tend to lose out in the struggle for survival. But if the habitat is altered by a natural cataclysm or, as happens ever more often, by man, then the hybrids may be better adapted to survive than either parent. This can be outlined as follows:

$$
\begin{array}{ccc}
\text{Strain A} & \times & \text{Strain B} \\
\text{(Adapted to} & & \text{(Adapted to} \\
\text{area I)} & & \text{area II)}
\end{array}
$$

Hybrids
(Cannot compete in area I or II,
but some offspring survive better
than A or B in altered area)

An entertaining example of this phenomenon has been described for two species of wild *Iris* growing on the Mississippi Delta. Although occurring near one another, the two species did not produce viable hybrids except on the land of one particular farmer. His land differed from that of his neighbors only in that he let his cows pasture on it without limit. The cows trampled down the grasses and turned the low-lying pasture into a quagmire which was ideal for hybrid forms of the iris (Fig. 21a.12). The sequel of the story runs thus: these beautiful hybrids were discovered by members of the New Orleans garden clubs who paid for the right to dig them up, soon removing all of the irises from the pasture. These incipient species did not die out however: they were transplanted and thrive today in gardens throughout Europe and America, where they may undergo sufficient further evolution to become new species (refer to Figure 21.12).

It is possible to picture how a similar situation may have occurred during Neolithic times when plants and animals were first domesticated by man. By creating disturbed environments, man may have provided habitats suitable to the survival of hybrid forms. If these proved more useful to him than the parent strains, it is easy to see how the latter may have died out while man tended and further improved by selection his domesticated hybrids.

Man has thus established a symbiotic relationship with his domesticated plants and animals. Neither is able to survive without the other. This has reached its highest level of refinement so far in the development of the high-yield grains of the Green Revolution. In parallel with the growth of medical knowledge which led to the population explosion, there has been an increase in understanding of the mechanisms of genetics, which has led to development of highly productive strains of wheat and rice. The new strains were developed to meet specific growth requirements, especially those encountered in tropical climates.

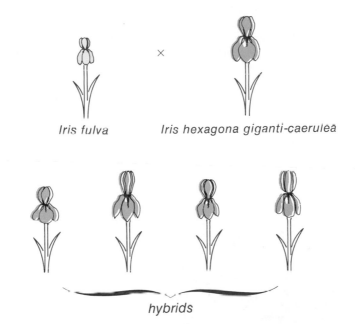

Iris fulva   *Iris hexagona giganti-caerulea*

hybrids

FIGURE 21a.12

Hybrids resulting from crosses between two wild iris species
Iris fulva is small, with unmarked bronze-colored flowers, while *Iris hexagona giganti-caerulea* is bright blue, with a yellow "throat" on the falls, outlined in white. The hybrids have a range of sizes, and vary in color from a deep wine to a very pale light blue.

First, the new strains had to be tough, able to withstand a variety of soils and climates. Second, they had to produce more grain per stalk if they were to provide more food. Third, in order not to collapse under the weight of their extra grain, the stalks had to be shorter and stockier. Fourth, the new grains had to be unaffected by day length, growing well at all seasons of the year and thus permitting year-round crops in temperate and tropical latitudes.

Although many scientists in many countries contributed to the development of the new strains of wheat and rice, two main centers achieved the final breakthroughs. The development of a high-yield wheat took place at the International Maize and Wheat Improvement Center in Mexico, sponsored by the Rockefeller Foundation. Under the Direction of Dr. Norman Borlaug, who received the Nobel Peace Prize in 1970 for his work, a new strain of wheat capable of increased yields and adapted to many soils and climates was produced (Fig. 21a.13). A similar development of genetically improved rice strains occurred at the International Rice Research Institute in the Philippines, sponsored by both the Rockefeller and Ford Foundations. The new grains have staved off for a time the famine which threatened much of the world. Borlaug himself, however, views the gains as only temporary unless human population growth is checked. It is important to realize that these man-made strains of grain, like other domesticated species, are totally dependent on man for their continued survival. As greater yields are achieved by genetic manipulation, so must greater energy subsidies be made by man to maintain his new strains. The "miracle" grains require irrigation and, above all, fertilizer, to survive; they also are less resistant to pests and disease, necessitating the use of pest-control programs.

We can thus conclude that man is speeding up the processes of evolution, hastening the extinction of some species and the evolution of others. However, the plants and animals which are favored by man are becoming ever more dependent on him for their continued existence. As man "puts more of his eggs into one basket" it would well repay him to be certain that his basket—the man-made environment he is creating in the world—won't suddenly collapse and destroy his food resources in the process. For not only does man tend to make simplified environments that are inherently unstable and more susceptible to stress than are complex natural systems; he also continues to place ever more stresses upon his environment in the form of pollution, plowing, chemical additives, weather and climate modification and so on.

FIGURE 21a.13

Dr. Norman Borlaug, holding stalks of regular wheat and one of his new high-yield strains. He received the Nobel Peace Prize in 1970 for his research. (*Wide World Photo,* from *Britannica* Book of the Year, 1971.)

SUMMARY

There is now sufficient fossil evidence to indicate that in the past million or so years man has evolved gradually from ape-men living in Africa. Upright gait, and the later increase in brain size, have become man's most distinguishing physical features. The gain in the cerebral cortex allowed finer discrimination of sensory input and refined motor movements as well as complex behavior such as language and abstract thought. There has probably been no increase in human intellectual potential in the last 100,000 years, and there is no basis for considering any modern race biologically superior to another. In fact, so-called racial differences constitute but a minute fraction of the total genes characteristic of man.

With the evolution of the human brain, man became a powerful selective force in the evolution of other species. Not only did he directly exterminate some species and protect others; he also altered environments, upsetting the previous balance between species and providing new habitats for hybrid variants. He has also transplanted organisms over long distances, introducing new selective forces wherever he went. More recently, his misuse of chemicals threatens the unintended extinction of several animal species, while his manipulation of the genetic constitution of certain plants has provided him new food resources. The danger to man lies in the great degree of reciprocal dependence that now exists between himself and the species he has favored. If he makes a mistake and causes the collapse of the simplified ecosystems he has created, it could well bring about his own downfall.

**READINGS AND
REFERENCES**

Brown, L. R. 1970. *Seeds of Change*. Praeger Publishers, New York. An account of the rapid development and adoption by farmers of high-yield strains of wheat and rice; and some of the new problems arising now that mass famine has been temporarily averted.

Dobzhansky, T. 1962. *Mankind Evolving: The Evolution of the Human Species*. Yale University Press, New Haven, Conn. An excellent general account of the history of mankind.

Ehrenfeld, D. W. 1970. *Biological Conservation*. Holt, Rinehart & Winston, New York. An able presentation of the modern conservationists' point of view.

Fox, R. 1968. Chinese have bigger brains than whites — Are they superior? The New York Times Magazine, June 30, 1968. A popular article pointing out that so-called racial differences in fact cut across the conventionally accepted racial classifications.

Howell, F. C. 1965. *Early Man*. Life Nature Library, Time, Incorporated, New York. A lively and well-illustrated account of the finds leading to our present understanding of human evolution.

Jensen, A. R. 1969. How much can we boost IQ and scholastic achievement? Harvard Educational Review. *39*:1–123. An attempt to assess the relative contributions of heredity and culture to intelligence. In this same volume are six articles critical of Jensen's conclusions, and Jensen's rebuttal to them.

King, J. C. 1971. *The Biology of Race*. Harcourt Brace Jovanovich, New York. A non-technical discussion of genetics, inheritance and the races of man.

Mangelsdorf, P. C. 1953. Wheat. Scientific American Offprint No. 25. The story of how wheat evolved from a weed into a useful cereal.

Martin, P. S. 1967. Pleistocene Overkill. Natural History, December 1967, pp. 32–38.

Martin, P. S., and Wright, H. E., Jr. 1967. *Pleistocene Extinctions: The Search for a Cause*. Yale University Press, New Haven, Conn. The above two references expound the hypothesis that man was responsible for extinction of great mammals at the end of the last Ice Age.

Moore, R. 1969. The search for mankind's ancestors. In *Evolution*. Time, Inc., New York. A highly readable account of the history of the search for man's origins.

Napier, J. 1967. The antiquity of human walking. Scientific American Offprint No. 1070. Changes in pelvic girdle and limbs preceded development of brain-hand relationship.

Thomas, W. L. and others (eds.) 1956. *Man's Role in Changing the Face of the Earth*. University of Chicago Press, Chicago. An excellent and highly recommended collection of articles from a symposium held at Stockholm in 1955—far ahead of its time! Especially appropriate to this chapter are:

Sauer, C. O. The agency of man on the Earth. Pp. 49–60.

Anderson, E. Man as a maker of new plants and new plant communities. Pp. 763–777.

Bates, M. Man as an agent in the spread of organisms. Pp. 788–804.

Young, L. B. (ed.) 1970. *Evolution of Man.* Oxford University Press, New York. A thoughtful series of readings about man's origins and his present evolution.

HUMAN REPRODUCTION

22

Internal fertilization and subsequent development of the young mammalian embryo are only possible because males and particularly females possess highly specialized reproductive systems. By providing a long period of postnatal care, mankind is able to produce offspring, which, although helpless at birth, have enormous potential for further development.

The current rapid rate of growth of the human population is cause for concern. Although human fertility has probably changed little over the millennia, increased survival of offspring through improved medical knowledge has recently brought about a great upsurge in the world's population (see Chapter 19a). In the past, environmental resistance has been a major controlling factor. As man learns to overcome that factor, he must also learn to control his own fertility if a population crisis is to be avoided. This chapter and the next will deal with the biological aspects of this problem. In this chapter we shall consider the physiology of human reproduction, and in the following, the various means available for controlling fertility.

FEMALE REPRODUCTIVE PHYSIOLOGY

We shall begin our discussion with the human female, since she exhibits to the greatest extent those hormonal feedback mechanisms which control fertility in both sexes.

ANATOMY OF THE FEMALE REPRODUCTIVE SYSTEM

The human baby, like those of other mammals, develops within the mother's body. The period of development, known as *gestation*, lasts about nine months, during which the infant receives oxygen and nutrients from the mother and utilizes the mother's body for excretion of most of its wastes. This close association between mother and fetus demands a highly specialized reproductive system, geared for internal fertilization of the egg and its subsequent development.

The anatomy of the female reproductive system is shown in Figure 22.1. Note that it is completely separate from the excretory system; female mammals are unique in this respect. The **urethra,** by which urine is expelled from the bladder, serves no other purpose in the female, whereas in the male it also carries seminal fluid during ejaculation.

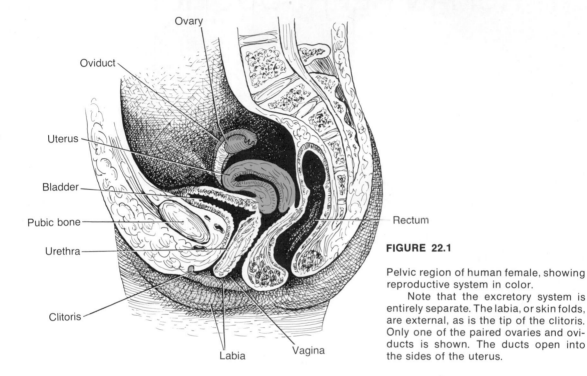

FIGURE 22.1

Pelvic region of human female, showing reproductive system in color.

Note that the excretory system is entirely separate. The labia, or skin folds, are external, as is the tip of the clitoris. Only one of the paired ovaries and oviducts is shown. The ducts open into the sides of the uterus.

The paired gonads—the **ovaries**—are located in the groin. **Ova** (eggs) are released from the ovaries into the body cavity, where they are quickly picked up by the **oviducts** or Fallopian tubes. In fact, the openings of the tubes are surrounded by finger-like projections which almost envelop the ovaries. The cilia on their walls create a current which insures that the ovum enters the tube. Muscular contractions of the oviduct hasten the transport of the ovum to the **uterus** or womb.

The uterus is a highly muscular organ about the size and shape of a pear. Its hollow interior is lined with a mucus-producing, vascularized epithelium called the **endometrium,** into which an egg, should it become fertilized, buries itself and where it grows into an embryo.

The lower end of the uterus, or **cervix,** is the region where uterine cancer most often develops in young women. In older women, cancers are more likely to develop in the upper regions of the uterus. It is from the cervix that "Pap" smears are taken and screened for malignant cells. Uterine cancer may be treated in several ways. If it is detected at an early stage, the malignant cells may be destroyed by cryoscopy (freezing); sometimes a radium needle is inserted to kill the tumor cells, after which the affected region is removed surgically. In more advanced cases, **hysterectomy,** or removal of the uterus, is necessary.

The uterine cervix projects into the **vagina,** an elastic, muscular canal which has three functions: The vagina is the receptacle for sperm ejaculated from the male penis; it contains mucus, secreted by the cervix, which is necessary to maintain it in a healthy condition; and it allows passage of the baby to the exterior at birth.

In front of the vaginal and urethral openings is the *clitoris,* a part of the female external genitalia. It is a highly sensitive structure, playing an important role in sexual receptivity. Anatomically it is homologous with the tip or *glans* of the male penis. On each side are two folds of tissue, the *labia.*

THE MENSTRUAL CYCLE

All female mammals exhibit some sort of regular reproductive cycles in which changes in the endometrial lining of the uterus are synchronized with the release of eggs from the ovaries. The endometrium must be in prime condition to receive and nourish a fertilized ovum, and apparently this condition cannot be maintained continuously. If conception does not occur, the old endometrium is lost and a new one grows in its place. In primates, including humans, the old endometrium is expelled with a rupture of blood vessels, resulting in menstrual flow, or **menstruation**. This occurs about once every 28 days, although the range among individuals is great; the flow of blood and sloughed tissue may last from three to seven days. The events occurring in the uterus, however, are secondary to those taking place in the ovary, which produces not only eggs but also hormones which regulate the changes occurring in the uterine lining.

The Ovarian Cycle. At birth, a human female already possesses all the ova she will ovulate, but they are in a non-fertilizable condition. These potential eggs have already developed from the germinal epithelium, and each is surrounded by a ring of supportive sister cells; together they form a *primordial follicle* (Fig. 22.2). Of the million or so follicles present in the ovaries at birth, only about 450 to 500 are destined to grow to maturity and be released during a lifetime. The rest degenerate. The release of an ovum, known as **ovulation,** occurs approximately once every 28 days, from puberty to menopause, or from about age 12 to 50.

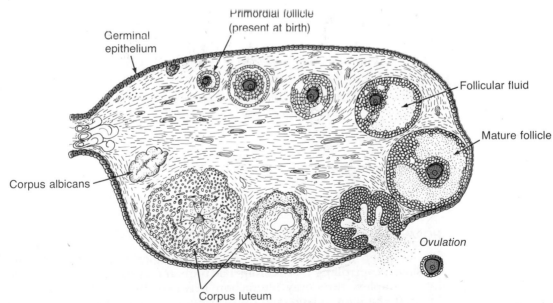

Schematic diagram of human ovary, showing sequential stages in ovarian cycle.
Each month one or the other ovary goes through such a cycle. Treatment of low-fertility women with hormones sometimes causes several follicles to develop simultaneously, leading to multiple births. (After Gorbman, A., and Bern, H. A. 1962. *Textbook of Comparative Endocrinology.* John Wiley and Sons, New York.)

FIGURE 22.2

The menstrual cycle can be conveniently divided into two phases. At the beginning of each menstrual cycle usually one follicle begins to grow, increasing in size for about two weeks during which the ovum enlarges. A cavity develops within the follicle, filled with a fluid containing **estrogen,** a female sex hormone produced by the follicle cells. This first two-week span constitutes Phase I.

When the follicle is mature, it ruptures, releasing the ovum, which passes through the oviducts to the uterus. Meantime, the collapsed follicle undergoes changes, and soon becomes a **corpus luteum** (Latin, "yellow body"). This persists for about two more weeks, during which it secretes two female hormones, estrogen and **progesterone.** If the ovum is not fertilized, the corpus luteum regresses and becomes scar tissue, the **corpus albicans** ("white body"). This is Phase II.

The Uterine Cycle. Meanwhile, the uterine lining also undergoes cyclical changes, which, as we shall see shortly, are controlled by the hormones produced in the ovary. The endometrium is composed of epithelial cells which form long, convoluted glands interspersed with numerous blood vessels. After a menstrual period, during Phase I, this lining is quite thin, but it soon begins to grow, and its tubular glands increase in size (Fig. 22.3). In about two weeks, just a few days after ovulation, the endometrium achieves maximum thickness and is ready to accept a fertilized ovum (Phase II). If pregnancy does not occur, however, menstruation ensues, and Phase I of a new cycle begins.

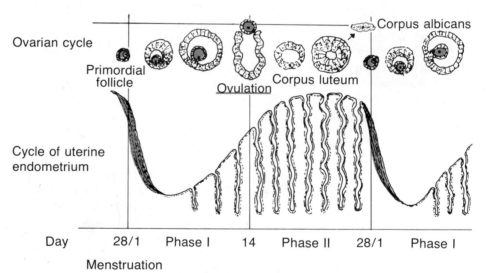

| Day | 28/1 | Phase I | 14 | Phase II | 28/1 | Phase I |

FIGURE 22.3   Menstruation

Synchronization of ovarian and uterine cycles.
Note that ovulation occurs just a few days before the endometrium reaches its maximal thickness.

HORMONAL CONTROL OF MENSTRUAL CYCLES

In order for a synchronized cyclic physiological process to occur, there must be a controlling mechanism. The human menstrual cycle is regulated by a complex, three-way interaction between the hypothalamus in the brain, the pituitary gland, and the ovaries; each produces hormones which stimulate the next gland in the triangular pathway (Fig. 22.4). Let us examine it in detail, since it is by interruption of this pathway that the Pill and several other chemical contraceptives exert their action.

Phase I. At the beginning of the cycle, during menstruation, one of the hypothalamic centers, as shown in our diagram, secretes a hormone called FSH-releasing factor, or **FRF**. This hormone passes via the blood to the anterior part of the pituitary, causing it to release a second hormone, a **gonadotrophic** ("gonad-feeding") hormone, stored in certain cells. This hormone, follicle-stimulating hormone or **FSH**, in turn causes one of the primordial ovarian follicles to develop. Finally, the follicle cells secrete estrogen, a fatty, steroid hormone related to cholesterol. Estrogen performs two important functions: it stimulates the growth of the uterine lining (Fig. 22.5), and it acts back on the hypothalamus.

FIGURE 22.4

Schematic diagram of the triangular hormonal system controlling menstrual cycles.
  1. Hypothalamus sequentially releases FRF and LF; these act on pituitary.
  2. Anterior pituitary releases first FSH, later LH; these act on ovary.
  3. Ovary first releases estrogen from the follicle, later releases estrogen and progesterone from corpus luteum; these act back on the hypothalamus. Note for future reference the hypothalamic nerve cells with enlarged endings terminating in posterior pituitary.

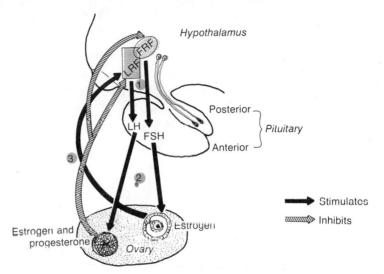

Phase II. Toward the end of Phase I there is a surge of estrogen. During this surge, estrogen is thought to *stimulate* the hypothalamic centers, especially the one which produces **LRF**, or LH-releasing factor (Fig. 22.4) LRF in turn causes a sudden release from the anterior pituitary of another gonadotrophic hormone, **LH**, or luteinizing hormone. It is thought that this sudden rise of LH in the blood causes ovulation to occur (Fig. 22.5).

LH then acts on the remnants of the follicle, converting it to the corpus luteum, whose cells produce both estrogen and progesterone. These two steroid hormones stimulate further glandular development of the endometrium, readying it for the implantation of an embryo. Estrogen and progesterone also act to inhibit the FRF and LRF centers of the hypothalamus, an example of **negative-feedback control.** If conception does not occur, hormone production by the corpus luteum drops off, initiating sloughing of the endometrium, or menstruation. The FRF center, released from steroid hormone inhibition, now begins a new cycle, starting with Phase I.

The cyclic nature of the production of hormones and their action on the ovary and uterus are summarized in Figure 22.5. *The main function of the menstrual cycle is to coordinate ovulation with receptivity of the uterine endometrium.*

The Pill, which contains steroid hormones, acts by suppressing the hormonal centers of the hypothalamus. Since neither development of the follicle nor ovulation can occur without the pituitary hormones released by FRF and LRF, pregnancy is impossible. The Pill blocks the entire triangular pathway. A pseudomenstruation is induced by sudden withdrawal of pills at the end of the month; this results in a decrease of steroid hormone levels in the blood, just as occurs before normal menstruation (see Fig. 22.5).

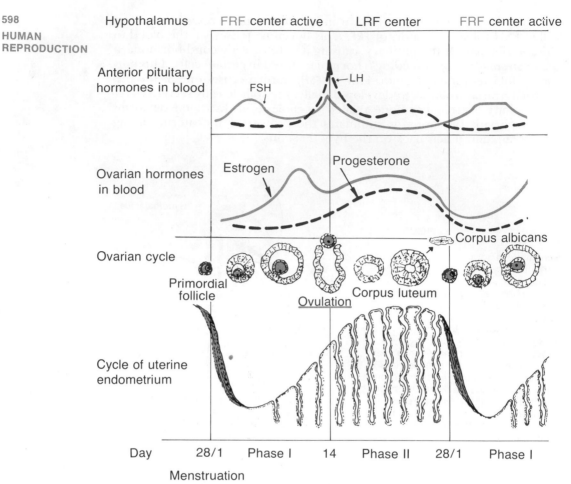

Hypothalamus · FRF center active · LRF center · FRF center active

Anterior pituitary hormones in blood · FSH · LH

Ovarian hormones in blood · Estrogen · Progesterone · Corpus albicans

Ovarian cycle · Primordial follicle · Ovulation · Corpus luteum

Cycle of uterine endometrium

Day · 28/1 · Phase I · 14 · Phase II · 28/1 · Phase I

Menstruation

FIGURE 22.5 A summary of the time sequence of events in the menstrual cycle.

Hormones in the hypothalamus affect blood levels of anterior pituitary hormones, which in turn act on the ovary.

Note the sharp surge of LH at the moment of ovulation.

OTHER ACTIONS OF STEROID HORMONES

In addition to its effect on the uterus, estrogen also increases contractility of the oviduct, hastening passage of the egg from ovary to uterus. During puberty, estrogen causes early closure of growing points in the bones, resulting in the average shorter stature of females than males. It also brings about the enlargement of the pelvic girdle or hips, a high-pitched voice and the development of the breasts (Fig. 22.6A). Estrogen facilitates water retention in the skin and is a component of expensive hormone cosmetics; it is doubtful, however, that the amounts of estrogen absorbed via the skin are sufficient to have much effect on wrinkles. Estrogen is also used to alleviate the symptoms of menopause experienced by some women, of which the most common are hot flashes, due to sudden dilation of blood vessels of the upper parts of the body, and a general feeling of irritability.

Progesterone also has effects on secondary sex characteristics, primarily on the breasts. The rise in this steroid hormone during Phase II of each cycle causes the breasts to enlarge, and during pregnancy, when progesterone levels are quite high, they grow even more. This growth is mainly due to glandular development at the end of the duct system (Fig. 22.6B).

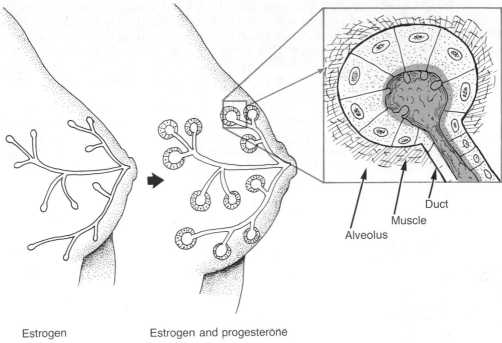

Estrogen   Estrogen and progesterone

A   B   C   **FIGURE 22.6**

Schematic diagram of breast development.

    A. During puberty, estrogen stimulates extensive duct growth.

    B. During pregnancy, high levels of progesterone cause development of secretory cells into clusters, or alveoli, at ends of ducts.

    C. Single alveolus enlarged, during lactation; milk collected in alveolar cavity is ejected when oxytocin-sensitized muscle contracts.

EVENTS FOLLOWING CONCEPTION

An ovum must be fertilized within about 24 hours after ovulation; otherwise it begins to degenerate. At this stage it is still within the oviduct, and it is here that fertilization occurs. The tiny sperm, which remain viable for only about 48 hours, must therefore swim rapidly from the vagina, through the uterus and far up the oviduct. They have only a few mitochondria to convert whatever food they meet on the way into useful energy to keep them swimming—a formidable task!

Within three to five days, the fertilized egg or zygote reaches the uterus, by which time several divisions have already occurred. Upon contacting the glandular endometrium, the tiny embryo becomes buried within it, being overgrown by uterine cells. This process is known as *implantation.*

Upon contact, the embryonic and maternal cells interact to form a dual organ, the *placenta,* which serves to nourish the embryo. Within the placenta, bloods of the mother and the embryo come into intimate contact, separated only by a thin layer of cells. Ordinarily, the baby's blood does not mix with that of the mother, and all molecules passing between the two circulations must diffuse across this thin capillary lining. Occasionally, however, small leaks develop which permit the exchange of blood cells. If the red blood cells of a baby carrying the Rh positive antigen on their surfaces pass into the circulation of its Rh negative mother, the mother's body reacts by forming antibodies against the "foreign" red blood cells of her baby. (Rh factors on erythrocytes, like other blood-group factors, are genetically inherited.) The antibody molecules pass back across the placenta, causing clumping of the baby's red blood cells (see Chapter 10 for discussion of antigen-antibody reactions). At birth, the baby's blood has insufficient oxygen-carrying capacity, and the infant dies unless a complete blood transfusion is given. Ordinarily, not enough antibodies are formed against a first Rh positive baby to do much harm, since the major mixing of the two circulations occurs at birth, but subsequent Rh positive babies born to a sensitized mother may be severely affected.

As the baby grows, the placenta enlarges, and at birth it occupies a large area of the uterine wall. The placenta functions to convey oxygen and nutrients to the developing fetus and removes most of its wastes; the connecting blood vessels are the arteries and veins of the **umbilical cord** (Fig. 22.7). Thus, the baby's heart pumps blood not only through its own organs but also, by way of this life-giving cord, to its placental circulation in the uterus.

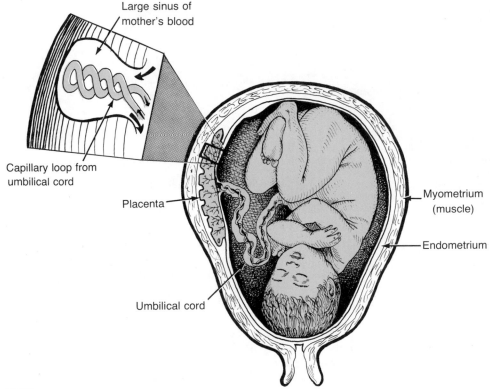

Large sinus of
mother's blood

Capillary loop from
umbilical cord

Placenta

Umbilical cord

Myometrium
(muscle)

Endometrium

FIGURE 22.7   Maternal-fetal circulatory relations.
Fetal blood passes via umbilical cord to placenta, where capillary loops are bathed in a sinus of maternal blood. Here nutrients, gases and wastes are exchanged. As seen in the inset, the two bloods almost never mix.

MAINTENANCE OF PREGNANCY

Just as menstrual cycles are maintained by hormones, so is their interruption during pregnancy a result of hormonal control. Soon after implantation, the placenta begins to parallel those functions of the pituitary gland related to ovarian cycles and starts to secrete large quantities of ovarian stimulating hormones, especially those which mimic the action of LH. Under the influence of these placental gonadotrophic hormones the corpus luteum continues to secrete large amounts of estrogen and progesterone (Fig. 22.8), which prevent the onset of menstruation. Phase II continues throughout gestation. Excess placental gonadotrophic hormones are excreted in the mother's urine and are the basis of pregnancy tests. The urine sample is mixed with blood from a rabbit immunized specifically against human placental gonadotrophins; if gonadotrophins are present in the urine specimen, they react with the antibodies and a precipitate forms (see Chapter 10).

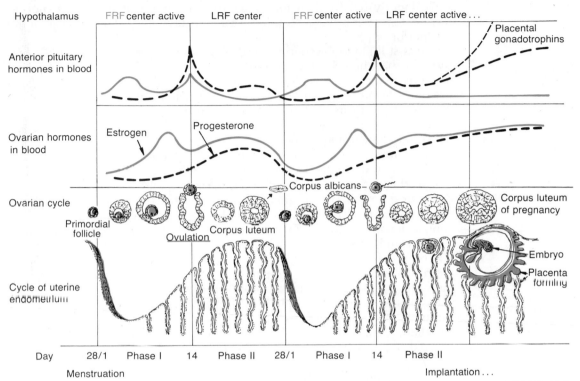

FIGURE 22.8

Maintenance of pregnancy.
    Compare the regular cycle, on the left, with the one in which implantation occurs, on the right. The placenta and pituitary secrete gonadotrophins which keep the corpus luteum active. While it secretes large amounts of estrogen and progesterone, menstruation does not occur. Later these hormones are produced by the placenta itself.

Later in pregnancy the placenta also takes over the job of secreting ovarian steroids, both estrogen and progesterone, which rise to high levels and maintain the uterus in its nutritive condition (Fig. 22.9). The placental steroids also induce further growth of the breasts in preparation for lacta-

FIGURE 22.9

Hormone production during pregnancy.
    At first placental gonadotrophins are produced to maintain the corpus luteum. Later the placenta, itself, takes over the latter's function, producing great quantities of female sex steroids. Note the sudden fall of steroids at parturition (birth).

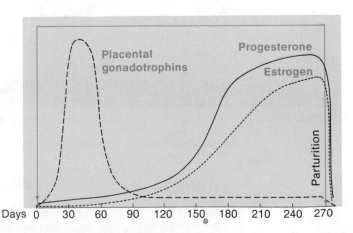

tion. The placental production of steroid hormones appears to be free of any of the usual feedback controls, and they totally block the cyclical functions of the hypothalamus, pituitary and ovaries.

BIRTH

Delivery of the human infant usually occurs about 280 days after the start of the last menstrual period, or 266 days after ovulation of the subsequently fertilized ovum. A handy way of calculating the expected birth date is to subtract 3 months and add 7 days to the date of onset of the last period.

It is still not altogether clear what initiates labor. Until recently it was thought that the release of **oxytocin**, a hormone produced in the hypothalamus and stored in the posterior pituitary (Fig. 22.4), was responsible for the onset of uterine contractions. Current opinion is that the intermittent release of oxytocin determines the strength and timing of contractions once labor is under way but does not trigger it initially. Possible factors which may cause labor to begin are (1) changes in ion concentration in the uterine muscle, especially calcium ion (Ca^{++}) and (2) physical distention of the uterine wall by the growing baby.

At the beginning of labor the cervix dilates, and the baby's head drops downward toward the vagina. This constitutes the first stage of labor, which often requires several hours. When dilation is complete, the uterine contractions become more frequent and forceful, expelling the infant rapidly through the vagina (Fig. 22.10). This is the second stage of labor. There are immediate changes in the baby's circulatory pattern—blood supply to the umbilical cord is cut off, and instead blood flows through the baby's lungs. Respiration begins almost immediately. This is the second stage of labor. Usually within about a half hour after birth there are renewed uterine contractions during which the placenta, or afterbirth, is expelled. After this third stage of labor, the uterus soon returns to its normal size. The stages of labor may thus be summarized as follows:

First stage: Onset of uterine contractions to full dilation of cervix.

Second stage: Full dilation of cervix to delivery of infant.

Third stage: Delivery of infant to expulsion of placenta.

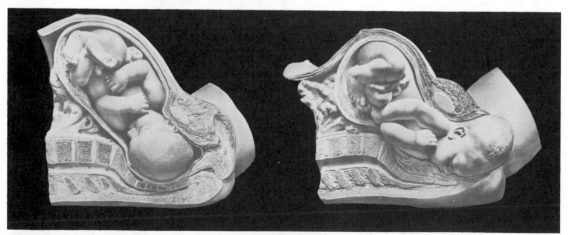

FIGURE 22.10 Models showing the second stage of labor. After dilation of the cervix and vaginal canal, rapid, forceful contractions of the uterine and abdominal musculature soon expel the infant. (Reproduced, with permission, from the *Birth Atlas,* published by Maternity Center Association, New York.)

LACTATION

The mammary glands, primed to their maximal growth by estrogen and progesterone during pregnancy, begin actually to secrete milk only about two days after birth. At this time the high blood levels of these ovarian hormones drop sharply (Fig. 22.9), apparently permitting the pituitary to begin releasing a hormone, **prolactin,** which induces milk production.

Release of the secreted milk is brought about by an interesting combination of neural and hormonal pathways. The suckling infant stimulates the sensory nerves in the nipple, which relay messages to the hypothalamus. This in turn causes release of oxytocin from the posterior pituitary. Oxytocin acts upon muscle cells lying around the alveoli, squeezing milk from the breasts (Fig. 22.6C). Since oxytocin also increases uterine contractions, suckling by the infant in the first days after birth may play a role in returning the uterus to its normal size. Lactation tends to cease if suckling is interrupted for several days.

In some but not all women, lactation suppresses menstruation and ovulation, although the hormonal pathways involved are not well understood. Lactation may thus have a contraceptive effect. This control system may be of importance among primitive peoples where nursing is essential to infant survival and may last for a year or more. If ovulation is suppressed, conception is automatically postponed until the previous child is weaned. This mechanism, however, should not be relied on as a means of contraception.

MALE REPRODUCTIVE PHYSIOLOGY

Having considered the events involved in female reproductive cycles and pregnancy, let us now turn to the other half of the picture, without which no pregnancies can occur. Physiologically speaking, the sexual function of the human male is to inseminate the female.

ANATOMY OF THE MALE REPRODUCTIVE SYSTEM

The male reproductive tract is designed to produce astronomical quantities of genetic information. Whereas the human female releases 500 eggs in a lifetime, the human male ejaculates 150 to 200 million sperm with each orgasm! Low sperm counts are associated with decreased male fertility. They may occur temporarily after illness or exhaustion, or because of either too frequent or too infrequent intercourse. In a few cases, a low count is due to permanent testicular damage.

The anatomy of the male urogenital system is shown in Figure 22.11. In the male, the urinary and genital tracts share a common duct to the exterior, the **urethra,** but it functions for only one purpose at a time. The paired gonads, or **testes** (singular, testis), develop in the body cavity of the male embryo, but shortly before birth they descend into the **scrotum.** The scrotal sacs are paired external extensions of the body cavity. Their function is to keep the testes cool, since sperm are heat sensitive and survive best at temperatures a few degrees below that of the rest of the body. Usually the connections between the body cavity and scrotal sacs remain tightly closed, but occasionally the intestine ruptures through on one or both sides, producing an inguinal hernia; this is relatively easily repaired surgically.

Sperm produced by each testis pass into a highly coiled storage duct, the **epididymis.** This duct connects with the urethra by way of a long tube known as the **vas deferens.** This "vas" is the one which is cut during a **vasectomy.** At the junction of the paired vasa deferentia with the urethra there are three sets of glands—the **seminal vesicles,** the **prostate gland** and **Cowper's glands.** All contribute nutrient secretions and mucus which are ejaculated with the sperm, together forming the **seminal fluid** or semen. These secretions help maintain and protect the sperm against the hostile environment of the female vagina. Two to three milliliters of seminal fluid are ejaculated through the urethra at each orgasm.

Sperm Capacitation. Although the sperm seem to undergo certain important physiological changes known as "ripening" during storage in the epididymis and passage through the male ducts, they are unable to fertilize an ovum in the condition in which they are released into the vagina. The seminal fluid contains a "decapacitation factor" which becomes diluted and neutralized by secretions in the female uterus. The sperm then undergo further physiological maturation, a process known as **capacitation**; this is essential for fertilization.

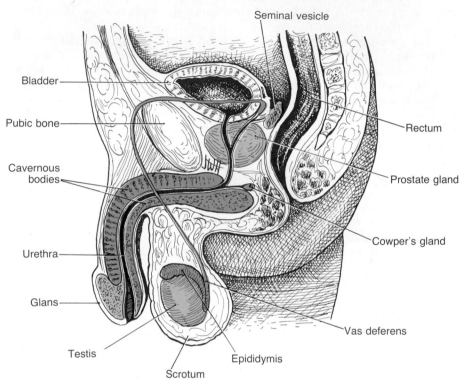

FIGURE 22.11

Pelvic region of human male, showing reproductive system in color.
    Sperm from testis, temporarily stored in epididymis, pass up vas deferens to enter urethra. Contributions to semen come from seminal vesicles, prostate and Cowper's glands. Erection occurs by filling of cavernous bodies with blood.

CONTROL OF SPERM PRODUCTION

The testes, or testicles, are composed of a system of highly coiled ducts, the **seminiferous tubules** (Fig. 22.12). Each tubule is lined by a germinal epithelium analogous to that in the female ovary. This epithelium is in almost continuous mitotic division, producing millions of potential sperm cells. These cells, in turn, undergo meiosis, achieving a haploid number of chromosomes. The male gametes then begin to lose cytoplasm, maturing into functional sperm (Fig. 22.13). This entire process depends upon secretion of FSH from the pituitary.

Between the seminiferous tubules of the testis there are clusters of **interstitial cells** that produce the male sex hormone, **testosterone**. Like the female sex hormones, testosterone is a steroid which controls the functioning of the male reproductive tract and determines secondary male sexual characteristics. The production of testosterone is controlled by the secretion of LH hormone from the pituitary.

Unlike the female, the human male does not undergo cyclical fluctuations of hypothalamic, pituitary or gonadal hormones; males are continuously fertile. Experiments on animals have shown, however, that there is a negative feedback between testicular steroids and the hypothalamus. If excess testosterone is injected, it inhibits the triangular hypothalamic-pituitary-testis pathway and the interstitial cells become inactive. We can thus conclude that, although there are no hormonal cycles in the male, the hormonal control of gamete formation follows a pathway similar to that in the female.

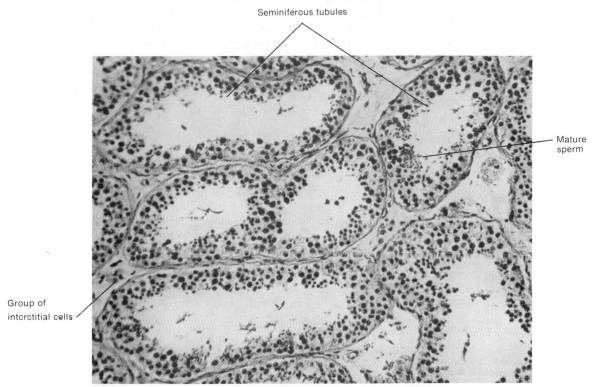

Seminiferous tubules

Mature sperm

Group of interstitial cells

Photograph of section of human testis. Note various stages of sperm development in seminiferous tubules, and small clusters of testosterone-secreting interstitial cells. (From Ham, A. W. 1969. *Histology.* J. B. Lippincott, Philadelphia.)

FIGURE 22.12

FIGURE 22.13

Stages in maturation of human sperm from haploid spermatocyte (left). Note condensation of DNA into head, formation of tail and loss of cytoplasm (RB = residual body, containing cast-off cytoplasm). Mature sperm (Sd$_2$) is shown in side and front view. Midpiece contains mitochondria.

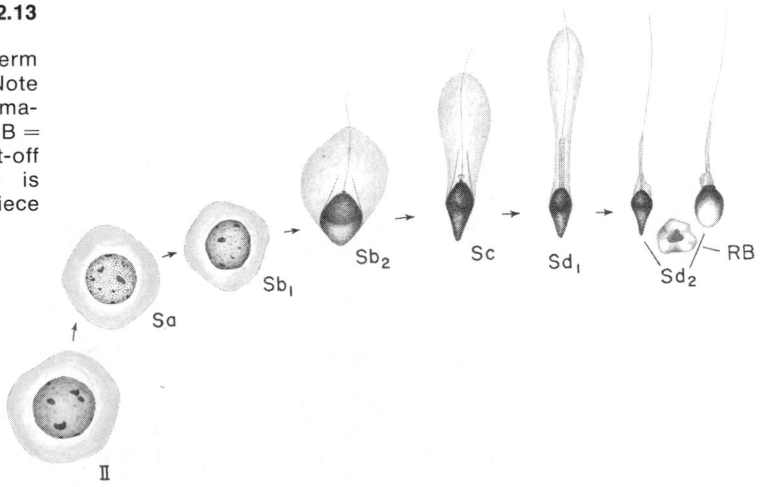

Sa

Sb$_1$

Sb$_2$

Sc

Sd$_1$

Sd$_2$

RB

II

Functions of male steroid hormone. Testosterone has been shown to play a similar role in the male to that of estrogen and progesterone in the female. During puberty, there is a gradual increase in testosterone levels in the blood which causes deepening of the voice, increased growth of facial and body hair, muscular and bone growth and delayed closure of the long bones—adolescent males thus grow for a longer period than females and hence are taller.

Testosterone is also responsible for maintaining the size and function of the male reproductive tract and its associated glands. Together with FSH, it stimulates spermatogenesis (sperm formation). Eunuchs (castrated males) or eunuchoids (males in which the testes have failed to descend into the scrotum) fail to produce adequate amounts of testosterone and hence do not exhibit the full range of male characteristics. Testosterone treatment of eunuchs and eunuchoids may restore secondary sexual characters but not, of course, sperm production.

It is of interest to note that, in both males and females, certain of those organs which are affected most by steroid hormones—uterus and mammary glands in the female and prostate gland in the male—are sites of high incidence of cancer. In some cases it is possible to treat prostate cancer with estrogens, since this hormone has an opposite effect to testosterone on prostate growth. It is still not known how sex hormones are related to development of cancer, although much research on this problem is in progress.

SEXUAL INTERCOURSE

In 1966, William H. Masters and Virginia E. Johnson, of the Reproductive Biology Research Foundation in St. Louis, published the first scientifically documented account of the physiological responses occurring in male and female human beings during sexual intercourse. Our résumé here is necessarily brief and is confined to those events which surround the orgasm.

The main results of sexual stimulation are an increase in muscle tension throughout the body and the development of *tumescence,* or swelling, of the sexually sensitive tissues—the nipples, vagina, labia and clitoris in the female, and the penis and scrotum in the male. All these tissues are highly vascularized; when stimulated, the veins constrict and the arteries dilate and the tissues become engorged with blood. This is particularly significant in the male penis, where the filling of its cavernous sinuses causes it to become erect and turgid; the penis is thus able to penetrate into the female vagina, which becomes dilated. Both the male penis and female cervix, together with glands located at the vaginal opening, secrete quantities of mucus as a lubricant.

At the peak of intercourse, the stimulation to both the male penis and female clitoris and vagina results in a sudden series of muscle contractions, especially of the genital tracts—the **orgasm.** Spontaneous contractions of the vagina and uterus occur in the female, and the ducts and glands of the male contract, ejaculating the seminal fluid. The contractions also force blood from the tumescent tissues, which then become flaccid. During an orgasm, there is virtually total involvement of the rest of the body as well, and especially of the nervous system.

WHY ARE SO MANY SPERM REQUIRED?

One may well wonder why, if a woman produces only 500 ova per lifetime, a man ejaculates 150,000,000 sperm with each orgasm. One might analogize the situation with 150,000,000 men trying to cross the Sahara on foot in search of an oasis with nothing but a quart of water each to sustain them. Only a few will be lucky enough to survive. Just so, a sperm on leaving the male body enters a hostile environment with virtually no resources on which to lean—only a few mitochondria in the midpiece and the seminal fluid, which is soon diluted. The female vagina contains quantities of organic acids to protect itself from fungal and bacterial infections. Although the resulting high acidity has a spermicidal (sperm killing) effect, sperm survival in an infected vagina is demonstrably lower than in a healthy vagina. Once past this barrier, the tiny sperm must yet find its way across the uterus and halfway up the oviduct before it encounters a fertilizable egg. Its only assistance comes from its single flagellum and from contractions of the uterine wall and oviduct. During its progress, the all-important process of capacitation occurs. The journey must be accomplished within 48 hours if the sperm is to remain alive.

Once having penetrated an ovum, however, a sperm cell causes irreversible changes. The egg membrane undergoes chemical alteration and cannot be penetrated by a second sperm. The nucleus of the successful sperm interacts with that of the egg and mitotic divisions begin, soon producing a minute embryo which will implant in the waiting wall of the uterus.

In Chapter 14 we outlined the major events in the early development of an animal embryo, and we have no space here to elaborate the myriad details known about the development of the tiny human embryo, not as large as a pinhead, into a human baby. It would require another textbook to do it justice, and several to which the interested reader may turn are listed at the end of this chapter.

Within the protective environment of the mother's womb, the infant develops according to its own unique set of genetic "blueprints," coded within the DNA of the particular egg and sperm that produced it. Its development is modified, however, by its maternal surroundings. The physical health and mental state of the mother continuously act upon the baby; malnutrition, illness and psychological disturbances may all have adverse effects. Mothers who smoke during pregnancy are known to produce small babies. The use of drugs or medications other than those recommended or prescribed by a doctor is to be avoided.

In but a few short months after fertilization, the human embryo develops into a viable organism not quite like any other that ever existed before or will ever exist again.

SUMMARY

The cyclical menstrual period in women insures that the uterine endometrium is in optimum condition to receive a fertilized egg. This is controlled by a three-way hormonal control system, which directs each of the two phases of the cycle, summed as follows:

Phase I

FRF → FSH → estrogen → endometrial
(hypothalamus) (anterior pituitary) (ovarian follicle) growth

Phase II

LRF → LH → estrogen → uterine
(hypothalamus) (anterior pituitary) progesterone glandular
(corpus luteum) development

The two phases are linked by the secretion of estrogen, which stimulates the LRF center in the hypothalamus. A sudden surge in the blood of LH about midcycle appears necessary for ovulation; this surge is inhibited by the female steroid hormones in the Pill. If implantation does not occur, both FRF and LRF centers are suppressed, followed by a fall in steroids; menstruation ensues.

If pregnancy occurs, the early placenta forms pituitary-like gonadotrophins which maintain steroid hormone production by the corpus luteum; later the placenta produces these steroids as well. The normal hormonal and menstrual cycles are thus blocked. The placenta, a combined fetal-maternal tissue, exchanges food and wastes between the two blood streams. After nine months, the steroid hormone levels fall, the uterine muscle undergoes physiological changes and labor begins. During gestation, alveolar growth of the mammary glands also takes place; following birth, suckling by the infant stimulates a neural-hormonal pathway which results in milk ejection.

The human male, in contrast, produces gametes continuously, and ejaculates about 150 million sperm at a time. This large number is necessary to insure sperm survival through the hostile vagina, the uterus and halfway up the oviduct, where fertilization takes place. The hormonal mechanisms in the male, although non-cyclic, are otherwise similar to those in the female.

**READINGS AND
REFERENCES**

Arey, L. B. 1965. *Developmental Anatomy.* 7th edition. W. B. Saunders Co., Philadelphia. One of the best standard texts on human embryology, including reproductive physiology and both normal and abnormal fetal development.

Masters, W. H. and Johnson, V. E. 1966. *Human Sexual Response.* Little, Brown & Co., Boston. A highly detailed, completely objective analysis of physiological responses of both sexes. Clears up many misconceptions about sexual activity.

Page, E. W., Villee, C. A., and Villee, D. B. 1972. *Human Reproduction.* W. B. Saunders Co., Philadelphia. A clearly written, up-to-date account of all biological aspects of human reproduction. Designed for beginning medical students, it is, however, free from obscure medical terminology.

Patten, B. M. 1968. *Human Embryology.* 3rd edition. Blakiston Co., Philadelphia. A detailed account of the development of organ systems in human embryos.

Turner, C. D. and Bagnara, J. T. 1971. *General Endocrinology.* 5th edition. W. B. Saunders Co., Philadelphia. Includes within its pages much about hormones involved in vertebrate and especially mammalian reproduction.

Villee, C. A. (ed.) 1961. *The Control of Ovulation.* Pergamon Press, Inc., New York. A series of papers dealing with this important subject, the partial understanding of which has made the Pill possible.

BIRTH CONTROL

Mankind is faced with two choices: either voluntary regulation of fertility, or a return to famine, disease and probably further war. Great advances in contraceptive methods have paralleled the medical advances which allowed the population explosion. Will mankind utilize the technology of birth control as he now utilizes that of medicine?

John E. Miller died at the age of 95 in a farmhouse near Cleveland. He was survived by five of his seven children, by 61 grandchildren, 338 great-grandchildren, and six great-great-grandchildren — a grand total of 410 descendants. Toward the end of his life, in the late 1950's, John Miller was being informed of the arrival of a new descendant on the average of once every 10 days. Had he lived another 10 years, they would have numbered over a thousand. His query regarding it all: "Where will they all find good farms?" With zero population growth, John Miller would have had about 15 descendants at his death, assuming they all survived him.

The present explosive growth of the world population has given rise to grave concern, not only by individuals but by governments and by agencies of the United Nations. Like that of any other organism, the size of the human population is determined by a balance between reproductive potential, or fertility, and environmental resistance, or mortality. In the past, famine, pestilence and war served as major regulators of population size (Fig. 22a.1). As these have declined without an accompanying decrease in fertility, the world population has increased. Until recently, emigration was an important means of alleviating crowded areas. Habitable regions which were relatively underpopulated, particularly the New World and Australia, served as areas of expansion for some of the excess population — often, however, at the expense of the low-density native populations. These frontiers are fast diminishing, and without control of reproduction there must, sooner or later, be a return of those former controlling factors which man has tried to escape — a death-rate solution. This is already happening in India, parts of South America and other areas where famine is widespread and increasing. This chapter is concerned with the alternative — a birth-rate solution.

FIGURE 22a.1

Durer's *The Four Horsemen*—death, war, pestilence and famine—from the Apocalypse of 1408. (Taken from *Six Centuries of Fine Prints* by Carl Zigrosser. © 1937 by Carl Zigrosser; © renewed 1965. Used by permission of Crown Publishers, Inc.)

Although certain biological factors—for instance, age, heredity, nutrition and general health—affect the fertility of individuals, the number of children produced in a population is more dependent upon cultural practices. There appears to have been little change in intrinsic human fertility with time, and thus changes in reproductive rate are primarily due to man-made control factors.

HISTORICAL METHODS

Not all cultures in the past have found it necessary to reproduce continuously at a maximum rate to maintain their numbers. In fact, over-population in areas of limited resources has often engendered a need for population control.

In some primitive societies, both infanticide and infant neglect have been practiced when the population threatened to outgrow food supplies. In hunting cultures, such as that of the Eskimos, female infants are selected against, since they do not contribute to the dangerous job of food gathering. Infanticide was also common in many of the Greek city-states. Abortion is another ancient method, although how successful it was in its aim before modern times is hard to assess.

Attempts to decrease fertility by cultural regulation have also been common. Taboos on intercourse are widespread. Postponing marriage is still common in Ireland, where late marriage is the rule and the percentage of celibate persons is high. Homosexuality as an outlet for sexual energy was not frowned upon in the Classical world and may have contributed partly to the decline in the population of Rome (but see also Chapter 10a). Lifelong celibacy for religious reasons has probably played little role in population control because of the small numbers involved. An exception was Tibet, where one male child in three was sent to a monastery (Fig. 22a.2).

A school for young Tibetan lamas who will remain celibate their entire lives. (T. S. Satyan Camera Press-PIX, from Petersen, W. 1969. *Population.* Macmillan, New York.)

Attempts were also made to decrease frequency of intercourse or prevent successful copulation by primitive operations. One tribe, recently described, practices clitoridectomy (removal of the clitoris) in young females to decrease sexual arousal. Australian bushmen make a second opening for the urethra at the rear of the base of the penis (urethrotomy) in volunteers to prevent their ejaculate from inseminating the female; as an inducement, such males are allowed access to all the tribe's females. Castration of male slaves, once practiced in many cultures, insured their inability to procreate.

A variety of less barbarous methods, which are still in use today, have also been practiced since prehistoric times. **Coitus interruptus,** or withdrawal of the penis before ejaculation, is perhaps the oldest and most common. The use of **spermicides** (sperm-killing agents), douches, sponges, and tampons also goes far back in history.

ATTITUDES TOWARD BIRTH CONTROL

There are a great many social and cultural factors involved in the acceptance and application of various birth control methods by different peoples. A few general observations about some of these factors may serve to introduce the interested student to the overall problem. On the whole, large families are more common in agrarian than industrialized cultures. Several reasons have been put forward to explain this. Rural people have been said to be less well educated and to have more conservative attitudes than urbanites; there are fewer diversions in rural areas to supplant procreation; and there is a need for extra hands. It is also probable that the desire for male heirs is strongest in the cultures of some underdeveloped countries, where sons are a form of old-age security. The rise of Christianity in Europe had a significant effect upon social attitudes toward sexual activity and procreation, replacing generally more liberal views with severe strictures about intercourse, contraception and abortion.

Today, however, there is an increasing demand, especially in developed countries, for effective birth control. The realization that population growth cannot continue at its present rate and the unwillingness of individuals to bear more children than they wish have given rise to cartoons like those of Figure 22a.3. In our discussion, however, we shall limit ourselves mainly to the biological aspects of various modern methods of birth control.

FIGURE 22a.3

A   B

Cartoons inspired by new attitudes toward birth control.
   A. Population limitation.
   B. Individual choice.
   (From Rubin, D. 1968. Modern Medicine *36*:76.)

CURRENTLY USED METHODS

Currently used methods of birth control fall into six major categories:
1. No chemical or mechanical devices
2. Local mechanical and chemical blocking devices
3. Intra-uterine devices
4. Hormonal contraceptives
5. Abortion
6. Sterilization

Each method has certain advantages and disadvantages. Aside from cultural and esthetic problems, and special problems of use discussed below, there are considerations as to the stage a couple has reached in attaining its desired family size. Newly married couples will often select quite different methods from couples who already have the desired number of children.

Table 22a.1 will give the reader a summary view of the main contraceptive methods used, how they act and their effectiveness in an average population. Effectiveness is usually measured in terms of "failures"—that is, pregnancies occurring in so many cycles. This is usually expressed as pregnancies per 100 woman-years of exposure (intercourse). Thus, five

pregnancies per 100 woman-years means that each year, five of every 100
women using a particular method will conceive. Likewise, 0.1 pregnancies
per 100 woman-years means that once in 10 years, one woman out of 100
using a certain method will become pregnant. It is important to bear in
mind that these statistics include human error as well as faults in the
method itself; this is implicit in the ranges given in the table.

613
BIRTH CONTROL

**TABLE 22a.1
MAJOR
CONTRACEPTIVE
METHODS AND
THEIR
EFFECTIVENESS††**

Method	Action	Approximate Failure Rate* (Pregnancies/100 Woman-Years)
Abstinence	No intercourse	0
Rhythm**	Intercourse in "safe" period	14–38 – Britain Up to 58 – U.S.
Coitus interruptus	Male withdrawal	15–23
Lactation	Nursing suppresses ovulation	No reliable data
Douche	Washing vagina	34–61
Condom	Intercepts sperm	7–17
Diaphragm and jelly	Intercepts and kills sperm	11–28
Vaginal chemicals	Spermicidal	9–40
IUD	Prevents implantation	3–8
Pill	Prevents ovulation and affects cervical mucus	0.1–2
Morning-after pill	Prevents implantation	No reliable data
Abortion	Removes fetus	0
Sterilization†	Prevents gamete release in male; prevents fertilization in female	About 0

*Ranges indicate best and worst estimates from clinical trials and population
studies.
**Abstinence required during one-third to one-half of menstrual cycle.
†Effectiveness of sterilization depends on method used.
††Data mainly from Peel, J. and Potts, M. 1969. *Textbook of Contraceptive Practice.* Cambridge University Press, New York.

CONTRACEPTION WITHOUT CHEMICAL OR MECHANICAL DEVICES

Several contraceptive practices fall within this category: abstinence,
rhythm, coitus interruptus, lactation, and douching. Several of these
require but little comment. Abstinence obviously is highly effective for
those individuals who elect to practice it. Douching is commonly practiced
in Catholic countries. The bidet, a low-standing washbasin designed for
flushing the vagina (modern models have a built-in fountain), is still commonly found in France and other countries which have come under the influence of French culture. In many countries, Coca Cola and other carbonated beverages are used as douches, primarily by teenagers; their
effectiveness is mainly due to the spermicidal effect of their high acidity.

Lactation is sometimes relied upon as a means of avoiding conception.
As we saw in the last chapter, in a proportion of nursing women menstrual
cycles may be absent for several months, and in some individuals ovulation
is also inhibited, making pregnancy impossible. Since the incidence of
this effect among nursing mothers is unknown and there is presently no
sure method of detecting it, it is a highly unreliable means of contraception
and, in any case, is but short-lived.

COITUS INTERRUPTUS

Male withdrawal before ejaculation can be a moderately effective
means of preventing pregnancy. Many colorful expressions have been

applied to it; those such as "getting off at Cottingham instead of going on through Beverley," used by inhabitants of the English city of Hull, may mislead unwary interviewers. Coitus interruptus is a somewhat more effective method of contraception than several other widely used techniques (Table 22a.1), and probably has no serious physiological side effects. The advice of an Indian government family planning manual is: "It is condemned by some doctors, but try it. You won't suddenly become a nervous wreck. If you notice bad side effects you can easily give it up." Failures among conscientious users of coitus interruptus are due to the fact that a few sperm are often present in the mucus secreted from the penis as a lubricant during sexual arousal, and these are transferred to the female vagina.

RHYTHM

Prevention of conception by having intercourse only on "safe" days of the month is the only method, other than abstinence, condoned by the Roman Catholic church. Since an egg is viable for only 24 hours, and sperm for about 48 hours, the maximum fertile period is three days. The snag is to determine precisely when a woman's fertile period is—that is, when ovulation occurs. In 1929, a Japanese and an Austrian independently discovered that, on the average, ovulation occurs 14 days *before* the next period. If a woman has very regular periods, there is no problem in estimating when she will ovulate each month. But most women show considerable variation in the length of their menstrual cycles—as much as 10 days or more—and herein lies the rub.

To determine the normal range of a woman's cycles, she should be observed by a trained worker for at least six months or, preferably, a year. This requires that she refrain from intercourse for that time, if she does not wish to become pregnant. Once the range of her cycles is determined, "safe" limits for intercourse can be calculated. The excluded time usually extends from the 10th to the 17th or 18th day of the cycle.

An alternative to this method is the use of basal temperature measurements: the basal body temperature of most women increases by about 0.6° Fahrenheit a day or so *after* ovulation (Fig. 22a.4). If a woman takes her temperature before getting out of bed each morning and records it on a chart, she can often judge when she has ovulated. She should then wait three days before having intercourse. Only if she has regular cycles, however, can she determine when to have intercourse *prior* to ovulation. The

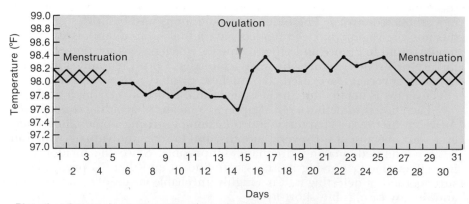

FIGURE 22a.4

Plot of early-morning temperature through one menstrual cycle.

Note the increase just after ovulation. Not all women have such distinctive rises, however. (After Peel, J., and Potts, M. 1969. *Textbook of Contraceptive Practice.* Cambridge University Press, New York.)

method has several drawbacks: many individuals are not able to read a thermometer accurately nor understand a graph; the temperature rise may be less obvious than shown in our figure; and fevers associated with minor colds may cause a "false rise," inviting the chance of pregnancy.

In the most favorable cases, the rhythm method will prohibit intercourse at least a third of the time; often the period of abstinence must be much longer. Its failure rate can be relatively high, and only intelligent and highly motivated couples are likely to find it significantly effective.

MECHANICAL OR CHEMICAL BLOCKING DEVICES

This category includes the use of mechanical methods, such as condoms and diaphragms, and of chemical spermicides which may be used in conjunction with the former. Their effectiveness is largely dependent upon the quality of the rubber product used, which is specifically regulated in some countries, and the faithfulness of the users in following directions.

THE CONDOM

It is probable that the **condom,** or penis sheath, as first used in sixteenth century Europe, was a device against the spread of venereal disease. At that time, cotton or linen cloths, soaked in lotions, were wrapped around the penis prior to intercourse. Its subsequent recognition as an insurance against pregnancy made the condom more popular, and by the eighteenth century they were being manufactured from animal intestines. The development of manufactured rubber in the following century greatly increased the popularity of condoms for contraceptive purposes. Improvements in manufacture, government controls on quality, and ease of purchase and use have all combined to make the condom the most widely used of all contraceptive devices in the Western world. When used with care it can be a highly effective agent (Table 22a.1). High failure rates with condoms are due largely to failure to use them regularly. Males, being less motivated than females to prevent pregnancy, tend to be poor contraceptive users.

THE DIAPHRAGM

Contraceptive methods which mechanically protect the cervix from penetration by sperm have had a long history. Leaves, halves of lemons and sponges made of wool or cotton, sometimes soaked in solutions with spermicidal properties, were all utilized. From these primitive methods the diaphragm or cervical cap eventually emerged (Fig. 22a.5). Made of rubber with a flexible, semi-rigid rim, the diaphragm acts as a mechanical barrier to the transmission of sperm from vagina to uterus. Used in conjunction with a spermicidal cream, it was the method most highly recommended for family planning prior to the birth control pill.

Problems with the diaphragm are need for instruction in use (Fig.

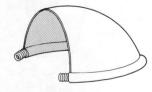

FIGURE 22a.5

Cervical diaphragms, cut in half to show two types of springs used to maintain shape.

Flat spring diaphragm   Coil spring diaphragm

22a.6), possibility of leakage, poor fitting (it must be fitted by a person well schooled in family planning to insure safety), need for replacement and careful hygiene during usage. To be maximally effective, a diaphragm must be used in conjunction with a chemical spermicide. Refitting after birth of a child is essential. The relative ineffectiveness of the diaphragm is probably due mainly to esthetic aversion, resulting from the necessity for action prior to intercourse. It can be highly successful, however, when used by strongly motivated women.

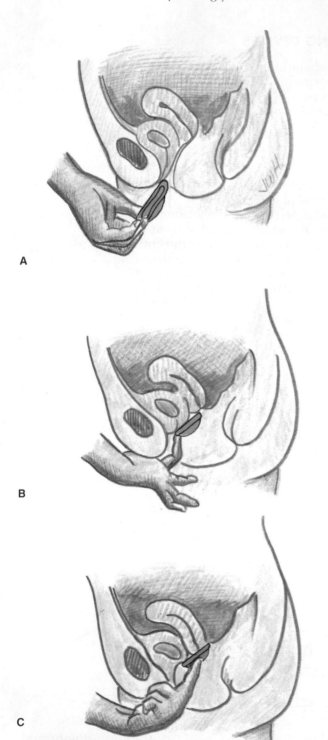

A

B

C

FIGURE 22a.6

Steps in insertion of a diaphragm.
A. The compressed diaphragm is slipped as far as possible into the vagina.
B. The front edge is pushed up under the pubic bone.
C. Testing to see that the cervix is completely covered.
(After Peel, J., and Potts, M. 1969. *Textbook of Contraceptive Practice.* Cambridge University Press, New York.)

CHEMICAL SPERMICIDES

Chemicals applied within the vagina probably were first used to physically impede the progress of sperm; such substances included gums, resins and vegetable products. In the late nineteenth century spermicidal agents came into use, one of the most widespread being quinine. Since then other, slightly more effective spermicides have been developed. All are now designed to be used in conjunction with a condom or diaphragm. They are provided as suppository tablets, creams, jellies or foams and must be applied shortly before intercourse to be effective. Their use has declined since introduction of the Pill, but further research should be undertaken to provide better protection for women unable to use oral contraceptives. The combination of a condom used by the male and spermicidal foam by the female can be 97 per cent effective in preventing conception.

SPONGES AND TAMPONS

These are primarily of historical interest. Designed to absorb sperm and deflect their passage from vagina to uterus, they have but marginal effectiveness in preventing contraception.

INTRA-UTERINE DEVICES

The IUD's, as they are now called, come in many shapes and sizes. Evidence of their use extends back into prehistory, although contemporary use is but a century old. Widespread acceptance of IUD's began only in the 1960's, with the advent of various stainless steel and plastic devices. A variety of these are shown in Figure 22a.7.

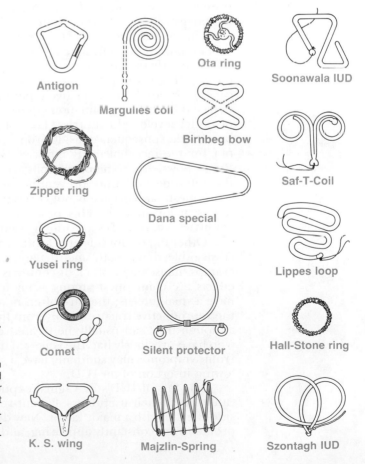

FIGURE 22a.7

Types of intrauterine devices.

All are flexible for insertion through the cervical canal by a doctor or trained assistant. Most usually have a thread or chain projecting into the vagina to detect for retention, and for removal if necessary. (After Peel, J., and Potts, M. 1969. *Textbook of Contraceptive Practice.* Cambridge University Press, New York.)

Antigon

Margulies coil

Ota ring

Soonawala IUD

Birnbeg bow

Zipper ring

Dana special

Saf-T-Coil

Yusei ring

Lippes loop

Comet

Silent protector

Hall-Stone ring

K. S. wing

Majzlin-Spring

Szontagh IUD

The devices shown in the figure are essentially two-dimensional and are flexible for easy insertion into the uterus. They prevent conception by inducing a mild inflammatory response in the endometrium. White blood cells infiltrate the uterine lining, producing a uterine environment unsuitable for implantation. A small amount of copper-containing wire on IUD's has been found to increase this inflammatory response and simultaneously to decrease failure rates. There is no indication of malignant changes in the uterus resulting from use of IUD's. IUD's are considerably more effective than the other methods we have so far considered (Table 22a.1).

ADVANTAGES OF IUD's

The intrauterine device has several advantages over previously described methods of contraception. It may be readily inserted without anesthesia into the uterus of women who have had children. For women who have not borne children anesthesia may be necessary, as their cervical canal is narrow. (Insertion requires a skilled operator, since perforation of the uterus is possible.) A string or "tail" (not always shown in Figure 22a.7) projects into the vagina, permitting easy removal and detection of possible loss during menstruation by the wearer. (The string, which may be felt by the penis, is bothersome to some males, however.) Once inserted, an IUD requires no further attention on the part of the wearer, other than to check that expulsion has not occurred. It is thus highly suitable for use among minimally educated or motivated people. IUD's are inexpensive to maintain and distribute.

DISADVANTAGES OF IUD's

Despite its great advantages and widespread use, the IUD is not at present an ideal solution for all women. It may be expelled unnoticed during menstruation. Examination of sanitary napkins and tampons and checking on the presence of the string can obviate this problem in educated women but are less effective among the illiterate.

If poorly qualified persons insert an IUD, the chances are greatly increased that it will perforate the uterine wall and enter the coelom (body cavity). Since this is a relatively painless phenomenon, it is usually only detected by x-ray examination, a service not widely available in rural areas of underdeveloped countries. Loss of an IUD into the body cavity would be of little consequence, aside from the possibility of conception, were it not for the fact that many types of IUD's have a "closed" configuration. Rings, bows, and some of the coils are capable of obstructing a segment of the intestine and thus preventing its normal peristaltic movements. If such intestinal blockage is not corrected, it can lead to conditions necessitating major surgery. This poses a serious problem in underdeveloped countries where IUD's are widely used and rural medical care is minimal.

Other important side effects often associated with today's IUD's are excessive bleeding—both menstrual and midcycle—increased backaches, cramps, nausea and other discomforts of severe menstruation. Such side effects are commonest among women who have borne no children and have a small uterus; they are often unable to tolerate IUD's. Infections of the reproductive tract resulting from the use of IUD's are not uncommon, although they can usually be treated effectively with antibiotics. Patients who have previously had the venereal disease gonorrhea, even though free from symptoms, may suddenly have a severe recurrence of the disease following insertion of an IUD.

The loss of IUD's by vaginal expulsion is common enough to significantly affect their usefulness. Patient-requested removal due to unpleasant side effects is also a major factor. New designs intended to circumvent these problems are constantly appearing, and greatly improved devices are likely to be available soon. The relatively low cost of manufacture and distribution of IUD's and their suitability for undereducated people make them still the choice contraceptive for family planning in many areas.

HORMONAL CONTRACEPTIVES

The recent availability of large quantities of inexpensive, synthetic female hormones has made the development of oral contraceptives possible. At present, the Pill is the only widely used hormonal contraceptive, but several improvements are likely in the near future.

THE PILL

Oral contraceptives consist of a sequence of 20 or 21 pills, one being taken each day. They contain synthetic estrogens and **progestogens** (the name given to artificial progesterones) which mimic the normal ovarian steroid hormones. Pill sequences are of two types. In one type, all pills are identical and contain both estrogens and progestogens; in the other type, the first 15 pills contain only estrogen, the remaining five contain both hormones. The effects are identical: the entire three-way hormonal control system of the menstrual cycle—hypothalamus, pituitary and ovary—is blocked by the synthetic hormones. The hormones suppress the release of LRF from the hypothalamus, which is thought to be necessary for triggering ovulation (see Chapter 22).

Advantages of the Pill. Taken regularly, the Pill is virtually 100 per cent effective (failure rates are about 0.1 per 100 woman-years). Forgetting to take the Pill is the major cause of failure among women who use it. Its convenience and esthetic desirability have made it an extremely popular method. Its use should be discontinued for a few months every two to three years to check that normal cycles are still possible. A normal menstrual period is one of the best indicators of good health in a woman.

Disadvantages of the Pill. The Pill is a drug, taken voluntarily by large numbers of healthy women over a long period of time. In this, it differs from all other drugs ever used, whose functions have been to cure or prevent diseases. Just like any other drug, oral contraceptives are bound to have undesirable side effects in a fraction of those using them and should only be taken under the continuous supervision of a doctor.

The most common side effects are merely unpleasant rather than dangerous. These include headache, nausea, dizziness, vomiting, excessive painful breast enlargement, midcycle or "breakthrough" bleeding, weight gain and occasional brown discoloration of the face. Often these side effects diminish with continued use of the Pill or are relieved by changing from one type to another. A few women, however, cannot tolerate the Pill.

A more serious objection raised against the Pill is that it has been implicated in **thromboembolism,** or obstructive blood clots, especially in lungs and heart. A study in Great Britain reported that the incidence of this disease was about 10 times higher in women taking the Pill than in other women of the same age. It is the estrogen in the Pill which is suspect, although no explanation of how it may be acting is yet forthcoming. The British study showed that the risk of death from thromboembolism to users of the Pill is about three per 100,000 women. The risk of death due to a normal pregnancy in the same population is 25 per 100,000. Deaths from illegal abortions are about twice as frequent, or 50 per 100,000 women. The woman using oral contraceptives thus has a lower risk of death than does a woman using nothing or using a less efficient method.

No increases in thromboembolism due to the Pill have been reported in the United States, although an increase in blood pressure has been shown to occur in a few women. Obviously women with a predisposition to circulatory disease should use an alternative form of contraception.

THE "MORNING-AFTER" PILL

The so-called "morning-after" Pill does not really live up to its name, for it must be taken for up to three days after intercourse to be effective. It contains enormous quantities of estrogens which produce overgrowth of the endometrium while the zygote is still in the oviduct, thus preventing its implantation. The side effects are severe: nausea, vomiting and excessive and irregular bleeding. If the "morning-after" Pill is taken after implantation has already occurred and fails to cause extrusion of the embryo, and if the fetus is a female, there is a high probability that when the child reaches puberty she will develop vaginal cancer. Excess estrogen thus has a latent carcinogenic effect on female embryos.

FUTURE HORMONAL CONTRACEPTIVES

Several future possibilities exist for improved contraception with drugs interfering with the normal cycle. One is continuous, low-level dosages of new types of synthetic progestogens, either taken as daily pills or implanted in tiny capsules in the muscle from which they gradually leach out. Precisely how these hormones act is unknown; in some but not all patients they suppress the LH peak thought to be necessary for ovulation (see Chapter 22); they may also act by altering the cervical mucus and making it more hostile to sperm. They are reasonably effective, with a failure rate of about 3 per 100 woman-years and do not cause thromboembolism since they contain no estrogens. The only side effects are somewhat irregular midcycle bleeding in about a third of the users. Clinical trials in the United States were stopped in 1970, however, after it was found that female dogs, which metabolize these progestogens quite differently from humans, develop precancerous mammary nodules.

Another possibility lies in drugs which specifically interfere with a single step in the cycle. One in the process of development—the so-called "Swedish abortion pill"—is a derivative of a chemical called diphenylethylene. It inhibits progesterone production by the corpus luteum and may also enhance excretion of the hormone from the body. As we saw in Chapter 22, the continued functioning of the corpus luteum is essential during the early stages of pregnancy, before the placental steroid hormones take over. One diphenylethylene pill is taken at the end of each month, or whenever a period is "late," to suppress corpus luteum function and induce menstruation. Safety tests are currently under way on this promising agent.

Another group of chemicals which hold some promise are the **prostaglandins.** These hormone-like substances are produced by all cells in the body. In relatively large amounts, they stimulate rhythmic uterine contractions, and when taken on the last day of a cycle will induce menstruation, even if implantation has occurred. They also appear to shorten the life of the corpus luteum so necessary for continuation of pregnancy. They are currently used clinically to induce abortion. The prostaglandins now in use produce nausea and diarrhea as side effects. There are many types of prostaglandins, however, and the search continues for one which will not produce either unpleasant or dangerous side effects.

ABORTION

The removal of the fetus, either spontaneously by miscarriage or intentionally, if it occurs during the first two-thirds of pregnancy, is known as

abortion. The fetus at this stage is incapable of independent survival. Our concern here is with induced abortions, and we shall consider only those means which are medically approved. Other methods are either extremely dangerous or essentially ineffective.

METHODS

During the first third of pregnancy, abortion can usually be carried out via the vagina and cervix. There are two main methods: dilatation and curettage, and uterine aspiration. Both require anesthesia and should take place under sterile conditions with the supervision of a doctor, usually in a hospital. In the first method, the cervix is dilated and forceps are used to remove the fetus and its membranes; the uterine wall is then gently scraped to remove the placenta. In uterine aspiration, a suction device is used (Fig. 22a.8). After dilatation, a hollow curette is inserted through the cervical canal, and the fetus, membranes and placenta are sucked out under vacuum. Complication from infection and excessive bleeding are minimal if the procedures are carried out by trained doctors. Some hospital-associated clinics in the United States now offer these types of abortion to healthy women on an out-patient, one day basis, with back-up measures immediately available in case of complications.

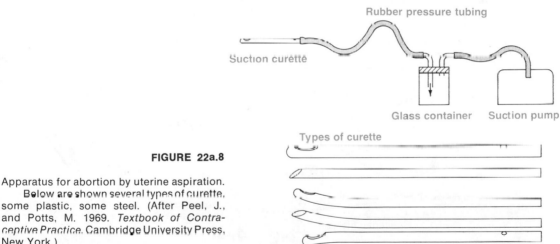

FIGURE 22a.8

Apparatus for abortion by uterine aspiration.
Below are shown several types of curette, some plastic, some steel. (After Peel, J., and Potts, M. 1969. *Textbook of Contraceptive Practice.* Cambridge University Press, New York.)

During the second three months of pregnancy two very different methods are commonly used to induce abortion. One is the injection of hypertonic solutions into the amniotic sac around the fetus, which induces labor within 8 to 72 hours. A small amount of amniotic fluid is removed by amniocentesis (see Chapter 15a) and replaced with an equal amount of strongly hypertonic salt solution which destroys the fetus. After the products of conception are expelled, dilatation and curettage is usually performed. The other method is ***hysterotomy,*** or opening of the uterus. Although sometimes carried out through the vagina, it is usually similar to the Caesarean section performed for difficult full-term deliveries and requires an abdominal operation. The uterus is opened surgically, and the fetus, membranes and placenta are removed. (Patients so treated must usually have any further children delivered by Caesarean section also.) Tubal ligation, or tying of the oviducts, can be done at the same time if the patient has requested sterilization. Both methods obviously pose some danger to the patient and should only be performed in hospitals by skilled surgeons.

ABORTION FOR BIRTH CONTROL

It is a simple fact that as the demand for birth control in a country rises, so does the demand for abortion. In eastern European countries such as Czechoslovakia and Hungary, however, where abortion has long been legal and is carried out at extremely low risk to the patient, the former high demand for abortion has declined as better contraceptives have become available. It would appear that, given a choice, women prefer not becoming pregnant to having a pregnancy terminated by abortion.

In all but a small number of states in this country, abortion is illegal except to save the mother's life. Interestingly enough, these strict anti-abortion laws, introduced early in the nineteenth century at a time when abortion was a highly dangerous procedure, were designed to protect the mother's life. The moral issues now being raised regarding repeal or reform of these laws were not part of their original intent! Whatever one's moral views, it is clear that women will seek abortions, and if legal means are unavailable, illegal means will be used, with all their attendant risks. If the experience in eastern Europe is a reliable guide, we can expect that as safe abortion becomes more available, it will be used not as a primary means of birth control, but as a back-up measure by women for whom contraceptives have failed.

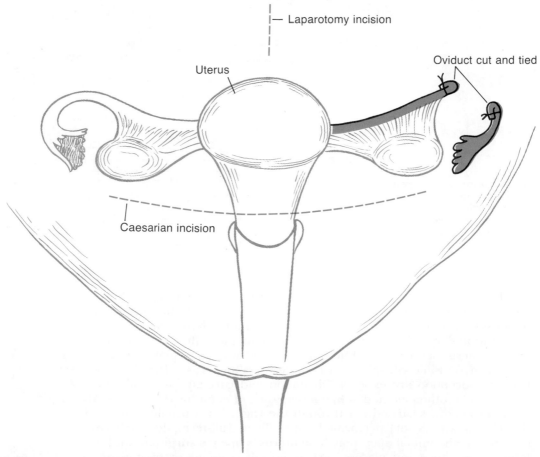

A

FIGURE 22a.9   Sterilization operations.
A. Female: Tubal ligation is carried out through an abdominal incision while patient is under general anesthesia.

Illustration continued on opposite page.

STERILIZATION

The most effective form of birth control, after total abstinence and abortion, is sterilization. It is a subject fraught with social and cultural overtones. In Harlem and other crowded urban areas in the United States, married women with large families are often refused sterilization in clinics, even when they plead for it. Some must have *seven* children before obtaining the operation. On the other hand, there are a few people who strongly advocate laws compelling involuntary sterilization of unwed mothers. It would seem more humane to sterilize those who request it, instead, while making certain first that the recipient gives her written consent to the operation and realizes that, to all intents and purposes, the operation is irreversible.

FEMALE STERILIZATION

Tying the oviducts, or **tubal ligation,** is the most common operation for sterilizing women and is most frequently done at the time of delivery of the last child (Fig. 22a.9A). It is generally carried out through the abdominal wall, but a recent new technique, known as **laparoscopy,** which involves a small opening through the umbilicus, is coming into use. Since hospitalization is minimal, this method is relatively inexpensive and also has the advantage of leaving no unsightly scar.

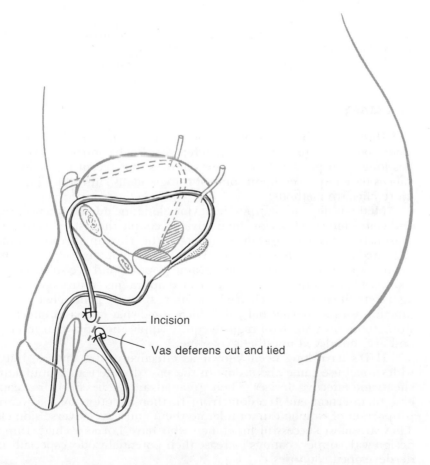

Incision

Vas deferens cut and tied

B

Continued
B. Male: Vasectomy is simply done in a brief operation with local anesthesia.

FIGURE 22a.9

Hysterectomy, or removal of the uterus, is sometimes used to effect sterilization, although it is usually performed because of a diseased condition. Likewise, ovariectomy (removal of the ovaries) is generally carried out only if they are unhealthy. Since the operation removes the natural sources of female hormones, it is not used primarily for sterilization.

MALE STERILIZATION

Except for castration, which is carried out only if the testes are severely diseased, the only practical method of sterilizing men is **vasectomy** or severance of the paired vasa deferentia (Fig. 22a.9B). Each duct is exposed through a small incision near the top of the scrotum, a short length is removed and the two ends are tied. Although sperm continue to be produced, they cannot pass across the cut ducts, and so they accumulate; they subsequently die and are phagocytosed by white blood cells. Note that the cut occurs *before* the junction of the prostate and seminal vesicles with the urethra; thus, during an orgasm, a male who has had a vasectomy ejaculates the same quantity of seminal fluid as before.

Occasional failures occur following both tubal ligations and vasectomies; the cut ends grow together, again permitting passage of gametes. Reversal of these sterilization operations cannot be counted upon; success is generally only 20 per cent. In the near future, reversible occlusion of the vasa deferentia by plastic plugs may become feasible. At the present, for persons who are absolutely certain they wish no more children, vasectomy and tubal ligation are extremely beneficial, for they permit totally normal sexual activity with unchanged potency but without the fear of pregnancy. It is likely that in the near future, as techniques for prolonged storage of human sperm are improved, a man will be able to store his sperm in a sperm bank before having a vasectomy. If the couple decides later to have a child, these sperm can be used for artificial insemination of the woman.

SUMMARY

If there is to be a humane solution to the population explosion it must come about by regulation of human fertility. Birth control in various forms has long been practiced in various cultures, but recent technological advances have improved upon some of these and also introduced newer and more effective methods.

Methods depending upon behavior alone include abstinence, rhythm and coitus interruptus. For the motivated couple the last of these can be a reasonably effective procedure, whereas the rhythm method cannot be regarded as reliable in most cases. With the development of rubber manufacture a century ago, condoms became popular and are still widely used today with moderate success. Cervical diaphragms, until about a decade ago, were the method of choice by women; they are generally used in conjunction with a spermicidal jelly which somewhat increases effectiveness. Both condoms and diaphragms require application prior to intercourse and are considered unesthetic by some.

IUD's have long been known as effective contraceptives, but their widespread use came about only in the past decade with manufacture of cheap and effective devices. Their great advantage lies in their cheapness, ease of insertion and freedom from further attention by the wearer. A proportion of women cannot tolerate them and have adverse side effects. They are most successful in women who have borne a child. Improved design with copper coatings increase their potential value, especially in underdeveloped countries.

Hormonal contraceptives which block the pituitary-ovarian pathway are highly effective in decreasing pregnancy rates when used consistently. As newer combinations of sex hormones are tried, many unwanted side effects of the Pill disappear. Other hormonal contraceptives are likely to be available in a few years which will afford protection without continuously blocking normal hormonal systems.

Safe abortion is now available in many areas at hospital-associated clinics. It is best carried out during the first three months of pregnancy and may require one day or less in a hospital. Past experience has shown that, where legal abortions are readily available, women tend to use this method only as a back-up measure when contraceptives have failed.

Sterilization is a highly effective means of birth control that should be considered irreversible by those who undertake it. Vasectomy in the male is a cheaper and less painful operation than tubal ligation in the female.

READINGS AND REFERENCES

Cuadros, A. and Hirsch, J. G. 1972. Copper on intrauterine devices stimulates leukocyte exudation. Science *175*:175–176. Data suggesting that IUD's work by increasing release of white blood cells into uterine cavity.

Family Planning Perspective *2*(3):2–3 June 1970. How IUD's prevent pregnancy in humans. An editorial discusses theories of this contraceptive method.

Family Planning Perspective *3*(2):62–63. April 1971. IUD's: T's, springs; how they protect. Further theories on action of IUD's.

Hardin, G. (ed.) 1964. *Population, Evolution, and Birth Control: A Collage of Controversial Readings.* W. H. Freeman, San Francisco. An interesting collection of attitudes and opinions in the last 150 years.

Peel, J. and Potts, M. 1969. *Textbook of Contraceptive Practice.* Cambridge University Press, New York. A comprehensive, accurate source, completely understandable to the lay reader. Highly recommended.

Petersen, W. 1969. *Population.* 2nd edition. Macmillan, New York. A sociological approach to human populations.

Rubin, A. (ed.) 1969. *Family Planning Today.* F. A. Davis Co., Philadelphia. A readable paperback.

Ryder, N. B. 1959. Fertility. Chapter 18, pp. 400–436, *in* Hauser, P. M. and Duncan, O. D. (eds.) *The Study of Population: An Inventory and Appraisal.* University of Chicago Press, Chicago. Discusses, among many other things, historical and social aspects of fertility regulation.

Scrimshaw, S. C. and Pasquariella, B. 1970. Variables associated with the demand for female sterilization in Spanish Harlem. Adv. Planned Parenthood 6:133–141. Describes the huge gap between demand and availability of sterilization.

Zipper, J. A., Tatum, H. J. *et al.* 1971. Contraception through the use of intrauterine metals. *I.* Copper as an adjunct to the 'T' device. American Journal of Obstetrics and Gynecology *109*:771–774. Observations on effectiveness of copper in contraceptive action of IUD's.

U.S. Forest Service photo

23

THE BIOLOGICAL FUTURE OF MAN

Man's great capacity for non-genetic, behavioral adaptiveness has placed him in his unique position in the biosphere. However, if he fails to make use of this adaptiveness to respond to the new environmental conditions he himself has created, then his own survival is in jeopardy.

Man, that very human animal, can never escape his biological nature nor his dependence on the biosphere he inhabits. If he survives the immediate future he will, as have all other organisms, go on evolving, both genetically and culturally. At the end of this chapter we shall speak of some of the potentials he holds for controlling his future evolution. But first let us summarize the biological problems that lie just ahead.

THE IMMEDIATE FUTURE: SURVIVAL

Like every other species of organism, past or present, man is a product of the forces of evolution—but man is uniquely different. For evolution has blessed him with the means for shaping his own further evolution and that of his surroundings. In the past, other species have become extinct because they were unable to adapt genetically to changing circumstances. To a great degree, man has overcome the need for genetic adaptation. Through innovation and experimentation he has the potential to adapt his behavior and modify his surroundings to meet the effects of change within but one or two generations.

As we have seen repeatedly in this text, however, there is great danger to man as a species in his enormous power over his environment. In human history, those who have exercised their power within society to their own benefit without regard for the health of the society from which they derived that power have eventually, by destroying the cultural integrity of the society, destroyed themselves. Just so, mankind as a part of the greater society of life on Earth is now endangering his fellow creatures, on whom directly or indirectly he depends. The real question is whether man *will* take advantage of his adaptability to alter his long-standing cultural attitudes and behavioral patterns which are no longer appropriate to the new circumstances his technology has created.

Consider, for example, human warfare. Whether war is, as some suggest, an inheritance from an aggressive ancestor or whether it is a *de novo* product of human history, it has long played a role in shaping human cultural evolution. But innovation, the product of human intelligence, has now put mankind in the ridiculous position of having created self-annihilistic means of killing. A sling-shot, a cannon or even a conventional bomb seldom boomeranged; but today's sophisticated and expensive nuclear weapons are as much a threat to the user as they are to the target. Whatever selective advantage, if any, warlike behavior may have had during man's evolutionary history, it certainly no longer serves such a function. The continued utilization of outmoded power politics by nation-states poses the greatest immediate threat to the survival of mankind. Adaptive changes in political behavior are of the utmost urgency. The following words were spoken by U Thant in 1969:

I do not wish to seem overdramatic, but I can only conclude from the information that is available to me as Secretary-General, that the Members of the United Nations have perhaps ten years left in which to subordinate their ancient quarrels and launch a global partnership to curb the arms race, to improve the human environment, to defuse the population explosion and to supply the required momentum to development efforts. If such a global partnership is not forged within the next decade, then I very much fear that the problems I have mentioned will have reached such staggering proportions that they will be beyond our capacity to control.

In a parallel way, many of man's other recent innovations are becoming more of a threat to his existence than a benefit to it. In attempting to fight and conquer his environment, his own physical suffering and his mortality, man has begun to create a world incapable of supporting him. Just like those fossil creatures whose genetic over-specialization caused their extinction, so now is man over-specializing in those technologies which serve him only at the moment. But, unlike the extinct dinosaurs, for example, who were felled not by their own effects on their environment but by forces beyond themselves, man's own technology is what is destroying his support system. It is for this reason that man, unlike the dinosaurs, does not *have* to become extinct—he does not *have* to destroy his environment. The causes of man-made environmental destruction are not genetic, and they do not have to be weeded out by the slow but irresistible forces of natural selection. If man continues to be behaviorally unadaptable, however, then just as surely as the genetically unadaptable dinosaurs died out, so will man. If man is to have a future he must learn to behave in accordance with biological laws. Of these, two are of primary importance.

BIOLOGICAL LAWS OF SURVIVAL

The First Law: A population cannot exceed the carrying capacity of its environment; N shall not exceed K (see Chapter 19). As man has modified his environment to increase its carrying capacity, K, so has his population size, N, increased to equal K. We are now in the third such upsurge of human population in response to a man-made increase in carrying capacity (refer to Figure 19a.1). But this time, owing to medical technology and lowered death rates, N threatens to overshoot K and so cause a population crash through famine, disease or nuclear war. Moreover, the global carrying capacity for man today is far more precarious than ever before, since it depends to a large degree on man-made ecosystems. Unless the technology for control of his own reproductive potential becomes as widely accepted by man as is the technology for control of his environment, the future of the human species is indeed bleak.

The Second Law: The productivity of an ecosystem is limited by the available energy. To increase the productivity and hence the carrying capacity of his environment, man has added energy subsidies—first in the form of his own labor, but more recently in the form of artificial fertilizers and other chemicals, widespread irrigation and the use of mechanical power. The genetically selected, productive strains of crops and cattle are, as we have seen, especially dependent on such energy subsidies, and most domesticated species could not survive in the world without them. Further genetic selection for increased productivity can only increase this dependence. Man now lives in a mutually obligatory symbiosis with his domesticated species (Chapter 21a).

On the whole, the large energy subsidies already pumped into agriculture have not been large enough to prevent environmental deterioration—man is harvesting too much energy, and not leaving enough behind for natural recycling (Chapters 17, 17a and 18). In many parts of the world, man's attempts to remove more from an ecosystem than he puts into it have resulted in total deterioration—the formation of deserts, lateritic soils, salt-poisoned fields, irreversibly eutrophied lakes and dust bowls. But the increased production of biological needs—adequate food, clothing and shelter—is only one cause of environmental deterioration. The rate of production of non-essential items is increasing even faster in some parts of the world. The environmental stress created by these new demands of man is actually *decreasing* the carrying capacity of his environment. A brief example may suffice to illustrate this:

Imagine an island with a limited amount of fertile land, populated by an agrarian people. The population grows to the point where the land just supplies food, shelter and clothing for all. Then someone discovers iron ore under the soil. The islanders, having heard of the advantages of washing machines, decide to provide one for each household. To smelt the ore and mold it into washing machines requires energy. The islanders therefore must grow trees on their cropland to provide fuel for smelting; productive land is lost and some of the population dies (K is reduced, and so N must also decrease). Once built, the washing machines require energy to run, and so the trees continue to be grown for fuel, and the island population remains diminished.

Although oversimplified, our island story is instructive. Both food production and affluence require energy, and if the total available energy—from the sun, from burning wood and fossil fuel and from nuclear power—remains constant, then an *increase* in affluence must mean a *decrease* in carrying capacity of the environment and hence a decrease in maximum population size.

THE LIMITS OF ENERGY USE

This brings us to the all-important question of how much extra energy man can safely introduce into the environment; how much can he increase his affluence without decreasing his population? As we saw in Chapter 7a, there is great disparity between the most optimistic and pessimistic scientific estimates as to the total amount of extra heat man can add to the Earth's heat budget before he significantly affects global climate. Let us take the most optimistic estimate,* which is that in a world of 20 billion,

---

*Weinberg, A. M. and Hammond, R. P. 1970. Limits to the use of energy. American Scientist *58*:412–418.

per capita energy production could rise to about twice that now consumed in the United States without seriously affecting the world's heat budget. As we have repeatedly seen, however, much of our environmental deterioration is the result of economic short cuts—of dumping wastes into the environment. Our present affluence is thus dependent on a continuing increase in environmental entropy—in other words, on pollution! Even if the *per capita* energy utilization in the United States were doubled, much of the increase would need to go toward cleaning up and restoring the environment in order for it to remain a habitat capable of supporting man. Thus, even using the most optimistic estimates for world energy production and assuming that world population does level off at around 20 billion people, the average world citizen would have a level of consumption not very much different from that of the *average* American today. It will be noted that such calculations take no account of the *local* effects of energy production, and especially of local thermal pollution; nor of the visual impact or safety of the giant nuclear power plants that would be needed; nor of the desirability of living in a world with six times more people than today.

Aside from its perhaps too optimistic estimate of the total extra heat man can safely add to the Earth's heat budget, however, this rosy projection assumes man will have the good sense to begin channeling some of his energy resources into reversing environmental deterioration. Even the production of energy itself will become less useful to man as an ever larger proportion of that energy is spent in dissipating waste heat generated in its own production. A rather less optimistic stable population size and level of affluence are projected by the authors of *The Limits to Growth*.* Using available data and computer programs to predict the outcome of interactions between population size, consumption of energy and resources, and environmental deterioration, this team from the Massachusetts Institute of Technology concluded that collapse can only be avoided with a world population but little larger than at present and with an average *per capita* consumption level about half that in the United States today.

One reason for the great discrepancy in these two projections is our lack of fundamental data about the environment. We just do not know (1) how much extra energy will be required to maintain it even in a barely stable condition and (2) whether this minimal support system will provide the quality of life necessary for a humane existence. Increased scientific information is needed urgently, not for further technological exploitation but for allowing us to make more accurate predictions about future dangers and how to avoid them (see Chapter 1a).

BRINGING ABOUT CHANGE

Although this is a book about biology, the subjects of sociology and political science are closely related to it in today's world. Therefore, a few words about how change in our cultural and behavioral patterns may occur are appropriate (see Chapters 1 and 1a, and the references listed after them).

The magnitude of the changes required is so great as to seem revolutionary—and therefore terrifying to some. "Revolution" today tends to connote armed political upheaval, yet we speak with pride of the Industrial Revolution, which has brought more permanent changes in the lives of men than any political revolution ever accomplished. What is now needed is a revolution *away* from needless consumption for its own sake and *toward* a new ethic for man as a partner with his environment.

---

*Meadows, D. H., Meadows, D. L., Randers, J. and Behrens, W. W., III. 1972. *The Limits to Growth*. Universe Books, Potomac Associates, New York.

The Role of the Individual. Although what each individual consumes may seem trivial, it may be instructive to consider one simple item you use—for example, plastic sandwich bags, cigarettes or deodorant spray—and list every conceivable way it and its container have affected and will affect the environment. Include raw materials, each step in its manufacture (don't forget the energy and resources for building and tooling the factory in the first place), advertising, the means by which the item was transported and sold to you, how you use it and, finally, what happens to it when you discard it. Then estimate how often you use this item, and multiply this figure by the number of others you think also use it. After that, you might begin a list of everything else you use and imagine the same computations for each item. The real cause of environmental deterioration soon becomes apparent! Such an exercise allows one to begin to put an environmental price tag on each item—which is far higher and far more meaningful than the price tag at the store.

If enough individuals become convinced of their own contributions to environmental deterioration, then an impact can be made through collective action—refusing to purchase unnecessary items, demanding alternative methods of transportation, eating foods lower on the food chain and so forth. This will come about, however, only through education. Perhaps only when local pollution becomes widely unacceptable will enough people be willing to understand the problem and accept their share of the responsibility.

The Role of Government. Democratic governments can bring about only those changes for which there is general support; they are responsive to political power. At the moment, the overwhelming demand of voters is for increased consumption; demands for a clean environment lag far behind. This has a subtle effect on public opinion, for the public tends to assume that its government will automatically solve environmental problems without any serious public pressure.

That this is not the case has already been demonstrated several times in our text. Limits set on environmental contaminants are seldom met; polluting industries are regularly granted "variances" (permission to exceed legal limits) by pollution control agencies. The asked-for advice of scientists goes unheeded when government policy is made. Environmental impact studies, submitted to the appropriate department in Washington by a public or private agency seeking federal funds for a project, seldom seriously consider alternative projects as they are required to do by law; as a result, the projects are often automatically given rubber-stamp approval. Monitoring agencies such as the Environmental Protection Agency and the Food and Drug Administration are badly underfunded and cannot possibly carry out their assigned tasks effectively.

To sum up, current government controls have a very minor impact on the problem of environmental deterioration. An elected government, no matter how enlightened, cannot carry out reforms for which there is insufficient popular support.

The Role of Technology. Many hope that somehow "improved" technology will allow a continued expansion of consumption while at the same time curing environmental problems. While it is true that such things as cleaner energy production and less-polluting cars will somewhat ease the present environmental impact of technology, these kinds of improvements will be minor compared to the total problem. The physical Laws of Thermodynamics (see Chapters 7 and 7a) can never be circumvented by any technology—either waste heat or waste materials are bound to be byproducts, and to clean up waste materials will result in more waste heat. To count on technological innovation for a solution is to invite disaster.

If mankind survives his present crises and arrives at a stable relationship with nature there will then open up great opportunities for his further cultural and even genetic evolution. As the English philosopher and economist John Stuart Mill wrote in 1857:

It is scarcely necessary to remark that a stationary condition of capital and population implies no stationary state of human improvement. There would be as much scope as ever for all kinds of mental culture, and moral and social progress; as much room for improving the Art of Living and much more likelihood of its being improved.

In the realm of biology, the most important immediate advances are likely to find application in improved medical care. In particular, we can expect improvements in the prevention and cure of diseases associated with aging, such as arthritis, arteriosclerosis and cancer. As we learn more about the functioning of the human brain new drugs may be developed for treating psychoses such as schizophrenia. The recent remarkable advances in the treatment of Parkinson's disease with L-DOPA point the way. We may look for improved treatments and control of infectious diseases, especially such age-old scourges of mankind as malaria and schistosomiasis. As noted in Chapter 22a, new, more effective chemicals for birth control are likely to become available soon. We can expect improved success in the field of organ transplants and in the treatment of autoimmune diseases as we increase our understanding of the way antibodies are synthesized and function. We can expect that the underlying biochemical causes of more and more genetic defects will be discovered and that treatments for at least some will be found. New means of screening potential carriers of various genetic diseases will be developed (see Chapter 15a). It is possible that one day we may learn how to induce the regeneration of lost limbs in humans, as occurs naturally in several species of amphibians.

As we learn more of the effects of prenatal and early postnatal experience in infants it may be possible to greatly increase the developmental potential of young children. The great importance of regular movement and physical contact to the development of normal behavior in young primates and humans has recently been uncovered (see Chapter 13). We have yet to fully assess the importance of environmental noise, crowding and lack of esthetic input on both the physiology and behavior of man, although studies in animals suggest they are of major importance.

Eventually biological research will make possible genetic selection of certain traits among offspring. Already, by amniocentesis (Chapter 15a), the sex of a fetus can be determined early in pregnancy, and the potential for abortion exists if the parents prefer a child of the opposite sex. Techniques may soon become available for separation of the X- and Y-bearing sperm in a man's ejaculate, thus allowing for artificial insemination of the woman with sperm containing the desired sex chromosomes. As sperm banks are developed (Chapter 22a) it will be possible to keep sperm of men with certain genetic traits for decades and use them for insemination at a later date. A couple desiring a child with particular traits could then choose the genetic father of the child, the other father becoming an adoptive parent. Eventually the sort of techniques now used for the transplantation of diploid adult nuclei into enucleated frog eggs (Chapter 14) may become feasible for human eggs. The unfertilized egg would have its nucleus removed and that of a selected person transplanted into it; then the egg would be implanted in the uterus of a woman, where it would undergo normal development.

Such potential for human selection of certain genetic traits is likely to be a reality in a few generations. Whether or not society will attempt to use this potential—and who will decide the great moral issues involved—are problems to be faced by man in the future. As with all man's knowledge, the power to control his own genetic destiny and guide his own evolution is neither good nor bad in itself. His decisions must be based on value judgments. Let us hope that when the time comes to make these decisions, man will have also acquired the necessary wisdom—the extremes lie between a highly efficient social machine, much like an ant colony, with little potential for further change, and a genetically, culturally and socially variable mixture with an infinite potential for forming new combinations and an ever changing future.

**READINGS
AND
REFERENCES**

(See also those listed after preceding chapters, especially Chapters 1 and 1a)

Darling, F. F. and Milton, J. P. (eds.) 1966. *Future Environments of North America.* The Natural History Press, Garden City, New York. An outstanding collection of papers dealing with conservation, social purpose, economics, regional planning and implementation of policies.

Handler, P. H. (ed.) 1970. *Biology and the Future of Man.* Oxford University Press, New York. A comprehensive survey of the present state of biology and its implications for the future, prepared by the Committee on Science and Public Policy of the National Academy of Sciences. This book is intended for the well-read layman, and should form part of the library of everyone concerned with the future of mankind.

Lappé, F. M. 1971. *Diet for a Small Planet.* Ballantine Books, Inc., New York. Recipes for obtaining a high protein diet while eating off the bottom of the food chain. See also other books in this series for practical suggestions to individuals.

Mumford, L. 1970. *The Myth of the Machine. The Pentagon of Power.* Harcourt Brace Jovanovich, Inc., New York. In this second volume of his study of modern technology, this foremost social philosopher shows how overdependence on the machine has led to loss of human individuality and creation of machine-oriented centers of power.

SOME BASIC CHEMISTRY

THE NATURE OF ATOMS

All matter is composed of minute particles known as **atoms.** More than 100 different types of atoms are known; these are the **elements**, such as oxygen (O), nitrogen (N) and so on. Basically, all atoms have a small, very heavy central region, the **nucleus.** Within the nucleus two types of extremely heavy particles are packed; **neutrons,** which have no charge, and **protons,** which have a positive (+) electrical charge. They are very nearly the same size. The neutrons provide mass which helps to keep the nucleus together, since mass at close distances produces highly attractive forces. Also, the neutrons prevent the like charges on the protons from repelling one another and causing the nucleus to fly apart. (Remember that opposite charges attract and like charges repel one another.)

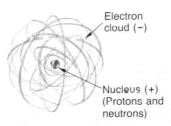

Electron cloud (−)

Nucleus (+)
(Protons and neutrons)

Whirling about the nucleus is a thin cloud of extremely small, very light **electrons,** each of which has a negative (−) electrical charge. In its elemental or uncombined state an atom has exactly equal numbers of protons and electrons and hence is electrically neutral. Each element has its own characteristic number of + and − charges. For example, H (hydrogen) has 1+ and 1− (1 proton and 1 electron); O (oxygen) has 8+ and 8− (8 protons and 8 electrons). This number is called the **atomic number** of an element and determines its chemical properties.

In addition to its atomic number, each element also has a characteristic weight. The units of weight are those of the heavy nuclear particles, since the light electrons contribute only negligibly. Thus the **atomic weight** is the sum of the number of neutrons and protons in the nucleus. In most of the lighter elements—those found in living organisms—the number of protons and neutrons are equal, and so the atomic weight is twice the atomic number; thus oxygen has an atomic weight of 16. An exception is hydrogen which has but one proton in its nucleus and no neutrons. The atomic weight and atomic number of this lightest element are both 1.

Variations occur in the nuclei of some elements, usually owing to the presence of an excess number of neutrons. A small proportion of all hydrogen atoms do, in fact, have a neutron in the nucleus and have an atomic weight of 2. This is heavy hydrogen, or **deuterium,** an **isotope** of hydrogen. Sometimes the imbalance of protons to neutrons causes the nucleus to be unstable, and it may spontaneously fragment, releasing particles and energy—a phenomenon known as **radioactivity.** Radioactivity is characteristic of all of the heavy elements, those with atomic numbers greater than 82, since they all have more neutrons than protons in their nuclei. An isotope of hydrogen containing two neutrons and with an atomic weight of 3 can be prepared artificially. This isotope, **tritium,** is radioactive. Another frequently used radioactive element is carbon 14 (^{14}C) which has two more neutrons than does the common stable isotope, carbon 12 (^{12}C).

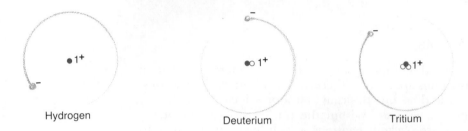

Hydrogen   Deuterium   Tritium

The electrons do not fly randomly around the nucleus, but tend to follow set paths which are determined by strict rules. The electrons have certain energy levels known as **orbitals,** and each orbital can be shared by only two electrons. Orbitals are arranged spatially in shells about the nucleus. The inner shell has but one orbital and thus can hold but two electrons; the next shell has four orbitals (not distinguished in the drawing) and so can hold eight electrons; the third shell has nine orbitals, and can hold 18 electrons. The very heaviest elements have seven shells made up of 52 different orbitals. In general, inner shells (lower energy levels) must be filled before electrons can take up positions in the outer shells (higher energy levels). Carbon and oxygen are examples of atoms found in living tissue.

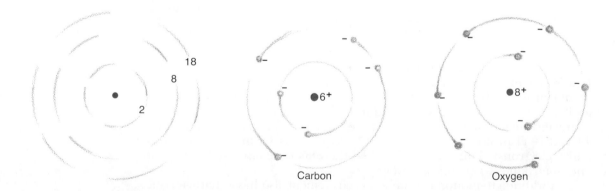

Carbon   Oxygen

CHEMICAL BONDS

It is a fact of nature that atoms "prefer" to have complete outer shells, and this is the basis of the chemical bonds formed between atoms to form compounds or molecules. Hydrogen, for example, would "like" to have one more electron, or else lose its electron and have none. Oxygen would "like" to gain two electrons and make a complete outer shell of eight.

Let us consider the example of H (hydrogen) and F (fluorine). All together, fluorine has nine electrons in two shells—two electrons in the inner shell and seven in the outer shell. Since the outer shell of fluorine has only seven electrons, it needs one more electron to make a complete set of eight. As just noted, hydrogen can either accept or give up an electron, making the number of electrons in its outer (and only) shell zero or two. When H and F combine they form a molecule, HF, or hydrogen fluoride, which is a gas. In this combined form, most of the time the electron of hydrogen is in orbit in fluorine's outer shell, and both H and F are "satisfied"—H has no electrons, and F has eight in its outer shell. Occasionally, however, a pair of electrons return to circle in the hydrogen shell, making a full shell of two, but temporarily leaving fluorine "unsatisfied." In the case of HF, then, the electrons are not shared equally between H and F—fluorine has them most of the time. *This sharing of electrons between two atoms is the basis of a chemical bond.*

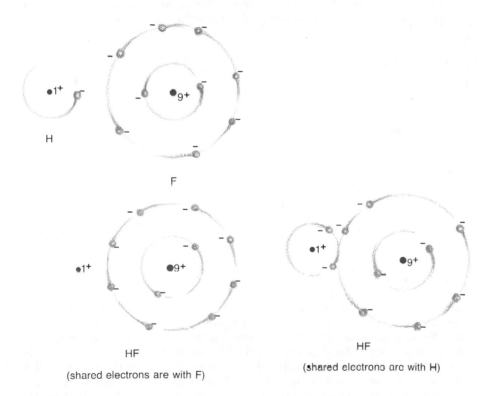

H

F

HF

(shared electrons are with F)

HF

(shared electrons are with H)

IONIC BONDS

There are several different types of chemical bonds whose nature is determined by the way in which the electrons are shared between the two atoms. In hydrogen fluoride gas, the electrons are mostly pulled to fluorine

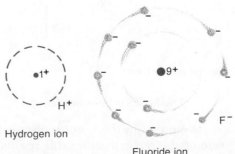

Hydrogen ion

Fluoride ion

because of the much greater + charge in its nucleus compared to that of hydrogen. (The reader will recall that opposite charges attract.) When the degree of electron sharing in a molecule is as unequal as it is in HF, an interesting thing happens when the molecule is dissolved in water; the two atoms separate, not as atoms, but as charged *ions.* Hydrogen loses its electron permanently to fluorine and becomes a positively charged hydrogen ion (H^+); fluorine, with one extra electron (nine protons in the nucleus, but 10 electrons) becomes negatively charged fluoride ion (F^-). This type of chemical bond, in which the electrons are so unequally shared that ions are formed when the molecule dissolves in water, is called an *ionic bond.*

POLAR BONDS

A second type of bond exists when two atoms do not share the electrons an equal proportion of the time but the inequality is not great enough to lead to ion formation. This is called a *polar bond,* an excellent example of which is found in water (H_2O). Oxygen has only six electrons in its

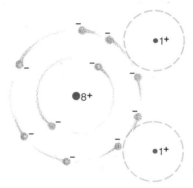

Water molecule (H_2O)

outer shell and therefore needs two electrons to fill that shell. Even though oxygen has an 8+ charge in its nucleus (almost as large as that of flourine's 9+), since it must attract *two* electrons to fill its outer shell (rather than just the one required by fluorine), its ability to hold each of those electrons tightly is much less than was the case for fluorine. The force of attraction of the 8+ charge must be divided between two electrons, one from each hydrogen atom. There is still a considerable tendency, however, for the shared electrons in water to spend most of their time in the outer shell of the oxygen atom rather than in the shells of the hydrogen atoms. Thus, although water molecules seldom form ions, more than half of the time the shared electrons are in orbit in the oxygen atom.

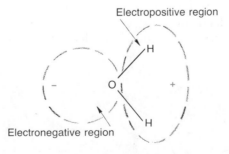

Polarity of a water molecule

Note that in the water molecule the hydrogen atoms are both on one side of the oxygen atom; this results in positively and negatively charged regions in the molecule. (If the hydrogen atoms were on opposite sides of the oxygen, there would be no regions of net + and net − charge, since the effects of the two hydrogen atoms would cancel one another.) Water is called a **polar molecule**. In most molecules containing oxygen, the oxygen atom attracts electrons from other atoms and is therefore **electronegative**. Hydrogen usually loses electrons and is **electropositive**.

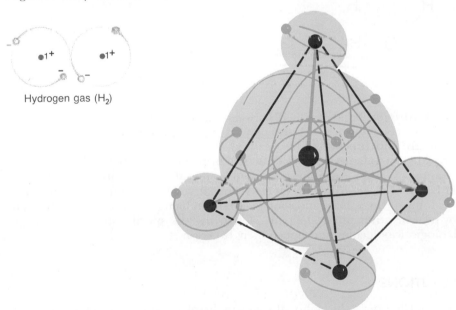

+ H —— O —— H +

Water is not like this.

COVALENT BONDS

A third type of chemical bond is one in which electrons are equally shared between the two atoms. These are called **covalent bonds.** Hydrogen gas (H_2) is one example; another is the carbon compound, methane (CH_4) shown in three dimensions. Note that the four bonds formed by carbon are directed to the four corners of a pyramid or tetrahedron. Both H_2 and CH_4 are non-polar molecules and have no electropositive or electronegative regions. They never ionize.

Hydrogen gas (H_2)

Methane (CH_4)

Carbon (C) is the major atom in all biological compounds. Since it has only four electrons in its outer shell, it could either give up all four electrons or take on four more to satisfy the condition of a full outer shell. In fact, it seems unable to "make up its mind" and does neither of these things. Instead it shares its four electrons with four from other atoms, making four covalent bonds in most molecules in which it occurs. Covalent bonds, such as those between two carbon atoms (C—C) or a carbon and a hydrogen atom (C—H), are always non-polar and electrically uncharged. When carbon (C) combines with electronegative atoms, such as O (oxygen), S (sulfur), N (nitrogen) and P (phosphorus), the latter tend to pull electrons away from carbon to a small extent, thus creating polar regions in biological molecues.

HYDROGEN BONDS

The major molecule in living systems is water. Because of their polar nature, water molecules interact with one another. We know very little about the exact nature of these interactions, but they are thought to be due to the formation of weak attractions between the electropositive H atoms and electronegative O atoms; the water molecules are oriented in a non-random fashion. These weak interactions are known as **hydrogen bonds,** or H-bonds.

Hydrogen bonds formed between various parts of a macromolecule such as an enzyme or DNA also play an important role in determining their structure and function. The weak bonds mentioned in Chapter 3, which provide the coiled secondary structure of proteins and which hold the two strands of DNA together in a double-helical configuration, are in fact H-bonds. These form between electronegative O atoms and electropositive H atoms of these large molecules. H-bonds, being weak, are easily broken by heat or other physical agents, destroying the specific structure of large biological molecules and hence their function.

SOLUTIONS

In Chapter 3 the ability of water to dissolve molecules containing ionic or polar bonds was explained in terms of the polar nature of water. Because fats contain very few polar regions, their solubility in water is quite small.

A solution is often characterized by the concentration of dissolved molecules it contains. There are many ways to express concentration; in describing blood constituents, for example, it is usual to speak of so many grams of a substance per liter of blood or, more commonly, milligrams per 100 milliliters. For instance, the concentration of human blood sugar normally ranges from 70 to 100 mg/100 ml blood.

MOLAR SOLUTIONS

A more useful expression, however, is one based on the number of molecules dissolved in a known volume, since in writing chemical reactions we are concerned with interactions between molecules rather than between weights of two or more substances. But molecules are awkward to deal with—a tiny drop of blood contains literally hundreds of trillions (10^{14}) of sugar molecules. Therefore we use the term **mole**. A mole of any substance contains 6×10^{23} molecules, and its weight, in grams, is the same as the **molecular weight** of the substance. Molecular weight is itself the sum of the atomic weights of all the atoms in a molecule. H_2O has a molecular weight of 18: $1 + 1 + 16$; glucose, $C_6H_{12}O_6$, has a molecular weight of 180: $(6 \times 12) + (12 \times 1) + (6 \times 16)$.

A molar solution, then, contains one mole of solute in one liter of total solution. A one molar glucose solution thus contains 180 grams of glucose in each liter of solution. This is written as 1M glucose. A 0.1M glucose solution contains 18 g/1; a 10^{-2}M glucose solution contains 1.8 g/1 and a 2M glucose solution contains 360 g/1.

EQUIVALENTS

In solutions of ions it is often useful to know not the number of moles but the number of + and − charges that are present, since charge plays an important part in the properties of ionic solutions. Electrical charges are measured in **equivalents,** and their concentration is denoted by the term equivalents per liter (eq/1). For molecules such as NaCl, which in solution forms but one + and one − charge for each molecule dissolved, the number of equivalents per liter is the same as the number of moles per liter. One mole of NaCl dissolved in a liter of solution results in 1 eq/1 of Na^+ and 1 eq/1 of Cl^-.

However, many salts exist in which one or both ions carry more than one charge—calcium chloride ($CaCl_2$) is an example. Each dissolved molecule of $CaCl_2$ yields one Ca^{++} ion and two Cl^- ions. Thus, a *one* molar solution of $CaCl_2$ contains *two* equivalents of both + and − charges. Magnesium sulfate ($MgSO_4$) forms Mg^{++} and $SO_4^=$ ions, and again, 1M $MgSO_4$ contains 2 eq/1 of positive and 2 eq/1 of negative ions.

ACIDS AND BASES

One of the most important uses of equivalents is in determining the neutralizing capacity of an **acid** or a **base**. An acid is any chemical which, on dissolving in water, yields H^+ as its positive ion. HF is thus an acid, since it produces H^+ and F^- ions. The acid HCl, secreted by the stomach, yields H^+ and Cl^-. Sulfuric acid, H_2SO_4, yields two H^+ ions when it dissociates; one mole of H_2SO_4 produces two equivalents of H^+. Many organic (carbon-containing) acids exist. Acetic acid, for example, has the molecular formula CH_3COOH and dissociates into H^+ and CH_3COO^- (acetate ion). A base, in contrast, is a chemical which yields hydroxyl or OH^- ions on dissolving in water. A common base is NaOH, which dissociates into Na^+ and OH^-.

Both H^+ and OH^- at high concentrations are strongly reactive. H^+, for example, is used to etch metals, and OH^- of caustic soda, sodium hydroxide (NaOH), is used to unplug grease-clogged drains. Both are damaging to living cells. Since many reactions in living cells produce acids, it is necessary for the body to neutralize these acids to prevent the buildup of too much H^+.

As noted previously, very few water molecules dissociate—about two in every billion—to form H^+ *and* OH^- ions. If we combine an acid and a base, then, there is a strong tendency for the H^+ and OH^- ions to recombine to form water; the acid and base neutralize one another.

$$H^+ + OH^- \longrightarrow H_2O$$

In estimating the ability of a base to neutralize an acid, therefore, we are interested not in the number of moles of each, but in the combining equivalents of H^+ and OH^- that are released into solution.

THE pH SCALE

The hydrogen ion concentration can thus be expressed in terms of equivalents of H^+ per liter of solution. A neutral solution, such as pure water, which has equal amounts of H^+ and OH^-, has 10^{-7} equivalents of H^+. This, however, tends to become awkward in use, and chemists have devised a simpler notation for expressing H^+ concentration, called the **pH**. By definition:

$$pH = \log \frac{1}{[H^+]}$$

where $[H^+] =$ hydrogen ion concentration. pH thus equals the logarithm of the reciprocal of the H^+ concentration.

This concept is most easily grasped by example. A solution containing 0.1 eq/1 of H^+ has a pH of 1 ($\log \frac{1}{0.1} = \log 10 = 1$); a solution containing 0.01 eq/1 of H^+ has a pH of 2 ($\log \frac{1}{0.01} = \log 100 = 2$) and so forth. Thus an *increase* of one pH unit signifies a tenfold *decrease* in hydrogen ion concentration.

A convenience achieved by the pH scale is that it tells us immediately the concentration of OH^-: the sum of pH and pOH (the log of the reciprocal of the OH^- concentration) is always 14; therefore, $pH = 14 - pOH$. Thus, a solution of pH 2 has a pOH of 12. Since pure water has an H^+ concentration of 10^{-7} eq/1, its pH is 7 and its pOH is 7; neither ion predominates, and water is therefore neutral. The pH of most cells lies between 6 and 8—they are not far from neutrality.

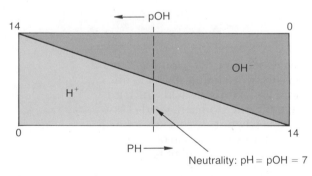

The reciprocal relationship between H^+ and OH^- concentrations, expressed in terms of pH and pOH

At this point we can extend our concept of a base as being not only a molecule which produces OH^- upon dissociation but any molecule capable of accepting or neutralizing hydrogen ions. Many proteins contain $-NH_2$ groups in the variable regions of their amino acids (see Chapter 3), and these groups are able to accept free hydrogen ions, forming a positively charged complex:

$$\boxed{\text{protein}} - NH_2 + H^+ \longrightarrow \boxed{\text{protein}} - NH_3^+$$

The hydrogen ion no longer exerts its acidic action. Such proteins are said to **buffer** the effects of excess H^+ produced during cellular reactions. The buffering action of plasma proteins is discussed in Chapter 10.

A REVIEW OF GRAPHS AND EXPONENTS

Many students find that they have forgotten some of the fundamentals of mathematics. Therefore, those concepts needed to understand this text are briefly reviewed here.

THE MEANING OF GRAPHS

A great deal of biological information is best expressed by means of graphs which give a visualization of trends that may be occurring.

FREQUENCY DISTRIBUTIONS

Among the things biologists are often interested in are **frequency distributions,** or the number of individuals in a population who possess a prop-

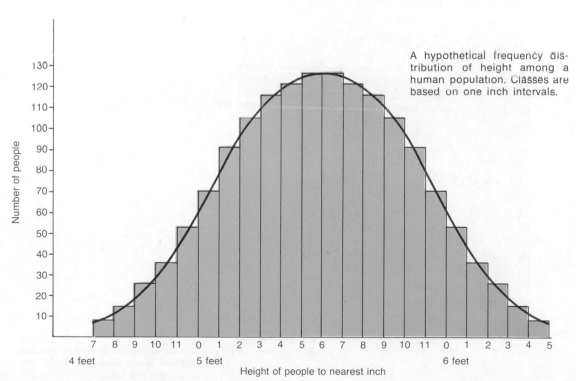

A hypothetical frequency distribution of height among a human population. Classes are based on one inch intervals.

Number of people

Height of people to nearest inch

4 feet   5 feet   6 feet

erty or character to a certain degree. The height of individual people is one such character (see Chapter 1a). If we had a sufficiently accurate ruler and could get all people to stand in exactly the same way while being measured, we would probably find that no two people have exactly the same height. But a list giving the height of every person, from the smallest to the largest, for example, would be useless: all it would tell us is what the mean and extreme heights of people are, but nothing of the way height is distributed in a population. In order to find this out, we must group people into categories, or classes — all those between, say 5'7" and 5'8" are put in one class, all those between 5'8" and 5'9" in another and so forth. By plotting the height classes against the number of people in each class we obtain a frequency distribution. Several of the curves shown in Chapter 1a are in fact derived from such frequency distributions. The student should reread the comments made there about the accuracy with which a curve is drawn to connect the actual data points.

POSITIVE AND NEGATIVE CORRELATION

In addition to frequency distributions, which may have the symmetrical, bell-shaped curve shown on the preceding page but may also have other shapes, biologists often plot data in other ways. One of the commonest types of graphs is that which relates one **variable** to another, graphically expressing cause and effect relationships.

In order to determine if two variables are related to one another, a biologist will select set numbers for one variable and measure the corresponding values for another variable. The factor chosen by the experimenter is called the **independent variable,** since its value is predetermined. The factor which is being measured in relation to the independent variable is called the **dependent variable.** An example is shown in the relationship between the age and reproductive capacity of the water flea, *Daphnia.*

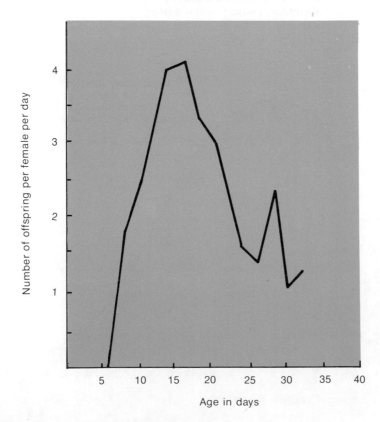

Relationship between age of adult female water fleas *(Daphnia)* and reproductive capacity.

In this case, we have arbitrarily selected age classes and estimated how reproductive capacity is related to age. Age class is the independent variable, since it has been pre-selected for study. Reproductive capacity is the dependent variable, since it is the factor that we are investigating *in relation to age*.

In the graph of age versus eggs produced per female, we can recognize that, up to 16 days of age, a female water flea increases her reproductive capacity with age. We can thus say that there is a **positive correlation** (upward-moving line) between age and reproductive capacity, up to a certain optimal age. But adult females older than 16 days of age produce fewer eggs than their slightly younger sisters; their reproductive capacity declines with age. For older female *Daphnia*, there is a **negative correlation** between age and reproductive capacity. If a horizontal line had been obtained, then there would have been no effect of age on reproductive capacity (female *Daphnia* of all ages would produce the same number of eggs), and we would conclude that the two variables were unrelated or, as biologists say, not correlated. These three possible types of relationships between two variables are shown on the following graph.

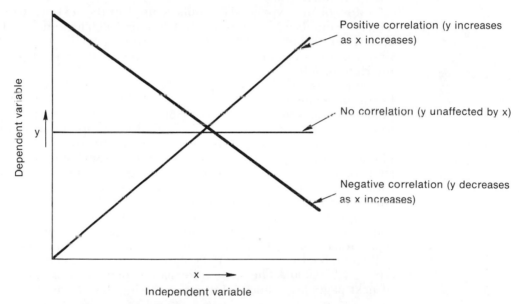

Three types of relationships between an independent variable, *x*, and a dependent variable, *y*.

Although it is usual to plot the independent variable on the horizontal axis and the dependent variable on the vertical axis, this is not always the case (see, for example, Figure 17a.6).

EXPONENTS

In biology, many numbers one meets are either very large or very small. It is possible, of course, to express these by the appropriate number of zeros, written before or after the decimal point ($1,000,000,000 = 1$ billion; $0.000000001 = 1$ billionth), but it is more convenient to use exponents.

A positive exponent is a direction to the reader to multiply the base number by itself a given number of times. Thus $2^4 = 2 \times 2 \times 2 \times 2 = 16$; the base number here is 2, the exponent is 4. A negative exponent says that the *reciprocal* of the number indicated is asked for. Thus:

$$2^{-4} = \frac{1}{2^4} = \frac{1}{2 \times 2 \times 2 \times 2} = \frac{1}{16}$$

As long as the base number is the same, when two exponential numbers are multiplied the answer is obtained by adding the exponents. Thus:

$$2^3 \qquad \times \qquad 2^4 \qquad = 2^7 \qquad (=128)$$
$$(2 \times 2 \times 2) \times (2 \times 2 \times 2 \times 2) = 128$$

Conversely, if an exponential number is divided by another, the exponents are subtracted:

$$2^3 \div 2^4 \qquad = 2^{-1} \left(= \frac{1}{2}\right)$$

$$\frac{(2 \times 2 \times 2)}{(2 \times 2 \times 2 \times 2)} \qquad = \frac{8}{16} = \frac{1}{2}$$

Note that $2^1 = 2$; $2^{-1} = \frac{1}{2}$; and $2^0 = 1$. The number zero cannot be expressed in exponential form.

Most of the exponential numbers used in the text are based on 10. Thus the number 5×10^5 is $5 \times 100,000$, or $500,000$, and 5×10^{-5} is 5×0.00001, or 0.00005. The base 10 is thus a very convenient notation, since it simply involves moving the decimal point the requisite number of digits to the right or left.

Exponential numbers are particularly useful in biology since many biological phenomena occur in an exponential fashion. Plotting data on exponential scales also allows one to present information which otherwise would be crammed together at one end of the graph in order to provide space for data at the other end. For example, the growth of cells is often plotted exponentially. The figure below, taken from Figure 8a.7, shows the effect of 100 parts per billion DDT on the growth of the single-celled alga *Cyclotella nana* ($\triangle$ = control; $\square$ = DDT-treated). Note that, although the curves for treated and control cultures do not appear at first glance to differ greatly, on taking into account the exponential scale of cell concentration, one can see that they are widely divergent. After five days of treatment, for instance, the DDT cultures have almost *ten times* fewer cells per milliliter (about $1.1 \times 10^5 = 110,000$) than do the untreated control cultures ($10^6 = 1,000,000$). The same information plotted on a linear scale (top of next page) loses much information during days 1 and 2.

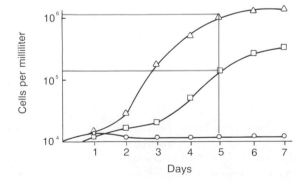

Effect of 100 ppb DDT on growth of the single-celled alga *Cyclotella nana*. Growth is estimated by counting the number of cells per milliliter of culture medium.

Another example of the use of exponentials is seen in Figure 17a.6A, which relates the yield of food harvested to amount of fertilizer applied. As countries use more fertilizer on their crops the yields go up, as one would expect. But the data are plotted on an exponential scale along both axes of

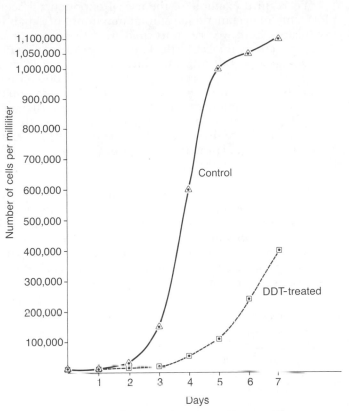

Effect of 100 ppb DDT on growth of *Cyclotella nana*, plotted on a linear scale.

the graph, which requires an understanding of exponents for interpretation. To estimate how much more fertilizer is required to double the yield, we select any point, say 1000 kg/hectare, as a starting point on the horizontal axis, and draw a vertical line to the curve (*a*). To produce this yield requires about 3.8 kg/hectare of fertilizer (see line *b* drawn to vertical axis). The increase in fertilizer required to double this yield (to achieve 2000 kg/hectare) is found by drawing a second vertical line (*c*) to the curve, and then projecting a second horizontal line (*d*) to the vertical axis, which reads about 20 kg/hectare of fertilizer. The increase of fertilizer needed to double the yield is thus $\frac{20}{3.8}$, or about 5 times.

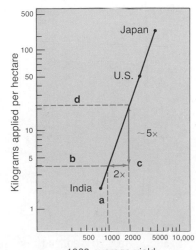

The effect of fertilizer on crop yields. Amount of fertilizer added is plotted on the vertical axis, crop yield on the horizontal axis.

Yet a third example of the use of exponents is seen in Figure 19.3, a plot of life expectancy against total maximum life span for three species of organisms. Here we are concerned with the mean or average life expectancy of an individual at birth. The average life expectancy, by definition, is the average age of death. If we start with 1000 newly born individuals, then the age at which 50 per cent, or 500, of these 1000 individuals are still alive will be shown by a horizontal line from the 500 mark on the scale. Note that half of the blacktail deer born have died before they reach 10 per cent of their life span, or at an age of about seven months. Half of the robins hatched die before they reach 20 per cent of the life span for robins, or by 17 months of age. Humans in developed countries, however, can expect at birth to live for 65 per cent of the total human life span of 110 years, or for 70 years.

By plotting the data on an exponential scale for survivors and on a percent scale for life span we can compare widely different populations of organisms without losing information. The same data plotted on a linear scale for survivors and an absolute time scale for life span would be extremely hard to comprehend, as shown in the accompanying diagram. It is thus apparent that much information is to be gained by transforming data onto scales which can be readily compared.

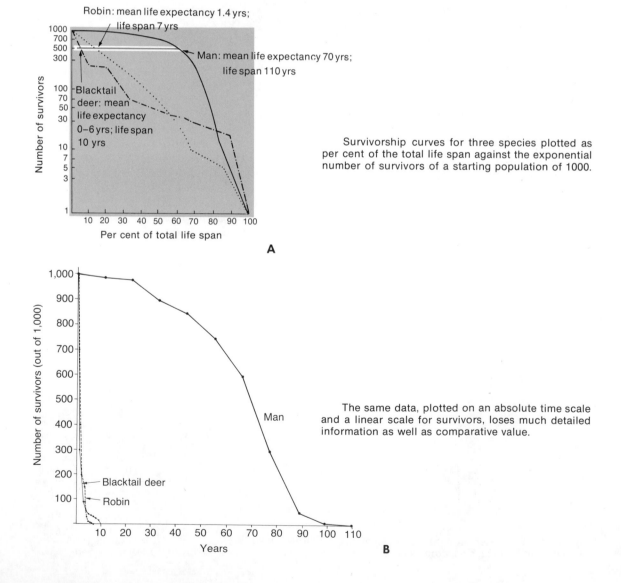

Survivorship curves for three species plotted as per cent of the total life span against the exponential number of survivors of a starting population of 1000.

The same data, plotted on an absolute time scale and a linear scale for survivors, loses much detailed information as well as comparative value.

THE CLASSIFICATION OF ORGANISMS

Agreement among biologists is still far from complete on the larger divisions into which organisms should be subdivided. The scheme presented here is modified from R. H. Whittaker (*Science 163*:150–160, 1969). It groups organisms into five large categories, the kingdoms.

KINGDOM MONERA

Single-celled organisms with no distinct nuclei.

Phylum Cyanophyta: bluc-green algae.
Phylum Schizophyta: bacteria
Subdivisions in this group include not only the true bacteria, but many bacteria-like organisms such as spirochaetes and rickettsia. For convenience the viruses are usually classified here.

KINGDOM PROTISTA

Single-celled or colonial organisms with a distinct nucleus; both photosynthetic and non-photosynthetic phyla are included.

Phylum Euglenophyta: chlorophyll-containing organisms; flagellated; inhabit mainly fresh water.
Phylum Chrysophyta: golden algae and diatoms; abundant in phytoplankton.
Phylum Pyrrophyta: dinoflagellates; most species marine; abundant in phytoplankton.
Phylum Xanthophyta: yellow-green algae; widespread but inconspicuous in fresh and marine waters.
Phylum Mastigophora: protozoans with one or more flagella; a few parasitic.
Phylum Sarcodina: protozoans moving by means of pseudopodia; a few parasitic.
Phylum Sporozoa: parasitic protozoans reproducing by spores.
Phylum Ciliata: protozoans which move by means of cilia.

KINGDOM PLANTAE

Organisms, usually multicellular, with rigid cell walls and chlorophyll.

Subkingdom Thallophyta: cells not differentiated into true tissues; do not form embryos.

Phylum Chlorophyta: true green algae; unicellular forms prominent in phytoplankton; multicellular green seaweeds.

Phylum Rhodophyta: red algae; conspicuous seaweeds.

Phylum Phaeophyta: brown algae; includes giant kelp.

Subkingdom Embryophyta: plants forming embryos; cells differentiated into tissues.

Phylum Bryophyta: plants without conducting tissues.

> *Class Hepaticae:* liverworts.
> *Class Anthocerotae:* hornworts.
> *Class Musci:* mosses.

Phylum Tracheophyta: plants with conducting tissues (true vascular plants).

> SUBPHYLUM PSILOPSIDA: plants lacking true leaves and roots; mostly fossil.
> SUBPHYLUM LYCOPSIDA: clubmosses and quillworts; leaves borne on unjointed stems.
> SUBPHYLUM SPHENOPSIDA: horsetails; true roots, reduced leaves, stems jointed.
> SUBPHYLUM PTEROPSIDA: ferns.
> Subphylum Spermopsida: seed plants.
>> *Class Gymnospermae:* conifers, cyclads and ginkgo; non-flowering seed plants with naked seeds borne on the surface of the cone scales.
>> *Class Angiospermae:* flowering plants; seeds enclosed in ovary.
>>> **Subclass Monocotyledoneae:** grasses, lilies and orchids; parallel-veined leaves.
>>> **Subclass Dicotyledoneae:** flowers, shrubs and trees with net-veined leaves.

KINGDOM FUNGI

Plant-like organisms; cells contain nuclei and a cell wall but no photosynthetic pigments.

Phylum Myxomycophyta: slime molds; found on decaying organic matter.

Phylum Eumycophyta: true fungi; feeding part of body usually of thread-like hyphae.

> *Class Phycomycetes:* algal fungi; bread molds and leaf molds.
> *Class Ascomycetes:* sac fungi; yeasts, mildews and cheese molds.
> *Class Basidiomycetes:* club fungi; mushrooms, rusts and smuts.
> Several other minor classes are also included.

KINGDOM ANIMALIA

Multicellular organisms without chlorophyll or cell walls; usually motile. (Several smaller phyla are omitted.)

Phylum Porifera: sponges.

Phylum Coelenterata: radially symmetrical marine animals with a central gastrovascular cavity.

> *Class Hydrozoa:* hydra-like animals; some colonial.
> *Class Scyphozoa:* jellyfish.
> *Class Anthozoa:* corals and sea anemones.

Phylum Ctenophora: comb jellies or sea walnuts, somewhat resembling jellyfish.

Phylum Platyhelminthes: flatworms.

> *Class Turbellaria:* free-living forms, mainly aquatic.
> *Class Trematoda:* parasitic flukes.
> *Class Cestoda:* parasitic tapeworms.

Phylum Nemertea: proboscis or ribbon worms; long, extensible, unsegmented bodies; mostly marine.

Phylum Aschelminthes: roundworms with eel-shaped bodies; many parasitic.

Phylum Rotifera: wheel animals; small, mostly freshwater.

Phylum Bryozoa: moss animals; encrusting forms, mostly marine.

Phylum Brachiopoda: lamp shells; shells somewhat resembling those of clams; marine.

Phylum Mollusca: unsegmented soft-bodied animals, usually with a shell secreted by mantle tissue.

> *Class Amphineura:* chitons; shell has eight plates; marine.
> *Class Gastropoda:* slugs, snails and abalones; if present, shell is coiled; marine, freshwater and terrestrial.
> *Class Pelecypoda:* clams, oysters and scallops; with a bi-valved shell; marine and freshwater.
> *Class Scaphopoda:* tooth shells; marine.
> *Class Cephalopoda:* octopuses and squids; a "head-foot," bearing tentacles; all marine.

Phylum Annelida: segmented worms.

> *Class Polychaeta:* bristle worms; parapodial flaps bear bristles on segments; mainly marine.
> *Class Oligochaeta:* earthworms and some freshwater species; no parapodia; bristles reduced.
> *Class Archiannelida:* primitive annelids without bristles or external segmentation.
> *Class Hirudinea:* leeches; no bristles or parapodia; flattened body with suckers at each end; mainly freshwater and terrestrial; some parasitic.

Phylum Onycophora: segmented animals with both annelid and arthropod features; only a few species, in tropics.

Phylum Arthropoda: joint-legged animals with a hard exoskeleton. (Extinct forms omitted.)

> SUBPHYLUM CHELICERATA: first pair of legs (on third segment) modified as pincers for feeding; no antennae.
>> *Class Xiphosura:* horseshoe crabs; marine.
>> *Class Arachnida:* spiders, scorpions, ticks and mites; all air-breathing; six paired appendages, four conspicuous for walking.
>> *Class Pycnogonida:* sea spiders.
>
> SUBPHYLUM MANDIBULATA: anterior segments externally fused; one or two pairs of antennae on head; next pair of appendages modified as mandibles for feeding.
>> *Class Crustacea:* barnacles, amphipods, isopods, copepods, lobsters and crabs; mainly aquatic; smaller forms important in zooplankton.
>> *Class Chilopoda:* centipedes; one pair of legs on all but first and last segments; terrestrial.
>> *Class Diplopoda:* millipedes; two pairs of legs on each external segment; terrestrial.
>> *Class Insecta:* insects; one pair of antennae on head; three pairs of walking legs; largest group of animals; mainly terrestrial. (Only commonest orders given.)

ORDER ORTHOPTERA: grasshoppers and cockroaches.

ORDER ISOPTERA: termites.

ORDER ODONATA: dragonflies and damsel flies.

ORDER ANAPLURA: lice.

ORDER HEMIPTERA: water boatmen, bedbugs and backswimmers.

ORDER HOMOPTERA: cicadas, aphids and scale insects.

ORDER COLEOPTERA: beetles, weevils.

ORDER LEPIDOPTERA: butterflies and moths.

ORDER DIPTERA: flies, mosquitos, gnats.

ORDER HYMENOPTERA: ants, bees, wasps and gallflies.

Phylum Echinodermata: animals having skin with calcareous, spine-bearing plates; larvae bilaterally symmetrical, but adults have secondary radial symmetry; all marine.

Class Asteroidea: starfish; arms broadly attached to body.

Class Ophiuroidea: brittle stars; long, narrow arms, sharply distinct from body.

Class Echinoidea: sea urchins and sand dollars; spines elongate and numerous.

Class Holothuroidea: sea cucumbers; soft-bodies; spiny plates greatly reduced.

Class Crinoidea: sea lilies and feather stars; cup-shaped adult is attached by stalk to substrate.

Phylum Hemichordata: acorn worms; proboscis connected by collar to long worm-like body; larvae resemble those of echinoderms.

Phylum Chordata: animals with body supported at some stage by a notochord; gill slits in pharynx; nerve cord hollow.

SUBPHYLUM UROCHORDATA: tunicates or sea squirts; adult attached; body sac-like, covered by cellulose test; marine.

SUBPHYLUM CEPHALOCHORDATA: lancelets: small, fish-like, sand-dwelling marine animals.

SUBPHYLUM VERTEBRATA: animals with nerve cord encased by vertebral column; brain well-developed, encased in cranium and associated with prominent sense organs; usually two pairs of limbs.

Class Agnatha: hagfish and lampreys; no jaws or paired fins.

Class Placodermi: spiny-skinned sharks; extinct forms with jaws and fins.

Class Chondrichthyes: sharks, rays and skates; fish with cartilaginous skeleton and dermal scales.

Class Osteichthyes: bony fish; skeleton of true bone; scales dermal.

Class Amphibia: frogs, toads and salamanders; skin without scales; larvae aquatic with gills; most adults semiterrestrial with lungs.

Class Reptilia: snakes, lizards, crocodiles and turtles; air-breathing; skin covered with epidermal scales.

Class Aves: birds; warm-blooded; forelimbs modified as wings; skin covered with feathers.

Class Mammalia: mammals; skin covered with hair; females with mammary glands.

Subclass Prototheria: duck-billed platypus and many extinct forms; egg-laying.

Subclass Metatheria: marsupials, including opossums, kangaroos and koala bears; young born in very immature state and complete development in mother's pouch.

Sublcass Eutheria: placental mammals; young develop in mother's uterus, being nourished via the placenta. (Extinct orders omitted.)

ORDER INSECTIVORA: hedgehogs, moles and shrews; primitive insect-eaters with teeth all similar.

ORDER CHIROPTERA: bats; forelimbs modified to support flight membrane.

ORDER CARNIVORA: dogs, cats, bears, minks, skunks, sea lions and seals; teeth sharp, with well-developed canines.

ORDER RODENTIA: squirrels, mice, beavers, porcupines; continuously growing sharp incisors for gnawing; canines absent.

ORDER LAGOMORPHA: rabbits and hares; rodent-like but with well-developed hindlimbs.

ORDER ARTIODACTYLA: cattle, deer, camels, pigs and hippopotamuses; even-toed ungulates.

ORDER PERISSODACTYLA: horses, tapirs and rhinoceroses; odd-toed ungulates with one functional digit.

ORDER PROBOSCIDEA: elephants; nose and upper lip modified as prehensile proboscis; upper incisors elongated as tusks.

ORDER EDENTATA: armadillos, sloths and anteaters; teeth reduced or absent.

ORDER CETACEA: whales, porpoises and dolphins; forelimbs modified as flippers; hindlimbs absent; tail broad and notched; head often pointed.

ORDER SIRENIA: manatees or sea cows; forelimbs modified as flippers; hindlimbs absent; tail not notched; head blunt.

ORDER PRIMATES: monkeys, apes and man; erect or partially erect posture; forelimb modified for grasping; large cerebrum.

GLOSSARY

abortion spontaneous or induced expulsion of the fetus during the first two-thirds of pregnancy

abscission the premature dropping of fruit and leaves by plants

absorption spectrum a plot of the amount of light energy absorbed by a substance in relation to the wavelength (color) of light

acetylcholine the neurotransmitter at neuromuscular junctions and at certain synapses in the brain, especially those related to cortical arousal

acid a chemical which yields H^+ as its positive ion on dissolving in water

acidic having a concentration of H^+ exceeding that of OH^- (refers to a solution)

actin a muscle protein composing the thin filaments of myofibrils; is involved in contraction

action potential the potential resulting from rapid influx of sodium into a cell following membrane depolarization; initiates a propagated impulse

action spectrum a plot of the amount of reaction occurring in a photoactivated system in relation to the wavelength (color) of light

activator a substance which facilitates the action of an enzyme

active site the region of an enzyme which binds specifically with its substrate molecule

active transport the movement of molecules across cell membranes in a direction opposite to their concentration or electrical gradients

adaptive coloration body coloration which increases the chance of survival

adaptive radiation the evolution of a diversity of body forms derived from one basic body plan

adrenal cortex the outer part of the adrenal gland, located above the kidney; this endocrine organ produces the cortical steroid hormones

adrenalin the hormone produced by the adrenal medulla; is released during an emergency and is thus known as the "fight-or-flight" hormone

adrenal medulla the inner region of the adrenal gland, an endocrine organ located above the kidney; secretes adrenalin

aerobic utilizing oxygen

aerosol a suspension of fine particles of dust and moisture in the air

agglutination the clumping of cells following reaction with antibodies

ALA δ-aminolevulinic acid, a precursor molecule in the synthesis of hemoglobin

aldosterone a steroid hormone from the adrenal cortex which increases the level of activity of sodium pumps in the kidney nephron and favors sodium retention

alkaline having a concentration of OH^- exceeding that of H^+ (refers to a solution)

alleles   two genes sharing the same locus on homologous chromosomes

allelopathy   the extrusion by a plant of noxious substances which inhibit the growth of other plant species

allergies   diseases resulting from hyperactive antibody production against innocuous foreign substances

all-or-none principle   the law which says that a nerve or muscle cell conducts an impulse either maximally or not at all

alveolus (pl. **alveoli**)   a grape-like sphere of cells surrounding a central space; especially, the thin-walled, air-filled alveoli of the lung

amino acids   small organic molecules having both an acid carboxyl group and a basic amino group; building blocks of proteins

ammonia   gaseous, toxic compound, NH_3, formed as a by-product of nitrogen metabolism; the main nitrogenous waste product of aquatic animals

amnesia   loss of the ability to transfer information to long-term memory storage, i.e., memory loss of recent events

amniocentesis   removal of a small sample of amniotic fluid through the mother's abdominal wall

amnion   the embryonic membrane immediately surrounding the fetus

amoebocytes   blood cells resembling small amoebae, occurring especially among invertebrates

amphetamines   drugs which mimic the action of noradrenalin; act as stimulants of the central nervous system

anaerobic   not utilizing oxygen

anemia   a condition characterized by decreased oxygen-carrying capacity of the blood, due to too few red blood cells, too little hemoglobin or the presence of faulty hemoglobin

angiosperms   flowering plants

annelids   segmented worms, including polychaetes, earthworms and leeches

anthers   the tips of the male sporangia in seed plants, where pollen is produced

antibody   a blood-borne protein synthesized in response to a foreign chemical which facilitates the inactivation of that foreign material

antidiuretic hormone   a hormone released by the posterior pituitary gland which increases water permeability of the collecting tubules in the kidney and hence favors water retention

antigen   a foreign chemical capable of inducing synthesis of a specific antibody

anti-metabolite   a chemical that structurally resembles a substrate and competes with it for the specific site on an enzyme

antitoxin   an antibody prepared by injecting animals with the antigen of a specific disease; used to treat disease

anus   the distal opening of the digestive tract through which feces are excreted

aquifer   an underground channel through which water may flow rapidly

arteriosclerosis   a disease in which the arterial walls lose their elasticity (hardening of the arteries)

arthropods   invertebrate animals with jointed appendages and a horny external covering, including insects, crustaceans and spiders

ascites   an accumulation of fluid in the abdominal cavity; is often due to liver disease

asexual reproduction   reproduction by mitosis or by budding

association neurons   small neurons with many synapses, located in the central nervous system; thought to play a role in behavioral modification

atom   the smallest unit an element can assume without losing its chemical properties

atomic number   the number of protons possessed by an atom of an element; also the number of electrons possessed by that atom

atomic weight   the number equal to the sum of the number of neutrons and protons in the nucleus of an atom

ATP   adenosine triphosphate, the molecule whose high energy bonds are immediately available for cell work

ATPases   enzymes capable of splitting ATP and releasing its energy for cell work

autoimmune disease   a disease in which the body produces antibodies against its own proteins

autotrophs   organisms capable of synthesizing their own food from inorganic precursors

auxin   a hormone produced by a plant's meristems, resulting in cell elongation

axon   the elongate process of a nerve cell which transmits information to other cells

barbiturates   a class of drugs which act as general depressants of the central nervous system

basal ganglia   areas of the brain located in the inner part of the cerebrum which serve as higher centers of motor coordination, emotional control and perhaps memory

base   a chemical which yields OH^- as its negative ion on dissolving in water

Benson-Calvin cycle   the complex cycle of reactions in photosynthesis in which CO_2 is added to phosphorylated intermediates present in the chloroplast; six turns of the cycle yields one glucose molecule

beri beri   a thiamin deficiency disease which affects the nervous system, resulting in paralysis

bilateral symmetry   exact correspondence of right and left halves so that both sides are identical (refers to body shape)

bile   a greenish fluid produced by the liver and emptying into the small intestine; contains bile salts, detoxified waste products and poisons, and the breakdown products of hemoglobin

bile salts   detergent-like substances that aid digestion by emulsifying fats in the gut

biogeochemical cycles   the cyclic processes by which elements pass continuously from the non living to the living to the non-living compartments of an ecosystem

biological magnification   the sequential concentration of a toxic chemical within a food chain

biological oxygen demand   the amount of organic matter in water measured by bacterial oxygen consumption

biomass   living matter

biome   a subdivision of the terrestrial environment characterized by a particular type of vegetation and associated organisms

biosphere   the entire Earth and all its living organisms

bivalves   molluscs with two shells, including clams and oysters

blastula   the hollow ball of cells formed early in the development of an animal embryo

blended inheritance   the inheritance of a genetic trait in which both alleles affect the phenotype

blood platelets   small blood cells which break down easily and initiate clotting by release of an enzyme, thrombokinase

blood filtrate   the non-cellular, protein-free fluid that is initially formed in the glomerulus of the kidney during urine formation

blue-green algae   primitive, chlorophyll-containing, single-celled organisms without a true nucleus

body burden the total amount of a toxic substance, such as lead, in the body

botulism a disease resulting from eating food containing botulinum toxin, a product of certain soil bacteria

bronchioles highly branched respiratory tubes of the lungs

bryophytes terrestrial plants without true leaves, roots or conducting systems; mosses and liverworts

buffer a chemical capable of combining with hydrogen ion and thus maintaining constant acidity in a solution

buffering capacity the ability of dissolved molecules, such as blood proteins, to combine with hydrogen ion and thus maintain a constant acidity

calorie the amount of energy required to raise the temperature of one gram of water one degree centigrade

cambium the region of cell division in a stem

capacitation the final maturation process of human sperm which occurs in the female reproductive tract

carbohydrases digestive enzymes that hydrolyze carbohydrates

carbon dioxide fixation the sequence of steps during photosynthesis in which glucose is synthesized from carbon dioxide, utilizing ATP and $NADP \cdot H_2$

carnivores organisms that feed on animals

carotenoids in plants, accessory light sensitive pigments, yellow to orange-red in color, which function during photosynthesis

carrier molecules protein molecules associated with cell membranes to facilitate movements of certain molecules in and out of cells

carrying capacity the maximum population size for a given species that can be supported by a particular ecosystem

catalyst an agent which speeds the rate of a chemical reaction without itself being changed

cellular differentiation the developmental process during which cells become chemically and anatomically specialized

cellular respiration the oxidation of organic molecules with simultaneous conversion of stored energy into useful energy

cellulose a glucose polymer found in plant cell walls

cell wall a thick, rigid layer external to the plasma membrane of plant cells

central nervous system the compact, central part of the nervous system (consisting in higher animals of the brain and spinal cord), as distinct from the more diffuse peripheral nervous system

centrioles a pair of short, rod-like structures forming the poles of the mitotic spindle in animal cells

centromere the constricted region of a chromosome at which two chromatids remain joined prior to mitosis

cephalization the evolutionary process by which the anterior part of the nervous system becomes predominant

cephalopods a group of molluscs including octopuses, squids, cuttlefish and the chambered nautilus

cercaria the final larval stage in the life cycle of the parasitic flatworm that causes schistosomiasis

cerebellum an outgrowth from the roof of the vertebrate brain, concerned mainly with motor coordination

cerebral cortex the evolutionarily most recent part of the brain; contains many association cells involved in behavioral modification; most highly developed in man

cerebrum the evolutionarily most advanced part of the brain, consisting of basal ganglia, old cortex and neocortex

cervix the lower end of the uterus, which opens into the vagina

chaparral the scrub biome characteristic of semi-arid regions

chelating agents chemicals capable of binding with multivalent ions such as calcium, magnesium and lead; used to treat lead poisoning

chemoreception the ability to detect chemical stimuli; the senses of taste and smell

chemosynthetic autotrophs bacteria capable of obtaining energy from the oxidation of inorganic matter such as Fe^{++} and H_2S

chiasma a visible point of reciprocal exchange between two chromatids during tetrad formation; a point of crossing-over

chitin a polysaccharide substance found in crustacean shells and insect cuticles

chitons marine molluscs with shells consisting of eight overlapping plates

chlorinated hydrocarbons synthetic organic molecules containing varying numbers of chlorine atoms

chlorophyll the photosensitive pigment which traps light energy in plant cells

chloroplast the organelle in green plant cells in which photosynthesis occurs

cholinesterase an enzyme which destroys the neurotransmitter acetylcholine in the synapse

chromatid one of two identical double helices of DNA in a chromosome which has duplicated itself in preparation for cell division; each chromatid becomes a daughter chromosome

chromatin stainable material in the nucleus associated with chromosomes

chromosomes elongate structures in the nucleus containing DNA and bearing hereditary information

ciliates protozoans bearing numerous cilia for locomotion

cilium (*pl.* **cilia**) a thread-like process extending from a cell which, by continuous beating, causes water to move past the cell; usually many occur on the same cell; used in protozoan locomotion

cirrhosis a liver disease in which normal cells are replaced by scar tissue; occurs in chronic alcoholism

cistron a region on the DNA molecule that is coded for synthesis of a particular protein

climax vegetation the major type of vegetation in a stable biome

clitoris a sexually sensitive organ in female mammals, located in primates just anterior to the urethral opening

cloaca the hindmost part of the gut of birds and reptiles, where water from both urine and feces is reabsorbed prior to their excretion

closed circulatory system a circulatory system of animals in which the blood is at all times confined within vessels

coacervates two-phase liquid systems in which large macromolecules become spontaneously concentrated in one of the phases

coelenterates radially symmetrical marine animals with a central gastrovascular cavity

coelom the body cavity of animals higher than the coelenterates

coenzymes small molecules which act in concert with an enzyme during the course of a reaction; many contain vitamins

cohesion theory a theory which explains the rise of sap as due to cohesive forces between water molecules in a continuous column; the driving force results from transpiration

coitus interruptus withdrawal of the penis before ejaculation

collecting tubule the last tubule through which urine passes during its formation in the kidney, and where its final concentration is determined

commensalism a symbiotic relationship which is beneficial to one species but without effect on the other

community   a group consisting of all the living organisms within a given ecosystem

complete proteins   proteins containing all 10 essential amino acids required by man

compound eyes   eyes found in insects and some crustaceans that are composed of many independent subunits, the ommatidia

concentration gradient   a non-uniform distribution of dissolved molecules

conditioned   exhibiting a conditioned response; a state in which the subject regularly responds in the same manner to a substitute stimulus as he (it) formerly did to the real one

condom   penis sheath; used for contraception

conjugation   a chemical reaction in which a potentially toxic molecule is rendered harmless by combination with a normal body chemical

convergent evolution   accidental similarity of structure in two organisms not phylogenetically related; brought about by a similarity in function

cooling towers   structures designed to discard waste heat from power plants into the atmosphere

copulation   physical union of male and female during which sperm are deposited in the female's reproductive tract.

corpus albicans   a small area of scar tissue in the ovary remaining at the site of a former corpus luteum

corpus luteum   a structure in the ovary derived from remnants of a ruptured follicle and secreting both estrogen and progesterone

cortex   the outer region of an organ such as the adrenal gland or kidney

cortical hormones   steroid hormones, produced by the adrenal cortex, which increase conversion of fats and proteins to glucose and affect sodium retention by the kidney

cortisone   an adrenal cortical hormone which helps convert fats and proteins into blood sugar and also reduces antibody-forming capacity; often used to treat severe autoimmune diseases

cotyledons   primitive leaves of the embryos of seed plants containing stored nutrients

courtship behavior   patterned behavior between a male and a female of the same species leading to either synchronized spawning or copulation

covalent bond   a chemical bond in which the shared electrons spend equal amounts of time with the two atoms

Cowper's glands   paired glands in the human male producing secretions that contribute to the seminal fluid

cranium   the bony case surrounding the brain of vertebrates

crossing-over   the reciprocal exchange of parts of homologous chromatids during tetrad formation in meiosis

crustaceans   the group of arthropods having a chitinous shell and paired jointed limbs; includes crabs, shrimps, amphipods, isopods and barnacles; small forms important in zooplankton

cryptic coloration   body coloration which helps to camouflage an organism

curare   a neurotoxin derived from vines native to South America; it blocks binding of acetylcholine to the post-synaptic membrane, resulting in paralysis

cuticle   a thick, often hard, protective covering on the outside surface of some plants and animals

cyanophytes   blue-green algae

cyclosis   movement of cytoplasmic contents in plant cells

cytochrome P450   a protein of the smooth endoplasmic reticulum to which molecules must bind before being detoxified

cytochromes   iron-containing protein molecules capable of electron transfer; found mainly in mitochondria and chloroplasts

cytokinins   plant hormones stimulating cell division
cytoplasm   the cell contents, excluding the nucleus
cytoplasmic streaming   movement of cytoplasmic contents in animal cells

dark reactions   the series of reactions during photosynthesis which can occur in the dark, when glucose is synthesized from CO_2, utilizing ATP and $NADP \cdot H_2$
Darwinism   Darwin's theory of biological evolution; states that new species arise through natural selection of variant individuals best adapted to survive
decomposers   organisms feeding on non-living organic matter, breaking it down into its basic components
denaturation   the physical disruption of the specific shape of a molecule
dendrites   nerve cell processes which receive stimuli and transmit them to the cell body; often highly branched
denitrification   the conversion of nitrate to gaseous nitrogen by certain soil bacteria
dependent variable   a property whose value depends upon the value of some other property
depolarization   a decrease in membrane potential across a cell membrane
detritus   non-living organic matter in soil, mud and elsewhere
deuterium ^2H, or heavy hydrogen; an isotope of hydrogen containing one neutron in its nucleus
diabetes mellitus   an insulin insufficiency disease characterized by high blood sugar levels and excess sugar in the urine
diaphragm   the major muscle involved in breathing, situated between the lung cavity and abdominal cavity; also, a contraceptive device that covers the cervix of the uterus
diatoms   single-celled algae with silicon-impregnated shells
differential centrifugation   a technique for separating components of ruptured cells by centrifuging them at different speeds
diffusion   the process that results in uniform distribution of matter or energy in a system
dinoflagellates   single-celled algae with two flagella; common in phytoplankton
dioxin   a teratogenic contaminant present in the herbicide 2,4,5-T
diploid (2N)   having a double set of chromosomes, one member of each pair coming from each parent (refers to cells)
direct effect   in radiation biology, the cell damage resulting from direct hits on macromolecules by radiant energy
displacement activity   inappropriate behavioral response utilized in a conflict situation when more appropriate responses are inhibited
DNA   deoxyribose nucleic acid; a polymer of nucleotides carrying genetic information
dominant   in genetics, producing a trait which is always expressed whether paired with an identical allele or not
dopamine   a neurotransmitter found in the human brain, especially in the substantia nigra
double helix   the double spiral of DNA threads
doubling time   the time required for a population to double its size
dust cells   phagocytic cells lining the alveoli of the lungs

ecdysone   an insect hormone favoring development of more adult characters at each molt
echinoderms   a group of marine animals characterized by spiny skins and radial symmetry; includes starfish, sea urchins and sea cucumbers
ecological succession   the temporal replacement of one set of organisms by another in an ecosystem
ecology   the science of the relationship of living organisms with their environment

ecosystem the sum total of interacting physical and biological features in a delimited area

ectoderm the outer layer of tissue in animal embryos, giving rise to skin and nerve tissue; also, the outer layer of coelenterates

ectoparasites parasitic organisms living on the outer body surface of the host

edema an excessive increase in fluids between cells; in the lung, the extra fluid coats the inner walls of the alveoli

effector that which effects a response; for example, a muscle, a gland or a stinging cell

electroencephalogram (EEG) the pattern of electrical brain waves observed by placing recording electrodes on the scalp

electromagnetic spectrum the entire range of non-particulate radiant energy, from short X-rays to long radio waves; includes visible light

electron a very light, negatively charged particle orbiting about the nucleus of an atom

electronegative tending to have an excess number of electrons

electron transport chain the series of reactions occurring on the mitochondrial cristae in which the iron of cytochromes is alternately oxidized and reduced and ATP is synthesized

electropositive tending to have a deficiency of electrons

element a substance composed entirely of one of the more than 100 different types of atoms

embryo a developing diploid organism, formed by repeated mitoses of the zygote

emphysema a disease condition of the lung in which breathing is impaired due to collapse of septa between alveoli and thickening of the bronchiolar walls

emulsifiers detergent-like molecules that function in dispersing fats

endocrine glands glands which produce and secrete chemical regulators, the hormones, into the blood

endoderm the inner layer of tissue in animal embryos, giving rise to the gut lining; also the inner layer of coelenterates

endometrium the glandular lining of the uterus

endoplasmic reticulum (ER) a system of tubules ramifying throughout the cytoplasm

endoskeleton a rigid, jointed structure within an animal against which the locomotory muscles exert their force

endosperm nutritive material contained in a seed

engram a permanent memory trace

entropy energy which is not available to do work; a measure of the randomness of a system

enzymes biological catalysts which speed the rate of a reaction; made of proteins

enzyme-substrate complex the close association of an enzyme with the molecule it acts on; an essential part of an enzyme-catalyzed process

environmental resistance the sum of factors external to an organism tending to cause its death

epidermis the outer layer of the body, composed of skin and its associated glands

epididymis a highly coiled duct for sperm storage; located beside the testis

epilimnion the upper layer of warm water in a lake, commonly occurring during summer stratification

epiphyte a plant living in a commensal relationship on another plant, deriving only support; nutrients are obtained from moisture and dust in the air

equivalent a unit expressing the amount of electrical charge in a solution of ions

erythrocytes   red blood cells; they are filled mainly with the respiratory pigment, hemoglobin

esophagus   a tube conducting food directly from mouth to stomach

essential amino acids   amino acids required in an organism's diet because it cannot synthesize them itself from other molecules

estrogen   the female sex hormone, produced both by the ovarian follicle and the corpus luteum

ethanol   the two-carbon alcohol contained in alcoholic beverages

ethology   the study of animal behavior under natural conditions

eutrophic   abundance of nutrients in an aqueous ecosystem, usually a lake

eutrophication   natural or man-made enrichment of water, leading to abundant plant growth

evapotranspiration   the combined evaporation of water from plants and soil

excitatory synapses   synapses at which an impulse results in depolarization of the postsynaptic membrane, often leading to an action potential

excited state   the state of a molecule in which one of its electrons has an excess of energy

excretion   removal from the body of toxic waste products

excretory tubule   the region of a kidney nephron in which urine volume is greatly reduced and where some molecules are removed from, and other molecules are added to, the urine

exoskeleton   an outer rigid body covering with appropriate joints, against which the locomotory muscles exert their force

extensor   a muscle which causes a limb or other extension to be moved away from the trunk of the body

extracellular digestion   digestion of food external to the cell

extrapyramidal pathway   the pathway in the brain by which modifying impulses from the cortex, basal ganglia and cerebellum reach motor neurons in the spinal cord

facilitation   the modification of a synaptic response as a result of prior impulses

FAD   flavin adenine dinucleotide; a hydrogen transfer molecule associated with flavoproteins

fatty acids   long-chain, water-insoluble hydrocarbons with a terminal carboxyl group

fertilization   the union of egg and sperm to form a zygote

fibrin   denatured fibrinogen; the protein which forms a blood clot

fibrinogen   a globular protein occurring in vertebrate blood plasma; plays a role in clotting

fibrous protein   an elongage protein, generally serving a structural function

filariasis   a disease, also called elephantiasis, caused by a roundworm that blocks lymph vessels

filter feeders   animals which sieve water to obtain suspended food particles

First Law of Thermodynamics   the physical law which states that in all processes, the total amount of energy remains constant

fixed action pattern   a behavioral response which more often than not follows a particular stimulus

flagellates   protozoans equipped with one or more flagella for locomotion

flagellum (*pl.* flagella)   a thread-like organelle extending from a cell; its whip-like beating propels the cell

flavoproteins   proteins located at the start of the electron transport chain whose coenzymes are capable of reversible oxidation and reduction

flexor   a muscle which causes a limb or other extension to be pulled toward the trunk of the body

flukes   parasitic flatworms causing several diseases in man and animals

fluorescence   re-emission of absorbed radiant energy as light, usually of a longer wavelength than the incident radiation

food vacuole   a membrane-bound cavity within a cell in which food is digested

food web   the total pattern by which energy (food) flows between trophic levels through an ecosystem

foraminifera   shelled, amoeba-like protozoans found in zooplankton

free energy   energy which is available to do work

free radical   an atom or molecule which contains an unpaired electron and is therefore highly reactive chemically

frequency distribution   a plot of the number of individuals in a population falling into certain categories or classes

FRF   FSH-releasing factor; a hormone produced in the hypothalamus and acting on the pituitary

FSH   follicle-stimulating hormone; a pituitary hormone causing growth of the ovarian follicle

gametes   the haploid sex cells, eggs and sperm

gametophyte   haploid (1N) stage of a plant life cycle

ganglion (*pl.* **ganglia**)   a cluster of interconnected nerve cells

gastrocoele   the single cavity, found in coelenterates, which functions as a gut

gastropods   a group of molluscs, often possessing a single, coiled shell, including snails, abalones and slugs; move by means of a muscle on the underside

gastrula   the stage of invagination to form a primitive gut during animal embryogenesis; two cell layers are present

Gause's competitive exclusion principle   the law which says that no two species fulfill identical functions in the same ecosystem

gene flow   the exchange of genetic information within a breeding population

gene pool   the sum total of all the genes within a population of organisms

genes   discrete units of inheritance controlling a specific trait

genetic code   the code whereby genetic information in triplets of nucleotides is converted into a specific sequence of amino acids in proteins

genetics   the study of inheritance

genotype   the genetic constitution of an individual as distinct from the outward appearance of inherited traits

geographic isolation   the separation of two subpopulations of a species by a geographic barrier

gestation   the period of intra-uterine development of mammalian embryos

gibberellins   plant hormones, produced by growing tips, which stimulate both cell division and cell elongation

gill slits   in the chordate group of animals, openings between the pharynx and exterior, used mainly for respiration

gizzard   a muscular region of the stomach in birds and certain invertebrates, such as the earthworm, that is specialized for grinding food

glans   the sexually sensitive tip of the penis in male mammals

globular protein   a protein with a specifically folded tertiary structure; many are enzymes

glucagon   a hormone produced by the pancreas which increases blood sugar levels

glucose   a common six-carbon sugar

gluteus maximus   the buttocks muscle, responsible for man's upright stature, especially during running and climbing

glycerol   a three-carbon alcohol; a component of fat molecules

glycogen   a glucose polymer; used for energy storage in animals

glycolysis   the series of cellular reactions by which a molecule of glucose is converted to two molecules of pyruvic acid

goiter   an enlargement of the thyroid gland in response to iodine deficiency

Golgi apparatus   a compact arrangement of cytoplasmic tubules in which proteins are processed

gonadotrophins   hormones produced by the pituitary and acting on the gonads

grana   stacks of membranes in chloroplasts, where chlorophyll is located

greenhouse effect   a heating of the atmosphere due to accumulation of gases (CO_2 and H_2O) which allow light energy to pass inward but do not transmit heat energy outward

gross primary productivity   the total photosynthetic rate of an ecosystem

ground state   the state of a molecule in which its electrons are all at their lowest energy levels

growth hormone   a hormone produced by the pituitary gland which affects growth rate and causes a rise in blood sugar

guano   phosphorus rich deposits of excreta left by shore birds

guard cells   in the leaves of plants, the pair of cells surrounding a stoma; capable of changing shape and hence the size of the stomatal opening

gymnosperms   bare-seeded plants, including cycads, ginkgo and conifers

habitat   the physical part of an ecosystem occupied by a species

habituation   loss of a reflex response in the continued absence of reinforcement

haploid (1N)   having a single set of chromosomes (refers to cells)

Hardy-Weinberg Law   the law of inheritance which says that in large, randomly mating populations the frequency of a gene remains unchanged from one generation to the next

heme   a special group within certain pigment molecules, such as hemoglobin, in which iron (Fe) is located

hemocyanin   a blue, copper-containing respiratory pigment occurring in molluscs and crustaceans

hemoglobin   a red, iron-containing respiratory pigment found in vertebrates and some invertebrates

hemophilia   an inherited disease in which the blood-clotting mechanism fails to function effectively

herbicides   chemicals used as weed-killers and defoliants

herbivores   animals feeding on plants

hermaphrodite   an animal producing both eggs and sperm in the same individual

heterogametic   possessing a dissimilar pair of sex chromosomes; in humans, the male is the heterogametic (XY) sex

heterotrophs   organisms feeding upon reduced organic molecules

heterozygous   having dissimilar alleles for a character

high energy bonds   chemial bonds which store about 8000 calories per mole

hirudineans   leeches

histamines   potent chemicals released from certain white blood cells during immune responses, causing swelling and inflammation

histones   protein molecules associated with chromosomes, which may suppress transcription of genetic information

holistic approach   a unified approach, encompassing all aspects of a problem

homeostasis   the tendency of an organism to maintain a constant internal environment

homologous chromosomes   matched pairs of chromosomes in a diploid cell

homozygous having identical alleles for a character

hookworm a small parasitic roundworm causing disease by attaching itself to the intestinal wall of its host

hormones chemical regulators produced by one organ and passing via the circulation to act on another

hydrogen bonds weak bonds formed between electropositive hydrogen atoms and electronegative atoms such as oxygen

hydrologic cycle the continuous cycling of water between oceans, air and land

hydrolysis splitting of a molecule by insertion of a molecule of water

hydrostatic skeleton a condition found in worms and many other soft-bodied invertebrates in which muscular contraction exerts pressure on the non-compressible body fluids to bring about changes in shape

hydroxylation a reaction in which an —OH group is added to a molecule, often making it readily susceptible to further modification

hydrozoans a subgroup of coelenterates, mostly colonial, including the Portuguese man-of-war

hyperosmotic having an osmotic pressure greater than that of a reference solution

hyperpolarization an increase in resting membrane potential, rendering the membrane less easily depolarized

hyphae fungal threads running through the substratum and used for obtaining food

hypolimnion the lower layer of cold water in a lake, commonly occurring during summer stratification

hypo-osmotic having an osmotic pressure less than that of a reference solution

hypothalamus the part of the brain most concerned with moderating behavior related to internal physiological states; "the seat of the emotions"

hypothesis a tentative theory to explain certain facts

hysterectomy surgical removal of the uterus

hysterotomy surgical opening of the uterus; may be used to effect an abortion

immunity the ability of the body to "remember" prior contact with an antigen and so to quickly synthesize antibodies on renewed contact

implantation the process by which a very young mammalian embryo becomes buried in the uterine endometrium

imprinting a behavioral modification occurring during a brief period in the life of a young animal when it identifies permanently with a parent object

independent variable in experiments, a property whose value is determined in advance by the experimenter

indirect effect in radiation biology, the cell damage resulting from reaction of the free radicals, H· and OH·, with macromolecules

indolacetic acid a growth hormone, or auxin, in plants

induction the initiation of enzyme synthesis in cells by the presence of substrate

inhibitor a substance which slows down or stops the action of an enzyme

inhibitory synapses synapses at which an impulse results in a hyperpolarization of the postsynaptic membrane, making an action potential more difficult to attain

innate that which is inborn; in the case of behavior, that which is not learned

insight learning learning which involves application of responses learned under somewhat different circumstances to a new situation

instars larval stages of insects

insulin a hormone produced by the pancreas which decreases blood sugar levels

integrated pest control the combined use of chemical, physical and biological approaches in the control of insects

intelligence quotient a measure of human intelligence; mental age relative to chronological age × 100

interferon a non-specific protein synthesized in response to viral infections

intermediary metabolism the set of chemical reactions in a cell by which proteins, fats and carbohydrates are converted into one another

interphase the state of a non-dividing cell

interstitial cells in the vertebrate male testis, cells lying between the seminiferous tubules and producing testosterone

intracellular digestion digestion of food within food vacuoles inside the cell

invertebrates all animals other than vertebrates; animals without backbones

ion an electrically charged atom or group of atoms

ionic bond a chemical bond in which one atom has the shared electrons far more often than the other

ion pumps transport systems, thought to be located within cell membranes, for moving ions against concentration gradients

irritability the property of all animal cells to respond rapidly to changes in their immediate environment

isosmotic having an osmotic pressure equal to that of a reference solution

isotopes forms of the same chemical element having dissimilar numbers of neutrons in their nuclei; such forms of an element differ only in their atomic weights

juvenile hormone a hormone found in insects which favors retention of larval characteristics at each molt

keratin a hard, water-impermeable protein found in skin, scales, feathers and hair of vertebrates

kilocalorie an energy unit, the so-called "large" calorie; 1000 times the standard calorie

Krebs cycle the series of cellular reactions within the mitochondrial matrix by which pyruvic acid is converted to carbon dioxide and coenzymes are reduced

kwashiorkor a protein deficiency disease of children

labia folds of tissue surrounding the openings of urethra and vagina in human females

lactic acid a three-carbon acid, the end product of anaerobic glycolysis in animals

lancelets fish-like, sand-dwelling marine animals belonging to the chordate group

laparoscopy abdominal incision through the umbilicus; used for performing tubal ligations in females

laterite a permanent rock-like crust developing in tropical soils containing iron and aluminum after exposure to air and sun

L-DOPA a precursor in the synthesis of the neurotransmitter dopamine; used to treat Parkinson's disease

learning in psychology, modification of behavior through prior experience

lethal mutations mutations whose effects are so severe that death ensues

leucocytes white blood cells of vertebrates which function in defense against foreign organisms

leucoplast   granules for starch storage in plant cells

LH   luteinizing hormone, a pituitary hormone perhaps responsible for ovulation and causing corpus luteum formation

lichens   complex organisms formed from the symbiotic association of an alga and a fungus

life expectancy   the mean age at death within a population

life span   the maximum length of life for members of a species

light reactions   the series of reactions during photosynthesis which require light; they result in synthesis of ATP, NADP·H$_2$, and O$_2$

limbic system   a diffuse region of the cerebrum, connected with the thalamus and hypothalamus; exerts control over emotional behavior and plays a role in learning and memory

linkage   a condition in which genes for two different traits are located on the same chromosome and so tend to be inherited together

lipases   digestive enzymes that hydrolyze lipids

lipid   a fat

locus (*pl.* **loci**)   the position on a chromosome at which the genes for a specific trait are located

locus coeruleus   a noradrenalin-producing center in the midbrain which is thought to control onset of REM sleep

long-term memory storage   the permanent storage of information in the nervous system

LRF   LH-releasing factor; a hormone produced in the hypothalamus and acting on the pituitary

lymph   the fluid filling the spaces between cells in animals with closed circulatory systems

lymph nodes   small organs scattered throughout the body and associated with the lymphatic system; they supply the circulatory system with white blood cells

lysis   the disintegration of a cell, especially that of bacterial cells, brought about by the action of antibodies

lysosomes   membrane-bound cell organelles containing hydrolytic or digestive enzymes

macromolecules   large organic molecules of complex structure

macronutrients   nutrients required in moderate to large amounts

mantle   in molluscs, a tissue which lies directly beneath and secretes the shell

mantle cavity   in molluscs, the space formed by a fold of mantle tissue and in which the gills are located

mechanoreception   the ability to detect changes in pressure

medulla   the evolutionarily oldest part of the vertebrate brain, controlling basic bodily functions; also, the central region of an organ such as the adrenal gland or kidney

meiosis   the sequence of divisions by which a diploid (2N) germ cell becomes a haploid (1N) gamete containing one of each pair of chromosomes

melanism   the genetic trait responsible for production of the black pigment, melanin

membrane potential   the electrical charge difference that exists between the outside and inside of virtually all cells

menstruation   the cyclic flow of blood due to sloughing of the uterine lining in female primates

meristems   growing points in a plant, where cell mitoses occur

mesoderm   the middle layer of tissue in animal embryos, giving rise to muscles, excretory organs, the blood-vascular system and gonads

mesoglea   in coelenterates, an intermediate, gelatinous layer of tissue; prominent in jellyfish

metabolic heat   waste heat generated during cellular reactions

metabolism   the sum total of all the chemical reactions carried out in the cells of an organism

methemoglobin   an oxidized form of hemoglobin which is no longer able to carry out its oxygen-carrying function

micronutrients   nutrients required in very small or trace amounts

microsomes   particles smaller than mitochondria and lysosomes, released from disrupted cells; mostly ribosomes and membrane fragments

microvilli   minute finger-like processes on the exposed surfaces of gut cells which greatly increase absorptive capacity

midbrain   the middle region of the vertebrate brain, primarily important in man for its ascending and descending fiber tracts

miracidium   the first larval stage in the life cycle of the parasitic flatworm that causes schistosomiasis

mitochondrion (*pl.* **mitochondria**)   cell organelle in which oxidative respiration takes place

mitosis (*pl.* **mitoses**)   cell division in which the hereditary material of the two daughter cells is identical to that of the parent cell

mitotic spindle   a cytoplasmic structure composed of delicate tubular fibers which attach to the chromosomes and appear to pull them to the spindle poles during mitosis

mole 6×10^{23} molecules of any compound

molecular weight   the sum, in grams, of the atomic weights of all the atoms in a molecule

molluscs   a group of invertebrate animals including snails, clams and octopuses

monoculture   the practice of planting large areas with a single crop

monomers   the repeating subunits of polymers

monosaccharides   simple sugars, such as glucose and ribose

motor unit   a group of muscle cells innervated by the same axon

m-RNA   messenger RNA; the molecule which carries genetic information from nucleus to cytoplasm

mucopolysaccharides   complex molecules of carbohydrate and protein characteristic of bacterial cell walls

mutagenic   causing biological mutations; mutagenic agents include radiant energy and some chemicals

mutations   genetic changes resulting from modification of DNA

mutualism   a symbiotic relationship beneficial to both species

myelin   a fatty material surrounding some nerve axons; formed from tightly wound membranes of supporting cells

myofibrils   rows of minute, overlapping filaments in muscle cells, composed of the contractile proteins actin and myosin

myogenic contractions   muscle contractions which occur spontaneously, without nervous excitation

myosin   a muscle protein composing the thick filaments of myofibrils; has ATPase activity and is involved in contraction

myxomatosis   a virus disease of rabbits characterized by edema of the mucous membranes and tumors of the skin

NAD   nicotine adenine dinucleotide; a hydrogen carrier molecule participating in cellular oxidation-reduction reactions

NAD·H₂   reduced nicotine adenine dinucleotide

NADP   nicotine adenine dinucleotide phosphate; a hydrogen carrier molecule participating in cellular oxidation-reduction reactions

NADP·H₂   reduced nicotine adenine dinucleotide phosphate

nasal salt glands   organs capable of secreting highly concentrated solutions of salt; found in certain sea birds

natality   the birth rate of a population

negative correlation   a relationship in which the dependent variable decreases as the independent variable is increased

negative feedback control a situation in which the end result of a process acts to slow down or prevent the further continuance of the process

negative reinforcement behavior modification in which punishment for a response discourages its repetition

nematodes roundworms ecologically important in soil and mud and medically significant in causing disease

neocortex the evolutionarily most recent part of the brain, concerned with learning, thinking and intelligence

neostigmine a cholinesterase inhibitor resembling physostigmine

nephridia excretory organs of lower animals

nephron a single functional unit of the vertebrate kidney

nerve net a diffuse nervous system, generally with conduction in both directions at synapses; characteristic of coelenterates

net primary productivity the rate of synthesis of new plant tissue in an ecosystem

neuromuscular junction a synapse between a motor neuron and a muscle cell

neurons nerve cells

neurotransmitter a chemical substance released from the end of an axon into a synapse and affecting the postsynaptic cell membrane

neutral having equal concentrations of H^+ and OH^- (refers to a solution)

neutron a heavy, uncharged particle found in the nuclei of atoms

niacin a B-vitamin which forms part of the hydrogen carrier molecules NAD and NADP

nitrification the two-step conversion of ammonia to nitrate; carried out by soil bacteria

nitrogen fixation the conversion of gaseous nitrogen to nitrate, especially by certain bacteria and blue-green algae

nondisjunction the failure of two chromatids to separate during cell division, resulting in one daughter cell with a chromosome deficiency and one with an excess

notochord a semi-rigid internal supporting rod characteristic of the chordate group of animals

nuclear envelope the double-layered membrane surrounding the nucleus of a cell

nucleic acids polymers of nucleotides; DNA and RNA

nucleolus a dense body within the cell nucleus, composed largely of RNA

nucleotides organic molecules consisting of a nitrogen-containing base, a 5-carbon sugar and phosphate; repeating subunits of nucleic acids

nucleus a large structure within a cell, containing the chromosomes; also, the central, very dense region of an atom

oligochaetes earthworms and a few freshwater and marine relatives

oligotrophic scarcity of nutrients in an aqueous ecosystem, usually a lake

ommatidia independent photoreceptor units in the compound eyes of insects and some crustaceans

open circulatory system a circulatory system of animals in which the blood is not continuously confined within vessels but flows freely between the cells

operant conditioning behavior modification brought about by reward or punishment for a response

orbital an energy level available to an electron within an atom or molecule

organ a structural part of an organism which carries out one or more specific functions

organelles small functionally specialized structures within cells

organic molecules molecules containing carbon-to-carbon bonds; characteristic of living organisms

organophosphates a group of phosphorus-containing organic molecules used as pesticides

orgasm   sudden contraction of muscles, mainly in the genital tract, leading to release of sexual tension; accompanied by ejaculation in the male

osmoconformer   an animal living in an environment of varying salinity which permits the salt concentration of its body fluids to vary with those of the medium

osmosis   the movement of water across a membrane in response to a concentration gradient

osmoregulation   the regulation of water and ion balance by organisms to maintain a constant internal environment despite changing external conditions

osmoregulators   animals living in environments with varying salinity which, at least partially, maintain an internal salt concentration different from that of their surroundings

osmotic pressure   the pressure that must be exerted on a solution to prevent influx of pure water across a selectively permeable membrane

ovum (*pl.* **ova**)   female gamete or egg

ovary   the female reproductive organ; produces ova

oviduct   the tube through which eggs pass from the ovary to the uterus

ovulation   release of an egg from the ovary

ovule   the base of the female sporangium in seed plants, where female gametophytes are formed

oxidation-reduction reactions   reactions in which one molecule loses hydrogen atoms or electrons (is oxidized) and another molecule simultaneously gains them (is reduced)

oxidative phosphorylation   the coupled process during which ATP is synthesized from energy released during an oxidation-reduction reaction

oxytocin   a hormone produced in the hypothalamus that plays a role in labor and in milk ejection

ozone   a highly oxidizing gas containing three atoms of oxygen per molecule

pancreas   in man, an organ lying behind the stomach that secretes digestive enzymes into the gut and the hormones insulin and glucagon into the blood

PAN's   peroxyacylnitrates; toxic products of photochemical smog formation

"paper factor"   a juvenile hormone analogue produced by a species of American fir tree which specifically interferes with development of one type of insect

parasitism   a symbiotic relationship beneficial to one species but harmful to the other

parthenogenesis   the development of an unfertilized haploid egg into a new individual

particulate radiation   tiny particles of vibrating matter released by the disintegration of radioactive elements

PCB's   polychlorinated biphenyls; synthetic organic molecules containing varying amounts of chlorine and widely used in industry

pellagra   a niacin-deficiency disease characterized by skin inflammation and diarrhea

peptide bond   the bond between two consecutive amino acids in a protein

peripheral nervous system   that part of the nervous system which connects, via nerves, to receptors and effectors located away from the central nervous system

peristaltic locomotion   movement brought about by sequential waves of contractions of circular and longitudinal muscles which exert forces on the hydrostatic skeleton

pH   the negative logarithm of the hydrogen ion concentration

phage   a virus capable of infecting bacterial cells

phagocytes   cells capable of engulfing food; especially amoebocytes and leucocytes which engulf foreign material

pharynx   the most anterior region of the gut, into which food is received

phenotype   the appearance of genetic traits in an individual

pheromone   a chemical used for communication between members of the same species

phloem   vascular tissue of plants, conducting organic molecules from sites of synthesis to sites of utilization

phosphagens   high energy storage molecules in muscle cells

photochemical smog   a type of air pollution characterized by high concentrations of ozone, nitrogen oxides, and irritants of mucus membranes; formed only in presence of sunlight

photolysis   the splitting of a molecule following the absorption of light energy

photon   the unit of light energy

photophosphorylation   the series of light-dependent reactions occurring in chloroplasts when ATP is synthesized during transfer of electrons through an electron transport chain

photoreceptors   sensory receptors responsive to light

photosynthesis   the series of reactions by which green plants utilize the energy of sunlight to convert carbon dioxide and water into organic molecules

phototropism   the orientation response of an organism to light

phycocyanin   a blue pigment characteristic of blue-green algae

phycoerythrin   an accessory photosensitive pigment occurring in certain algae

phylum   a major taxonomic subdivision within a kingdom

physostigmine   a substance extracted from the seed of an African bean-like plant; acts as a cholinesterase inhibitor

phytoplankton   single-celled plants living in the upper layers of oceans and fresh waters

pica   a craving for unnatural foods, including the habit in young children of chewing and swallowing non-food items

pistil   the stalk of the female sporangium in seed plants, where pollination occurs

pith   the central zone of a stem, serving for nutrient storage

pituitary gland   the "master" endocrine gland of the vertebrate body, located beneath the hypothalamus of the brain; it produces several hormones, most of which act on other endocrine glands

placenta   the organ in the uterine wall formed in pregnancy from both embryonic and maternal tissues, which serves to nourish the embryo

plasma   the non-cellular fraction of blood, a clear, yellowish fluid

plasma membrane   the bounding membrane of a cell

plasmolysis   in plants, the shrinking away of cell contents from cell walls during water deprivation

polar bond   a chemical bond in which the shared electrons tend to spend more time with one atom than with the other

polar molecule   a molecule having electropositive and electronegative regions

pollen   the minute male gametophyte stage of seed plants

pollination   transfer of pollen to the female sporangium in seed plants

polychaetes   bristle worms with parapodial flaps on each segment; mainly marine

polyculture   the agricultural practice of growing several crops side by side

polymers   large molecules composed of many similar, repeating subunits

polypeptide   a polymer consisting of amino acid subunits joined by peptide bonds

polyploidy   possession of more then two sets of chromosomes (3N, 4N, 5N, etc.)

polysaccharides   macromolecules composed of many repeating sugar units

polyunsaturated fats   fat molecules whose fatty acids bear two or more double bonds (C=C) along their length; they are essential to the function of cell membranes

pons   the connecting "bridge" of fiber tracts between the medulla and higher brain centers

population   the total of all individuals belonging to the same species within a given ecosystem and so sharing the same gene pool

positive correlation   a relationship in which the value of the dependent variable increases as the value of the independent variable is increased

positive reinforcement   behavior modification in which reward for a response encourages its repetition

postsynaptic membrane   the membrane on the "receiving" side of a synapse

potential energy   stored energy capable of doing work

primary production   the initial synthesis of organic molecules from inorganic precursors by green plants

primordial follicle   a potential ovum before it has begun to grow, surrounded by a layer of supporting cells

productivity   the rate at which energy is trapped and/or stored as biomass by an entire ecosystem or by one of its component parts

progesterone   the female sex hormone produced by the corpus luteum; essential during early stages of pregnancy

progestogens   synthetic progesterones

prolactin   a hormone produced by the pituitary which induces milk production

prostaglandins   ubiquitous chemicals in the body which in large amounts stimulate uterine contractions; have possible future contraceptive uses

prostate gland   a gland of the human male reproductive tract whose secretions contribute to the seminal fluid

proteases   digestive enzymes that hydrolyze proteins

proteins   large polymers composed of amino acid subunits

prothrombin   a protein in blood plasma which is converted to thrombin by the enzyme thrombokinase; it functions in blood clotting

protobionts   the hypothetical precursors of the first living organisms

proton   a heavy, positively charged particle found in the nuclei of atoms

protozoa   single-celled organisms with nuclei, without chlorophyll, and usually motile; sometimes considered as primitive animals

puffing   the swelling of specific regions of DNA on chromosomes of certain insects during various stages of development

pupa   the intermediate stage between larva and adult in some insects, during which dormancy may occur

pyramidal pathway   the pathway in the brain by which motor impulses from the cortex are relayed directly to the spinal cord

radial symmetry   identical configuration (especially, of body shape) in all directions about one central point or axis

radioactivity   the property of nuclear disintegration characteristic of isotopes with unstable nuclei

radioactive disintegration   the breakdown of an unstable atomic nucleus with the release of energy, often as tiny, fast-moving particles

rapid-eye-movement (REM) sleep   paradoxical sleep during which muscles are more relaxed than usual but brain waves indicate cortical arousal; dreaming is thought to occur at this time

receptor   an organ capable of detecting energy (stimuli) in the environment

recessive   in genetics, producing a trait which can only be expressed when paired with an identical allele

recycling the continuous process by which mineral nutrients pass from non-living to living to non-living compartments in an ecosystem

reducing atmosphere an atmosphere devoid of molecular oxygen

reducing smog a type of air pollution produced mainly from burning coal and oil of high sulfur content; characterized by the presence of sulfur dioxide

reductionist approach a piecemeal approach, considering only one aspect of a problem at a time

reflex arc a direct nervous pathway from receptor to central nervous system to effector

refractory not responsive; refers especially to a state in which a cell, particularly a nerve or muscle, is less responsive to stimuli

rem *r*oentgen *e*quivalent for *m*an; a unit of radiation that indicates the amount of biological damage

Renshaw cell a nerve cell in the spinal column of vertebrates; it receives stimuli from a collateral branch of a motor nerve axon and in turn sends inhibitory impulses to the cell body of the motor neuron, thus acting as a negative feedback control

repressor a substance on a chromosome, possibly histone, which prevents a cistron from directing synthesis of messenger RNA

reproductive isolation the inability of two long-separated subpopulations of a species to produce viable offspring

reproductive potential the maximum number of offspring an individual can produce

reserpine a drug which tends to deplete brain stores of all three monoamines; once used for treatment of manic-depression, but may induce symptoms of Parkinson's disease

respiration in cells, the oxidation of organic molecules with simultaneous conversion of stored energy into useful energy

respiratory pigments colored protein molecules in the blood of animals which are capable of binding reversibly with molecular oxygen (O_2)

resting potential the membrane potential of an unexcited muscle or nerve cell

reticular activating system that part of the reticular formation in the midbrain which controls cortical arousal

reticular formation a diffuse network of ascending and descending fibers in the midbrain; acts as a sensory filter for incoming information

reverberating loops diffuse control systems in the brain by which incoming information is continuously cycled in circular pathways

rhizoids undifferentiated root-like structures occurring in primitive plants

riboflavin a B-vitamin required as part of the hydrogen transfer molecule FAD

ribosomes in the cell, minute slotted granules composed mainly of r-RNA, on which protein synthesis takes place

rickets a vitamin D deficiency disease characterized by deformed bones

RNA ribose nucleic acid; a polymer of nucleotides functioning during transcription of genetic information

root hairs extensions of single root cells which penetrate particles of sand and clay to absorb water and mineral nutrients

rough-ER endoplasmic reticulum whose outer surface is covered with ribosomes

r-RNA ribosomal RNA, occurring in the ribosomes

salivary glands glands associated with the pharynx, often secreting both lubricant (mucus) and digestive enzymes

sample that fraction of a population chosen for study

schistosomiasis debilitating disease caused by flatworms of the genus *Schistosoma*

sclerophyll tissue   the hardened external layers on leaves of some desert plant species

scrotum   the sac external to the body cavity in which the testes of mammals are located

scurvy   a vitamin C deficiency disease characterized by sore, bleeding gums and failure of wounds to heal properly

Second Law of Thermodynamics   the physical law that states that all systems tend toward uniformity and that to restore order requires the addition of free energy from outside the system

segmentation   regular repetition of structures along the length of the body

selective permeability   a property of membranes such that only certain molecules can pass through

seminal fluid   the total ejaculate of the male, including sperm, nutrients and mucus

seminal vesicles   in lower animals, organs for sperm storage; in mammals, glands synthesizing components of seminal fluid

seminiferous tubules   highly coiled ducts in the male testis which produce sperm

sensory cells   modified nerve cells in sensory organs, capable of transducing various forms of energy into action potentials

septa   thin muscular partitions subdividing the body cavity

sessile   permanently attached to the substratum; said of animals, such as some coelenterates, that are not free-moving

sex chromosome   a chromosome which determines sex and whose mate in one or the other sex is not entirely homologous

sex-linked   sexually inherited; refers to traits whose loci are on the sex chromosomes

sexual reproduction   reproduction resulting from union of haploid gametes to form a diploid cell, the zygote

short-term memory storage   the temporary storage of information in the central nervous system

sieve plates   the perforated end walls of phloem cells

siphons   in bivalves, extensions of the mantle cavity through which respiratory water currents flow

smooth-ER   endoplasmic reticulum devoid of ribosomes

sodium pump   a transport system thought to be located on cell membranes for moving sodium ions out of cells

solar flux   the rate at which the sun's energy falls on the Earth

solute   a dissolved substance

solvent   a liquid capable of dissolving another substance

spawning   the release of gametes from the body, usually into water

species   a group of organisms whose individual members are distinct from all other types of organisms and are capable of interbreeding only with one another

species diversity   the number of different species found in an ecosystem

specificity   the unique shape of a molecule which allows it to join in lock-and-key fashion only with molecules of reciprocal shape

spermatozoa   male gametes or sperm

spermicides   sperm-killing agents

spiracles   the external openings of the tracheolar air ducts in the cuticle of insects, millipedes and centipedes

sporocyst   an asexual reproductive stage in the life cycle of the parasitic flatworm that causes schistosomiasis

sporozoans   parasitic protozoans, including gregarines and the organism causing malaria

spleen   an organ located beside the stomach, which acts as a reservoir for red blood cells and a site of formation of white blood cells

sporophyte   diploid (2N) stage of plant life cycle

stable state   the state of an ecosystem in which photosynthesis exactly equals respiration and the system is neither increasing nor decreasing in biomass

stamen   the stalk of the male sporangium in seed plants

starch   a glucose polymer used for energy storage in plants

stimulus   anything causing a response, such as environmental information perceived by an animal

stomach   a part of the gut in which both grinding and preliminary digestion of food may occur

stoma (*pl.* **stomata**)   openings on leaves through which oxygen, carbon dioxide and water vapor are exchanged with the air

stratosphere   the upper atmosphere, above seven miles, in which conditions become relatively stable

substantia nigra   the "black substance"; a region of the midbrain many of whose neurons send dopamine-mediated axons to the basal ganglia to depress motor inhibition from that region

substrate   the chemical on which an enzyme acts

successional stage   the temporary community composition of an unstable ecosystem

supernatant   that which remains in suspension following centrifugation

survivorship curves   curves relating survival of individuals within a population to the life span for that species

sustainable yield   the maximum amount of a natural population that may be harvested without depleting the supply

symbiosis   the very close association of two species of organisms

synapse   the junction between a nerve cell and another cell; synapses generally transmit information in but one direction

taiga   the boreal coniferous forest biome

temperature inversion   a meteorological condition in which cool air near the ground is prevented from rising by an overlying layer of warm air

template   that which serves as a model or mold for the manufacture of something

teratogenic defect   a defect produced in an animal embryo through toxic effects of a chemical present at a critical period of organ development

testis   the male reproductive organ; produces sperm

testosterone   the male sex hormone responsible for maintenance of the reproductive tract and secondary sex characteristics

tetrads   a structure formed by the joining along their length of two doubled homologous chromosomes in preparation for meiosis

thalamus   the terminus of the ascending fiber tracts from the spinal cord in the human brain; an important relay center

thallophytes   simple, undifferentiated plants such as algae and fungi

thallus   an undifferentiated leaf-like structure occurring in primitive plants

thermal pollution   the addition of waste heat, generated mainly by power plants, into the environment; usually refers to water-cooled systems.

thiamin   vitamin B_1; a constituent of the coenzyme involved in cellular reactions where carbon dioxide is removed

threshold potential   the critical level of depolarization at which membrane permeability to sodium suddenly increases

thrombin   the product of the action of thrombokinase on the plasma protein prothrombin; thrombin catalyzes the conversion of fibrinogen to fibrin during blood clotting

thromboembolism   obstruction of a blood vessel by a blood clot

thrombokinase   an enzyme released into the blood from damaged blood platelets or tissue cells, and which initiates clotting

thromboses   spontaneously forming blood clots which may damage vital organs if they block small blood vessels

thyroid an endocrine gland overlying the larynx in the throat; it secretes thyroxin

thyroxin a hormone produced by the thyroid gland; implicated in control of basal metabolism and of blood sugar levels

tissue a group of cells performing the same function

tonus a state of continuous tension in a muscle, developed by the sequential contraction of different combinations of motor units

tracheae the air tubes that together constitute the branched system of respiratory ducts found in insects, millipedes and centipedes

tracheophytes plants with true conducting tissues; include ferns and seed plants

translocation the movement of organic solutes throughout plants

transpiration the evaporation of water from plant leaves

trial-and-error learning learning by error; often requires positive or negative reinforcement

trichinosis a disease caused by eating pork infected with the larvae of a species of roundworm

triploid containing three sets of chromosomes (3N); characteristic of the endosperm of flowering plants

tritium ^{3}H, or radioactive hydrogen; an isotope of hydrogen containing two neutrons in its nucleus

t-RNA transfer RNA; the molecule which brings an amino acid to the site of protein synthesis and codes it into the correct sequence

trophic level a step in a food chain

troposphere the lower atmosphere, up to seven miles, in which conditions are variable

tubal ligation surgical cutting and tying of the oviducts to effect sterilization in the female

tumescence swelling of tissues, especially of sexually sensitive tissues during sexual arousal

tundra the arctic-alpine biome

tunicates sea squirts, primitive members of the chordate group

turbidity the cloudiness of a transparent substance due to suspended particles

turgor distention of plant cells resulting from internal osmotic pressure

turnover time the time required for all of a chemical present in one compartment of an ecosystem to flow through that compartment

umbilical cord the connecting link between the mammalian embryo and its placenta

upwelling the upward movement of deep ocean waters, especially near certain coasts

urea a small nitrogen-containing molecule; normally a waste product of protein metabolism that is excreted, but retained by sharks and rays to maintain internal osmotic pressure

uric acid a by-product of nitrogen metabolism, especially in insects, reptiles and birds; has low solubility in water and is excreted mainly as a solid

urethra the canal by which urine is expelled from the bladder

uterus womb; the organ in which an embryo develops

vacuoles fluid-filled cavities within cells, especially the large vacuoles of plant cells

vagina the canal leading from uterus to exterior in female mammals

variable a measurable property which is capable of assuming a range of values

vas deferens the duct through which sperm pass from the epididymis to the urethra

vasectomy surgical cutting of the vasa deferentia to effect male sterility

vectors   organisms which carry the agents of plant or animal diseases from one host to another

vertebrates   animals with backbones

vitamins   organic molecules required in small quantities in an organism's diet because it cannot synthesize them itself; many function as coenzymes

water expulsion vesicle   a vacuole occurring in some protozoans which alternately fills and empties, expelling excess water

water table   the upper level of ground water in the soil

water vascular system   a subdivision of the coelom unique to echinoderms, which functions in movement of their tube feet

wavelength   the reciprocal of the frequency at which a propagated wave vibrates, times its speed; usually used to define the colors of light

X-chromosome   in humans, the sex chromosome possessed homozygously by females; the larger of the two types of sex chromosomes

xylem   vascular tissue of plants conducting water and minerals upward

Y-chromosome   in humans, the sex chromosome which determines maleness; the smaller of the two types of sex chromosomes

yolk   nutrients stored in an egg for use by the developing embryo

zooplankton   small, suspended aquatic animals subsisting mainly on phytoplankton

zygote   the diploid (2N) cell formed by union of two haploid (1N) gametes

INDEX

Page numbers in *italics* refer to illustrations; page numbers followed by (t) refer to tables.

i

A	adenine
A	argon
A	normal hemoglobin subunit
ACh	acetylcholine
ADP	adenosine diphosphate
ALA	δ-aminolevulinic acid
ALAS	δ-aminolevulinic acid synthetase
Arg	arginine
ATP	adenosine triphosphate
B	boron
$B_4O_7^=$	borate ion
C	carbon
^{14}C	carbon 14 (radioactive)
C	cytosine
Ca	calcium
Ca^{++}	calcium ion
CH_4	methane
$[CH_2O]$	a shorthand notation for the carbohydrates
$C_6H_{12}O_6$	glucose
Cl	chlorine
Cl^-	chloride ion
CO	carbon monoxide
CO_2	carbon dioxide
COMT	catechol-O-methyltransferase
CN	cyanide
Cu	copper
Cu^{++}	copper ion
D	dopamine
2,4-D	2, 4 dichlorophenoxyacetic acid
DDT	dichlorodiphenyltrichloro ethane
DNA	deoxyribonucleic acid
e^-	free electron
FAD	flavin adenine dinucleotide
$FAD \cdot H_2$	reduced FAD
Fe	iron
Fe^{++}	reduced iron ion
Fe^{+++}	oxidized iron ion
FMN	flavin mononucleotide
$FMN \cdot H_2$	reduced FMN
FRF	FSH-releasing factor
FSH	follicle stimulating hormone
G	guanine
H	hydrogen
H^+	hydrogen ion
H_2S	hydrogen sulfide
$H \cdot$	hydrogen radical (free H atom)
HC	hydrocarbon
H_2CO_3	carbonic acid
HCO_3^-	bicarbonate ion
H_2O	water
5-HT	5-hydroxytryptamine
^{131}I	iodine-131 (radioactive)
IAA	indoleacetic acid
K	potassium
K^+	potassium ion
L-DOPA	l-dihydroxyphenylalanine
Leu	leucine
LH	luteinizing hormone
LRF	LH-releasing factor
LSD	lysergic acid diethylamide
Lys	lysine
MAO	monoamine oxidase
Mg	magnesium
Mn	manganese
Mn^{++}	manganese ion
m-RNA	messenger RNA
N	nitrogen
N_2	molecular nitrogen
Na	sodium
Na^+	sodium ion
NA	noradrenalin
NAD	nicotinamide adenine dinucleotide
$NAD \cdot H_2$	reduced NAD
NADP	NAD phosphate
$NADP \cdot H_2$	reduced NADP
NH_3	ammonia
NH_4^+	ammonium ion
NO	nitric oxide
NO_2	nitrogen dioxide
N_2O	nitrous oxide